EXCEL 2024/MICROSOFT 365
PROGRAMMING BY EXAMPLE

EXCEL 2024/MICROSOFT 365 PROGRAMMING BY EXAMPLE

Julitta Korol

MERCURY LEARNING AND INFORMATION
Boston, Massachusetts

MERCURY LEARNING AND INFORMATION
121 High Street, 3rd Floor
Boston, MA 02110
info@merclearning.com

J. Korol. *Excel 2024 / Microsoft 365 Programming by Example.*
ISBN: 978-1-50152-415-8

Library of Congress Control Number: 2025931826
242526321   This book is printed on acid-free paper in the United States of America.

To those who believe in self-education.
It simply astounds me that the language I learned so
many years ago is still so useful today.

CONTENTS

PART II WORKING WITH THE FILE SYSTEM 273

Chapter 10 Manipulating Files and Folders with VBA 275

Chapter 11 Automating File System Tasks with
Windows Script Host and FileSystemObject..............291

PART IV ENHANCING USER EXPERIENCE 427

PART VIII WORKING TOGETHER: VBA, HTML, XML, AND THE REST API   905

Chapter 25 Using HTML and XML in Excel 2024 907

ACKNOWLEDGMENTS

First, I'd like to express my gratitude to the editorial team at DeGruyter and Mercury Learning and Information for working extremely hard to bring this book to print. A sincere thank-you to David Pallai for offering me the opportunity to update this book to the new 2024 version. We've been through so many versions together and I am so grateful for all his support and guidance during the past decade we've been working together on this and other titles. Together with Jennifer Blaney, David and I have kept this book updated for the new generation of Excel power users and VBA developers. Jen continues to be my go-to person now at DeGruyter. She has numerous years of publishing expertise, and she's a great manager. With so many other staff members at the DeGruyter she tirelessly steered this book to publication from beginning to end. I also want to thank my copyeditor and compositor for their efforts in meticulous proofreading, manuscript editing and page layout.

Finally, I'd like to acknowledge readers like you who cared enough to post reviews of the previous editions of this book. Your invaluable feedback has helped me improve the quality of this work by including the material that matters to you most. Please continue to inspire me with your ideas and suggestions. And please forgive me if some of your suggestions didn't make it into this book. I simply ran out of pages!

INTRODUCTION

If you ever wanted to open a new worksheet without using built-in commands or create a custom, fully automated form to gather data and store the results in a worksheet, you've picked up the right book. This book shows you what's doable with Microsoft® Excel® 2024 beyond the standard user interface. This book's purpose is to teach you how to delegate many time-consuming and repetitive tasks to Excel by using its built-in language, VBA (Visual Basic for Applications). By executing special commands and statements and using several of Excel's built-in programming tools, you can work smarter than you ever thought possible. I will show you how.

When I first started programming in Excel (circa 1990), I was working in a sales department, and it was my job to calculate sales commissions and send the monthly and quarterly statements to our sales representatives spread all over the United States. As this was a very time-consuming and repetitive task, I became immensely interested in automating the whole process. In those days, it wasn't easy to get started in programming on your own. There weren't as many books written on the subject; all I had was the built-in documentation, which was hard to read. Nevertheless, I succeeded; my first macro worked like magic. It automatically calculated our salespeople's commissions and printed out nicely formatted statements. While the computer was busy performing the same tasks repeatedly, I was free to deal with other more interesting projects.

Many years have passed since that day, and Excel is still working like magic for me and a great number of other people who have taken the time to familiarize themselves with its programming interface. If you'd like to join these people and have Excel do magical things for you as well, this book provides an easy step-by-step introduction to VBA and other technologies that work nicely with Microsoft

Excel. Besides this book, there is no extra cost to you; all the tools you need are built into Excel. If you have not yet discovered them, *Excel 2024 / Microsoft 365 Programming by Example* will lead you through the process of creating your first macros, VBA procedures, Excel tables and charts, power queries, and XML documents from start to finish. Along the way, you'll find detailed, practical "how-to" examples and plenty of illustrations. The book's approach is to learn by doing. There's no better way than step by step. Simply turn on the computer, open this book, launch Microsoft Excel, and do all the guided Hands-On exercises. Before you get started, allow me to give you a short overview of the things you'll be learning as you progress through this book.

Excel 2024 / Microsoft 365 Programming by Example is divided into 8 parts (26 chapters) that progressively introduce you to programming Excel 2024 as well as controlling other applications with Excel.

Part I introduces you to VBA, the programming language for Microsoft Excel. In this part of the book, you'll acquire the fundamentals of VBA that you will use over and over again in building real-life spreadsheet applications.

PART I CONSISTS OF THE FOLLOWING NINE CHAPTERS:

Chapter 1—Excel Macros—A Quick Start in Excel VBA Programming
In this chapter, you'll learn how you can introduce automation into your Excel worksheets by simply using the built-in macro recorder. You'll learn about different phases of macro design and execution. You will also learn about macro security.

Chapter 2—Excel Programming Environment—A Quick Overview of Its Tools and Features
In this chapter, you'll learn almost everything you need to know about working with the Visual Basic Editor window, commonly referred to as VBE. Some of the programming tools that are not covered here are discussed and used in Chapter 9.

Chapter 3—Excel VBA Fundamentals—A Quick Reference to Writing VBA Code
In this chapter, you will be introduced to the basic VBA concepts, such as the Microsoft Excel object model and its objects, properties, and methods. You will also learn concepts that allow you to store various pieces of information in variables for later use.

Chapter 4—Excel VBA Procedures—A Quick Guide to Writing Function Procedures
In this chapter, you'll learn how to write and execute function procedures. You will also learn how to pass additional information to your procedures before they are

run. You will be introduced to working with some useful built-in functions and methods that allow you to interact with your VBA procedure users.

Chapter 5—Adding Decisions to Excel VBA Programs—A Quick Introduction to Conditional Statements
In this chapter, you'll learn how to control your program flow with several different decision-making statements.

Chapter 6—Adding Repeating Actions to Excel VBA Programs—A Quick Introduction to Looping Statements
In this chapter, you'll learn how you can repeat certain groups of statements using procedure loops.

Chapter 7—Storing Multiple Values in Excel VBA Programs—A Quick Introduction to Working with Arrays
In this chapter, you'll learn about the concept of static and dynamic arrays, which you can use for holding various values. You will also learn about built-in array functions.

Chapter 8—Keeping Track of Multiple Values in Excel VBA Programs—A Quick Introduction to Creating and Using Collections
In this chapter, you'll learn how to create and use your custom collection object to track and manipulate data in your VBA procedures.

Chapter 9—Excel Tools for Testing and Debugging—A Quick Introduction to Testing VBA Programs
In this chapter, you will begin using built-in debugging tools to test your programming code and trap errors.

The above nine chapters will give you the fundamental techniques and concepts you will need to continue your Excel VBA learning path. The skills obtained in *Excel VBA Primer* are very portable. They can be utilized in programming other Microsoft 365 applications that also use VBA as their native programming language, such as Access, Word, PowerPoint, Outlook, and others.

In Part II of this book, you will discover various methods of working with the file system.

While VBA offers numerous built-in functions and statements for working with the file system, you can also perform file and folder manipulation tasks via objects and methods included in the Windows Script Host installed by default on computers running the Windows operating system. Additionally, you can directly manipulate files and their contents via the low-level file I/O (input/output) functions.

PART II CONSISTS OF THE FOLLOWING THREE CHAPTERS:

Chapter 10—Manipulating Files and Folders with VBA
In this chapter, you'll learn about numerous VBA statements used to work with Windows files and folders.

Chapter 11—Automating File System Tasks with Windows Script Host and FileSystemObject
In this chapter, you'll learn how the Windows Script Host works together with VBA and allows you to get information about files and folders.

Chapter 12—Direct Manipulation of Files and Their Contents
In this chapter, you'll learn how to get in direct contact with your data by using the process known as low-level file I/O. You will also learn about various types of file access.

The VBA programming language goes beyond Excel. It can be used to program any application that supports this language. In Part III, you will learn how to use VBA to interact with other applications.

PART III CONSISTS OF THE FOLLOWING TWO CHAPTERS:

Chapter 13—Office 365 Automation with Excel VBA
In this chapter, you'll learn about Office automation, linking and embedding, and working with Word, Outlook, PowerPoint, and Access. Most of the time will be spent using Access, which provides data for many Excel applications. You will learn how to connect to an Access database and use various data retrieval methods.

Chapter 14—Using PowerShell with Excel VBA
This chapter delves into PowerShell—a robust, task-oriented scripting language and command-line shell developed by Microsoft. By integrating PowerShell with Excel VBA, you can perform a wide range of system-level tasks on both local and remote Windows systems. If you've ever worked with DOS commands, you'll find some familiarity here, but PowerShell offers far more flexibility, functionality, and power. You'll learn how to launch PowerShell from Excel VBA and pass data between both these applications.

In Part IV, we will focus on various ways of enhancing user experience. You will be introduced to many events that occur in Excel workbooks. You'll learn how to work with dialog boxes and user forms and format worksheets for display and printing. We also spend lots of time in programming context menus and ribbon customizations.

PART IV CONSISTS OF THE FOLLOWING FOUR CHAPTERS:

Chapter 15—Utilizing Event-Driven Programming
In this chapter, you'll learn about the types of events that can occur when you are running VBA procedures in Excel. You will gain a working knowledge of writing event procedures and handling various types of events.

Chapter 16—Working with Dialog Boxes and User Forms
In this chapter, you'll learn about working with Excel's built-in dialog boxes and various controls for designing user-friendly forms. This chapter has a hands-on user form application you will build from scratch.

Chapter 17—Formatting Worksheets for Display and Printing
In this chapter, you'll learn how to perform basic formatting tasks with VBA for numbers, text, date, columns, rows, and cell ranges. You will also perform advanced formatting tasks with VBA using conditional formatting and applying visual features such as data bars, color scales, icon sets, shapes, and sparklines. You'll learn how to produce consistent-looking worksheets by using document themes and styles. Finally, you will format your worksheets for printing and emailing.

Chapter 18—Context Menu Programming and Ribbon Customizations
In this chapter, you'll learn how to add custom options to Excel's built-in context (shortcut) menus and how to work programmatically with the Ribbon interface and Backstage view.

In Part V, you will work with Excel tools for data analysis. You will gain experience in programming advanced Excel features such as Excel tables, PivotTables, and PivotCharts, and get introduced to the Power Query feature, which allows you to create powerful queries that simplify data import and transformation.

PART V CONSISTS OF THE FOLLOWING THREE CHAPTERS:

Chapter 19—Using and Programming Excel Tables
In this chapter, you'll learn how to work with Excel tables. You will learn how to retrieve information from an Access database, convert it into a table, and enjoy database-like functionality in the spreadsheet. You will also learn how tables are exposed through Excel's object model and manipulated via VBA.

Chapter 20—Programming PivotTables and PivotCharts
In this chapter, you'll learn how to use VBA to work with powerful Microsoft Excel objects that are used for data analysis: PivotTables, PivotCharts, and slicers. You

will learn how to manipulate these objects to quickly produce reports that allow you and your users to easily examine large amounts of data pulled from an Excel worksheet range or from an external data source such as an Access database.

Chapter 21—Getting and Transforming Data
In this chapter, you will be introduced to data import, transformation, and shaping features available in the Get & Transform section of Excel 2024's Data tab. You will work with Query Editor and Advanced Editor and learn formulas and functions written in the M expression language while bringing together data from various sources.

While VBA provides a very comprehensive object model for automating worksheet tasks, some of the processes and operations that you may need to program are an integral part of the Windows operating system and cannot be controlled via VBA.

In Part VI, you will learn how to take charge of a programming environment. You'll begin by using VBA with VBA projects, modules, and procedures. Next, you will be introduced to the Windows API library of functions that will come to your rescue when you need to overcome the limitations of the native VBA library.

PART VI CONSISTS OF THE FOLLOWING TWO CHAPTERS:

Chapter 22—Programming the Visual Basic Editor (VBE)
In this chapter, you'll learn how to use numerous objects, properties, and methods from the Microsoft Visual Basic for Applications Extensibility object library to control the VBE to gain full control over Excel.

Chapter 23—Calling Windows API Functions from VBA
In this chapter, you will be introduced to the Windows API library. After learning basic Windows API concepts, you will be shown how to declare and utilize API functions from VBA.

As Excel VBA procedures grow more complex, managing scattered and disorganized code becomes increasingly challenging. Copying code across projects and constantly updating procedures to accommodate new requirements can lead to a tangled, unmanageable mess. Before you go down this route, you should know that Excel has a special feature known as a *class module* that allows you to create code that is self-contained and reusable, which we will cover in Part VII.

PART VII CONSISTS OF THE FOLLOWING CHAPTER:

Chapter 24—Creating Classes in VBA
In this chapter, you will learn how coding in a standalone class module can help you organize your code into more manageable objects that can easily be reused and adjusted when necessary. Here, you will learn about creating and using custom objects, declaring class members, defining class properties, and writing property procedures and class methods. You'll also learn how to use event procedures in a class module. These advanced topics will be covered in a custom application that you will build from scratch.

With the Internet and intranets, worksheet data becomes easily accessible and shareable around the clock. Excel not only allows you to capture data from the Web but also enables you to publish it seamlessly. Part VIII will introduce you to integrating Excel VBA with Hypertext Markup Language (HTML) and Extensible Markup Language (XML), as well as leveraging a modern Web service known as the REST API.

PART VIII CONSISTS OF THE FOLLOWING TWO CHAPTERS:

Chapter 25—Using HTML and XML in Excel 2024
In this chapter, you'll learn how to create hyperlinks and publish HTML files using VBA. You will also learn how to use XML with Excel. You'll learn about enhanced XML support in Excel 2024 and the many objects and technologies that are used to process XML documents.

Chapter 26—Excel and REST APIs
In this chapter, you will explore several external libraries that will help you build more advanced VBA applications. You'll learn about a Dictionary object and how it compares to the native VBA Collection object. You'll learn about regular expressions and how to use them to extract data. Finally, you will be introduced to the JSON format and learn the basics of making HTTP GET requests to pull data from a RESTful Web service.

INTENDED AUDIENCE

This book is designed for Excel users who want to expand their knowledge and learn what can be accomplished with Excel beyond the provided user interface.

Consider this book as a sort of private course that you can attend in the comfort of your office or home. Some courses have prerequisites, and this is no exception. *Excel 2024 / Microsoft 365 Programming by Example* does not explain how to select options from the ribbon or use shortcut keys. The book assumes that you can easily locate in Excel the options that are required to perform any of the tasks already pre-programmed by the Microsoft team. With the basics already mastered, this book will take you to the next learning level where your custom requirements and logic are rendered into the language that Excel can understand. Let your worksheets perform magical things for you and let the fun begin.

THE COMPANION FILES

The example files for all the hands-on activities in this book are available in the companion files included with this book. All companion files for this title are available by visiting the website for the book at *sciendo.com/book/9781501524158* and by clicking on the "COMPANION FILES" tab. Digital versions of this title are available at Amazon.com and other digital vendors.

Chapter 2

LIST OF TABLES

EXCEL VBA PRIMER

Part **I**

Excel VBA Primer is divided into nine chapters that progressively introduce you to programming Microsoft Excel using the 2024 version of the product. These chapters present the fundamental techniques and concepts that you need to master before you can take further steps in Excel programming.

Chapter 1

EXCEL MACROS—
A QUICK START IN
EXCEL VBA
PROGRAMMING

Visual Basic for Applications (VBA) is the programming language built into Microsoft Excel® and other Microsoft® 365® applications such as Excel, Outlook, Word, PowerPoint, and Access. With VBA, you can streamline repetitive tasks and processes, create custom functions, and enhance Excel's capabilities. Millions of companies worldwide use Microsoft 365 and its VBA language to enhance productivity, automate a broad range of tedious tasks, and even build entire custom applications. VBA's simplicity, accessibility, and seamless integration with all Microsoft 365 applications make it a valuable tool for a wide range of users who need to automate tasks across Excel, Word, and other applications, without leaving the Office ecosystem.

In this chapter, you will acquire the fundamentals of VBA by using the Excel macro recording feature and the Visual Basic Editor environment to examine and edit the VBA code behind the recorded macro.

MACROS AND VBA

Macros are programs that store a series of commands. When you create a macro, you simply combine a sequence of keystrokes into a single command that you can later "play back." Because macros can reduce the number of steps required to complete tasks, using macros can significantly decrease the time you spend creating, formatting, modifying, and printing your worksheets.

You can create macros by using Microsoft Excel's built-in recording tool (macro recorder) or you can write them from scratch using Visual Basic Editor, a special development environment built into Excel. You can combine recorded macros with your own programming code to create unique VBA applications that meet your everyday needs.

Microsoft Excel comes with dozens of built-in, time-saving features that allow you to work faster and smarter. Before you decide to automate a worksheet task with a recorded macro or programming code written from scratch, make sure there is not already a built-in feature that you can use to perform that task. Consider writing your own VBA code or recording a macro when you find yourself performing the same series of actions multiple times or when Excel does not provide a built-in tool to do the job.

Just by learning how to handle Excel's macro recorder and use basic VBA statements to enhance your macros, you'll be able to automate any part of your worksheet.

For example, you can automate data entry by recording a macro that enters headings in a worksheet or replaces column titles with new labels. Adding a little bit of conditional logic to your VBA code will allow you to automatically check duplicate entries in a specified range of your worksheet. With a macro, you can quickly apply formatting to several worksheets, as well as combine different formats, such as fonts, colors, borders, and shading. Macros will save you keystrokes when it comes to setting print areas, margins, headers, and footers, and selecting special options for printouts.

Excel Macro-Enabled File Formats

When you record a macro or write your own programming code in Excel, your workbook will contain a VBA project, a feature that requires that you save the workbook in one of the following macro-enabled file formats:

- Excel macro-enabled workbook (`.xlsm`)
- Excel binary workbook (`.xlsb`)
- Excel macro-enabled template (`.xltm`)

These macro-enabled file types can be selected from the File Type list when you save your workbook. Note that the familiar `.xlsx` file format applies only to macro-free workbooks.

Macro Security Settings

Because macros can contain malicious code designed to put a virus on a user's computer, it is important to understand the different security settings that are available in Excel. It is also critical that you run up-to-date antivirus software on your computer and only open workbook files from trusted sources. The default macro security setting is to disable all macros with notification, as shown in Figure 1.1.

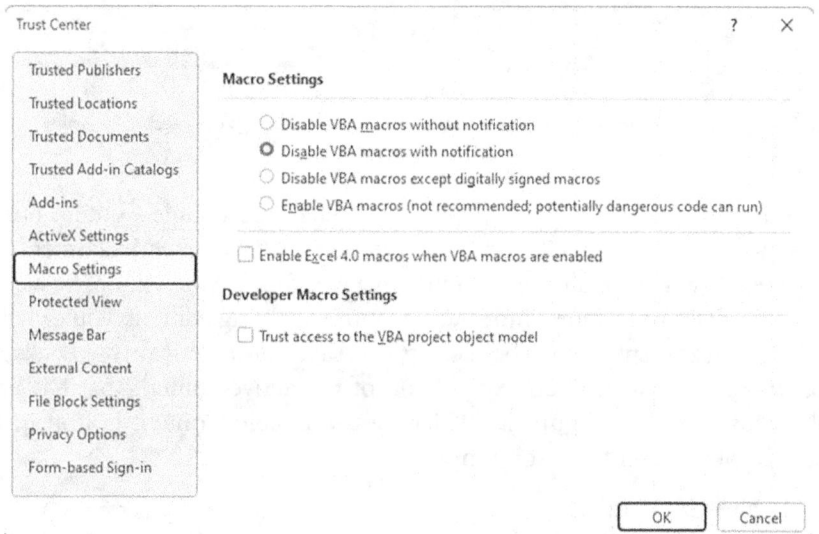

FIGURE 1.1.   The Macro Settings options in the Trust Center allow you to control how Excel should deal with macros when they are present in an open workbook. To open the Trust Center's Macro Settings, choose File | Options | Trust Center | Trust Center Settings and click the Macro Settings link shown in the left pane of the Trust Center dialog.

Note that VBA macros are the macros you create using the Excel built-in language, VBA. You will be working with these macros throughout this book. In contrast, Excel 4.0 macros are legacy Excel macros, also known as XLM 4.0 macros, which were introduced in 1992. They are still in Excel for backward compatibility reasons. Using these macros is discouraged as they can hide malicious code in Excel formulas.

If VBA macros are present in a workbook you are trying to open, you will receive a security warning message just under the ribbon, as shown in Figure 1.2.

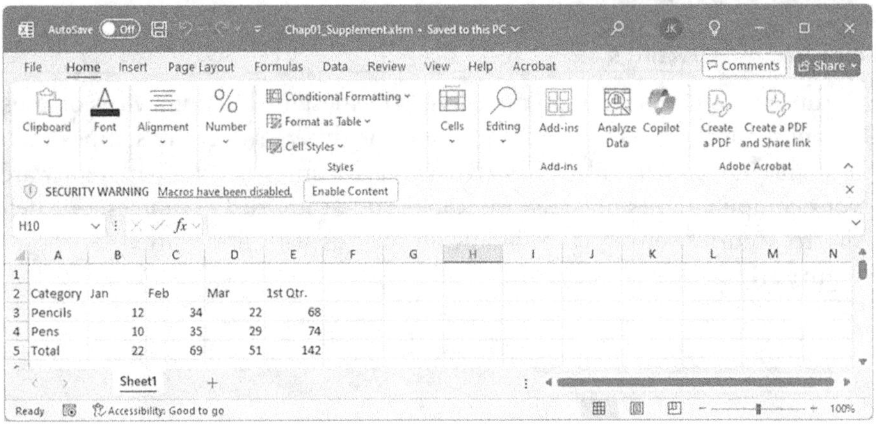

FIGURE 1.2. Upon opening a workbook with VBA macros, Excel brings up a security warning message.

To use the disabled components, you should click the Enable Content button on the message bar. This will add the workbook to the Trusted Documents list in your registry. The next time you open this workbook, you will not be alerted to macros. If you need more information before enabling content, you can click the message text displayed in the security message bar to activate the Backstage view, where you will find an explanation of the active content that has been disabled, as shown in Figure 1.3. Clicking the Enable Content button in the Backstage view presents two options:

- Enable Content:

  This option provides the same functionality as the Enable Content button in the security message bar. This will enable all the content and make it a trusted document.

- Advanced Options:

  This option brings up the Microsoft Office Security Options dialog shown in Figure 1.4. This dialog provides options for enabling content for the current session only.

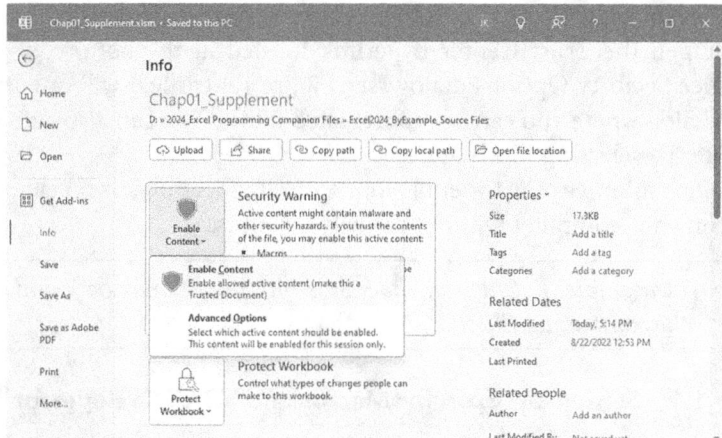

FIGURE 1.3. The Info option in the Backstage view of Excel offers detailed information about the active content within the workbook. It also provides options to either make the document trusted or enable active content solely for the current session.

FIGURE 1.4. Disabled macros can be enabled for the current session in the Microsoft Office Security Options dialog that is activated when you choose Advanced Options in the Security Warning dropdown (see Figure 1.3).

ENABLING THE DEVELOPER TAB IN EXCEL

To simplify working with macro-enabled workbooks while following this book's exercises, you can permanently trust your workbooks containing recorded macros or VBA code by placing them in a folder on your local drive designated as trusted. The Open the Trust Center hyperlink located at the bottom of the Microsoft Office Security Options dialog (see Figure 1.4 earlier) will open the Trust Center dialog where you can set up a trusted folder. You can also activate the Trust Center by selecting File | Options.

Let's take a few minutes now to set up your Excel application so you can run VBA macros on your computer without security prompts.

NOTE	*Please note that files for the hands-on exercise may be found in the companion files.*

Hands-On 1.1   Setting Up Excel for Macros and VBA Development

1. Create a folder on your hard drive named `C:\VBAExcel2024_ByExample`.
2. Launch Excel and choose Options.
3. In the Excel Options dialog, click Customize Ribbon.
4. In the Main Tabs list on the right-hand side, select Developer, as illustrated in Figure 1.5, and click OK.
   The Developer tab should now be visible in the ribbon.

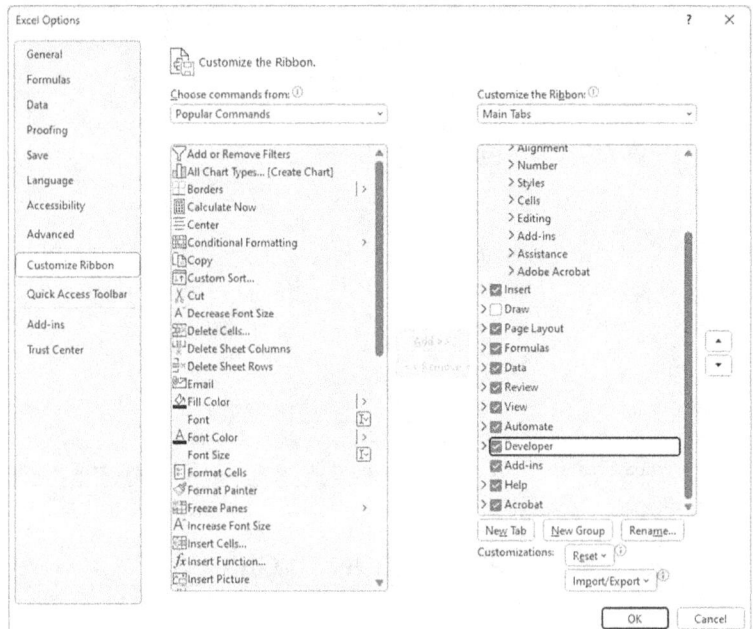

FIGURE 1.5. To enable the Developer tab on the ribbon, use the Excel Options dialog and select Customize Ribbon.

5. In the Code group of the Developer tab on the ribbon, click the Macro Security button, as shown in Figure 1.6.
The Trust Center dialog appears as depicted in Figure 1.1 earlier.

6. In the left pane of the Trust Center dialog, click Trusted Locations.
The Trusted Locations dialog already shows several predefined trusted locations that were created when you installed Excel. For this book, we will add a custom location to this list.

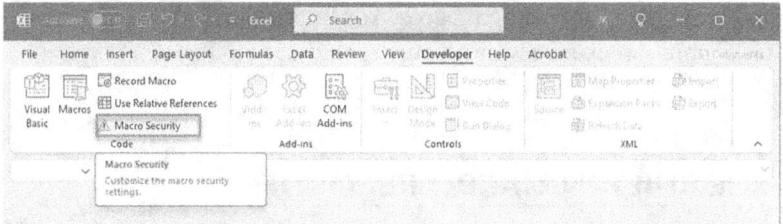

FIGURE 1.6.   Use the Macro Security button in the Code group on the Developer tab to customize the macro security settings.

7. On the Trusted Locations screen of the Trust Center dialog, click the Add new location… button.

8. In the Path text box, type the name of the folder you created in Step 1 of this hands-on exercise, as shown in Figure 1.7.

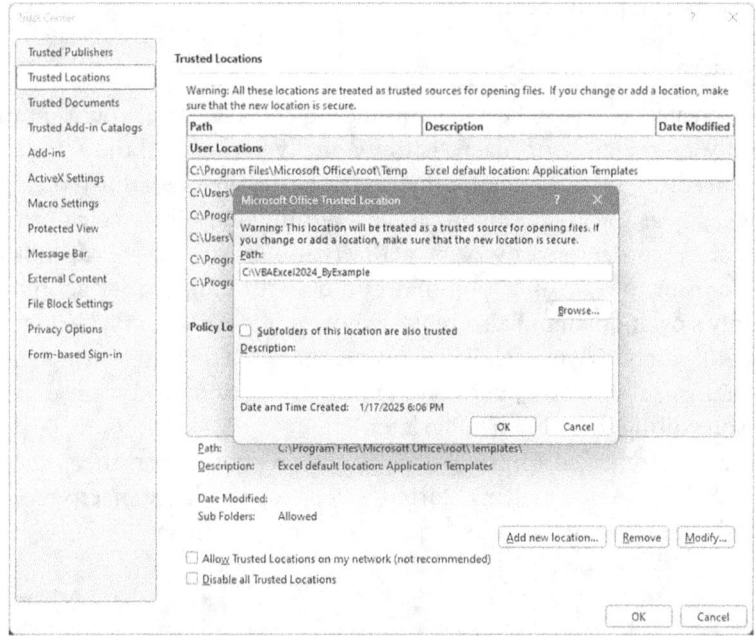

FIGURE 1.7.   Designating a Trusted Location folder for this book's programming examples.

9. Click OK to close the Microsoft Office Trusted Location dialog.
   The Trusted Locations list in the Trust Center now includes the `C:\ VBAExcel2024_ByExample` folder as a trusted location. Files placed in a trusted location can be opened without being checked by the Trust Center security feature.

10. Click OK to close the Trust Center dialog box.

Your Excel application is now set up for easy macro and VBA development as well as opening files containing macros. You should save all the files created in the book's hands-on exercises into your trusted `C:\VBAExcel2024_ByExample` folder.

USING THE BUILT-IN MACRO RECORDER

In this section, we will go through the process of recording several short macros that perform data entry and formatting tasks in an Excel worksheet. You will learn how to plan your macros, record your keystrokes, edit and improve your recorded macro code, run your macros, and learn basic troubleshooting techniques that will get you back on track if you encounter errors while running your macros. You will also learn how to save your macros, rename them, combine them, and print them.

Planning a Macro

Before you create a macro, take a few minutes to consider what you want to do. The easiest way to plan your macro is to manually perform all the actions that the macro needs to do. As you enter the keystrokes, write them down on a piece of paper exactly as they occur. Don't leave anything out. Like a voice recorder, Excel's macro recorder records every action you perform. If you do not plan your macro prior to recording, you may end up with unnecessary actions that will not only slow it down but also require more editing later to make it work as intended. Although it's easier to edit a macro than it is to erase unwanted passages from a voice recording, performing only the actions you want to record will save you editing time and trouble later.

Suppose you are asked to programmatically create the worksheet depicted in Figure 1.8. No worries. Getting started is very easy with the macro recorder. Let's begin by identifying the tasks required to complete this worksheet.

Task 1	Insert a new sheet into a workbook and call it Employee Wages.
Task 2	Enter column headings into the first row of the worksheet and apply the required formatting (column size, font styles, and so on).
Task 3	Enter employees' data (Full Name, Hourly Rate, and Hours Worked).
Tasks 4 and 5	Enter formulas to fill in the employees' First and Last Name columns.
Task 6	Enter formulas to calculate employees' total wages.
Task 7	Apply formatting to the completed worksheet.

Instead of recording one macro to complete your assignment, you will create a separate macro for each task. This approach will give you a chance to learn how to combine code from several simpler macros and how to create a master macro. Let's get started.

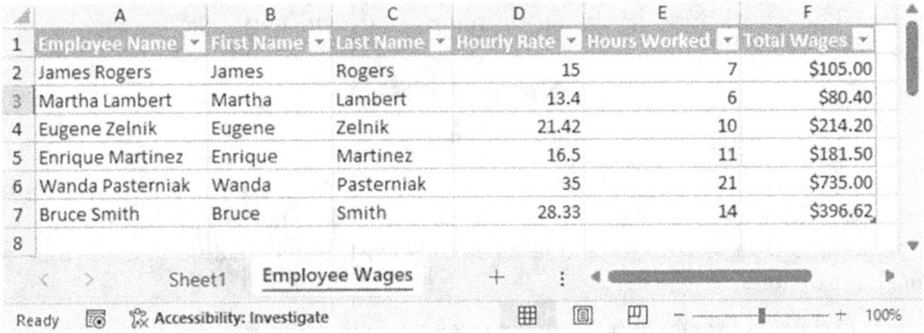

	A	B	C	D	E	F
1	Employee Name	First Name	Last Name	Hourly Rate	Hours Worked	Total Wages
2	James Rogers	James	Rogers	15	7	$105.00
3	Martha Lambert	Martha	Lambert	13.4	6	$80.40
4	Eugene Zelnik	Eugene	Zelnik	21.42	10	$214.20
5	Enrique Martinez	Enrique	Martinez	16.5	11	$181.50
6	Wanda Pasterniak	Wanda	Pasterniak	35	21	$735.00
7	Bruce Smith	Bruce	Smith	28.33	14	$396.62
8						

Sheet1   Employee Wages   +

Ready   Accessibility: Investigate   100%

FIGURE 1.8.   This sample worksheet will be created and formatted as shown with the help of the Excel built-in macro recorder.

⊙ Hands-On 1.2   Getting Ready for Macro Recording

1. Open a new workbook and save it as `Chap01.xlsm` in your trusted `VBAExcel2024_ByExample` folder.
Remember that you must save the file in the macro-enabled file format (`.xlsm`) to allow for storing macros. Keep this file open as you will use it to record all the macros in this chapter.

Recording a Macro

Before you record a macro, you need to decide whether you want to record the positioning of the active cell. If you want the macro to always start in a specific location on the worksheet, turn on the macro recorder first and then select the cell you want to start in. If the location of the active cell does not matter, select a single cell first and then turn on the macro recorder.

⊙ Hands-On 1.3   Inserting and Naming a Worksheet (Macro Task 1)

1. Choose Developer | Record Macro.
2. In the Record Macro dialog box, enter Insert_NewSheet for the macro name, as shown in Figure 1.9. Do not dismiss this dialog box until you are instructed to do so.

About Macro Names

If you forget to enter a name for the macro, Excel assigns a default name, such as *Macro1*, *Macro2*, and so on. Macro names can contain letters, numbers, and the underscore character, but the first character must be a letter. For example, *Report1* is a correct macro name, while *1Report* is not. Spaces are not allowed. If you want a space between the words, use the underscore.

---

3. Select This Workbook in the Store macro in dropdown.

Storing Macros

Excel allows you to store macros in three locations:

❑ Personal Macro Workbook—Macros stored in this location will be available each time you work with Excel. You can find the personal macro workbook in the XLStart folder. If the Personal Macro Workbook doesn't already exist, Excel creates it the first time you select this option.

❑ New Workbook—Excel will place the macro in a new workbook.

❑ This Workbook—The macro will be stored in the workbook you are currently using.

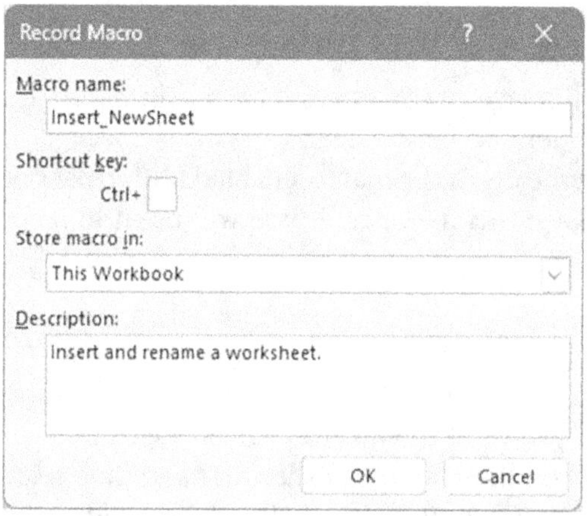

Record Macro   ?   ✕

Macro name:

Insert_NewSheet

Shortcut key:

Ctrl+ [ ]

Store macro in:

This Workbook   ⌄

Description:

Insert and rename a worksheet.

OK   Cancel

FIGURE 1.9.   When you record a new macro, you must provide a name for your macro. In the Record Macro dialog box, you can also supply a shortcut key, a storage location, and a description for your macro.

4. In the Description box, enter the following text: Insert and rename a worksheet.
5. Choose OK to close the Record Macro dialog box.

   The Stop Recording button shown in Figure 1.10 appears in the status bar, meaning the workbook is now in recording mode. Do not click this button until you are instructed to do so.

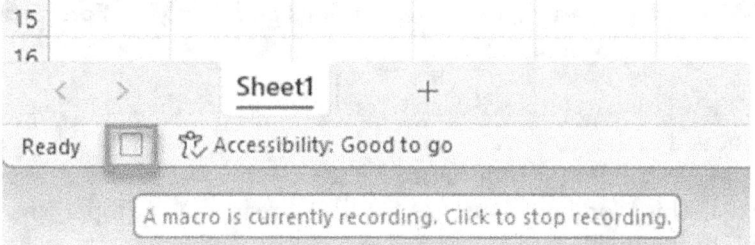

FIGURE 1.10.   The Stop Recording button depicted as a small square image in the Excel status bar indicates that the macro recording mode is active.

The Stop Recording button remains in the status bar while you record your macro. Only the actions finalized by pressing Enter or clicking OK are recorded. If you press the Esc key or click Cancel before completing the entry, the macro recorder does not record that action.

6. Add a new sheet to the current workbook. You can do this by either right-clicking the Sheet1 tab and choosing Insert | Worksheet | OK or simply clicking the plus button to the right of the Sheet1 tab.
7. Double-click the inserted sheet tab and rename it Employee Wages. Press Enter to record this change.
8. Click the Stop Recording button in the status bar or choose View | Macros | Stop Recording.

   When you stop the macro recorder, the status bar displays a button that allows you to record another macro (see Figure 1.11).

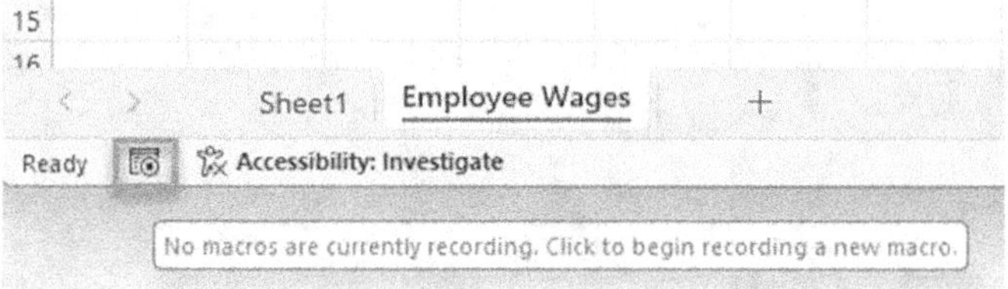

15				
16				
<	>	Sheet1	Employee Wages	+
Ready	🔲	🏃 Accessibility: Investigate		

No macros are currently recording. Click to begin recording a new macro.

FIGURE 1.11.   The Excel status bar with the macro recording button turned off.

You have now recorded your first macro. Excel has written all the necessary statements to execute the actions you performed. Let's continue recording all the remaining actions to complete the tasks that we defined earlier. After that, you will have a chance to review the recorded macro code and try out your macros.

(⊙)   **Hands-On 1.4   Inserting Column Headings and Applying Formatting (Macro Task 2)**

1. Choose View | Macros | Record Macro (or you may click the Begin Recording button located in the Excel status bar).
2. Enter Insert_Headings for your macro name.
3. Ensure that This Workbook is selected in the Store macro in dropdown.
4. Click OK to exit the Record Macro dialog.
   Excel turns on the macro recorder, as shown earlier in Figure 1.10. All your Excel actions from now on will be recorded.
5. Select cell A1 and enter the first heading: Employee Name.
6. Press the Tab key to move to cell B1 and enter First Name.
7. Enter the remaining headings in cells C1:F1 (Last Name, Hourly Rate, Hours Worked, and Total Wages).
8. Select cell range A1:F1 and apply the bold formatting to the selection by pressing the B button in the Font group of the ribbon's Home tab.
9. With the range A1:F1 still selected, choose Home | Cells | Format | Autofit Column Width.
10. Click the Stop Recording button in the status bar, as shown in Figure 1.10, or choose View | Macros | Stop Recording.

You have just recorded your second macro. The Employee Wages worksheet should now have the required headings in row 1 formatted in bold font style and auto-sized.

Using Relative or Absolute References in Macros

The Excel macro recorder can record your actions using absolute or relative cell references (see Figure 1.12):

☐ To have your macro execute the recorded action in a specific cell, no matter what cell is selected during the execution of the macro, use absolute cell addressing. Absolute cell references have the following form: $A$1, $C$5, etc. By default, the Excel macro recorder uses absolute references. Before you begin to record a new macro, make sure the Use Relative References option is not selected when you click the Macros button, as shown in Figure 1.12.

☐ To have your macro perform the action in any cell, be sure to select the Use Relative References option before you choose the Record Macro option. Relative cell references have the following form: A1, C5, etc. The Excel macro recorder will continue to use relative cell references until you exit Microsoft Excel or click the Use Relative References option again.

☐ During the process of recording your macro, you may use both methods of cell addressing. For example, you may select a specific cell (e.g., $A$4), perform an action, and then choose another cell relative to the selected cell (e.g., C9, which is located five rows down and two columns to the right of the currently active cell, $A$4). Relative references automatically adjust when you copy them, and absolute references don't.

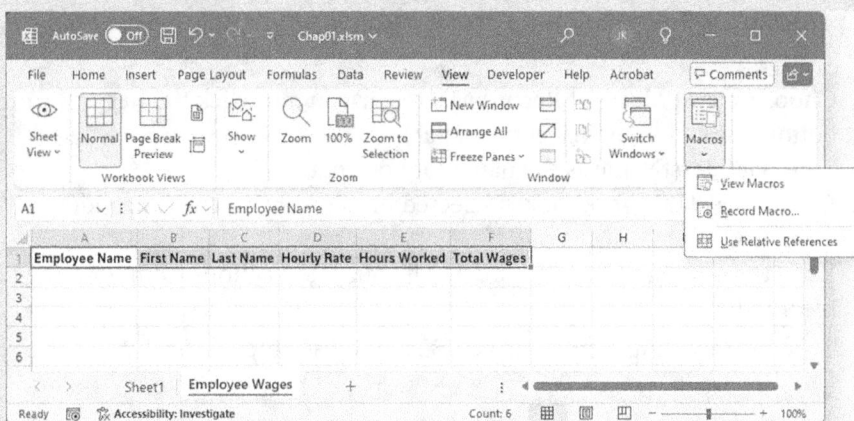

FIGURE 1.12.   The Excel macro recorder can record your actions using absolute or relative cell references. To choose between relative or absolute referencing in your macros, use the Macros dropdown on the ribbon's View tab.

⊙ Hands-On 1.5   Entering Employee Data (Macro Task 3)

1. Choose View | Macros | Record Macro (or you may click the Begin Recording button located in the Excel status bar).
2. Enter Insert_EmployeeData as the name for your macro.
3. Ensure that This Workbook is selected in the Store macro in dropdown.
4. Click OK.

   Excel will turn on the macro recorder. All your Excel actions from now on will be recorded.

   Enter employee data in columns A, D, and E, as shown in Figure 1.8. Leave the First Name, Last Name, and Total Wages columns blank as they will be filled in later. If you make any typos while entering data, don't be afraid to correct them.
5. Click the Stop Recording button in the status bar, as shown in Figure 1.10, or choose View | Macros | Stop Recording.

You have just recorded your third macro. The static data entry has been completed. We will now proceed to record macros that use formulas to fill the remaining columns of the worksheet.

⊙ Hands-On 1.6   Entering Formulas to Fill in Employee First Name (Macro Task 4)

1. Choose View | Macros | Record Macro (or you may click the Begin Recording button, located in the Excel status bar).
2. Enter Get_FirstName as the name for your macro.
3. Ensure that This Workbook is selected in the Store macro in dropdown.
4. Click OK.

   Excel turns on the macro recorder. All your Excel actions from now on will be recorded.
5. Type the following formula in cell B2 and press Enter:

   ```
   =LEFT(A2,FIND(" ", A2)-1)
   ```

6. Copy the formula down to cells B3:B7 by dragging the selection handle in the bottom-right corner of cell B2.

   Excel will fill in the first names of all employees.
7. Click the Stop Recording button in the status bar, as shown in Figure 1.11, or choose View | Macros | Stop Recording.

You have just recorded a macro that makes use of a formula to retrieve employee first names from their full name. The next macro will populate the last name column using another formula.

⊙ **Hands-On 1.7 Entering Formulas to Fill in Employee Last Name (Macro Task 5)**

1. Choose View | Macros | Record Macro (or you may click the Begin Recording button located in the Excel status bar).
2. Enter Get_LastName as the name for your macro.
3. Ensure that This Workbook is selected in the Store macro in dropdown.
4. Click OK.
   Excel turns on the macro recorder. All your Excel actions from now on will be recorded.
5. Enter the following formula in cell C2:

   ```
   =RIGHT(A2,LEN(A2)-FIND(" ", A2))
   ```

6. Copy the formula down to cells C3:C7 by dragging the selection handle in the bottom-right corner of cell C2.
   Excel will fill in the last names of all employees.
7. Click the Stop Recording button in the status bar as shown in Figure 1.10 or choose View | Macros | Stop Recording.

You have just recorded a macro that makes use of a formula to retrieve employees' last names from their full names. We have one more column to fill in before we can apply the final formatting to this worksheet.

⊙ **Hands-On 1.8 Entering Formulas to Calculate Employee Total Wages (Macro Task 6)**

1. Choose View | Macros | Record Macro (or you may click the Begin Recording button located in the status bar).
2. Enter CalculateWages as the name for your macro.
3. Ensure that This Workbook is selected in the Store macro in dropdown.
4. Click OK.
   Excel will turn on the macro recorder. All your Excel actions from now on will be recorded.
5. Select cells F2:F7 and type the formula shown below. Press Ctrl+Enter to ensure that the formula is entered into the selected range, F2:F7.

   ```
   =D2*E2
   ```

6. Apply the Currency format to cells F2:F7 by choosing Home | Number | Currency.
7. Click the Stop Recording button in the Excel status bar or choose View | Macros | Stop Recording.

In the next macro, you will complete the worksheet by applying the desired formatting.

(◉) Hands-On 1.9   Applying Table Format (Macro Task 7)

1. Choose View | Macros | Record Macro (or you may click the Begin Recording button located in the Excel status bar).
2. Enter FormatTable as the name for your macro.
3. Ensure that This Workbook is selected in the **Store macro in** dropdown.
4. Click OK.
   Excel will turn on the macro recorder. All your Excel actions from now on will be recorded.
5. Select all data in the Employee Wages worksheet (range A1:F7) and choose Home | Format as a Table. Select any of the predefined table styles.
   Excel will display the Create Table dialog asking you for the location of the data for your table. Because we have already selected the data we want, including the heading row, Excel should pick up the correct range, $A$1:$F$7.
6. Click OK to close the Create Table dialog box.
7. Select cell A1.
8. Click the Stop Recording button or choose View | Macros | Stop Recording.

You have now completed recording a set of macros that create and format a worksheet. Let's find and examine the macro code that Excel has written for us.

Locating the Macro Code

Before you can modify your macro, you must find the location where the macro recorder placed its code. As you recall, when you turned on the macro recorder, you selected This Workbook for the location. To find the location of your macros, you will use the Macro dialog box, as instructed in Hands-On 1.10.

(◉) Hands-On 1.10   Examining the Macro Code

1. Choose View | Macros | View Macros.
   You should see all seven macros you recorded earlier (see Figure 1.13).

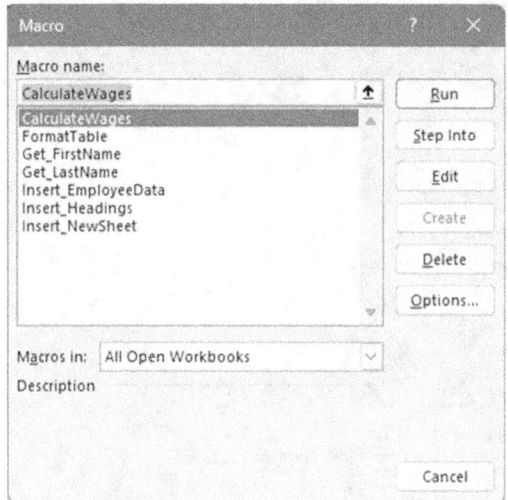

FIGURE 1.13.   In the Macro dialog box, you can select a macro to run, debug (Step Into), edit, or delete. You can also set options for your macro.

2. Select the Insert_NewSheet macro name and click the Edit button.
   Excel will open a special window called Visual Basic Editor (VBE), as shown in Figure 1.14. This window is your VBA programming environment. This is where you will find the code of your recorded macros and where you can write your VBA programming code from scratch.

> ### Switching Between the Excel Application Window and the VBE Window
>
> By using the keyboard shortcut Alt+F11, you can quickly switch between the Microsoft Excel application window and the VBE window. Take a moment and try switching between both windows. When you are done, ensure that you are back in the VBE window.

3. Close the VBE window by using the key combination Alt+Q or choosing File | Close and Return to Microsoft Excel.
   Don't worry if the VBE window seems a bit confusing right now. As you work with the recorded macros and start writing your own VBA procedures from scratch, you will become familiar with all the elements of the VBE screen.

4. In the Microsoft Excel application window, choose Developer | Visual Basic to switch again to the programming environment.

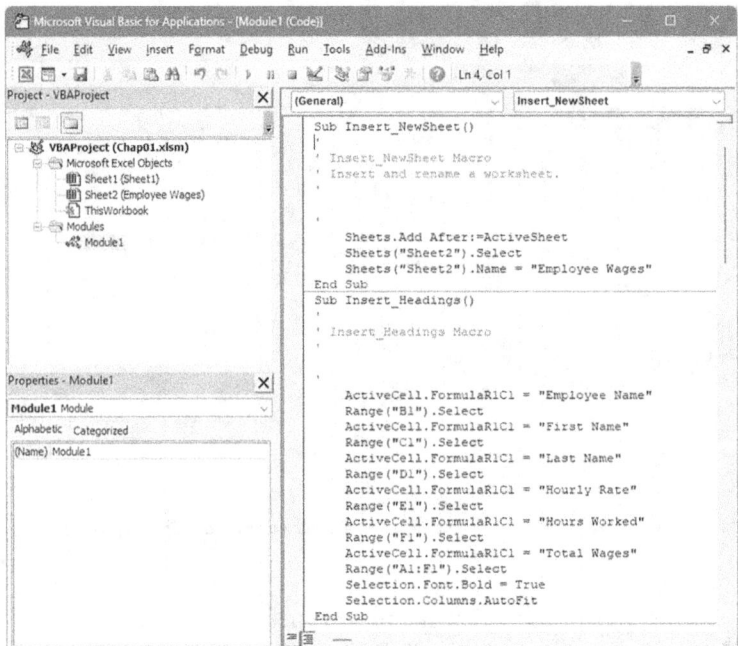

FIGURE 1.14. The VBE window is used for editing macros as well as writing your own programming code in the VBA language.

A Brief Introduction to the VBE Window

If you examine Figure 1.14, you will notice that the menu bar and toolbar in the VBE window look different from those in the Microsoft Excel window. As you can see, there is no ribbon interface. VBE uses the old Excel-style menu bar and toolbar, which provide tools required for programming and testing your recorded macros as well as your own VBA code. As you work through the individual chapters of this book, you will feel very comfortable using these tools.

The main part of the VBE window is a docking surface for various windows, which you will find extremely useful while creating and testing your VBA procedures. In Figure 1.14, you can see three windows: the Project Explorer window, the Properties window, and the Code window.

The Project Explorer window that appears in the left panel shows a Microsoft Excel Objects folder with the current workbook sheets and an open Modules folder. Excel records your macro actions in special worksheets called Module1, Module2, and so on. These modules are stored in the Modules folder. Later in this book, you will also use modules to write your own programming code from scratch. A module resembles a blank document in Microsoft Word. In Figure 1.14, it shows the VBA code that was generated by the Excel macro recorder.

The Properties window displays the properties of the object that is currently selected in the Project Explorer window. In Figure 1.14, the Module1 object is selected in the Project - VBAProject window, and therefore, the Properties - Module1 window displays the properties of Module1. Notice that the only available property for the module is the Name property. You can use this property to change the name of Module1 to a more meaningful name.

A Macro or a Procedure?

A macro is a series of commands or functions recorded with the help of a built-in macro recorder or entered manually in a Visual Basic module. The term "macro" is often replaced with the broader term "procedure." Although the words can be used interchangeably, many programmers prefer "procedure." While macros allow you to mimic keyboard actions, true procedures can also execute actions that cannot be performed using the mouse, keyboard, or menu options. In other words, procedures are more complex macros that incorporate language structures found in traditional programming languages. You will learn about these structures later in this book.

REVIEWING MACRO CODE

The Module1 (Code) window (see Figure 1.14) displays the code of all macros you recorded earlier. Note that the following code may not exactly match the code in your Code window. Excel records all the actions while the recorder is on, so you may see more or fewer statements, recorded depending on your own actions.

```
Sub Insert_NewSheet()
'
' Insert_NewSheet Macro
' Insert and rename a worksheet
'


'
    Sheets.Add After:=ActiveSheet
    Sheets("Sheet2").Select
    Sheets("Sheet2").Name = "Employee Wages"
End Sub

Sub Insert_Headings()
'
```

```vba
' Insert_Headings Macro
'

'
    Range("A1").Select
    ActiveCell.FormulaR1C1 = "Employee Name"
    Range("B1").Select
    ActiveCell.FormulaR1C1 = "First Name"
    Range("C1").Select
    ActiveCell.FormulaR1C1 = "Last Name"
    Range("D1").Select
    ActiveCell.FormulaR1C1 = "Hourly Rate"
    Range("E1").Select
    ActiveCell.FormulaR1C1 = "Hours Worked"
    Range("F1").Select
    ActiveCell.FormulaR1C1 = "Total Wages"
    Range("A1:F1").Select
    Selection.Font.Bold = True
    Selection.Columns.AutoFit
End Sub

Sub Insert_EmployeeData()
'
' Insert_EmployeeData Macro
'

'
    Range("A2").Select
    ActiveCell.FormulaR1C1 = "James Rogers"
    Range("D2").Select
    ActiveCell.FormulaR1C1 = "15"
    Range("E2").Select
    ActiveCell.FormulaR1C1 = "7"
    Range("A3").Select
    ActiveCell.FormulaR1C1 = "Martha Lambert"
    Range("D3").Select
    ActiveCell.FormulaR1C1 = "13.4"
    Range("E3").Select
    ActiveCell.FormulaR1C1 = "6"
    Range("A4").Select
    ActiveCell.FormulaR1C1 = "Eugene Zelnik"
    Range("D4").Select
    ActiveCell.FormulaR1C1 = "21.42"
    Range("E4").Select
    ActiveCell.FormulaR1C1 = "10"
```

```
    Range("A5").Select
    ActiveCell.FormulaR1C1 = "Enrique Martinez"
    Range("D5").Select
    ActiveCell.FormulaR1C1 = "16.5"
    Range("E5").Select
    ActiveCell.FormulaR1C1 = "11"
    Range("A6").Select
    ActiveCell.FormulaR1C1 = "Wanda Pasterniak"
    Range("D6").Select
    ActiveCell.FormulaR1C1 = "35"
    Range("E6").Select
    ActiveCell.FormulaR1C1 = "21"
    Range("A7").Select
    ActiveCell.FormulaR1C1 = "Bruce Smith"
    Range("D7").Select
    ActiveCell.FormulaR1C1 = "28.33"
    Range("E7").Select
    ActiveCell.FormulaR1C1 = "14"
    Range("A8").Select
End Sub

Sub Get_FirstName()
'
' Get_FirstName Macro
'

'
    Range("B2").Select
    ActiveCell.FormulaR1C1 = "=LEFT(RC[-1],
        FIND("" "", RC[-1])-1)"
    Range("B2").Select
    Selection.AutoFill Destination:=Range("B2:B7"),
Type:=xlFillDefault
    Range("B2:B7").Select
End Sub
Sub Get_LastName()
'
' Get_LastName Macro
'

'
    Range("C2").Select
    ActiveCell.FormulaR1C1 = "=RIGHT(RC[-2],
        LEN(RC[-2])-FIND("" "", RC[-2]))"
    Range("C2").Select
```

```
        Selection.AutoFill Destination:=Range("C2:C7"),
            Type:=xlFillDefault
        Range("C2:C7").Select
End Sub
Sub CalculateWages()
'
' CalculateWages Macro
'

'
        Range("F2:F7").Select
        Selection.FormulaR1C1 = "=RC[-2]*RC[-1]"
        Selection.NumberFormat = "$#,##0.00"
End Sub
Sub FormatTable()
'
' FormatTable Macro
'

'
        Range("A1:F7").Select
        Application.CutCopyMode = False
        Application.CutCopyMode = False
        ActiveSheet.ListObjects.Add(xlSrcRange,
            Range("$A$1:$F$7"), , xlYes).Name = "Table1"
        Range("Table1[#All]").Select
        ActiveSheet.ListObjects("Table1").TableStyle =
            "TableStyleMedium4"
        Range("Table1[[#Headers],[Employee Name]]").Select
End Sub
```

For now, let's focus on finding answers to two questions:

- How do you read the macro code?
- How can you edit macros?

Notice that each macro code you recorded is located between Sub and End Sub. These words are known as keywords. You read the code line by line from top to bottom. You can edit the recorded macros by deleting or modifying existing code or typing new instructions in the Code window.

Macro Comments

Look again at the recorded macro code. The lines that begin with a single quote denote comments. By default, comments appear in green. When the macro code is executed, Visual Basic ignores the comment lines. Comments are often placed within the macro code to document the meaning of certain lines that aren't obvious. Comments can also be used to temporarily disable certain blocks of code that you don't want to execute. This is often done while testing and troubleshooting your macros.

Let's add a comment to the CalculateWages macro to make the code easier to understand.

⊙ Hands-On 1.11   Adding Comments to the Macro Code

1. Make sure that the VBE screen shows the Code window with the `CalculateWages` macro.
2. Click after `Range("F2:F7").Select` and press Enter.
3. In the empty line you just created, type the following comment. Be sure to start with a single quote:

   ```
   ' Apply Currency Format
   ```

4. Press Ctrl+S to save the changes in `Chap01.xlsm` or choose File | Save Chap01. xlsm.

Macro Statements

All macro procedures begin with the keyword `Sub` and end with the keywords `End Sub`. The `Sub` keyword is followed by the macro name and a set of parentheses. Between the keywords `Sub` and `End Sub` are statements that Visual Basic executes each time you run your macro. Visual Basic reads the lines from top to bottom, ignoring the statements preceded with a single quote (see the information about comments), and stops when it reaches the keywords `End Sub`. Notice that the recorded macro contains many periods. The periods appear in almost every line of code and are used to join various elements of the VBA language. How do you read the instructions written in this language? They are read from the right side of the last period to the left. Here are a few statements from the `Insert_Headings` macro and a description of what they mean:

Code Segment	Description
`Range("B1").Select`	Select cell B1.
`ActiveCell.FormulaR1C1 = "Employee Name"`	Let the formula of the active cell be "Employee Name."
`Range("A1:F1").Select`	Select cells A1 to F1.
`Selection.Font.Bold = True`	Applies bold format to all selected cells.

`Selection.Columns.AutoFit`	Extend the selection width so that all entries fit in all selected columns.

Cleaning Up the Macro Code

As you review and analyze your macro code line by line, you may notice that Excel recorded other information that you didn't intend to include. For example, if you used the Font dialog box to apply bold formatting to the heading cells in your `Insert_Headings` macro, in addition to setting the font style to bold, Excel also recorded the current state of other options on the Font tab—strikethrough, superscript, subscript, outline font, shadow, underline, theme color, tint and shade, and theme font, as shown in the following:

```
With Selection.Font
     .Name = "Calibri Light"
     .FontStyle = "Bold"
     .Size = 9
     .Strikethrough = False
     .Superscript = False
     .Subscript = False
     .OutlineFont = False
     .Shadow = False
     .Underline = xlUnderlineStyleNone
     .ThemeColor = xlThemeColorLight1
     .TintAndShade = 0
     .ThemeFont = xlThemeFontMajor
End With
```

When you use dialog boxes, Excel always records all the settings. These additional instructions make your macro code longer and more difficult to understand. Therefore, when you finish recording your macro, it is a good idea to go over the recorded statements and delete the unnecessary lines of code. In the previous code snippet, different font settings are applied to the selection of cells. This is done with the special block of code that begins with the keyword `With` and ends with the keywords `End With`. Assume you just wanted to change the font name, style, and size of the selected cells. In this case, you can simply delete all the other settings that were recorded, and you will be left with the following code:

```
With Selection.Font
     .Name = "Calibri Light"
     .FontStyle = "Bold"
     .Size = 9
End With
```

This change makes your macro code easier to understand as only the settings you selected are shown. Notice that each setting in the `With...End With` block begins with a period. If you wanted to list each setting separately, you would write them as:

```
Selection.Font.Name = "Calibri Light"
Selection.Font.FontStyle = "Bold"
Selection.Font.Size = 9
```

By using the `With...End With` block, however, you simply write the repeating code once, which saves you time and speeds up your macros. Simply move the repeating code `Selection.Font` to the right of the `With` keyword and end the entire block with `End With`.

As you work more with the Excel macro recorder and learn more about VBA statements, you will be able to make your recorded macros much cleaner.

EXECUTING YOUR MACRO

You can run your macros from either the Microsoft Excel window or the VBE window. When you execute a macro from the VBE screen, Visual Basic executes the macro behind the scenes. You can't see when Visual Basic performed a specific action. To watch Visual Basic at work, you must run your macro from the Macro dialog box or arrange your screen in such a way that the Microsoft Excel and VBA windows can be viewed at the same time. A split screen or two monitors attached to your computer will help you greatly in the development work when you need to observe actions performed by the programming code.

After you create a macro, you should run it at least once to make sure it works correctly. Later in this chapter, you will learn other ways to run macros, but for now, let's use the Macro dialog box.

⊙ **Hands-On 1.12   Running a Macro Using the Macro Dialog Box**

1. Make sure that the `Chap01.xlsm` workbook is open.
2. Delete the Employee Wages worksheet so you can start from scratch.
3. Choose View | Macros | View Macros.
4. In the Macro dialog box, click the Insert_NewSheet macro name.
5. Click Run to execute the macro.

   The `Insert_NewSheet` macro inserts a blank worksheet and, all of a sudden, the VBE window appears with the `Sheets("Sheet2").Select` line of code highlighted in yellow (see Figure 1.15).

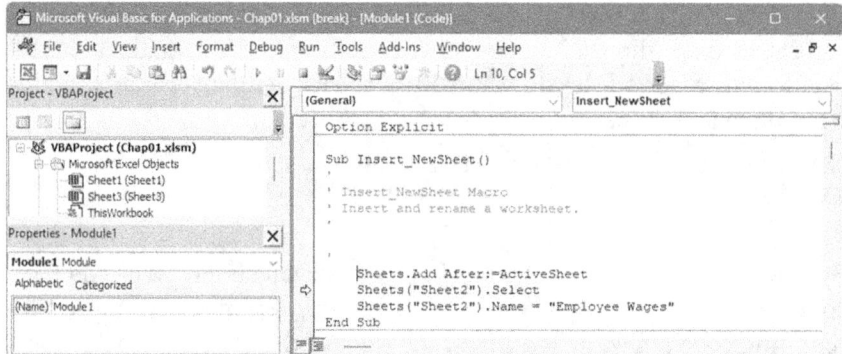

FIGURE 1.15.　When Visual Basic encounters an error while running your macro, it stops the execution, enters break mode, and highlights the line of code that triggered the error.

By highlighting the problematic line, Visual Basic helps you quickly identify the source of the issue, allowing you to diagnose and fix the error. In this example, Visual Basic cannot find Sheet2 in the workbook, as we deleted it in Step 2 of this hands-on exercise. With Sheet2 missing, Visual Basic cannot proceed, causing the entire execution process to halt until the problem is resolved. The VBE Module code window enters break mode, as indicated in the title bar of the VBE application window. While in break mode, you can make code adjustments that will allow you to continue. Let's fix this issue in the next step.

6. Delete the highlighted line of code and change the last line of code before the `End Sub` keywords to `ActiveSheet.Name = "Employee Wages"`.

7. Drag the yellow arrow to the updated statement you just typed so that the entire line is highlighted, as shown in Figure 1.16, and press F5 to continue the execution of the `Insert_NewSheet` macro.

```
(General)                              Insert_NewSheet

    Option Explicit

    Sub Insert_NewSheet()
    '
    ' Insert_NewSheet Macro
    ' Insert and rename a worksheet.
    '

    '
        Sheets.Add After:=ActiveSheet
        ActiveSheet.Name = "Employee Wages"
    End Sub
```

FIGURE 1.16.　In break mode, you can resolve the issue that triggered the error by deleting the problematic line of code and entering the corrected code. Afterward, move the highlight to the line where the code execution should resume.

> ### How to Move the Highlight in Break Mode
>
> After correcting the code that caused the error, you need to set the next statement.
>
> To do this, click on the line of code where you want execution to resume. Then, right-click and select Set Next Statement from the context menu.
>
> Alternatively, you can use the keyboard shortcut. Place the cursor on the desired line and press Ctrl+F9.
>
> This will move the yellow highlight to the line you've selected, indicating that the code execution will resume from that point when you continue running the macro.

Now, you should have the Employee Wages worksheet, and you can proceed to run other macros.

8. Choose View | Macros | View Macros.
9. In the Macro dialog box, click the Insert_Headings macro name.
10. Click Run to execute the macro.
11. Run the remaining macros in the order they were created: `Insert_EmployeeData`, `Get_FirstName`, `Get_LastName`, `CalculateWages`, and `FormatTable`.

After running all the macros, you should see the completed and formatted Employee Wages worksheet.

Testing and Debugging a Macro

As you've seen in the previous section, quite often, you will notice that your macro does not perform as expected the first time you run it. Perhaps, during the macro recording, you selected the wrong font or forgot to change the cell color, or maybe you just realized it would be better to include an additional step. Don't panic. Excel makes it possible to modify the macro without forcing you to go through the tedious process of recording your keystrokes again.

When you modify a recorded macro, it is quite possible that you will introduce some errors. For example, you may delete an important line of code, or you may inadvertently remove or omit a necessary period. To make sure that your macro continues to work correctly after your modifications, you need to run it again.

You can run a macro directly from the VBE screen by clicking anywhere within the macro code. Then, click the Run button, which is located in the toolbar at the top of the VBE window and is represented by a green triangle icon. Alternatively, you can press the F5 key on your keyboard.

Various errors can pop up during macro execution. For example, if your macro code references a non-existent worksheet or workbook, Visual Basic will generate the `Run time error '9': Subscript out of range` message (see Figure 1.17) when running this line of code: `Sheets("Sheet2").Select`.

Instead of pressing the End button to end your macro, you can click the Debug button in the message box so you can correct your macro code right away. When you press Debug, you will be placed in the Code window's break mode, as you've seen in the previous section. The yellow highlighter will indicate the problematic line of code.

Although you can edit code in break mode, some edits may prevent the code from continuing execution. VBA may need to reset your project before allowing you to proceed. This reset occurs because certain changes to the code can affect the overall structure or flow of the project, making it necessary to restart the execution environment. When this happens, the project will be reset, and you'll need to run the macro again from the beginning to test the changes. This ensures that the updates are correctly integrated, and that the macro operates as expected.

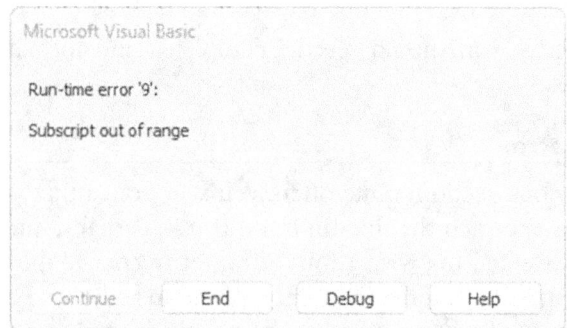

Microsoft Visual Basic

Run-time error '9':

Subscript out of range

| Continue | End | Debug | Help |

FIGURE 1.17.   VBE displays an error message when it encounters an error while executing a macro. The error may be caused by an incorrect statement or incorrect setup of the worksheet environment prior to running a macro.

Saving and Renaming a Macro

The macros you recorded in this chapter are stored in a Microsoft Excel workbook. All macros will be automatically saved when you save the workbook.

⊙ **Hands-On 1.13   Saving Macros and Running Macros from Another Workbook**

1. Save and close the `Chap01.xlsm` workbook.

2. Open a brand-new workbook and press Alt+F8 to open the Macro dialog box. Notice that there is no trace of your macros in the Macro dialog box. If you'd like to run the macros you recorded earlier in this chapter in another workbook, you need to open the file that stores these macros.

3. Close the Macro dialog box and save the open workbook file as `Chap01.xlsx` in your trusted `C:\VBAExcel2024_ByExample` folder. You will not have any macros in this workbook, so saving it in Excel's default file format will work just fine.

4. Open the `C:\VBAExcel2024_ByExample\Chap01.xlsm` workbook file.

5. Activate Sheet1 in the `Chap01.xlsx` workbook.

6. Press Alt+F8 to activate the Macro dialog box.
   Notice that the Macro dialog now displays macros in all open workbooks. Each macro name is preceded by the name of the workbook where it is located.

7. Run each of the macros listed in this dialog box in the order you have recorded them.
   Your macros will go to work again. You should end up with the Employee Wages worksheet formatted to your liking.

8. Close the `Chap01.xlsx` workbook file. Do not save the changes. Do not close the `Chap01.xlsm` workbook file. We will work with it in the next section.

Changing the Macro Name

When you add additional actions to your macro, you may want to change the macro name to better indicate its purpose. The name of the macro should communicate its function as clearly as possible. To change the macro name, you don't need to press a specific key. In the Code window, simply delete the old macro name and enter the new name following the Sub keyword.

Printing Macro Code

If you want to document your macros or perhaps study the macro code when you are away from the computer, you can print your macros. You can print the entire module sheet where your macro is stored or indicate a selection of lines to print. Let's print the entire module sheet that contains your macros.

⊙ Hands-On 1.14   Printing Macro Code

1. Switch to the VBE window and double-click Module1 in the Project Explorer window to activate the module containing your macros.

2. Choose File | Print.

3. In the Print - VBAProject dialog box, the Current Module option button should be selected.

4. Click OK to print the entire module sheet.

5. If you'd like to print only a certain block of programming code, perform the following steps:

   - In the module sheet, highlight the code you want to print.
   - Choose File | Print.
   - In the Print - VBAProject dialog box, the Selection option button should be selected.
   - Click OK to print the highlighted code.

IMPROVING YOUR RECORDED MACROS

After you record your macro, you may realize that you'd like the macro to perform additional tasks. Adding new instructions to the macro code is not very difficult if you are already familiar with the VBA language. In most situations, however, you can do this more efficiently when you delegate the extra tasks to the macro recorder. You may argue that Excel records more instructions than are necessary. One thing is for sure, however—the macro recorder does not make mistakes. If you want to add additional instructions to your macro using the macro recorder, you must record a new macro, copy the sections you want, and paste them into the correct location in your original macro.

At times, you may need to modify your macro code by removing some statements. Before you start deleting unnecessary lines of code, think of how you can use the comment feature that you've recently learned about. You can comment out the unwanted lines and run the macro with the commented code. If the VBE does not generate errors, you can safely delete the commented lines. By following this path, you will never find yourself recording the same keystrokes more than once. If the macro does not perform correctly, you can remove the comments from the lines that may be needed after all.

When you create macros with the Excel macro recorder, you can quickly learn the VBA equivalents for the Excel commands and dialog box settings. Then, you can look up the meaning and the usage of these commands in on-line help. It's obvious that the more instructions VBA needs to read, the slower your macro will execute. Eliminating extraneous commands will speed up your macro.

> ### Including Additional Instructions in Your Macro
>
> To include additional instructions in the existing macro, add empty lines in the required places of the macro code by pressing Enter and typing in the necessary Visual Basic statements. If the additional instructions are keyboard actions or menu commands, you may use the macro recorder to generate the necessary code and then copy and paste these code lines into the original macro.

Want to add more improvements to your macro? How about a message to notify you when Visual Basic has finished executing the last macro line? This sort of action cannot be recorded, as Excel does not have a corresponding ribbon command or shortcut menu option. By using the VBA language, however, you can add new instructions to your macro by hand. Let's see how this is done.

⊙ Hands-On 1.15   Adding Visual Basic Statements to the Recorded Macro Code

1. In the Code window containing the code of the `FormatTable` macro, click in front of the `End Sub` keywords and press Enter to add an empty line.
2. Click in the empty line and type the following statement:

```
MsgBox "Your worksheet is ready."
```

3. Press Ctrl+S to save the changes made in your macro code.

When you run this macro next time around, you should see a message box with your programmed message text. You must click the OK button in the message box to discard this message. `MsgBox` is one of the most frequently used built-in VBA functions. We will discuss it in more detail in a later chapter.

CREATING A MASTER MACRO

In this chapter, you've recorded several macros that required you execute them in the order they were recorded. Instead of running your macros one by one, it is more convenient to have one master macro that will perform all the required tasks automatically in the correct order. Let's see how this is done in the next hands-on exercise.

⊙ Hands-On 1.16   Creating a Master Macro Procedure

1. Switch to the Microsoft Visual Basic for Application window and click VBAProject (Chap01.xlsm) in the Project Explorer window.

2. Choose Insert | Module to add a new module to the selected VBAProject (Chap01.xlsm).
3. In the Properties window, select Module2 next to the (Name) property and rename it MasterProcedure.
4. In the Code window, enter the following procedure below the Option Explicit statement. This statement, which appears at the top of the VBA module before any other code, is used in VBA to enforce variable declarations before they are used in code. This topic will be covered in detail as we progress through this book.

```
Sub CreateEmployeeWorksheet()
    Insert_NewSheet
    Insert_Headings
    Insert_EmployeeData
    Get_FirstName
    Get_LastName
    CalculateWages
    FormatTable
End Sub
```

5. Press Ctrl+S to save the changes.
6. Choose File | Close and Return to Microsoft Excel.
7. In the Microsoft Excel window, choose File | New | Blank workbook.
8. Choose View | Macros | View Macros to display the Macro dialog box.
9. Select the CreateEmployeeWorksheet macro name and click Run. Excel will run your code and display a message box that you added in the previous hands-on exercise.
10. Click OK to dismiss the message box.
11. Close the Excel workbook you just created without saving it.

This hands-on exercise demonstrated how easy it is to combine standalone macros into a master macro. All you need to do is list the macro names on separate lines between the Sub and End Sub keywords and name your macro. You could also copy all the code of the recorded macros into a new macro; however, this will make the macro code more difficult to troubleshoot. It is much easier to understand and work with shorter macros. When referencing macro names in other macros, any misspelling of a macro name will cause a compile error, Sub or Function not defined, when you attempt to run your macro.

Later, you will learn about different types of errors and techniques that will allow you to test your macros using Excel built-in tools.

VARIOUS METHODS OF RUNNING MACROS

You have learned two methods of running macros. You can run a macro from the VBE screen or a Macro dialog box in the Microsoft Excel application window.

On the VBE screen, you can run your macro in one of the following ways:

- Press F5 on the keyboard.
- Choose Run | Run Sub/UserForm.
- Choose Tools | Macros
- Click the Run Sub/UserForm (F5) button on the Standard toolbar, as shown in Figure 1.18.

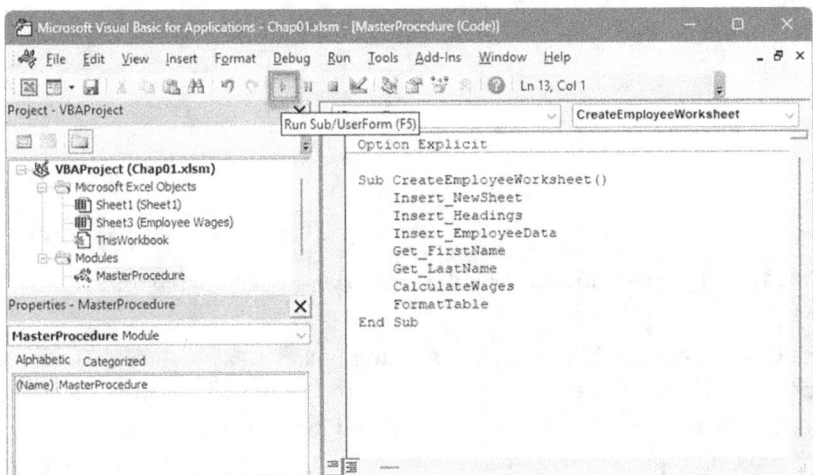

FIGURE 1.18.   The Visual Basic code can be run from the toolbar button on the VBE screen.

Now, let's learn about other methods of macro execution that will allow you to run your macros using a keyboard shortcut, toolbar button, or worksheet button. Let's get started.

Running the Macro Using a Keyboard Shortcut

A popular method to run a macro is by using an assigned keyboard shortcut. It is much faster to press Ctrl+Shift+I than it is to activate the macro from the Macro dialog box. Before you can use the keyboard shortcut, you must assign it to your macro.

⊙ **Hands-On 1.17   Assigning a Macro to a Keyboard Shortcut**

1. In the Excel application window, press Alt+F8 to open the Macro dialog box.
2. In the list of macros, click the CreateEmployeeWorksheet macro, and then choose the Options button.
3. When the Macro Options dialog box appears, the cursor is in the Shortcut key text box.
4. Hold down the Shift key and press the letter I on the keyboard. Excel records the keyboard combination as Ctrl+Shift+I. The result is shown in Figure 1.19.

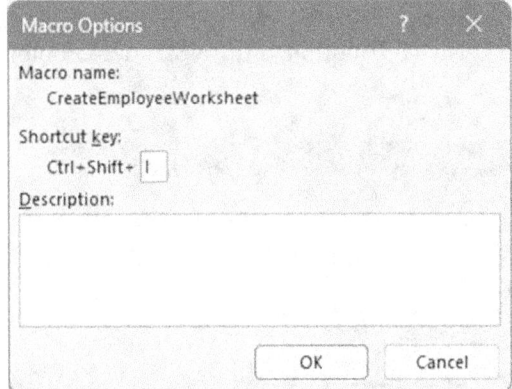

FIGURE 1.19.   Using the Macro Options dialog box, you can assign a keyboard shortcut for running a macro.

5. Click OK to close the Macro Options dialog box.
6. Click Cancel to close the Macro dialog box and return to the worksheet.
7. To run your macro using the newly assigned keyboard shortcut, open a new workbook and press Ctrl+Shift+I. Your macro will go to work and your worksheet will be ready to use.
8. Close the workbook with the employee worksheet you just created without saving it.

***

Avoid Shortcut Conflicts

If you assign a keyboard shortcut to your macro that conflicts with a Microsoft Excel built-in shortcut, Excel will run your macro if the workbook containing the macro code is currently open.

***

Running the Macro from the Quick Access Toolbar

You can add your own buttons to the built-in Quick Access toolbar.

⊙ Hands-On 1.18   Running a Macro from the Quick Access Toolbar

1. In the Microsoft Excel window, click the Customize Quick Access Toolbar button (the downward-pointing arrow in the title bar) and choose More Commands…, as shown in Figure 1.20.
   The Excel Options dialog box will appear with the page titled Customize the Quick Access Toolbar.

2. In the Choose commands from drop-down list box, select Macros.

3. Select CreateEmployeeWorksheet in the list box on the left-hand side.

4. Click the Add >> button to move the `CreateEmployeeWorksheet` macro to the list box on the right-hand side.
   The current selections are shown in Figure 1.21.

5. To change the button image for your macro, click the Modify button.

6. In the button gallery, select any button you like and click OK.

7. After closing the gallery window, make sure that the image to the left of the macro name has changed.

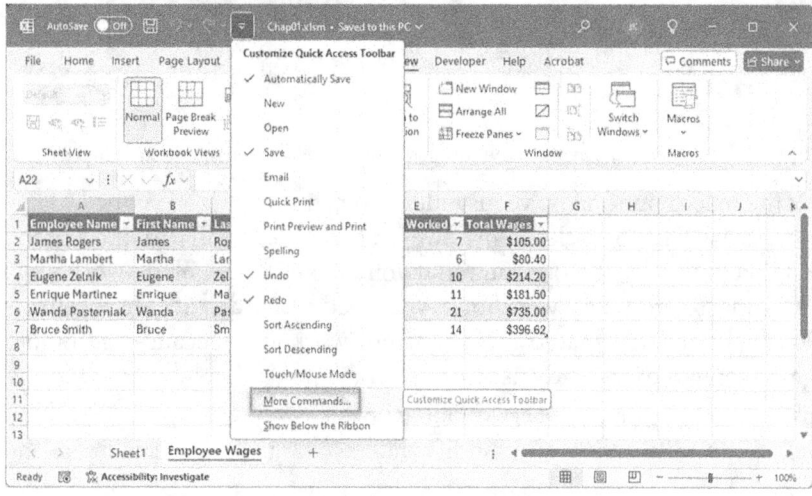

FIGURE 1.20.   Adding a new button to the Quick Access toolbar (step 1).

8. Click OK to close the Excel Options dialog.
   You should now see a new button on the Quick Access toolbar, as shown in Figure 1.22. This button will be available for any open workbook.

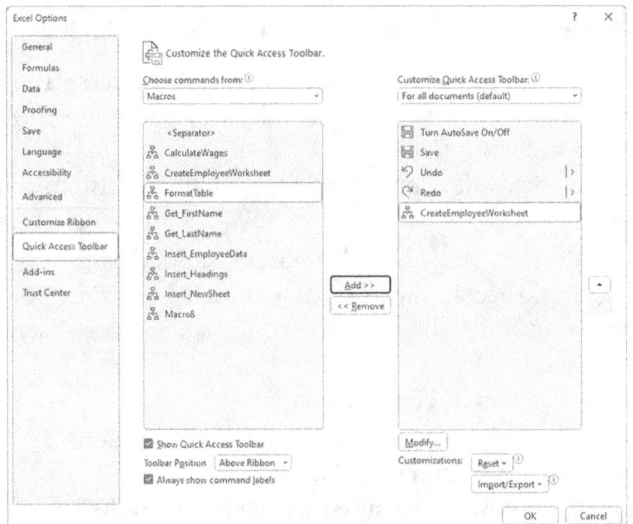

FIGURE 1.21.   Adding a new button to the Quick Access toolbar (step 2).

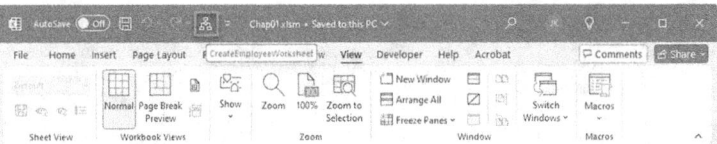

FIGURE 1.22.   A custom button placed on the Quick Access toolbar will run the specified macro.

9. Click the macro button you've just added to run the macro assigned to it. Your macro will go to work again; however, this time, it will run into a problem. Recall that previously before you ran it you opened a new blank workbook. To run this macro from any workbook, you need to modify it. Microsoft Visual Basic cannot run the macro in the current workbook because the Employee Wages sheet name is already used.

10. Click the End button in the error dialog box.

11. Switch to the VBE screen and modify the `Insert_NewSheet` macro, as shown in Figure 1.23.

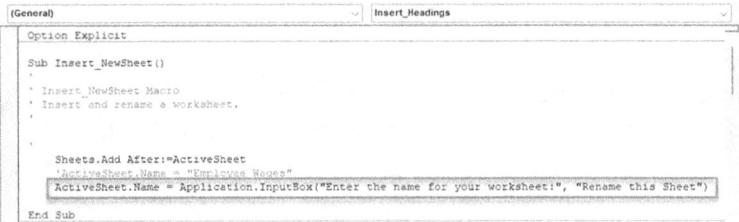

FIGURE 1.23.   The recorded Insert_NewSheet macro was modified to correct issues encountered during its execution.

To allow the user to name the sheet during the macro execution you can use the Excel built-in `InputBox` method:

```
ActiveSheet.Name = Application.InputBox("Enter the name for
your worksheet:", "Rename this Sheet")
```

Notice that we have commented out the previous line of code, so it does not trigger an error.

12. Save the workbook and return to the Microsoft Excel window.

13. Click the macro button on the Quick Access toolbar (see Figure 1.22).

Excel will add a new worksheet to the active workbook and prompt you for the name of the worksheet.

14. Enter any name for the newly created worksheet that does not conflict with an existing name and click OK.

NOTE	*If you click the Cancel button instead of typing in the name for the worksheet, Visual Basic will run into an issue and you will see* `Application-defined or object-defined run time error 1004`. *Click End to close the error message and you will be returned to the Microsoft Excel application window. Manually delete the empty sheet that was added to the workbook and execute the macro again—this time, enter the name for the sheet when prompted. You will learn how to handle the Cancel button later on when we get to this topic.* *After you supply the worksheet name, Visual Basic continues to execute the remaining macros in your master procedure. The execution fails again when the program reaches the* `FormatTable` *procedure. What's wrong with this macro code? It worked perfectly well when you recorded it. Oftentimes, issues with recorded code arise with the named ranges. A line of the* `FormatTable` *procedure assigns the name to the table range. Because you are running the master procedure inside the workbook where the specified table name already exists, Visual Basic throws the error* `Select method of range class failed`. *All table names within the workbook must be unique. For your code to run correctly, you must revise the* `FormatTable` *procedure.*

15. Click the Debug button in the error message dialog, and Visual Basic will highlight the line of code it cannot execute.

16. Exit break mode by choosing Run | Reset.

40 EXCEL 2024/MICROSOFT 365 PROGRAMMING BY EXAMPLE

17. Modify the `FormatTable` procedure as shown in Figure 1.24.

```
(General)                                    FormatTable
Sub FormatTable()
'
' FormatTable Revised Macro


'
Dim strTableName As String
strTableName = InputBox("Enter the name for your table:", "Name your table range")

    ActiveSheet.ListObjects.Add(xlSrcRange, Range("$A$1:$F$7"), , xlYes).Name = strTableName
    ActiveSheet.ListObjects(strTableName).TableStyle = "TableStyleMedium4"
    Range("A1").Select
    MsgBox "Your worksheet is ready."
End Sub
```

FIGURE 1.24. The recorded macro FormatTable was modified to correct issues encountered during its execution.

The first line of code in the revised procedure uses the VBA `Dim` statement to declare the `strTableName` variable to hold the name of the table supplied by the `InputBox` function on the next line. You will learn about variables and their types, declarations, and assignments in Chapter 3. The third line creates a new list object and assigns it a name stored in the `strTableName` variable. Every time you run the procedure and are prompted for a table name, you must enter a unique name.

After adding and assigning a name to the table object, the macro again refers to the `strTableName` variable to assign a predefined formatting style to the table. The procedure then selects cell A1 in the active worksheet and displays a message to the user.

18. After making changes to the `FormatTable` procedure, save your code and return to the Microsoft Excel application window.
19. Run the procedure again by clicking the button on the Quick Access toolbar. Provide the values for the sheet name and the table name when prompted. The master procedure should now run as expected.

Running the Macro from a Worksheet Button

Sometimes, it makes the most sense to place a macro button right on the worksheet, where it cannot be missed.

Hands-On 1.19 Running a Macro from a Button Placed on a Worksheet

1. Copy the `Chap01_Supplement.xlsm` workbook from the companion files to your `C:\VBAExcel2024_ByExample` folder.
2. Open the copied workbook file in Excel.

3. Choose Developer | Insert. The Forms toolbar appears, as shown in Figure 1.25.

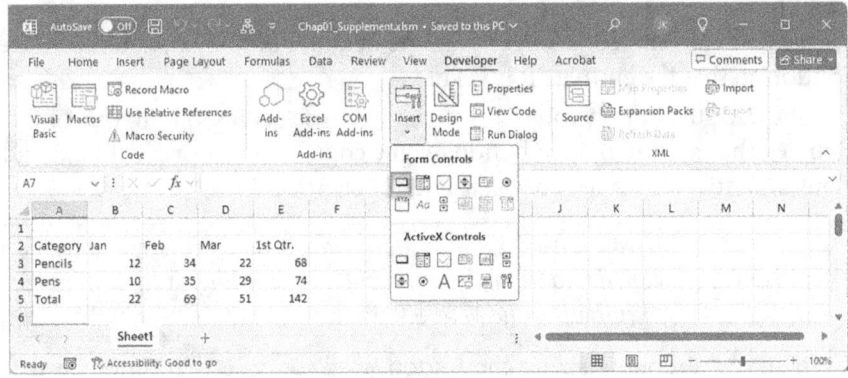

FIGURE 1.25.   Adding a button to a worksheet.

4. In the Form Controls area, click the first image, which represents a button.
5. Click anywhere in the empty area of the worksheet. When the Assign Macro dialog box appears, choose WhatsInACell macro and click OK.
6. Excel will create a button with the default label `Button 1`. To change the button's label, double-click inside the button, delete the default text, and type Format Cells. If the text does not fit, do not worry; you will resize the button in the next step. When the button is selected, it looks like the one shown in Figure 1.26. If the selection handles are not displayed, right-click Button 1 on the worksheet and choose Edit Text on the context menu. Select the default text and enter the new label.

Button 1		⋮ ⨯ ✓	f_x ⌄			
	A	B	C	D	E	F
1						
2	Category	Jan	Feb	Mar	1st Qtr.	
3	Pencils		12	34	22	68	
4	Pens		10	35	29	74	
5	Total		22	69	51	142	
6						
7						
8			Format Cells			
9						
10						
11						

Sheet1   +

FIGURE 1.26.   A button with an attached macro.

7. When you're done renaming the button, click outside the button to exit the edit mode.

   If the text you entered is longer than the default button text, you can resize the button so that the entire text is visible.

8. Right-click the button you've just renamed to select it, point to one of the tiny circles that appear on the button's right edge, and drag right to expand the button until you see the complete entry, Format Cells.

NOTE	*If you left-click the button inadvertently, there is nothing you can do to stop the macro from running. You can resize the button after the macro has run.*

9. When you're done resizing the button, click outside the button to exit the selection mode.

10. To run your macro, click the button you just created.

    Your macro will go to work, and your worksheet will be formatted as shown in Figure 1.27.

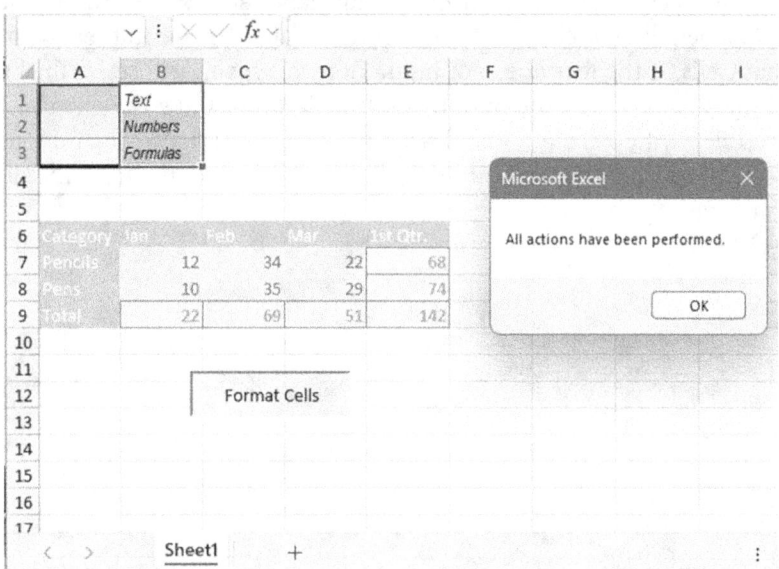

FIGURE 1.27. This worksheet was formatted with a macro attached to the Format Cells button.

Let's remove the formatting you just applied by running the `RemoveFormats` macro.

11. Click OK to close the message box and then press Alt+F8 to open the Macro dialog box. Select the RemoveFormats macro and click the Run button. Custom formatting has now been removed.

12. On your own, create another button on this worksheet that will be used for running the `RemoveFormats` macro.

13. Save your workbook with a different file name so that the original workbook can be reused again in case you'd like to revisit the button creation process.

NOTE	*The code of the* `WhatsInACell` *and* `RemoveFormat` *macros in this practice workbook was created by the built-in macro recorder while executing a series of commands via Excel ribbon options.*

You can also run macros from a hyperlink, or a button placed in the ribbon. These techniques will be introduced later in this book because they require an understanding of the advanced topic of ribbon customization.

SUMMARY

In this chapter, you learned how to create macros by recording your cell selections and data entry in the Microsoft Excel application window. You also explored how to view, read, and modify the recorded macros in the VBE window. Additionally, you experimented with various methods of running macros. This chapter also pointed out macro security issues that you should be aware of when opening workbooks containing macro code. Moreover, you got a glimpse of methods used in testing and debugging your macros.

The next chapter focuses on various tools available in the VBE window to assist you in programming and troubleshooting your VBA code, whether recorded by the macro recorder or written from scratch.

Chapter 2

EXCEL PROGRAMMING ENVIRONMENT—A QUICK OVERVIEW OF ITS TOOLS AND FEATURES

Now that you know how to record, run, and edit macros, let's spend some time in the VBE window and become familiar with its features. With the tools located in the VBE window, you can:

- Write your own VBA procedures.
- Create custom forms.
- View and modify object properties.
- Test VBA procedures and locate errors.

The VBE window can be accessed in the following ways:

- Choose Developer | Visual Basic.
- Choose Developer | View Code.
- Press Alt+F11.

UNDERSTANDING THE PROJECT EXPLORER WINDOW

The Project Explorer window displays a hierarchical list of currently open projects and their elements. A VBA project can contain the following elements:

- Worksheets
- Charts
- `ThisWorkbook`—The workbook where the project is stored
- Modules—Special sheets where programming code is stored
- Classes—Special modules that allow you to create your own objects
- Forms
- References to other projects

With the Project Explorer, you can manage your projects and easily move between projects that are loaded into memory. You can activate the Project Explorer window in one of three ways:

- From the View menu by selecting Project Explorer
- From the keyboard by pressing Ctrl+R
- From the Standard toolbar by clicking the Project Explorer button, as shown in Figure 2.1

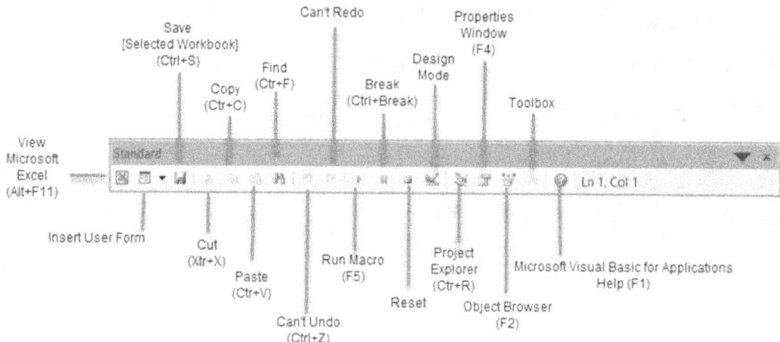

FIGURE 2.1. Buttons on the Standard toolbar provide a quick way to access many of the VBE features.

The Project Explorer window contains three buttons, as shown in Figure 2.2. The first button from the left (View Code) displays the Code window for the selected module. The middle button (View Object) displays either the selected sheet in the Microsoft Excel Objects folder or a form located in the Forms folder.

The button on the right (Toggle Folders) hides and/or activates the display of folders in the Project Explorer window.

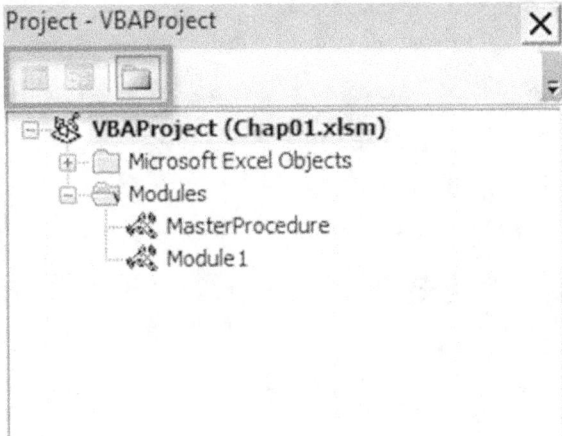

FIGURE 2.2.   The Project Explorer window displays a list of currently open projects.

UNDERSTANDING THE PROPERTIES WINDOW

The Properties window allows you to review and set properties of various objects in your project. The name of the currently selected object is displayed in the Object box located just below the Properties window title bar. For example, Figure 2.3 displays the properties of the `Sheet3` object. The properties of the object can be viewed alphabetically or by category by clicking the appropriate tab.

- Alphabetic tab—Lists all properties for the selected object alphabetically. You can change the property setting by selecting the property name and typing or selecting the new setting.
- Categorized tab—Lists all properties for the selected object by category. You can collapse the list so that you see the categories, or you can expand a category to see the properties. The plus sign (+) icon to the left of the category name indicates that the category list can be expanded. The minus sign (–) indicates that the category is currently expanded.

The Properties window can be accessed in three ways:

- From the View menu by selecting Properties Window
- From the keyboard by pressing F4
- From the toolbar by clicking the Properties Window button

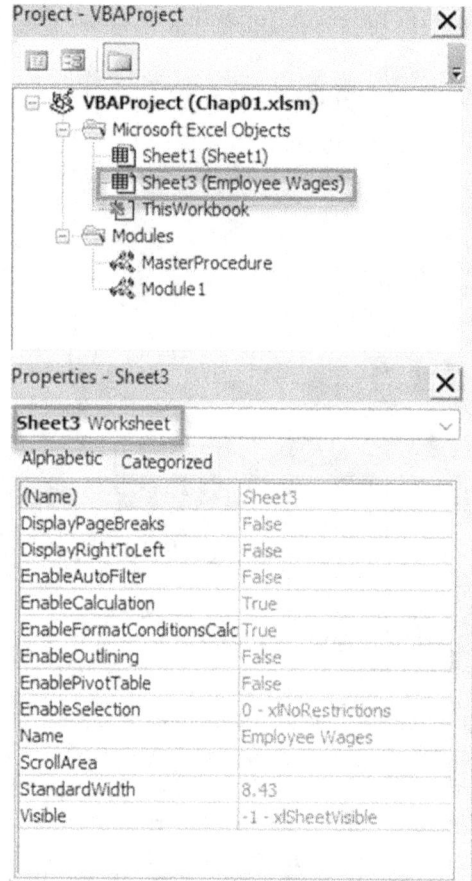

FIGURE 2.3. The Properties window displays the settings for the object currently selected in the Project Explorer.

UNDERSTANDING THE CODE WINDOW

The Code window is used for writing VBA code as well as viewing and modifying the code of recorded macros and existing VBA procedures. Each module can be opened in a separate Code window. There are several ways to activate the Code window:

- From the Project Explorer window, choose the appropriate UserForm or module, and click the View Code button.
- From the menu bar, choose View | Code.
- From the keyboard, press F7.

In Figure 2.4, you will notice, at the top of the Code window, two drop-down list boxes that allow you to move quickly within the Visual Basic code. In the Object box on the left side of the Code window, you can select the object whose code you want to view. The box on the right side of the Code window lets you quickly choose a procedure or event procedure to view. When you open this box, the names of all procedures located in a module are sorted alphabetically. If you select a procedure in the Procedure box, the cursor will jump to the first line of this procedure.

By dragging the split bar shown in Figure 2.4 down to a selected position in the Code window, you can divide the Code window into two panes. You can then view different sections of a long procedure or a different procedure in each pane. This two-pane display in the Code window is often used for copying or cutting and pasting sections of code between procedures of the same module. To return to the one-window display, simply drag the split bar to the top of the Code window.

FIGURE 2.4. The Visual Basic Code window has several elements that make it easy to locate procedures and review the VBA code.

At the bottom left of the Code window, there are two icons. The Procedure view icon displays one procedure at a time in the Code window. To select another procedure, use the Procedure drop-down list. The Full Module view icon displays all the procedures in the selected module. Use the vertical scroll bar to scroll through the module's code.

The margin indicator bar is used by VBE to display helpful indicators during editing and debugging. If you'd like to take a quick look at some of these indicators, skim through Chapter 9, Excel Tools for Testing and Debugging.

SETTING THE VBE OPTIONS

There are several other windows that are frequently used in the Visual Basic environment.

Figure 2.5 displays the list of windows that can be docked in the VBE window. You will learn how to use some of these windows in Chapter 3 (Object Browser and Immediate Window) and Chapter 9 (Locals Window and Watch Window). The Options dialog can be accessed via the Tools menu.

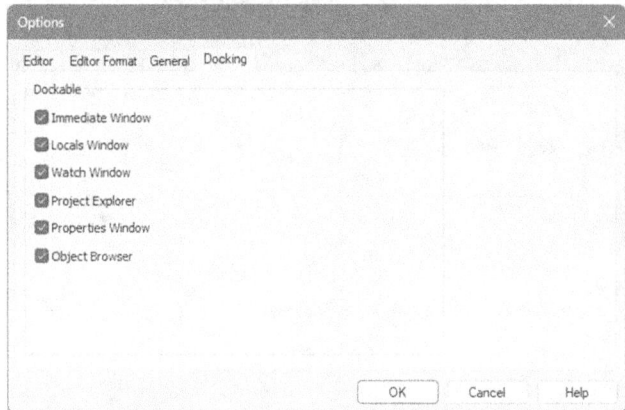

FIGURE 2.5.   The Docking tab in the Options dialog box allows you to choose which windows you want to be dockable on the VBE screen.

SYNTAX AND PROGRAMMING ASSISTANCE

Figure 2.6 shows the Edit toolbar in the VBE window with several buttons that let you enter correctly formatted VBA instructions with speed and ease. If the Edit toolbar isn't currently docked in the VBE window, you can turn it on by choosing View | Toolbars | Edit.

Writing procedures in Visual Basic requires that you use hundreds of built-in instructions and functions. Because most people cannot memorize the correct syntax of all the instructions that are available in VBA, the IntelliSense®

technology provides you with syntax and programming assistance on demand when entering instructions. You can have special windows pop up and guide you through the process of creating correct VBA code.

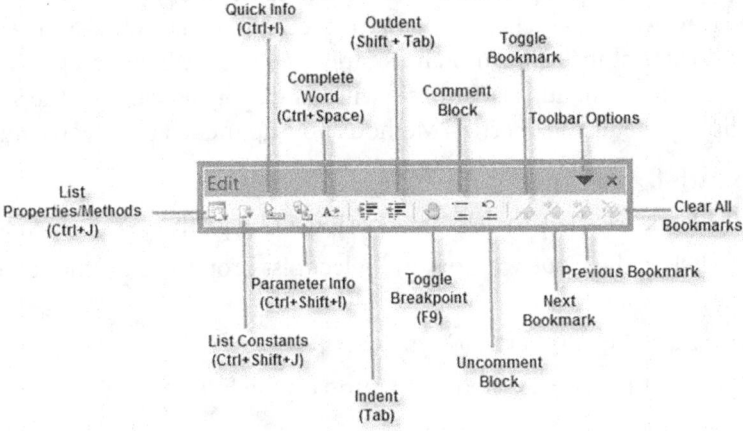

FIGURE 2.6. Buttons located on the Edit toolbar make it easy to write and format VBA instructions.

List Properties/Methods

Each object can contain several properties and methods. When you enter the name of the object and a period that separates the name of the object from its property or method in the Code window, a pop-up menu may appear. This menu lists the properties and methods available for the object that precedes the period, as shown in Figure 2.7. To turn on this automated feature, choose Tools | Options. In the Options dialog box, click the Editor tab, and make sure the Auto List Members checkbox is selected.

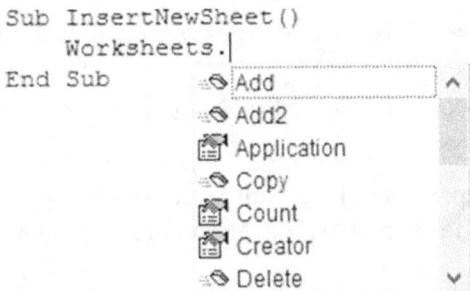

```
Sub InsertNewSheet()
    Worksheets.|
End Sub
```

FIGURE 2.7. While you are entering the VBA instructions, Visual Basic suggests properties and methods that can be used with the object.

To choose an item from the pop-up menu, start typing the name of the property or method that you want to select. When Excel highlights the correct item name, press Enter to insert the item into your code and start a new line. If you want to continue writing instructions on the same line, press the Tab key instead. You can also double-click the item to insert it in your code. To close the pop-up menu without inserting an item, simply press Esc. When you press Esc to remove the pop-up menu, Visual Basic will not display it again for the same object. To display the List Properties/Methods pop-up menu again, you can:

- Press Ctrl+J.
- Use the Backspace key to delete the period and type the period again.
- Right-click in the Code window and select List Properties/Methods from the shortcut menu.
- Choose Edit | List Properties/Methods.
- Click the List Properties/Methods button on the Edit toolbar.

List Constants

A *constant* is a value that indicates a specific state or result. Excel has many predefined, built-in constants. You will learn about constants, their types, and their usage in Chapter 3.

Let's say you want your program to turn on the Page Break Preview of your worksheet. In the Microsoft Excel application window, the View tab lists four types of workbook views:

- Normal View is the default view for most tasks in Excel.
- Page Layout View allows you to view the document as it will appear on the printed page.
- Page Break Preview allows you to see where pages will break when the document is printed.
- Custom Views allows you to save the set of display and print settings as a custom view.

The first three view options are represented by a built-in constant. Microsoft Excel constant names begin with the characters "xl." As soon as you enter the following instruction:

```
ActiveWindow.View =
```

a pop-up menu will appear with the names of valid constants for the View property, as shown in Figure 2.8.

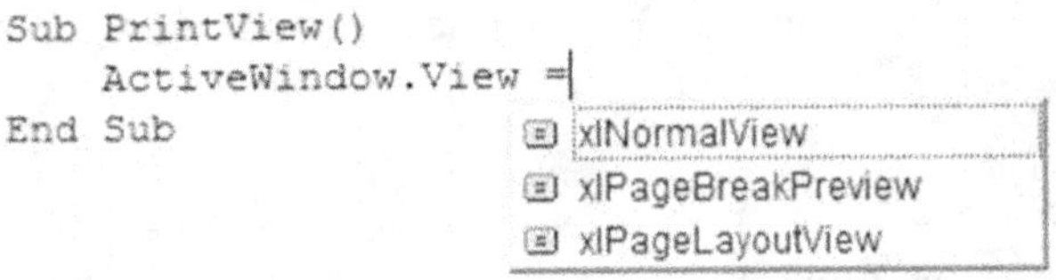

```
Sub PrintView()
    ActiveWindow.View =|
End Sub                    ☰ xlNormalView
                           ☰ xlPageBreakPreview
                           ☰ xlPageLayoutView
```

FIGURE 2.8.   The List Constants pop-up menu displays a list of constants that are valid for the property entered.

To work with the List Constants pop-up menu, use the same techniques as for the List Properties/Methods pop-up menu outlined in the preceding section. The List Constants pop-up menu can be activated by pressing Ctrl+Shift+J or clicking the List Constants button on the Edit toolbar.

Parameter Info

If you've had a chance to work with Excel worksheet functions, you already know that many functions require one or more arguments (or *parameters*). For example, here's the syntax for the most common worksheet function:

```
SUM(number1,number2, ...)
```

where `number1, number2,` ... are 1 to 30 arguments that you can add up.

Like Excel built-in functions, VBA methods may require one or more arguments. If a method requires an argument, you can see the names of required and optional arguments in a tooltip box that appears just below the cursor as soon as you type the beginning parenthesis, as illustrated in Figure 2.9. In the tooltip, the current argument is displayed in bold. When you supply the first argument and enter the comma, Visual Basic displays the next argument in bold. Optional arguments are surrounded by square brackets, [ ].

You can open the Parameter Info tooltip using the keyboard. To do this, enter the method or function name, follow it with the left parenthesis, and press Ctrl+Shift+I. You can also click the Parameter Info button on the Edit toolbar or choose Edit | Parameter Info.

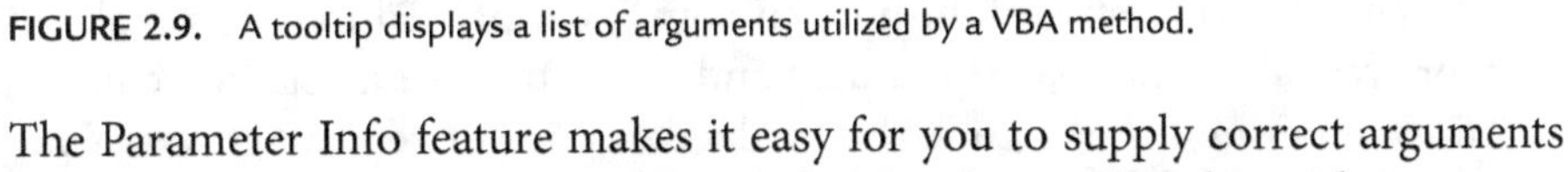

```
Sub PrintView()
    ActiveWindow.View = xlPageBreakPreview
    ActiveWorkbook.SaveAs
End Sub          SaveAs([Filename], [FileFormat], [Password], [WriteResPassword], [ReadOnlyRecommended], [CreateBackup], [AccessMode As
                XlSaveAsAccessMode = xlNoChange], [ConflictResolution], [AddToMru], [TextCodepage], [TextVisualLayout], [Local])
```

FIGURE 2.9.   A tooltip displays a list of arguments utilized by a VBA method.

The Parameter Info feature makes it easy for you to supply correct arguments to a VBA method. In addition, it reminds you of two other things that are very important for the method to work correctly: the order of the arguments and the required data type of each argument. You will learn about data types in Chapter 3.

Quick Info

When you select an instruction, function, method, procedure name, or constant in the Code window and then click the Quick Info button on the Edit toolbar (or press Ctrl+I), Visual Basic displays the syntax of the highlighted item, as well as the value of a constant, as depicted in Figure 2.10. The Quick Info feature can be turned on or off using the Options dialog box. To use the feature, click the Editor tab and choose the Auto Quick Info option.

```
Sub PrintView()
    ActiveWindow.View = xlPageBreakPreview
    ActiveWorkbook.SaveA xlPageBreakPreview = 2
End Sub
```

FIGURE 2.10.   The Quick Info feature displays a list of arguments required by a selected method or function, a value of a selected constant, or the type of the selected object or property.

Complete Word

Another way to increase the speed of writing VBA procedures in the Code window is with the Complete Word feature. As you enter the first few letters of a keyword and press Ctrl+Spacebar or click the Complete Word button on the Edit toolbar, Visual Basic will fill in the remaining letters by completing the keyword entry for you. For example, when you enter the first four letters of the keyword Application (Appl) in the Code window and press Ctrl+Spacebar, Visual Basic will complete the rest of the word, and in the place of "Appl," you will see the entire word "Application."

Indent/Outdent

If the Auto Indent option is turned on, you can automatically indent the selected lines of code by the number of characters specified in the Tab Width text box. The default entry for Auto Indent is four characters. You can easily change this setting via the Options dialog box; choose Tools | Options and click the Editor tab.

When you indent certain lines in your VBA procedures, you make them more readable and easier to understand. Indenting is especially recommended for entering lines of code that make decisions or repeat actions. You will learn how to create these kinds of Visual Basic instructions in Chapters 5 and

6. Let's spend a few minutes learning how to apply the indent and outdent features to the lines of code in the `WhatsInACell` macro that you worked with in Chapter 1.

⊙ **Hands-On 2.1 Indenting/Outdenting Visual Basic Code**

1. Open the `Chap01_Supplement.xlsm` workbook that you worked with in Chapter 1.
2. Press Alt+F11 to switch to the VBE window.
3. Choose View | Toolbars | Edit to gain access to the Editing toolbar. If the toolbar pops up in the middle of the screen, double-click its title bar to get it docked at the top of the VBE window.
4. In the Project Explorer window, select the Chap01_Supplement.xlsm VBA project and activate `Module1`, which contains the code of the `WhatsInACell` macro.
5. Select the block of code located between the keywords `With` and `End With`:

```
With Selection.Font
    .Name = "Arial Narrow"
    .FontStyle = "Italic"
    .Size = 10
End With
```

6. Click the Indent button on the Edit toolbar or press Tab on the keyboard. The selected block of instructions will move four spaces to the right if you are using the default setting in the Tab Width box in the Options dialog box (Editor tab).
7. Click the Outdent button on the Edit toolbar or press Shift+Tab to return the selected lines of code to the previous location in the Code window.
8. Close the `Chap01_Supplement.xlsm` workbook.
9. The Indent and Outdent options are also available from the Edit menu.

Comment Block/Uncomment Block

In Chapter 1, you learned that a single quote placed at the beginning of a line of code denotes a comment. Not only do comments make it easier to understand what the procedure does but they are also very useful in testing and trouble-shooting VBA code.

For example, when you execute your code, it may not run as expected. Instead of deleting the lines that may be responsible for the problems you encountered, you may want to skip those lines of code for now and return to them later.

By placing a single quote at the beginning of the line you want to avoid, you can continue checking the other parts of your procedure.

- To comment out a few lines of code, simply select the lines and click the Comment Block button on the Edit toolbar.
- To turn the commented code back into VBA instructions, select the lines and click the Uncomment Block button on the Edit toolbar.

If you don't select text and click the Comment Block button, the single quote is added only to the line of code where the cursor is currently located.

USING THE OBJECT BROWSER

You can move easily through the myriad of VBA elements and features by examining the capabilities of the Object Browser. To access the Object Browser, use any of the following methods in the VBE window:

- Press F2.
- Choose View | Object Browser.
- Click the Object Browser button on the toolbar.

The Object Browser allows you to browse through the objects that are available to your VBA procedures, as well as view their properties, methods, and events. With the aid of the Object Browser, you can move quickly between procedures in your own VBA projects, as well as search for objects and methods across installed object-type libraries.

The Object Browser window is divided into three sections, as illustrated in Figure 2.11. The top of the window displays the Project/Library drop-down list box with the names of all libraries and projects that are available to the currently active VBA project.

A *library* is a special file that contains information about the objects in an application. New libraries can be added via the References dialog box (Tools | References). The entry for <All Libraries> lists the objects of all libraries that are installed on your computer. When you select the library called Excel, you will see only the names of the objects that are exclusive to Microsoft Excel. In contrast to the Excel library, the VBA library lists the names of all the objects in Visual Basic for Applications.

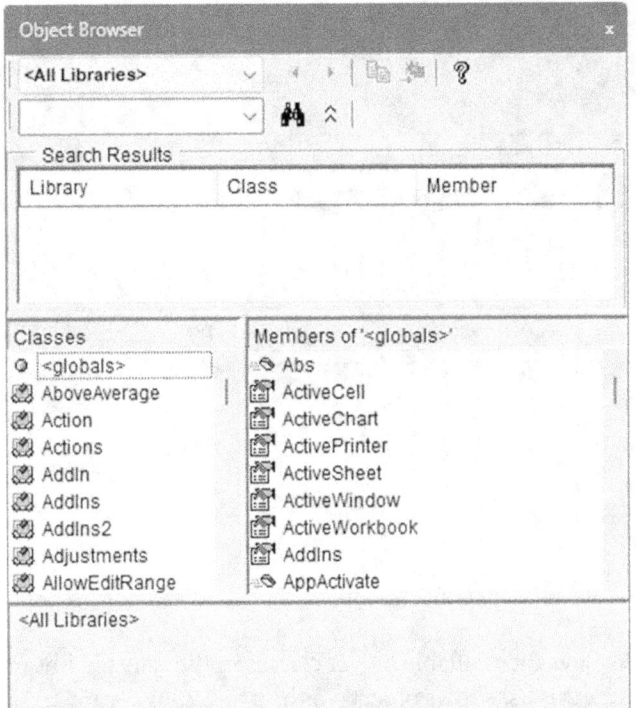

FIGURE 2.11.   The Object Browser window allows you to browse through all the objects, properties, and methods available to the current VBA project.

Below the Project/Library drop-down list box is a Search text box that you can use to quickly find information in a library. This field remembers the last four items for which you searched. To find only whole words, you can right-click anywhere in the Object Browser window and choose Find Whole Word Only from the context menu.

The Search Results section of the Object Browser displays the library, class, and member elements that met the criteria entered in the Search text box, as shown in Figure 2.12.

When you type the search text and click the Search button (the binoculars icon), Visual Basic expands the Object Browser dialog box to show the Search Results area. You can hide or show the search results by clicking the button located to the right of the Search button.

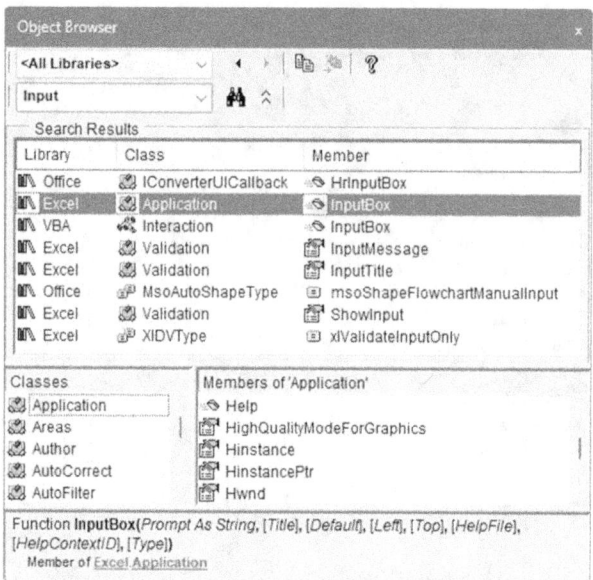

FIGURE 2.12.   Searching for answers in the Object Browser.

The Classes list box displays the available object classes in the selected library. If you select a VBA project, this list shows objects in the project. In Figure 2.12, the Excel Application object class is selected. When you highlight a class, the list on the right-hand side (Members) shows the properties, methods, and events available for that class. By default, members are listed alphabetically. You can, however, organize the Members list by group type (properties, methods, or events) using the Group Members command from the Object Browser context menu.

If you select a VBA project in the Project/Library list box, the Members list box will list all the procedures available in this project. To examine the code of a procedure, simply double-click its name. If you select a VBA library, you will see a list of Visual Basic built-in functions and constants. If you need more information on the selected class or a member, click the question mark button at the top of the Object Browser window.

The bottom of the Object Browser window displays a code template area with the definition of the selected member. If you click the hyperlink text in the code template, you can quickly jump to the selected member's class or library in the Object Browser window. Text displayed in the code template area can be copied to the Windows clipboard and then pasted to a Code window. If the Code window is visible while the Object Browser window is open, you can save time by dragging the highlighted code template and dropping it into the Code window.

You can easily adjust the size of the various sections of the Object Browser window by dragging the dividing horizontal and vertical lines.

Now that you've discovered the Object Browser, you may wonder how you can put it to use in VBA programming. Let's assume that you placed a text box in the middle of your worksheet. How can you make Excel move this text box so that it is positioned in the top left-hand corner of the sheet? Hands-On 2.2 should provide the answer to this question.

Hands-On 2.2 Writing a VBA Procedure to Move a Text Box on the Worksheet

1. Open a new workbook.
2. Choose Insert | Text | Text Box.
3. Draw a box in the middle of the sheet and enter any text, as shown in Figure 2.13.
4. Select any cell outside the text box area.
5. Press Alt+F11 to activate the Visual Basic Editor window.
6. Choose Insert | Module to add a new module sheet.
7. In the Properties window, enter the new name for this module: Manipulations.
8. Choose View | Object Browser or press F2.
9. In the Project/Library list box, click the drop-down arrow and select the Excel library.
10. Enter textbox as the search text in the Search box, as shown in Figure 2.14, and then click the Search button. Make sure you don't enter a space in the search string.

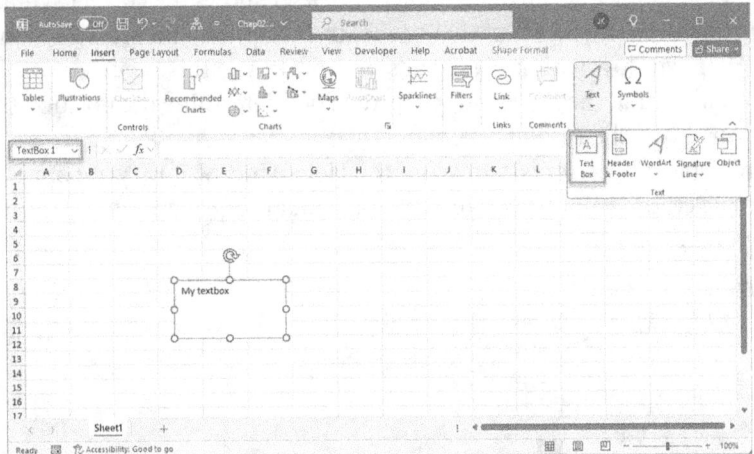

FIGURE 2.13.   Excel displays the name of the inserted object in the Name box above the worksheet.

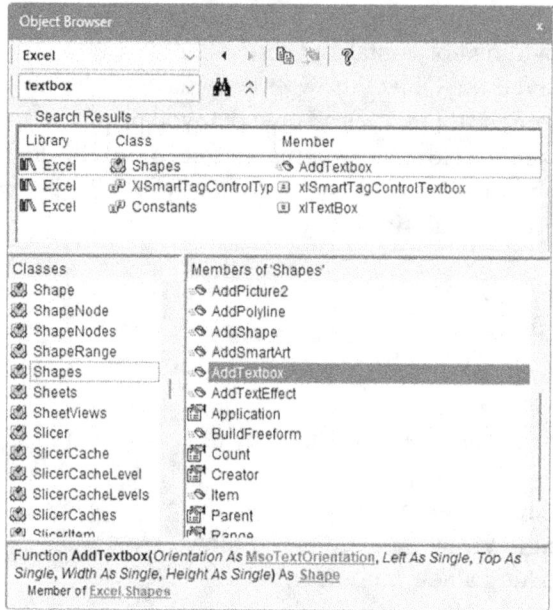

FIGURE 2.14. Using the Object Browser window, you can find the appropriate VBA instructions for writing your own procedures.

11. Visual Basic searches the Excel library and displays the search results. It appears that the Shapes object shown in Figure 2.14 is in control of our text box operations. Looking at the Members list, you can quickly determine that the AddTextbox method is used for adding a new text box to a worksheet. The code template at the bottom of the Object Browser shows the correct syntax for using this method. If you select the AddTextbox method and press F1, you will see the Help window with more details on how to use this method. The Help window tells us that the Left and Top properties determine the position of the text box in a worksheet.

12. Close the Object Browser window and the Help window if they are open.

13. Double-click the Manipulations module and enter the MoveTextBox procedure, as shown here:

```
Sub MoveTextBox()
    With ActiveSheet.Shapes("TextBox 1")
            .Select
            .Left = 0
            .Top = 0
    End With
End Sub
```

The `MoveTextBox` procedure selects `TextBox 1` in the collection of `Shapes` objects. `TextBox 1` is the default name of the first text box object placed in the worksheet. Each time you add a new object to your worksheet, Excel assigns a new number (index) to it. Instead of using the object name, you can refer to the member of a collection by its index. For example, instead of:

```
With ActiveSheet.Shapes("TextBox 1")
```

enter:

```
With ActiveSheet.Shapes(1)
```

14. Choose Run | Run Sub/UserForm to execute this procedure.
15. Press Alt+F11 to switch to the Microsoft Excel application window.
16. The text box should be positioned in the top left-hand corner of the worksheet.
17. Save the workbook file as `Chap02.xlsm`. Keep this file open as you will continue to work with it in Hands-On 2.3.

Let's manipulate another object with Visual Basic.

Hands-On 2.3 Writing a VBA Procedure to Move a Circle on the Worksheet

1. Place a small circle in the same worksheet where you originally placed the text box in Hands-On 2.2. Use the Oval shape in the Basic Shapes area of the Insert | Illustrations | Shapes tool. Hold down the Shift key while drawing on the worksheet to create a perfect circle.
2. Click outside the circle to deselect it.
3. Press Alt+F11 to activate the VBE screen.
4. In the Manipulations module's Code window, write a VBA procedure that will place the circle inside the text box. Keep in mind that Excel assigns numbers to its objects consecutively. The first object is assigned the number 1, the second object the number 2, and so on. The type of object—whether it is a text box, a circle, or a rectangle—does not matter.
5. The `MoveCircle` procedure shown here demonstrates how to move a circle to the top left-hand corner of the active worksheet:

```
Sub MoveCircle()
    With ActiveSheet.Shapes(2)
        .Select
        .Left = 0
        .Top = 0
    End With
End Sub
```

Moving a circle is like moving a text box or any other object placed in a worksheet. Notice that instead of referring to the circle by its name, the procedure uses the object's index.

6. Run the `MoveCircle` procedure.
7. Press Alt+F11 to return to the Microsoft Excel window.
8. The circle should now appear at the top of the text box.

Locating VBA Procedures with the Object Browser

In addition to locating objects, properties, and methods, the Object Browser is a handy tool for locating and accessing procedures written in various VBA projects. The Hands-On 2.4 exercise demonstrates how you can quickly find which procedures are stored in the selected project.

Hands-On 2.4   Using Object Browser to Locate VBA Procedures

1. In the Object Browser, select VBAProject from the Project/Library drop-down list, as shown in Figure 2.15.

   The left side of the Object Browser displays the names of objects that are included in the selected project. The Members list box on the right shows the names of all the available procedures.

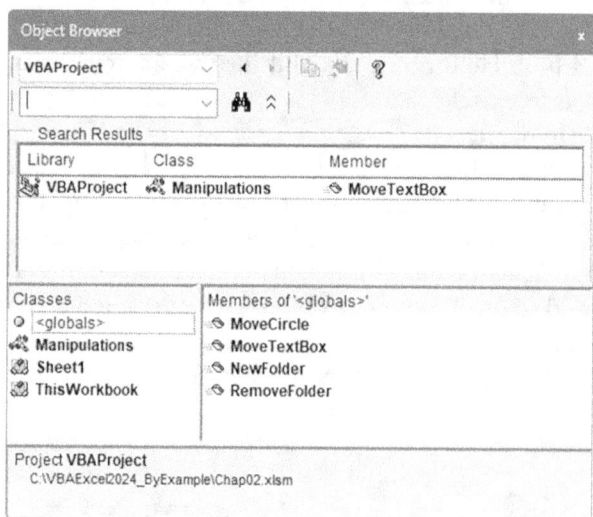

FIGURE 2.15. The Object Browser lists all the procedures available in a VBA project.

2. In the Members list, double-click the MoveCircle procedure.
3. Excel will locate the selected procedure in the Code window.

USING THE VBA OBJECT LIBRARY

In the previous examples, you used the properties of objects that are members of the Shapes collection in the Excel Objects library. While the Excel library contains objects specific to using Microsoft Excel, the VBA Objects library provides access to many built-in VBA functions that are general in nature. They allow you to manage files, set the date and time, interact with users, convert data types, deal with text strings, or perform mathematical calculations.

In the following Hands-On 2.5 exercise, you will use one of the built-in VBA functions to create a new Windows subfolder without leaving Excel.

Hands-On 2.5   Writing a VBA Procedure to Create a Folder in Windows

1. Press Alt+F11 to return to the `Manipulations` module, in which you entered the `MoveTextBox` and `MoveCircle` procedures.
2. On a new line, type the name of the new procedure: Sub NewFolder().
3. Press Enter. Visual Basic will enter the ending keywords `End Sub`.
4. Press F2 to activate the Object Browser.
5. Click the drop-down arrow in the Project/Library list box and select VBA.
6. Enter file as the search text in the Search box and press the Search button.
7. Scroll down in the Members list box and highlight the MkDir method, as shown in Figure 2.16.
8. Click the Copy button (the middle button in the top row) in the Object Browser window to copy the selected method name to the Windows clipboard.
9. Return to the Manipulations Code window and paste the copied instruction inside the procedure NewFolder.
10. Enter a space, followed by "C:\Study". Be sure to enter the name of the entire path in quotes. The `NewFolder` procedure should look like this:

```
Sub NewFolder()
    MkDir "C:\Study"
End Sub
```

11. Position the insertion point within the code of the `NewFolder` procedure and choose Run | Run Sub/UserForm to execute the `NewFolder` procedure.
    When you run the `NewFolder` procedure, Visual Basic creates a new folder on drive C. To see the folder, activate Windows Explorer. After creating a new folder, you may realize that you don't need it after all. Although you could easily delete the folder while in Windows Explorer, how about getting rid of it programmatically? The Object Browser displays many other methods that are

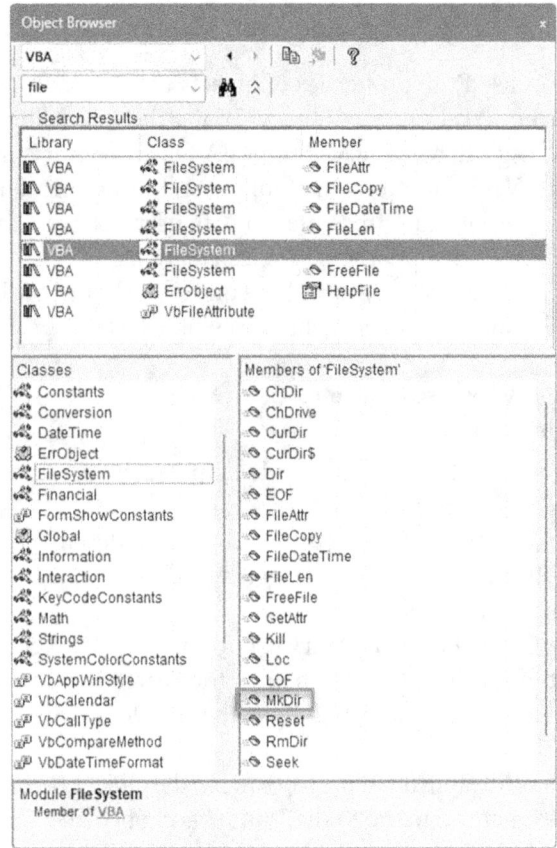

FIGURE 2.16. When writing procedures from scratch, consult the Object Browser for the names of the built-in VBA functions.

useful for working with folders and files. The RmDir method is just as simple to use as the MkDir method.

12. To remove the Study folder from your hard drive, you could replace the MkDir method with the RmDir method and then rerun the NewFolder procedure. Let's write a new procedure called RemoveFolder in the Manipulations Code window, as shown here:

```
Sub RemoveFolder()
    RmDir "C:\Study"
End Sub
```

The RmDir method allows you to remove unwanted folders from your hard disk.

13. Position the insertion point within the code of the `RemoveFolder` procedure and choose Run | Run Sub/UserForm to execute the `RemoveFolder` procedure.

14. Check Windows Explorer to see that the `Study` folder has gone.

USING THE IMMEDIATE WINDOW

The Immediate window is used for trying out various instructions, functions, and operators present in the Visual Basic language before using them in your own VBA procedures. It is a great tool for experimenting with your new language.

The Immediate window allows you to type VBA statements and test their results immediately without having to write a VBA procedure. It is like a scratch pad. Use it to try out your statements. If the statement produces the expected result, you can copy the statement from the Immediate window into your procedure (or you can drag it right onto the Code window if it is visible).

The Immediate window can be moved anywhere on the VBE screen, or it can be docked so that it always appears in the same area of the screen. The docking setting can be turned on and off on the Docking tab in the Options dialog box (Tools | Options).

- To quickly access the Immediate window, press Ctrl+G while on the VBE screen.
- To close the Immediate window, click the Close button in the top right-hand corner of the window.

Before you start creating full-fledged VBA procedures (this awaits you in the next chapter!), begin with some warm-up exercises to build up your VBA vocabulary. How can you do this quickly and painlessly? How can you try out some of the newly learned VBA statements? Here is a short, interactive language exercise: Enter a simple VBA instruction in the Immediate window and Excel will check it out and display the result on the next line.

⊙ Hands-On 2.6   Entering and Executing VBA Statements in the Immediate Window

1. On the VBE window, choose View | Immediate Window.

2. Arrange the screen so that both the Microsoft Excel window and the Visual Basic window are placed horizontally or vertically side by side, as presented in Figure 2.17, or use a setup with two monitors displaying Excel windows on separate screens.

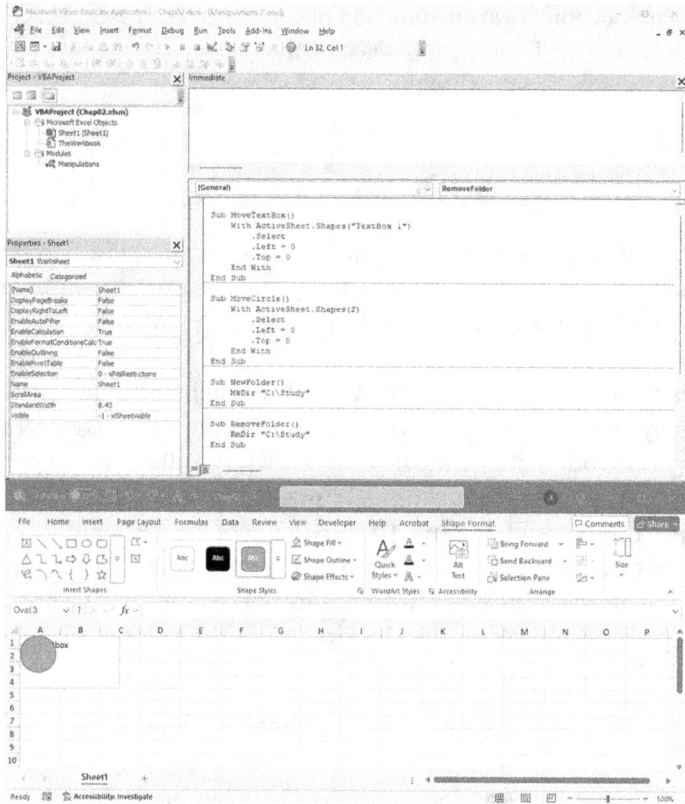

FIGURE 2.17. By positioning the Microsoft Excel and Visual Basic windows side by side, you can watch the execution of the instructions entered in the Immediate window.

3. On the VBE screen, press Ctrl+G to activate the Immediate window.

4. In the Immediate window, type the following instruction and press Enter:

```
Worksheets.Add
```

When you press the Enter key, Visual Basic gets to work. If you entered the foregoing VBA statement correctly, VBA adds a new sheet in the current workbook. The Sheet2 tab at the bottom of the workbook should now be highlighted.

5. In the Immediate window, type another VBA statement and be sure to press Enter when you're done:

```
Range("A1:A4").Select
```

As soon as you press Enter, Visual Basic highlights cells A1, A2, A3, and A4 in the active worksheet.

6. Enter the following instruction in the Immediate window:

```
[A1:A4].Value = 55
```

When you press Enter, Visual Basic places the number 55 in every cell of the specified range, A1:A4. This statement is an abbreviated way of referring to the `Range` object. The full syntax is more readable:

```
Range("A1:A4").Value = 55
```

7. Enter the following instruction in the Immediate window:

```
Selection.ClearContents
```

When you press Enter, VBA will delete the results of the previous statement from the selected cells. Cells A1:A4 are now empty.

8. Enter the following instruction in the Immediate window:

```
ActiveCell.Select
```

9. When you press Enter, Visual Basic will make cell A1 active.
10. Figure 2.18 shows all the instructions entered in the Immediate window in this exercise. Every time you pressed the Enter key, Excel executed the statement on the line where the cursor was located. If you want to execute the same instruction again, click anywhere in the line containing the instruction and press Enter.

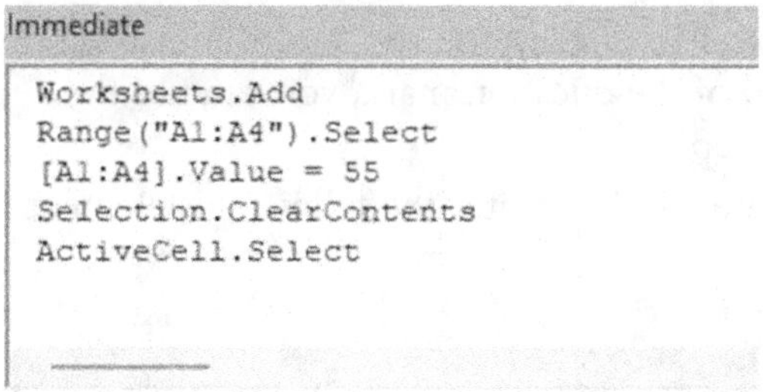

```
Immediate
Worksheets.Add
Range("A1:A4").Select
[A1:A4].Value = 55
Selection.ClearContents
ActiveCell.Select
```

FIGURE 2.18. Instructions entered in the Immediate window are executed as soon as you press the Enter key.

Obtaining Information in the Immediate Window

So far, you have used the Immediate window to perform actions. These actions could have been performed manually by clicking the mouse in various areas of the worksheet and entering data.

Instead of simply performing actions, the Immediate window also allows you to ask questions. Suppose you want to find out which cells are currently selected, what the value of the active cell is, what the name of the active sheet is, or what the number of the current window is. When working in the Immediate window, you can easily get answers to these and other questions.

In the preceding exercise, you entered several instructions.

Let's return to the Immediate window to ask some questions. Excel remembers the instructions entered into the Immediate window even after you close this window. Note that the contents of the Immediate window are automatically deleted when you exit Microsoft Excel but, at any time, you can delete the contents of the Immediate window by selecting it and pressing the Delete key.

⊙ Hands-On 2.7   Obtaining Information in the Immediate Window

1. Click the mouse in the second line of the Immediate window where you previously entered the instruction `Range("A1:A4").Select`.
2. Press Enter to have Excel reselect cells A1:A4.
3. Click on the new line of the Immediate window, enter the following question, and press Enter:

```
?Selection.Address
```

When you press Enter, Excel will not select anything in the worksheet. Instead, it will display the result of the instruction on a separate line in the Immediate window. In this case, Excel returns the absolute address of the cells that are currently selected ($A$1:$A$4).

The question mark (?) tells Excel to display the result of the instruction in the Immediate window. Instead of the question mark, you can use the `Print` keyword, as shown in the next step.

4. On a new line in the Immediate window, enter the following statement and press Enter:

```
Print ActiveWorkbook.Name
```

Excel enters the name of the active workbook on a new line in the Immediate window.

How about finding the name of the application?

5. On a new line in the Immediate window, enter the following statement and press Enter:

```
?Application.Name
```

Excel will reveal its full name: Microsoft Excel.

The Immediate Window can also be used for quick calculation.

6. On a new line in the Immediate window, enter the following statement and press Enter:

```
?12/3
```

Excel shows the result of the division on the next line. What if you want to know the result of 3+2 and 12*8 right away?

Instead of entering these instructions on separate lines, you can enter them on one line, as in the following example:

```
?3+2:?12*8
```

Notice the colon separating the two blocks of instructions. When you press the Enter key, Excel displays the results 5 and 96 on separate lines in the Immediate window.

The following lists all the instructions you entered in the Immediate window, including Excel's answers to your questions:

```
Worksheets.Add
Range("A1:A4").Select
[A1:A4].Value = 55
Selection.ClearContents
ActiveCell.Select
?Selection.Address
$A$1:$A$4
Print ActiveWorkbook.Name
Chap02.xlsm
?Application.Name
Microsoft Excel
?12/3
 4
?3+2:?12*8
 5
 96
```

To delete the instructions from the Immediate window, make sure that the selection point is in the Immediate window, press Ctrl+A to highlight all the lines, and then press the Delete key.

WORKING WITH WORKSHEET CELLS AND RANGES

When you are ready to write your own VBA procedure to automate a worksheet task, you will most likely begin searching for instructions that allow you to manipulate worksheet cells. You will need to know how to select cells, how to enter data in cells, how to assign range names, how to format cells, and how to move, copy, and delete cells. Although these tasks can be easily performed with the mouse or keyboard, mastering these techniques in VBA requires little practice. You must use the Range object to refer to a single cell, a range of cells, a row, or a column. There are three properties that allow you to access the Range object: the Range property, the Cells property, and the Offset property.

Using the Range Property

The Range property returns a cell or a range of cells. The reference to the range must be in an A1 style and in quotation marks (for example, "A1"). The reference can include the range operator, which is a colon (for example, "A1:B2"), or the union operator, which is a comma (for example, "A5", "B12").

Hands-On 2.8   Using the Range Property to Select Worksheet Cells

To render this into VBA:	Enter this in the Immediate window:
Select a single cell (e.g., A5).	`Range("A5").Select`
Select a range of cells (e.g., A6:A10).	`Range("A6:A10").Select`
Select several non-adjacent cells (e.g., A1, B6, C8).	`Range("A1, B6, C8").Select`
Select several non-adjacent cells and cell ranges (e.g., A11:D11, C12, D3).	`Range("A11:D11, C12, D3").Select`

Using the Cells Property

You can use the Cells property to return a single cell. When selecting a single cell, this property requires two arguments. The first argument indicates the row number and the second one is the column number. Arguments are entered in parentheses. When you omit arguments, Excel selects all the cells in the active worksheet. Let's try out a couple of statements in Hands-On 2.9.

Hands-On 2.9   Using the Cells Property to Select Worksheet Cells (Part I)

To render this into VBA:	Enter this in the Immediate window:
Select a single cell (e.g., A5).	`Cells(5, 1).Select`
Select a range of cells (e.g., A6:A10).	`Range(Cells(6, 1), Cells(10, 1)).Select`
Select all cells in a worksheet.	`Cells.Select`

Notice how you can combine the `Range` property and the `Cells` property:

```
Range(Cells(6, 1), Cells(10, 1)).Select
```

In this example, the first `Cells` property returns cell A6, while the second one returns cell A10. The cells returned by the `Cells` properties are then used as a reference for the `Range` object. As a result, Excel will select the range of cells where the top cell is specified by the result of the first `Cells` property and the bottom cell is defined by the result of the second `Cells` property.

A worksheet is a collection of cells. You can also use the `Cells` property with a single argument that identifies a cell's position in the collection of a worksheet's cells. Excel numbers the cells in the following way: cell A1 is the first cell in a worksheet, cell B1 is the second one, cell C1 is the third one, and so on. Cell 16384 is the last cell in the first worksheet row. Now let's write some practice statements in Hands-On 2.10.

Hands-On 2.10   Using the Cells Property to Select Worksheet Cells (Part II)

To render this into VBA:	Enter this in the Immediate window:

Select cell A1.	`Cells(1).Select`
	or
	`Cells.Item(1).Select`

Select cell C1.	`Cells(3).Select`
	or
	`Cells.Item(3).Select`

Select cell XFD.	`Cells(16384).Select`
	or
	`Cells.Item(16384).Select`

Notice that the word `Item` is a property that returns a single member of a collection. Because an item is the default member for a collection, you can refer to a worksheet cell without explicitly using the `Item` property.

Now that you've discovered two ways to select cells (`Range` property and `Cells` property), you may wonder why you should bother using the more complicated `Cells` property. It's obvious that the `Range` property is more readable; after all, you used the `Range` references in Excel formulas and functions long before you decided to learn about VBA. Using the `Cells` property is more convenient, however, when it comes to working with cells as a collection. Use this property to access all the cells or a single cell from a collection.

Using the Offset Property

Another very flexible way to refer to a worksheet cell is with the `Offset` property. Quite often when automating worksheet tasks, you may not know exactly where a specific cell is located. How can you select a cell whose address you don't know? The answer is to have Excel select a cell based on an existing selection.

The `Offset` property calculates a new range by shifting the starting selection down or up a specified number of rows. You can also shift the selection to the right or left a specified number of columns. In calculating the position of a new range, the `Offset` property uses two arguments. The first argument indicates the row offset and the second one is the column offset. Let's try out some examples in Hands-On 2.11.

(•) Hands-On 2.11   Selecting Cells Using the Offset Property

To render this into VBA:	Enter this in the Immediate window:
Select a cell located one row down and three columns to the right of cell A1.	`Range("A1").Offset(1, 3).Select`
Select a cell located two rows above and one column to the left of cell D15.	`Range("D15").Offset(-2, -1).Select`
Select a cell located one row above the active cell. If the active cell is in the first row, you will get an error message.	`ActiveCell.Offset(-1, 0).Select`

In the first example, Excel selects cell D2. As soon as you enter the second example, Excel chooses cell C13.

If cells A1 and D15 are already selected, you can rewrite the first two statements in the following way:

```
Selection.Offset(1, 3).Select
Selection.Offset(-2, -1).Select
```

Notice that the third example in the practice table displays zero (0) in the position of the second argument. Zero entered as a first or second argument of the `Offset` property indicates a current row or column. The instruction `ActiveCell.Offset(-1, 0).Select` will cause an error if the active cell is located in the first row.

Using the Resize Property

When working with the `Offset` property, you may occasionally need to change the size of a selection of cells. Suppose that the starting selection is A5:A10. How about shifting the selection two rows down and two columns to the right and then changing the size of the new selection? Let's say the new selection should highlight cells C7:C8. The `Offset` property can take care of only the first part of this task. The second part requires another property. Excel has a special `Resize` property. You can combine the `Offset` property with the `Resize` property to answer the foregoing question. Before you combine these two properties, let's proceed to Hands-On 2.12 to learn how you can use them separately.

Hands-On 2.12   Writing a VBA Statement to Resize a Selection of Cells

1. Arrange the screen so that the Microsoft Excel window and the Visual Basic window are side by side.
2. Activate the Immediate window and enter the following instructions:

```
Range("A5:A10").Select
Selection.Offset(2, 2).Select
Selection.Resize(2, 4).Select
```

The first instruction selects range A5:A10. Cell A5 is an active cell. The second instruction shifts the current selection to cells C7:C12. Cell C7 is located two rows below the active cell A5 and two columns to the right of A5. Now the active cell is C7.

The last instruction resizes the current selection. Instead of range C7:C12, cells C7:F8 are selected.

Like the `Offset` property, the `Resize` property takes two arguments. The first argument is the number of rows you intend to include in the selection, and the second argument specifies the number of columns. Hence, the instruction `Selection.Resize(2, 4).Select` resizes the current selection to two rows and four columns.

The last two instructions can be combined in the following way:

```
Selection.Offset(2, 2).Resize(2, 4).Select
```

In this statement, the `Offset` property calculates the beginning of a new range, the `Resize` property determines the new size of the range, and the `Select` method selects the specified range of cells.

Recording a Selection of Cells

By default, the macro recorder selects cells using the `Range` property. If you turn on the macro recorder and select cell A2, enter any text, and select cell A5, you will see the following lines of code in the VBE window:

```
Range("A2").Select
ActiveCell.FormulaR1C1 = "text"
Range("A5").Select
```

You can have the macro recorder use the `Offset` property if you tell it to use relative references. To do this, click View | Macros | Use Relative References, and then choose Record Macro. The macro recorder produces the following lines of code:

```
ActiveCell.Offset(-1, 0).Range("A1").Select
ActiveCell.FormulaR1C1 = "text"
ActiveCell.Offset(3, 0).Range("A1").Select
```

When you record a procedure using the relative references, the procedure will always select a cell relative to the active cell. The first and third lines in this set of instructions reference cell A1, even though nothing was said about cell A1. As you remember from Chapter 1, the macro recorder has its own way of getting things done. To make things simpler, you can delete the reference to `Range("A1")`:

```
ActiveCell.Offset(-1, 0).Select
ActiveCell.FormulaR1C1 = "text"
ActiveCell.Offset(3, 0).Select
```

After recording a procedure using the relative reference, make sure Use Relative References is not selected if your next macro does not require the use of relative addressing.

Using the End Property

If you must quickly access certain remote cells in your worksheet, you may already be familiar with the following keyboard shortcuts: End+up arrow, End+down arrow, End+left arrow, and End+right arrow. In VBA, you can use the `End` property to quickly move to remote cells. Let's move around the worksheet by writing the statements listed in Hands-On 2.13.

⊙ Hands-On 2.13   Selecting Cells Using the End Property

To render this into VBA:	Enter this in the Immediate window:
Select the last cell in any row.	`ActiveCell.End(xlToRight).Select`
Select the last cell in any column.	`ActiveCell.End(xlDown).Select`

To render this into VBA:	Enter this in the Immediate window:
Select the first cell in any row.	`ActiveCell.End(xlToLeft).Select`
Select the first cell in any column.	`ActiveCell.End(xlUp).Select`

Notice that the `End` property requires an argument that indicates the direction you want to move. Use the following Excel built-in direction enumeration constants to jump in the specified direction: `xlToRight`, `xlToLeft`, `xlUp`, and `xlDown`.

Moving, Copying, and Deleting Cells

In the process of developing a new worksheet model, you often find yourself moving and copying cells and deleting cell contents. VBA allows you to automate these worksheet editing tasks with three simple-to-use methods: `Cut`, `Copy`, and `Clear`. Let's do some hands-on exercises to get familiar with the most frequently used worksheet operations.

Hands-On 2.14   Moving, Copying, and Deleting Cells

To render this into VBA:	Enter this in the Immediate window:
Move the contents of cell A5 to cell A4.	`Range("A5").Cut` `Destination:=Range("A4")`
Copy a formula from cell A3 to cells D5:F5.	`Range("A3").Copy` `Destination:=Range("D5:F5")`

Delete the contents of cell A4.	`Range("A4").Clear`
	or
	`Range("A4").Cut`

Notice that the first two methods listed in the table use a special argument called `Destination`. This argument specifies the address of a cell or a range of cells where you want to place the cut or copied data. In the last example, the `Cut` method is used without the `Destination` argument to remove data from the specified cell.

The `Clear` method deletes everything from the specified cell or range, including any applied formats and cell comments. If you want to be specific about what you delete, use the following methods:

- `ClearContents`—Clears only data from a cell or range of cells
- `ClearFormats`—Clears only applied formats
- `ClearComments`—Clears all cell comments from the specified range
- `ClearNotes`—Clears notes and sound notes from all the cells in the specified range

- `ClearHyperlinks`—Removes all hyperlinks from the specified range
- `ClearOutline`—Clears the outline for the specified range

WORKING WITH ROWS AND COLUMNS

Excel uses the `EntireRow` and `EntireColumn` properties to select the entire row or column. Let's now write the statements in Hands-On 2.15 to quickly select entire rows and columns.

(⊙) Hands-On 2.15   Selecting Entire Rows and Columns

To render this into VBA:	Enter this in the Immediate window:
Select an entire row where the active cell is located.	`Selection.EntireRow.Select`
Select an entire column where the active cell is located.	`Selection.EntireColumn.Select`

When you select a range of cells, you may want to find out how many rows or columns are included in the selection. Let's have Excel count rows and columns in `Range("A1:D15")`.

1. Type the following VBA statement in the Immediate window and press Enter:

    `Range( "A1:D15 ").Select`

    If the Microsoft Excel window is visible, Visual Basic will highlight the range A1:D15 when you press Enter.

2. To find out how many rows are in the selected range, enter the following statement:

    `?Selection.Rows.Count`

    As soon as you press Enter, Visual Basic displays the answer on the next line. Your selection includes 15 rows.

3. To find out the number of columns in the selected range, enter the following statement:

    `?Selection.Columns.Count`

    As soon as you press Enter, Visual Basic tells you that the selected `Range("A1:D15")` occupies the width of four columns.

4. In the Immediate window, position the cursor anywhere within the word Rows or Columns and press F1 to find out more information about these useful properties.

Obtaining Information about the Worksheet

How big is an Excel worksheet? How many columns and rows does it contain? If you ever forget the details, use the `Count` property, as shown in Hands-On 2.16.

(◉) **Hands-On 2.16   Counting Rows and Columns**

To render this into VBA:	Enter this in the Immediate window:
Find out the total number of rows in an Excel worksheet.	`?Rows.Count`
Find out the total number of columns in an Excel worksheet.	`?Columns.Count`

A Microsoft Excel worksheet has 1,048,576 rows and 16,384 columns.

ENTERING DATA AND FORMATTING CELLS

The information entered in a worksheet can be text, numbers, or formulas. To enter data in a cell or range of cells, you can use either the `Value` property or the `Formula` property of the `Range` object.

- Using the `Value` property:

```
ActiveSheet.Range("A1:C4").Value = "=4 * 25"
```

- Using the `Formula` property:

```
ActiveSheet.Range("A1:C4").Formula = "=4 * 25"
```

In both examples, cells A1:C4 display 100—the result of the multiplication 4 * 25. Let's proceed to some practice in Hands-On 2.19.

(◉) **Hands-On 2.17   Using VBA Statements to Enter Data in a Worksheet**

To render this into VBA:	Enter this in the Immediate window:
Enter in cell A5 the following text: Amount Due	`Range("A5").Formula = "Amount Due"`

To render this into VBA:	Enter this in the Immediate window:

Enter the number 123 in cell D21.	`Range("D21").Formula = 123`
	or
	`Range("D21").Value = 123`

Enter in cell B4 the following formula: = D21 * 3	`Range("B4").Formula = "=D21 * 3"`

Returning Information Entered in a Worksheet

In some Visual Basic procedures, you will undoubtedly need to return the contents of a cell or a range of cells. Although you can use either the `Value` or `Formula` property, this time, the two `Range` object's properties are not interchangeable.

- The `Value` property displays the result of a formula entered in a specified cell. If, for example, cell A1 contains a formula = 4 * 25, then the instruction

    `?Range( "A1 ").Value`

    will return the value of `100`.

- If you want to display the formula instead of its result, you must use the `Formula` property:

    `?Range( "A1 ").Formula`

Excel will display the formula (= 4 * 25) instead of its result (100).

Finding Out About Cell Formatting

A frequent worksheet task is applying formatting to a selected cell or a range. Your VBA procedure may need to find out the type of formatting applied to a worksheet cell. To retrieve the cell formatting, use the `NumberFormat` property:

`?Range("A1").NumberFormat`

Upon entering the foregoing statement in the Immediate window, Excel displays the word `General`, which indicates that no special formatting was applied to the selected cell. To change the format of a cell to dollars and cents using VBA, enter the following instruction:

`Range("A1").NumberFormat = "$#,##0.00"`

If you enter 125 in cell A1 after it has been formatted using this code, cell A1 will display $125.00. You can look up the available format codes in the Format Cells dialog box in the Microsoft Excel application window, as shown in Figure 2.19.

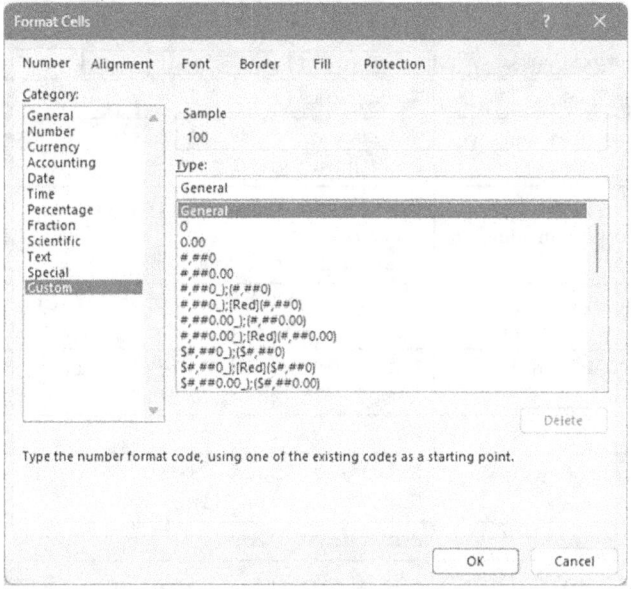

FIGURE 2.19.   You can apply different formatting to selected cells and ranges using format codes, as displayed in the Custom category in the Format Cells dialog box. To quickly bring up this dialog box, press the ALT, H, F, and M keys one at a time.

WORKING WITH WORKBOOKS AND WORKSHEETS

Let's move up one level and learn how you can control a single workbook, as well as an entire collection of workbooks. You cannot prepare a new worksheet if you don't know how to open a workbook. You cannot remove a workbook from the screen if you don't know how to close it. You cannot work with an existing workbook if you don't know how to open it. These important tasks are handled by the following VBA methods: `Add`, `Open`, and `Close`. The next series of drills in Hands-On 2.18 and 2.19 will give you the language skills necessary for dealing with workbooks and worksheets.

◉ Hands-On 2.18   Working with Workbooks

To render this into VBA:	Enter this in the Immediate window:
Open a new workbook.	`Workbooks.Add`
Find out the name of the first workbook.	`?Workbooks(1).Name`

To render this into VBA:	Enter this in the Immediate window:
Find out the number of open workbooks.	`?Workbooks.Count`
Activate the second open workbook.	`Workbooks(2).Activate`
Close the `Chap01.xlsm` workbook and save the changes.	`Workbooks("Chap01.xlsm").Close` `SaveChanges:=True`

Open the `Chap01.xlsm` workbook. Type the correct path to the file location on your computer.	`Workbooks.Open "C:\VBAEx-`
	`cel2024_ByExample\`
	`Chap01_ExcelPrimer.xlsm"`

Activate the `Chap01.xlsm` workbook.	`Workbooks("Chap01.xlsm").Acti-` `vate`
Save the active workbook as NewChap.xlsm.	`ActiveWorkbook.SaveAs File-` `name:= "NewChap.xlsm"`
Close the first workbook.	`Workbooks(1).Close`
Close the active workbook without saving recent changes to it.	`ActiveWorkbook.Close` `SaveChanges:=False`
Close all open workbooks.	`Workbooks.Close`

If you worked through the last example in Hands-On 2.18, all workbooks are now closed. Before you experiment with worksheets, make sure you have opened a new workbook.

When you deal with individual worksheets, you must know how to add a new worksheet to a workbook, select a worksheet or a group of worksheets, name a worksheet, and copy, move, and delete worksheets. In the VBA language, each of these tasks is handled by a special method or property.

Hands-On 2.19   Working with Worksheets

To render this into VBA:	Enter this in the Immediate window:
Add a new worksheet.	`Worksheets.Add`
Find out the name of the first worksheet.	`?Worksheets(1).Name`
Select a sheet named Sheet3.	`Worksheets(3).Select`
Select sheets 1, 3, and 4.	`Worksheets(Array(1,3,4)).Select`
Activate a sheet named Sheet1.	`Worksheets("Sheet1").Activate`
Move Sheet2 before Sheet1.	`Worksheets("Sheet2").Move` `Before:=Worksheets("Sheet1")`
Rename worksheet Sheet2 to Expenses.	`Worksheets("Sheet2").Name = "Expenses"`

To render this into VBA:	Enter this in the Immediate window:
Find out the number of worksheets in the active workbook.	`?Worksheets.Count`
Remove the worksheet named Expenses from the active workbook.	`Worksheets("Expenses").Delete`

Notice the difference between the `Select` and `Activate` methods:

- The `Select` and `Activate` methods can be used interchangeably if only one worksheet is selected.

- If you select a group of worksheets, the `Activate` method allows you to decide which one of the selected worksheets is active. As you know, only one worksheet can be active at a time.

Sheets Other than Worksheets

In addition to worksheets, the collection of workbooks contains chart sheets. To add a new chart sheet to your workbook, use the `Add` method:

`Charts.Add`

To count the chart sheets, use:

`?Charts.Count`

WORKING WITH WINDOWS

When you work with several Excel workbooks and need to compare or consolidate data or you want to see different parts of the same worksheet, you are bound to use the options available in the Excel ribbon's View tab: New Window and Arrange All.

In Hands-On 2.20, you will learn how to work with Windows using VBA.

(•) Hands-On 2.20   Working with Windows

To render this into VBA:	Enter this in the Immediate window:
Show the active workbook in a new window.	`ActiveWorkbook.NewWindow`
Display on screen all open workbooks.	`Windows.Arrange`
Activate the second window.	`Windows(2).Activate`
Find out the title of the active window.	`?ActiveWindow.Caption`
Change the active window's title to My Window.	`ActiveWindow.Caption = "My Window"`

When you display windows on the screen, you can decide how to arrange them. The `Arrange` method has many arguments, as shown in Table 2.1. The argument that allows you to control the way the windows are positioned on your screen is called `ArrangeStyle`. If you omit the `ArrangeStyle` argument, all windows are tiled.

TABLE 2.1. Arguments of the Arrange method of the Windows object.

Constant	Value	Description
xlArrangeStyleTiled	1	Windows are tiled (the default value).
xlArrangeStyleCascade	7	Windows are cascaded.
xlArrangeStyleHorizontal	2	Windows are arranged horizontally.
xlArrangeStyleVertical	3	Windows are arranged vertically.

Instead of the names of constants, you can use the value equivalents shown in Table 2.1.

To cascade all windows, use the following VBA instruction:

```
Windows.Arrange ArrangeStyle:=xlArrangeStyleCascade
```

or simply:

```
Windows.Arrange ArrangeStyle:=7
```

WORKING WITH THE EXCEL APPLICATION OBJECT

The Excel `Application` object serves as the highest-level object in the Excel object model. It represents the entire Excel application and provides numerous properties, methods, and events to control and interact with Excel. Let's write some statements that use the `Application` object.

⊙ Hands-On 2.21 Working with the Excel Application Object

To render this into VBA:	Enter this on one line in the Immediate window:
Open a workbook located at the specified path.	`Application.Open "C:VBAExcel2024_ByExample\ Chap01.xlsm"`
Display a message in the Excel status bar.	`Application.StatusBar = "Pre- paring your environment..."`
Recalculate all open workbooks.	`Application.Calculate`

To render this into VBA:	Enter this on one line in the Immediate window:
Run the specified macro within the Excel environment.	`Application.Run "NewFolder"`
Check the name of the active application.	`?Application.Name`
Change the title of the Excel application to My Application.	`Application.Caption = "My Application"`
Change the title of the Excel application back to Microsoft Excel.	`Application.Caption = "Microsoft Excel"`
Find out what operating system you are using.	`?Application.OperatingSystem`
Find out the name of a person or firm to whom the application is registered.	`?Application.OrganizationName`
Find out the name of the folder where the Excel executable file (`Excel.exe`) resides.	`?Application.Path`
Quit working with Microsoft Excel.	`Application.Quit`

SUMMARY

This chapter provided an overview of the VBE development environment and its various tools. You learned many essential VBA terms and practiced them by writing and executing single statements from the Immediate window.

In the next chapter, you will discover how to store data for later use in variables. Additionally, you will delve into VBA data types and constants.

Chapter 3

Excel VBA Fundamentals— A Quick Reference to Writing VBA Code

In programming, just as in life, certain things need to be done at once while others can be put off until later. When you postpone a task, you may enter it in your mental or paper "to do" list and classify it by its type or importance. When you delegate the task or finally get around to doing it yourself, you cross it off the list. This chapter shows you how your VBA procedures can memorize important pieces of information for use in later statements or calculations. You will learn how a procedure can keep a "to do" entry in a variable, how variables are declared, and how they relate to data types and constants.

EXCEL OBJECTS, PROPERTIES, AND METHODS

You can create procedures that control many features of Microsoft Excel using VBA. VBA can also be used to control other applications. The power of VBA comes from its ability to control and manage various objects. What is an object, though?

An *object* is an entity you can control with VBA. Examples of objects in Excel include workbooks, worksheets, ranges in a worksheet, charts, and toolbars. Excel contains a multitude of objects that you can manipulate in different ways, and these objects are organized in a hierarchy. Some objects may contain other objects. For example, Microsoft Excel is an `Application` object. The `Application` object contains other objects, such as workbooks and command bars. The `Workbook` object may contain other objects, such as worksheets or charts.

In this chapter, you will learn how to control the following Excel objects: `Range`, `Window`, `Worksheet`, `Workbook`, and `Application`. You will begin by learning about the `Range` object, as you can't do much work in spreadsheets without knowing how to manipulate ranges of cells.

Certain objects look alike. For example, if you open a new workbook and examine its worksheets, you won't see any differences. A group of similar objects is called a *collection*. A `Worksheets` collection includes all worksheets in a workbook. Collections are also objects.

In Microsoft Excel, the most frequently used collections are:

- `Workbooks` collection—represents all currently open workbooks.
- `Worksheets` collection—represents all the `Worksheet` objects in the specified or active workbook. Each `Worksheet` object represents a worksheet.
- `Sheets` collection—represents all the sheets in the specified or active workbook. The `Sheets` collection can contain `Chart` or `Worksheet` objects.
- `Windows` collection—represents all the `Window` objects in Microsoft Excel. The `Windows` collection for the `Application` object contains all the windows in the application, whereas the `Windows` collection for the `Workbook` object contains only the windows in the specified workbook.

When you work with collections, you can perform the same action on all the objects in the collection. Each object has some characteristics that allow you to describe the object. In Visual Basic, the object's characteristics are called *properties*. For example, a `Workbook` object has a `Name` property, and the `Range` object has such properties as `Column`, `Font`, `Formula`, `Name`, `Row`, `Style`, and `Value`.

The object properties can be set. When you set an object's property, you control its appearance or its position. Object properties can take on only one specific value at any one time. For example, the active workbook can't be called two different names at the same time.

Understanding that some properties can also be objects is one of the most challenging aspects of VBA. Let's consider the `Range` object. You can change the appearance of a selected range of cells by setting the `Font` property. The font, however, can have a different name (Times New Roman, Arial, etc.), different size (10, 12, 14, etc.), and different style (bold, italic, underline, etc.). These are `Font` properties. Since `Font` has properties, it is also considered an object.

Properties are great for changing the appearance of an object, but how can you control its actions? This is where another important concept comes in: *methods*. Objects have methods, which are actions you want the object to perform. One of the most crucial Visual Basic methods is the `Add` method, used to add a new workbook or worksheet. Various objects can utilize different methods. For example, the `Range` object has methods to clear cell contents (`ClearContents`), clear formats (`ClearFormats`), and clear both contents and formats (`Clear`). Other methods allow objects to be selected, copied, or moved.

Methods can have optional parameters that specify how the method is to be carried out. For instance, the `Workbook` object has a `Close` method, used to close any open workbook. If there are changes to the workbook, Microsoft Excel will prompt you to save them. You can use the `Close` method with the `SaveChanges` parameter set to `False` to close the workbook and discard any changes that have been made to it, as in the following example:

```
Workbooks("Chap01.xlsm").Close SaveChanges:=False
```

MICROSOFT EXCEL OBJECT MODEL

When you learn new things, theory provides the necessary background, but how do you really know what's where? All the available Excel objects, along with their properties and methods, can be looked up in the online Excel object model reference. You can access it by selecting Help | Microsoft Visual Basic for Applications Help in the VBE window.

Figure 3.1 illustrates the Excel object model reference in the online help. You can access this page via the following link:

https://docs.microsoft.com/en-us/office/vba/api/overview/Excel/object-model

Objects are listed alphabetically for easy perusal, and when you click the object, you will see object subcategories listing the object's properties, methods, and events. Reading the object model reference is a great way to learn about Excel objects and collections of objects. The time you spend here will pay big dividends later when you need to write complex VBA procedures from scratch. A good way to get started is to always look up objects that you come across in Excel programming texts or example procedures.

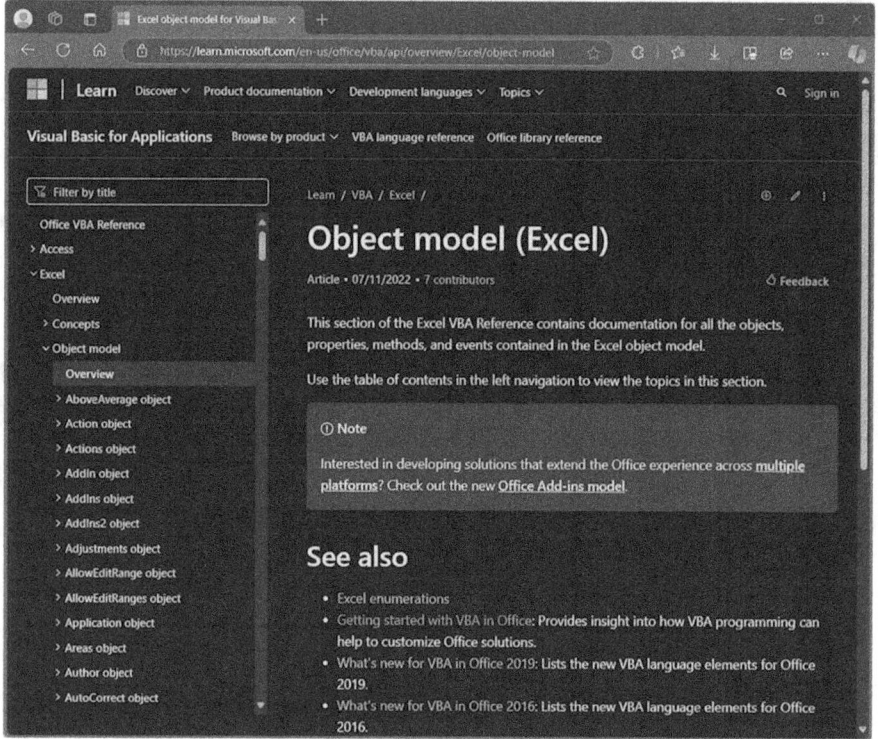

FIGURE 3.1. In your VBA programming work, always refer to the Excel object model reference, which contains documentation for all the objects, properties, methods, and events contained in the Excel object model.

Now, take a few minutes to familiarize yourself with the main Excel object—`Application`. This object allows you to specify application-level properties and execute application-level methods. Recall that, in Chapter 2, you saw several examples of VBA statements that use the `Application` object.

WRITING SIMPLE AND COMPLEX VBA STATEMENTS

Now that you know about the basic elements of VBA (objects, properties, and methods), it's time to begin using them. How do you combine objects, properties, and methods into correct language structures? Every language has grammar rules that people follow to make themselves understood. Whether you communicate in English, Spanish, French, or another language, you apply certain rules to your writing and speech. In programming, we use the term *syntax* to specify language rules. You can look up the syntax of each object, property, or method in the online help or in the Object Browser window, which was discussed in Chapter 2.

To make sure Excel always understands what you mean, just stick to the following rules.

Rule #1: Referring to the property of an object

If the property does not have arguments, the syntax is as follows:

```
Object.Property
```

`Object` is a placeholder. It is where you should place the name of the actual object that you are trying to access. `Property` is also a placeholder. Here, you place the name of the object's characteristics. For example, to refer to the value entered in cell A4 on your worksheet, you can write the following instruction:

```
Range("A4").Value
```

Notice that there is a period between the name of the object and its property.

To access the property of an object that is contained within several other objects, you must include the names of all objects in turn, separated by the dot operator, as shown here:

```
ActiveSheet.Shapes(2).Line.Weight
```

This example references the `Weight` property of the `Line` object and refers to the second object in the collection of `Shapes` located in the active worksheet.

Some properties require one or more arguments. For example, when using the `Offset` property, you can select a cell relative to the active cell. The `Offset` property requires two arguments. The first argument indicates the row number (`rowOffset`), and the second one determines the column number (`columnOffset`). For instance:

```
ActiveCell.Offset(3, 2)
```

Assuming the active cell is A1, `Offset(3,  2)` will reference the cell located three rows down and two columns to the right of cell A1. In other words, cell C4 is referenced.

Because the arguments placed within parentheses are often difficult to understand, it's common practice to precede the value of the argument with its name, as in the following example:

```
ActiveCell.Offset(rowOffset:=3, columnOffset:=2)
```

Notice that a colon and an equals sign must always follow the named arguments. When you use the named arguments, you can list them in any order. The foregoing instruction can also be written as follows:

```
ActiveCell.Offset(columnOffset:=2, rowOffset:=3)
```

The revised instruction does not change the meaning; you are still referencing cell C4, assuming that A1 is the active cell. If you transpose the arguments in a statement that does not use named arguments, however, you will end up referencing another cell. For example, the statement `ActiveCell.Offset(2,  3)` will reference cell D3 instead of C4.

Rule #2: Changing the property of an object

Here is the syntax:

```
Object.Property = Value
```

`Value` is a new value that you want to assign to the property of the object. The value can be:

- A number—In the following example, 25 is entered in cell A4:

  ```
  Range("A4").Value = 25
  ```

- Text entered in quotes—In the following statement, we specify a new font name for the active cell:

  ```
  ActiveCell.Font.Name =   "Times New Roman "
  ```

- A logical value (`True` or `False`)—The following instruction applies bold formatting to the active cell:

  ```
  ActiveCell.Font.Bold = True
  ```

Rule #3: Returning the current value of the object property

Here is the syntax:

```
Variable = Object.Property
```

`Variable` is the name of the storage location where Visual Basic will store the specified property setting. You will learn about variables later in this chapter. For instance, the following statement:

```
CellValue = Range("A4").Value
```

instructs Visual Basic to save the current value of cell A4 in the variable named `CellValue`.

Rule #4: Referring to the object's method

If the method does not have arguments, the syntax is as follows:

```
Object.Method
```

The `Object` is a placeholder. It is where you should place the name of the actual object that you are trying to access. The `Method` is also a placeholder. Here, you place the name of the action you want to perform on the object. For example, to clear the contents in cell A4, use the following instruction:

```
Range("A4").ClearContents
```

If the method requires arguments, the syntax is as follows:

```
Object.Method (argument1, argument2, ... argumentN)
```

For example, using the `GoTo` method, you can quickly select any range in a workbook. The syntax of the `GoTo` method is shown here:

```
Object.GoTo(Reference, Scroll)
```

The `Reference` argument is the destination cell or range. The `Scroll` argument can be set to `True` to scroll through the window or to `False` to not scroll through the window. For example, the following VBA statement selects cell P100 in `Sheet1` and scrolls through the window:

```
Application.GoTo _
    Reference:=Worksheets("Sheet1").Range("P100"), _
    Scroll:=True
```

The foregoing instruction did not fit on one line, so it was broken into sections using the special line continuation character (the underscore), described in the next section.

Suppose you want to delete the contents of cell A4. To do this manually, you would select cell A4 and press the Delete key. To perform the same operation using VBA, you first need to find out how to make Excel select the appropriate cell. Cells in a worksheet can be accessed using the `Range` object. Visual

Basic does not have a `Delete` method for deleting contents of cells. Instead, you should use the `ClearContents` method, as in the following example:

```
Range( "A4 ").ClearContents
```

Notice the dot operator between the name of the object and its method. This instruction removes the contents of cell A4. How do you make Excel delete the contents of cell A4 located in the first sheet of the `Chap01.xlsm` workbook, however? Let's also assume that you have several workbooks open in Excel. To ensure that you don't delete the contents of cell A4 from the wrong workbook or worksheet, you must write detailed instruction:

```
Application.Workbooks("Chap01.xlsm")
.Worksheets("Sheet1").Range("A4").ClearContents
```

The foregoing instruction should be written on one line and read from right to left as follows: clear the contents of cell A4, which is part of a range located in a worksheet named `Sheet1` contained in a workbook named `Chap01.xlsm`, which, in turn, is part of the Excel application. Be sure to include the letter "s" at the end of the collection names: `Workbooks` and `Worksheets`.

All references to the names of workbooks, worksheets, and cells must be enclosed in quotation marks.

Breaking Up Long VBA Statements

When you start writing complete VBA procedures from scratch, you will need to know how to break up a long VBA statement into two or more lines to make your procedure more readable. VBA has a special line continuation character that can be used at the end of a line to indicate that the next line is a continuation of the previous one, as in the following example:

```
Selection.PasteSpecial _
   Paste:=xlValues, _
   Operation:=xlMultiply, _
   SkipBlanks: =False, _
  Transpose:=False
```

The line continuation character is the underscore (_). It must be preceded by a single space.

You can use the line continuation character in your code before or after:

- Operators, for example: `&`, `+`, `Like`, `NOT`, and `AND`
- A comma
- An equals sign
- An assignment operator (`:=`)

Note that you cannot use the line continuation character between a colon and an equals sign.

For example, the following use of the continuation character is invalid:

```
Selection.PasteSpecial Paste: _
    =xlValues, Operation: _
    =xlMultiply, SkipBlanks: _
    =False, Transpose: _
    =False
```

Also, you may not use the line continuation character within text enclosed in quotes. For example, the following usage of the underscore is invalid:

```
MsgBox "To continue the long instruction, use the _
    line continuation character."
```

Instead, break up this statement as follows:

```
MsgBox "To continue the long instruction, use the " & _
    "line continuation character."
```

SAVING RESULTS OF VBA STATEMENTS

In Chapter 2, while working in the Immediate window, you experimented with several Visual Basic instructions to retrieve information. For instance, by entering ?Rows.Count, you discovered that a worksheet contains 1,048,576 rows. When writing Visual Basic statements outside of the Immediate window, however, you cannot use the question mark. Instead, you can store the results of Visual Basic instructions in variables. Since variables can hold various types of data, the next section will introduce you to VBA data types. Understanding the basics of data types will make it easier to work with variables.

INTRODUCING DATA TYPES

When you create Visual Basic procedures, you have a purpose in mind, such as the manipulation of data. As your procedures will need to handle different kinds of information, you should understand how Visual Basic stores data.

The *data type* determines how the data is stored in the computer's memory. For example, it can be stored as a number, text, date, object, and so on. If you forget to tell Visual Basic the type of your data, it assigns the Variant data type. The Variant type can figure out on its own what kind of data is being manipulated and then take on that type. Visual Basic common data types are listed in Table 3.1.

In addition to the built-in data types, you can define your own data types. Data types vary in how much space they take up in the computer's memory, making some more expensive than others. To conserve memory and enhance the performance of your VBA procedures, it's important to select a data type that uses the fewest bytes while still being able to handle the required data.

TABLE 3.1.   Common VBA data types.

Data Type (Name)	Size (Bytes)	Description
Boolean	2	Stores a value of True (0) or False (–1).
Byte	1	A number in the range of 0 to 255.
Integer	2	A number in the range of –32,768 to 32,767. The type declaration character for Integer is the percent sign (%).
Long (Long integer)	4	A number in the range of –2,147,483,648 to 2,147,483,647. The type declaration character for Long is the ampersand (&).
LongLong (LongLong integer)	8	Stored as a signed 64-bit (8-byte) number ranging in value from –9,223,372,036,854,775,808 to 9,223,372,036,854,775,807. The type declaration character for LongLong is the caret (^). LongLong is a valid declared type only on 64-bit platforms.
LongPtr (Long integer on 32-bit systems; LongLong integer on 64-bit systems)	4 on 32-bit 8 on 64-bit	Numbers ranging in value from –2,147,483,648 to 2,147,483,647 on 32-bit systems; –9,223,372,036,854,775,808 to 9,223,372,036,854,775,807 on 64-bit systems. Using LongPtr enables writing code that can run in both 32-bit and 64-bit environments.
Single (single-precision floating-point)	4	Single-precision floating-point real number ranging in value from –3.402823E38 to –1.401298E–45 for negative values and from 1.401298E–45 to 3.402823E38 for positive values. The type declaration character for Single is the exclamation point (!).
Double (double-precision floating-point)	8	Double-precision floating-point real number in the range of –1.79769313486231E308 to –4.94065645841247E–324 for negative values and 4.94065645841247E–324 to 1.79769313486231E308 for positive values. The type declaration character for Double is the number sign (#).
Currency (scaled integer)	8	(Scaled integer) monetary values used in fixed-point calculations: –922,337,203,685,477.5808 to 922,337,203,685,477.5807. The type declaration character for Currency is the at sign (@).

Data Type (Name)	Size (Bytes)	Description

Decimal	14	96-bit (12-byte) signed integer scaled by a variable power of 10. The power of 10 scaling factor specifies the number of digits to the right of the decimal point and ranges from 0 to 28.
		 With no decimal point (scale of 0), the largest value is +/–79,228,162,514,264,337,593,543,950,335.
		 With 28 decimal places, the largest value is +/–7.9228162514264337593543950335. The smallest non-zero value is +/–0.0000000000000000000000000001.
		 You cannot declare a variable to be of type Decimal. You must use the Variant data type. Use the `CDec` function to convert a value to a decimal number:
		 `Dim numDecimal As Variant`
		 `numDecimal = CDec(0.02 * 15.75 * 0.0006)`

Date	8	Date from January 1, 100, to December 31, 9999, and times from 0:00:00 to 23:59:59. Date literals must be enclosed within number signs (#)—for example, `#January 1, 2022#`
String (variable-length)	10 bytes + string length	A variable-length string can contain up to approximately 2 billion characters. The type declaration character for String is the dollar sign ($).
String (fixed-length)	Length of string	A fixed-length string can contain 1 to approximately 65,400 characters.
Object	4	Object variable used to refer to any Excel object. Use the `Set` statement to declare a variable as an Object.
Variant (with numbers)	16	Any numeric value up to the size of a Double.
Variant (with characters)	22 bytes + string length	Any valid non-numeric data type in the same range as for a variable-length string.

User-Defined (using Type)	One or more elements	A data type you define using the `Type` statement. User-defined data types can contain one or more elements of a data type, an array, or a previously defined user-defined type—for example:
		 `Type custInfo`
		 `  custFullName as String`
		 `  custTitle as String`
		 `  custBusinessName as String`
		 `  custFirstOrderDate as Date`
		 `End Type`

Excel 2024 introduces some new data types. The *Stocks* data type allows you to enter real-time financial data directly into your spreadsheets. Another new data type is *Geography*, which provides geographical information for locations. These new data types can be very useful for enhancing data analysis and visualization capabilities in Excel.

NOTE	*For more information about data types see the online help at https://docs.microsoft.com/en-us/office/vba/language/reference/user-interface-help/data-type-summary.*

USING VARIABLES

A *variable* is simply a name used to refer to an item of data. Each time you want to remember the result of a VBA instruction, think of a name that will represent it. For example, if the number 1,048,576 must remind you of the total number of rows in a worksheet—a crucial piece of information when importing external data into Excel—you could choose a name such as `AllRows`, `NumOfRows`, or `TotalRows`.

Variable names can contain characters, numbers, and some punctuation marks, except for the following: !, #, $, %, &, @, and ,. The name of a variable cannot begin with a number or contain spaces. If you want the name to include more than one word, use the underscore (_) as a separator.

Although a variable can contain up to 254 characters, it's best to use short and simple names to save typing time when referring to the variable in your Visual Basic procedure. Visual Basic doesn't differentiate between uppercase or lowercase letters in variable names, but most programmers prefer using lowercase. For variable names consisting of multiple words, you might use title case, such as `NumOfRows` or `First_Name`.

Reserved Words Can't Be Used for Variable Names

You can use any label you want for a variable name, except for the reserved VBA words. Visual Basic statements and certain other words that have a special meaning in VBA cannot be used as names of variables. For example, words such as Name, Len, Empty, Local, Currency, or Exit will generate an error message if used as a variable name.

Meaningful Variable Names

Give variables names that can help you remember their roles. Some programmers use a prefix to identify the type of variable. A variable name that begins with "str" (for example, `strName`) can be quickly recognized within the code of your procedure as the one holding the Text data type.

How to Create Variables

You can create a variable by declaring it with a special command or using it in a VBA statement. When you declare your variable, you make Visual Basic aware of the variable's name and data type. This is called *explicit variable declaration*. There are several advantages to explicit variable declaration:

- Explicit variable declaration speeds up the execution of your procedure. Because Visual Basic knows the data type, it reserves only as much memory as is necessary to store the data.

- It makes your code easier to read and understand because all the variables are listed at the very beginning of the procedure.

- It helps prevent errors caused by misspelled variable names. Visual Basic automatically corrects the variable name based on the spelling used in the variable declaration.

If you don't let Visual Basic know about the variable prior to using it, you are implicitly telling VBA that you want to create this variable. Variables declared implicitly are automatically assigned the Variant data type (see Table 3.1 in the previous section). Although implicit variable declaration is convenient (it allows you to create variables on the fly and assign values without knowing in advance the data type of the values being assigned), it can cause several problems, as outlined here:

- If you misspell a variable name in your procedure, Visual Basic may display a runtime error or create a new variable. You are guaranteed to waste some time troubleshooting problems that could have been easily avoided had you declared your variable at the beginning of the procedure.

- Because Visual Basic does not know what type of data your variable will store, it assigns it a Variant data type. This causes your procedure to run slower because Visual Basic must check the data type every time it deals with your variable. Because Variant can store any type of data, Visual Basic must reserve more memory to store your data.

How to Declare Variables

You declare a variable with the `Dim` keyword. `Dim` stands for *dimension*. The `Dim` keyword is followed by the name of the variable and the variable type.

Suppose you want your VBA procedure to display the age of an employee. Before you can calculate the age, you must tell the procedure the employee's date of birth. To do this, you declare a variable called `DateOfBirth`, as follows:

```
Dim DateOfBirth As Date
```

Notice that the `Dim` keyword is followed by the name of the variable (`DateOf-Birth`). This name can be anything you choose, but it cannot be one of the reserved VBA keywords. You must specify the type of data the variable will hold using the `As` keyword after the variable name, followed by the desired data type. The Date data type tells Visual Basic that the variable `DateOfBirth` will store a date.

To store the employee's age, declare the `age` variable as follows:

```
Dim age As Integer
```

The `age` variable will store the number of years between today's date and the employee's date of birth. Because `age` is displayed as a whole number, this variable has been assigned the Integer data type.

You may also want your procedure to keep track of the employee's name, so you declare another variable to hold the employee's first and last name:

```
Dim FullName As String
```

Because the word "Name" is on the VBA list of reserved words, using it in your VBA procedure would guarantee an error. To hold the employee's full name, call the variable `FullName` and declare it as the String data type, because the data it will hold is text.

Declaring variables is regarded as a good programming practice because it makes programs easier to read and helps prevent certain types of errors.

Informal Variables

Variables that are not explicitly declared with `Dim` statements are said to be implicitly declared. These variables are automatically assigned Variant data type. They can hold numbers, strings, and other types of information.

You can create a variable by simply assigning some value to a variable name anywhere in your VBA procedure. For example, you will implicitly declare a variable in the following way:

```
DaysLeft = 100
```

Now that you know how to declare your variables, let's look at a complete VBA procedure that uses them:

```
Sub AgeCalc()
    ' variable declaration
    Dim FullName As String
    Dim DateOfBirth As Date
    Dim age As Integer

    ' assign values to variables
    FullName = "John Smith"
    DateOfBirth = #01/03/1981#

    ' calculate age
    age = Year(Now())-Year(DateOfBirth)

    ' print results to the Immediate window
    Debug.Print FullName & " is " & age & " years old."
End Sub
```

The variables are declared at the beginning of the procedure in which they are going to be used. While, in this procedure, the variables are declared on separate lines, you can declare several variables on the same line, separating each variable name with a comma, as shown here:

```
Dim FullName As String, DateOfBirth As Date, age As Integer
```

Notice that the `Dim` keyword appears only once at the beginning of the variable declaration line.

When Visual Basic executes the variable declaration statements, it creates the variables with the specified names and reserves memory space to store their values. Then, further in the procedure, specific values are assigned to these variables.

To assign a value to a variable, begin with a variable name followed by an equals sign. The value entered to the right of the equals sign is the data you want to store in the variable. The data you enter here must be of the type determined by the variable declaration. Text data should be surrounded by quotation marks, and dates by # characters.

Using the data supplied by the `DateOfBirth` variable, Visual Basic calculates the age of an employee and stores the result of the calculation in the `age` variable. Then, the full name of the employee as well as the age is printed in the Immediate window using the instruction `Debug.Print`. When the Visual Basic procedure has been executed, you must view the Immediate window to see the results.

Let's see what happens when you declare a variable with the incorrect data type. The purpose of the following procedure is to calculate the total number of rows in a worksheet and then display the results in a dialog box:

```
Sub HowManyRows()
  Dim NumOfRows As Integer

  NumOfRows = Rows.Count

  MsgBox "The worksheet has " & NumOfRows & " rows."
End Sub
```

A wrong data type can cause an error. In the foregoing procedure, when Visual Basic attempts to write the result of the Rows.Count statement to the variable NumOfRows, the procedure fails, and Excel displays the message Run-time error 6: Overflow. This error results from selecting an invalid data type for that variable. The number of rows in a spreadsheet does not fit the Integer data range. To correct the problem, you should choose a data type that can accommodate a larger number:

```
Sub HowManyRows2()
  Dim NumOfRows As Long

  NumOfRows = Rows.Count
  MsgBox "The worksheet has " & NumOfRows & " rows."
End Sub
```

You can also correct the problem caused by the assignment of the wrong data type in the first example by deleting the variable type (As Integer). When you rerun the procedure, Visual Basic will assign to your variable the Variant data type. Although Variants use up more memory than any other variable type and slow down the speed at which your procedures run (because Visual Basic must do extra work to check the Variant's context), when it comes to short procedures, the cost of using Variants is barely noticeable.

What Is the Variable Type?

You can quickly find out the type of variable used in your procedure by right-clicking the variable name and selecting Quick Info from the context menu.

Concatenation

You can combine two or more strings to form a new string. The joining operation is called *concatenation*. You have seen examples of concatenated strings in the foregoing `AgeCalc` and `HowManyRows2` procedures. Concatenation is represented by an ampersand character (`&`).

For instance, `"His name is " & FirstName` will produce the following string: His name is John. The name of the person is determined by the contents of the `FirstName` variable. Notice that there is an extra space between "is" and the ending quotation mark: `"His name is "`.

Concatenation of strings also can be represented by a plus sign (+). Many programmers, however, prefer to restrict the plus sign to operations on numbers to eliminate ambiguity.

Specifying the Data Type of a Variable

If you don't specify the variable's data type in the `Dim` statement, you end up with an *untyped* variable. Untyped variables in VBA are always Variant data types. It's highly recommended that you create typed variables. When you declare a variable of a certain data type, your VBA procedure runs faster because Visual Basic does not have to stop analyzing the Variant variable to determine its type.

Visual Basic can work with many types of numeric variables. Integer variables can hold only whole numbers from –32,768 to 32,767. Other types of numeric variables are Long, Single, Double, and Currency. Long variables can hold whole numbers in the range –2,147,483,648 to 2,147,483,647. Unlike the Integer and Long variables, the Single and Double variables can hold decimals. String variables are used to refer to text. When you declare a variable of the String data type, you can tell Visual Basic how long the string should be—for instance:

```
Dim extension As String * 3
```

declares a fixed-length String variable named `extension` that is three characters long. If you don't assign a specific length, the String variable will be dynamic. This means that Visual Basic will make enough space in computer memory to handle whatever amount of text is assigned to it.

After declaring a variable, you can only store in it the type of information specified in the declaration statement. Assigning string values to numeric variables or numeric values to string variables results in the "Type mismatch" error or causes Visual Basic to modify the value. For example, if your variable is declared to hold whole numbers and you assign a decimal value to it, Visual Basic will disregard the decimals and use only the whole part of the number.

When you run the `MyNumber` procedure shown here, Visual Basic modifies the data to fit the variable's data type (Integer), and instead of `23.11`, the variable ends up holding a value of `23`:

```
Sub MyNumber()
  Dim myNum As Integer
  myNum = 23.11
  MsgBox myNum
End Sub
```

If you don't declare a variable with a `Dim` statement, you can still designate a type for it by using a special character at the end of the variable name. To declare the `FirstName` variable to hold a text string, you can append the dollar sign to the variable name, like this:

```
Dim FirstName$
```

This declaration is the same as `Dim FirstName As String`. Type declaration characters are shown in Table 3.2.

TABLE 3.2. Type declaration characters.

Data Type	Character
Integer	%
Long	&
Single	!
Double	#
Currency	@
String	$

Notice that type declaration characters can be used only with six data types. To use type declaration, simply append the required character to the end of the variable name.

In the `AgeCalc2` procedure shown below, we use two type declaration characters from Table 3.2.

```
Sub AgeCalc2()
  ' variable declarations
  Dim FullName$
  Dim DateOfBirth As Date
  Dim age%

  ' assign values to variables
  FullName$ = "John Smith"
  DateOfBirth = #1/3/1981#
```

```
' calculate age
age% = Year(Now()) - Year(DateOfBirth)

' print results to the Immediate window
     Debug.Print FullName$ & " is " & age% & " years old."
End Sub
```

Declaring Typed Variables

The variable type can be indicated by the As keyword or a type symbol (see Table 3.2). If you don't add the symbol type or the As command followed by a space and the data type name, the variable will be the default data type—Variant.

Assigning Values to Variables

You've now mastered naming and declaring variables, and we can move on to exploring how to effectively use and manipulate these variables within your VBA procedures to make your code more dynamic and powerful. In Hands-On 3.1, you will create and assign specific values to variables.

NOTE	*Please note files for the "Hands-On" exercise can be found in the companion files.*

◉ Hands-On 3.1   Writing a VBA Procedure with Variables

1. Create a new workbook and save it as C:\VBAExcel2024_ByExample\Chap03. xlsm. Be sure to save it as an Excel macro-enabled workbook (*.xlsm).
2. Activate the VBE window and, in the Project Explorer window, select the new project, VBAProject (Chap03.xlsm).
3. Choose Insert | Module to add a new module.
4. When Module1 is selected, use the Properties window to change its name to Variables.
5. In the Code window, enter the CalcCost procedure shown here:

```
Sub CalcCost()
  slsPrice = 35
  slsTax = 0.085

  Range("A1").Formula = "The cost of calculator"
  Range("A4").Formula = "Price"
  Range("B4").Formula = slsPrice
  Range("A5").Formula = "Sales Tax"
  Range("A6").Formula = "Cost"
  Range("B5").Formula = slsPrice * slsTax
```

```
    cost = slsPrice + (slsPrice * slsTax)

    With Range("B6")
      .Formula = cost
      .NumberFormat = "0.00"
    End With

    strMsg = "The calculator total is $" & cost & "."
    Range("A8").Formula = strMsg
End Sub
```

The foregoing procedure calculates the cost of purchasing a calculator using the following assumptions: the price of a calculator is $35 and the sales tax equals 8.5%.

The procedure uses four variables: `slsPrice`, `slsTax`, `cost`, and `strMsg`. Because none of these variables have been explicitly declared, they all have the same data type—Variant. The variables `slsPrice` and `slsTax` were created by assigning some values to variable names at the beginning of the procedure. The `cost` variable was assigned a value that is a result of a calculation: `slsPrice + (slsPrice * slsTax)`.

The cost calculation uses the values supplied by the `slsPrice` and `slsTax` variables. The `strMsg` variable prepares a text message. This message is then entered as a complete sentence in a worksheet cell. When you assign values to variables, place an equals sign after the name of the variable. After the equals sign, you must enter the value of the variable. This can be a number, a formula, or text surrounded by quotation marks. While the values assigned to the variables `slsPrice`, `slsTax`, and `cost` are easily understood, the value stored in the `strMsg` variable is a little more involved. Let's examine the contents of the `strMsg` variable:

```
strMsg = "The calculator total is $ " & cost & "."
```

- The string `"The calculator total is "` is surrounded by quotation marks. Notice that there is extra space before the ending quotation marks.
- The dollar sign inside the quotes is used to denote the Currency data type. Because the dollar symbol is a character, it is surrounded by quotes.
- The `&` character allows another string or the contents of a variable to be appended to the string. The `&` character must be used every time you want to append a new piece of information to the previous string.

- The `cost` variable is a placeholder. The actual cost of the calculator will be displayed here when the procedure runs.
- The `&` character attaches yet another string.
- The period is surrounded by quotes. When you require a period at the end of a sentence, you must attach it separately when it follows the name of the variable.

Variable Initialization

When Visual Basic creates a new variable, it initializes the variable. Variables assume their default value. Numerical variables are set to zero (0), Boolean variables are initialized to `False`, String variables are set to the empty string (`""`), and Date variables are set to December 30, 1899.

Now, let's execute the `CalcCost` procedure.

6. Position the cursor anywhere within the `CalcCost` procedure and choose Run | Run Sub/UserForm.

   When you run this procedure, Visual Basic may display the following message: "Compile error: Variable not defined." If this happens, click OK to close the message box. Visual Basic will select the `slsPrice` variable and highlight the name of the `CalcCost` procedure. The title bar displays "Microsoft Visual Basic – Chap03.xlsm [break]." The Visual Basic break mode allows you to correct the problem before you continue.

7. For now, exit the break mode by choosing Run | Reset.

8. Now, go to the top of the Code window and comment out or delete the statement `Option Explicit` that appears on the first line.

   The `Option Explicit` statement requires that all variables used within this module be formally declared. You will learn about this statement in the next section.

9. When the `Option Explicit` statement is removed from the Code window, choose Run | Run Sub/UserForm to rerun the procedure. This time, Visual Basic executes the code without objections.

10. After the procedure has finished executing, press Alt+F11 to switch to Microsoft Excel.

    The result of the procedure should match Figure 3.2.

A1			f_x	The cost of calculator		
	A	B	C	D	E	F
1	The cost of calculator					
2						
3						
4	Price	35				
5	Sales Tax	2.975				
6	Cost	37.98				
7						
8	The calculator total is $37.975.					
9						

FIGURE 3.2. The VBA procedure can enter data and calculate results in a worksheet.

Cell A8 displays the contents of the strMsg variable. Notice that the cost entered in cell B6 has two decimal places, while the cost in strMsg displays three decimals. To display the cost of a calculator with two decimal places in cell A8, you must apply the required format not to the cell but to the cost variable itself.

VBA has special functions that allow you to change the format of data. To change the format of the cost variable, you can use the Format function. This function has the following syntax:

```
Format(expression, format)
```

where expression is a value or variable that you want to format, and format is the type of format you want to apply.

11. In the VBE window, select the entire code of the CalcCost procedure and copy and paste it below the current procedure on the first empty line. Add some spacing between the two procedures by pressing Enter two times after the first procedure End Sub keywords.

12. Change the name of the copied procedure to CalcCost_Modified.

13. Change the calculation of the cost variable in the CalcCost procedure:

```
cost = Format(slsPrice + (slsPrice * slsTax), "0.00")
```

14. Replace the With...End With block of instructions with the following:

```
Range("B6").Formula = cost
```

15. Replace the statement Range("B5").Formula = slsPrice * slsTax with the following instruction:

```
Range( "B5 ").Formula = Format((slsPrice * slsTax),  "0.00 ")
```

16. Rerun the modified procedure.

After running the procedure, the text displayed in cell A8 shows the cost of the calculator formatted with two decimal places, and the Sales Tax value is also shown with two decimal places.

After trying out the `CalcCost` procedure, you may wonder why you should bother declaring variables if Visual Basic can handle undeclared variables so well. The `CalcCost` procedure is very short, so you don't need to worry about how many bytes of memory will be consumed each time Visual Basic uses the Variant variable. In short procedures, however, it is not the memory that matters but the mistakes you are bound to make when typing variable names. What will happen if the second time you use the `cost` variable, you omit the "o" and refer to it as `cst`?

```
Range( "B6 ").Formula = cst
```

Also, what will you end up with if, instead of `slsTax`, you use the word `Tax` in the formula?

```
Cost = Format(slsPrice + (slsPrice * Tax), "0.00")
```

The result of the `CalcCost` procedure after introducing these two mistakes is shown in Figure 3.3.

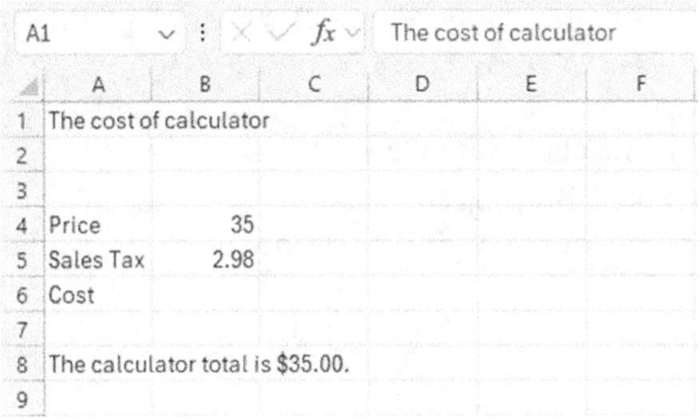

A1		⌄ ⋮ ✕ ✓ _fx_ ⌄	The cost of calculator			
◢	A	B	C	D	E	F
1	The cost of calculator					
2						
3						
4	Price	35				
5	Sales Tax	2.98				
6	Cost					
7						
8	The calculator total is $35.00.					
9						

FIGURE 3.3. Misspelling variable names will produce incorrect results.

Notice that, in Figure 3.3, cell B6 does not show a value because Visual Basic does not find the assignment statement for the `cst` variable. Because Visual Basic does not know the sales tax, it displays the price of the calculator (see cell A8) as the total cost. Visual Basic does not guess. It simply does what you tell it to do. This brings us to the next section, which explains how to make sure this kind of error doesn't occur.

NOTE	*If you have made changes in the variable names as described earlier, be sure to replace the names of the variables* cst *and* tax *with* cost *and* slsTax *in the appropriate lines of the VBA code before you continue.*

Forcing Declaration of Variables

Visual Basic has the Option Explicit statement, which automatically reminds you to formally declare all your variables. This statement must be entered at the top of each of your modules. The Option Explicit statement will cause Visual Basic to generate an error message when you try to run a procedure that contains undeclared variables, as demonstrated in Hands-On 3.1.

 Hands-On 3.2   Writing a VBA Procedure with Explicitly Declared Variables

This exercise requires the prior completion of Hands-On 3.1.

1. Return to the Code window where you entered the CalcCost and CalcCost_ Mofified procedures.
2. At the top of the module window (in the first line), type Option Explicit and press Enter, or uncomment this statement if it's already there.
   Excel will display the statement in blue.
3. Run the CalcCost_Modified procedure. Visual Basic displays the error message Compile error: Variable not defined.
4. Click OK to exit the message box.
   Visual Basic highlights the name of the variable slsPrice.
5. Choose Run | Reset to reset the VBA project.
6. Enter the following declarations at the beginning of the CalcCost_Modified procedure:

```
' declaration of variables
Dim slsPrice As Currency
Dim slsTax As Single
Dim cost As Currency
Dim strMsg As String
```

The revised CalcCost_Modified procedure is shown here:

```
Sub CalcCost()
   ' declaration of variables
   Dim slsPrice As Currency
   Dim slsTax As Single
```

```
Dim cost As Currency
Dim strMsg As String

slsPrice = 35
slsTax = 0.085

Range("A1").Formula = "The cost of calculator"
Range("A4").Formula = "Price"
Range("B4").Formula = slsPrice
Range("A5").Formula = "Sales Tax"
Range("A6").Formula = "Cost"
Range("B5").Formula = Format((slsPrice * slsTax), "0.00")
cost = Format(slsPrice + (slsPrice * slsTax), "0.00")

Range("B6").Formula = cost
strMsg = "The calculator total is $" & cost & "."
Range("A8").Formula = strMsg
End Sub
```

7. Rerun the procedure to ensure that Excel no longer displays the error.

Option Explicit in Every Module

To automatically include `Option Explicit` in every new module you create, follow these steps:

- Choose Tools | Options.
- Make sure that the Require Variable Declaration checkbox is selected in the Options dialog box (Editor tab).
- Choose OK to close the Options dialog box.

From now on, every new module you insert into your VBA project will come with the `Option Explicit` statement on line 1. If you want to require variables to be explicitly declared in a previously created module, you must enter the `Option Explicit` statement manually by editing the module yourself.

 `Option Explicit` forces formal (explicit) declaration of all variables in a module. One big advantage of using `Option Explicit` is that any mistyping of the variable name will be detected at compile time (when Visual Basic attempts to translate the source code to executable code).

 If included, the `Option Explicit` statement must appear in a module before any VBA procedures.

Understanding the Scope of Variables

Variables can have different ranges of influence in a VBA procedure. The term *scope* defines the availability of a variable to the same procedure, other procedures, and other VBA projects.

Variables can have the following three levels of scope in VBA:

- Procedure-level scope
- Module-level scope
- Project-level scope

Procedure-Level (Local) Variables

From this chapter, you already know how to declare a variable by using the `Dim` keyword. The position of the `Dim` keyword in a module determines the scope of a variable. Variables declared with the `Dim` keyword placed within a VBA procedure have a *procedure-level scope*.

Procedure-level variables are frequently referred to as *local variables*. Local variables can be used only in the procedure in which they were declared. Undeclared variables always have a procedure-level scope. A variable's name must be unique within its scope. This means that you cannot declare two variables with the same name in the same procedure. You can use the same variable name in different procedures, however. In other words, the `CalcCost` procedure can have the `slsTax` variable, and the `ExpenseRep` procedure in the same module can have its own variable called `slsTax`. Both variables are independent of each other.

Module-Level Variables

Local variables help save computer memory. As soon as the procedure ends, the variable dies and Visual Basic returns the memory space used by the variable to the computer. In programming, however, you often want the variable to be available to other VBA procedures after the procedure in which the variable was declared has finished running. This situation requires that you change the scope of a variable. Instead of a procedure-level variable, you should declare a module-level variable. This is done by placing the `Dim` keyword at the top of the module sheet before any procedures (just below the `Option Explicit` keyword). For instance, to make the `slsTax` variable available to any other procedure in the `Variables` module, declare the `slsTax` variable in the following way:

```
Option Explicit
Dim slsTax As Single

Sub CalcCost()
...Instructions of the procedure...
End Sub
```

In the foregoing example, the `Dim` keyword is located at the top of the module, below the `Option Explicit` statement. To demonstrate how module-level variables work, we need another procedure that also uses the `slsTax` variable.

⊙ Hands-On 3.3   Writing a VBA Procedure with a Module-Level Variable

1. In the Code window, cut the declaration line Dim slsTax As Single in the Variables module from the CalcCost_Modified procedure and paste it at the top of the module sheet just below the Option Explicit statement.
2. In the same module, where the CalcCost_Calculated procedure is located, enter the code of the ExpenseRep procedure, as shown here:

```
Sub ExpenseRep()
    Dim slsPrice As Currency
    Dim cost As Currency

    slsPrice = 55.99

    cost = slsPrice + (slsPrice * slsTax)
    MsgBox slsTax
    MsgBox cost
End Sub
```

The ExpenseRep procedure declares two Currency type variables: slsPrice and cost. The slsPrice variable is then assigned a value of 55.99. The slsPrice variable is independent of the slsPrice variable that is declared within the CalcCost_Modified procedure.

The ExpenseRep procedure calculates the cost of a purchase. The cost includes the sales tax stored in the slsTax variable. Because the sales tax is the same as the one used in the CalcCost_Modified procedure, the slsTax variable has been declared at the module level.

3. Run the ExpenseRep procedure.
   Because you have not yet run the CalcCost_Modified procedure, Visual Basic does not know the value of the slsTax variable, so it displays zero in the first message box.
4. Click OK to dismiss the message boxes generated by the CalcCost_Modified procedure and then run the CalcCost procedure.
   After Visual Basic executes the CalcCost_Modified procedure, the contents of the slsTax variable equals 0.085. If slsTax were a local variable, the contents of this variable would be empty upon the termination of the CalcCost procedure.

   When you run the CalcCost_Modified procedure, Visual Basic erases the contents of all the variables except for the slsTax variable, which was declared at a module level.

5. Run the ExpenseRep procedure again.
   Notice that, now, the first message box displays the value of slsTax.

Private Variables

When you declare variables at a module level, you can use the `Private` keyword instead of the `Dim` keyword. For example:

```
Private slsTax As Single
```

Private variables are accessible only to the procedures within the module where they were declared. These variables should be declared at the top of the module, following the `Option Explicit` statement. This ensures that they are recognized throughout the module and helps to maintain code clarity and organization.

Project-Level Variables

Module-level variables that are declared with the `Public` keyword (instead of `Dim`) have project-level scope. This means that they can be used in any VBA module. When you want to work with a variable in all the procedures in all the open VBA projects, you must declare it with the `Public` keyword. For example:

```
Option Explicit
Public slsTax As Single

Sub CalcCost()
  ...procedure statements...
End Sub
```

Notice that the `slsTax` variable is now declared at the top of the module with the `Public` keyword to make it available to any other procedure in the VBA project.

Keeping the Project-Level Variable Private

To prevent a project-level variable's contents from being referenced outside its project, you can use the `Option Private Module` statement at the top of the module sheet, just below the `Option Explicit` statement and before the declaration line. For example:

```
Option Explicit
Option Private Module
Public slsTax As Single

Sub CalcCost()
... procedure statements...
End Sub
```

Lifetime of Variables

In addition to scope, variables have a lifetime. The *lifetime* of a variable determines how long a variable retains its value. Module-level and project-level variables preserve their values as long as the project is open. Visual Basic, however, can reinitialize these variables if required by the program's logic. Local variables declared with the Dim statement lose their values when a procedure has finished. Local variables have a lifetime while a procedure is running, and they are reinitialized every time the program is run. Visual Basic allows you to extend the lifetime of a local variable by changing the way it is declared.

Finding a Variable Definition

When you find an instruction in a VBA procedure that assigns a value to a variable, you can quickly locate the definition of the variable by selecting the variable name and pressing Shift+F2 or choosing View | Definition. Visual Basic will jump to the variable declaration line. Press Ctrl+Shift+F2 or choose View | Last Position to return your mouse pointer to its previous position.

Determining a Data Type of a Variable

You can find out the type of variable by using one of the VBA built-in functions. The VarType function returns an integer indicating the type of a variable. Table 3.3 displays the VarType function's syntax and the values it returns. Let's try using the VarType function in the Immediate window.

Hands-On 3.4   Using the Built-In VarType Function

1. In the VBE window, choose View | Immediate Window.
2. Type the following statements that assign values to variables:
```
age = 18
birthdate = #1/1/1981#
firstName = "John"
```

3. Now, ask Visual Basic what type of data each of the variables holds:
```
?VarType(age)
```

When you press Enter, Visual Basic returns 2. As shown in Figure 3.4, number 2 represents the Integer data type. If you type:
```
?VarType(birthdate)
```

Visual Basic returns 7 for Date. If you make a mistake in the variable name (e.g., type birthday instead of birthdate), Visual Basic returns zero (0).

If you type:

```
?VarType(firstName)
```

Visual Basic tells you that the value stored in the variable `firstName` is a String type (8).

TABLE 3.3.   Return values of the VarType function.

Constant	Value	Description
vbEmpty	0	Empty (uninitialized)
vbNull	1	Null (no valid data)
vbInteger	2	Integer
vbLong	3	Long Integer
vbSingle	4	Single-precision floating-point number
vbDouble	5	Double-precision floating-point number
vbCurrency	6	Currency value
vbDate	7	Date value
vbString	8	String
vbObject	9	Object
vbError	10	Error value
vbBoolean	11	Boolean value
vbVariant	12	Variant (used only with arrays of Variants)
vbDataObject	13	Data access object
vbDecimal	14	Decimal value
vbByte	17	Byte value
vbLongLong	20	LongLong integer (valid on 64-bit platforms only)
vbUserDefinedType	36	Variants that contain user-defined types
vbArray	8192	Array (always added to another constant when returned by this function)

USING CONSTANTS

The contents of a variable can change while your procedure is executing. If your procedure needs to refer to unchanged values repeatedly, you should use constants. A *constant* is like a named variable that always refers to the same value. Visual Basic requires that you declare constants before you use them. Declare constants by using the `Const` statement, as in the following examples:

```
Const dialogName = "Enter Data" As String
Const slsTax = 8.5
Const ColorIdx = 3
```

A constant, like a variable, has a scope. To make a constant available within a single procedure, declare it at the procedure level, just below the name of the procedure. For example:

```
Sub WedAnniv()
  Const Age As Integer = 25
  MsgBox (Age)
End Sub
```

If you want to use a constant in all the procedures of a module, use the `Private` keyword in front of the `Const` statement. For example:

```
Private Const driveLetter As String = "C:"
```

The `Private` constant has to be declared at the top of the module, just before the first `Sub` statement. To make a constant available to all modules in the work-book, use the `Public` keyword in front of the `Const` statement. For example:

```
Public Const NumOfChars As Integer = 255
```

The `Public` constant must be declared at the top of the module, just before the first `Sub` statement.

When declaring a constant, you can use any one of the following data types: Boolean, Byte, Integer, Long, Currency, Single, Double, Date, String, or Variant. Like variables, several constants can be declared on one line if separated by commas. For example:

```
Const Age As Integer = 25, City As String = "Denver"
```

Using constants makes your VBA procedures more readable and easier to maintain. For example, if you refer to a certain value several times in your procedure, use a constant instead of the value. This way, if the value changes (for example, the sales tax goes up), you can simply change the value in the declaration of the `Const` statement instead of tracking down every occurrence of that value.

Built-In Constants

Both Microsoft Excel and VBA have a long list of predefined constants that do not need to be declared. These built-in constants can be looked up using the Object Browser. Let's proceed with Hands-On 3.5, where we will open the Object Browser to look at the list of Excel constants.

Hands-On 3.5   Viewing Excel Constants in the Object Browser

1. In the VBE window, choose View | Object Browser.
2. In the Project/Library list box, click the drop-down arrow and select Excel.
3. Enter constants as the search text in the Search box and press Enter or click the Search button. Visual Basic shows the result of the search in the Search Results area.
4. Scroll down in the Classes list box to locate and then select Constants, as shown in Figure 3.4. The right side of the Object Browser window displays a list of all built-in constants that are available in the Microsoft Excel Objects library. Notice that the names of all the constants begin with the prefix xl.

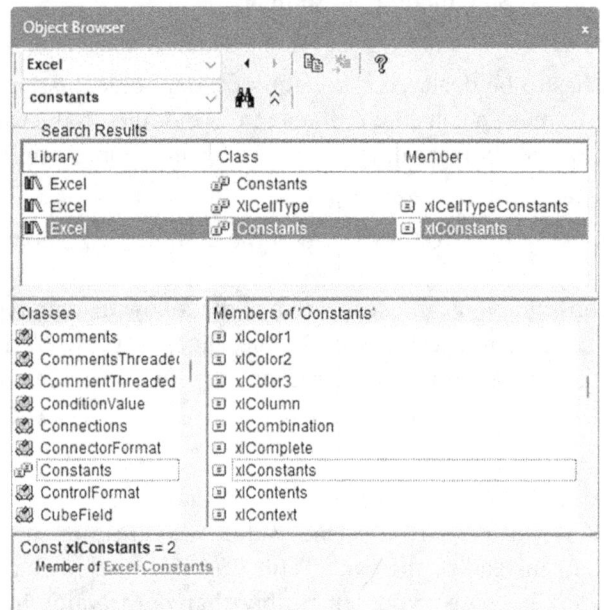

FIGURE 3.4.   Use the Object Browser to look up any built-in constant.

5. To look up VBA constants, choose VBA in the Project/Library list box (see Figure 3.5). Notice that the names of the VBA built-in constants begin with the prefix vb.

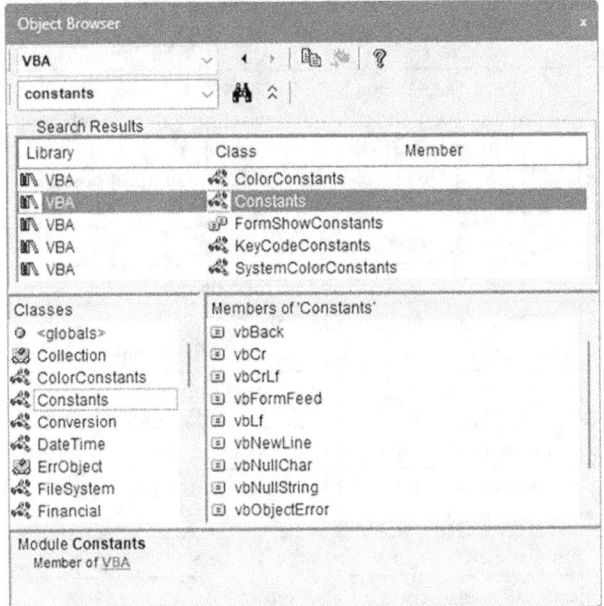

FIGURE 3.5.   The names of VBA constants begin with the prefix "vb."

CONVERTING BETWEEN DATA TYPES

While VBA handles a lot of data type conversion automatically in the background, it also provides several data conversion functions (see Table 3.4) that allow you to convert one data type to another.

TABLE 3.4.   VBA data type conversion functions.

Conversion Function	Return Type	Description
CBool	Boolean	Any valid string or numeric expression.
CByte	Byte	0 to 255.
CCur	Currency	−922,337,203,685,477.5808 to 922,337,203,685,477.5807.
CDate	Date	Any valid date expression.
CDbl	Double	−1.79769313486231E308 to −4.94065645841247E-324 for negative values; 4.94065645841247E-324 to 1.79769313486232E308 for positive values.

(Contd.)

Conversion Function	Return Type	Description
CDec	Decimal	+/–79,228,162,514,264,337,593,543,950,335 for zero-scaled numbers—that is, numbers with no decimal places. For numbers with 28 decimal places, the range is +/–7.9228162514264337593543950335. The smallest possible nonzero number is 0.0000000000000000000000000001.
CInt	Integer	–32,768 to 32,767; fractions are rounded.
CLng	Long	–2,147,483,648 to 2,147,483,647; fractions are rounded.
CLngLng	LongLong	–9,223,372,036,854,775,808 to 9,223,372,036,854,775,807; fractions are rounded. (Valid on 64-bit platforms only.)
CLngPtr	LongPtr	–2,147,483,648 to 2,147,483,647 on 32-bit systems; –9,223,372,036,854,775,808 to 9,223,372,036,854,775,807 on 64-bit systems. Fractions are rounded for 32-bit and 64-bit systems.
CSng	Single	–3.402823E38 to –1.401298E-45 for negative values; 1.401298E-45 to 3.402823E38 for positive values.
CStr	String	Returns for CStr depend on the Expression argument.

If Expression is	CStr returns
Boolean	A string containing True or False
Date	A string containing a date in the short date format of your system
Null	A runtime error
Empty	A zero-length string ("")
Error	A string containing the word Error followed by the error number
Other numeric	A string containing the number

Conversion Function	Return Type	Description
Cvar	Variant	Same range as Double for numerics. Same range as String for non-numeric.

The conversion functions should be used in situations where you want to show the result of an operation as a specific data type rather than the default data type. For example, instead of showing the result of your calculation as an integer or single- or double-precision number, you may want to use the CCur function to force currency arithmetic, as in the following example procedure:

```
Sub ShowMoney()
  'declare variables of two different types
  Dim myAmount As Single
  Dim myMoneyAmount As Currency
```

```
    myAmount = 345.34

    myMoneyAmount = CCur(myAmount)
    Debug.Print "Amount = $" & myMoneyAmount
End Sub
```

When using the CCur function, currency options are recognized depending on the locale setting of your computer. The same holds true for the CDate function. By using this function, you can ensure that the date is formatted according to the locale setting of your system.

Use the IsDate function to determine whether a return value can be converted to date or time:

```
Sub ConvertToDate()
    'assume you have entered  Jan 1 2021 in cell A1
    Dim myEntry As String
    Dim myRangeValue As Date

    myEntry = Sheet2.Range("A1").Value
    If IsDate(myEntry) Then
    myRangeValue = CDate(myEntry)
    End If
    Debug.Print myRangeValue
End Sub
```

In cases where you need to round the value to the nearest even number, you will find the CInt and CLng functions quite handy, as demonstrated in the following procedure:

```
Sub ShowInteger()
    'declare variables of two different types
    Dim myAmount As Single
    Dim myIntAmount As Integer

    myAmount = 345.64

    myIntAmount = CInt(myAmount)
    Debug.Print "Original Amount = " & myAmount
    Debug.Print "New Amount = " & myIntAmount
End Sub
```

As you can see in the code of the foregoing procedures, the syntax for the VBA conversion functions is as follows:

```
conversionFunctionName(variablename)
```

where variablename is the name of a variable, a constant, or an expression (such as x + y) that evaluates to a specific data type.

⊙ **Hands-On 3.6   Using Data Type Conversion Functions in VBA**

1. Select Insert | Module to insert a new module into your VBA project (`Chap03.xslm`).
2. Use the Properties window to rename the module to `DataTypeConversion`.
3. Enter the code of the procedures introduced in this section: `ShowMoney`, `ConvertToDate`, and `ShowInteger`.
4. Insert a new worksheet into the current workbook and enter Jan 1 2025 in cell A1.
5. Run each procedure and check the results in the Immediate window.

USING STATIC VARIABLES IN VBA PROCEDURES

A variable declared with the `Static` keyword is a special type of local variable. Static variables are declared at the procedure level. Unlike local variables declared with the `Dim` keyword, static variables do not lose their contents when the program is not in its procedure. For example, when a VBA procedure with a static variable calls another procedure, after Visual Basic executes the statements of the called procedure and returns to the calling procedure, the static variable will still retain the original value. The `CostOfPurchase` procedure shown in Hands-On 3.7 demonstrates the use of the static variable named `allPurchase`. Notice how this variable keeps track of the running total.

⊙ **Hands-On 3.7   Writing a VBA Procedure with a Static Variable**

1. In the Code window of the `Variables` module, write the following procedure:

```
Sub CostOfPurchase()
  ' declare variables
  Static allPurchase
  Dim newPurchase As String
  Dim purchCost As Single

  newPurchase = InputBox("Enter the cost of a purchase:")
  purchCost = CSng(newPurchase)
  allPurchase = allPurchase + purchCost

  ' display results
  MsgBox "The cost of a new purchase is: " & newPurchase
  MsgBox "The running cost is: " & allPurchase
End Sub
```

The foregoing procedure begins with declaring a static variable named `allPurchase` and two other local variables: `newPurchase` and `purchCost`. The `InputBox` function used in this procedure displays a dialog box and waits for the user to enter the value. After inputting the value and clicking OK, Visual Basic assigns this value to the variable `newPurchase`.

The `InputBox` function is discussed in detail in Chapter 4. Because the result of the `InputBox` function is always a string, the `newPurchase` variable was declared as the String data type. You can't, however, use strings in mathematical calculations. That's why the next instruction uses a type conversion function (`CSng`) to translate the text value into a numeric variable of the Single data type. The `CSng` function requires one argument—the value you want to translate. To find out more about the `CSng` function, position the insertion point anywhere within the word `CSng` and press F1. The number obtained as the result of the `CSng` function is then stored in the variable `purchCost`.

The next instruction, `allPurchase = allPurchase + purchCost`, adds to the current purchase value the new value supplied by the `InputBox` function.

2. Position the cursor anywhere within the `CostOfPurchase` procedure and press F5. When the dialog box appears, enter a number. For example, enter 100 and click OK or press Enter. Visual Basic displays the message "The cost of a new purchase is: 100." Click OK in the message box. Visual Basic displays the second message, "The running cost is: 100."

3. When you run this procedure for the first time, the content of the `allPurchase` variable is the same as the content of the `purchCost` variable.

4. Rerun the same procedure. When the input dialog appears, enter another number. For example, enter 50 and click OK or press Enter. Visual Basic displays the message "The cost of a new purchase is: 50." Click OK in the message box. Visual Basic displays the second message, "The running cost is: 150."

5. When you run the procedure the second time, the value of the static variable is increased by the new value supplied in the dialog box. You can run the `CostOfPurchase` procedure as many times as you want. The `allPurchase` variable will keep the running total for as long as `VBAProject(Chap03.xlsm)` is open.

USING OBJECT VARIABLES IN VBA PROCEDURES

The variables that we discussed in the preceding sections are used to store data. Storing data is the main reason for using "normal" variables in your VBA procedures. In addition to the normal variables that store data, there are special variables that refer to Visual Basic objects. These variables are called *object variables*. In Chapter 2, you worked with several objects in the Immediate window. Now, you will learn how you can represent an object with the object variable.

Object variables don't store data; instead, they tell where the data is located. For example, with the object variable, you can tell Visual Basic that the data is in cell E10 of a worksheet. Object variables make it easy to locate data. When writing VBA procedures, you often need to write long instructions, such as:

```
ActiveSheet.Range(Cells(1, 1), Cells(10, 5).Select
```

Instead of using long references to the object, you can declare an object variable that will tell Visual Basic where the data is located. Object variables are declared similarly to the variables you already know. The only difference is that after the As keyword, you enter the word Object as the data type. For example:

```
Dim myRange As Object
```

The foregoing statement declares the object variable named myRange.

Well, it's not enough to declare the object variable. You also must assign a specific value to the object variable before you can use this variable in your procedure. Assign a value to the object variable by using the Set keyword. The Set keyword must be followed by the equals sign and the value that the variable will refer to. For example:

```
Set myRange = ActiveSheet.Range(Cells(1, 1), Cells(10, 5))
```

This statement assigns value to the object variable myRange. This value refers to cells A1:E10 on the active sheet. If you omit the word Set, Visual Basic will respond with an error message—"Run-time error 91: Object variable or With block variable not set."

Again, it's time to see a practical example.

⊙ Hands-On 3.8 Writing a VBA Procedure with Object Variables

1. In the Code window of the Variables module, write the following procedure:

```
Sub UseObjVariable()
  Dim myRange As Object
  Sheets.Add
  Set myRange = ActiveSheet.Range(Cells(1, 1), _
    Cells(10, 5))
  myRange.BorderAround Weight:=xlMedium

  With myRange.Interior
```

```
        .ColorIndex = 6
        .Pattern = xlSolid
    End With

    Set myRange = ActiveSheet.Range(Cells(12, 5), _
        Cells(12, 10))
    myRange.Value = 54

    Debug.Print IsObject(myRange)
End Sub
```

Let's examine the code of the `UseObjVariable` procedure line by line. The procedure begins with the declaration of the object variable `myRange`. The next statement sets the object variable `myRange` to the range A1:E10 on the active sheet. From now on, every time you want to reference this range, instead of using the entire object's address, you'll use the shortcut—the name of the object variable. The purpose of this procedure is to create a border around the worksheet range A1:E10. Instead of writing a long instruction:

```
ActiveSheet.Range(Cells(1, 1), _
    Cells(10, 5)).BorderAround Weight:=xlMedium
```

you can take a shortcut by using the name of the object variable:

```
myRange.BorderAround Weight:=xlMedium
```

The next series of statements changes the color of the selected range of cells (A1:E10). Again, you don't need to write long instructions to reference the object that you want to manipulate. Instead of the full object name, you can use the `myRange` object variable.

The next statement assigns a new reference to the object variable `myRange`. Visual Basic forgets the old reference, and the next time you use `myRange`, it refers to another range (E12:J12).

After the number 54 is entered into the new range (E12:J12), the procedure shows you how you can make sure that a specific variable is of the Object type. The instruction `Debug.Print IsObject(myRange)` will enter `True` in the Immediate window if `myRange` is an object variable. `IsObject` is a VBA function that indicates whether a specific value represents an object variable.

2. Position the cursor anywhere within the `UseObjVariable` procedure and press F5.

The Advantages of Using Object Variables

❑ They can be used instead of the actual object.

❑ They are shorter and easier to remember than the actual values to which they point.

❑ You can change their meaning while your procedure is running.

Using Specific Object Variables

The object variable can refer to any type of object. Because Visual Basic has many types of objects, it's a good idea to create object variables that refer to a specific object to make your programs faster and more readable. For instance, in the `UseObjVariable` procedure (see the previous section), instead of the generic object variable (`Object`), you can declare the `myRange` object variable as a `Range` object:

```
Dim myRange As Range
```

If you want to refer to a specific worksheet, then you can declare the `Worksheet` object:

```
Dim mySheet As Worksheet
Set mySheet = Worksheets("Marketing")
```

When the object variable is no longer needed, you can assign `Nothing` to it. This frees up memory and system resources:

```
Set mySheet = Nothing
```

SUMMARY

This chapter introduced several crucial VBA concepts, including data types, variables, and constants. You learned how to declare different types of variables and specify their data types. You also explored the difference between variables and constants. Additionally, we discussed the scope and lifetime of variables, the advantages of using object variables, and emphasized the importance of the `Option Explicit` statement. With this knowledge of variables and their usage, you can now create VBA procedures to manipulate data in more meaningful ways.

In the next chapter, you will expand your VBA knowledge by writing custom function procedures. Additionally, you will learn about built-in functions that will allow your VBA procedures to interact with users.

Chapter 4

EXCEL VBA PROCEDURES—

A QUICK GUIDE TO WRITING FUNCTION PROCEDURES

Earlier in this book, you learned that a procedure is a group of instructions that allows you to accomplish specific tasks when your program runs. As you progress through this book, you will get acquainted with the following types of VBA procedures:

- *Subroutine procedures* (*subroutines*) perform some useful tasks but don't return any values. They begin with the keyword Sub and end with the keywords End Sub. Subroutines can be recorded with the macro recorder or written from scratch in the VBE window.

- *Function procedures* (*functions*) perform specific tasks that return values. They begin with the keyword Function and end with the keywords End Function. Function procedures can be executed from a subroutine or accessed from a worksheet just like any Excel built-in function.

- *Property procedures* are used with custom objects. Use them to set and get the value of an object's property or set a reference to an object. Property procedures are an advanced feature of VBA and are covered later in this book.

In this chapter, you will learn how to create and execute function procedures. You will also explore the critical role variables play in passing values to both subroutines and functions. As you advance through this chapter, you will get an in-depth look at two of the most versatile and useful built-in functions in VBA: `MsgBox` and `InputBox`. The `MsgBox` function will help you master how to display message boxes to interact with users dynamically, while the `InputBox` function will provide you with the tools to obtain user input effectively.

UNDERSTANDING FUNCTION PROCEDURES

With the hundreds of functions already built into Excel, you can perform a wide variety of calculations automatically. There will be times, however, when you may require a custom calculation or functionality. With VBA, you can quickly fulfill this special need by writing a function procedure. Among the reasons for creating custom VBA functions are the following:

- Analyze data and perform calculations.
- Modify data and report information.
- Take specific action based on supplied or calculated data.

Creating a Function Procedure

Like Excel functions, function procedures perform calculations and return values. The best way to learn about functions is to create one, so let's get started. After setting up a new VBA project, you will create a simple function procedure that sums two values.

NOTE	*Please note that files for the "Hands-On" projects can be found in the companion files.*

⊙ Hands-On 4.1   Writing a Simple Function Procedure

1. Open a new Excel workbook and save it as `C:\VBAExcel2024_ByExample\Chap04.xlsm`.
2. Switch to the VBE window and select VBAProject (Chap04.xlsm).
3. In the Properties window, change the name of the project name to `ProcAndFunctions`.

4. Select the ProcAndFunctions (Chap04.xlsm) project in the Project Explorer window and choose Insert | Module.

5. In the Properties window, change the Module1 name to `Sample1`.

6. In the Project Explorer window, highlight Sample1 and click anywhere in the Code window. Choose Insert | Procedure. The Add Procedure dialog box appears.

7. In the Add Procedure dialog box, make the entries shown in Figure 4.1:

   - Name: SumItUp
   - Type: Function
   - Scope: Public

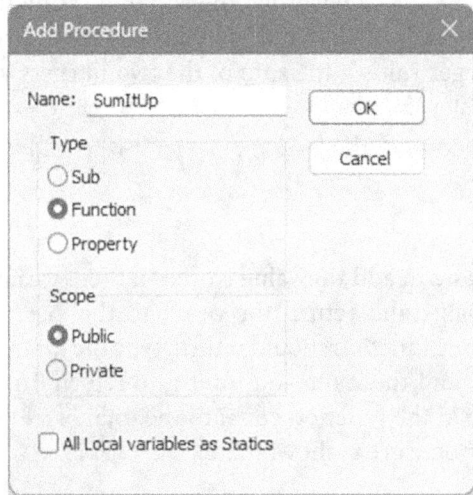

FIGURE 4.1.   When you use the Add Procedure dialog box, Visual Basic automatically creates the procedure type you choose.

8. Click OK to close the Add Procedure dialog box. Visual Basic enters an empty function procedure that looks like this:

```
Public Function SumItUp()

End Function
```

9. Modify the function declaration as follows:

```
Public Function SumItUp(m As Integer, n As Integer) As Integer

End Function
```

The first statement declares the name of the function procedure. The `Public` keyword indicates that the function is accessible to all other procedures in all modules. The `Public` keyword is optional. Notice the keyword `Function` followed by the name of the function (`SumItUp`) and a pair of parentheses. In the parentheses, you will list the data items that the function will use in the calculation. Every function procedure ends with the `End Function` statement. This function is designed to add two integer values. Instead of hardcoding the values, you can make the function more flexible by passing values as variables. By doing this, your custom function will be able to add any two integers that you specify at runtime. Each of the passed-in variables m and n represents a value, which you will provide when you call the function. Notice that the function is declared with the `As Integer` keyword at the end, indicating that the function will return an integer. This means that when you call the SumItUp function, the result will be an integer value – the sum of the two integers you pass to the function (see the next step).

10. Type the following statement between the `Public Function` and `End Function` statements:

```
SumItUp = m + n
```

This statement instructs Visual Basic to add the value stored in the n variable to the value stored in the m variable and return the result to the `SumItUp` function. To specify the value that the function should return, type the function name followed by the equals sign and the value you want it to return. In the foregoing statement, set the name of the function equal to the total of m + n. The completed custom function procedure is shown here:

```
Public Function SumItUp(m As Integer, n As Integer) As Integer
    SumItUp = m + n
End Function
```

11. To test this function, open the Immediate window, type the following statement, and press Enter to execute:

```
SumItUp(5, 8)
```

12. The function will return 13, which is an integer.

About Function Names

Function names should suggest the role that the function performs and must conform to the rules for naming variables (see Chapter 3).

Scoping VBA Procedures

In the previous chapter, you learned that the variable's scope determines which modules and procedures can access it. Similarly, VBA procedures have scope, dictating whether they can be called by procedures in other modules.

By default, all VBA procedures are public, meaning they can be called by other procedures in any module. Because procedures are public by default, you can omit the `Public` keyword if you prefer. If you use the `Private` keyword, however, your procedure will only be accessible to other procedures within the same module and not to those in other modules.

VARIOUS METHODS OF RUNNING FUNCTION PROCEDURES

A function procedure can be executed from the Immediate window for testing purposes (see Step 11 in Hands-On 4.1). It can also be used in a worksheet formula or called from another procedure. In the following sections, you will learn advanced techniques for executing functions.

Running a Function Procedure from a Worksheet

A custom function procedure is like an Excel built-in function. If you don't know the exact name of the function or its arguments, you can use the Formula palette to help enter the required function in a worksheet, as shown in Hands-On 4.2.

Hands-On 4.2   Executing a Function Procedure from within an Excel Worksheet

1. Switch to the Microsoft Excel window and select any cell.
2. Click the Insert Function (fx) button on the Formula bar.
3. Excel displays the Insert Function dialog box. The lower portion of the dialog box displays an alphabetical listing of all the functions in the selected category.
4. In the category drop-down box, select User Defined. In the function name box, locate and select the SumItUp function that was created in Hands-On 4.1.
5. When you highlight the name of the function in the function name box (Figure 4.2), the bottom part of the dialog box displays the function's syntax: SumItUp(m,n).

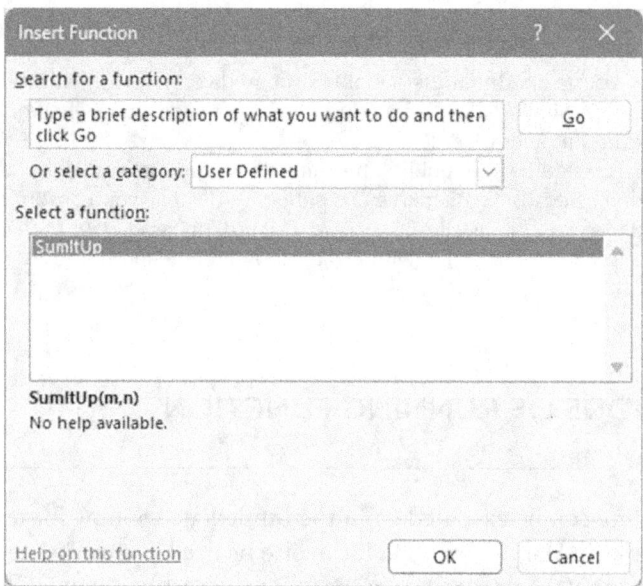

FIGURE 4.2. VBA custom function procedures are listed under the User Defined category in the Insert Function dialog box. They also appear in the list of all Excel built-in functions when you select All in the category dropdown.

6. Select the SumItUp function and click OK to begin writing a formula.

7. The Function Arguments dialog box appears, as shown in Figure 4.3. This dialog displays the name of the function and each of its arguments: m and n.

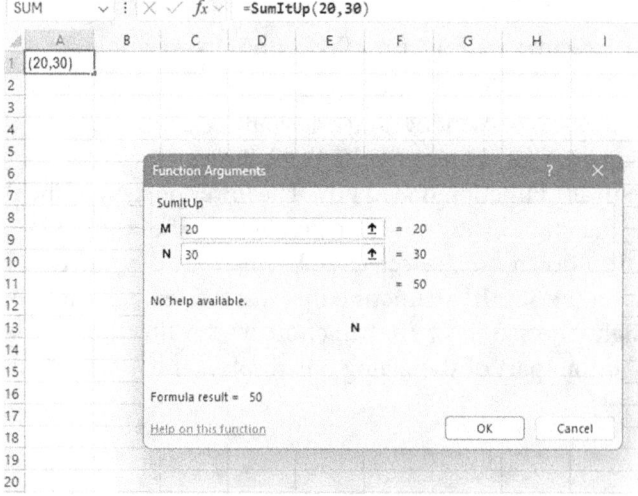

FIGURE 4.3. The Formula palette feature is helpful in entering any worksheet function, whether built-in or custom-made with VBA programming.

8. Enter the values for the arguments as shown in Figure 4.3 or enter your own values.

9. As you type the values in the argument text boxes, Excel displays the values you entered and the current result of the function. Because both arguments (m and n) are required, the function will return an error if you skip either one of the arguments.

10. Click OK to exit the Function Arguments dialog.
    Excel enters the SumItUp function in the selected cell and displays its result.

11. To edit the function, select the cell that displays the function's result and click the Insert Function (fx) button to access the Function Arguments dialog box. Enter new values for the function's m and n arguments and click OK.

NOTE	*To edit the arguments' values directly in the cell, double-click the cell containing the function and make the necessary changes. You may also set up the SumItUp function to perform calculations based on the values entered in cells. To do this, in the Function Arguments dialog box shown in Figure 4.3, enter cell references instead of values. For example, enter C1 for the m argument and C2 for the n argument. When you click OK, Excel will display zero (0) as the result of the function. On the worksheet, enter the values in cells C1 and C2 and your custom function will recalculate the result just like any other built-in Excel function.*

Running a Function Procedure from Another VBA Procedure

To execute a custom function, write a VBA subroutine and call the function when you need it. The following procedure calls the SumItUp function and prints the result of the calculation to the Immediate window.

(◉) **Hands-On 4.3   Executing a Function from a VBA Procedure**

1. In the same module where you entered the code of the SumItUp function procedure, enter the RunSumItUp procedure.

```
Sub RunSumItUp()
    Dim m As Integer, n As Integer
    m = 37
    n = 34

    Debug.Print SumItUp(m, n)
    MsgBox "Open the Immediate Window to see the result."
End Sub
```

Notice how the foregoing subroutine uses one `Dim` statement to declare the `m` and `n` variables. These variables will be used to feed the data to the function. The next two statements assign the values to those variables. Next, Visual Basic calls the `SumItUp` function and passes it the values stored in the `m` and `n` variables. When the function procedure statement `SumItUp = m + n` is executed, Visual Basic returns to the `RunSumItUp` subroutine and uses the `Debug.Print` statement to print the function's result to the Immediate window. Finally, the `MsgBox` function informs the user where to look for the result. You can find more information about using the `MsgBox` function later in this chapter.

2. Place the mouse pointer anywhere within the `RunSumItUp` procedure and press F5 to run it.

ENSURING AVAILABILITY OF YOUR CUSTOM FUNCTIONS

Your custom VBA function is available only while the workbook where the function is stored is open. If you close the workbook, the function is no longer available. To make sure your custom VBA functions are available every time you work with Microsoft Excel, you can do one of the following:

- Store your functions in the Personal macro workbook.
- Save the workbook with your custom VBA function in the `XLStart` folder.
- Set up a reference to the workbook containing your custom functions.

PASSING ARGUMENTS TO FUNCTION PROCEDURES

Procedures (both subroutines and functions) often take arguments. *Arguments* are one or more values needed for a procedure to do something. They are entered within parentheses. Multiple arguments are separated with commas.

Having used Excel for a while, you already know that Excel's built-in functions can produce different results based on the values you supply to them. For example, if cells A4 and A5 contain the numbers 5 and 10, respectively, the `Sum` function `=SUM(A4:A5)` will return `15`, unless you change the values entered in the specified cells. Just like you can pass any values to Excel's built-in functions, you can pass any values to custom VBA procedures.

Let's explore passing values from a subroutine procedure to the `SumItUp` function. We will write a sub-procedure that collects the user's first and last name. Next, we will call the `SumItUp` function to get the sum of characters in a person's first and last name.

⊙ Hands-On 4.4 Passing Arguments to Functions (Example 1)

1. Type the following NumOfCharacters subroutine in the same module (Sample1) where you entered the SumItUp function:

```
Sub NumOfCharacters()
  Dim f As Integer
  Dim l As Integer

  f = Len(InputBox("Enter first name:"))
  l = Len(InputBox("Enter last name:"))
  MsgBox SumItUp(f,l)
End Sub
```

2. Place the mouse pointer within the code of the NumOfCharacters procedure and press F5. Visual Basic displays the input box prompting you for the first name. This box is generated by the following function: InputBox("Enter first name:"). For more information on the use of this function, see the section titled "Getting to Know the InputBox Function" later in this chapter.

3. Enter any name, and press Enter or click OK.
   Visual Basic takes the text you entered and supplies it as an argument to the Len function. The Len function calculates the number of characters in the supplied text string. Visual Basic places the result of the Len function in the f variable for further reference. After that, Visual Basic displays the next input box, this time asking for the last name.

4. Enter any last name, and press Enter or click OK.
   Visual Basic passes the last name to the Len function to get the number of characters. Then, that number is stored in the l variable. What happens next? Visual Basic encounters the MsgBox function, which tells it to display the result of the SumItUp function. Because the result is not yet ready, however, Visual Basic jumps quickly to the SumItUp function to perform the calculation using the values saved earlier in the f and l variables. Inside the function procedure, Visual Basic substitutes the m argument with the value of the f variable and the n argument with the value of the l variable. Once the substitution is done, Visual Basic adds up the two numbers and returns the result to the SumItUp function.

   There are no more tasks to perform inside the function procedure, so Visual Basic returns to the subroutine and provides the SumItUp function's result as an argument to the MsgBox function. Now, a message appears on the screen displaying the total number of characters.

5. Click OK to exit the message box.

You can run the `NumOfCharacters` procedure as many times as you'd like, each time supplying different first and last names.

To pass a specific value from a function to a subroutine, assign the value to the name of the function. For example, the `NumOfDays` function shown here passes the value of 7 to the subroutine `DaysInAWeek`:

```
Function NumOfDays()
    NumOfDays = 7
End Function

Sub DaysInAWeek()
    MsgBox "There are " & NumOfDays & " days in a week."
End Sub
```

Specifying Argument Types

In the preceding section, you learned that functions perform some calculations based on data received through their arguments. When you declare a function procedure, you list the names of arguments inside a set of parentheses. Argument names are like variables. Each argument name refers to whatever value you provide at the time the function is called. When a subroutine calls a function procedure, it passes the required arguments as variables to it. Once the function does something, the result is assigned to the function name. Notice that the function procedure's name is used as if it were a variable.

Like variables, functions can have types. The result of your function procedure can be String, Integer, Long, and so on. To specify the data type for your function's result, add the keyword `As` and the name of the desired data type to the end of the function declaration line—for example:

```
Function MultiplyIt(num1, num2) As Integer
```

Let's look at an example of a function that returns an Integer number, although the arguments passed to it are declared as Single data types in a calling subroutine.

⊙ Hands-On 4.5 Passing Arguments to Functions (Example 2)

1. Add a new module to the `ProcAndFunctions` (Chap04.xlsm) project and change the module's name to `Sample2`.
2. Activate the `Sample2` module and enter the `HowMuch` subroutine, as shown here:

```
Sub HowMuch()
  Dim num1 As Single
```

```
    Dim num2 As Single
    Dim result As Single

    num1 = 45.33
    num2 = 19.24

    result = MultiplyIt(num1, num2)
    MsgBox result
End Sub
```

3. Enter the `MultiplyIt` function procedure below the `HowMuch` subroutine:

```
Function MultiplyIt(num1, num2) As Integer
    MultiplyIt = num1 * num2
End Function
```

Because the values stored in the variables `num1` and `num2` are not whole numbers, you may want to assign the Integer data type to the result of the function to ensure that the result is a whole number. If you don't assign the data type to the `MultiplyIt` function's result, the `HowMuch` procedure will display the result in the data type specified in the declaration line of the result variable. Instead of `872`, the result of the multiplication will be `872.1492`.

4. Run the `HowMuch` procedure.

How about passing different values each time you run the procedure? Instead of hardcoding the values to be used in the multiplication, you can use the `InputBox` function to ask the user for the values at runtime—for example:

```
num1 = InputBox("Enter a number:")
```

The `InputBox` function is discussed in detail in a later section of this chapter.

Passing Arguments by Reference and by Value

In VBA, when you pass arguments to a function or subroutine, you can pass them either by reference (`ByRef`) or by value (`ByVal`). The difference between these two methods lies in how the arguments are treated inside the function or subroutine:

- By reference (`ByRef`):

  When you pass the arguments by reference, you are passing a memory address of the variable. This means that any changes made to the arguments inside the function or subroutine will directly affect the original value. You should use `ByRef` when you need the function or subroutine to modify the original variable or when you want to avoid making a copy of a large dataset.

Let's look at the following example where the original variable value passed to a function from a subroutine procedure is changed by the function:

```
Sub SendByRef(ByRef x As Integer)
    x = x + 10
End Sub

Sub CallSendByRef()
    Dim num As Integer
    Dim orgNum As Integer
    num = 10
    orgNum = num
    SendByRef num
    MsgBox "The num variable was originally " & orgNum & _
        " and was changed by the function to " & num & "."
End Sub
```

When you run the `CallSendByRef` procedure, the value of the `num` variable will be changed to `20`.

- By value (`ByVal`):

When you pass an argument by value, you are passing a copy of the variable. This means that any changes made to the argument inside the function or subroutine procedure will not affect the original variable. Use `ByVal` when you want to ensure that the original variable remains unchanged by the function or subroutine procedure.

Let's look at the following example where the original variable value passed to a function from a subroutine procedure is not changed by the function:

```
Sub SendByVal(ByVal x As Integer)
    x = x + 10
End Sub

Sub CallSendByVal()
    Dim num As Integer
    Dim orgNum As Integer
    num = 10
    orgNum = num
    SendByVal num
    MsgBox "The num variable was originally " & orgNum & _
        " and came back from the function as " & num & "."
End Sub
```

When you run the `CallSendByVal` procedure, the value of the `num` variable will be the same as the original value (`10`).

To summarize, `ByRef` allows the function to modify the original value, while `ByVal` ensures that the original value remains unchanged. The default in VBA when passing arguments to a function or a subroutine is by reference (`ByRef`). This means that unless you explicitly specify otherwise, VBA passes arguments by reference, allowing the called procedure to modify the actual variables passed to it.

Using Optional Arguments

At times, you may want to supply an additional value to a function. Let's say you have a function that calculates the price of a meal per person. Sometimes, however, you'd like the function to perform the same calculation for a group of two or more people. To indicate that a procedure argument is not always required, precede the name of the argument with the `Optional` keyword. Arguments that are optional always come at the end of the argument list, following the names of all the required arguments.

Optional arguments must always be the Variant data type, meaning that you can't specify the optional argument's type by using the `As` keyword.

Let's write the `Avg` function to calculate the average of two or three numbers.

⊙ Hands-On 4.6   Writing Functions with Optional Arguments

1. Add a new module to the `ProcAndFunctions` (`Chap04.xlsm`) project and change the module's name to `Sample3`.
2. Activate the `Sample3` module and enter the function procedure `Avg` shown here:

```
Function Avg(num1 As Single, num2 As Single, _
    Optional num3) As Single
    Dim totalNums As Integer

    totalNums = 3

    If IsMissing(num3) Then
        num3 = 0
        totalNums = totalNums - 1
    End If

    Avg = (num1 + num2 + num3) / totalNums
End Function
```

Let's take a few minutes to analyze the `Avg` function. This function takes two required arguments (`num1` and `num2`) and one optional argument (`num3`). It returns a Single data type, which is used for floating-point numbers. A variable `totalNums` is declared to keep track of the number of arguments provided. The variable `totalNums` is initially set to `3`, assuming that the optional third argument (`num3`) will be provided. The `IsMissing` function checks whether num3 is provided. If num3 is missing, num3 is set to `0` and `totalNums` is decreased by one because only two arguments are provided. The sum of num1, num2, and num3 is divided by `totalNums` to calculate the average. The result is assigned to the function's name, `Avg`, which returns the average value.

The `IsMissing` built-in VBA function allows you to determine whether the optional argument was supplied. This function returns the logical value `true` if the third argument is not supplied, and it returns `false` when the third argument is given. The `IsMissing` function is used here with the decision-making statement `If…Then`. (See Chapter 5 for a detailed description of decision-making statements used in VBA.) If the num3 argument is missing (`IsMissing`), then (`Then`) Visual Basic uses a zero for the value of the third argument (num3 = 0) and reduces the value stored in the argument `totalNums` by one (`totalNums = totalNums – 1`).

3. Now call this function from the Immediate window like this:

```
?Avg(2,3)
```

As soon as you press Enter, Visual Basic displays the result: `2.5`.

If you enter the following:

```
?Avg(2,3,5)
```

Visual Basic displays the result as `3.333333`.

As you've seen, the `Avg` function allows you to calculate the average of two or three numbers. You decide which values and how many values (two or three) you want to average. When you start typing the values for the function's arguments in the Immediate window, Visual Basic displays the name of the optional argument enclosed in square brackets.

How else can you run the `Avg` function? On your own, run this function from a worksheet. Make sure you run it with two and then with three arguments.

TESTING A FUNCTION PROCEDURE

To test whether a custom function does what it was designed to do, write a simple subroutine that will call the function and display its result. In addition, the subroutine should show the original values of arguments. This way, you'll be able to quickly determine when the values of arguments were altered. If the function procedure uses optional arguments, you'll also need to check those situations in which the optional arguments may be missing.

LOCATING BUILT-IN FUNCTIONS

VBA comes with numerous built-in functions. These functions can be looked up in Visual Basic online help:

https://docs.microsoft.com/en-us/office/vba/Language/Reference/functions-visual-basic-for-applications

Take, for example, the `MsgBox` or `InputBox` function. One of the features of a good program is its interaction with the user. When you work with Microsoft Excel, you interact with the application by using various dialog boxes. When you make a mistake, a dialog box comes up and displays a message informing you of the error. When you write your own procedures, you can also inform users about an unexpected error or the result of a specific operation. You do this with the help of the `MsgBox` function. So far, you have seen a simple implementation of this function. In the next section, you will find out how to control the way your message looks. You will also learn how to get information from the user with the `InputBox` function.

GETTING TO KNOW THE MSGBOX FUNCTION

The `MsgBox` function that you have used thus far was limited to displaying a message to the user in a simple one-button dialog box. You have closed the message box by clicking the OK button or pressing the Enter key. You can create a simple message box by following the `MsgBox` function name with the text enclosed in quotation marks. In other words, to display the message "The procedure completed successfully.", you write the following statement:

```
MsgBox "The procedure completed successfully."
```

You can quickly try out the foregoing instruction by entering it in the Immediate window. When you type this instruction and press Enter, Visual Basic displays the message box shown in Figure 4.4.

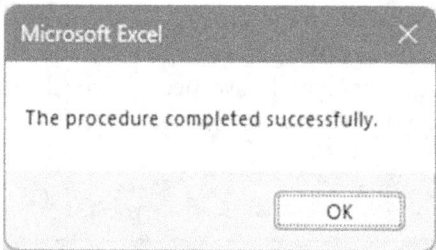

FIGURE 4.4. To display a message to the user, place the text you want to display as the argument of the MsgBox function.

The MsgBox function allows you to use other arguments that make it possible to set the number of buttons that should be available in the message box or change the title of the message box from the default, Microsoft Excel. You can also assign your own help topic.

The syntax of the MsgBox function is as follows:

```
MsgBox (prompt [, buttons] [, title], [, helpfile, context])
```

Notice that while the MsgBox function has five arguments, only the first one, prompt, is required. The arguments listed in square brackets are optional. When you enter a long text string for the prompt argument, Visual Basic decides how to break the text, so it fits the message box. Let's do some exercises in the Immediate window to learn various text formatting techniques.

Hands-On 4.7   Formatting Text for Display in the MsgBox Function

1. Enter the following instruction in the Immediate window. Be sure to enter the entire text string on one line, and then press Enter:

```
MsgBox "All processes completed successfully. Now connect the
external storage device to your computer. The following procedure
will copy the workbook file to the attached device."
```

As soon as you press Enter, Visual Basic shows the resulting dialog box (Figure 4.5).

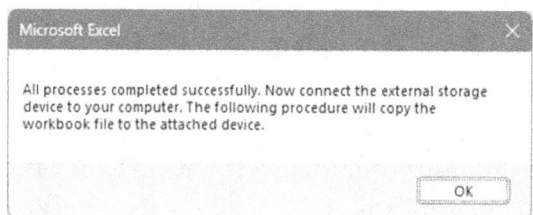

FIGURE 4.5. This long message will look more appealing when you take the text formatting into your own hands.

When you write a VBA procedure that requires long messages, you can break your message text into several lines using the VBA Chr function. The Chr function takes one argument (a number from 0 to 255), and it returns a character represented by this number. For example, Chr(13) returns a carriage return character (this is the same as pressing the Enter key), and Chr(10) returns a linefeed character (useful for adding spacing between the text lines). For example, take a look at the following subroutine procedure:

```
Sub LongTextMessage()
    MsgBox "All processes completed successfully. " & Chr(13) _
    & "Now connect the external storage device to " & Chr(13) _
    & "your computer. The following procedure " & Chr(13) _
    & "will copy the workbook file to the attached device."
End Sub
```

Figure 4.6 depicts the message box after running the LongTextMessage procedure.

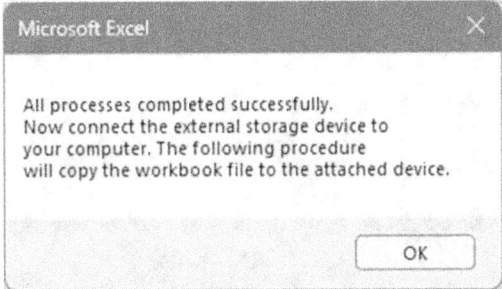

FIGURE 4.6.   You can break a long text string into several lines by using the Chr(13) function.

Note that you must surround each text fragment with quotation marks. The Chr(13) function indicates a place where you'd like to start a new line. The string concatenation character (&) is used to add a carriage return character to a concatenated string. Quoted text embedded in a text string requires an additional set of quotation marks, as shown in the revised subroutine below:

```
Sub LongTextMessage_Modified()
    MsgBox "All processes completed successfully. " & Chr(13) _
    & "Now connect the external storage device to " & _
        Chr(13) & "your computer. " & _
    "The following procedure ""TestProc()""" & Chr(13) & _
        "will copy the workbook file to the attached device."
End Sub
```

When you enter exceptionally long text messages on one line, it's easy to make a mistake. As you recall, Visual Basic has a special line continuation character (an underscore, _) that allows you to break a long VBA statement into several lines. Unfortunately, the line continuation character cannot be used in the Immediate window.

2. Add a new module to the `ProcAndFunctions` `(Chap04.xlsm)` project and change the module's name to `Sample4`.

3. Activate the `Sample4` module and enter the `LongTextMessage` and `LongTextMessage_Modified` subroutines as shown earlier. Be sure to precede each line continuation character (_) with a single space.

4. Execute each procedure.
Notice that the text entered on several lines is more readable and the code is easier to maintain.

To improve the readability of your message, you may want to add more spacing between the text lines by including blank lines. To do this, use two `Chr(13)` or two `Chr(10)` functions, as shown in the following step.

5. Enter the following `LongTextMessage_Modified2` procedure and run it:

```
Sub LongTextMessage_Modified2()
    MsgBox "All processes completed successfully. " & _
        Chr(10) & Chr(10) _
    & "Now connect the external storage device " & _
        Chr(13) & Chr(13) & "to your computer." & _
    " The following procedure ""TestProc()""" & _
        Chr(10) & Chr(10) _
    & "will copy the workbook file to the attached device."
End Sub
```

Figure 4.7 displays the message box generated by the `LongTextMessage2` procedure.

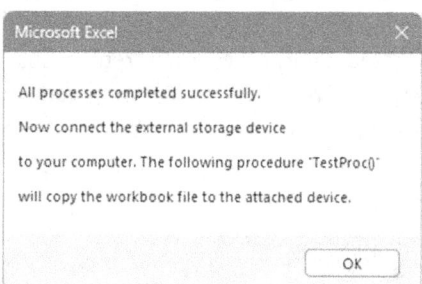

FIGURE 4.7. You can increase the readability of your message by increasing the spacing between the selected text lines.

Now that you've mastered the text formatting techniques, let's take a closer look at the next argument of the `MsgBox` function. Although the `buttons` argument is optional, it is frequently used. The `buttons` argument specifies how many and what types of buttons you want to appear in the message box. This argument can be a constant or a number, as shown in Table 4.1. If omitted, the resulting message box includes only the OK button, as you've seen in the preceding examples.

TABLE 4.1.   Settings for the MsgBox buttons argument.

Constant	Value	Description
Button settings		
vbOKOnly	0	Displays only an OK button. This is the default.
vbOKCancel	1	OK and Cancel buttons.
vbAbortRetryIgnore	2	Abort, Retry, and Ignore buttons.
vbYesNoCancel	3	Yes, No, and Cancel buttons.
vbYesNo	4	Yes and No buttons.
vbRetryCancel	5	Retry and Cancel buttons.
Icon settings		
vbCritical	16	Displays the Critical Message icon.
vbQuestion	32	Displays the Question Message icon.
vbExclamation	48	Displays the Warning Message icon.
vbInformation	64	Displays the Information Message icon.
Default button settings		
vbDefaultButton1	0	The first button is the default.
vbDefaultButton2	256	The second button is the default.
vbDefaultButton3	512	The third button is the default.
vbDefaultButton4	768	The fourth button is the default.
Message box modality		
vbApplicationModal	0	The user must respond to the message before continuing to work on the current application.
vbSystemModal	4096	All applications are suspended until the user responds to the message box.
Other MsgBox display settings		
vbMsgBoxHelpButton	16384	Adds a Help button to the message box.
vbMsgBoxSetForeground	65536	Specifies the message box window as the foreground window.
vbMsgBoxRight	524288	Text is right aligned.
vbMsgBoxRtlReading	1048576	Text appears as right-to-left reading on Hebrew and Arabic systems.

When should you use the `buttons` argument? Suppose you want the user of your procedure to respond to a question with Yes or No. Your message box may then require two buttons. If a message box includes more than one button, one of them is the default button. When the user presses Enter, the default button is selected automatically. Because you can display various types of messages (critical, warning, or information), you can visually indicate the importance of the message by including in the `buttons` argument the graphical representation (icon) for the chosen message type.

In addition to the type of message, the `buttons` argument can include a setting to determine whether the message box must be closed before a user activates another application. It's quite possible that the user may want to switch to another program or perform another task before responding to the question posed in your message box.

If the message box is application modal (`vbApplication Modal`), the user must close the message box before continuing to use your application. On the other hand, if you want to suspend all the applications until the user responds to the message box, you must include the `vbSystemModal` setting in the `buttons` argument.

The `buttons` argument settings are divided into the following five groups:

- Button settings
- Icon settings
- Default button settings
- Message box modality
- Other `MsgBox` display settings

Only one setting from each group can be included in the `buttons` argument.

To create a `buttons` argument, you can add up the values for each setting you want to include. For example, to display a message box with two buttons (Yes and No), the question mark icon, and the No button as the default button, look up the corresponding values in Table 4.1 and add them up. You should arrive at 292 (4 + 32 + 256).

Let's go back to the Immediate window for more testing of the capabilities of the `MsgBox` function.

Hands-On 4.8 Using the MsgBox Function with Arguments (Example 1)

1. To quickly see the message box using the calculated message box argument, enter the following statement in the Immediate window, and press Enter:

```
MsgBox "Do you want to proceed?", 292
```

The resulting message box is shown in Figure 4.8.

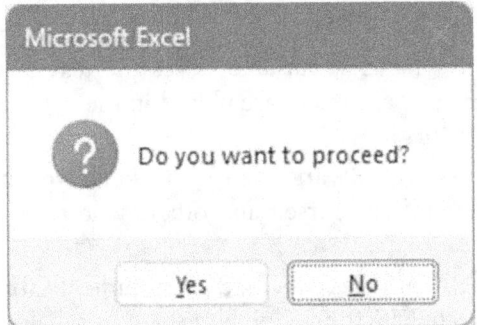

FIGURE 4.8.   You can specify the number of buttons to include in the message box by using the optional buttons argument.

When you derive the `buttons` argument by adding up the constant values, your procedure becomes less readable. There's no reference table where you can check the hidden meaning of 292. To improve the readability of your `MsgBox` function, it's better to use the constants instead of their values.

2. Now, enter the following revised statement on one line in the Immediate window and press Enter:

```
MsgBox "Do you want to proceed?", vbYesNo + vbQuestion +
    vbDefaultButton2
```

This statement (which must be entered on one line) produces the same result as shown in Figure 4.8 and is more readable.

The following example demonstrates how to use the `buttons` argument inside the Visual Basic procedure.

Hands-On 4.9   Using the MsgBox Function with Arguments (Example 2)

1. Add a new module to the `ProcAndFunctions` (Chap04.xlsm) project and change the module's name to `Sample5`.

2. Activate the `Sample5` module, enter the `MsgYesNo` subroutine shown here, and then run it:

```
Sub MsgYesNo()
  Dim question As String
  Dim myButtons As Integer
```

```
    question = "Do you want to open a new workbook?"
    myButtons = vbYesNo + vbQuestion + vbDefaultButton2

    MsgBox question, myButtons
End Sub
```

In the foregoing subroutine, the `question` variable stores the text of your message. The settings for the `buttons` argument are placed in the `myButtons` variable.

By specifying the names of the `buttons` argument's constants, you make your procedure easier to understand for yourself and others who may work with this procedure in the future.

The `question` and `myButtons` variables are used as arguments for the `MsgBox` function. `No` button is selected. It's the default button for this dialog box specified by the `vbDefaultButton2` setting. If you press Enter, Excel removes the `MsgBox` from the screen. Nothing happens because your procedure does not have any more instructions following the `MsgBox` function. To change the default button, use the `vbDefaultButton1` setting instead.

The third argument of the `MsgBox` function is `title`. While this is also an optional argument, it's very handy, as it allows you to create procedures that don't provide visual clues to the fact that you programmed them with Microsoft Excel. By using this argument, you can set the title bar of your message box to any text you want.

3. Suppose you want the `MsgYesNo` procedure to display in its title the text "New workbook." The following `MsgYesNo2` procedure demonstrates the use of the `title` argument:

```
Sub MsgYesNo2()
    Dim question As String
    Dim myButtons As Integer
    Dim myTitle As String

    question = "Do you want to open a new workbook?"
    myButtons = vbYesNo + vbQuestion + vbDefaultButton2
    myTitle = "New workbook"

    MsgBox question, myButtons, myTitle
End Sub
```

The text for the `title` argument is stored in the variable `myTitle`. If you don't specify the value for the `title` argument, Visual Basic displays the default text, Microsoft Excel.

Notice that the arguments are listed in the order determined by the MsgBox function. If you would like to list the arguments in any order, you must precede the value of each argument with its name. For example:

```
MsgBox title:=myTitle, prompt:=question, buttons:=myButtons
```

The last two optional arguments—helpfile and context—are used by programmers who are experienced with using help files in the Windows environment.

The helpfile argument indicates the name of a special help file that contains additional information you may want to display to your VBA procedure user. When you specify this argument, the Help button will be added to your message box.

Returning Values from the MsgBox Function

When you display a simple message box dialog with one button, clicking the OK button or pressing the Enter key removes the message box from the screen. When the message box has more than one button, however, your procedure should detect which button was pressed so you can take appropriate action further in your procedure code. To do this, you must save the result of the message box in a variable. Table 4.2 shows values that the MsgBox function returns.

TABLE 4.2.   Values returned by the MsgBox function.

Button Selected	Constant	Value
OK	vbOK	1
Cancel	vbCancel	2
Abort	vbAbort	3
Retry	vbRetry	4
Ignore	vbIgnore	5
Yes	vbYes	6
No	vbNo	7

Let's revise the MsgYesNo2 procedure to show which button the user has chosen.

Hands-On 4.10   Using the MsgBox Function with Arguments (Example 3)

1. Activate the Sample5 module and enter the MsgYesNo3 subroutine as shown here:

```
Sub MsgYesNo3()
    Dim question As String
```

```
    Dim myButtons As Integer
    Dim myTitle As String

    Dim myChoice As Integer

    question = "Do you want to open a new workbook?"
    myButtons = vbYesNo + vbQuestion + vbDefaultButton2
    myTitle = "New workbook"
    myChoice = MsgBox(question, myButtons, myTitle)

    MsgBox myChoice
End Sub
```

In the foregoing procedure, we assigned the result of the `MsgBox` function to the variable `myChoice`. Notice that when you return a value from the `MsgBox` function, the arguments of the `MsgBox` function must be listed in parentheses:

```
myChoice = MsgBox(question, myButtons, myTitle)
```

2. Run the `MsgYesNo3` procedure.

When you run the `MsgYesNo3` procedure, a two-button message box is displayed. When you click on the Yes button, the statement `MsgBox myChoice` displays the number 6. When you click the No button, the number 7 is displayed (see the corresponding numbers in Table 4.2).

MsgBox Function with or Without Parentheses?

Use parentheses around the `MsgBox` function's argument list when you want to use the result returned by the function. By listing the function's arguments without parentheses, you tell Visual Basic that you want to ignore the function's result. Most likely, you will want to use the function's result when the `MsgBox` contains more than one button.

GETTING TO KNOW THE INPUTBOX FUNCTION

The `InputBox` function displays a dialog box with a message that prompts the user to enter data (see Figure 4.9 in Hands-On 4.11). This dialog box has two buttons—OK and Cancel. When you click OK, the `InputBox` function returns the information entered in the text box. When you select Cancel, the function returns an empty string ("").

The syntax of the `InputBox` function is as follows:

```
InputBox(prompt [, title] [, default] [, xpos] [, ypos]
    [, helpfile, context])
```

The first argument, `prompt`, is the text message that you want to display in the dialog box. Long text strings can be entered on several lines by using the `Chr(13)` or `Chr(10)` functions (see the examples of using the `MsgBox` function earlier in this chapter). All the remaining `InputBox` arguments are optional.

The second argument, `title`, allows the changing of the default dialog box title. The default title is Microsoft Excel.

The third argument of the `InputBox` function, `default`, shows a default value in the text box. If you omit this argument, an empty edit box is displayed.

The following two arguments, `xpos` and `ypos`, can be used to specify the exact position where the dialog box should appear on the screen. If you omit these arguments, the box appears in the middle of the current window. The `xpos` argument determines the horizontal position of the dialog box from the left edge of the screen. When omitted, the dialog box is centered horizontally. The `ypos` argument determines the vertical position from the top of the screen. If you omit this argument, the dialog box is positioned vertically approximately one-third of the way down the screen. Both `xpos` and `ypos` are measured in special units called *twips*. One twip is equivalent to approximately 0.0007 inches.

The last two arguments, `helpfile` and `context`, are used in the same way as the corresponding arguments of the `MsgBox` function discussed earlier in this chapter.

⊙ Hands-On 4.11   Using the InputBox Function (Example 1)

1. Add a new module to the `ProcAndFunctions` (`Chap04.xlsm`) project and change the module's name to `Sample6`.
2. Activate the `Sample6` module and enter the `Informant` subroutine shown here:

```
Sub Informant()
    InputBox prompt:="Enter your place of birth:" & Chr(13) _
        & " (e.g., Boston, Great Falls, etc.) "
End Sub
```

This procedure displays a dialog box with two buttons, as shown in Figure 4.9. The input prompt is displayed on two lines.

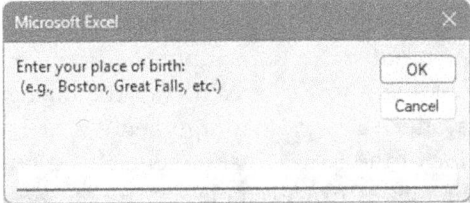

FIGURE 4.9. A dialog box generated by the Informant subroutine.

As with the MsgBox function, if you plan on using the data entered by the user in the dialog box, you should store the result of the InputBox function in a variable.

3. Type the Informant2 procedure shown here to assign the result of the InputBox function to the variable town:

```
Sub Informant2()
  Dim myPrompt As String
  Dim town As String

  Const myTitle = "Enter data"

  myPrompt = "Enter your place of birth:" & Chr(13) _
      & "(e.g., Boston, Great Falls, etc.)"
  town = InputBox(myPrompt, myTitle)

  MsgBox "You were born in " & town & ".", , "Your response"
End Sub
```

Notice that, this time, the arguments of the InputBox function are listed within parentheses. Parentheses are required if you want to use the result of the InputBox function later in your procedure. Notice that the Informant2 subroutine uses a constant to specify the text to appear in the title bar of the dialog box. Because the constant value remains the same throughout the execution of your procedure, you can declare the input box title as a constant. If you'd rather use a variable, however, you still can.

When you run a procedure using the InputBox function, the dialog box generated by this function always appears in the same area of the screen. To change the location of the dialog box, you must supply the xpos and ypos arguments, as explained earlier.

4. Run the Informant2 procedure.
5. To display the dialog box in the top left-hand corner of the screen, modify the InputBox function in the Informant2 procedure as follows and then run it:

```
town = InputBox(myPrompt, myTitle, , 1, 200)
```

Notice that the argument myTitle is followed by two commas. The second comma marks the position of the omitted default argument. The next two arguments determine the horizontal and vertical position of the dialog box. If you omit the second comma after the myTitle argument, Visual Basic will use the number 1 as the value of the default argument. If you precede the values of arguments by their names (for example, prompt:=myPrompt, `title:=myTitle, xpos:=1, ypos:=200`), you won't have to remember to place a comma in the place of each omitted argument.

What will happen if you enter a number instead of the name of a town? Because users often supply incorrect data in an input dialog box, your procedure must verify that the supplied data type can be used in further data manipulations. The `InputBox` function itself does not provide a facility for data validation. To validate user input, you must learn additional VBA instructions, which are presented in the next chapter.

Determining and Converting Data Types

The result of the `InputBox` function is always a string. If the user enters a number, the string value the user entered should be converted to a numeric value before your procedure can use this number in mathematical computations. Visual Basic can convert values from one data type to another.

NOTE	*Refer to Chapter 3 for more information about using the* `VarType` *function to determine the data type of a variable and common data type conversion functions.*

Let's try out a procedure that suggests what type of data the user should enter by supplying a default value in the `InputBox` dialog.

Hands-On 4.12 Using the InputBox Function (Example 2)

1. Activate the `Sample6` module in the `ProcAndFunctions` (`Chap04.xlsm`) project and enter the following `AddTwoNums` procedure:

```
Sub AddTwoNums()
  Dim myPrompt As String
  Dim value1 As String
  Dim value2 As Integer
  Dim mySum As Single

  Const myTitle = "Enter data"

  myPrompt = "Enter a number:"
  value1 = InputBox(myPrompt, myTitle, 0)
  value2 = 2
  mySum = value1 + value2

  MsgBox "The result is " & mySum & _
      " (" & value1 & " + " & CStr(value2) + ")", _
```

```
                vbInformation, "Total"
End Sub
```

The `AddTwoNums` procedure displays the dialog box shown in Figure 4.10. Notice that this dialog box has two special features that are obtained by using the `InputBox` function's optional `title` and `default` arguments. Instead of the default title "Microsoft Excel," the dialog box displays a text string defined by the contents of the `myTitle` constant. The zero entered as the default value in the edit box suggests that the user enter a number instead of text. Once the user provides the data and clicks OK, the user's input is assigned to the variable `value1`.

```
value1 = InputBox(myPrompt, myTitle, 0)
```

2. Run the `AddTwoNums` procedure, supply any number when prompted, and then click OK.

FIGURE 4.10.   To suggest that the user enter a specific type of data, you may want to provide a default value in the edit box.

The data type of the variable `value1` is String.

3. You can check the data type easily if you follow the foregoing instruction in the procedure code with this statement:

```
MsgBox VarType(value1)
```

When Visual Basic runs the foregoing line, it will display a message box with the number 8. Recall from Chapter 3 (Table 3.3) that this number represents the String data type.

The statement `mySum = value1 + value2` adds the value stored in the `value2` variable to the user's input and assigns the result of the calculation to the variable `mySum`. Because the `value1` variable's data type is String, prior to using this variable's data in the computation, Visual Basic goes to work behind the scenes to perform the data type conversion. Visual Basic understands the need for conversion. Without it, the two incompatible data types (String and

Integer) would generate a type mismatch error. The procedure ends with the MsgBox function displaying the result of the calculation and showing the user how the total was derived. Notice that the value2 variable must be converted from an Integer to a String data type using the CStr function to display it in the message box:

```
MsgBox "The result is " & mySum & _
    " (" & value1 & " + " & CStr(value2) + ")", _
    vbInformation, "Total"
```

Define a Constant

To ensure that all title bars in a particular VBA procedure display the same text, assign the title text to a constant. By doing this, you will save time by not having to type the title text more than once.

USING THE INPUTBOX METHOD

In addition to the built-in InputBox VBA function, there is also the Excel Input-Box method. If you activate the Object Browser window, type "inputbox" in the search box, and press Enter, you will see two occurrences of InputBox—one in the Excel library and the other in the VBA library, as shown in Figure 4.11.

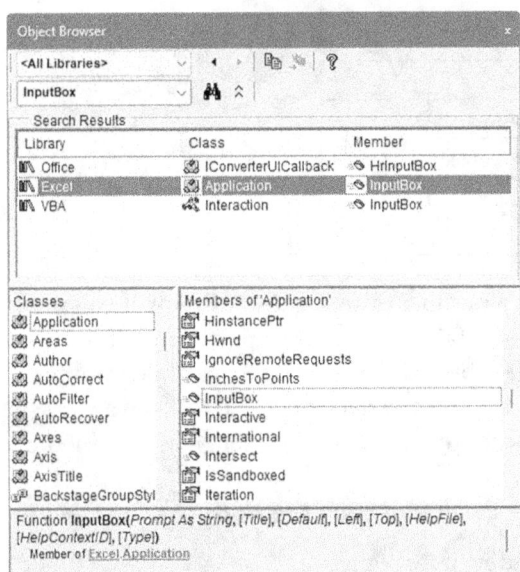

FIGURE 4.11. Don't forget to use the Object Browser when researching Visual Basic functions and methods.

The `InputBox` method available in the Microsoft Excel library has a slightly different syntax from the `InputBox` function that was covered earlier in this chapter. Its syntax is:

```
expression.InputBox(prompt, [title], [default], [left], [top], _
        [helpfile], [helpcontextID], [type])
```

All bracketed arguments are optional. The `prompt` argument is the message to be displayed in the dialog box, `title` is the title for the dialog box, and `default` is a value that will appear in the text box when the dialog box is initially displayed.

The `left` and `top` arguments specify the position of the dialog box on the screen. The values for these arguments are entered in points. Note that one point equals 1/72 inch. The arguments `helpfile` and `helpcontextID` identify the name of the help file and the specific number of the help topic to be displayed when the user clicks the Help button.

The last argument of the `InputBox` method, `type`, specifies the return data type. If you omit this argument, the `InputBox` method will return a string. The values of the `type` argument are shown in Table 4.3.

TABLE 4.3.   Data types returned by the InputBox method.

Value	Type of Data Returned
0	A formula
1	A number
2	A string (text)
4	A logical value (`True` or `False`)
8	A cell reference, as a `Range` object
16	An error value (for example, `#N/A`)
64	An array of values

You can allow the user to enter a number or text in the edit box if you use 3 for the `type` argument. This value is obtained by adding up the values for a number (1) and a string (2), as shown in Table 4.3. The `InputBox` method is quite useful for VBA procedures that require a user to select a range of cells in a worksheet.

Let's look at an example procedure that uses the Excel `InputBox` method.

Hands-On 4.13   Using the Excel InputBox Method

1. Close the Object Browser window if it is open.

2. In the `Sample7` module, enter the following `WhatRange` procedure:

```
Sub WhatRange()
   Dim newRange As Range
   Dim tellMe As String

   tellMe = "Use the mouse to select a range:"
   Set newRange = Application.InputBox(prompt:=tellMe, _
      Title:="Range to format", _
      Type:=8)
   newRange.NumberFormat = "0.00"
   newRange.Select
End Sub
```

The `WhatRange` procedure begins with a declaration of an object variable—`newRange`. As you recall from Chapter 3, object variables point to the location of the data. The range of cells that the user selects is assigned to the object variable `newRange`. Notice the keyword `Set` before the name of the variable:

```
Set newRange = Application.InputBox(prompt:=tellMe, _
   Title:="Range to format", _
   Type:=8)
```

The `Type` argument (`Type:=8`) enables the user to select any range of cells. When the user highlights the cells, the next instruction changes the format of the selected cells:

```
newRange.NumberFormat = "0.00"
```

The last instruction selects the range of cells that the user highlighted.

3. Press Alt+F11 to activate the Microsoft Excel Application window, and then press Alt+F8 to activate the Macro dialog box.

4. Scroll down in the macro name list box, select the WhatRange procedure, and click the Run button.
Visual Basic displays a dialog box prompting the user to select a range of cells in the worksheet.

5. Use the mouse to select any cells you want. Figure 4.12 shows how Visual Basic enters the selected range reference in the edit box as you drag the mouse to select the cells.

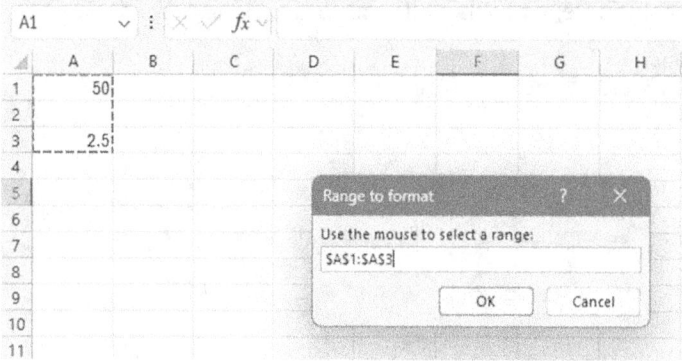

FIGURE 4.12. By using Excel's InputBox method, you can get the range address from the user.

6. When you're done selecting cells, click OK in the dialog box.
   The selected range is now formatted. To check this out, enter a whole number in any of the selected cells. The number should appear formatted with two decimal places.

7. Rerun the procedure, and when the dialog box appears, click Cancel.
   When you click the Cancel button or press Esc, Visual Basic displays an error message—Run-time error '424': Object Required. When you click the Debug button in the error dialog box, Visual Basic will highlight the line of code that caused the error. Because you don't want to select anything when you cancel the dialog box, you must find a way to ignore the error that Visual Basic displays. Using a special statement, `On Error GoTo labelname`, you can take a detour when an error occurs. This instruction should be placed just below the variable declaration lines. `Labelname` can be any word you want, except for a Visual Basic keyword. If an error occurs, Visual Basic will jump to the specified label, as shown in Step 10 ahead.

8. Choose Run | Reset to cancel the procedure you were running.

9. Modify the `WhatRange` procedure so it looks like the `WhatRange2` procedure shown here:

```
Sub WhatRange2()
   Dim newRange As Range
   Dim tellMe As String

   On Error GoTo VeryEnd

   tellMe = "Use the mouse to select a range:"
   Set newRange = Application.InputBox(prompt:=tellMe, _
```

```
        Title:="Range to format", _
        Type:=8)
    newRange.NumberFormat = "0.00"
    newRange.Select

    VeryEnd:
End Sub
```

10. Run the modified procedure and click Cancel as soon as the input box appears. Notice that, this time, the procedure does not generate the error when you cancel the dialog box. When Visual Basic encounters an error, it jumps to the `VeryEnd` label placed at the end of the procedure. The statements placed between `On Error Goto VeryEnd` and the `VeryEnd` labels are ignored. In Chapter 9, you will find other examples of handling errors in your VBA procedures.

Subroutines and Functions: Which Should You Use?

Create a subroutine when...	Create a function when...
You want to perform some actions.	You want to perform a simple calculation more than once.
You want to get input from the user.	You must perform complex computations.
You want to display a message on the screen.	You must call the same block of instructions more than once.
	You want to check whether a certain expression is `True` or `False`.

SUMMARY

In this chapter, you learned the difference between subroutine procedures, which perform actions, and function procedures, which return values. While you can create subroutines by recording or typing code directly into the Visual Basic module, function procedures cannot be recorded because they can take arguments. You must write them manually. You learned how to pass arguments to functions and determine the data type of a function's result. You increased your repertoire of VBA keywords with the `ByVal`, `ByRef`, and `Optional` keywords. You also learned how, with the help of parameters, subprocedures can pass values back to the calling procedures. After working through this chapter, you should be able to create some custom functions of your own that are suited to your specific needs. You should also be able to interact easily with your procedure users by employing the `MsgBox` and `InputBox` functions as well as the Excel `InputBox` method.

The next chapter will introduce you to decision-making. You will learn how to change the course of your VBA procedure based on the results of the conditions that you supply.

Chapter 5

ADDING DECISIONS TO EXCEL VBA PROGRAMS—A QUICK INTRODUCTION TO CONDITIONAL STATEMENTS

VBA, like other programming languages, offers statements that allow you to include decision points in your VBA procedures. In programming, all decisions are based on supplied answers. If the answer is positive, the procedure executes a specified block of instructions. If the answer is negative, the procedure executes another block of instructions or simply doesn't do anything.

In this chapter, you will learn how to use VBA conditional statements to alter the flow of your program. Conditional statements are often referred to as "control structures," as they give you the ability to control the flow of your VBA procedure by skipping over certain statements and "branching" to another part of the procedure.

RELATIONAL AND LOGICAL OPERATORS

You make decisions in your VBA procedures by using conditional expressions inside special control structures. A *conditional expression* is an expression that uses one of the relational operators listed in Table 5.1, one of the logical operators listed in Table 5.2, or a combination of both. When Visual Basic encounters a conditional expression in your program, it evaluates the expression to determine whether it is true or false.

TABLE 5.1.   Relational operators in VBA.

Operator	Description
=	Equal to
<>	Not equal to
>	Greater than
<	Less than
>=	Greater than or equal to
<=	Less than or equal to

TABLE 5.2.   Logical operators in VBA.

Operator	Description
AND	All conditions must be true before an action can be taken.
OR	At least one of the conditions must be true before an action can be taken.
NOT	Used for negating a condition. If a condition is true, NOT makes it false. If a condition is false, NOT makes it true.

USING THE IF...THEN STATEMENT

The simplest way to get some decision-making into your VBA procedure is to use the If...Then statement. Suppose you want to choose an action depending on a condition. You can use the following structure:

```
If condition Then statement
```

For example, to delete a blank row from a worksheet, first check whether the active cell is blank. If the result of the test is True, go ahead and delete the entire row that contains that cell:

```
If ActiveCell = "" Then Selection.EntireRow.Delete
```

If the active cell is not blank, Visual Basic will ignore the statement following the Then keyword.

Sometimes you may want to perform several actions when the condition is true. Although you could add other statements on the same line by separating them with colons, your code will look clearer if you use the multiline version of the If...Then statement, as shown here:

```
If condition Then
   statement1
   statement2
```

```
    statementN
End If
```

For example, to perform some actions when the value of the active cell is greater than 50, you can write the following block of instructions:

```
If ActiveCell.Value > 50 Then
  MsgBox "The exact value is " & ActiveCell.Value
  Debug.Print ActiveCell.Address & ": " & ActiveCell.Value
End If
```

In this example, the statements between the `Then` and `End If` keywords are not executed if the value of the active cell is less than or equal to 50. Notice that the `If...Then` block statement must end with the keywords `End If`.

How does Visual Basic make a decision? It evaluates the condition it finds between the `If...Then` keywords. Let's try to evaluate the following condition: `ActiveCell.Value > 50.`

NOTE | *Please note that files for the "Hands-On" projects can be found in the companion files.*

Hands-On 5.1   Evaluating Conditions in the Immediate Window

1. Open a new Microsoft Excel workbook.
2. Select any cell in a blank worksheet, enter 50, and then reselect that cell.
3. Switch to the VBE window.
4. Activate the Immediate window.
5. Type the following statement and press Enter when you're done:

   ```
   ? ActiveCell.Value > 50
   ```

   When you press Enter, Visual Basic writes the result of this test—`False`. When the result of the test is `False`, Visual Basic will not bother to read the statement following the `Then` keyword in your code. It will simply go on to read the next line of your procedure if there is one. If there are no more lines to read, the procedure will end.

6. Now change the operator to less than or equal to, and have Visual Basic evaluate the following condition:

   ```
   ? ActiveCell.Value <= 50
   ```

   This time, the test returns `True`, and Visual Basic will jump to whatever statement or statements it finds after the `Then` keyword.

7. Close the Immediate window.

Now, let's use the If...Then statement in a VBA procedure.

> **Hands-On 5.2   Writing a VBA Procedure with a Simple If...Then Statement**

1. Open a new workbook and save it as `C:\VBAExcel2024_ByExample\Chap05.xlsm`.
2. Switch to the VBE screen and rename the VBA project `Decisions`.
3. Insert a new module in the `Decisions (Chap05.xlsm)` project and rename this module `IfThen`.
4. In the `IfThen` module, enter the following procedure:

```
Sub SimpleIfThen()
  Dim weeks As String
  weeks = InputBox("How many weeks are in a year?", "Quiz")
  If weeks <> 52 Then MsgBox "Try Again"
End Sub
```

The `SimpleIfThen` procedure stores the user's answer in the variable named `weeks`. The variable's value is then compared to the number 52. If the result of the comparison is `True` (that is, if the value stored in the variable `weeks` is not equal to 52), Visual Basic will display the message "Try Again."

5. Run the `SimpleIfThen` procedure and enter a number other than 52.
6. Rerun the `SimpleIfThen` procedure and enter 52.
   When you enter the correct number of weeks, Visual Basic does nothing. The procedure simply ends. It would be nice to display a message when the user guesses right.
7. Enter the following instruction on a separate line before the `End Sub` keywords:

```
If weeks = 52 Then MsgBox "Congratulations!"
```

8. Run the `SimpleIfThen` procedure again and enter 52.
   When you enter the correct answer, Visual Basic does not execute the statement `MsgBox "Try Again."` When the procedure is executed, the statement to the right of the `Then` keyword is ignored if the result from evaluating the supplied condition is `False`.

   A VBA procedure can call another procedure, including calling itself.

9. Modify the first `If` statement in the SimpleIfThen procedure as follows:

```
If weeks <> 52 Then MsgBox "Try Again": SimpleIfThen
```

We added a colon and the name of the `SimpleIfThen` procedure to the end of the existing If...Then statement. If the user enters the incorrect answer, they will see a message; after dismissing the message, the input box will appear again, giving them another chance to supply the correct answer. If they click Cancel, they will have to deal with the unfriendly error message "Type mismatch." You saw in the previous chapter how to use the `On Error GoTo labelname` statement to get around the error, at least temporarily until you learn more about error handling in Chapter 9. For now, you may want to revise your `SimpleIfThen` procedure as follows:

```
Sub SimpleIfThen()
  Dim weeks As String
  On Error GoTo VeryEnd
  weeks = InputBox("How many weeks are in a year:", "Quiz")
  If weeks <> 52 Then MsgBox "Try Again": SimpleIfThen
  If weeks = 52 Then MsgBox "Congratulations!"
  VeryEnd:
End Sub
```

10. Run the `SimpleIfThen` procedure a few times by supplying incorrect answers. The error trap that you added to your procedure will allow you to quit gracefully without seeing the "Type mismatch" error.

Two Formats for the If...Then Statement

The `If...Then` statement has two formats—single line and multiline.

The single-line format is good for short or simple statements such as:

```
If secretCode <> 01W01 Then MsgBox "Access denied"
```

In this example, Visual Basic displays the message "Access denied" if the value of `secretCode` is not equal to `01W01`.

In the example shown below, Visual Basic sets the value of the `alpha` variable to `True` and the `beta` variable to `False` when the `secretCode` variable is equal to `01W01`:

```
If secretCode = 01W01 Then alpha = True : beta = False
```

Notice that the second statement to be executed is separated from the first by a colon.

The multiline `If...Then` statement is clearer when there are more statements to be executed when the condition is true or when the statement to be executed is extremely long, as in the following example:

```
If ActiveSheet.Name = "Sheet1" Then
  ActiveSheet.Move after:= Sheets(Worksheets.Count)
End If
```

Here, Visual Basic examines the active sheet name. If it is `Sheet1`, the condition `ActiveSheet.Name = "Sheet1"` is true, and Visual Basic proceeds to execute the line following the `Then` keyword. As a result, the active sheet is moved to the last position in the workbook.

If Block Instructions and Indenting

To make the `If` blocks easier to read and understand, use indentation. Compare the following:.

`If condition Then`
`action`
`End If`
	`If condition Then`
	`    action`
	`End If`

In the `If...Then` block statement on the right, you can easily see where the block begins and where it ends.

USING THE IF...THEN...ELSE STATEMENT

By adding the `Else` clause to the simple `If...Then` statement, you can direct your procedure to the appropriate statement depending on the result of the test. The `If...Then...Else` statement has two formats—single line and multiline. The single-line format is as follows:

```
If condition Then statement1 Else statement2
```

The statement following the `Then` keyword is executed if the condition is true, and the statement following the `Else` clause is executed if the condition is false—for example:

```
If Sales > 5000 Then Bonus = Sales * 0.05 Else MsgBox "No Bonus"
```

If the value stored in the variable `Sales` is greater than 5000, Visual Basic will calculate the bonus using the following formula: `Sales * 0.05`. If the variable `Sales` is not greater than 5000, however, Visual Basic will display the message "No Bonus."

The `If...Then...Else` statement should be used to decide which of the two actions to perform.

When you need to execute more statements when the condition is true or false, it's better to use the multiline format of the `If...Then...Else` statement:

```
If condition Then
    statements to be executed if condition is True
Else
    statements to be executed if condition is False
End If
```

Notice that the multiline (block) `If...Then...Else` statement ends with the `End If` keywords. Use the indentation shown in the previous section to make this block structure easier to read. Here's a code example that uses the preceding syntax:

```
If ActiveSheet.Name = "Sheet1" Then
  ActiveSheet.Name = "My Sheet"
  MsgBox "This sheet has been renamed."
Else
  MsgBox "This sheet name is not default."
End If
```

If the condition (`ActiveSheet.Name = "Sheet1"`) is true, Visual Basic will execute the statements between `Then` and `Else` and ignore the statement between `Else` and `End If`. If the condition is false, Visual Basic will omit the statements between `Then` and `Else` and execute the statement between `Else` and `End If`.

Let's look at the complete procedure example.

Hands-On 5.3   Writing a VBA Procedure with an If…Then…Else Statement

1. Insert a new module into the `Decisions` (`Chap05.xlsm`) project.
2. Change the module name to `IfThenElse`.
3. Enter the following `WhatTypeOfDay` procedure and then run it:

```
Sub WhatTypeOfDay()
  Dim response As String
  Dim question As String
  Dim strmsg1 As String, strmsg2 As String
  Dim myDate As Date

  question = "Enter any date in the format mm/dd/yyyy:" _
    & Chr(13)& " (e.g., 11/22/2024)"
  strmsg1 = "weekday"
  strmsg2 = "weekend"
  response = InputBox(question)
  myDate = Weekday(CDate(response))
  If myDate >= 2 And myDate <= 6 Then
    MsgBox strmsg1
  Else
    MsgBox strmsg2
  End If
End Sub
```

The preceding procedure asks the user to enter any date. The user-supplied string is then converted to the Date data type with the built-in `CDate` function. The `Weekday` function converts the date into an integer that indicates the day of the week. The day of the week constants are listed in Table 5.3. The integer is stored in the variable `myDate`. The conditional test is performed to check whether the value of the variable `myDate` is greater than or equal to 2 (`>=2`) and less than or equal to 6 (`<=6`). If the result of the test is `True`, the user is told that the supplied date is a weekday; otherwise, the program announces that it's a weekend.

4. Run the procedure from the Visual Basic window. Run it a few times, each time supplying a different date. Check the Visual Basic answers against your desktop or wall calendar.

TABLE 5.3.   Values returned by the built-in Weekday function.

Constant	Value
vbSunday	1
vbMonday	2
vbTuesday	3
vbWednesday	4
vbThursday	5
vbFriday	6
vbSaturday	7

USING THE IF...THEN...ELSEIF STATEMENT

Quite often, you will need to check the results of several different conditions. To join a set of `If` conditions together, you can use the `ElseIf` clause. By using the `If...Then...ElseIf` statement, you can supply more conditions to evaluate than is possible with the `If...Then...Else` statement discussed earlier. Here's the syntax of the `If...Then...Else` statement:

```
If condition1 Then
   statements to be executed if condition1 is True
ElseIf condition2 Then
   statements to be executed if condition2 is True
ElseIf condition3 Then
   statements to be executed if condition3 is True
ElseIf conditionN Then
   statements to be executed if conditionN is True
Else
   statements to be executed if all conditions are False
End If
```

The `Else` clause is optional; you can omit it if there are no actions to be executed when all conditions are false. Your procedure can include any number of `ElseIf` statements and conditions. The `ElseIf` clause always comes before the

`Else` clause. The statements in the `ElseIf` clause are executed only if the condition in this clause is true.

Let's look at the following code example:

```
If ActiveCell.Value = 0 Then
  ActiveCell.Offset(0, 1).Value = "zero"
ElseIf ActiveCell.Value > 0 Then
  ActiveCell.Offset(0, 1).Value = "positive"
ElseIf ActiveCell.Value < 0 Then
  ActiveCell.Offset(0, 1).Value = "negative"
End if
```

This example checks the value of the active cell and enters the appropriate label (`zero`, `positive`, or `negative`) in the adjoining column. Notice that the `Else` clause is not used. If the result of the first condition (`ActiveCell.Value = 0`) is `False`, Visual Basic jumps to the next `ElseIf` statement and evaluates its condition (`ActiveCell.Value > 0`). If the value is not greater than zero, Visual Basic skips to the next `ElseIf` and the condition `ActiveCell.Value < 0` is evaluated.

Let's see how the `If...Then...ElseIf` statement works in a complete procedure.

⊙ **Hands-On 5.4   Writing a VBA Procedure with an**
If...Then...ElseIf Statement

1. Insert a new module into the current VBA project.
2. Rename the module `IfThenElseIf`.
3. Enter the following `WhatValue` procedure:

```
Sub WhatValue()
  Range("A1").Select
  If ActiveCell.Value = 0 Then
    ActiveCell.Offset(0, 1).Value = "zero"
  ElseIf ActiveCell.Value > 0 Then
    ActiveCell.Offset(0, 1).Value = "positive"
  ElseIf ActiveCell.Value < 0 Then
    ActiveCell.Offset(0, 1).Value = "negative"
  End If
End Sub
```

Because you need to run the `WhatValue` procedure several times to test each condition, let's have Visual Basic assign a temporary keyboard shortcut to this procedure.

4. Open the Immediate window and type the following statement:

```
Application.OnKey "^+y", "WhatValue"
```

When you press Enter, Visual Basic runs the OnKey method, which assigns the WhatValue procedure to the key sequence Ctrl+Shift+Y. This keyboard shortcut is only temporary—it will not work when you restart Microsoft Excel. To assign the shortcut key to a procedure, use the Options button in the Macro dialog box accessed from Developer | Macros in the Microsoft Excel window.

5. Now, switch to the Microsoft Excel window and activate Sheet1.
6. Type 0 (zero) in cell A1 and press Enter. Then press Ctrl+Shift+Y.
7. Visual Basic calls the WhatValue procedure and enters zero in cell B1.
8. Enter any number greater than zero in cell A1 and press Ctrl+Shift+Y.
   Visual Basic again calls the WhatValue procedure. Visual Basic evaluates the first condition, and because the result of this test is False, it jumps to the ElseIf statement. The second condition is true, so Visual Basic executes the statement following Then and skips over the next statements to the End If. Because there are no more statements following End If, the procedure ends. Cell B1 now displays the word positive.
9. Enter any number less than zero in cell A1 and press Ctrl+Shift+Y.
   This time, the first two conditions return False, so Visual Basic goes to examine the third condition. Because this test returns True, Visual Basic enters the word negative in cell B1.
10. Enter any text in cell A1 and press Ctrl+Shift+Y.
    Visual Basic's response is positive. This is not a satisfactory answer, however. You may want to differentiate between positive numbers and text by displaying the word text. To make the WhatValue procedure smarter, you need to learn how to make more complex decisions by using nested If...Then statements.

NESTED IF...THEN STATEMENTS

You can make more complex decisions in your VBA procedures by placing an If...Then or If...Then...Else statement inside another If...Then or If...Then...Else statement.

Structures in which an If statement is contained inside another If block are referred to as nested If statements. The following TestConditions procedure is a revised version of the WhatValue procedure created in the previous section. The WhatValue procedure has been modified to illustrate how nested If...Then statements work.

```
Sub TestConditions()
    Range("A1").Select
```

```
If IsEmpty(ActiveCell) Then
    MsgBox "The cell is empty."
Else
    If IsNumeric(ActiveCell.Value) Then
        If ActiveCell.Value = 0 Then
            ActiveCell.Offset(0, 1).Value = "zero"
        ElseIf ActiveCell.Value > 0 Then
            ActiveCell.Offset(0, 1).Value = "positive"
        ElseIf ActiveCell.Value < 0 Then
            ActiveCell.Offset(0, 1).Value = "negative"
        End If
    Else
        ActiveCell.Offset(0, 1).Value = "text"
    End If
End If
End Sub
```

To make the `TestConditions` procedure easier to understand, each `If…Then` statement is shown with different formatting. You can now clearly see that the procedure uses three `If…Then` blocks.

The first `If` block (in bold) checks whether the active cell is empty. If this is true, the message is displayed, and Visual Basic skips over the `Else` part until it finds the matching `End If`. This statement is located just before the `End Sub` keywords. If the active cell is not empty, the `IsEmpty(ActiveCell)` condition returns `False`, and Visual Basic runs the single underlined `If` block following the `Else` formatted in bold. This (underlined) `If…Then…Else` statement is said to be nested inside the first `If` block (in bold). This statement checks whether the value of the active cell is a number. Notice that this is done with the help of another built-in function—`IsNumeric`. If the value of the active cell is not a number, the condition is false, so Visual Basic jumps to the statement following the underlined `Else` and enters `text` in cell B1. If the active cell contains a number, however, Visual Basic runs the double-underlined `If` block, evaluating each condition and making the appropriate decision.

The first `If` block (in bold) is called the outer `If` statement. This outer statement contains two inner `If` statements (single and double underlined).

USING THE SELECT CASE STATEMENT

To avoid complex nested `If` statements that are difficult to follow, you can use the `Select Case` statement instead. The syntax of this statement is as follows:

```
Select Case testexpression
```

```
   Case expressionlist1
statements if expressionlist1 matches testexpression
   Case expressionlist2
statements if expressionlist2 matches testexpression
   Case expressionlistN
statements if expressionlistN matches testexpression
   Case Else
statements to be executed if no values match testexpression
End Select
```

You can place any number of `Case` clauses to test between the keywords `Select Case` and `End Select`. The `Case Else` clause is optional. Use it when you expect that there may be conditional expressions that return `False`. In the `Select Case` statement, Visual Basic compares each `expressionlist` with the value of `testexpression`.

Here's the logic behind the `Select Case` statement. When Visual Basic encounters the `Select Case` clause, it makes note of the value of `testexpression`. Then, it proceeds to test the expression following the first `Case` clause. If the value of this expression (`expressionlist1`) matches the value stored in `testexpression`, Visual Basic executes the statements until another `Case` clause is encountered and then jumps to the `End Select` statement. If, however, the expression tested in the first `Case` clause does not match `testexpression`, Visual Basic checks the value of each `Case` clause until it finds a match. If none of the `Case` clauses contain the expression that matches the value stored in `testexpression`, Visual Basic jumps to the `Case Else` clause and executes the statements until it encounters the `End Select` keywords. Notice that the `Case Else` clause is optional. If your procedure does not use `Case Else` and none of the `Case` clauses contain a value matching the value of `testexpression`, Visual Basic jumps to the statements following `End Select` and continues executing your procedure.

Let's look at an example of a procedure that uses the `Select Case` statement. In Chapter 4, you learned quite a few details about the `MsgBox` function, which allows you to display a message with one or more buttons. You also learned that the result of the `MsgBox` function can be assigned to a variable. Using the `Select Case` statement, you can now decide which action to take based on the button the user pressed in the message box.

⊙ Hands-On 5.5   Writing a VBA Procedure with a Select Case Statement

1. Insert a new module into the current VBA project.
2. Rename the new module `SelectCase`.

3. Enter the following `TestButtons` procedure:

```
Sub TestButtons()
   Dim question As String
   Dim bts As Integer
   Dim myTitle As String
   Dim myButton As Integer

   question = "Do you want to open a new workbook?"
   bts = vbYesNoCancel + vbQuestion + vbDefaultButton1
   myTitle = "New Workbook"
   myButton = MsgBox(prompt:=question, _
      buttons:=bts, _
      title:=myTitle)
   Select Case myButton
      Case 6
            Workbooks.Add
      Case 7
            MsgBox "You can open a new book manually later."
      Case Else
            MsgBox "You pressed Cancel."
   End Select
End Sub
```

The first part of the `TestButtons` procedure displays a message with three buttons: Yes, No, and Cancel. The value of the button selected by the user is assigned to the variable `myButton`. If the user clicks Yes, the variable `myButton` is assigned the `vbYes` constant or its corresponding value—6. If the user selects No, the variable `myButton` is assigned the constant `vbNo` or its corresponding value—7. Lastly, if Cancel is pressed, the contents of the variable `myButton` will equal `vbCancel`, or 2. The `Select Case` statement checks the values supplied after the `Case` clause against the value stored in the variable `myButton`. When there is a match, the appropriate `Case` statement is executed.

The `TestButtons` procedure will work the same if you use the constants instead of button values:

```
Select Case myButton
   Case vbYes
      Workbooks.Add
   Case vbNo
      MsgBox "You can open a new book manually later."
   Case Else
      MsgBox "You pressed Cancel."
End Select
```

You can omit the `Else` clause. Simply revise the `Select Case` statement as follows:

```
Select Case myButton
  Case vbYes
    Workbooks.Add
  Case vbNo
    MsgBox "You can open a new book manually later."
  Case vbCancel
    MsgBox "You pressed Cancel."
End Select
```

4. Run the `TestButtons` procedure three times, each time selecting a different button.

Using Is with the Case Clause

Sometimes a decision is made based on a relational operator, listed in Table 5.1, such as whether the test expression is greater than, less than, or equal to. The `Is` keyword lets you use a conditional expression in a `Case` clause. The syntax for the `Select Case` clause using the `Is` keyword is shown here:

```
Select Case testexpression
  Case Is condition1
    statements if condition1 is True
  Case Is condition2
    statements if condition2 is True
  Case Is conditionN
    statements if conditionN is True
End Select
```

Although using `Case Else` in the `Select Case` statement isn't required, it's always a good idea to include one, just in case the variable you are testing has an unexpected value. The `Case Else` statement is a good place to put an error message. For example, let's compare some numbers:

```
Select Case myNumber
  Case Is <=10
    MsgBox "The number is less than or equal to 10."
  Case 11
    MsgBox "You entered eleven."
  Case Is >=100
    MsgBox "The number is greater than or equal to 100."
  Case Else
    MsgBox "The number is between 12 and 99."
End Select
```

Assuming that the variable `myNumber` holds `120`, the third `Case` clause is true, and the only statement executed is the one between the `Case Is >=100` and the `Case Else` clause.

Specifying a Range of Values in a Case Clause

In the preceding example, you saw a simple `Select Case` statement that uses one expression in each `Case` clause. Often, however, you may want to specify a range of values in a `Case` clause. Do this by using the `To` keyword between the values of expressions, as in the following example:

```
Select Case unitsSold
    Case 1 To 100
      Discount = 0.05
    Case Is <= 500
      Discount = 0.1
    Case 501 To 1000
      Discount = 0.15
    Case Is > 1000
      Discount = 0.2
End Select
```

Let's analyze the preceding `Select Case` block with the assumption that the variable `unitsSold` currently holds the value `99`. Visual Basic compares the value of the variable `unitsSold` with the conditional expression in the `Case` clauses. The first and third `Case` clauses illustrate how to use a range of values in a conditional expression by using the `To` keyword. Because `unitsSold` equals `99`, the condition in the first `Case` clause is true; thus, Visual Basic assigns the value `0.05` to the variable `Discount`. How about the second `Case` clause, which is also true? Although it's obvious that 99 is less than or equal to 500, Visual Basic does not execute the associated statement `Discount = 0.1`. The reason for this is that once Visual Basic locates a `Case` clause with a true condition, it doesn't bother to look at the remaining `Case` clauses. It jumps over them and continues to execute the procedure with the instructions that may be following the `End Select` statement.

Specifying Multiple Expressions in a Case Clause

You may specify multiple conditions within a single `Case` clause by separating each condition with a comma, as shown in the following code example:

```
Select Case myMonth
    Case "January", "February", "March"
      Debug.Print myMonth & ": 1st Qtr."
```

```
   Case "April", "May", "June"
      Debug.Print myMonth & ": 2nd Qtr."
   Case "July", "August", "September"
      Debug.Print myMonth & ": 3rd Qtr."
   Case "October", "November", "December"
      Debug.Print myMonth & ": 4th Qtr."
End Select
```

Multiple Conditions with the Case Clause

The commas used to separate conditions within a `Case` clause have the same meaning as the `OR` operator used in the `If` statement. The `Case` clause is true if at least one of the conditions is true.

What Is Nesting?

Nesting means placing one type of control structure inside another control structure. You will see more nesting examples when we discuss looping structures in Chapter 7.

VBA PROCEDURE WITH MULTIPLE CONDITIONS

The `SimpleIfThen` procedure that you worked with earlier evaluated only a single condition in the `If...Then` statement. This statement, however, can take more than one condition.

To specify multiple conditions in an `If...Then` statement, use the logical operators `AND` and `OR` (listed in Table 5.2 at the beginning of this chapter). Here's the syntax with the `AND` operator:

```
If condition1 AND condition2 Then statement
```

In the preceding syntax, both `condition1` and `condition2` must be true for Visual Basic to execute the statement to the right of the `Then` keyword. For example:

```
If sales = 10000 AND salary < 45000 Then SlsCom = Sales
* 0.07
```

In the above example:

```
Condition1 sales = 10000
Condition2 salary < 45000
```

When AND is used in the conditional expression, both conditions must be true before Visual Basic can calculate the sales commission (SlsCom). If either of these conditions is false or both are false, Visual Basic ignores the statement after Then.

When it's good enough to meet only one of the conditions, you should use the OR operator. Here's the syntax:

```
If condition1 OR condition2 Then statement
```

The OR operator is more flexible. Only one of the conditions has to be true before Visual Basic can execute the statement following the Then keyword.

Let's look at this example:

```
If dept = "S" OR dept = "M" Then bonus = 500
```

In this example, if at least one condition is true, Visual Basic assigns 500 to the bonus variable. If both conditions are false, Visual Basic ignores the rest of the line.

Now, let's write a complete procedure. Suppose you can get a 10% discount if you purchase 50 units of a product, each priced at $7.00. The IfThenAnd procedure demonstrates the use of the AND operator.

⊙ Hands-On 5.6   Writing a VBA Procedure with Multiple Conditions

1. Enter the following procedure in the IfThen module of the Decisions (Chap05.xlsm) project:

```
Sub IfThenAnd()
  Dim price As Single
  Dim units As Integer
  Dim rebate As Single

  Const strmsg1 = "To get a rebate you must buy an additional "
  Const strmsg2 = "Price must equal $7.00"

  units = Range("B1").Value
  price = Range("B2").Value

  If price = 7 AND units >= 50 Then
     rebate = (price * units) * 0.1
     Range("A4").Value = "The rebate is: $" & rebate
  End If

  If price = 7 AND units < 50 Then
     Range("A4").Value = strmsg1 & 50 - units & " unit(s)."
  End If
```

```
If price <> 7 AND units >= 50 Then
   Range("A4").Value = strmsg2
End If

If price <> 7 AND units < 50 Then
   Range("A4").Value = "You didn't meet the criteria."
End If
End Sub
```

The `IfThenAnd` procedure just shown has four `If...Then` statements that are used to evaluate the contents of two variables: `price` and `units`. The `AND` operator between the keywords `If...Then` allows more than one condition to be tested. With the `AND` operator, all conditions must be true for Visual Basic to run the statements between the `Then...End If` keywords. Because the `IfThenAnd` procedure is based on the data entered in worksheet cells, it's more convenient to run it from the Microsoft Excel window.

2. Switch to the Microsoft Excel application window and choose Developer | Macros.
3. In the Macro dialog box, select the IfThenAnd macro and click the Options button.
4. While the cursor is blinking in the Shortcut key box, press Shift+I to assign the shortcut key Ctrl+Shift+I to your macro, and then click OK to exit the Macro Options dialog box.
5. Click Cancel to close the Macro dialog box.
6. Enter the sample data in a worksheet, as shown in Figure 5.1.

⁄	A	B	C
1	units	300	
2	price	7	
3			

FIGURE 5.1. Sample test data in a worksheet.

7. Press Ctrl+Shift+I to run the `IfThenAnd` procedure.
8. Change the values of cells B1 and B2 so that every time you run the procedure, a different `If...Then` statement is true.

CONDITIONAL LOGIC IN FUNCTION PROCEDURES

To get more practice with the `Select Case` statement, let's use it in a function procedure. As you recall from Chapter 4, function procedures allow you to return the function result to a subroutine. Suppose a subroutine must display a discount based on the number of units sold. You can get the number of units from the user and then run a function to determine which discount applies.

⊙ Hands-On 5.7   Writing a Function Procedure with a Select Case Statement

1. Enter the following subroutine in the SelectCase module:

```
Sub DisplayDiscount()
  Dim unitsSold As Integer
  Dim myDiscount As Single
  unitsSold = InputBox("Enter the number of units sold:")
  myDiscount = GetDiscount(unitsSold)
  MsgBox myDiscount
End Sub
```

2. In the same module, enter the following function procedure:

```
Function GetDiscount(unitsSold As Integer)
  Select Case unitsSold
    Case 1 To 200
      GetDiscount = 0.05
    Case Is <= 500
      GetDiscount = 0.1
    Case 501 To 1000
      GetDiscount = 0.15
    Case Is > 1000
      GetDiscount = 0.2
  End Select
End Function
```

3. Place the cursor anywhere within the code of the DisplayDiscount procedure and press F5 to run it. Run the procedure several times, entering values to test each Case statement.

The DisplayDiscount procedure passes the value stored in the variable unitsSold to the GetDiscount function. When Visual Basic encounters the Select Case statement, it checks whether the value of the first Case clause expression matches the value stored in the unitsSold parameter. If there is a match, Visual Basic assigns a 5% discount (0.05) to the function name and then jumps to the End Select keywords. Because there are no more statements to execute inside the function procedure, Visual Basic returns to the calling procedure—DisplayDiscount. Here, it assigns the function's result to the variable myDiscount. The last statement displays the value of the retrieved discount in a message box.

SUMMARY

Conditional statements let you control the flow of your procedure. By testing the truth of a condition, you can decide which statements should be run and which should be skipped over. In other words, instead of running your procedure from top to bottom, line by line, you can execute only certain lines.

With so many conditional statements, you may be wondering which one you should use in your VBA procedure. Here are a few guidelines:

- If you need to supply only one condition, the simple `If...Then` statement is the best choice.
- If you need to decide which of two actions to perform, use the `If...Then... Else` statement.
- If your procedure requires two or more conditions, use the `If...Then... ElseIf` or `Select Case` statements.
- If your procedure has a great number of conditions, use the `Select Case` statement. This statement is more flexible and easier to comprehend than the `If...Then...ElseIf` statement.

Some decisions must be repeated. For example, you may want to repeat the same actions for each cell in a worksheet or each sheet in a workbook. The next chapter teaches you how to perform the same steps repeatedly.

Chapter 6

ADDING REPEATING ACTIONS TO EXCEL VBA PROGRAMS—
A QUICK INTRODUCTION TO LOOPING STATEMENTS

Now that you've learned how conditional statements can give your VBA procedures decision-making capabilities, it's time to go a step further. Not all decisions are easy. Sometimes, you will need to perform several statements multiple times to arrive at a certain condition. On other occasions, however, after you've reached the decision, you may need to run the specified statements as long as a condition is true or until a condition becomes true. In programming, performing repetitive tasks is called *looping*. VBA has various looping structures that allow you to repeat a sequence of statements several times. In this chapter, you will learn how to loop through your VBA code.

INTRODUCING LOOPING STATEMENTS

A loop is a programming structure that causes a section of program code to execute repeatedly. VBA provides several structures to implement loops in your procedures: `Do...While`, `Do...Until`, `For...Next`, `For...Each`, and `While...Wend`.

UNDERSTANDING DO...WHILE AND DO...UNTIL LOOPS

Visual Basic has two types of `Do` loop statements that repeat a sequence of statements either as long as or until a certain condition is true. The `Do...While` loop lets you repeat an action as long as a condition is true. This loop has the following syntax:

```
Do While condition
  statement1
  statement2
  statementN
Loop
```

When Visual Basic encounters this loop, it first checks the truth value of the condition. If the condition is false, the statements inside the loop are not executed. Visual Basic will continue to execute the program with the first statement after the `Loop` keyword. If the condition is true, the statements inside the loop are run one by one until the `Loop` statement is encountered. The `Loop` statement tells Visual Basic to repeat the entire process, as long as the testing of the condition in the `Do...While` statement is true. Let's now see how you can put the `Do...While` loop to good use in Microsoft Excel.

In Chapter 5, you learned how to make a decision based on the contents of a cell. Let's take it a step further and see how you can repeat the same decision for a number of cells. Our task is to apply bold formatting to any cell in a column, as long as it's not empty.

NOTE	*Please note, the files for the "Hands-On" projects can be found in the companion files.*

⊚ Hands-On 6.1   Writing a VBA Procedure with a Do...While Statement

1. Open a new workbook and save it as `C:\VBAExcel2024_ByExample\Chap06.xlsm`.
2. Switch to the VBE screen and change the name of the new project to `Repetition`.
3. Insert a new module into the `Repetition` project and change its name to `DoLoops`.
4. Enter the following procedure in the `DoLoops` module:

```
Sub ApplyBold()
  Do While ActiveCell.Value <> ""
    ActiveCell.Font.Bold = True
```

```
        ActiveCell.Offset(1, 0).Select
    Loop
End Sub
```

5. Press Alt+F11 to switch to the Microsoft Excel application window, activate `Sheet1`, and then enter any data (text or numbers) in cells A1:A7.
6. When finished with the data entry, select cell A1.
7. Choose Developer | Macros. In the Macro dialog box, double-click the ApplyBold procedure (or highlight the procedure name and click Run).

   When you run the `ApplyBold` procedure, Visual Basic first evaluates the condition in the `Do While` statement—`ActiveCell.Value <>""`. The condition says: Perform the following statements as long as the value of the active cell is not an empty string (`""`). Because you have entered data in cell A1 and made this cell active (see Steps 5 and 6), the first test returns `True`, so Visual Basic executes the statement `ActiveCell.Font.Bold = True`, which applies the bold formatting to the active cell. Next, Visual Basic selects the cell in the next row (the `Offset` property was discussed in Chapter 3). Because the statement that follows is the `Loop` keyword, Visual Basic returns to the `Do While` statement and again checks the condition. If the newly selected active cell is not empty, Visual Basic repeats the statements inside the loop. This process continues until the contents of cell A8 are examined. Because this cell is empty, the condition is false, so Visual Basic skips the statements inside the loop. Because there are no more statements to execute after the `Loop` keyword, the procedure ends.

Let's look at another `Do...While` loop example.

The `Do...While` loop has an alternative syntax that lets you test the condition at the bottom of the loop in the following way:

```
Do
    statement1
    statement2
    statementN
Loop While condition
```

When you test the condition at the bottom of the loop, the statements inside the loop are executed at least once. Let's take a look at an example:

```
Sub SignIn()
    Dim secretCode As String
    Do
        secretCode = InputBox("Enter your secret code:")
        If secretCode = "sp1045" Then Exit Do
```

```
    Loop While secretCode <> "sp1045"
End Sub
```

Notice that by the time the condition is evaluated, Visual Basic has already executed the statements one time. In addition to placing the condition at the end of the loop, the SignIn procedure shows how to exit the loop when a condition is reached using the Exit Do statement.

Another handy loop, Do...Until, allows you to repeat one or more statements until a condition becomes true. In other words, Do...Until repeats a block of code as long as something is false. Here's the syntax:

```
Do Until condition
   statement1
   statement2
   statementN
Loop
```

Using the preceding syntax, you can now rewrite the previous ApplyBold procedure in the following way:

```
Sub ApplyBold2()
   Do Until IsEmpty(ActiveCell)
     ActiveCell.Font.Bold = True
     ActiveCell.Offset(1, 0).Select
   Loop
End Sub
```

The first line of this procedure says to perform the following statements until the first empty cell is reached. As a result, if the active cell is not empty, Visual Basic executes the two statements inside the loop. This process continues as long as the condition IsEmpty(ActiveCell) evaluates to False. Because the ApplyBold2 procedure tests the condition at the beginning of the loop, the statements inside the loop will not run if the first cell is empty. You will get the chance to try out this procedure in the next section.

Like the Do...While loop, the Do...Until loop has a second syntax that lets you test the condition at the bottom of the loop:

```
Do
   statement1
   statement2
   statementN
Loop Until condition
```

If you want the statements to be executed at least once, place the condition on the line with the Loop statement, no matter what the value of the condition.

Let's write a procedure that deletes empty sheets from a workbook.

◉ Hands-On 6.2   Writing a VBA Procedure with a Do...Until Statement

1. Enter the `DeleteBlankSheets` procedure, as shown here, in the `DoLoops` module that you created earlier:

```
Sub DeleteBlankSheets()
    Dim myRange As Range
    Dim shcount As Integer
    shcount = Worksheets.Count
    Do
        Worksheets(shcount).Select
        Set myRange = ActiveSheet.UsedRange
        If myRange.Address = "$A$1" And _
            Range("A1").Value = "" Then
            Application.DisplayAlerts = False
            Worksheets(shcount).Delete
            Application.DisplayAlerts = True
        End If
        shcount = shcount - 1
    Loop Until shcount = 1
End Sub
```

2. Press Alt+F11 to switch to the Microsoft Excel window and manually insert three new worksheets into the current workbook. In one of the sheets, enter text or a number in cell A1. On another sheet, enter some data in cells B2 and C10. Do not enter any data on the third inserted sheet.

3. Run the `DeleteBlankSheets` procedure.

When you run this procedure, Visual Basic deletes the selected sheet whenever two conditions are true—the `UsedRange` property address returns cell A1 and cell A1 is empty. The `UsedRange` property applies to the `Worksheet` object and contains every non-empty cell on the worksheet, as well as all the empty cells that are among them. For example, if you enter something in cells B2 and C10, the used range is $B$2:$C$10. If you later enter data in cell A1, the used range will be $A$1:$C$10. The used range is bounded by the farthest upper-left and farthest lower-right non-empty cell on a worksheet.

Because the workbook must contain at least one worksheet, the code is executed until the variable `shcount` equals one. The statement `shcount = shcount - 1` makes sure that the `shcount` variable is reduced by one each time the statements in the loop are executed. The value of `shcount` is initialized at the beginning of the procedure with the following statement:

```
Worksheets.Count
```

When deleting sheets, Excel normally displays the confirmation dialog box. If you'd rather not be prompted to confirm the deletion, use the following statement:

```
Application.DisplayAlerts = False
```

When you are finished, turn the system messages back on using this statement:

```
Application.DisplayAlerts = True
```

Counters

A *counter* is a numeric variable that keeps track of the number of items that have been processed. The `DeleteBlankSheets` procedure just shown declares the variable `shcount` to keep track of sheets that have been processed. A counter variable should be initialized (assigned a value) at the beginning of the program. This ensures that you always know the exact value of the counter before you begin using it. A counter can be incremented or decremented by a specified value. See other examples of using counters with the `For...Next` loop later in this chapter.

AVOIDING INFINITE LOOPS

If you don't design your loop correctly, you get an infinite loop—a loop that never ends. You will not be able to stop the procedure by using the Esc key. The following procedure causes the loop to execute endlessly because the programmer forgot to include the test condition:

```
Sub SayHello()
  Do
    MsgBox "Hello."
  Loop
End Sub
```

To stop the execution of the infinite loop, you must press Ctrl+Break. When Visual Basic displays the message box that says "Code execution has been interrupted," click End to end the procedure.

Note that the Break key, often labeled as Pause/Break, is typically located in the upper-right area of the keyboard, near the Scroll Lock (ScrLk) key. Not all modern keyboards have it. If yours doesn't, you might need to use an alternative method to interrupt code in VBA. On my Dell laptop, the Break key is not present as a standalone key, and to simulate it, I have to press Fn+Ctrl+B.

EXECUTING A PROCEDURE LINE BY LINE

When you run procedures that use looping structures, it's sometimes hard to see whether the procedure works as expected. Occasionally, you'd like to watch the procedure execute in slow motion so that you can check the logic of the program. Let's examine how Visual Basic allows you to execute a procedure line by line.

⊙ **Hands-On 6.3   Executing a Procedure Line by Line**

1. Insert a new sheet into the current workbook and enter any data in cells A1:A5.
2. Select cell A1 and choose Developer | Macros.
3. In the Macro dialog box, select the ApplyBold procedure and click the Step Into button.
   The VBE screen will appear with the name of the procedure highlighted in yellow, as shown in Figure 6.1. Notice the yellow arrow in the margin indicator bar of the Code window.

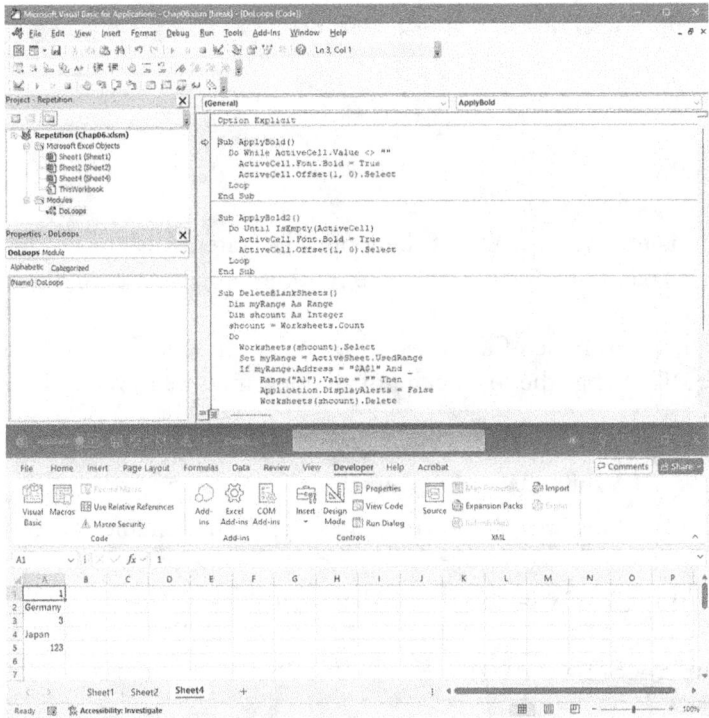

FIGURE 6.1. Watching the procedure code execute line by line.

4. Arrange the screens side by side, as shown in Figure 6.1.
5. Make sure cell A1 is selected and that it contains data.
6. Click the title bar in the Visual Basic window to move the focus to this window, and then press F8. The yellow highlight in the Code window jumps to this line:

```
Do While ActiveCell.Value <> ""
```

7. Continue pressing F8 while watching both the Code window and the worksheet window.

NOTE	*You will find more information related to stepping through VBA procedures in Chapter 9.*

UNDERSTANDING THE WHILE...WEND LOOP

The While...Wend loop is functionally equivalent to the Do...While loop. This statement is a carryover from earlier versions of Microsoft Basic and is included in VBA for backward compatibility. The loop begins with the keyword While and ends with the keyword Wend. Here's the syntax:

```
While condition
   statement1
   statement2
   statementN
Wend
```

The condition is tested at the top of the loop. The statements are executed as long as the given condition is true. Once the condition is false, Visual Basic exits the loop.

Let's look at an example of a procedure that uses the While...Wend looping structure. We will change the row height of all non-empty cells in a worksheet.

Hands-On 6.4 Writing a VBA Procedure with a While...Wend Statement

1. Insert a new module into the current VBA project. Rename the module WhileLoop.
2. Enter the following procedure in the WhileLoop module.

```
Sub ChangeRHeight()
   While ActiveCell <> ""
     ActiveCell.RowHeight = 28
     ActiveCell.Offset(1, 0).Select
```

```
   Wend
End Sub
```

3. Switch to the Microsoft Excel window and enter some data in cells B1:B4 of any worksheet.

4. Select cell B1 and choose Developer | Macros.

5. In the Macro dialog, select the ChangeRHeight procedure and click Run.

The ChangeRHeight procedure sets the row height to 28 when the active cell is not empty. The next cell is selected by using the Offset property of the Range object. The statement ActiveCell.Offset(1, 0).Select tells Visual Basic to select the cell that is located one row below (1) the active cell and in the same column (0).

UNDERSTANDING THE FOR...NEXT LOOP

The For...Next loop is used when you know how many times you want to repeat a group of statements. The syntax of a For...Next loop looks like this:

```
For counter = start To end [Step increment]
   statement1
   statement2
   statementN
Next [counter]
```

The code in the brackets is optional. counter is a numeric variable that stores the number of iterations. start is the number at which you want to begin counting, and end indicates how many times the loop should be executed. For example, if you want to repeat the statements inside the loop five times, use the following For statement syntax:

```
For counter = 1 To 5
   Your statements go here
Next
```

When Visual Basic encounters the Next keyword, it will go back to the beginning of the loop and execute the statements inside the loop again, as long as counter hasn't reached the value in end. As soon as the value of counter is greater than the number entered after the To keyword, Visual Basic exits the loop. Because the variable counter automatically changes after each execution of the loop, sooner or later, the value stored in counter will exceed the value specified value. By default, every time Visual Basic executes the statements inside the loop, the value of the variable `counter` is increased by one. You can change this default setting by using the `Step` clause. For example, to increase the variable `counter` by three, use the following statement:

```
For counter = 1 To 5 Step 3
  Your statements go here
Next counter
```

When Visual Basic encounters the preceding instruction, it executes the statements inside the loop twice. The first time in the loop, `counter` equals `1`. The second time in the loop, `counter` equals `4` (3 + 1). After the second time inside the loop, `counter` equals `7` (4 + 3). This causes Visual Basic to exit the loop. Note that the `Step` increment is optional and isn't specified unless it's a value other than `1`. You can also place a negative number after `Step`. Visual Basic will then decrement this value from `counter` each time it encounters the `Next` keyword. The name of the variable (`counter`) after the `Next` keyword is also optional. It's good programming practice, however, to make your `Next` keywords explicit by including `counter`.

How can you use the `For...Next` loop in a Microsoft Excel worksheet? Suppose, in your sales report, you'd like to include only products that were sold in a particular month. When you imported data from a Microsoft Access table, you also got rows with the sold amount equal to zero. How can you quickly eliminate those "zero" rows? Although there are many ways to solve this problem, let's see how you can handle it with a `For...Next` loop.

Hands-On 6.5   Writing a VBA Procedure with a For...Next Statement

1. In the Visual Basic window, insert a new module into the current project and rename it `ForNextLoop`.
2. Enter the following procedure in the `ForNextLoop` module:

```
Sub DeleteZeroRows()
  Dim totalR As Integer
  Dim r As Integer

  Range("A1").CurrentRegion.Select
  totalR = Selection.Rows.Count
  Range("B2").Select

  For r = 1 To totalR - 1
    If ActiveCell = 0 Then
        Selection.EntireRow.Delete
        totalR = totalR - 1
```

```
    Else
            ActiveCell.Offset(1, 0).Select
    End If
  Next r
End Sub
```

Let's examine the `DeleteZeroRows` procedure line by line. The first two statements calculate the total number of rows in the current range and store this number in the variable `totalR`. Next, Visual Basic selects cell B2 and encounters the `For` keyword. Because the first row of the worksheet contains the column headings, decrease the total number of rows by one (`totalR - 1`). Visual Basic will need to execute the instructions inside the loop six times. The conditional statement (`If...Then...Else`) nested inside the loop tells Visual Basic to make a decision based on the value of the active cell. If the value is equal to zero, Visual Basic deletes the current row and reduces the value of `totalR` by one. Otherwise, the condition is false, so Visual Basic selects the next cell. Each time Visual Basic completes the loop, it jumps to the `For` keyword to compare the value of `r` with the value of `totalR - 1`.

3. Switch to the Microsoft Excel window and insert a new worksheet. Enter the data shown here:

	A	B
1	Product Name	Sales (in Pounds)
2	Apples	120
3	Pears	0
4	Bananas	100
5	Cherries	0
6	Blueberries	0
7	Strawberries	160

4. Choose Developer | Macros.
5. In the Macro dialog, select the DeleteZeroRows procedure and click Run. When the procedure ends, the sales worksheet will not include products that were not sold.

Paired Statements

`For` and `Next` must be paired. If one is missing, Visual Basic generates the following error message: "For without Next."

UNDERSTANDING THE FOR EACH...NEXT LOOP

When your procedure needs to loop through all of the objects of a collection or all of the elements in an array (arrays will be covered in Chapter 7), the `For Each...Next` loop should be used. This loop does not require a `counter` variable. Visual Basic can figure out on its own how many times the loop should be executed.

Let's take, for example, a collection of worksheets. To remove a worksheet from a workbook, you must first select it and then choose Home | Cells | Delete | Delete Sheet. To leave only one worksheet in a workbook, you need to use the same command several times, depending on the total number of worksheets. Because each worksheet is an object in a collection of worksheets, you can speed up the process of deleting worksheets by using the `For Each...Next` loop. This loop looks like the following:

```
For Each element In Group
  statement1
  statement2
  statementN
Next [element]
```

In the preceding syntax, `element` is a variable to which all the elements of an array or collection will be assigned. This variable must be of the Variant data type for an array and an Object data type for collection. `Group` is the name of a collection or an array.

Let's use the `For Each...Next` loop to remove some worksheets.

Hands-On 6.6 Writing a VBA Procedure with a For Each...Next Statement

1. Insert a new module into the current project and rename it `ForEachNextLoop`.
2. Type the following procedure in the `ForEachNextLoop` module:

```
Sub RemoveSheets()
  Dim mySheet As Worksheet

  Application.DisplayAlerts = False

  Workbooks.Add
  Sheets.Add After:=ActiveSheet, Count:=3

  For Each mySheet In Worksheets
    If mySheet.Name <> "Sheet1" Then
```

```
        ActiveWindow.SelectedSheets.Delete
    End If
Next mySheet

  Application.DisplayAlerts = True
End Sub
```

Visual Basic will open a new workbook, add three new sheets after the default `Sheet1` (`ActiveSheet`), and proceed to delete all the sheets except for `Sheet1`. Notice that the variable `mySheet` represents an object in a collection of worksheets. Therefore, this variable has been declared as being of the specific object data type `Worksheet`. The first instruction, `Application.DisplayAlerts = False`, makes sure that Microsoft Excel does not display alerts and messages while the procedure is running. The `For Each...Next` loop steps through each worksheet and deletes it as long as it is not `Sheet1`. When the procedure ends, the workbook has only one sheet—`Sheet1`.

3. Position the insertion point anywhere within the `RemoveSheets` procedure code and press F5 to run it.

EXITING LOOPS EARLY

Sometimes, you may not want to wait until the loop ends on its own. A user may have entered the wrong data, a procedure may have encountered an error, or perhaps the task has been completed and there's no need to do additional looping. You can leave the loop early without reaching the condition that normally terminates it. Visual Basic has two types of `Exit` statements:

- The `Exit For` statement is used to end either a `For...Next` or a `For Each...Next` loop early.

- The `Exit Do` statement immediately exits any of the VBA `Do` loops.

The following procedure demonstrates how to use the `Exit For` statement to leave the `For Each...Next` loop early.

⊙ **Hands-On 6.7   Writing a VBA Procedure with an Early Exit from a For Each...Next Statement**

1. Enter the following procedure in the `ForEachNextLoop` module:

```
Sub EarlyExit()
  Dim myCell As Variant
```

```
Dim myRange As Range

Set myRange = Range("A1:H10")
For Each myCell In myRange
    If myCell.Value = "" Then
        myCell.Value = "empty"
    Else
        Exit For
    End If
Next myCell
End Sub
```

The `EarlyExit` procedure examines the contents of each cell in the specified range—A1:H10. If the active cell is empty, Visual Basic enters the text `empty` in the active cell. When Visual Basic encounters the first non-empty cell, it exits the loop.

2. Open a new workbook and enter a value in any cell within the specified range—A1:H10.
3. Choose Developer | Macros.
4. In the Macro dialog, select the EarlyExit procedure and click Run.

USING A DO...WHILE STATEMENT

The next example procedure demonstrates how to display today's date and time in Microsoft Excel's status bar for 10 seconds.

(⊙) **Hands-On 6.8   Writing a VBA Procedure with a Do...While Statement**

1. Enter the following procedure in the `DoLoops` module:

```
Sub TenSeconds()
  Dim stopme

  stopme = Now + TimeValue("00:00:10")

  Do While Now < stopme
    Application.DisplayStatusBar = True
    Application.StatusBar = Now
  Loop

  Application.StatusBar = False
End Sub
```

In the `TenSeconds` procedure, the statements inside the `Do...While` loop will be executed as long as the time returned by the `Now` function is less than the value of the variable called `stopme`. The variable `stopme` holds the current time plus 10 seconds. (See the online help for other examples of using the built-in `TimeValue` function.)

The statement `Application.DisplayStatusBar` tells Visual Basic to turn on the status bar display. The next statement places the current date and time in the status bar. While the time is displayed for 10 seconds, the user cannot work with the system (the mouse pointer turns into an hourglass). After the 10 seconds are over (that is, when the condition `Now < stopme` evaluates to `True`), Visual Basic leaves the loop and executes the statement after the `Loop` keyword. This statement returns the default status bar message "Ready."

2. Press Alt+F11 to switch to the Microsoft Excel application window.
3. Choose Developer | Macros. In the Macro dialog box, double-click the TenSeconds macro name (or highlight the macro name and click Run). Observe the date and time displayed in the status bar. The status bar should return to "Ready" after 10 seconds.

USING LOOPS AND CONDITIONALS

Let's combine the looping statements and some conditional logic to write a procedure that checks whether a certain sheet is part of a workbook.

Hands-On 6.9   Writing a VBA Procedure with Loops and Conditionals

1. Enter the following procedures in a new module:

```
Sub IsSuchSheet(strSheetName As String)
    Dim mySheet As Worksheet
    Dim counter As Integer

    counter = 0

Workbooks.Add
Sheets.Add After:=ActiveSheet, Count:=3
    For Each mySheet In Worksheets
        If mySheet.Name = strSheetName Then
            counter = counter + 1
            Exit For
        End If
```

```
    Next mySheet

    If counter = 1 Then
        MsgBox strSheetName & " exists."
    Else
        MsgBox strSheetName & " was not found."
    End If
End Sub

Sub FindSheet()
   Call IsSuchSheet("Sheet4")
End Sub
```

2. The `IsSuchSheet` procedure uses the `Exit For` statement to ensure that we exit the loop as soon as the sheet name passed in the procedure argument is found in the workbook. The `FindSheet` procedure is used to show you how to call one procedure from another.
3. To execute the `IsSuchSheet` procedure, run the `FindSheet` procedure.

SUMMARY

In this chapter, you learned how to use procedure loops to repeat certain groups of statements. While working with several types of looping statements, you saw how each loop performs repetitions in a slightly different way. As you gain programming experience, you'll find it easier to choose the appropriate flow control structure for your task.

The next chapter will show you how arrays are used in VBA to work with larger sets of data.

Chapter 7

STORING MULTIPLE VALUES IN EXCEL VBA PROGRAMS—
A QUICK INTRODUCTION TO WORKING WITH ARRAYS

In previous chapters, you worked with VBA procedures that used variables to hold specific information about an object, property, or value. For each single value that you wanted your procedure to manipulate, you declared a variable, but what if you have a series of values? If you had to write a VBA procedure to deal with larger amounts of data, you would have to create enough variables to handle all the data. Can you imagine the nightmare of storing currency exchange rates in your program for all the countries in the world? To create a table to hold the necessary data, you'd need at least three variables for each country: country name, currency name, and exchange rate. Fortunately, Visual Basic has a way to get around this problem. By clustering the related variables together, your VBA procedures can manage a large amount of data with ease. In this chapter, you'll learn how to manipulate lists and tables of data with arrays.

UNDERSTANDING ARRAYS

An *array* is a special type of variable that represents a group of similar values that are of the same data type (String, Integer, Currency, Date, etc.). The two most common types of arrays are one-dimensional arrays (lists) and two-dimensional arrays (tables).

One-Dimensional Arrays

A one-dimensional array is sometimes referred to as a *list*. A shopping list, a list of the days of the week, and an employee list are examples of one-dimensional arrays or, simply, numbered lists. Each element in the list has an index value that allows access to that element.

For example, in the following illustration, we have a one-dimensional array of six elements indexed from 0 to 5:

(0)	(1)	(2)	(3)	(4)	(5)

You can access the third element of this array by specifying index (2). By default, the first element of an array is indexed zero. You can change this behavior by using the `Option Base 1` statement or by explicitly coding the lower bound of your array, as explained further in this chapter.

All elements of the array must be of the same data type. In other words, one array cannot store both strings and integers. The following are two examples of one-dimensional arrays: a one-dimensional array called `cities` that is populated with text (String data type—$) and a one-dimensional array called `lotto` that contains six lottery numbers stored as integers (Integer data type—%).

A one-dimensional array: `cities$`

cities(0)	Baltimore
cities(1)	Atlanta
cities(2)	Boston
cities(3)	Washington
cities(4)	New York
cities(5)	Trenton

A one-dimensional array: `lotto%`

lotto(0)	25
lotto(1)	4
lotto(2)	31
lotto(3)	22
lotto(4)	11
lotto(5)	5

As you can see, the contents assigned to each array element match the Array type. If you want to store values of different data types in the same array, you must declare the array as Variant. You will learn how to declare arrays in the next section.

Two-Dimensional Arrays

A two-dimensional array may be thought of as a table or matrix. The position of each element in a table is determined by its row and column numbers.

For example, an array that holds the yearly sales for each product your company sells has two dimensions (the product name and the year). The following is a diagram of an empty two-dimensional array:

(0,0)	(0,1)	(0,2)	(0,3)	(0,4)	(0,5)
(1,0)	(1,1)	(1,2)	(1,3)	(1,4)	(1,5)
(2,0)	(2,1)	(2,2)	(2,3)	(2,4)	(2,5)
(3,0)	(3,1)	(3,2)	(3,3)	(3,4)	(3,5)
(4,0)	(4,1)	(4,2)	(4,3)	(4,4)	(4,5)
(5,0)	(5,1)	(5,2)	(5,3)	(5,4)	(5,5)

You can access the first element in the second row of this two-dimensional array by specifying indexes (1, 0). Following are two examples of a two-dimensional array: an array named `yearlyProductSales@` that stores yearly product sales using the Currency data type (@) and an array named `exchange` (of the Variant data type) that stores the name of the country, its currency, and the U.S. dollar exchange rate.

A two-dimensional array: `yearlyProductSales@`

Walking Cane (0,0)	$25,023 (0,1)
Pill Crusher (1,0)	$64,085 (1,1)
Electric Wheelchair (2,0)	$345,016 (2,1)
Folding Walker (3,0)	$85,244 (3,1)

A two-dimensional array: `exchange`

Japan (0,0)	Japanese Yen (0,1)	108.83 (0,2)
Australia (1,0)	Australian Dollar (1,1)	1.28601 (1,2)
Canada (2,0)	Canadian Dollar (2,1)	1.235 (2,2)
Norway (3,0)	Norwegian Krone (3,1)	6.4471 (3,2)
Europe (4,0)	Euro (4,1)	0.816993 (4,2)

In these examples, the `yearlyProductSales@` array can hold a maximum of 8 elements (4 rows * 2 columns = 8) and the `exchange` array will allow a maximum of 15 elements (5 rows * 3 columns = 15).

Three-Dimensional Arrays

Although VBA arrays can have up to 60 dimensions, most people find it difficult to picture dimensions beyond 3D. A three-dimensional array is a collection of elements organized in three dimensions: rows, columns, and layers, where a "layer" refers to a set of elements that share the same position in one of the three indices. You can think of it as a series of tables stacked on top of each other. Each layer is essentially a 2D table with rows and columns, and the third dimension indicates the layer (or table) number you're accessing. For example, to declare a 3D array, you specify the number of elements in each dimension:

```
Dim my3DArray(1 to 3, 1 to 3, 1 to 3) As Integer
```

In this example, `my3DArray` is a 3-dimensional array of integers. The first dimension (rows) ranges from 1 to 3. The second dimension (columns) ranges from 1 to 3. The third dimension (layers) ranges from 1 to 3. This declaration will create a 3x3x3 array that contains 27 elements in total. Each element in the array can be accessed using its corresponding indices in the three dimensions.

NOTE	*A practical example of using a 3D array was developed specifically for this chapter with the assistance of Microsoft Copilot and is available in the companion file—see the* `ArraysDemo_3D.xlsm` *macro-enabled workbook. This workbook contains a separate sheet with quarterly sales for three different salespeople. The* `ProcessThreeDimArray` *procedure in this workbook reads data from Sheets 2 through 4, storing it in a 3D array variable* `myArray`. *The procedure then sums the unit sales values across each layer (sheet) for each corresponding cell and stores the results in the* `totalArray` *variable. Prior to writing the summed values to the destination sheet, we call the* `TransferHeadings` *procedure to prepare the headings for the Final Qtr1 Sales worksheet. This example demonstrates how you can manage complex data structures in Excel VBA.*

Declaring Arrays

Because an array is a variable, you must declare it in a similar way as you declare other variables by using the `Dim`, `Private`, or `Public` keywords. Let's look at some examples:

Array Declaration (One-Dimensional)	Description
`Dim cities(5) as String`	Declares a 6-element array, indexed 0 to 5
`Dim lotto(1 to 6) as String`	Declares a 6-element array, indexed 1 to 6
`Dim supplies(2 to 11)`	Declares a 10-element array, indexed 2 to 11
`Dim myIntegers(-3 to 6)`	Declares a 10-element array, indexed −3 to 6 (the lower bound of an array can be 0, 1, or negative)
`Dim dynArray() as Integer`	Declares a variable-length array whose bounds will be determined at runtime (see examples later in this chapter)

Array Declaration (Two-Dimensional)	Description
`Dim exchange(4,2) as Variant`	Declares a two-dimensional array (five rows by three columns)
`Private yearlyProductSales (3, 1) as Currency`	Declares a two-dimensional array (four rows by two columns)
`Public my2DArray(1 to 3, 1 to7) as Single`	Declares a two-dimensional array (three rows indexed 1 to 3 by seven columns indexed 1 to 7)

Array Declaration (Three-Dimensional)	Description
`Dim exchange(2, 1 to 6, 4) as Variant`	Declares a three-dimensional array (the first dimension has three elements, the second dimension has six elements indexed 1 to 6, and the third dimension has five elements)

Note that the last part of the array declaration is the definition of the data type that the array will hold. It can be Integer, Long, Single, Double, Variant, Currency, String, Boolean, Byte, or Date.

What Are Array Bounds?

Array bounds define the range of indices you use to access elements within an array. You can specify lower and upper bounds for each dimension of the array. For example, the `lotto` array shown earlier is a one-dimensional array with bounds from 1 to 6. So, it has 6 elements, accessed using indices 1 through 6. For fixed-length arrays, the array bounds are listed in parentheses following the variable name. If a variable-length or dynamic array is being declared, the variable name is followed by an empty pair of parentheses. Array bounds are discussed in detail in the next section.

When you declare an array, Visual Basic automatically reserves enough memory space. The amount of memory allocated depends on the array's size and data type. When you declare a one-dimensional array named `lotto` with six elements, Visual Basic sets aside 12 bytes—2 bytes for each element of the array (recall that the size of the Integer data type is 2 bytes, and hence 2 * 6 = 12). The larger the array, the more memory space is required to store the data. Because arrays can eat up a lot of memory and impact your computer's performance, it's recommended that you declare arrays with only as many elements as you think you'll use.

What Is an Array Variable?

An array is a group of variables that have a common name. While a typical variable can hold only one value, an array variable can store many individual values. You refer to a specific value in the array by using the array name and an index number.

Subscripted Variables

The numbers inside the parentheses of the array variables are called *subscripts*, and each individual variable is called a subscripted variable or element. For example, `cities(5)` is the sixth subscripted variable (element) of the array `cities()`.

Array Upper and Lower Bounds

By default, VBA assigns zero (0) to the first element of the array. Therefore, number 1 represents the second element of the array, number 2 represents the third, and so on. With numeric indexing starting at 0, the one-dimensional array `cities(5)` contains six elements numbered from 0 to 5. If you prefer arrays to start at 1, you can override this by using the `Option Base 1` statement at the beginning of your module before any `Sub` statements.

If you don't specify `Option Base 1` in a procedure that uses arrays, VBA assumes that the statement `Option Base 0` is to be used and begins indexing your array's elements at 0.

If you'd rather not use the `Option Base 1` statement and still have the array indexing start at a number other than 0, you must specify the bounds of an array when declaring the array variable.

The bounds of an array are its lowest and highest indices. Let's look at the following example:

```
Dim cities(3 To 6) As Integer
```

The preceding statement declares a one-dimensional array with four elements. The numbers enclosed in parentheses after the array name specify the lower (3) and upper (6) bounds of the array. The first element of this array is indexed 3, the second 4, the third 5, and the fourth 6. Notice the keyword `To` between the lower and upper indices.

By using the `LBound` and `UBound` functions, you can dynamically find the lower and upper bounds of an array, as discussed in a later section of this chapter.

Initializing and Filling an Array

After declaring an array, you must assign values to its elements. This is often referred to as "initializing," "filling," or "populating" an array. The three methods you can use to load data into an array are discussed in this section.

Filling an Array Using Individual Assignment Statements

Assume you want to store the names of your six favorite cities in a one-dimensional array named `cities`. After declaring a fixed-size array with the `Dim` statement:

```
Dim cities(5) as String
```

or

```
Dim cities$(5)
```

you can assign values to the array variable like this:

```
cities(0) = "Baltimore"
cities(1) = "Atlanta"
cities(2) = "Boston"
cities(3) = "San Diego"
cities(4) = "New York"
cities(5) = "Denver"
```

Filling an Array Using the Array Function

VBA's built-in function `Array` returns an array of Variants. Because Variant is the default data type, the `As Variant` clause is optional in the array variable declaration:

```
Dim cities() As Variant
```

or

```
Dim cities()
```

Notice that we didn't specify the number of elements between the parentheses.

Next, use the `Array` function, as shown here, to assign values to your `cities` array:

```
cities = Array("Baltimore", "Atlanta", "Boston", "San Diego",
    "New York", "Denver")
```

When using the `Array` function for array population, the lower bound of an array is 0 or 1 and the upper bound is 5 or 6, depending on the setting of `Option Base` (see the previous section titled "Array Upper and Lower Bounds").

Filling an Array Using a For...Next Loop

You can use the `For...Next` loop discussed in the previous chapter to populate an array. Let's look at the example procedure here:

```
Sub LoadArrayWithIntegers()
  Dim myIntArray(1 To 10) As Integer
  Dim i As Integer

  ' Initialize random number generator
  Randomize

  ' Fill the array with 10 random numbers between 1 and 100
  For i = 1 To 10
    myIntArray(i) = Int((100 * Rnd) + 1)
  Next

  ' Print array values to the Immediate window
  For i = 1 To 10
    Debug.Print myIntArray(i)
  Next
End Sub
```

The preceding procedure uses a `For...Next` loop to fill `myIntArray` with 10 random numbers between 1 and 100. The second loop is used to print out the values from the array. Notice that the procedure uses the `Rnd` function to generate a random number. This function returns a value less than 1 but greater than or equal to 0. You can try it out in the Immediate window by entering:

```
x=rnd
?x
```

Before calling the `Rnd` function, the `LoadArrayWithIntegers` procedure uses the `Randomize` statement to initialize the random number generator. To become more familiar with the `Randomize` statement and `Rnd` function, be sure to follow up with the Excel online help.

USING A ONE-DIMENSIONAL ARRAY

Having learned the basics of array variables, let's write a couple of procedures to make arrays a part of your VBA skill set. The procedure in Hands-On 7.1 uses a one-dimensional array to programmatically display a list of six North American cities.

NOTE	*Please note that the files for the "Hands-On" projects can be found in the companion files.*

Hands-On 7.1   Using a One-Dimensional Array

1. Open a new workbook and save it as C:\VBAExcel2024_ByExample\Chap07. xlsm.
2. Switch to the Microsoft VBE window and rename the VBA project Arrays.
3. Insert a new module into the Arrays (Chap07.xlsm) project and rename this module StaticArrays.
4. In the StaticArrays module, enter the following FavoriteCities procedure:

```
' start indexing array elements at 1
Option Base 1

Sub FavoriteCities()
    ' now declare the array
    Dim cities(6) As String

    ' assign the values to array elements
    cities(1) = "Baltimore"
    cities(2) = "Atlanta"
    cities(3) = "Boston"
    cities(4) = "San Diego"
    cities(5) = "New York"
    cities(6) = "Denver"

    ' display the list of cities
    MsgBox cities(1) & Chr(13) & cities(2) & Chr(13) _
        & cities(3) & Chr(13) & cities(4) & Chr(13) _
        & cities(5) & Chr(13) & cities(6)
End Sub
```

Before the FavoriteCities procedure begins, the default indexing for an array is changed. Notice that the position of the Option Base 1 statement is at the top of the module window before the Sub statement. The array cities() of the String data type is declared with six elements, and values are specified for each element. The last statement uses the MsgBox function to display the list of cities. When you run this procedure in Step 5, the city names will appear on separate lines in the message box, as shown in Figure 7.1. You can change the order of the displayed data by switching the index values.

FIGURE 7.1. You can display the elements of a one-dimensional array with the MsgBox function.

5. Position the insertion point anywhere within the procedure code and press F5 to run the FavoriteCities procedure.
6. On your own, modify the FavoriteCities procedure so that it displays the names of the cities in the reverse order (from 6 to 1).

USING A TWO-DIMENSIONAL ARRAY

The following procedure creates a two-dimensional array that will hold the country name, currency name, and exchange rate for three countries.

Hands-On 7.2   Storing Data in a Two-Dimensional Array

1. In the StaticArrays module, enter the following procedure:

```
Sub Exchange()
    Dim t As String
    Dim r As String
    Dim Ex(3, 3) As Variant

    t = Chr(9)  ' tab
    r = Chr(13) ' Enter

    Ex(1, 1) = "Japan"
    Ex(1, 2) = "Yen"
    Ex(1, 3) = 104.57
    Ex(2, 1) = "Mexico"
    Ex(2, 2) = "Peso"
    Ex(2, 3) = 11.2085
    Ex(3, 1) = "Canada"
    Ex(3, 2) = "Dollar"
    Ex(3, 3) = 1.2028
```

```
    MsgBox "Country " & t & t & "Currency" & t & t & "per US$" _
        & r & r _
        & Ex(1, 1) & t & t & Ex(1, 2) & t & t & Ex(1, 3) & r _
        & Ex(2, 1) & t & t & Ex(2, 2) & t & t & Ex(2, 3) & r _
        & Ex(3, 1) & t & t & Ex(3, 2) & t & t & Ex(3, 3) & r & r _
        & "* Sample Exchange Rates for Demonstration Only", , _
        "Exchange"
End Sub
```

2. Run the Exchange procedure.

   When you run the Exchange procedure, you will see a message box with the exchange information presented in three columns, as shown in Figure 7.2.

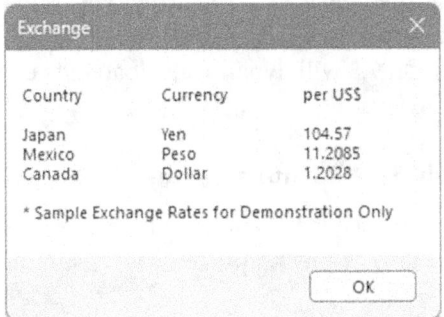

FIGURE 7.2.   The text displayed in a message box can be custom formatted.

USING A DYNAMIC ARRAY

The arrays introduced thus far in this chapter were static. A *static array* (or fixed-size array) is an array of a specific size. Use a static array when you know in advance how big the array should be. The size of the static array is specified in the array's declaration statement.

For example, the statement Dim Fruits(9) As String declares a static array called Fruits that is made up of 10 elements (assuming you have not changed the default indexing to 1). What if you're not sure how many elements your array will contain? If your procedure depends on user input, the number of user-supplied elements might vary every time the procedure is executed. How can you ensure that the array you declare is not a waste of memory?

After declaring an array, VBA sets aside enough memory to accommodate the array. If you declare an array to hold more elements than what you need, you'll end up wasting valuable computer resources. The solution to this problem is to make your arrays dynamic.

A *dynamic array* is an array whose size can change. You use a dynamic array when the array size is determined each time the procedure is run. A dynamic array is declared by placing empty parentheses after the array name:

```
Dim Fruits() As String
```

To allocate or reallocate storage space for dynamic arrays, you must use the ReDim statement. For example, initially, you may want to hold five fruits in the array:

```
Redim Fruits(1 To 5)
```

The ReDim statement redimensions arrays as the code of your procedure executes and informs Visual Basic about the new size of the array. This statement can be used several times in the same procedure.

The example procedure in Hands-On 7.3 will dynamically load data entered in a worksheet into a one-dimensional array.

Hands-On 7.3   Loading Worksheet Data into an Array

1. Insert a new module into the Arrays project and rename it DynamicArrays.
2. In the DynamicArrays module, enter the following procedure:

```
Sub LoadArrayFromWorksheet()
   Dim myDataRng As Range
   Dim myArray() As Variant
   Dim cnt As Integer
   Dim i As Integer
   Dim cell As Variant
   Dim r As Integer
   Dim last As Integer

   Set myDataRng = ActiveSheet.UsedRange

   ' get the count of nonempty cells (text and numbers only)
   last = myDataRng.SpecialCells(xlCellTypeConstants, 3).Count

   If IsEmpty(myDataRng) Then
       MsgBox "Sheet is empty."
       Exit Sub
   End If

   ReDim myArray(1 To last)

   i = 1
```

```
' fill the array from worksheet data
' reformat all numeric values
For Each cell In myDataRng
    If cell.Value <> "" Then
        If IsNumeric(cell.Value) Then
            myArray(i) = Format(cell.Value, "$#,#00.00")
        Else
            myArray(i) = cell.Value
        End If
        i = i + 1
    End If
Next
' print array values to the Immediate window
For i = 1 To last
    Debug.Print myArray(i)
Next
Debug.Print "Items in the array: " & UBound(myArray)
End Sub
```

3. Switch to the Microsoft Excel application window of the `Chap07.xlsm` workbook and enter some data in the active sheet. For example, enter your favorite fruits in cells A1:B6 and numbers in cells D1:D9.

4. Choose Developer | Macros. In the Macro dialog box, choose LoadArray-FromWorksheet, and click Run.

   When the procedure is completed, check the data in the Immediate window. You should see the entries you typed in the worksheet. The numeric data should appear formatted with the currency format.

When resizing a dynamic array, you can use the `Preserve` keyword with the `ReDim` statement to retain the existing data. The following VBA procedure demonstrates the declaration, allocation, and resizing of a dynamic array:

```
Sub myDynArraay()
    ' Declare dynamic array
    Dim dynArray() As Integer
    Dim i As Integer

    ' Initial allocation
    ReDim dynArray(1 To 3)

    ' Assign values to the array
    For i = 1 To 3
        dynArray(i) = i + 5
    Next i
```

```
' Resize the array while preserving the existing data
ReDim Preserve dynArray(1 To 6)

' Assign values to the newly allocated elements
For i = 4 To 6
    dynArray(i) = i * 5
Next i

' Print the values of the array to the Immediate Window
For i = 1 To 6
    Debug.Print "dynArray(" & i & ") = " & dynArray(i)
Next i
End Sub
```

USING ARRAY FUNCTIONS

You can manipulate arrays with five built-in VBA functions: Array, IsArray, Erase, LBound, and UBound. The following sections demonstrate the use of each of these functions in VBA procedures.

The Array Function

The Array function allows you to create an array during code execution without having to dimension it first. This function always returns an array of Variants. Using the Array function, you can quickly place a series of values in a list.

The CarInfo procedure shown here creates a fixed-size, one-dimensional, three-element array called auto.

⊙ Hands-On 7.4   Using the Array Function

1. Insert a new module into the current project and rename it Array_Function.
2. Enter the following CarInfo procedure:

```
Option Base 1

Sub CarInfo()
  Dim auto As Variant
  auto = Array("Ford", "Black", "2024")
  MsgBox auto(2) & " " & auto(1) & ", " & auto(3)
  auto(2) = "4-door"
  MsgBox auto(2) & " " & auto(1) & ", " & auto(3)
End Sub
```

3. Run the CarInfo procedure.

The IsArray Function

Using the `IsArray` function, you can test whether a variable is an array. The `IsArray` function returns either `True` if the variable is an array or `False` if it's not an array. Here's an example.

⊙ Hands-On 7.5   Using the IsArray Function

1. Insert a new module into the current project and rename it `IsArray_Function`.
2. Enter the code of the `IsThisArray` procedure, as shown here:

```
Sub IsThisArray()
    ' declare a dynamic array
    Dim sheetNames() As String
    Dim totalSheets As Integer
    Dim counter As Integer

    ' count the sheets in the current workbook
    totalSheets = ActiveWorkbook.Sheets.Count

    ' specify the size of the array
    ReDim sheetNames(1 To totalSheets)

    ' enter and show the names of sheets
    For counter = 1 To totalSheets
        sheetNames(counter) = _
            ActiveWorkbook.Sheets(counter).Name
        MsgBox sheetNames(counter)
    Next counter

    ' check if this is indeed an array
    If IsArray(sheetNames) Then
        MsgBox "The sheetNames variable is an array."
    End If
End Sub
```

3. Run the `IsThisArray` procedure.

The Erase Function

When you want to remove the data from an array, you should use the `Erase` function. This function deletes all the data held by static or dynamic arrays. In addition, the `Erase` function reallocates all the memory assigned to a dynamic array. If a procedure has to use the dynamic array again, you must use the `ReDim` statement to specify the size of the array.

The following example shows how to erase the data from the array `cities`.

Hands-On 7.6   Using the Erase Function

1. Insert a new module into the current project and rename it `Erase_Function`.
2. Enter the code of the `FunCities` procedure shown here:

```
' start indexing array elements at 1
Option Base 1

Sub FunCities()
' declare the array
Dim cities(1 To 5) As String

' assign the values to array elements
cities(1) = "Las Vegas"
cities(2) = "Orlando"
cities(3) = "Atlantic City"
cities(4) = "New York"
cities(5) = "San Francisco"

' display the list of cities
  MsgBox cities(1) & Chr(13) & cities(2) & Chr(13) _
    & cities(3) & Chr(13) & cities(4) & Chr(13) _
    & cities (5)
  Erase cities

' show all that were erased
MsgBox cities(1) & Chr(13) & cities(2) & Chr(13) _
  & cities(3) & Chr(13) & cities(4) & Chr(13) _
  & cities (5)
End Sub
```

After the `Erase` function deletes the values from the array, the `MsgBox` function displays an empty message box.

3. Run the `FunCities` procedure.

The LBound and UBound Functions

The `LBound` and `UBound` functions return whole numbers that indicate the lower bound and upper bound of an array.

Hands-On 7.7   Using the LBound and UBound Functions

1. Insert a new module into the current project and rename it `L_and_UBound_Function`.

2. Enter the code of the `FunCities2` procedure shown here:

```
Sub FunCities2()
  ' declare the array
  Dim cities(1 To 5) As String

  ' assign the values to array elements
  cities(1) = "Las Vegas"
  cities(2) = "Orlando"
  cities(3) = "Atlantic City"
  cities(4) = "New York"
  cities(5) = "San Francisco"

  ' display the list of cities
  MsgBox cities(1) & Chr(13) & cities(2) & Chr(13) _
    & cities(3) & Chr(13) & cities(4) & Chr(13) _
    & cities (5)
  ' display the array bounds
  MsgBox "The lower bound: " & LBound(cities) & Chr(13) _
    & "The upper bound: " & UBound(cities)
End Sub
```

3. Run the `FunCities2` procedure.

TROUBLESHOOTING ERRORS IN ARRAYS

When working with arrays, it's easy to make a mistake. If you try to assign more values than there are elements in the declared array, VBA will display the error message "Subscript out of range," as shown in Figure 7.3.

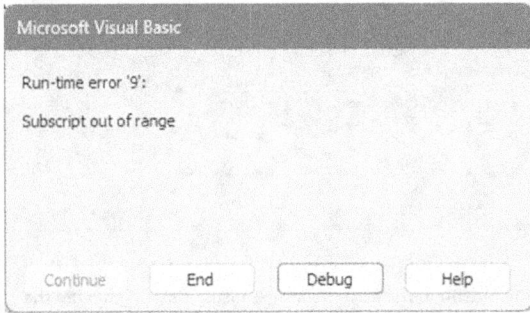

FIGURE 7.3. This error was caused by an attempt to access a nonexistent array element.

Suppose you declared a one-dimensional array that consists of six elements and you are trying to assign a value to the seventh element. When you run the procedure, Visual Basic can't find the seventh element, so it displays the error message. When you click the Debug button, Visual Basic will highlight the line of code that caused the error.

To fix the "Subscript out of range" error, you should begin by looking at the array's declaration statement. Once you know how many elements the array should hold, it's easy to figure out that the culprit is the index number that appears in the parentheses in the highlighted line of code. In the example shown in Figure 7.4, once we replace the line of code `cities(7) = "Denver"` with `cities(6) = "Denver"` and press F5 to resume the procedure, the procedure will run as intended.

Another frequent error you may encounter while working with arrays is "Type mismatch." To avoid this error, keep in mind that each element of an array must be of the same data type. If you attempt to assign to an element of an array a value that conflicts with the data type of the array declared in the `Dim` statement, you'll obtain the "Type mismatch" error during code execution. To hold values of different data types in an array, declare the array as a Variant.

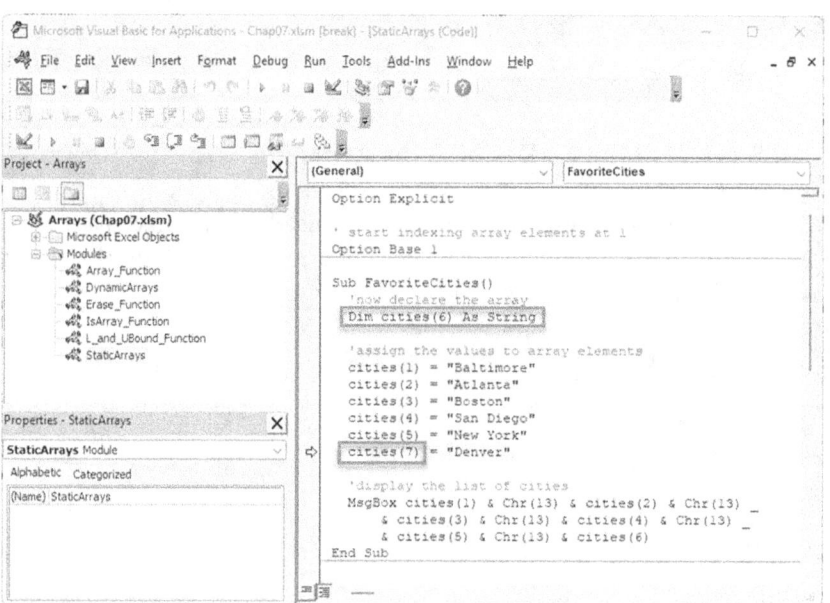

FIGURE 7.4. When you click the Debug button in the error message, Visual Basic highlights the statement that triggered the error.

USING THE PARAMARRAY KEYWORD

You can pass values between subroutines or functions as required or optional arguments. If an argument isn't strictly necessary for the procedure to execute, prefix its name with the keyword Optional. Sometimes, however, you don't know in advance how many arguments you need. A classic example is addition: you may want to add two numbers, but later, you might want to add 3, 10, or even 15 numbers.

By using the keyword ParamArray, you can pass an array consisting of any number of elements to your subroutines and function procedures.

The following AddMultipleArgs function will add up as many numbers as you require. This function begins with the declaration of an array, myNumbers. Notice the use of the ParamArray keyword. The array must be declared as an array of the type Variant, and it must be the last argument in the procedure definition.

Hands-On 7.8 Passing an Array to Procedures Using the ParamArray Keyword

1. Insert a new module into the current project and rename it ParameterArrays.
2. In the ParameterArrays module, enter the following AddMultipleArgs function procedure:

```
Function AddMultipleArgs(ParamArray myNumbers() As Variant)
    Dim mySum As Single
    Dim myValue As Variant
    For Each myValue in myNumbers
      mySum=mySum+myValue
    Next
    AddMultipleArgs = mySum
End Function
```

3. To try out the AddMultipleArgs function, activate the Immediate window and type the following instruction:

```
?AddMultipleArgs(1, 23.24, 3, 24, 8, 34)
```

When you press Enter, Visual Basic returns the total of all the numbers in the parentheses: 93.24. You can supply an unlimited number of arguments. To add more values, enter additional values inside the parentheses and press Enter. Notice that each function argument must be separated by a comma.

DATA ENTRY WITH AN ARRAY

Earlier in this chapter, you learned how to use various `Array` functions. The following procedure demonstrates how a simple `Array` function can speed up data entry.

> **Hands-On 7.9 Using the Array Function to Enter Headings in a Worksheet**

1. Insert a new module into the current project and rename it `DataEntry_withArray`.
2. In the `EnterData_Array` module, enter the following `ColumnHeads` procedure:

```
Sub ColumnHeads()
  Dim heading As Variant
  Dim cell As Range
  Dim i As Integer
  i = 0
  heading = Array("First Name", "Last Name", _
    "Position", "Salary")
  Workbooks.Add

  For Each cell In Range("A1:D1")
    cell.Formula = heading(i)
  i = i + 1
  Next

  Columns("A:D").Select
  Selection.Columns.AutoFit
  Range("A1").Select
End Sub
```

3. Switch to the Microsoft Excel window and use the Macros dialog to run the `ColumnHeads` procedure.

SORTING AN ARRAY WITH EXCEL

We all find it easier to work with sorted data. Some operations on arrays, such as finding maximum and minimum values, require that the array is sorted. Once it is sorted, you can find the maximum value by assigning the upper bound index to the sorted array, as in the following:

```
y = myIntArray(UBound(myIntArray))
```

The minimum value can be obtained by reading the first value of the sorted array:

```
x = myIntArray(1)
```

So, how can you sort an array? This section demonstrates how to use Excel to get your array data into the sorted order. An easy way to sort an array is by copying your array values to a new worksheet and then using the Excel built-in Sort function. After completing the sort, you can load your sorted values back into a VBA array. This technique is the simplest since you can use a macro recorder to get your Sort statement started for you. With a large array, it is also faster than the classic bubble sort routine that is commonly used with arrays.

⊙ Hands-On 7.10 Using Excel to Sort a VBA Array

1. Insert a new module into the current project and rename it SortArray_ withExcel.
2. In the SortArray_withExcel module, enter the following SortArrayWithExcel procedure:

```
Sub SortArrayWithExcel()
  Dim myIntArray() As Integer
  Dim i As Integer
  Dim x As Integer
  Dim y As Integer
  Dim r As Integer
  Dim myDataRng As Range

  'initialize random number generator
  Randomize

  ReDim myIntArray(1 To 10)

  ' Fill the array with 10 random numbers between 1 and 100
  For i = 1 To 10
    myIntArray(i) = Int((100 * Rnd) + 1)
    Debug.Print "aValue" & i & ":" & vbTab & myIntArray(i)
  Next

  'write array to a worksheet
  Worksheets.Add

  r = 1    'row counter
  With ActiveSheet
```

```
For i = 1 To 10
  Cells(r, 1).Value = myIntArray(i)
  r = r + 1
Next i
  End With

  'Use Excel Sort to order values in the worksheet
  Set myDataRng = ActiveSheet.UsedRange

  With ActiveSheet.Sort
  .SortFields.Clear
  .SortFields.Add Key:=Range("A1"), _
    SortOn:=xlSortOnValues, Order:=xlAscending, _
    DataOption:=xlSortNormal
  .SetRange myDataRng
   .Header = xlNo
  .MatchCase = False
  .Apply
  End With

  'free the memory used by array by using Erase statement
  Erase myIntArray

  ReDim myIntArray(1 To 10)

  'load sorted values back into an array

  For i = 1 To 10
  myIntArray(i) = ActiveSheet.Cells(i, 1).Value
  Next

  'write out sorted array to the Immediate Window

  i = 1
  For i = 1 To 10
  Debug.Print "aValueSorted: " & myIntArray(i)
  Next

  'find minimum and maximum values stored in the array
  x = myIntArray(1)
  y = myIntArray(UBound(myIntArray))
  Debug.Print "Min value=" & x & vbTab; "Max value=" & y
End Sub
```

The `SortArrayWithExcel` procedure populates a dynamic array with 10 random Integer values and prints out this array to an Immediate window and a new worksheet. Next, the values entered in the worksheet are sorted in ascending order using the Excel `Sort` object.

The `Sort` statements have been generated by the macro recorder and then modified for this procedure's needs. Once sorted, the `Erase` statement is used to free the memory used by the dynamic array. Before reloading the array with the sorted values, the procedure redeclares the array variable using the `ReDim` statement. The last statements in the procedure demonstrate how to retrieve the minimum and maximum values from the array variable.

3. Switch to the Microsoft Excel window and run the `SortArrayWithExcel` procedure. Compare with the results in the Immediate window.

Using Bubble Sort to Sort the Array

Bubble sort is a custom sorting algorithm that steps through the list of items, compares adjacent elements, and swaps them if they are in the wrong order. This process is repeated until the list is sorted. Start by declaring and initializing your array with sample data. Then, start with the first element and compare it with the next element. If the first element is greater than the second, swap them. Move to the next pair of elements and repeat this process. Continue until all the elements have been processed. Use `For...Next` loops to swap adjacent elements if they are out of order.

The `BubbleSort` procedure shown below was provided with the assistance of Microsoft Copilot. It has been corrected to avoid the "Type mismatch" error due to the assignment of the incorrect data type for the `arr` variable. This version of `BubbleSort` ensures that the array is properly declared as a Variant to handle different data types without errors. While AI assistance is welcome for providing initial guidance and error-checking, always review and test the code to ensure it's free of errors and meets your specific use case and requirements.

```
Sub BubbleSort()
    Dim arr() As Variant
    Dim temp As Integer
    Dim i As Integer, j As Integer
    Dim swapped As Boolean
    Dim n As Integer

    ' Initialize the array with sample data
    arr = Array(34, 7, 23, 32, 5, 62, 32)

    n = UBound(arr)
```

```
    ' Bubble sort algorithm
    For i = 0 To n - 1
        swapped = False
        For j = 0 To n - i - 1
            If arr(j) > arr(j + 1) Then
                temp = arr(j)
                arr(j) = arr(j + 1)
                arr(j + 1) = temp
                swapped = True
            End If
        Next j
        If Not swapped Then Exit For
    Next i

    ' Print the sorted array in the Immediate Window
    For i = 0 To n
        Debug.Print arr(i)
    Next i
End Sub
```

Bubble sort gets its name from the way the algorithm works—it sorts elements like bubbles rising to the top in water.

USING CHATGPT WITH EXCEL

In the previous section, you learned how to sort an array with the bubble sort algorithm. This type of sort is easy to implement and understand but it is not very suitable for large datasets. A more efficient algorithm compared to bubble sort is the quick sort algorithm. If your data processing needs require fast performance, I suggest you explore the topic of using quick sort. Use your favorite AI tool to get more insight into this topic or review the provided chat example in the companion files—see *Chapter 7 – Using Chat GPT with Excel*.

Using ChatGPT with Excel can greatly enhance your productivity while improving your Excel skills. By crafting VBA procedures or getting assistance with writing complex formulas, ChatGPT or your favorite AI tool can provide quick guidance to many topics of interest that can't be covered in detail in technical books due to space constraints. By incorporating ChatGPT into your Excel routine you can get real-time help in obtaining solutions to complex and difficult problems. Learn how to craft prompts so the AI can help you with various tasks. Giving it longer commands with all sorts of instructions can often provide you with useful information that you can refine to suit your needs.

SUMMARY

In this chapter, you learned how to use arrays in complex VBA procedures that require many variables. You worked with examples of procedures that demonstrated how to declare and use a one-dimensional array (list) and a two-dimensional array (table). Additionally, you saw how to sum data across multiple worksheets using a 3D array and learned about sorting arrays with Excel and by using a custom bubble sort algorithm.

You explored the difference between static and dynamic arrays and practiced using five built-in VBA functions that are frequently used with arrays: `Array`, `IsArray`, `Erase`, `LBound`, and `UBound`. You were also introduced to statements and keywords frequently used with arrays: `ReDim`, `ReDim Preserve`, and `ParamArray`.

In the next chapter, you will learn how to use collections instead of arrays to manipulate large amounts of data.

## Chapter 8

KEEPING TRACK OF MULTIPLE VALUES IN EXCEL VBA PROGRAMS—

A QUICK INTRODUCTION TO CREATING AND USING COLLECTIONS

In the previous chapter, you learned how arrays are used to manipulate multiple items quickly and easily. Instead of creating several variables to keep track of your data, you only need to declare one variable. Using arrays instead of defining individual variables saves you from writing many lines of repetitive code. As you have seen so far, in programming, there are many ways of performing the same task. The method you use depends on your needs. So it is with storing multiple values. In addition to arrays, you can maintain your data while your program is running by using a special type of object—a collection. Like arrays, collections are used for grouping variables.

Because collections have built-in properties and methods that allow you to add, remove, and count their elements, they make working with multiple data items much easier than arrays. Collections can also be used to hold objects. In this chapter, we will focus on learning the basic skills of using collections for tracking and maintaining data in your VBA procedures.

WORKING WITH BUILT-IN COLLECTIONS

A set of similar objects is known as a collection. In Microsoft Excel, for example, all open workbooks belong to the `Workbooks` collection, and all the sheets in a workbook are members of the `Worksheets` collection. Collections are objects that contain other objects. No matter what collection you want to work with, you can do the following:

- Refer to a specific object in a collection by using an index value. For example, to refer to the second object in the `Worksheets` collection, use either of the following statements:

```
Worksheets(2).Select
```

or

```
Worksheets("Sheet2").Select
```

- Determine the number of items in the collection by using the `Count` property. For example, after entering the following statement in the Immediate window:

```
?Worksheets.Count
```

VBA returns the total number of worksheets in the current workbook.

- Use the `Add` method to insert new items into the collection. For example, after entering the following statement in the Immediate window:

```
Worksheets.Add
```

VBA inserts a new worksheet into the current workbook. The `Worksheets` collection now contains one more item.

- Use the `For Each... Next` loop to cycle through every object in the collection.

Suppose you opened a workbook containing five worksheets with the following names: `Daily wages`, `Weekly wages`, `Monthly wages`, `Yearly salary`, and `Bonuses`. To delete the worksheets that contain the word "wages" in the name, you could write the following procedure:

```
Sub DeleteSheets()
   Dim ws As Worksheet
   Application.DisplayAlerts = False
       For Each ws In Worksheets
       If InStr(ws.Name, "wages") Then
```

```
        ws.Delete
    End If
    Next
    Application.DisplayAlerts = True
End Sub
```

The statement `Application.DisplayAlerts = False` is used to suppress some prompts and messages that Excel displays while the code is running. In this case, we want to suppress the confirmation message that Excel displays when worksheets are deleted. The `InStr` function is very useful for string comparisons as it allows you to find one string within another. The statement `InStr(ws.Name, "wages")` tells Excel to determine whether the worksheet name (stored in the `ws` object variable) contains the string of characters `wages`.

Creating Your Own Collection

`Collection` is an object; it is a data type in VBA. To create a user-defined collection, begin by declaring an object variable of the `Collection` type:

```
Dim collection_name As Collection
Set collection_name = New Collection
```

or

```
Dim collection_name As New Collection
```

Notice the `New` keyword. VBA uses this keyword to create a new instance of an object. `collection_name` is the name of your collection. You can use any name provided it is not one of the reserved words that Excel uses for its own collections or other internal operations.

You can define more than one collection; however, each collection you define must have a distinct name so you can easily reference it in your code. Collections can be defined at the top of the standard module or within a procedure. They can also be defined in class modules, which are often used in more advanced Excel VBA programming. Unlike arrays, collections do not require you to predefine their size. Once you define the object variable of the `Collection` type, you are ready to begin adding items to your collection.

Adding Objects to a Custom Collection

After you've declared the `Collection` object, you can use the `Add` method to insert new items into the collection. The `Add` method looks like this:

```
object.Add item[, key, before, after]
```

You are required to specify only the object and the item. All arguments in the square brackets are optional. `object` is the collection name. This is the same name that was used in the declaration of the `Collection` object. `item` is the object that you want to add to the collection. For example, if the name of your collection is `colFruits`, you can add new items to it like this:

```
colFruits.Add "Apple"
colFruits.Add "Pear"
colFruits.Add "Strawberry"
colFruits.Add "Blueberry"
colFruits.Add "Orange"
colFruits.Add "Peach"
```

Although other arguments are optional, they are quite useful. It's important to understand that the items in a collection are automatically assigned numbers starting from 1. They can also, however, be assigned a unique key value. Instead of accessing a specific item with an index (1, 2, 3, and so on), you can assign a key for that object. For instance, if you are creating a collection of custom sheets, you could use a sheet name as a key. To identify an individual in a collection of students or employees, you could use their ID numbers as a key. For example, here's how you can add `James Allen` to the `colPeople` collection, using his employee ID as a key:

```
Dim colPeople As New Collection
colPeople.Add "James Allen", Key:="123456"
```

To specify the position of the object in the collection, use either a `before` or `after` argument. The `before` argument is the object before which the new object is added. The `after` argument is the object after which the new object is added.

For example, to add `Kiwi` to the `colFruits` collection so that it appears after the second item, use the following statement:

```
colFruits.Add "Kiwi", , , 2
```

or

```
colFruits.Add "Kiwi", After:=2
```

Notice that if you are not using the named argument `After`, you must place a comma for each of the preceding optional arguments that you are not specifying.

To enter `Cherry` in the first position, use the named `Before` argument:

```
colFruits.Add "Cherry", Before:=1
```

By using the optional `Before` or `After` arguments, you can easily add elements in any position.

Each element of a collection can be a different data type. Please note that arrays can support different data types only if they are defined as variants. To store a date item in your collection, use the following statement:

```
colFruits.Add #12/10/2021#, Key:="InvoiceDate"
```

To store a number in your collection, the following statement can be used:

```
colFruits.Add 100.99, Key:="InvoiceTotal"
```

Determining the Number of Items in Your Collection

Use the `Count` property to find out the number of items in your collection. To determine the current number of fruit items in `colFruits`, write the following statement:

```
Debug.Print colFruits.Count
```

Accessing Items in a Collection

To refer to a specific item in your collection, use its index or key value.

For example, to find out the names of the first collection item, use this statement:

```
Debug.Print colFruits.Item(1)
```

Because the `Item` method is a default method of the collection, you may omit it from the statement, as shown here:

```
Debug.Print colFruits(1)
```

If you know the item key, then you can quickly retrieve it like this:

```
Debug.Print colPeople("123456")
```

By using the key to access a collection item you can go to it directly without the need to iterate through all the items. Excel VBA does not provide a built-in method to check if the Key exists, but you can write your own function to return a Boolean value of True if the key exists. Here's how you would do it:

```
Function KeyExists(colName As Collection, _
        key As String) As Boolean
    On Error GoTo EndHere
    IsObject (colName.Item(key))
    KeyExists = True
```

```
EndHere:
End Function
```

IsObject is a built-in VBA function that returns True if the passed-to-it expression (i.e., key name) represents an object variable. The statement OnError GoTo EndHere tells Excel to jump to the line EndHere: if the result of the IsObject function is False. The error-handling statements introduced here are covered in detail in Chapter 9.

Removing Items from a Collection

Removing an item from your custom collection is as easy as adding an item. To remove an item, use the Remove method in the following format:

```
object.Remove index
```

object is the name of the custom collection that contains the object you want to remove. index is an expression specifying the position of the object in the collection. To remove the third fruit item from colFruits, you simply write the following statement:

```
colFruits.Remove 3
```

Collections are reindexed automatically when an item is removed. Therefore, to remove all items from a custom collection, you can use 1 for the Index argument, as in the following example:

```
Do While colFruits.Count > 0
  colFruits.Remove Index:=1
Loop
```

Another way to remove all objects from your collection is by using the For Each...Next loop. For example, to remove all objects from colFruits, use the following looping structure:

```
Dim m As Variant

For Each m in colFruits
  colFruits.Remove 1
Next
```

Note that the control variable used in the For Each...Next loop must be declared as Variant or Object. Because collections are reindexed, the preceding statement will remove the first item of the collection on each iteration. After the loop is completed, colFruits should have zero items. To be sure, however, use the Count property to find out:

```
Debug.Print colFruits.Count
```

You can also remove all items from a collection by setting the `Collection` object variable to a new collection, like this:

```
Set colFruits = New Collection
```

Updating Items in a Collection

When you add an item to your collection that has a basic data type such as String, Integer, Long, Currency, or Date, your collection will be read-only, meaning you won't be able to change the value of the item. Excel will display an error if you try to assign a value to an existing item in your collection:

```
' this statement will produce Run-time error 424 'Object required'
  colFruits(4) = "Cranberry"
```

Therefore, if your procedure requires that the values be updated, you should group your items into an array. The only time that the collection is updatable is when it references objects, and this is a subject for the advanced VBA.

Let's proceed with the first "Hands-On" project in this chapter, where you will use collections in VBA procedures.

> **NOTE** *Please note that the files for the "Hands-On" projects can be found in the companion files.*

Hands-On 8.1 Creating and Manipulating a Custom Collection (Example 1)

1. Start Excel and create a new macro-enabled workbook named `Chap08.xlsm` in your `C:\VBA2024_ByExample` folder.
2. Press Alt+F11 to switch to the VBE window.
3. Choose Insert | Module to add a new standard module.
   In the Module1 Code window, enter the following `WorkWith_Collection`, `Display_Items`, and `KeyExists` procedures:

```
Sub WorkWith_Collection()
    Dim colFruits As New Collection
    Dim itm As Variant
    Dim strColItems As String

    colFruits.Add "Apple"
    colFruits.Add "Pear"
    colFruits.Add "Strawberry"
    colFruits.Add "Blueberry"
    colFruits.Add "Orange"
```

```
    colFruits.Add "Peach"
    colFruits.Add "Kiwi", , , 2
    colFruits.Add "Mango", , 5
    colFruits.Add "Cherry", Before:=1
    colFruits.Add 100.99, Key:="InvoiceTotal"
    colFruits.Add #12/10/2021#, Key:="InvoiceDate"

    Debug.Print "Total Items in colFruits: " & colFruits.Count

    'call a procedure to display all items in the collection
    Display_Items colFruits, itm

    colFruits.Remove 3
    Debug.Print "New Total Items in colFruits: " & colFruits.Count

    For Each itm In colFruits
        strColItems = strColItems & ", " & itm
    Next

    ' remove a comma and a space from the beginning of
    ' the strColItems variable
    strColItems = Mid(strColItems, 3, Len(strColItems))
    Debug.Print strColItems

    'Find if keys exist and if not display a message
    'and go to the next line

    If KeyExists(colFruits, "InvoiceDate") And _
        KeyExists(colFruits, "InvoiceTotal") Then
    Debug.Print "Invoiced on: " & colFruits("InvoiceDate") & _
        vbCrLf & "Total: " & colFruits("InvoiceTotal")
    Else
        MsgBox "Provided key(s) not found."
    End If

    ' Remove all items from collection one by one
    For Each itm In colFruits
        colFruits.Remove 1
    Next
    Debug.Print "Total Items in colFruits: " & colFruits.Count

End Sub
```

```vba
Sub Display_Items(col As Collection, myItm As Variant)

    For Each myItm In col
        Debug.Print myItm
    Next
End Sub

Function KeyExists(colName As Collection, _
        key As String) As Boolean

    On Error GoTo EndHere

    IsObject (colName.Item(key))
    KeyExists = True

EndHere:
End Function
```

4. Position the pointer anywhere within the `WorkWith_Collection` procedure and choose Run | Run Sub/UserForm to execute it.
5. Press Ctrl+G to open the Immediate Window and check the output of the procedure.
   The WorkWith_Collection procedure performs various operations on the declared `Collection` object variable `colFruits`. If you plan on using the same collection in other procedures in the same module, you will need to move its variable declaration statement to the top of the module.

Returning a Collection from a Function

Like arrays, collections can be used as parameters or return values to functions or subroutine procedures. In Hands-On 8.2, you will collect entries from the user via the VBA `InputBox` function and store them in an array. Next, you will pass that array to a function and return a collection with the same items.

Hands-On 8.2 Creating and Manipulating a Custom Collection (Example 2)

1. Choose Insert | Module to add a new standard module to the current VBA project.
2. In the Code window, enter the code shown as follows.
   Notice that the `allItems` variable is declared at the top of the module (above all the procedure code). This placement will make this variable available to all the procedures in this module:

```vba
Dim allItems As String

Sub ShowCollItems()
    Dim coll As Collection
    Dim myArray As Variant
    Dim itm As Variant

    ' get items from user input
    If AskForItems <> "" Then
        Debug.Print allItems
        ' extract items from the user input string (allItems)
        ' and place them in an array
        myArray = Split(allItems, ",")
       Debug.Print "Array has " & UBound(myArray) + 1 & " items."

        ' call function to create a collection from the array
        Set coll = CreateCollection(myArray)

        ' iterate through the collection to display each item
        For Each itm In coll
            Debug.Print itm
        Next
       Debug.Print "Total items in the collection: " & coll.Count
    End If
End Sub

Function AskForItems() As String
  allItems = InputBox("Enter your items separated by a comma", _
        "Demo - Get User Input", _
        "item1, item2")

  If allItems = "" Then
    AskForItems = ""
  Else
    AskForItems = allItems
  End If
End Function

Function CreateCollection(arrMyItems As Variant) As Collection
    Dim coll As New Collection
    Dim i As Integer
    For i = 0 To UBound(arrMyItems)
        coll.Add arrMyItems(i)
    Next i
    'Return a collection
```

```
      Set CreateCollection = coll
End Function
```

3. Position the pointer anywhere within the `ShowCollItems` procedure and choose Run | Run Sub/UserForm to execute it.

4. Press Ctrl+G to open the Immediate window and check the output of the procedure.

Let's review the previous code. The main procedure `ShowCollItems` declares three variables that we need for working with the array and collection. Notice that you don't need to use the `New` keyword to declare the collection because the collection is created inside the `CreateCollection` function. The declared object variable `coll` will be assigned a collection received from this function. In addition to the `CreateCollection` function, the `ShowCollItem` procedure relies on the following functions: `AskForItems` and `Split`.

The custom function procedure `AskForItems` populates the `allItems` string variable (declared at the top of the module) with the values obtained from the user via the VBA `InputBox` function. The user is requested to input values as a comma-delimited string (each item must be separated by a comma). If the user does not enter any values and presses Cancel instead, the function will return an empty string to the calling procedure (`ShowCollItems`). If items are entered, then the entire string will be returned. If the string returned from the `AskForItems` function is not empty, we continue running the statements within the `If` block. If the string is empty, the procedure ends.

Within the `If` block, we print the contents of the `allItems` variable to the Immediate window. Next, we use the built-in VBA `Split` function to extract the values from the `allItems` string using a comma delimiter and return an array (`myArray`). We use the `UBound` function to find the number of items in the array. Because, by default, arrays are zero-based, we need to add `1` to the count to get the correct number of items.

Refer to the previous chapter on arrays if you'd like to add code here to list the items in the `myArray` variable.

The next set of statements focuses on creating a collection. To do this, we set the `coll` variable to the result obtained from the `CreateCollection` function. We call the `CreateCollection` function and pass it the `myArray` variable. Notice that `myArray` is a variant and the `CreateCollection` function was defined to expect the parameter of the `Variant` type. Inside the `CreateCollection` function, we start by declaring the `coll` variable of the `Collection` type. We also need a counter (`i`) to loop through the items of the array variable that we passed to this function. Note that the parameter name that the `CreateCollection` function expects can be any name you define.

Using the `For...Next` loop, we loop through the items of the array starting from zero and add each array item to the collection. When finished, we pass the entire collection to the calling procedure—`ShowCollItems`. The `coll` variable is an object variable, so we need to use the `Set` keyword to return it from the function:

```
'Return a collection
Set CreateCollection = coll
```

Now we are back again in the `ShowCollItems` procedure—this time, returning a collection. To view the collection items, we use the `For each...Next` loop to print each item to the Immediate window. Note that the `itm` iterator must be defined as `Variant`. The procedure ends by printing the total number of collection items.

Using Custom and Built-In Collections Together

Let's apply our knowledge of custom collections to working with the Excel built-in collection of worksheets. As worksheets can contain various objects, you may need to collect different values stored within them. The `GetNotes` procedure in Hands-On 8.3 declares a custom collection object named `colNotes`. We will use this collection to store notes that you insert in various worksheets of the active workbook.

___ **NOTE**	*In Excel for Microsoft 365, notes are like the classic yellow sticky notes that are used to make annotations or reminders about the data in your worksheets. These were previously known as comments. Comments, on the other hand, are now threaded and intended for discussions. They include a reply box, so you can have conversations directly in your workbook. When people reply to a comment, you can see all the responses connected together, forming a thread or conversation. In the following Hands-On you will work with notes.*

⊙ **Hands-On 8.3   Using a Custom Collection Object**

1. Right-click any cell in `Sheet1` of the `Chap8.xlsm` workbook and choose New Note from the context menu. You can also enter a note by choosing Review | Notes | New Note. Type any text you want for your note. Click outside the

note box to exit the edit mode. Add two new sheets to the workbook. Use the same technique to enter two notes in Sheet2. Enter different text for each note. Add a note in any cell on Sheet3. You should now have four notes in three worksheets.

2. Click the File tab and choose Options. In the Excel Options window's General section, in the area named Personalize your copy of Microsoft Office, you should see a text box with your name. Delete your name, enter Joan Smith, and then click OK.

3. Enter one note anywhere on Sheet2 and one note anywhere on Sheet3. These notes should be automatically stamped with Joan Smith's name. When you're done entering the note text, return to the Excel Options window and change the Username text box entry back to the way it was (your name).

4. Switch to the VBE and add a new module to the current project.

5. Use the Properties box to rename the module. Enter MyCollection next to the Name property.

6. In the MyCollection module, enter the GetNotes procedure, as shown here:

```vba
Sub GetNotes()
    Dim sht As Worksheet
    Dim colNotes As New Collection
    Dim myNote As Comment
    Dim i As Integer
    Dim t As Integer
    Dim strName As String

    strName = InputBox("Enter author's name:")
    For Each sht In ThisWorkbook.Worksheets
      sht.Select
      i = ActiveSheet.Comments.Count
      For Each myNote In ActiveSheet.Comments
          If myNote.Author = strName Then
            MsgBox myNote.Text
            If colNotes.Count = 0 Then
              colNotes.Add Item:=myNote, key:="first"
            Else
              colNotes.Add Item:=myNote, Before:=1
            End If
          End If
      Next
      t = t + i
    Next
    If colNotes.Count <> 0 Then MsgBox colNotes("first").Text
```

```
    MsgBox "Total notes in workbook: " & t & Chr(13) & _
    "Total notes in collection: " & colNotes.Count
    Debug.Print "Notes by " & strName
  For Each myNote In colNotes
    Debug.Print Mid(myNote.Text, Len(myNote.Author) + 2, _
      Len(myNote.Text))
  Next
End Sub
```

The preceding procedure begins by declaring the custom collection object called `colNotes`. Next, the procedure prompts for an author's name and then loops through all the worksheets in the active workbook to locate this author's notes. Only notes entered by the specified author are added to the custom collection. It is important to note that we are working here with the note feature, which is compatible with previous versions of Excel.

To use the new comments feature using Excel VBA, you must declare `myNote` as `CommentThreaded` and make other changes in the procedure.

The procedure assigns a key to the first note and then adds the remaining notes to the collection by placing them before the note that was added last (notice the use of the `before` argument). If the collection includes at least one note, the procedure displays a message box with the text of the note that was identified with the special `key` argument. Notice how the `key` argument is used in referencing an item in a collection. The procedure then prints the text of all the notes included in the collection to the Immediate window.

Text functions (`Mid` and `Len`) are used to get only the text of the note without the author's name. Next, the total number of notes in a workbook and the total number of notes in the custom collection are returned by the `Count` property.

7. Run the `GetNotes` procedure twice each time, supplying a different name of the author (your name and `Joan Smith`). Check the procedure results in the Immediate window.

Let's modify the `GetNotes` procedure that you prepared in the previous Hands-On. At the end of the procedure, we'll display the contents of the items that are currently in the `colNotes` collection one by one and ask the user whether the item should be removed from the collection. If, for any reason, your name does not appear in the notes you entered after you've saved your workbook, enter `Author` for the name of the author.

To resolve the issue of the author's name being sometimes removed from your notes, review the options available from File | Info. Choose Check for Issues next to the Inspect Workbook section and select Inspect Document to launch Document Inspector. Make sure that Comments is not selected.

⊙ Hands-On 8.4 Removing Items from the Custom ColNotes Collection

1. Add the following line to the declaration section of the GetNotes procedure:

```
Dim response As Integer
```

This statement declares the variable called response. You will use this variable to store the result of the MsgBox function.

2. Locate the following statement in the GetNotes procedure:

```
Debug.Print Mid(myNote.Text, Len(myNote.Author) + 2, _
    Len(myNote.Text))
```

Enter the following block of instructions below that statement:

```
response = MsgBox("Remove this note?" & Chr(13) _
    & Chr(13) & myNote.Text, vbYesNo + vbQuestion)
    If response = 6 Then
        colNotes.Remove Index:=myID
    End If
```

3. Enter the following statements at the end of the procedure before the End Sub keywords:

```
Debug.Print "These notes remain in the collection:"
For Each myNote in colNotes
    Debug.Print Mid(myNote.Text, Len(myNote.Author) + 2, _
        Len(myNote.Text))
Next
```

The revised GetNotes procedure, named GetNotes_Modified, is shown here. The procedure removes the specified notes from the custom collection. It does not delete the notes from the worksheets:

```
Sub GetNotes_Modified()
  Dim sht As Worksheet
  Dim colNotes As New Collection
  Dim myNote As Comment
  Dim i As Integer
  Dim t As Integer
  Dim strName As String
  Dim response As Integer

    strName = InputBox("Enter author's name:")
    For Each sht In ThisWorkbook.Worksheets
      sht.Select
      i = ActiveSheet.Comments.Count
```

```
            For Each myNote In ActiveSheet.Comments
         If myNote.Author = strName Then
         MsgBox myNote.Text
         If colNotes.Count = 0 Then
            colNotes.Add Item:=myNote, key:="first"
         Else
            colNotes.Add Item:=myNote, Before:=1
         End If
       End If
     Next
   t = t + i
 Next
 If colNotes.Count <> 0 Then MsgBox colNotes("first").Text

    MsgBox "Total notes in workbook: " & t & Chr(13) & _
    "Total notes in collection:" & colNotes.Count
    Debug.Print "Notes by " & strName

    For Each myNote In colNotes
     Debug.Print Mid(myNote.Text, Len(myNote.Author) + 2, _
       Len(myNote.Text))
     response = MsgBox("Remove this note?" & Chr(13) _
       & Chr(13) & myNote.Text, vbYesNo + vbQuestion)
     If response = 6 Then
       colNotes.Remove index:=1
     End If
    Next

    MsgBox "Total notes in workbook: " & t & Chr(13) & _
    "Total notes in collection: " & colNotes.Count
    Debug.Print "These notes remain in the collection:"

    For Each myNote In colNotes
       Debug.Print Mid(myNote.Text, Len(myNote.Author) + 2, _
       Len(myNote.Text))
    Next
 End Sub
```

4. Run the `GetNotes_Modified` procedure and remove one of the notes displayed in the message box.

You can delete all notes from the workbook using the following code:

```
Sub DeleteWorkbookNotes()
  Dim myComment As Comment
  Dim sht As Worksheet
```

```
For Each sht In ThisWorkbook.Worksheets
    For Each myComment In sht.Comments
        myComment.Delete
    Next
  Next
End Sub
```

COLLECTIONS VERSUS ARRAYS

As you have seen so far, both collections and arrays provide a very convenient way of storing and manipulating groups of similar items. Most people find collections easier to use and master than arrays. Before you decide which grouping structure you should use for storing items in your program, however, examine your needs. Arrays are usually faster and more convenient to use if you know the number of items you are going to store ahead of time. If the number of elements varies and you often need to add and remove elements, collections may be more efficient to use.

Let's do some feature comparison—collections versus arrays:

- Custom collections you create use 1 by default as a first element. Arrays by default are zero-based. You need to use the Option Base 1 statement to force the numbering of array items to start at position 1.

- You don't need to specify the size of your collection upfront as collections are dynamically allocated. Arrays, on the other hand, require that you define their size, and if you need to change the array size further in your procedure, you must use the ReDim keyword. Each time you re-dimension the array, Excel takes up more resources.

- It's very easy to add or remove items from a collection with the Add and Remove methods. With arrays, before you can add or remove items, you need to find the size of the array by using its upper and lower bounds.

- You can add new items to a collection in any position. To perform the same task using arrays, you must write more code.

- Collections can store items of different data types. Arrays can only store items of different data types when they are declared as Variant.

- You can use a For and For Each loop to access items in a collection, while with arrays, you must first set and verify the upper and lower bounds to iterate through the items.

- Collections allow you to use keys to access a particular item directly, while arrays don't provide this feature.

SUMMARY

This chapter introduced you to the concept of collections in Excel VBA, demonstrating their creation, use, and how they compare to arrays. Collections provide a flexible way to manage groups of related items, which is essential for efficient coding. You learned how to create and initialize collections by using the `Collection` object and the `Set` statement. Next, you manipulated collections by adding, accessing, and removing items, and used the `Count` property to return the number of items in the collection. Additionally, you learned how to build an array and convert it into a collection. We compared collections versus arrays and stated that the choice between them depends on the specific needs of the task you want to accomplish.

As your procedures become more complex, you will need to start using special tools for tracing errors, which are covered in the next chapter.

Chapter **9**

Excel Tools for Testing and Debugging—A Quick Introduction to Testing VBA Programs

It does not take much time for an error to creep into your VBA procedure. The truth is that no matter how careful you are, it is rare that all your VBA procedures will work correctly the first time. There are three types of errors in VBA: syntax errors, logic errors, and runtime errors.

This chapter introduces you to the VBE tools that are available for you to use in the process of analyzing the code of your VBA procedures and locating the source of errors.

TESTING VBA PROCEDURES

Because most of the procedures we created earlier were quite short, finding errors wasn't very difficult. Locating the source of errors in longer and more complex procedures, however, is often more tedious and time-consuming. Fortunately, VBE provides a set of handy tools that can make the process of tracking down your VBA problems easier, faster, and less frustrating.

Bugs are errors in computer programs and *debugging* is the process of locating and fixing those errors by stepping through the code of your procedure or checking the values of variables.

When testing your VBA procedure, use the following guidelines:

- To analyze your procedure, step through your code one line at a time by pressing F8 or choosing Debug | Step Into.
- To locate an error in a specific place in your procedure, use a breakpoint.
- To monitor the value of a variable or expression used by your procedure, add a watch expression.
- To get to sections of code that interest you, set up a bookmark to jump quickly to the desired location.

Each of these guidelines is demonstrated in a hands-on scenario later in this chapter.

STOPPING A PROCEDURE

While testing your VBA procedure, you may want to halt its execution. This can be done by pressing Ctrl+Break or Ctrl+Pause on your keyboard while the procedure is running. VBA will then display the message shown in Figure 9.1. VBA also offers other methods of stopping your procedure. When you stop your procedure, you enter what is called *break mode*.

To enter break mode, do one of the following:

- Press the Ctrl+Break or Ctrl+Pause key combination.
- Set one or more breakpoints.
- Insert the `Stop` statement into your procedure code.
- Add a watch expression.

A break occurs when the execution of your VBA procedure is suspended. Visual Basic remembers the values of all variables and the statement from which the execution of the procedure should resume when the user decides to continue by clicking Run Sub/UserForm on the toolbar (or the Run menu option with the same name), or by clicking the Continue button in the dialog box.

The error dialog box shown in Figure 9.1 informs you that the procedure was halted. The buttons in this dialog are described in Table 9.1.

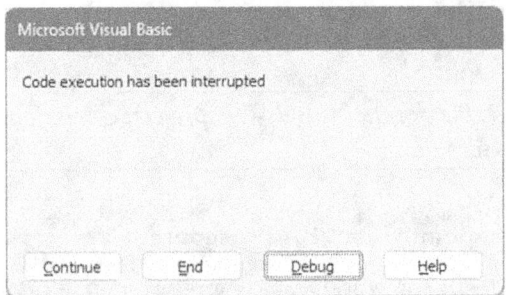

Microsoft Visual Basic

Code execution has been interrupted

Continue   End   Debug   Help

FIGURE 9.1.   This message appears when you press Ctrl+Break or Ctrl+Pause while your VBA procedure is running.

TABLE 9.1.   Error dialog buttons.

Continue	Click this button to resume code execution. This button will be grayed out if an error is encountered.
End	Click this button if you do not want to troubleshoot the procedure at this time. VBA will stop code execution.
Debug	Click this button to enter break mode. The Code window will appear, and VBA will highlight the line at which the procedure execution was suspended. You can examine, debug, reset, or step through the code.
Help	Click this button to view the online help that explains the cause of this error message.

You can prevent application users from halting your procedure by including the following statement in the procedure code:

```
Application.EnableCancelKey = xlDisabled
```

USING BREAKPOINTS

If you know more or less where you can expect a problem in the code of your procedure, suspend code execution on a given line by pressing F9. This will set a breakpoint on that line. When VBA gets to that line while running your procedure, it will immediately display the Code window. At this point, you can step through the procedure code line by line by pressing F8 or choosing Debug | Step Into.

To see how this works, let's look at the following scenario. Assume that during the execution of the ChangeCode procedure in Hands-On 9.1, the following line of code could get you in trouble:

```
ActiveCell.FormulaR1C1 "=VLOOKUP(RC[1],Codes.xlsx!R1C1:R6C2,2)"
```

NOTE	*Please note that the files for the "Hands-On" projects can be found in the companion files.*

⊚ Hands-On 9.1   Setting Breakpoints in a VBA Procedure

1. Copy the Chap09.xlsm and Codes.xlsx workbooks from the companion files to your C:\VBAExcel2024_ByExample folder.
2. Start Microsoft Excel and open both of these files (Chap09.xlsm and Codes.xlsx) from the C:\VBAExcel2024_ByExample folder.
3. Examine the data in both workbooks. It should look like Figures 9.2 and 9.3.

⊿	A	B	C	D
1	Teacher	Position	Amount	Code
2	Ann Marie Smith	A	6500	227.163-23-220
3	Barbara Kaufman	A	6500	227.163-14-100
4	John Frederick	A	6500	211.163-23-330
5	Katherine Stein	B	6300	211.163-23-330
6	Christine Martin	B	6300	211.163-23-330
7	Mark O'Brian	B	6300	211.163-23-220
8	Jorge Rodriguez	B	6300	227.163-11-100
9				

FIGURE 9.2.   The data entered in column D of this worksheet will be used to look up the equivalent code from column B in the Codes.xlsx workbook (see Figure 9.3).

4. Close the Codes.xlsx workbook. Leave the other file open.
5. With Chap09.xlsm active, switch to the VBE window.
6. In Project Explorer, open the Modules folder in the Debugging (Chap09.xlsm) project and double-click the Breaks module.

The Breaks module code window lists the following ChangeCode procedure:

⊿	A	B
1	211.163-23-220	65
2	211.163-23-330	73
3	211.163-28-330	78
4	227.163-11-100	67
5	227.163-14-100	62
6	227.163-23-220	66
7		

FIGURE 9.3. The ChangeCode procedure uses this code table for lookup purposes.

```
Sub ChangeCode()
  Workbooks.Open Filename:="C:\VBAExcdl2024_ByExample\Codes.xlsx"
  Windows("Chap09_ExcelPrimer.xlsm").Activate
  Columns("D:D").Insert Shift:=xlToRight
  Range("D1").Formula = "Simple Code"
  Columns("D:D").SpecialCells(xlBlanks).Select
  ActiveCell.FormulaR1C1 = "=VLookup(RC[1],Codes.xlsx!R1C1:R6C2,2)"
  Selection.FillDown
    With Columns("D:D")
      .EntireColumn.AutoFit
      .Select
    End With
  Selection.Copy
  Selection.PasteSpecial Paste:=xlValues
  Rows("1:1").Select
    With Selection
      .HorizontalAlignment = xlCenter
      .VerticalAlignment = xlBottom
      .Orientation = xlHorizontal
    End With
  Workbooks("Codes.xlsx").Close
End Sub
```

7. In the ChangeCode procedure, click anywhere on the line containing the following statement:

```
ActiveCell.FormulaR1C1 = "=VLookup(RC[1], _
    Codes.xlsx!R1C1:R6C2,2)"
```

8. Set a breakpoint by pressing F9 (or choosing Debug | Toggle Breakpoint or clicking on the margin indicator to the left of the line).

When you set the breakpoint, Visual Basic displays a red circle in the margin. At the same time, the line that has the breakpoint is indicated as white text on a red background, as in Figure 9.4. The color of the breakpoint can be changed on the Editor Format tab in the Options dialog box (Tools menu).

9. Press F5 to run the ChangeCode procedure.

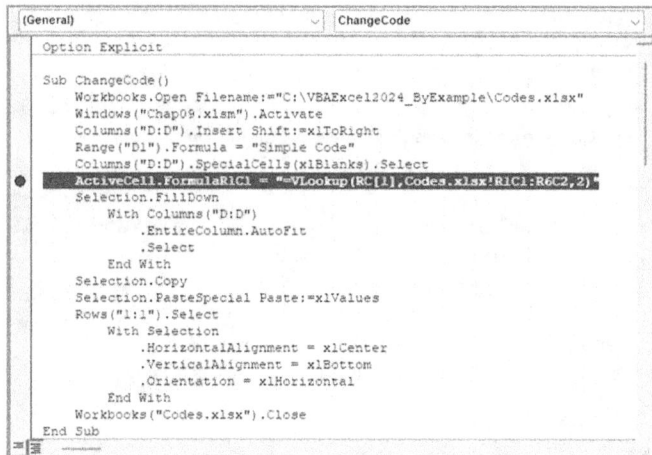

```
(General)                                    ChangeCode

Option Explicit

Sub ChangeCode()
    Workbooks.Open Filename:="C:\VBAExcel2024_ByExample\Codes.xlsx"
    Windows("Chap09.xlsm").Activate
    Columns("D:D").Insert Shift:=xlToRight
    Range("D1").Formula = "Simple Code"
    Columns("D:D").SpecialCells(xlBlanks).Select
    ActiveCell.FormulaR1C1 = "=VLookup(RC[1],Codes.xlsx!R1C1:R6C2,2)"
    Selection.FillDown
        With Columns("D:D")
            .EntireColumn.AutoFit
            .Select
        End With
    Selection.Copy
    Selection.PasteSpecial Paste:=xlValues
    Rows("1:1").Select
        With Selection
            .HorizontalAlignment = xlCenter
            .VerticalAlignment = xlBottom
            .Orientation = xlHorizontal
        End With
    Workbooks("Codes.xlsx").Close
End Sub
```

FIGURE 9.4. The line of code where the breakpoint is set is displayed in the color specified on the Editor Format tab in the Options dialog box.

```
(General)                                    ChangeCode

Option Explicit

Sub ChangeCode()
    Workbooks.Open Filename:="C:\VBAExcel2024_ByExample\Codes.xlsx"
    Windows("Chap09.xlsm").Activate
    Columns("D:D").Insert Shift:=xlToRight
    Range("D1").Formula = "Simple Code"
    Columns("D:D").SpecialCells(xlBlanks).Select
    ActiveCell.FormulaR1C1 = "=VLookup(RC[1],Codes.xlsx!R1C1:R6C2,2)"
    Selection.FillDown
        With Columns("D:D")
            .EntireColumn.AutoFit
            .Select
        End With
    Selection.Copy
    Selection.PasteSpecial Paste:=xlValues
    Rows("1:1").Select
        With Selection
            .HorizontalAlignment = xlCenter
            .VerticalAlignment = xlBottom
            .Orientation = xlHorizontal
        End With
    Workbooks("Codes.xlsx").Close
End Sub
```

FIGURE 9.5. When Visual Basic encounters a breakpoint, it displays the Code window and indicates the current statement.

When you run the procedure, Visual Basic will execute all the statements until it encounters the breakpoint. Figure 9.5 shows the yellow arrow in the margin to the left of the statement at which the procedure was suspended, and the statement inside a box with a yellow background. The arrow and the box indicate the current statement or the statement that is about to be executed. If the current statement also contains a breakpoint, the margin displays both indicators overlapping one another (the circle and the arrow).

While in break mode, you can change code, add new statements, execute the procedure one line at a time, skip lines, set the next statement, use the Immediate window, and more. When Visual Basic is in break mode, all the options on the Debug menu are available.

If you change certain code, VBA will prompt you to reset the project by displaying the following error message: "This action will reset your project, proceed anyway?" You can click OK to stop the program's execution and proceed with editing your code or click Cancel to delete the new changes and continue running the code from the point at which it was suspended.

10. Press F5 (or choose Run Sub/UserForm) to continue running the procedure. Visual Basic leaves break mode and continues to run the procedure statements until it reaches the end of the procedure. When the procedure finishes executing, Visual Basic does not automatically remove the breakpoint. Notice that the line of code with the VLookup function is still highlighted in red.

In this example, you have set only one breakpoint, but Visual Basic allows you to set any number of breakpoints in your VBA procedure. In this way, you can suspend and continue the execution of your procedure as you please. You can analyze the code of your procedure and check the values of variables while execution is suspended. You can also perform various tests by typing statements in the Immediate window.

11. Remove the breakpoint by choosing Debug | Clear All Breakpoints or pressing Ctrl+Shift+F9 or clicking on the red circle in the margin area to remove the breakpoint.

All the breakpoints are removed.

NOTE	*If you had set several breakpoints in a given procedure and would like to remove only one or some of them, click on the line containing the breakpoint that you want to remove and press F9 (or choose Debug | Clear Breakpoint or simply click the red dot in the margin). You should clear the breakpoints when they are no longer needed. The breakpoints are automatically removed when you close the file.*

12. Switch to the Microsoft Excel application window and notice that a new column with the looked-up codes, like the one in Figure 9.6, was added on Sheet1 of the Chap09.xlsm workbook.

	A	B	C	D	E
1	Teacher	Position	Amount	Simple Code	Code
2	Ann Marie Smith	A	6500	66	227.163-23-220
3	Barbara Kaufman	A	6500	62	227.163-14-100
4	John Frederick	A	6500	73	211.163-23-330
5	Katherine Stein	B	6300	73	211.163-23-330
6	Christine Martin	B	6300	73	211.163-23-330
7	Mark O'Brian	B	6300	65	211.163-23-220
8	Jorge Rodriguez	B	6300	67	227.163-11-100

FIGURE 9.6.   This worksheet was modified by the ChangeCode procedure in Hands-On 9.1.

When to Use a Breakpoint

Consider setting a breakpoint if you suspect that your procedure never executes a certain block of code. In break mode, you can quickly find out the contents of the variable at the cursor in the Code window by holding the mouse pointer over it.

For example, in the VarValue procedure shown in Figure 9.7, the breakpoint has been set on the Workbooks(strName).Activate statement. When Visual Basic encounters this statement, the Chap09.xlsm [break] window appears. Because Visual Basic has already executed the statement that stores the name of ActiveWorkbook in the variable strName, you can quickly find out the value of this variable by resting the mouse pointer over the variable name. The name of the variable and its current value appear in a tooltip frame.

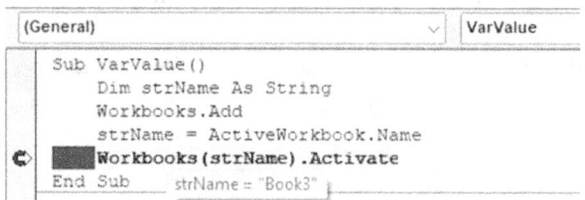

```
(General)                                        VarValue

    Sub VarValue()
        Dim strName As String
        Workbooks.Add
        strName = ActiveWorkbook.Name
        Workbooks(strName).Activate
    End Sub      strName = "Book3"
```

FIGURE 9.7.   In break mode, you can find out the value of a variable by resting the mouse pointer on that variable.

_____ **NOTE**	*To show the values of several variables used in a procedure at once, you should use the Locals window, which is discussed later in this chapter.*

USING THE IMMEDIATE WINDOW IN BREAK MODE

Once the procedure execution is suspended and the Code window appears, you can activate the Immediate window and type VBA instructions to find out, for instance, which cell is currently active or ask for the name of the active sheet. You can also use the Immediate window to change the contents of variables in order to correct values that may be causing errors.

Figure 9.8 shows the suspended `ChangeCode` procedure and the Immediate window with the questions that were asked of Visual Basic while in break mode.

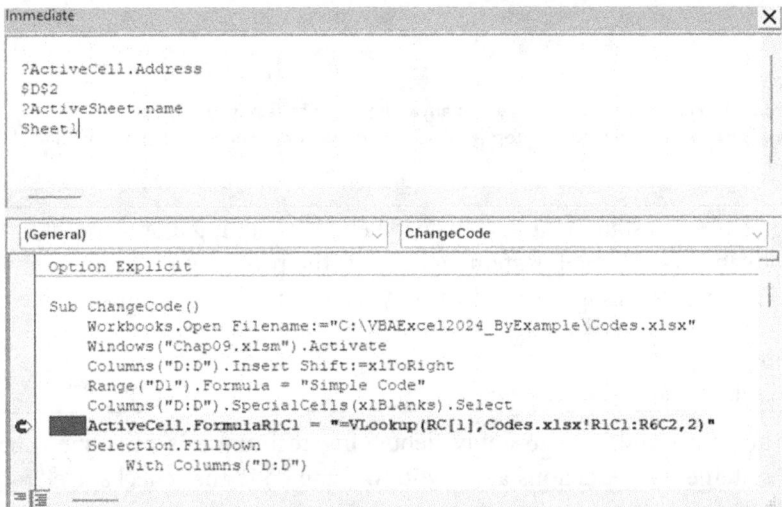

FIGURE 9.8. When the code execution is suspended, you can find the values of your variables and execute additional commands by entering appropriate statements in the Immediate window.

USING THE STOP AND ASSERT STATEMENTS

Sometimes, you won't be able to test your procedure right away. If you set up your breakpoints and then close the file, Excel will remove your breakpoints, and the next time you are ready to test your procedure, you'll have to begin by setting up breakpoints again.

Using the Stop Statement

To postpone the task of testing your procedure until you reopen the file, insert a `Stop` statement into your code wherever you want to halt a procedure. Figure 9.9 shows a `Stop` statement placed before the `For Each...Next` loop. Visual Basic will suspend the execution of the `StopExample` procedure when it encounters the `Stop` statement. The screen will display the Code window in break mode.

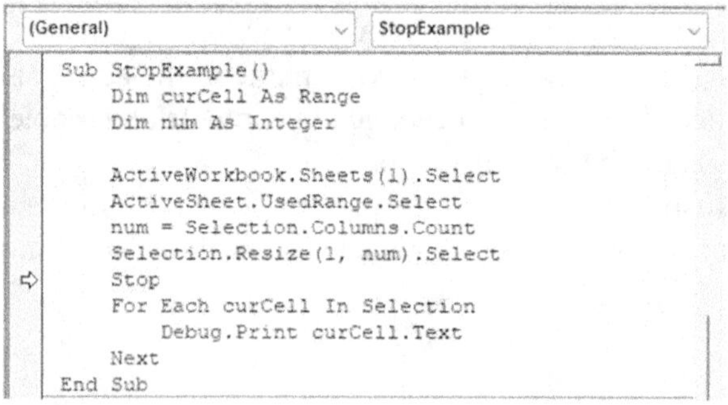

```
(General)                          ∨    StopExample                   ∨

    Sub StopExample()
        Dim curCell As Range
        Dim num As Integer

        ActiveWorkbook.Sheets(1).Select
        ActiveSheet.UsedRange.Select
        num = Selection.Columns.Count
        Selection.Resize(1, num).Select
⇨       Stop
        For Each curCell In Selection
            Debug.Print curCell.Text
        Next
    End Sub
```

FIGURE 9.9.   You can insert a Stop statement anywhere in the code of your VBA procedure. The procedure will halt when it gets to the Stop statement, and the Code window will appear with the line highlighted.

Although the `Stop` statement has the same effect as setting a breakpoint, it has one disadvantage—all `Stop` statements stay in the procedure until you remove them. When you no longer need to stop your procedure, you must locate and remove all the `Stop` statements.

Using the Debug.Assert Statement

A very powerful and easy-to-apply debugging technique is utilizing `Debug.Assert` statements. Assertions allow you to write code that checks itself while running. By including assertions in your programming code, you can verify that a particular condition or assumption is true.

Assertions give you immediate feedback when an error occurs. They are great for detecting logic errors early in the development phase instead of hearing about them later from your end users. The fact that your procedure ran on your system without generating an error does not mean that there are no bugs in that procedure. Don't assume anything—always test for the validity of expressions and variables in your code.

The `Debug.Assert` statement takes any expression that evaluates to `True` or `False` and activates break mode when that expression evaluates to `False`. The syntax for `Debug.Assert` is shown here:

```
Debug.Assert condition
```

where `condition` is a VBA code or expression that returns `True` or `False`. If `condition` evaluates to `False` or `0` (zero), VBA will enter break mode. For example, when running the following looping structure, the code will stop executing when the variable i equals 50:

```
Sub TestDebugAssert()
    Dim i As Integer

    For i = 1 To 100
       Debug.Assert i <> 50
    Next
End Sub
```

Keep in mind that Debug.Assert does nothing if the condition is False or zero. The execution simply stops on that line of code and the VBE screen opens with the line containing the false statement highlighted so that you can start debugging your code. You may need to write an error handler to handle the identified error. Error-handling techniques are discussed later in this chapter.

Debug.Assert differs from the Stop statement in its conditional aspect; it will stop your code only under specific conditions. Conditional breakpoints can also be set by using the Watches window (see the next section).

After you have debugged and tested your code, comment out or remove the Debug.Assert statements from your final code. The easiest way to do this is to use Edit | Replace on the VBE screen:

- To comment out the statements, enter Debug.Assert in the Find What box. In the Replace With box, enter an apostrophe followed by Debug. Assert.

- To remove the Debug.Assert statements from your code, enter Debug. Assert in the Find What box. Leave the Replace With box empty but be sure to mark the Use Pattern Matching checkbox.

USING THE WATCHES WINDOW

Many errors in procedures are caused by variables that assume unexpected values. If a procedure uses a variable whose value changes in various locations, you may want to stop the procedure and check the current value of that variable.

Visual Basic offers a special Watches window that allows you to keep an eye on variables or expressions while your procedure is running. To add a watch expression to your procedure, perform the following:

1. In the Code window, select the variable whose value you want to monitor.
2. Choose Debug | Add Watch.

The screen will display the Add Watch dialog box, as shown in Figure 9.10. The Add Watch dialog box contains three sections, which are described in Table 9.2.

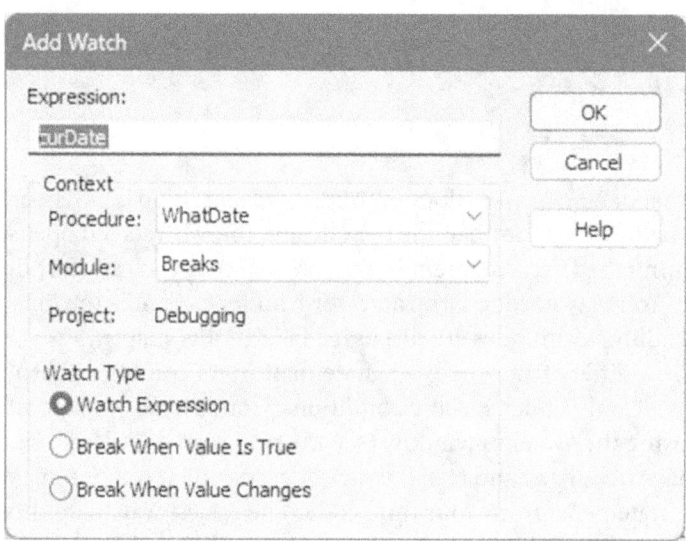

FIGURE 9.10. The Add Watch dialog box allows you to define conditions that you want to monitor while a VBA procedure is running.

TABLE 9.2. Add Watch dialog options.

Expression	Displays the name of a variable that you have highlighted in your procedure. If you opened the Add Watch dialog box without selecting a variable name, type the name of the variable you want to monitor in the Expression text box.
Context	In this section, you should indicate the name of the procedure that contains the variable and the name of the module where this procedure is located.
Watch Type	Specifies how to monitor the variable. If you choose the Watch Expression option button, you will be able to read the value of the variable in the Watches window while in break mode. If you choose Break When Value Is True, Visual Basic will automatically stop the procedure when the variable evaluates to `True` (nonzero). The last option button, Break When Value Changes, stops the procedure each time the value of the variable or expression changes.

You can add a watch expression before running a procedure or after the execution of your procedure has been suspended. The difference between a breakpoint and a watch expression is that the breakpoint always stops a procedure in a specified location and the watch stops the procedure only when the specified condition (Break When Value Is True or Break When Value Changes) is met.

Watches are extremely useful when you are not sure where the variable is being changed. Instead of stepping through many lines of code to find the location where the variable assumes the specified value, you can simply put a watch expression on the variable and run your procedure as normal. Let's see how this works.

⊙ Hands-On 9.2   Watching the Values of VBA Expressions

1. The `Breaks` module in the `Chap09.xlsm` workbook lists the following `WhatDate` procedure:

```
Sub WhatDate()
  Dim curDate As Date
  Dim newDate As Date
  Dim x As Integer

  curDate = Date
  For x = 1 To 365
    newDate = Date + x
  Next
End Sub
```

The `WhatDate` procedure uses the `For…Next` loop to calculate the date that is x days in the future. If you run this procedure, you won't get any results unless you insert the following instruction in the code of the procedure:

```
MsgBox "In " & x & " days, it will be " & NewDate
```

In this example, however, you don't care to display the individual dates, day after day. What if all you want to do is stop the program when the value of the variable x reaches `211`?

In other words, what date will it be 211 days from now? To get the answer, you could insert the following statement into your procedure:

```
If x = 211 Then MsgBox "In " & x & " days it will be " & NewDate
```

Introducing new statements into your procedure just to get an answer about the value of a certain variable when a specific condition occurs will not always be viable.

Instead of adding `MsgBox` or other `Debug` statements to your procedure code that you will later need to delete, you can use the Watches window and avoid extra code maintenance. If you add watch expressions to the procedure, Visual Basic will stop the `For…Next` loop when the specified condition is met, and you'll be able to check the values of the desired variables.

2. In the VBE screen, choose View | Watch Window.
An empty Watches window should appear. This window is divided into four areas: Expression, Value, Type, and Context. The Expression column will list all the watch expressions that you'll set in this Hands-On. The Value column will list the value of the expression at the time of transition into break mode. The Type column will list the expression type, and the Context column will list the context of the watch expression.

Let's set a few watches for the WhatDate procedure.

3. Choose Debug | Add Watch.

4. In the Expression text box, enter the following expression: x = 211. In the Context section, choose WhatDate from the Procedure combo box and Breaks from the Module combo box if they are not already selected. In the Watch Type section, select the Break When Value Is True option button.

5. Click OK to close the Add Watch dialog box.
You have now added your first watch expression.

Visual Basic places your expression x = 211 in the first line of the Watches window. Notice that the Value column will show an <Out of context> entry. This entry will change as you run your procedure. Now, let's add another expression for tracking the current date.

6. In the Code window, position the insertion point anywhere within the name of the curDate variable.

7. Choose Debug | Add Watch and click OK to set up the default watch type as Watch Expression.
Notice that a new line is added to the Watches window with the curDate variable in the Expression column. Let's add another variable to the Watches window.

8. In the Code window, position the insertion point anywhere within the name of the newDate variable.

9. Choose Debug | Add Watch and click OK to set up the default watch type as Watch Expression.
Notice that newDate now appears in the Expression column of the Watches window.

After performing the preceding steps, the WhatDate procedure contains the following three watches:

```
x = 211—Break When Value is True
curDate—Watch Expression
newDate—Watch Expression
```

10. Position the insertion point anywhere inside the code of the `WhatDate` procedure, and press F5.

Figure 9.11 shows the Watches window when Visual Basic stops the procedure when x equals `211`.

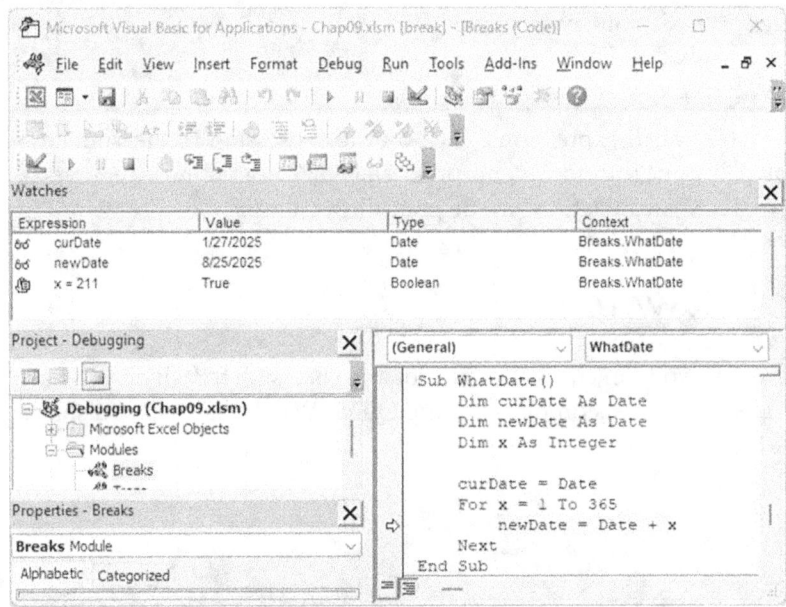

FIGURE 9.11. Using the Watches window.

Notice that the value of the variable x in the Watches window is the same as the value that you specified in the Add Watch dialog. In addition, the Watches window shows the value of both variables—`curDate` and `newDate`. The procedure is in break mode. You can press F5 to continue or you can ask another question, such as "What date will it be in 277 days?" The next step shows how to do this.

11. Choose Debug | Edit Watch and enter the following expression: x = 277.

12. Click OK to close the Edit Watch dialog box.

Notice that the Watches window now displays a new value for the expression. x is now `False`.

13. Press F5 to continue running the procedure.

The procedure stops again when the value of x equals `277`. The value of `curDate` is the same; however, the `newDate` variable now contains a new value date that is 277 days from now. You can change the value of the expression again or finish running the procedure.

14. Press F5 to finish running the procedure.

When your procedure is running and a watch expression has a value, the Watches window displays the value of the watch expression. If you open the Watches window after the procedure has finished, you will see <out of context> instead of the variable values. In other words, when the watch expression is out of context, it does not have a value.

Removing Watch Expressions

To remove the watch expressions, click on the expression in the Watches window that you want to remove and press Delete. You may now remove all the watch expressions you defined in the preceding example.

USING QUICK WATCH

In break mode, you can check the value of an expression for which you have not defined a watch expression by using the Quick Watch dialog box displayed in Figure 9.12.

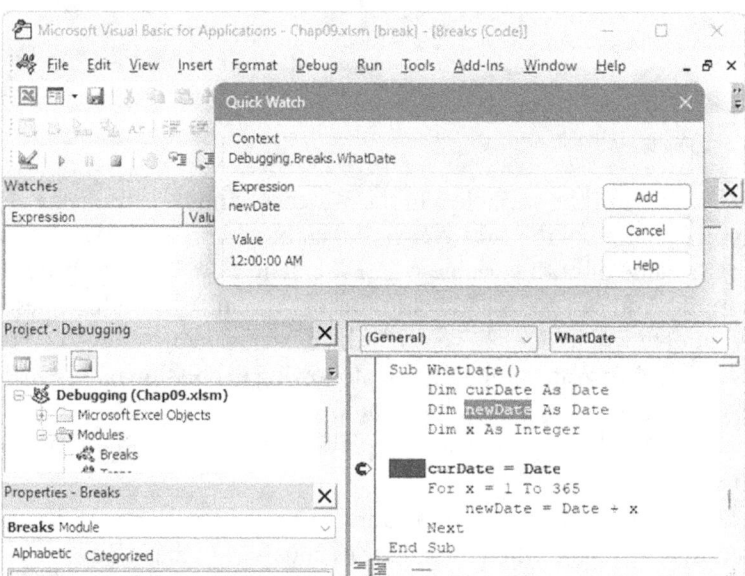

FIGURE 9.12. The Quick Watch dialog box shows the value of the selected expression in a VBA procedure.

The Quick Watch dialog box can be accessed in one of the following ways:

- While in break mode, position the insertion point anywhere inside the name of a variable or expression you wish to watch.
- Choose Debug | Quick Watch.
- Press Shift+F9.

The Add button in the Quick Watch dialog box allows you to add the expression to the Watches window.

⊙ Hands-On 9.3   Using the Quick Watch Dialog Box

1. Make sure that the `WhatDate` procedure you entered in the previous Hands-On exercise does not contain any watch expressions that we defined earlier. See the section called "Removing Watch Expressions" for instructions on how to remove a watch expression from the Watches window.
2. In the `WhatDate` procedure, position the insertion point on the name of the variable `x`.
3. Choose Debug | Add Watch.
4. Enter the following expression: `x = 50`.
5. Choose the Break When Value Is True option button and click OK.
6. Run the `WhatDate` procedure.
   Visual Basic will suspend procedure execution when `x` equals `50`. Notice that the Watches window does not contain the `newDate` or `curDate` variables. To check the values of these variables, you can position the mouse pointer over the appropriate variable name in the code window, or you can invoke the Quick Watch dialog box.
7. In the Code window, position the mouse pointer inside the `newDate` variable and press Shift+F9 or choose Debug | Quick Watch.
   The Quick Watch dialog shows the name of the expression and its current value.
8. Click Cancel to return to the Code window.
9. In the Code window, position the mouse pointer inside the `curDate` variable and press Shift+F9.
   The Quick Watch dialog now shows the value of the variable `curDate`.
10. Click Cancel to return to the Code window.
11. Press F5 to continue running the procedure.
12. In the Watches window, highlight the line containing the expression `x = 50` and press Delete to remove it.
13. Close the Watches window.

THE LOCALS AND CALL STACK WINDOWS

To keep an eye on all the declared variables and their current values, choose View | Locals Window before you run the procedure. Figure 9.13 shows a list of variables and their corresponding values in the Locals window, which is displayed while Visual Basic is in break mode. Notice that the Locals window contains three columns.

The Expression column displays the names of variables that are declared in the current procedure. The first row displays the name of the module preceded by the plus sign. When you click the plus sign, you can check whether any variables have been declared at the module level.

For class modules, which are discussed later in this book, the system variable Me will be defined. For standard modules, the first variable is the name of the current module. Note that the global variables and variables in other projects are not accessible from the Locals window.

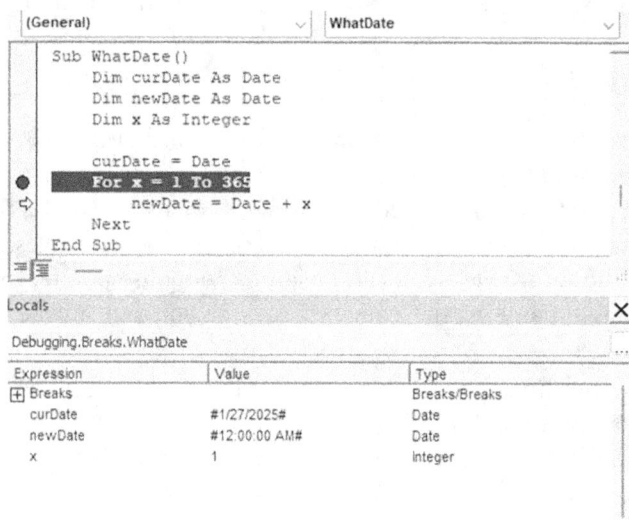

FIGURE 9.13.   The Locals window displays the current values of all the declared variables in the current VBA procedure.

The second column shows the current values of variables. In this column, you can change the value of a variable by clicking it and typing the new value. After changing the value, press Enter to save the change. You can also press Tab, Shift+Tab, or the up or down arrows, or click anywhere within the Locals window after you've changed the variable value. The third column displays the type of declared variable.

⊚ Hands-On 9.4   Using the Locals and Call Stack Windows

1. Choose View | Locals Window.
2. Click anywhere inside the WhatDate procedure and press F8.

   By pressing F8, you place the procedure in break mode. The Locals window displays the name of the current module, the local variables, and their beginning values.
3. Press F8 a few more times while keeping an eye on the Locals window.

   Notice that the Locals window also contains a button with three dots. This button opens the Call Stack dialog box shown in Figure 9.14, which displays a list of all active procedure calls. An active procedure call is a procedure that is started but not completed. You can also activate the Call Stack dialog box by choosing View | Call Stack. This option is only available in break mode.

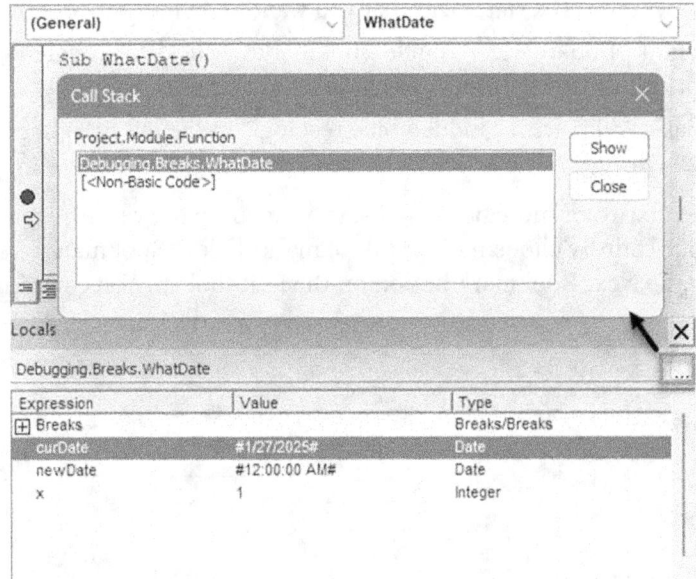

FIGURE 9.14.   The Call Stack dialog box displays a list of the procedures that are started but not completed. Clicking the three dots button in the Locals window activates the Call Stack dialog box.

The Call Stack dialog box is especially helpful for tracing nested procedures. Recall that a nested procedure is a procedure that is being called from within another procedure. If a procedure calls another, the name of the called procedure is automatically added to the Calls list in the Call Stack dialog box. When Visual Basic has finished executing the statements of the called procedure, the procedure name is automatically removed from the Call Stack dialog box. You can use the Show button in the Call Stack dialog box to display the statement that calls the next procedure listed in the dialog box.

4. Press F5 to continue running the `WhatDate` procedure.
5. Close the Locals window.

NAVIGATING WITH BOOKMARKS

In the process of analyzing or reviewing your VBA procedures, you will often find yourself jumping to certain areas of code. Using the built-in bookmark feature, you can easily mark the spots in your code that you want to navigate between.

To set up a bookmark:

- Click anywhere in the statement that you want to define as a bookmark.
- Choose Edit | Bookmarks | Toggle Bookmark or click the Toggle Bookmark button on the Edit toolbar, as illustrated in Figure 9.15.
- Visual Basic will place a rounded blue rectangle in the left margin beside the statement.

Once you've set up two or more bookmarks, you can jump between the marked locations of your code by choosing Edit | Bookmarks | Next Bookmark or simply by clicking the Next Bookmark button on the Edit toolbar. You can remove

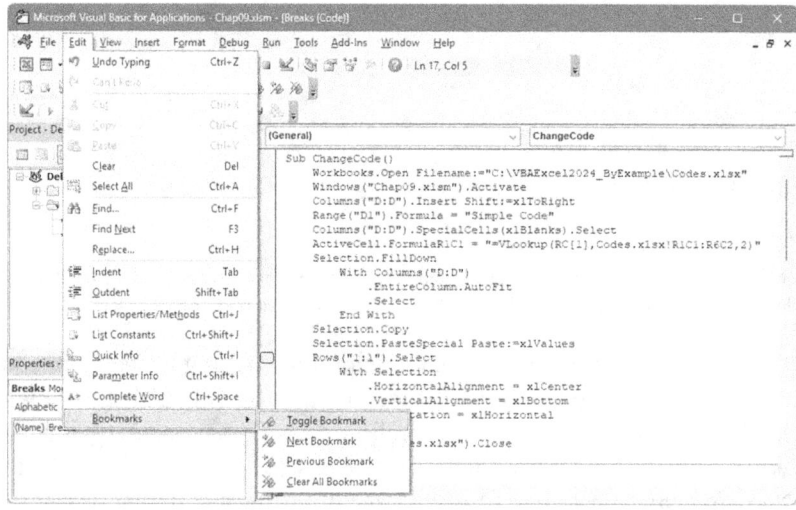

FIGURE 9.15. Set a bookmark to quickly jump between often-used sections of your procedures.

bookmarks at any time by choosing Edit | Bookmarks | Clear All Bookmarks or by clicking the Clear All Bookmarks button on the Edit toolbar. To remove a single bookmark, click anywhere in the bookmarked statement and choose Edit | Bookmarks | Toggle Bookmark or click the Toggle Bookmark button on the Edit toolbar.

TRAPPING ERRORS

No one writes bug-free programs the first time. When you write VBA procedures, you must determine how your program will respond to errors. Many unexpected errors happen during the runtime.

For example, your procedure may try to give a workbook the same name as an open workbook. Runtime errors are often discovered by users who attempt to do something that the programmer has not anticipated. If an error occurs when the procedure is running, Visual Basic displays an error message and the procedure is stopped. Most often, the error message that VBA displays is quite cryptic to the user. You can prevent users from seeing many runtime errors by including error-handling code in your VBA procedures. This way, when Visual Basic encounters an error, instead of displaying a default error message, it will show a much friendlier and more comprehensive error message.

In programming, mistakes and errors are not the same thing. A mistake, such as a misspelled or missing statement, a misplaced quote or comma, or assigning a value of one type to a variable of a different (and incompatible) type, can be removed from your program through proper testing and debugging. Even though your code may be free of mistakes, this does not mean that errors will not occur. An error is the result of an event or an operation that doesn't work as expected. For example, if your VBA procedure accesses a particular file on a disk and someone has deleted this file or moved it to another location, you'll get an error no matter what. An error prevents the procedure from carrying out a specific task.

To implement error handling, place the `On Error` statement in your procedure. This statement tells VBA what to do if an error occurs while your program is running. VBA uses the `On Error` statement to activate your error-handling procedure that will trap runtime errors. Depending on the type of procedure, you can exit the error trap by using one of the following statements: `Exit Sub`, `Exit Function`, `Exit Property`, `End Sub`, `End Function`, or `End Property`. You should write an error-handling routine for each procedure.

Table 9.3 shows how the `On Error` statement can be used.

TABLE 9.3. On Error statement options.

`On Error GoTo Label`	Denotes a label to jump to when an error occurs. This label marks the beginning of the error-handling code. An error handler is a routine for trapping and responding to errors in your application. The label must appear in the same procedure as the `On Error` statement.
`On Error Resume Next`	When a runtime error occurs, Visual Basic ignores the line that caused the error and does not display an error message but continues the procedure with the next line.
`On Error GoTo 0`	Turns off error trapping in a procedure. When VBA runs this statement, errors are detected but not trapped within the procedure.

Using the Err Object

Your error-handling code can utilize various properties and methods of the `Err` object. For example, to check which error occurred, check the value of `Err.Number`. The `Number` property of the `Err` object will tell you the value of the last error that occurred, and the `Description` property will return a description of the error. You can also find the name of the application that caused the error by using the `Source` property of the `Err` object (this is very helpful when your procedure launches other applications).

After handling the error, use the `Err.Clear` statement to reset `Err.Number` back to zero.

To test your error-handling code, use the `Raise` method of the `Err` object. For example, to raise the "Disk not ready" error, use the following statement:

```
Err.Raise 71
```

The `OpenToRead` procedure in Hands-On 9.5 demonstrates the use of the `Resume Next` and `Error` statements, as well as the `Err` object.

⊙ Hands-On 9.5 Writing a VBA Procedure with Error-Handling Code

1. Review the code of the `OpenToRead` procedure in the `Traps` module of the `Chap09.xlsm` workbook:

```
Sub OpenToRead()
Dim myFile As String
Dim myChar As String
Dim myText As String
Dim FileExists As Boolean

FileExists = True
```

```
On Error GoTo ErrorHandler

myFile = InputBox("Enter the name of file to open:")
Open myFile For Input As #1
If FileExists Then
' loop until the end of file (EOF)
    Do While Not EOF(1)
       ' get one character
       myChar = Input(1, #1)
       ' store in the variable myText
       myText = myText + myChar
    Loop
    Debug.Print myText
    ' close the file
    Close #1
End If
Exit Sub

ErrorHandler:
  FileExists = False
  Select Case Err.Number
    Case 76
    MsgBox "The path you entered cannot be found."
    Case 53
    MsgBox "This file can't be found on the " & _
        "specified drive."
    Case 75
      Exit Sub
    Case Else
      MsgBox "Error " & Err.Number & " :" & _
        Error(Err.Number)
    Exit Sub
  End Select
  Resume Next
End Sub
```

The OpenToRead procedure reads the contents of the user-supplied text file character by character. When the user enters a filename, various errors can occur. For example, the filename or the path may be wrong, or the user may try to open a file that is already open. To trap these errors, the error-handling routine at the end of the OpenToRead procedure uses the Number property of the Err object.

There are several methods you can use in VBA for reading a text file. In this example, to read data from a text file, the procedure uses the Windows low-level file I/O (Input/Output) method. To open the file for reading, we use the Open statement, like this:

```
Open myFile For Input As #1
```

Here's the general syntax of the Open statement, followed by an explanation of each component:

```
Open pathname For mode[Access access][lock] As [#]filenumber
    [Len=reclength]
```

The Open statement has three required arguments: pathname, mode, and filenumber. pathname is the name of the file you want to open. The filename may include the name of a drive and folder.

- mode is a keyword that determines how the file was opened. Use Input mode to read the file, Output to write to a file by overwriting any existing file, and Append to write to a file by adding to any existing information.

- The optional Access clause can be used to specify permissions for the file (Read, Write, or Read Write).

- The optional lock argument determines which file operations are allowed for other processes. For example, if a file is open in a network environment, lock determines how other people can access it. The following lock keywords can be used: Shared, Lock Read, Lock Write, or Lock Read Write.

- filenumber is a number from 1 to 511. This number is used to refer to the file in subsequent operations. You can obtain a unique file number using the Visual Basic built-in FreeFile function.

- The last element of the Open statement, reclength, specifies the buffer size (total number of characters) for sequential (text) files, or the record size for random-access files (text files where data is stored in records of equal length and fields separated by commas).

If the specified file exists, the procedure uses the Do...While loop to tell Visual Basic to execute the statements inside the loop until the end of the file has been reached. The end of the file is determined by the result of the EOF function. The Input function is used to return the specified number of characters:

```
myChar = Input(1, #1)
```

#1 is the file number that was used in the process of opening the file with the Open statement.

Each character being read is stored in the `myChar` variable. Next, the `myChar` variable is appended to the `myText` variable, like this:

```
myText = myText + myChar
```

The procedure then writes the contents of the `myText` variable to the Immediate window using the `Debug.Print` statement. When the file has been read, we must close it using the `Close` statement:

```
Close #1     ' close the file
```

The `Err` object contains information about runtime errors. If an error occurs while the procedure is running, the statement `Err.Number` will return the error number.

If errors 76, 53, or 75 occur, Visual Basic will display user-friendly messages stored inside the `Select...Case` block and then proceed to the `Resume Next` statement, which will send it to the line of code following the one that caused the error.

If another error occurs, Visual Basic will return its error code (`Err.Number`) and error description (`Error (Err.Number)`).

At the beginning of the procedure, the variable `FileExists` is set to `True`. This way, if the program doesn't encounter an error, all the instructions inside the `If FileExists Then` block will be executed. If VBA encounters an error, however, the value of the `FileExists` variable will be set to `False` (see the first statement in the error-handling routine just below the `ErrorHandler` label). This way, Visual Basic will not cause another error while trying to read a file that caused the error on opening.

Notice the `Exit Sub` statement before the `ErrorHandler` label. Put the `Exit Sub` statement just above the error-handling routine because you don't want Visual Basic to carry out the error handling if there are no errors.

To test the `OpenToRead` procedure and better understand error trapping, we will need a text file (see Step 3).

2. Use Windows Notepad to prepare a text file. Enter any text you want in this file. When done, save the file as `C:\VBAExcel2024_ByExample\Readme.txt`.

3. Run the `OpenToRead` procedure three times in step mode by using the F8 key, each time supplying one of the following:

   - Name of the `C:\VBAExcel2024_ByExample\Readme.txt` file
   - Filename that does not exist on drive C
   - Path that does not exist on your computer (e.g., `K:\Test`)

Setting Error Trapping Options in a VBA Project

You can specify the error-handling settings for your current Visual Basic project by choosing Tools | Options and selecting the General tab (shown in Figure 9.16).

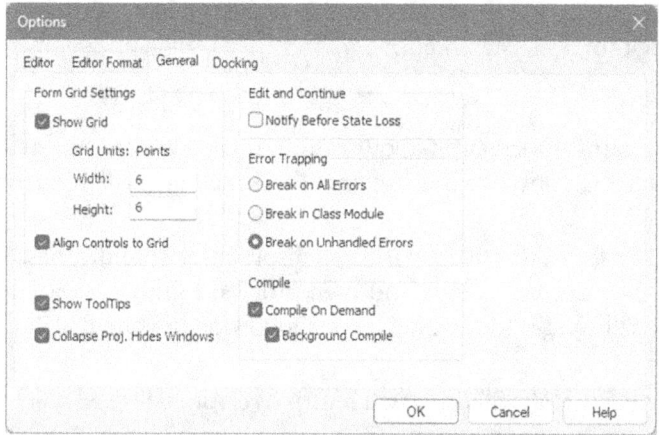

FIGURE 9.16.   Setting the Error Trapping options in the Options dialog box will affect all instances of Visual Basic started after you change the setting.

The Error Trapping area on the General tab determines how errors are handled in the Visual Basic environment. The following options are available:

- Break on All Errors:

  This setting will cause Visual Basic to enter the break mode on any error, whether an error handler is active or whether the code is in a class module (class modules are covered later in this book).

- Break in Class Module:

  This setting will trap any unhandled error in a class module. Visual Basic will activate a break mode when an error occurs and will highlight the line of code in the class module that produced this error.

- Break on Unhandled Errors:

  This setting will trap errors for which you have not written an error handler. The error will cause Visual Basic to activate a break mode. If the error occurs in a class module, the error will cause Visual Basic to enter break mode on the line of code that invoked the offending procedure of the class.

STEPPING THROUGH VBA PROCEDURES

Stepping through the code means running one statement at a time. This allows you to check every line in every procedure that is encountered. To start stepping through a procedure from the beginning, place the insertion point anywhere inside the code of your procedure and choose Debug | Step Into or press F8.

Figure 9.17 shows the Debug menu, which contains several options that allow you to execute a procedure in step mode. When you run a procedure one statement at a time, Visual Basic executes each statement until it encounters the End Sub keywords. If you don't want Visual Basic to step through every statement, you can press F5 at any time to run the rest of the procedure without stepping through it.

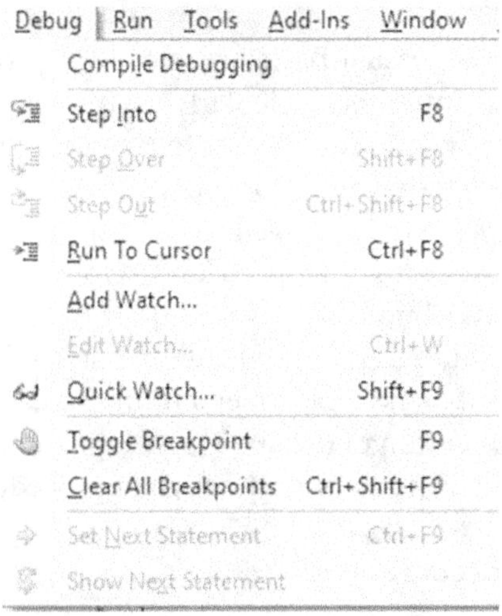

Debug	Run	Tools	Add-Ins	Window	
	Compile Debugging				
🖳	Step Into			F8	
🖳	Step Over		Shift+F8		
🖳	Step Out		Ctrl+Shift+F8		
🖳	Run To Cursor		Ctrl+F8		
	Add Watch...				
	Edit Watch...		Ctrl+W		
66	Quick Watch...		Shift+F9		
🖐	Toggle Breakpoint		F9		
	Clear All Breakpoints	Ctrl+Shift+F9			
⇨	Set Next Statement		Ctrl+F9		
🔲	Show Next Statement				

FIGURE 9.17. The Debug menu offers many commands for stepping through VBA procedures.

Let's step through a VBA procedure line by line.

⊙ Hands-On 9.6 Stepping Through a VBA Procedure

1. Place the insertion point anywhere inside the code of the procedure whose execution you wish to trace. For example, use the OpenToRead procedure you prepared in Hands-On 9.5.
2. Press F8 or choose Debug | Step Into.
   Visual Basic executes the current statement, automatically advances to the next statement, and suspends execution. While in break mode, you can activate

the Immediate window, Watches window, or Locals window to see the effect of a particular statement on the values of variables and expressions. If the procedure you are stepping through calls other procedures, you can activate the Call Stack window to see which procedures are currently active.

3. Press F8 again to execute the selected statement.
   After executing this statement, Visual Basic will select the next statement, and the procedure execution will be halted again.

4. Continue stepping through the procedure by pressing F8, or press F5 to continue the code execution without stopping.
   You can also choose Run | Reset to stop the procedure at the current statement without executing the remaining statements.

Stepping Over a Procedure and Running to Cursor

When you step over procedures (Shift+F8), Visual Basic executes each procedure as if it were a single statement. This option is particularly useful if a procedure contains calls to other procedures and you don't want to step into these procedures because they have already been tested and debugged, or you want to focus only on the new code that has not yet been debugged.

Suppose that the current statement in `MyProcedure` (see Hands-On 9.7) calls the `SpecialMsg` procedure. If you choose Debug | Step Over (Shift+F8) instead of Debug | Step Into (F8), Visual Basic will quickly execute all the statements inside the `SpecialMsg` procedure and select the next statement in the calling procedure (`MyProcedure`). During the execution of the `SpecialMsg` procedure, Visual Basic continues to display the Code window with the current procedure.

Hands-On 9.7   Stepping over a Procedure

1. In the Breaks Module Code window, locate the following procedures:

```
Sub MyProcedure()
  Dim strName As String

  Workbooks.Add
  strName = ActiveWorkbook.Name
  ' choose Step Over to avoid stepping through the
  ' lines of code in the called procedure - SpecialMsg
  SpecialMsg strName
  Workbooks(strName).Close
End Sub
```

```
Sub SpecialMsg(n As String)
  If n = "Book2" Then
    MsgBox "You must change the name."
  End If
End Sub
```

2. Add a breakpoint at the following statement:

```
SpecialMsg strName
```

3. Place the insertion point anywhere within the code of `MyProcedure`, and press F5 to run it.

Visual Basic halts execution when it reaches the breakpoint.

4. Press Shift+F8 or choose Debug | Step Over.

Visual Basic quickly runs the `SpecialMsg` procedure and advances to the statement immediately after the call to the `SpecialMsg` procedure.

5. Press F5 to finish running the procedure without stepping through its code.

6. Remove the breakpoint you set in Step 2.

Stepping over a procedure is particularly useful when you don't want to analyze individual statements inside the procedure that is being called.

Another command on the Debug menu, Step Out (Ctrl+Shift+F8), is used when you step into a procedure and then decide that you don't want to step all the way through it. When you choose this option, Visual Basic will execute the remaining statements in this procedure in one step and proceed to activate the next statement in the calling procedure.

In the process of stepping through a procedure, you can switch between the Step Into, Step Over, and Step Out options. The option you select depends on which code fragment you wish to analyze at a given moment.

The Debug menu's Run to Cursor (Ctrl+F8) command lets you run your procedure until the line you have selected is encountered. This command is useful if you want to stop the execution before a large loop or intend to step over a called procedure.

Now, let's suppose you want to execute MyProcedure to the line that calls the `SpecialMsg` procedure.

7. Click inside the statement `SpecialMsg strName`.

8. Choose Debug | Run to Cursor.

Visual Basic will stop the execution of the `MyProcedure` code when it reaches the specified line.

9. Press Shift+F8 to step over the `SpecialMsg` procedure.

10. Press F5 to execute the remaining statements in the procedure.

Setting the Next Statement

At times, you may want to rerun previous lines of code in the procedure or skip over a section of code that is causing trouble. In each of these situations, you can use the Set Next Statement option on the Debug menu. When you halt the execution of a procedure, you can resume the procedure from any statement you want. Visual Basic will skip the execution of the statements between the selected statement and the statement where execution was suspended.

Suppose that in `MyProcedure` (see the code of this procedure in the preceding section), you set a breakpoint on the statement calling the `SpecialMsg` procedure. To skip the execution of the `SpecialMsg` procedure, you can place the insertion point inside the statement `Workbooks (strName).Close` and press Ctrl+F9 (or choose Debug | Set Next Statement).

Keep in mind that you can't use the Set Next Statement option unless you have suspended the execution of the procedure.

While skipping lines of code can be very useful in the process of debugging your VBA procedures, it should be done with care. When you use the Next Statement option, you tell Visual Basic that this is the line you want to execute next. All lines in between are ignored. This means that certain things that you may have expected to occur don't happen, which can lead to unexpected errors.

Showing the Next Statement

If you are not sure from which statement the execution of the procedure will resume, you can choose Debug | Show Next Statement and Visual Basic will place the cursor on the line that will run next. This is particularly useful when you have been looking at other procedures and are not sure where execution will resume. The Show Next Statement option is available only in break mode.

Stopping and Resetting VBA Procedures

At any time, while stepping through the code of a procedure in the Code window, you can:

- Press F5 to execute the remaining instructions without stepping through.
- Choose Run | Reset to finish the procedure without executing the remaining statements.

When you reset your procedure, all the variables lose their current values. Numeric variables assume the initial value of zero, variable-length strings are initialized to a zero-length string (""), and fixed-length strings are filled with the character represented by the ASCII character code 0 or Chr(0). Variant variables are initialized to Empty, and the value of object variables is set to Nothing.

TERMINATING A PROCEDURE BASED ON A CONDITION

You may recall that in Chapter 1 (see Hands-On 1.20), we ran into an error while executing the Insert_NewSheet macro. We modified this macro to prompt the user for the sheet name using the Excel InputBox method. To make this macro error-proof, however, we need to ensure that the macro will not fail if the user clicks Cancel or enters a space or several blank spaces for the worksheet name. Let's address this problem now that you have more Excel VBA knowledge under your belt. Here is the Insert_NewSheet procedure as we modified it in Chapter 1:

```
Sub Insert_NewSheet()
'
' Insert_NewSheet Macro
' Insert and rename a worksheet.
'

    Sheets.Add After:=ActiveSheet
    ActiveSheet.Name = Application.InputBox _
      ("Enter the name for your worksheet:", "Rename This Sheet")
End Sub
```

The InputBox method is a member of the Excel Application object, and this requires that you precede its name with the name of the object (Application). Note that the following code uses the line continuation character (an underscore) to break up the long statement. Refer to Chapter 4 for more details on using this method and its arguments.

One of the arguments we absolutely must add to the InputBox method to get the expected results with the Cancel button is called type and it specifies the return data type. When the user clicks Cancel, the Application.InputBox method returns False. Therefore, we need to introduce conditional logic to test for the return type. We also need to prevent the user from feeding us blank spaces for the sheet name. By now, you should be familiar with writing VBA conditional statements. Conditional logic will allow you to make many enhancements to your recorded macro code. Let's look at the revised Insert_NewSheet procedure:

```
Sub Insert_NewSheet()
'
' Insert_NewSheet Macro revised
' Insert and rename a worksheet
'
    Dim userInput As Variant

    userInput = Application.InputBox _
      ("Enter the name for your worksheet:", _
        "Rename This Sheet", , , , , 2)
    If userInput = False Then
        MsgBox ("You pressed the Cancel button." & _
            "The procedure will terminate.")
        sFlag = True
        Exit Sub
    ElseIf userInput = "" Or Trim(userInput) = "" Then
        MsgBox "Please enter the sheet name or & _
            press Cancel to exit."
        Insert_NewSheet

    Else
        Sheets.Add After:=ActiveSheet
        ActiveSheet.Name = userInput
    End If
End Sub
```

Note that we will now store the user-supplied sheet name in the `userInput` variable. This variable is declared as a `Variant` data type because the `InputBox` method can return different types of data, and we want Excel to handle it for us. We begin by asking the user for the input. First, we must define the message that the user will see. Next, we specify the text that appears in the title bar of the dialog box. We don't care about the five arguments that follow, so we are using commas as their placeholders (or you can forgo commas when you specify the names for your arguments, as shown in Chapter 4's procedures). Recall that you can use named arguments to make your methods easier to understand. What we care about is the last argument. The value of `2` specifies that we expect to get a string (text).

Once we fill in the `userInput` variable, it's time for the `If` statements. If the contents of the variable are `False`, then we want to display a message to the user and terminate the procedure. You already know that you can exit early from a VBA procedure by using the `Exit Sub` statement. Before terminating the procedure, however, you may want to store some vital pieces of information in additional variables. In the case of the `Insert_NewSheet` procedure, we need to remember that we exited the procedure so we don't run other procedures that may depend on this one. Recall that our `Insert_NewSheet` procedure is part of a larger master procedure, which will also need to be terminated. The `sFlag` variable would hold a Boolean value of `True` if the user clicked Cancel and `False` otherwise. You are free to choose names for your variables. Notice that the `sFlag` variable is not declared anywhere in the `Insert_NewSheet` procedure. Since we must use it also in the `CreateEmployeeWorksheet` master procedure, we need a project-level scope declaration.

Recalling from Chapter 3 that public variables can be used in any module, here is the perfect opportunity to utilize them. Figure 9.18 shows the revised `CreateEmployeeWorksheet` procedure from Chapter 1. Notice the declaration of the `sFlag` variable at the top of the module. The first line of code in the procedure makes sure that `sFlag` is set to `False` when we start. When the `Insert_NewSheet` procedure has finished running, `sFlag` will be `True` if the user clicked Cancel. Again, we can use the `Exit Sub` statement to stop further code execution. If `sFlag` is `False`, we will continue with the remaining statements.

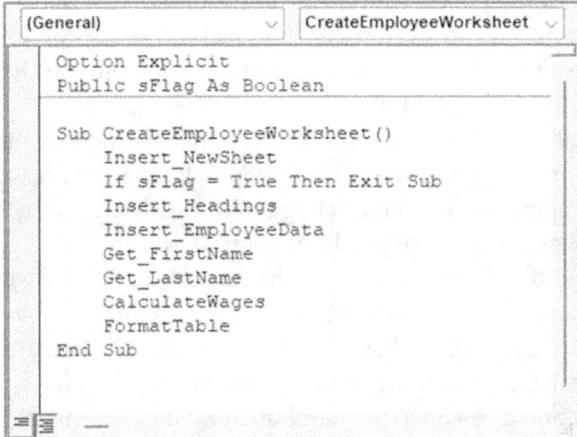

```
(General)                    CreateEmployeeWorksheet

    Option Explicit
    Public sFlag As Boolean

    Sub CreateEmployeeWorksheet()
        Insert_NewSheet
        If sFlag = True Then Exit Sub
        Insert_Headings
        Insert_EmployeeData
        Get_FirstName
        Get_LastName
        CalculateWages
        FormatTable
    End Sub
```

FIGURE 9.18. The revised CreateEmployeeWorksheet procedure uses a public variable of the Boolean data type.

Note that the `Insert_NewSheet` procedure also checks in the `Elseif` clause whether the user clicked OK without supplying any data or entered one or more spaces. The VBA `Trim` function removes the leading and trailing spaces from a supplied text string. If the value of the `userInput` variable is an empty string (`""`), then we display a message to the user and call the procedure again. This way, the user has a chance to either enter the required data or click Cancel.

Finally, if everything looks good, then we execute the statements in the `Else` clause. A new worksheet is inserted after the current sheet and is renamed with the text stored in the `userInput` variable.

(•) Hands-On 9.8   Working with All the Debugging Tools

The revised `Chap01.xlsm` file can be found in the companion files (`Chap01_Revised.xlsm`). Be sure to run the `CreateEmployeeWorksheet` master procedure at least three times to check all the conditions used in the `Insert_NewSheet` procedure.

SUMMARY

In this chapter, you learned how to trap errors and test your VBA procedures to make sure they run as intended. You debugged your code by stepping through it using breakpoints and watches. You discovered how to use the Immediate window in break mode and how the Locals window helps monitor the variable values. The Call Stack dialog box also proved useful for tracking your progress in complex programs. You explored numerous debugging options from the Debug menu, with practical examples of their application.

By leveraging these built-in debugging tools, you can quickly identify problem spots in your VBA procedures and functions. Spend time familiarizing yourself with the Debug menu options and debugging tools discussed in this chapter, as mastering debugging can save you hours of trial and error.

Congratulations on completing the first nine chapters of this book, which introduced you to the essentials of Excel VBA programming. You've covered a lot of ground, from understanding the basics of VBA syntax and structure to writing procedures and functions, working with conditional and looping statements, creating and manipulating collections and arrays, and even mastering error handling and debugging techniques.

In the chapters that follow, we will delve into more advanced VBA topics that will enhance your programming skills and allow you to build more sophisticated applications.

Part **II**

WORKING WITH
THE FILE SYSTEM

VBA provides numerous functions and statements for file system operations. Additionally, the `FileSystemObject` object from the Microsoft scripting library allows you to perform common file tasks such as reading, writing, and managing files and directories.

In this part, you will explore various methods of working with files and folders, including how to programmatically open, read, and write text files.

Chapter **10** *MANIPULATING FILES AND FOLDERS WITH VBA*

I n the course of your work, you've surely accessed, created, renamed, copied, and deleted hundreds of files and folders; however, you've probably never performed these tasks programmatically. This chapter will introduce you to VBA functions and statements that specifically deal with files and folders and allow you to:

- Find out the name of the current folder (CurDir function).
- Change the name of a file or folder (Name function).
- Check whether a file or folder exists on a disk (Dir function).
- Find out the date and time a file was last modified (FileDateTime function).
- Get the size of a file (FileLen function).
- Check and change file attributes (GetAttr and SetAttr functions).
- Change the default folder or drive (ChDir and ChDrive statements).
- Create and delete a folder (MkDir and RmDir statements).
- Copy and delete a file or folder (FileCopy and Kill statements).

MANIPULATING FILES AND FOLDERS

This section discusses a set of VBA functions used to perform operations on files and folders.

Finding Out the Name of the Active Folder

When working with files, you often need to find out the name of the current folder. You can get this information easily with the `CurDir` function, which looks like this:

```
CurDir([drive])
```

Note that `drive` is an optional argument. If you omit `drive`, VBA uses the current drive. The `CurDir` function returns a file path as `Variant`. To return the path as `String`, use `CurDir$` (where `$` is the type declaration character for a string).

To see this function in action, let's perform a couple of exercises in the Immediate window.

NOTE	*Please note that the files for the "Hands-On" projects can be found in the companion files.*

(●) Hands-On 10.1   Using the CurDir Function

1. Open a new workbook and save it as `Chap10.xlsm` in your `C:\VBAExcel2024_ByExample` folder.
2. Switch to Microsoft VBE and press Ctrl+G to activate the Immediate window. Type the following statement and press Enter:

```
?CurDir
```

When you press Enter, Visual Basic displays the name of the current folder. For example:

```
C:\VBAExcel2024_ByExample
```

3. If you have a second disk drive, you can find out the current folder on drive D, as follows:

```
?CurDir("D:\")
```

NOTE	*Supplying a letter for a drive that does not exist generates the following error message: "Run-time error '68': Device unavailable."*

4. To store the name of the current disk drive in a variable called `myDrive`, type the following statement and press Enter:

```
myDrive = Left(CurDir$,1)
```

When you press Enter, Visual Basic stores the letter of the current drive in the variable `myDrive`. Notice how the `CurDir$` function is used as the first argument of the `Left` function. The `Left` function tells Visual Basic to extract the leftmost character from the string returned by the `CurDir$` function and store it in the `myDrive` variable.

5. To check the contents of the variable `myDrive`, type the following statement and press Enter:

```
?myDrive
```

6. To return the letter of the drive followed by the colon, type the following instructions, pressing Enter after each line:

```
myDrive = Left(CurDir$,2)
?myDrive
```

Changing the Name of a File or Folder

To rename a file or folder, use the `Name` function, as follows:

```
Name old_pathname As new_pathname
```

`old_pathname` is the current path and name of a file or folder that you want to rename. `new_pathname` specifies the new path and name of the file or folder.

Using the `Name` function, you can move a file from one folder to another (you can't move a folder). Here are some precautions to consider while working with the `Name` function:

- The filename in `new_pathname` cannot refer to an existing file. The following statement replaces the filename:

```
Name "C:\Reports.txt" As "C:\Reports2024.txt"
```

However, if the file `c:\Reports.txt` already exists on drive C, Visual Basic generates the following error message: "File already exists." Similarly, the "File not found" error message will appear if the file you are trying to rename does not exist.

- If `new_pathname` already exists and it's different from `old_pathname`, the `Name` function moves the specified file to a new folder and changes its name, if necessary:

```
Name "C:\Reports.txt" As "D:\Reports2024.txt"
```

However, if the `Reports.txt` file doesn't exist in the root directory on drive D, Visual Basic moves the `c:\Reports.txt` file to the specified drive without renaming the file.

- If `new_pathname` and `old_pathname` refer to different directories and both supplied filenames are the same, the `Name` function moves the specified file to a new location without changing the filename:

```
Name "D:\Reports2024.txt" As "C:\VBAExcel2024_ByExample\
Reports2024.txt"
```

Renaming an Open File

You must close an open file before attempting to rename it. Also, note that the filename cannot contain the wildcard characters (* or ?).

Checking the Existence of a File or Folder

The `Dir` function, which returns the name of a file or folder, has the following syntax:

```
Dir[(pathname[, attributes])]
```

Notice that both arguments of the `Dir` function are optional. `pathname` is the name of a file or folder. You can use one of the constants or values in Table 10.1 for the `attributes` argument:

TABLE 10.1.   File attributes.

Constant	Value	Attribute Name
vbNormal	0	Normal
vbHidden	2	Hidden
vbSystem	4	System
vbVolume	8	Volume label
vbDirectory	16	Directory or folder

The `Dir` function is often used to check whether a file or folder exists on a disk. If a file or folder does not exist, the empty string (`""`) is returned.

Let's try out the `Dir` function in several exercises in the Immediate window.

⊙ Hands-On 10.2   Using the `Dir` Function

1. In the Immediate window, type the following statement and press Enter:

```
?Dir("C:\", vbNormal)
```

As you press Enter, Visual Basic returns the name of the first file in the specified folder. A normal file (`vbNormal`) is any file that does not have a hidden, volume label, directory, folder, or system file attribute.

2. To return the names of other files in the current directory, type the `Dir` function without an argument and press Enter:

```
?Dir
```

3. Enter the following instructions in the Immediate window and examine their results as you press Enter:

```
myfile = Dir("C:\", vbHidden)
?myfile
myfile = Dir
?myfile
```

Here the `Dir` function gets the name of the first file in the `c:\` directory that has the `Hidden` attribute. The `vbHidden` constant specifies that only hidden files should be returned. The result is assigned to the variable `myfile`. The `myfile = Dir` line calls the `Dir` function again but without any arguments. When `Dir` is called without arguments, it continues the previous search started by the initial `Dir` call and retrieves the name of the next file or directory that matches the original criteria (in this case, the next hidden file in `c:\`).

4. Type the following instruction on one line in the Immediate window and press Enter:

```
If Dir("C:\stamp.bat") = "" Then Debug.Print "File not found."
```

Because the `stamp.bat` file doesn't exist on drive C, Visual Basic prints the message "File not found" to the Immediate window.

The `Dir` function allows you to use the wildcards in the specified pathname— an asterisk (`*`) for multiple characters and a question mark (`?`) for a single character. For example, to find all Control Panel files in the `WINDOWS\System32` folder, you can look for all the MSC files, as shown below (the lines in italics show what Visual Basic might return as you call the `Dir` function):

```
?Dir("C:\WINDOWS\System32\*.msc", vbNormal)
azman.msc
```

```
?dir
certlm.msc
?dir
certmgr.msc
```

Let's write complete VBA procedures that use the `Dir` function. How about writing out the names of files in the specified directory to the Immediate window and a worksheet? We'll make our output consistent by using the `LCase$` function, which causes the names of files to appear in lowercase.

Hands-On 10.3   Using the Dir Function in a Procedure

1. In the Project Explorer window, rename the VBA project `FileMan_VBA`.
2. Insert a new module into the `FileMan_VBA (Chap10.xlsm)` project, and rename it `DirFunction`.
3. Enter the `MyFiles` procedure in the Code window, as shown below:

```
Sub MyFiles()
    Dim myfile As String
    Dim mpath As String
    Dim myPrompt As String

    myPrompt = "Enter pathname, "
    myPrompt = myPrompt & "e.g. C:\VBAExcel2024_ByExample"
    mpath = InputBox(myPrompt)
    If Right(mpath, 1) <> "\" Then mpath = mpath & "\"

    myfile = Dir(mpath & "*.*")
    If myfile <> "" Then Debug.Print "Files in the " & _
        mpath & " folder:"
    Debug.Print LCase$(myfile)
    If myfile = "" Then
        MsgBox "No files found."
        Exit Sub
    End If
    Do While myfile <> ""
        myfile = Dir
        Debug.Print LCase$(myfile)
    Loop
End Sub
```

The `MyFiles` procedure shown above asks the user for the pathname.

If the path does not end with a backslash, the `Right` function appends the backslash to the end of the `pathname` string. Next, Visual Basic looks for all the files (*) in the specified path. If there are no files, a message is displayed. If files exist, the filenames are written to the Immediate window.

4. Run the `MyFiles` procedure.

5. To output the filenames to a worksheet, enter the following `GetFiles` procedure in the same `DirFunction` module:

```
Sub GetFiles()
    Dim myfile As String
    Dim nextRow As Integer

    nextRow = 1
    With Worksheets("Sheet1").Range("A1")
        myfile = Dir("C:\VBAExcel2024_ByExample\*.*", vbNormal)
        .Value = myfile
        Do While myfile <> ""
            myfile = Dir
            .Offset(nextRow, 0).Value = myfile
            nextRow = nextRow + 1
        Loop
    End With
End Sub
```

6. Run the `GetFiles` procedure.

The `GetFiles` procedure obtains the names of files located in the specified directory of drive C and writes each filename to a worksheet.

Retrieving the Date and Time the File Was Modified

If your procedure must check when a file was last modified, use the `FileDate-Time` function in the following form:

`FileDateTime(pathname)`

`pathname` is a string that specifies the file you want to work with. The `pathname` may include the drive and folder where the file resides. The function returns the date and timestamp for the specified file. The date and time format depends on the regional settings selected in the Windows Control Panel.

Let's practice using this function in the Immediate window.

⦿ Hands-On 10.4   Using the FileDateTime Function

1. Enter the following statement in the Immediate window:

`?FileDateTime("C:\VBAExcel2024_ByExample\Chap10.xlsm")`

When you press Enter, Visual Basic returns the date and timestamp in the following format:

```
2/3/2025 7:00:19 PM
```

2. To return the date and time separately, use the `FileDateTime` function as an argument of the `DateValue` or `TimeValue` functions. For instance, enter the following statements on one line in the Immediate window:

```
?DateValue(FileDateTime("C:\VBAExcel2024_ByExample\Chap10.xlsm"))
?TimeValue(FileDateTime("C:\VBAExcel2024_ByExample\Chap10.xlsm"))
```

The `Date` function returns the current system date as it is set in the Date and Time Properties dialog box accessed in the Windows Control Panel.

Retrieving the Size of a File (the FileLen Function)

To check the size of a file, use the `FileLen` function in the following form:

```
FileLen(pathname)
```

The `FileLen` function returns the size of a specified file in bytes. If the file is open, Visual Basic returns the size of the file when it was last saved.

Returning and Setting File Attributes (the GetAttr and SetAttr Functions)

Files and folders can have attributes such as read-only, hidden, system, and archive. To find out the attributes of a file or folder, use the `GetAttr` function, which returns an integer that represents the sum of one or more of the constants shown in Table 10.2. This function requires one argument, which is the name of the file or folder you want to work with:

```
GetAttr(pathname)
```

TABLE 10.2. File and folder attributes.

Constant	Value	Attribute
vbNormal	0	Normal (other attributes are not set)
vbReadOnly	1	Read-only (file or folder can't be modified)
vbHidden	2	Hidden (file or folder isn't visible under normal setup)
vbSystem	4	System file
vbDirectory	16	The object is a directory
vbArchive	32	Archive (the file has been modified since it was last backed up)

To determine whether a file has any of the attributes shown in Table 10.2, we will use the AND operator to compare the result of the GetAttr function with the value of the constant. If the function returns a nonzero value, it means that the file or folder specified in the pathname has the attribute for which you are testing.

◉ Hands-On 10.5   Returning File Attributes with the GetAttr Function

1. Insert a new module into the FileMan_VBA (Chap10.xlsm) project and rename it GetAttrFunction.

2. Enter the following GetAttributes procedure and run it:

```
Sub GetAttributes()
    Dim attr As Integer
    Dim msg As String
    Dim strFileName As String

    strFileName = InputBox("Enter the complete file name:", _
        "Drive\Folder\Filename")
    If strFileName = "" Then Exit Sub
    attr = GetAttr(strFileName)

    msg = ""

    If attr And vbReadOnly Then msg = msg & "Read-Only (R)"
    If attr And vbHidden Then msg = msg & Chr(10) & "Hidden (H)"
    If attr And vbSystem Then msg = msg & Chr(10) & "System (S)"
    If attr And vbArchive Then msg = msg & Chr(10) & "Archive (A)"
    MsgBox msg, , strFileName
End Sub
```

The opposite of the GetAttr function is the SetAttr function, which allows you to set the attributes for files or folders that are closed. Its syntax is:

```
SetAttr pathname, attributes
```

pathname is a string that specifies the file or folder that you want to work with. The second argument, attributes, is one or more constants that specify the attributes you want to set. See Table 10.2 previously for the list of available constants.

Suppose you have a file called C:\stamps.txt and you want to set two attributes: read-only and hidden. You can do this with the SetAtrr function, as shown in the next exercise.

⦿ Hands-On 10.6 Setting File Attributes with the SetAttr Function

1. To set the file attributes, type the following instruction in the Immediate window and press Enter (replace `C:\stamps.txt` with the name of a file that exists on your disk):

```
SetAttr "C:\stamps.txt", vbReadOnly + vbHidden
```

2. To find out what attributes were set in Step 1, type the following instruction in the Immediate window and press Enter (check the returned value against Table 10.2):

```
?GetAttr("C:\stamps.txt")
```

Changing the Default Folder or Drive (the ChDir and ChDrive Statements)

You can easily change the default folder by using the `ChDir` statement, as follows:

```
ChDir pathname
```

In the statement above, `pathname` is the name of the new default folder and may include the name of the disk drive. If `pathname` doesn't include a drive designation, the default folder will be changed on the current drive. The current drive will not be changed. Suppose the default folder is `C:\WINDOWS`. The statement:

```
ChDir "D:\MyFiles"
```

changes the default folder to `D:\MyFiles`; however, the current drive is still drive `C`.

To change the current drive, use the `ChDrive` statement in the following format:

```
ChDrive drive
```

The `drive` argument specifies the letter of the new default drive.

For instance, to change the default drive to drive `D` or `E`, use the following statements:

```
ChDrive "D"
ChDrive "E"
```

If you refer to a nonexistent drive, you will get the message "Device unavailable."

Creating and Deleting Folders (the MkDir and RmDir Statements)

You can create a new folder using the following syntax of the `MkDir` statement:

```
MkDir pathname
```

`pathname` specifies the new folder you want to create. If you don't include the name of the drive, Visual Basic will create a new folder on the current drive.

To delete a folder you no longer need, use the `RmDir` function. This function has the following syntax:

```
RmDir pathname
```

`pathname` specifies the folder you want to delete and may include the drive name. If you omit the name of the drive, Visual Basic will delete the folder on the current drive if a folder with the same name exists. Otherwise, Visual Basic will display the error message "Path not found."

Let's run through some examples in the Immediate window.

Hands-On 10.7   Creating and Deleting Folders with the MkDir and RmDir Statements

1. Type the following instruction in the Immediate window and press Enter to create a folder called `Mail` on drive `C`:

```
MkDir "C:\Mail"
```

2. To change the default folder to `C:\Mail`, enter the following statement and press Enter:

```
ChDir "C:\Mail"
```

3. To find out the name of the active folder, enter the following statement and press Enter:

```
?CurDir
```

4. To delete the `C:\Mail` folder that was created in Step 1, enter the following statements and press Enter:

```
ChDir "C:\"
RmDir "C:\Mail"
```

RmDir Removes Empty Folders

You cannot delete a folder if it still contains files. You should first delete the files with the `Kill` statement (discussed later in this chapter).

Copying Files (the FileCopy Statement)

To copy files between folders, use the `FileCopy` statement shown below:

```
FileCopy source, destination
```

The first parameter of this statement, `source`, specifies the name of the file that you want to copy. The name may include the drive in which the file resides. The second parameter, `destination`, is the name of the destination file and may include the drive and folder designation. Both parameters are required.

Suppose you want to copy a user-specified file specified to the `C:\Abort` folder.

Hands-On 10.8 demonstrates how to do this.

⊙ Hands-On 10.8   Copying Files with the FileCopy Statement

1. Insert a new module into the `FileMan_VBA`   (`Chap10.xlsm`) project and rename it `FileCopyAndKill`.
2. In the module's Code window, enter the following `CopyToAbortFolder` procedure:

```vba
Sub CopyToAbortFolder()
    Dim folder As String
    Dim source As String
    Dim dest As String
    Dim msg1 As String
    Dim msg2 As String
    Dim p As Integer
    Dim s As Integer
    Dim i As Long

    On Error GoTo ErrorHandler

    folder = "C:\Abort"
    msg1 = "The selected file is already in this folder."
    msg2 = "was copied to"
    p = 1
    i = 1
    ' get the name of the file from the user
    source = Application.GetOpenFilename
    ' don't do anything if cancelled
    If source = "False" Then Exit Sub
    ' get the total number of backslash & _
            characters "\" in the source
    ' variable's contents
    Do Until p = 0
      p = InStr(i, source, "\", 1)
      If p = 0 Then Exit Do
      s = p
      i = p + 1
```

```
Loop
' create the destination filename
dest = folder & Mid(source, s, Len(source))
  ' create a new folder with this name
  MkDir folder
  ' check if the specified file already exists in the
  ' destination folder
  If Dir(dest) <> "" Then
    MsgBox msg1
  Else
  ' copy the selected file to the C:\Abort folder
    FileCopy source, dest
    MsgBox source & " " & msg2 & " " & dest
  End If
  Exit Sub
ErrorHandler:
  If Err = "75" Then
    Resume Next
  End If
  If Err = "70" Then
    MsgBox "You can't copy an open file."
  Exit Sub
  End If
End Sub
```

The procedure CopyToAbortFolder uses the Excel Application object's GetOpenFilename method to get the name of the file from the user. This method opens the built-in Open dialog box where you can choose any file, in any directory, and on any disk drive. If the dialog box is canceled, Visual Basic returns False and the procedure ends. When the user selects the file and clicks Open, the filename is saved in the source variable.

For the purpose of copying, you'll only need the filename (without the path), so the Do...Until loop determines the position of the last backslash (\) in the file stored in the source variable, the first argument of the FileCopy statement. The second argument of the FileCopy statement, the dest variable, holds the string obtained by concatenating the name of the destination folder C:\Abort with the user-specified filename preceded by a backslash (\).

The MkDir function creates a new folder called C:\Abort if it doesn't exist on drive C. If the folder already exists, Visual Basic will need to handle error 75 in the ErrorHandler code.

When Visual Basic encounters the Resume Next statement, it will continue to execute the procedure from the instruction following the instruction that caused the error. This means that the statement `MkDir folder` won't be executed.

3. Next, the procedure checks whether the selected file already exists in the destination folder. If it's `True`, the user will get the message stored in the variable `msg1`. If the file does not exist and is not currently open, Visual Basic will copy the file to the specified folder and notify the user with the appropriate message. If the file is open, Visual Basic will encounter runtime error `70` and will run the corresponding instructions in the `ErrorHandler` code section of the procedure.

4. Run the `CopyToAbortFolder` procedure several times, each time selecting files from different folders.

5. Try to copy a file that was copied before by this procedure to the `C:\Abort` folder.

6. Try to copy an open file while using the `CopyToAbortFolder` procedure.

7. Run the procedure `MyFiles` prepared earlier in this chapter to write the contents of the `Abort` folder to the Immediate window.

NOTE	*Do not delete the* `C:\Abort` *folder and its files. You'll delete them in the next section using a VBA procedure.*

Deleting Files (the Kill Statement)

You already know from one of the earlier sections in this chapter that you can't delete a folder if it still contains files. To delete the files from any folder, use the following `Kill` statement:

```
Kill pathname
```

`Pathname` specifies the names of one or more files that you want to delete. Optionally, `pathname` may include the drive and folder name where the file resides. To enable the quick deletion of files, you can use the wildcard characters (`*` or `?`) in the `pathname` argument.

You can't delete a file that is open. If you worked through the exercises in the preceding section, your hard drive now contains the folder `C:\Abort` with several files. Let's write a VBA procedure to dispose of this folder and its files.

(⊙) Hands-On 10.9   Deleting Files with the Kill Statement

1. Insert a new module into the `FileMan_VBA(Chap10.xlsm)` project and rename it `KillStatement`.

2. Enter the code of the `RemoveMe` procedure, as shown below:

```
Sub RemoveMe()
   Dim folder As String
   Dim myFile As String

   ' assign the name of folder to the folder variable
   ' notice the ending backslash "\"
   folder = "C:\Abort\"
   myFile = Dir(folder, vbNormal)

   Do While myFile <> ""
     Kill folder & myFile
     myFile = Dir
   Loop
   RmDir folder
End Sub
```

3. Run the `RemoveMe` procedure. When the procedure ends, check Windows Explorer to see that the `Abort` folder was removed.

Obtaining Information about Recent Files

Excel has a `RecentFiles` property that can be used to return a `RecentFiles` collection that represents the list of recently used files. Let's try it out in the Immediate window:

```
?Application.RecentFiles(1).Name
\VBAExcel2024_ByExample\Chap10.xlsm
?Application.RecentFiles.Count
50
```

The first statement in the above code snippet returns the name of the most recent file. The second statement displays the total count of recent files.

SUMMARY

In this chapter, you explored VBA functions and statements that enable you to interact with the file system. You learned how to manage files and folders using various functions.

For instance, the `CurDir` function helps you obtain the name of the current folder. Additionally, you discovered how to use the `GetAttr` and `SetAttr` functions to check and modify file attributes. Furthermore, you practiced creating, copying, and deleting files and folders with the `MkDir`, `FileCopy`, `RmDir`, and `Kill` statements. Lastly, you retrieved information about recent files using the properties and methods of the Excel `RecentFiles` object.

In the next chapter, you'll learn how working with the `FileSystemObject` (FSO) object and Windows Script Host (WSH) can significantly enhance your ability to manage and automate file system tasks.

11 AUTOMATING FILE SYSTEM TASKS WITH WINDOWS SCRIPT HOST AND FILESYSTEMOBJECT

There is a hidden gem in your computer called Windows Script Host (WSH). This powerful tool allows you to create small programs that can control the Windows operating system and its applications, as well as retrieve information from the system. WSH is an ActiveX control that enables you to create scripts for both simple or complex operations—tasks that were once only possible with batch (.bat) files.

WSH is not a standalone language but rather a scripting environment that can execute scripts written in various scripting languages such as VBScript and JScript. A *script* is essentially a set of commands that can be executed automatically. You can run these scripts directly from the command prompt using the Command Script Host (Cscript.exe) or from Windows using WSH (Wscript.exe).

In the following sections, you'll discover how WSH can work in tandem with VBA.

USEFUL OPERATIONS YOU CAN PERFORM WITH WSH

WSH operates within its own object hierarchy, providing a structured way to interact with various components of the Windows operating system. Before you start writing VBA procedures that utilize WSH objects, it's essential to understand the range of tasks you'll be able to control. With WSH, you can:

- Manipulate files and folders. You can create, delete, move, copy, and manage files and directories.
- Read and write text files. This includes accessing and modifying the contents of text files, supporting both sequential and random access.
- Access drives to retrieve information about disk drives, such as available space or file system type.
- Read, set, and manipulate environment variables that can affect system behavior and application settings.
- Execute programs and scripts. You can run executable programs, batch files, and other scripts from within your VBA code.
- Perform network operations, such as interacting with network resources and performing network-related tasks.
- Perform registry operations, such as reading from and writing to the Windows registry to manage system and application settings.
- Retrieve user, domain, and computer names.
- Launch other applications.
- Display dialog boxes and retrieve user input.
- Create and manage shortcuts on your Windows desktop.

REFERENCING THE MICROSOFT SCRIPTING RUNTIME LIBRARY

Microsoft Scripting Runtime is a library that allows you to work with the file system and manipulate text files, folders, and drives. It's part of the WSH environment and provides objects and methods to perform file operations that aren't natively supported by VBA.

To explore various features of WSH, let's begin by creating a reference to Microsoft Scripting Runtime.

> **NOTE** *Please note that the files for the "Hands-On" projects can be found in the companion files.*

Hands-On 11.1   Controlling Objects with WSH

1. Open a new workbook and save it as `C:\VBAExcel2024_ByExample\Chap11.xlsm`.

2. Switch to the VBE window and choose Tools | References. Click the checkbox next to Microsoft Scripting Runtime, as in Figure 11.1, then click OK to close the References dialog box.

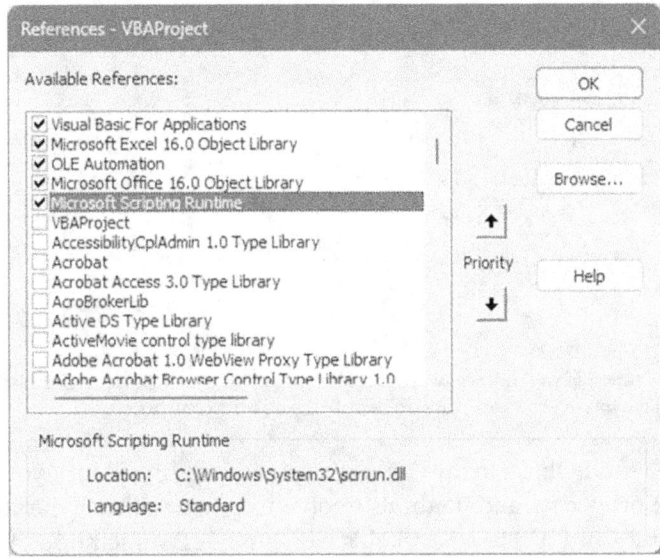

FIGURE 11.1.   Creating a reference to Microsoft Scripting Runtime.

3. Press F2 to open the Object Browser.

4. In the <All Libraries> combo box, choose Scripting. You will see a list of objects that are part of the WSH library, as shown in Figure 11.2.

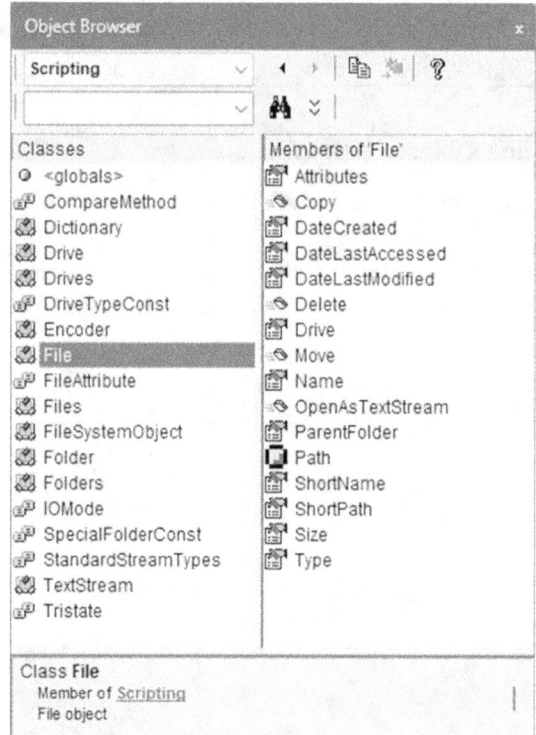

FIGURE 11.2. After establishing a reference to Microsoft Scripting Runtime, the Object Browser displays objects that allow you to work with disks, folders, files, and their contents.

By selecting FileSystemObject in the Object Browser's Scripting library, you will be able to locate properties and methods to obtain answers to such questions as "On which disk can I locate a particular file?" (GetDrive method), "What is the extension of a filename?" (GetExtensionName method), and "Does this folder or file exist on a given drive?" (FolderExists and FileExists methods).

5. Close the Object Browser.

FINDING INFORMATION ABOUT FILES

WSH exposes an object called FileSystemObject. This object has several methods for working with the file system. In the following procedure, we will obtain some information about a specific file.

⊙ Hands-On 11.2   Using WSH to Obtain File Information

1. In the VBE window, activate the Properties window and change the name of `VBAProject(Chap11.xlsm)` to `FileMan_WSH`.
2. Insert a new module into the `FileMan_WSH` project and rename it `WSH`.
3. In the Code window, enter the following `FileInfo` procedure (you may need to change the path to the Windows folder to make it run on your computer):

```
Sub FileInfo()
  Dim objFs As Object
  Dim objFile As Object
  Dim strMsg As String

  Set objFs = CreateObject("Scripting.FileSystemObject")
  Set objFile = objFs.GetFile("C:\WINDOWS\System.ini")
  strMsg = "File name: " & _
      objFile.Name & vbCrLf
  strMsg = strMsg & "Disk: " & _
  objFile.Drive & vbCrLf
  strMsg = strMsg & "Date Created: " & _
      objFile.DateCreated & vbCrLf
  strMsg = strMsg & "Date Modified: " & _
      objFile.DateLastModified & vbCrLf
  MsgBox strMsg, , "File Information"
End Sub
```

The `FileInfo` procedure shown above uses the `CreateObject` VBA function to create an instance of an automation object (`Scripting.FileSystemObject`), enabling you to use its methods and properties to interact with the file system:

```
Dim objFs As Object
Set objFs = CreateObject("Scripting.FileSystemObject")
```

The `FileSystemObject` is assigned to the variable `objFs`. In the next statement, the `GetFile` method of the `FileSystemObject` is used to get a reference to a specific file, in this case, `System.ini` located in the `C:\WINDOWS` directory:

```
Set objFile = objFs.GetFile("C:\WINDOWS\System.ini")
```

The file reference is then assigned to the variable `objFile`.

The `File` object has many properties that you can read. For example, the statement `objFile.Name` returns the full name of the file. The statement `objFile.Drive` returns the drive name where the file is located. The statements `objFile.DateCreated` and `objFile.DateLastModified` return the date the file was created and when it was last modified. This procedure can be modified easily to return the type of file, its attributes, and the name of the parent folder. Try to modify this procedure on your own by adding the following instructions to the code: `objFile.Type`, `objFile.Attributes`, `objFile.ParentFolder`, and `objFile.Size`. Check the Object Browser for other things you can learn about the file by referencing the `File` object.

4. Run the `FileInfo` procedure.

Accessing the FileSystemObject

There are two ways to access the `FileSystemObject` in VBA. You can use the `CreateObject` function, as was demonstrated in Hands-On 11.2, or you can set a reference to the Microsoft Scripting Runtime library (see Hands-On 11.1). Each method has its own advantages and disadvantages.

Using the CreateObject Function

The `CreateObject` function gives you more flexibility, allowing you to dynamically create instances of various objects without needing to set up library references in advance. Your VBA code can be more portable and less dependent on specific library versions.

When you use the `CreateObject` function to access the `FileSystemObject`, you set it up as a generic type `Object`, like this:

```
Dim objFs As Object
Set objFs = CreateObject("Scripting.FileSystemObject")
```

This method uses *late binding*, meaning that the actual type of `objFs` is determined at runtime. As a result, you will not have access to *early binding*, such as IntelliSense (auto-completions, parameter info, and object browsing) and compile-time error checking (which detects and reports errors in the code before the program is actually run). Late binding can be slower than early binding because the object type is resolved during runtime.

Using a Library Reference

When you set up a reference to Microsoft Scripting Runtime in the References dialog box (Tools | References in the VBE screen), you use early binding like this:

```
Dim objFs As FileSystemObject
Set objFs = New FileSystemObject
```

With early binding, the object type is known at compile time, which allows you to take advantage of IntelliSense. It also enables compile-time error checking so you can catch syntax errors such as missing semicolons, mismatched parentheses, incorrect keywords, and so on before your VBA procedure is executed. Early binding can be faster than late binding since the object is resolved at compile time. The dependency on setting the reference to Microsoft Scripting Runtime, however, can make your code less portable. You must make sure that the necessary references are set up on each system where your code runs.

Methods and Properties of the FileSystemObject Object

The `FileSystemObject` provides a number of methods, some of which are shown below with code examples.

- `FileExists` returns `True` if the specified file exists, as follows:

```
Sub FileExists()
  Dim objFs As Object
  Dim strFile As String
  Set objFs = CreateObject("Scripting.FileSystemObject")
  strFile = InputBox("Enter the full name of the file: ")
  If objFs.FileExists(strFile) Then
    MsgBox strFile & " was found."
  Else
    MsgBox "File does not exist."
  End If
End Sub
```

- `GetFile` returns a `File` object.
- `GetFileName` returns the filename and path.
- `GetFileVersion` returns the file version.
- `CopyFile` creates a copy of a file, as follows:

```
Sub CopyFile()
  Dim objFs As Object
  Dim strFile As String
  Dim strNewFile As String

  strFile = "C:\Hello.txt"
  strNewFile = "C:\VBAExcel2024_ByExample\Hello.txt"

  Set objFs = CreateObject("Scripting.FileSystemObject")
  objFs.CopyFile strFile, strNewFile
  MsgBox "A copy of the specified file was created."
  Set objFs = Nothing
End Sub
```

- MoveFile moves a file.

- DeleteFile deletes a file, as follows:

```
Sub DeleteFile()
  ' This procedure uses early binding and requires
  ' a reference to the Microsoft Scripting Runtime
  ' Object Library (use Tools | References
  ' in the VBE window)
  Dim objFs As FileSystemObject
  Set objFs = New FileSystemObject

  objFs.DeleteFile "C:\VBAExcel2024_ByExample\Hello.txt"
  MsgBox "The requested file was deleted."
End Sub
```

- DriveExists returns True if the specified drive exists, as follows:

```
Function DriveExists(disk)
  Dim objFs As Object
  Dim strMsg As String
  Set objFs = CreateObject("Scripting.FileSystemObject")
  If objFs.DriveExists(disk) Then
    strMsg = "Drive " & UCase(disk) & " exists."
  Else
    strMsg = UCase(disk) & " was not found."
  End If
  DriveExists = strMsg
' run this function from the worksheet
' by entering the following in any cell :
' =DriveExists("E:\")
End Function
```

- GetDrive returns a Drive object, as follows:

```
Sub DriveInfo()
  Dim objFs As Object
  Dim objDisk As Object
  Dim infoStr As String
  Dim strDiskName As String
  strDiskName = InputBox("Enter the drive letter:", _
      "Drive Name", "C:\")

  Set objFs = CreateObject("Scripting.FileSystemObject")
  Set objDisk=objFs.GetDrive(objFs.GetDriveName(strDiskName))
  infoStr = "Drive: " & UCase(strDiskName) & vbCrLf
  infoStr = infoStr & "Drive letter: " & _
      UCase(objDisk.DriveLetter) & vbCrLf
```

```
    infoStr = infoStr & "Drive Type: " & _
        objDisk.DriveType & vbCrLf
    infoStr = infoStr & "Drive File System: " & _
        objDisk.FileSystem & vbCrLf
    infoStr = infoStr & "Drive SerialNumber: " & _
        objDisk.SerialNumber & vbCrLf
    infoStr = infoStr & "Total Size in Bytes: " & _
        FormatNumber(objDisk.TotalSize / 1024, 0) & _
            " Kb" & vbCrLf
    infoStr = infoStr & "Free Space on Drive: " & _
        FormatNumber(objDisk.FreeSpace / 1024, 0) & _
            " Kb" & vbCrLf
    MsgBox infoStr, vbInformation, "Drive Information"
End Sub
```

- GetDriveName returns a string containing the name of a drive or network share, as follows:

```
Function DriveName(disk) As String
    Dim objFs As Object
    Dim strDiskName As String

    Set objFs = CreateObject("Scripting.FileSystemObject")
    strDiskName = objFs.GetDriveName(disk)
    DriveName = strDiskName
' run this function from the Immediate window
' by entering ?DriveName("C:\")
End Function
```

- FolderExists returns True if the specified folder exists, as follows:

```
Sub DoesFolderExist()
    Dim objFs As Object
    Set objFs = CreateObject("Scripting.FileSystemObject")
    MsgBox objFs.FolderExists("C:\Program Files")
End Sub
```

- GetFolder returns a Folder object, as follows:

```
Sub FilesInFolder()
    Dim objFs As Object
    Dim objFolder As Object
    Dim objFile As Object

    Set objFs = CreateObject("Scripting.FileSystemObject")
    Set objFolder = objFs.GetFolder("C:\")

    Workbooks.Add
```

```
    For Each objFile In objFolder.Files
        With ActiveCell
            .Formula = objFile.Name
            .Offset(0, 1).Range("A1").Formula = objFile.Type
            .Offset(1, 0).Range("A1").Select
        End With
    Next
    Columns("A:B").AutoFit
End Sub
```

- `GetSpecialFolder` returns the path to the operating system folders, as follows:

  - 0—Windows folder

  - 1—System folder

  - 2—Temp folder

```
Sub SpecialFolders()
    Dim objFs As Object
    Dim strWindowsFolder As String
    Dim strSystemFolder As String
    Dim strTempFolder As String

    Set objFs = CreateObject("Scripting.FileSystemObject")
    strWindowsFolder = objFs.GetSpecialFolder(0)
    strSystemFolder = objFs.GetSpecialFolder(1)
    strTempFolder = objFs.GetSpecialFolder(2)

    MsgBox strWindowsFolder & vbCrLf _
        & strSystemFolder & vbCrLf _
        & strTempFolder, vbInformation + vbOKOnly, _
        "Special Folders"
End Sub
```

- `CreateFolder` creates a folder, as follows:

```
Sub MakeNewFolder()
    Dim objFs As Object
    Dim objFolder As Object
    Set objFs = CreateObject("Scripting.FileSystemObject")
    Set objFolder = objFs.CreateFolder("C:\TestFolder")
    MsgBox "A new folder named " & _
        objFolder.Name & " was created."
End Sub
```

- `CopyFolder` creates a copy of a folder, as follows:

```
Sub MakeFolderCopy()
  Dim objFs As FileSystemObject
  Set objFs = New FileSystemObject
  If objFs.FolderExists("C:\TestFolder") Then
    objFs.CopyFolder "C:\TestFolder", "C:\FinalFolder"
    MsgBox "The folder was copied."
  End If
End Sub
```

- `MoveFolder` moves a folder.

- `DeleteFolder` deletes a folder, as follows:

```
Sub RemoveFolder()
  Dim objFs As Object
  Dim objFolder As Object
  Set objFs = CreateObject("Scripting.FileSystemObject")

  If objFs.FolderExists("C:\TestFolder") Then
    objFs.DeleteFolder "C:\TestFolder"
    MsgBox "The folder was deleted."
  End If
End Sub
```

- `CreateTextFile` creates a text file (see the example procedure later in this chapter).

- `OpenTextFile` opens a text file, as follows:

```
Sub ReadTextFile()
    Dim objFs As Object
    Dim objFile As Object
    Dim strContent As String
    Dim strFileName As String

    strFileName = "C:\VBAExcel2024_ByExample\Vacation.txt"
    Set objFs = CreateObject("Scripting.FileSystemObject")
    Set objFile = objFs.OpenTextFile(strFileName)
    Do While Not objFile.AtEndOfStream
        strContent = strContent & objFile.ReadLine & vbCrLf
    Loop

    objFile.Close
    Set objFile = Nothing
    ActiveWorkbook.Sheets(1).Select
    Range("A1").Formula = strContent
```

```
        Columns("A:A").Select
        With Selection
            .ColumnWidth = 62.43
            .Rows.AutoFit
        End With
    End Sub
```

The above example demonstrates how to retrieve the contents of a text file. The `objFile.AtEndOfStream` property is a property of the `TextStream` object, which is part of the Microsoft Scripting Runtime library. This property checks whether the current position in the text stream is at the end of the stream, which means that there is no more data to read from the file. This property is useful when reading a file line by line in a loop, as it allows you to know when you've reached the end of the file and should stop reading. In this example, the `Do While Not objFile.AtEndOfStream` loop continues to read lines from the file until the end of the stream is reached, indicated by `objFile.AtEndOfStream` being `True`.

The `FileSystemObject` has only one property. The `Drives` property returns a reference to the collection of drives. Using this property, you can create a list of drives on a computer, as shown below:

```
Sub DrivesList()
  Dim objFs As Object
  Dim colDrives As Object
  Dim strDrive As String
  Dim Drive As Variant

  Set objFs = CreateObject("Scripting.FileSystemObject")
  Set colDrives = objFs.Drives

  For Each Drive In colDrives
    strDrive = "Drive " & Drive.DriveLetter & ": "
    Debug.Print strDrive
  Next
End Sub
```

Properties of the File Object

With the `File` object, you can access all the properties of a specified file. The following lines of code create a reference to the `File` object:

```
Set objFs = CreateObject("Scripting.FileSystemObject")
Set objFile = objFs.GetFile("C:\My Documents\myFile.doc")
```

You will find an example of using the `File` object in the `FileInfo` procedure that was created earlier in this chapter.

These are the properties of the `File` object:

- `Attributes`—Returns file attributes (compare this property to the `GetAttr` VBA function explained in Chapter 10, "Manipulating Files and Folders with VBA").
- `DateCreated`—File creation date.
- `DateLastAccessed`—File last-access date.
- `DateLastModified`—File last-modified date.
- `Drive`—Drive name followed by a colon.
- `Name`—Name of the file.
- `ParentFolder`—Parent folder of the file.
- `Path`—Full path of the file.
- `Size`—File size in bytes (compare this property to the `FileLen` VBA function introduced in Chapter 10).
- `Type`—File type. This is the text that appears in the Type column in Windows Explorer, e.g., configuration settings, application, and shortcut.

Properties of the Folder Object

The `Folder` object provides access to all of the properties of a specified folder. The following lines of code create a reference to the `Folder` object:

```
Set objFs = CreateObject("Scripting.FileSystemObject")
Set objFolder = objFs.GetFolder("C:\My Documents")
```

The `Folder` object has the following properties:

- `Attributes`—Folder attributes
- `DateCreated`—Folder creation date
- `Drive`—Returns the drive letter of the folder where the specified folder resides
- `Files`—Collection of files in the folder

```
Sub CountFilesInFolder()
    Dim objFs As Object
    Dim strFolder As String
    Dim objFolder As Object
    Dim objFiles As Object

    strFolder = InputBox("Enter the folder name:")
```

```
      If Not IsFolderEmpty(strFolder) Then
        Set objFs = CreateObject ("Scripting.FileSystemObject")
        Set objFolder = objFs.GetFolder(strFolder)
        Set objFiles = objFolder.Files
        MsgBox "The number of files in the folder " & _
           strFolder & " = " & objFiles.Count
      Else
         MsgBox "Folder " & strFolder & " has 0 files."
      End If
    End Sub
```

The above procedure calls the `IsFolderEmpty` function (see the next code example).

- `IsRootFolder`—Returns `True` if the folder is the root folder
- `Name`—Name of the folder
- `ParentFolder`—Parent folder of the specified folder
- `Path`—Full path to the folder
- `Size`—Folder size in bytes

```
Function IsFolderEmpty(myFolder)
  Dim objFs As Object
  Dim objFolder As Object

  Set objFs = CreateObject("Scripting.FileSystemObject")
  Set objFolder = objFs.GetFolder(myFolder)
  IsFolderEmpty = (objFolder.Size = 0)
End Function
```

- `SubFolders`—Collection of subfolders in the folder
- `Type`—Folder type, e.g., file folder or Recycle Bin

Properties of the Drive Object

The `Drive` object provides access to the properties of the specified drive on a computer or a server. The following lines of code create a reference to the `Drive` object:

```
Set objFs = CreateObject("Scripting.FileSystemObject")
Set objDrive = objFs.GetDrive("C:\")
```

The `Drive` object has the following properties:

- `AvailableSpace`—Available space in bytes
- `FreeSpace`—Same as `AvailableSpace`

- `DriveLetter`—Drive letter (without the colon)
- `DriveType`—Type of drive:
    - `0`—Unknown
    - `1`—Removable
    - `2`—Fixed
    - `3`—Network
    - `4`—CD-ROM
    - `5`—RAM disk

```
Sub Removable_DriveLetter()
    Dim objFs As Object
    Dim colDrives As Object
    Dim Drive As Object
    Dim counter As Integer
    Const Removable = 1

    Set objFs = CreateObject("Scripting.FileSystemObject")
    Set colDrives = objFs.Drives
    counter = 0
    For Each Drive In colDrives
        If Drive.DriveType = Removable Then
            counter = counter + 1
            Debug.Print "The Removable Drive:
                " & Drive.DriveLetter
        End If
    Next
    MsgBox "There are " & counter & " Removable drives."
End Sub
```

- `FileSystem`—File system such as FAT, NTFS, or CDFS
- `IsReady`—Returns `True` if the appropriate media (CD-ROM) is inserted and ready for access:

```
Function IsCDROMReady(strDriveLetter)
  Dim objFs As Object
  Dim objDrive As Object

  Set objFs = CreateObject("Scripting.FileSystemObject")
  Set objDrive = objFs.GetDrive(strDriveLetter)

  IsCDROMReady = (objDrive.DriveType = 4) And _
      objDrive.IsReady = True
```

```
' run this function from the Immediate window
' by entering: ?IsCDROMReady("D:")
End Function
```

- `Path`—Path of the root folder
- `SerialNumber`—Serial number of the drive
- `TotalSize`—Total drive size in bytes

CREATING A TEXT FILE USING WSH

WSH offers three methods for creating text files: `CreateTextFile`, `OpenText-File`, and `OpenAsTextStream`.

Creating a Text File Using the CreateTextFile Method

The syntax of the `CreateTextFile` method, the method's arguments, and an example procedure are shown below:

```
CreateTextFile object.CreateTextFile(filename[,
   overwrite[, unicode]])
```

- `object` is the name of the `FileSystemObject` or the `Folder` object, and `filename` is a string expression that specifies the file to create.
- `overwrite` (optional) is a Boolean value that indicates whether you can overwrite an existing file. The value is `True` if the file can be overwritten and `False` if it can't be overwritten. If omitted, existing files are not overwritten.
- `unicode` (optional) is a Boolean value that indicates whether the file is created as a Unicode or ASCII file. The value is `True` if the file is created as a Unicode file and `False` if it's created as an ASCII file. If omitted, an ASCII file is assumed.

```
Sub CreateFile_Method1()
  Dim objFs As Object
  Dim objFile As Object

  Set objFs = CreateObject("Scripting.FileSystemObject")
  Set objFile = objFs.CreateTextFile( _
          "C:\VBAExcel2024_ByExample\Phones.txt", True)
  objFile.WriteLine ("Margaret Kubiak: 212-338-8778")
  objFile.WriteBlankLines (2)
  objFile.WriteLine ("Robert Prochot: 202-988-2331")
```

```
    objFile.Close
End Sub
```

The above procedure creates a text file to store the names and phone numbers of two people. Because there is a Boolean value of `True` in the position of the `overwrite` argument, the `Phones.txt` file will be overwritten if it already exists in the specified folder.

Creating a Text File Using the OpenTextFile Method

The syntax of the `OpenTextFile` method, the method's arguments, and an example procedure are shown below:

```
OpenTextFile object.OpenTextFile(filename[, iomode[,
      create[, format]]])
```

`object` is the name of the `FileSystemObject` and `filename` is a string expression that identifies the file to open.

`iomode` (optional) is a Boolean value that indicates whether a new file can be created if the specified filename doesn't exist. The value is `True` if a new file is created and `False` if it isn't created. If omitted, a new file isn't created. The `iomode` argument can be one of the following constants:

- `ForReading` (1)
- `ForWriting` (2)
- `ForAppending` (8)

`create` (optional) is a Boolean value that indicates whether a new file can be created if the specified filename doesn't exist. The value is `True` if a new file is created and `False` if it isn't created. If omitted, a new file isn't created.

`format` (optional) is one of three `Tristate` values used to indicate the format of the opened file. If omitted, the file is opened as ASCII:

- `TristateTrue`—Open the file as ASCII.
- `TristateFalse`—Open the file as Unicode.
- `TristateUseDefault`—Open the file using the system default.

The following procedure creates a text file using OpenTextFile method:

```
Sub CreateFile_Method2()
  Dim objFs As Object
  Dim objFile As Object

  Const ForWriting = 2
```

```
  Set objFs = CreateObject("Scripting.FileSystemObject")
  Set objFile = _
  objFs.OpenTextFile("C:\VBAExcel2024_ByExample\Shopping.txt", _
      ForWriting, True)

  objFile.WriteLine ("Bread")
  objFile.WriteLine ("Milk")
  objFile.WriteLine ("Strawberries")
  objFile.Close
End Sub
```

Creating a Text File Using the OpenAsTextStream Method

The syntax of the `OpenAsTextStream` method, the method's arguments, and an example procedure are shown below:

```
OpenAsTextStream object.OpenAsTextStream([iomode, [format]])
```

`object` is the name of the `File` object.

`iomode` (optional) indicates input/output mode. This can be one of three constants:

- `ForReading` (1)
- `ForWriting` (2)
- `ForAppending` (8)

`format` (optional) is one of three `Tristate` values used to indicate the format of the opened file. If omitted, the file is opened as ASCII:

- `TristateTrue`—Open the file as ASCII.
- `TristateFalse`—Open the file as Unicode.
- `TristateUseDefault`—Open the file using the system default.

The following procedure creates a text file using OpenAsTextStream method:

```
Sub CreateFile_Method3()
  Dim objFs As Object
  Dim objFile As Object
  Dim objText As Object
  Const ForWriting = 2
  Const ForReading = 1
  Const TristateUseDefault = -2

  Set objFs = CreateObject("Scripting.FileSystemObject")
  objFs.CreateTextFile "New.txt"
  Set objFile = objFs.GetFile("New.txt")
  Set objText = objFile.OpenAsTextStream(ForWriting, _
      TristateUseDefault)
```

```
   objText.Write "Wedding Invitation"
   objText.Close
   Set objText = objFile.OpenAsTextStream(ForReading, _
      TristateUseDefault)
   MsgBox objText.ReadLine
   objText.Close
End Sub
```

PERFORMING OTHER OPERATIONS WITH WSH

WSH makes it possible to manipulate any `Automation` object installed on your computer. In addition to accessing the file system through `FileSystemObject`, you can perform such tasks as handling WSH and ActiveX objects, mapping and unmapping printers and remote drives, manipulating the registry, creating Windows and Internet shortcuts, and accessing the Windows NT Active Directory service.

The WSH object model is made of the following three main objects:

- `WScript`
- `WshShell`
- `WshNetwork`

In the following sections, we'll use the `WshShell` object to write procedures that start other applications and create shortcuts.

Running Other Applications

Suppose you want to start up Windows Notepad from your VBA procedure. The procedure that follows shows you how easy it is to run an application using the `WshShell` object. If you'd rather launch the built-in calculator, just replace the name of the `Notepad` application with `Calc`.

⊚ Hands-On 11.3   Running Other Applications Using the WSH Object

1. Insert a new module into the `FileMan_WSH`   (`Chap11.xlsm`) project and rename it `WSH_Additional`.
2. Enter the `RunNotepad` procedure in the `WSH_Additional` module's Code window, as shown below:

```
Sub RunNotepad()
   Dim WshShell As Object
   Set WshShell = CreateObject("WScript.Shell")
   WshShell.Run "Notepad"
```

```
    Set WshShell = Nothing
End Sub
```

The above procedure begins by declaring and creating a WshShell object. The next statement uses the Run method to run the required application.

Using the same concept, it is easy to run Windows utility applications such as Calculator or Explorer:

```
WshShell.Run "Calc"
WshShell.Run "Explorer"
```

The last line in the procedure destroys the WshShell object because it is no longer needed.

```
Set WshShell = Nothing
```

3. Execute the RunNotepad procedure.

Instead of launching an empty application window, you can start your application with a specific document, as shown in the following procedure:

```
Sub OpenTxtFileInNotepad()
    Dim WshShell As Object
    Set WshShell = CreateObject("WScript.Shell")
    WshShell.Run "Notepad C:\VBAExcel2024_ByExample\Vacation.txt"
    Set WshShell = Nothing
End Sub
```

If the specified file cannot be found, Visual Basic will prompt you to create the file. If the path contains any spaces, the pathname must be enclosed in double quotes, or the Run method will raise a runtime error. For example, to open the C:\My Files\my text file.txt file in Notepad, use the following statement:

```
WshShell.Run "Notepad " & """C:\My Files\my text file.txt"""
```

or use the ANSI (American National Standards Institute) equivalent for double quote marks, as shown below:

```
WshShell.Run "Notepad " & Chr(34) & _
"C:\My Files\my text file.txt" & Chr(34)
```

The above statement uses the VBScript function Chr to convert an ANSI value to its character equivalent. 34 is the ANSI value for double quotes.

To open a Web page, pass the Web page address to the Run method, as in the following:

```
Sub Launch_MSN()
Dim WshShell As Object
```

```
Set WshShell = CreateObject("WScript.Shell")

WshShell.Run "https://msn.com"
Set WshShell = Nothing
End Sub
```

The following statement invokes a Control Panel:

```
WshShell.Run "Control.exe"
```

The Control Panel has various property sheets. The following statement will open the System property page with the Hardware tab selected:

```
WshShell.Run "Control.exe Sysdm.cpl, ,2"
```

The parameters after the name of the Control Panel property page (identified by the `.cpl` extension) specify which page is selected. To select the General tab on this property page, replace 2 with 0 (zero).

Note that the `Run` method has two optional arguments that allow you to specify the window style and whether the system should wait until the executed process completes. For example, you can launch Notepad in the minimized view as shown below:

```
WshShell.Run "Notepad", vbMinimizedFocus
```

Because the second optional parameter is not specified, the `Run` method executes the command (`Notepad.exe`) and immediately terminates the process. If the second parameter is set to `True`, the `Run` method will create a new process, execute the command, and wait until the process terminates:

```
WshShell.Run "Notepad", vbMinimizedFocus, True
```

If you run the above statement with the second parameter set to `True`, you will not be able to work with Excel until you close Notepad.

Obtaining Information About Windows

Windows stores various kinds of information in environment variables. You can use the `Environment` property of the `WshShell` object to access these variables. Depending on which version of the operating system you are using, the environment variables are grouped into `System`, `User`, `Volatile`, and `Process` categories. You can use the name of the category as the index of the `Environment` property.

The following procedure demonstrates how to retrieve the values of several environment variables from the `Process` category:

```
Sub ReadEnvVar()
  Dim WshShell As Object
  Dim objEnv As Object

  Set WshShell = CreateObject("WScript.Shell")
  Set objEnv = WshShell.Environment("Process")

  Debug.Print "Path=" & objEnv("PATH")
  Debug.Print "System Drive=" & objEnv("SYSTEMDRIVE")
  Debug.Print "System Root=" & objEnv("SYSTEMROOT")
  Debug.Print "Windows folder=" & objEnv("Windir")
  Debug.Print "Operating System=" & objEnv("OS")
  Set WshShell = Nothing
End Sub
```

Retrieving Information About the User, Domain, or Computer

You can use the properties of the WshNetwork object of WSH to retrieve the user's name, domain name, or computer name, as shown in the procedure below:

```
Sub GetUserDomainComputer()
  Dim WshNetwork As Object
  Dim myData As String

  Set WshNetwork = CreateObject("WScript.Network")
  myData = myData & "Computer Name: " _
      & WshNetwork.ComputerName & vbCrLf
  myData = myData & "Domain: " _
      & WshNetwork.UserDomain & vbCrLf
  myData = myData & "User Name: " _
      & WshNetwork.UserName & vbCrLf

  MsgBox myData
End Sub
```

Creating Shortcuts

When you start distributing your VBA applications, users might request that you automatically place a shortcut to your application on their desktop. VBA does not provide a way to create Windows shortcuts. Luckily for you, you now know how to work with WSH, and you can use its Shell object to create shortcuts to applications or Web sites without any user intervention.

The WshShell object exposes the CreateShortcut method, which you can use in the following way:

```
Set myShortcut = WshShell.CreateShortcut(pathname)
```

`pathname` is a string indicating the full path to the shortcut file. All shortcut files have the `.lnk` extension, which must be included in the pathname.

The `CreateShortcut` method returns a `Shortcut` object that exposes a number of properties and one method.

The `TargetPath` property is the path to the shortcut's executable:

```
WshShell.TargetPath = ActiveWorkbook.FullName
```

The `WindowStyle` property identifies the window style used by a shortcut:

- 1—Normal window
- 3—Maximized window
- 7—Minimized window

```
WshShell.WindowStyle = 1
```

The `Hotkey` property is a keyboard shortcut—for example, Alt+F, Shift+G, Ctrl-Shift+Z, and so on:

```
WshShell.Hotkey = "Ctrl+Alt+W"
```

The `IconLocation` property is the location of the shortcut's icon. Because icon files usually contain more than one icon, you should provide the path to the icon file followed by the index number of the desired icon in this file. If not specified, Windows uses the default icon for the file:

```
WshShell.IconLocation = "notepad.exe, 0"
```

The `Description` property contains a string value describing a shortcut:

```
WshShell.Description = "Mercury Learning Web Site"
```

The `WorkingDirectory` property identifies the working directory used by a shortcut:

```
strWorkDir = WshShell.SpecialFolders("Desktop")
WshShell.WorkingDirectory = strWorkDir
```

- `Save` is the only method of the `Shortcut` object. After using the `CreateShortcut` method to create a `Shortcut` object and set the `Shortcut` object's properties, the `CreateShortcut` method must be used to save the `Shortcut` object to disk.

Creating a shortcut is a three-step process:

1. Create an instance of a `WshShortcut` object.
2. Initialize its properties (shown above).
3. Save it to disk with the `Save` method.

The following Hands-On creates a `WshShell` object and uses the `CreateShortcut` method to create two shortcuts: a shortcut to the active Microsoft Excel workbook file and an Internet shortcut to the De Gruyter Brill Web site. Both shortcuts are placed on the user's desktop.

(⊙) **Hands-On 11.4   Creating Shortcuts Using the WshShell Object**

1. In the `WSH_Additional` module created in the previous Hands-On exercise, enter the `CreateShortcut` procedure, as shown below:

```
Sub CreateShortcut()
    ' this script creates two desktop shortcuts
    Dim WshShell As Object
    Dim objShortcut As Object
    Dim strWebAddr As String

    strWebAddr = "https://www.DeGruyterBrill.com"

    Set WshShell = CreateObject("WScript.Shell")

    ' create an Internet shortcut
    Set objShortcut = WshShell.CreateShortcut(WshShell. _
        SpecialFolders("Desktop") & "\DeGruyterBrill.url")
    With objShortcut
        .TargetPath = strWebAddr
        .Save
    End With

    ' create a file shortcut
    ' you cannot create a shortcut to unsaved workbook file
    Set objShortcut = WshShell.CreateShortcut(WshShell. _
        SpecialFolders("Desktop") & "\" &_
            ActiveWorkbook.Name & ".lnk")
    With objShortcut
        .TargetPath = ActiveWorkbook.FullName
        .Description = "Discover De Gruyter Brill"
        .WindowStyle = 7
        .Save
    End With

    Set objShortcut = Nothing
    Set WshShell = Nothing
End Sub
```

2. The above procedure uses the `SpecialFolders` property of the `WshShell` object to return the path to the Windows desktop.

3. Run the `CreateShortcut` procedure.

4. Switch to your desktop and click the `DeGruyterBrill` shortcut.

5. Close the active workbook file and test the shortcut to this file, which you should now have on your desktop.

Using the SpecialFolders Property

The `SpecialFolders` property allows you to access the paths of various special folders on your machine. The listed folders are `AllUsersDesktop`, `AllUsersPrograms`, `AllUsersStartMenu`, `AllUsersStartup`, `Desktop`, `Favorites`, `Fonts`, `MyDocuments`, `NetHood`, `PrintHood`, `Programs`, `Recent`, `SendTo`, `StartMenu`, `Startup`, and `Templates`. If the requested folder is not available, the property will return an empty string.

Listing Shortcut Files

The following procedure prints to the Immediate window the list of all shortcut files found on the desktop. This procedure uses the `InStrRev` function to check whether the file is a shortcut. This function has the same syntax as the `InStr` function, except that it returns the position of an occurrence of one string within another, from the end of the string:

```
Sub ListShortcuts()
    Dim objFs As Object
    Dim objFolder As Object
    Dim wshShell As Object
    Dim strLinks As String
    Dim s As Variant
    Dim f As Variant
    Dim i As Integer

    Set wshShell = CreateObject("WScript.Shell")
    Set objFs = CreateObject("Scripting.FileSystemObject")
    strLinks = ""

    i = 1
    For Each s In wshShell.SpecialFolders
        Set objFolder = objFs.GetFolder(s)
        strLinks = strLinks & objFolder.Name _
            & " Shortcuts:" & vbCrLf

        If objFolder.Name = "Desktop" Then
            For Each f In objFolder.Files
                If InStrRev(UCase(f), ".LNK") Then
```

```
                    strLinks = strLinks & f.Name & vbCrLf
                ' List the shortcut name in the worksheet
                Cells(i, 1).Value = f.Name
                i = i + 1
                End If
            Next
        End If
        Exit For
    Next
    Debug.Print strLinks
End Sub
```

SUMMARY

In this chapter, you learned how, by integrating `FileSystemObject` with WSH, you can automate a wide range of file system tasks efficiently. Whether you need simple file operations or more complex automation, leveraging these tools can significantly enhance your programming capabilities.

In the next chapter, you will learn how to use VBA to work with three types of files: sequential, random access, and binary.

12 DIRECT MANIPULATION OF FILES AND THEIR CONTENTS

In addition to opening files within a particular application, VBA contains special statements that allow you to directly manipulate files and read their contents. This chapter will put you in direct contact with your data by introducing you to the process known as *low-level file I/O* (input/output). Low-level file I/O involves directly interacting with the bytes and characters of a file, giving you greater control and precision over your data handling tasks. By mastering these techniques, you'll be able to efficiently manage file operations, from simple text file manipulations to more complex file processing.

FILE ACCESS TYPES

There are three types of files used by a computer:

- *Sequential access files* are files where data is retrieved in the same order as it is stored, such as files stored in the CSV format (comma-delimited text), TXT format (text separated by tabs), or PRN format (text separated by spaces). Sequential access files are often used for writing text files such as error logs, configuration settings, and reports. Sequential access files have the following modes: Input, Output, and Append. The mode specifies how you can work with a file after it has been opened. In a sequential access file, the records must be accessed in the order they occur in the file. This means that before you can access the third record, you must first access record number 1 and then record number 2, and so on.

- *Random-access files* are text files where data is stored in records of equal length and fields separated by commas. In a random-access file, data is stored in records that can be accessed without having to read every record preceding it. Random-access files have only one mode: `Random`.
- *Binary access files* allow you to read from and write to files at the byte level. This is particularly useful when dealing with non-text data such as images, videos, and other binary formats. Binary access is more flexible and efficient compared to other file types such as sequential access or random-access files, especially for larger datasets and complex file structures. Binary files can only be accessed in a `Binary` mode.

WORKING WITH SEQUENTIAL FILES

Your computer hard drive contains hundreds of sequential files. Configuration files, error logs, HTML files, and all sorts of plain text files are all sequential files. These files are stored on disk as a sequence of characters.

In a sequential access file, the beginning of a new text line is indicated by two special characters: the carriage return and the linefeed. When you work with sequential files, start at the beginning of the file and move forward character by character, line by line, until you encounter the end of the file. Sequential access files can be easily opened and manipulated by just about any text editor.

Opening and Reading Sequential Files

Let's take one of the sequential files already present on your computer and read its contents with VBA straight from the VBE window. You can read any other text file you want. To read data from a file, open the file with the `Open` statement. Here's the general syntax of this statement, followed by an explanation of each component:

```
Open pathname For mode[Access access] [lock] As [#]filenumber
    [Len=reclength]
```

The `Open` statement has three required arguments: `pathname`, `mode`, and `file-number`.

`pathname` is the name of the file you want to open. The filename may include the name of a drive and folder.

`mode` is a keyword that determines how the file was opened. Sequential files can be opened in one of the following modes: `Input`, `Output`, or `Append`. Use

`Input` to read the file, `Output` to write to a file by overwriting any existing file, and `Append` to write to a file by adding to any existing information.

The optional `Access` clause can be used to specify permissions for the file (`Read`, `Write`, or `Read Write`).

The optional `lock` argument determines which file operations are allowed for other processes. For example, if a file is open in a network environment, `lock` determines how other people can access it. The following `lock` keywords can be used: `Shared`, `Lock Read`, `Lock Write`, or `Lock Read Write`.

`filenumber` is a number from 1 to 511. This number is used to refer to the file in subsequent operations. You can obtain a unique file number using the Visual Basic built-in `FreeFile` function.

The last element of the `Open` statement, `reclength`, specifies the buffer size (total number of characters) for sequential files, or the record size for random-access files.

Taking the preceding into consideration, to open `Readme.txt` or any other sequential file to read its data, use the following instruction:

```
Open "C:\VBAExcel2024_ByExample\Readme.txt" For Input As #1
```

If a file is opened for input, it can only be read from. After you open a sequential file, you can read its contents with the `Line Input #` or `Input #` statements or by using the `Input` function, as shown in the next section. When you use sequential access to open a file for input, the file must already exist.

Reading a File Line by Line

To read the contents of a sequential file line by line, use the following `Line Input #` statement:

```
Line Input #filenumber, variableName
```

`#filenumber` is the file number that was used in the process of opening the file with the `Open` statement. `variableName` is a `String` or `Variant` variable that will store the line being read. The statement `Line Input #` reads a single line in an open sequential file and stores it in a variable. Bear in mind that the `Line Input #` statement reads the sequential file one character at a time until it encounters a carriage return (`Chr(13)`) or a carriage return linefeed sequence (`Chr(13) & Chr(10)`). These characters are omitted from the text retrieved in the reading process.

The `ReadMe` procedure that follows demonstrates how to use the `Open` and `Line Input #` statements to read the contents of the `Readme.txt` file line by line. Apply the same method for reading other sequential files.

> **NOTE** *Please note that the files for the "Hands-On" projects can be found in the companion files.*

Hands-On 12.1 Reading File Contents with the Open and Line Input # Statements

1. Open a new workbook and save it as `C:\VBAExcel2024_ByExample\Chap12.xlsm`.
2. Switch to the VBE and use the Properties window to rename `VBAProject (Chap12.xlsm)` `FileMan_IO`.
3. Insert a new module in the `FileMan_IO` project and rename it `SeqFiles`.
4. In the `SeqFiles` module's Code window, enter the `ReadMe` procedure shown below:

```
Sub ReadMe(strFileName As String)
   Dim rLine As String
   Dim i As Integer

   ' line number

   i = 0

   On Error GoTo ExitHere
   Open strFileName For Input As #1

   ' stay inside the loop until the end of file is reached
   Do While Not EOF(1)
      i = i + 1
      Line Input #1, rLine
      MsgBox "Line " & i & " in " & strFileName & " reads: " _
         & Chr(13) & Chr(13) & rLine
   Loop
   MsgBox i & " lines were read."
   Close #1
   Exit Sub
ExitHere:
   MsgBox "File " & strFileName & " could not be found."
End Sub
```

The `ReadMe` procedure opens the specified text file in the `Input` mode as file number `1` in order to read its contents. If the specified file cannot be opened (because it may not exist), Visual Basic jumps to the label `ExitHere` and displays a message box.

After the file is successfully opened, the `Do...While` loop tells Visual Basic to execute the statements inside the loop until the end of the file has been reached. The end of the file is determined by the result of the `EOF` (end of file) function. The `EOF` function returns a logical value of `True` if the next character to be read is past the end of the file. Notice that the `EOF` function requires one argument—the number of the open file you want to check. This is the same number used in the `Open` statement. Use the `EOF` function to ensure that Visual Basic doesn't read past the end of the file.

The `Line Input #` statement stores each line's contents in the variable `rLine`. Next, a message is displayed showing the line number and its contents. Visual Basic exits the `Do...While` loop when the result of the `EOF` function is `True`. Before VBA ends the procedure, two more statements are executed. A message is displayed with the total number of lines that have been read, and the file is closed.

5. To run the procedure, open the Immediate window, type the following statement, and press Enter to execute:

```
ReadMe "C:\VBAExcel2024_ByExample\Readme.txt"
```

Reading Characters from Sequential Files

Suppose that your procedure needs to check how many commas appear in a file. Instead of reading entire lines, you can use the `Input` function to return the specified number of characters. Next, the `If` statement can be used to compare the obtained character against the one you are looking for. Before you write a procedure that does that, let's review the syntax of the `Input` function:

```
Input(number, [#]filenumber)
```

Both arguments of the `Input` function are required; `number` specifies the number of characters you want to read, and `filenumber` is the same number that the `Open` statement used to open the file. The `Input` function returns all the characters being read, including commas, carriage returns, end-of-file markers, quotes, and leading spaces.

⊙ Hands-On 12.2   Reading Characters from Sequential Files

1. Enter the following `CountChar` procedure in the `SeqFiles` module:

```
Sub CountChar(strFileName As String, srchChar As String)
    Dim counter As Integer
    Dim char As String
```

```
      counter = 0
      Open strFileName For Input As #1

      Do While Not EOF(1)
         char = Input(1, #1)
          If char = srchChar Then
              counter = counter + 1
          End If
      Loop
      If counter <> 0 Then
          MsgBox "Characters (" & srchChar & ") found: " & counter
      Else
          MsgBox "The specified character (" & srchChar & _
              ") has not been found."
      End If
      Close #1
   End Sub
```

2. To run the procedure, open the Immediate window, type the following statement, and press Enter to execute:

```
CountChar "C:\VBA2024_ByExample\Readme.txt", "."
```

You should get a message that the `Readme.txt` file contains 1 period.

3. Run the procedure again after replacing the period character with any other character you'd like to find.

The `Input` function allows you to return any character from the sequential file. If you use the Visual Basic `LOF` (length of file) function as the first argument of the `Input` function, you'll be able to quickly read the contents of the sequential file without looping through the entire file.

The `LOF` function returns the number of bytes in a file. Each byte corresponds to one character in a text file.

The following `ReadAll` procedure shows how to read the contents of a sequential file to the Immediate window:

```
Sub ReadAll(strFileName As String)
   Dim all As String

   Open strFileName For Input As #1
   all = Input(LOF(1), #1)
   Debug.Print all
   Close #1
End Sub
```

4. To execute the above procedure, open the Immediate window, type the following statement, and press Enter:

```
ReadAll "C:\VBAExcel2024_ByExample\Readme.txt"
```

Instead of printing the file contents to the Immediate window, you can read it into a text box placed in a worksheet, like the one in Figure 12.1. Let's take a few minutes to write this procedure.

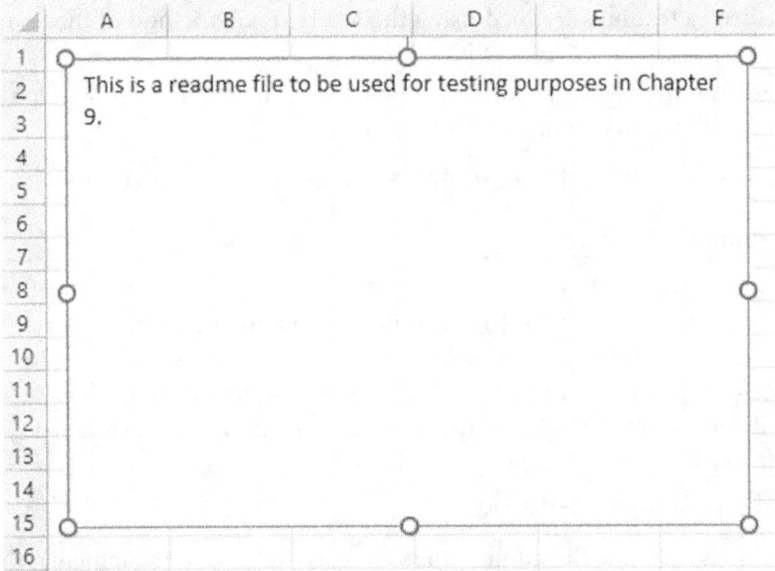

FIGURE 12.1.   The contents of a text file are displayed in a text box placed in an Excel worksheet. This Readme.txt file was prepared in Chapter 9 (see Hands-On 9.5).

Hands-On 12.3   Printing File Contents to a Worksheet Text Box

1. Enter the WriteToTextBox procedure below in the SeqFiles module:

```
Sub WriteToTextBox(strFileName As String)
   Dim sh As Worksheet
   Set sh = ActiveSheet

   On Error GoTo CloseFile:

   Open strFileName For Input As #1
   sh.Shapes.AddTextbox(msoTextOrientationHorizontal, _
      10, 10, 300, 200).Select
```

```
   Selection.Characters.Text = Input(LOF(1), #1)
CloseFile:
   Close #1
End Sub
```

The statement `On Error GoTo CloseFile` activates error trapping. If an error occurs, the program will jump to the error-handling routine that follows the `CloseFile` label. The statement `Close #1` will be executed, whether or not the program encounters an error. Notice that before the file contents are placed in a worksheet, a text box is added using the `AddTextbox` method of the `Shapes` object.

To execute the above procedure, open the Immediate window, type the following statement, and press Enter:

```
WriteToTextBox "C:\VBAExcel2024_ByExample\Readme.txt"
```

Reading Delimited Text Files

In some text files (usually, files saved in CSV, TXT, or PRN format), data entered on each line of text is separated (or delimited) with a comma, tab, or space character. These types of files can be read faster with the `Input #` statement instead of the `Line Input #` statement introduced earlier in this chapter. The `Input #` statement allows you to read data from an open file into several variables. This statement's syntax is shown below:

```
Input #filenumber, variablelist
```

`filenumber` is the same file number that was used in the `Open` statement. `variablelist` is a comma-separated list of variables that you will want to use to store the data being read. Note that you cannot use arrays or object variables. You can, however, use a user-defined variable (explained later in this chapter). An example of a sequential file with comma-delimited values is shown below:

```
Smith,John,15
Malloney,Joanne,28
Ikatama,Robert,15
```

Note that, in this example, there are no spaces before or after the commas. To read text formatted in this way, you must specify one variable for each item of data: last name, first name, and age. Let's try it out.

⊙ Hands-On 12.4 Reading a Comma-Delimited (CSV) File with the Input # Statement

1. Open a new workbook and enter the data shown in the worksheet in Figure 12.2:

◢	A	B	C	D
1	Smith	John	15	
2	Malloney	James	28	
3	Ikatama	Robert	15	
4				

FIGURE 12.2. You can create a comma-delimited file from an Excel workbook.

2. Click the File tab on the ribbon, then click Save As. Switch to the `C:\` `VBAExcel2024_ByExample` folder. In the Save As type drop-down box, select CSV (Comma delimited) (*.csv). Change the filename to `Winners.csv` and click Save.

3. Excel will display a warning message that some features will be lost if you save the file as CSV. Close the warning message.

NOTE	*The CSV file type does not support workbooks that contain multiple sheets. If you add more sheets to the workbook and try to save the workbook as a comma-delimited file (CSV), Excel displays a warning message.*

4. Close the `Winners.csv` file. Click No if prompted to save changes.

5. Activate the `Chap12.xlsm` workbook and switch to the VBE.

6. In the Project Explorer window, double-click the SeqFiles module in the `FileMan_IO (Chap12.xlsm)` project and enter the `Winners` procedure, as shown below:

```
Sub Winners()
   Dim lname As String
   Dim fname As String
   Dim age As Integer

   Open "C:\VBAExcel2024_ByExample\Winners.csv" For Input As #1
   Do While Not EOF(1)
       Input #1, lname, fname, age
       MsgBox lname & ", " & fname & ", " & age
   Loop
   Close #1
End Sub
```

The above procedure opens the `Winners.csv` file for input and sets up a `Do…While` loop that runs through the entire file until the end of the file is reached. The `Input #1` statement is used to write the contents of each line of text into three variables: `lname`, `fname`, and `age`. Then, a message box displays the contents of these variables. The procedure ends by closing the `Winners.csv` file.

7. Run the `Winners` procedure.

Now that you've mastered reading sequential files, let's explore the process of adding new data to these files.

Writing Data to Sequential Files

To write data to a sequential file, open the file in `Append` or `Output` mode. The following are the differences between these two modes:

- `Append` mode—Use `Append` to add data to the end of an existing text file. For example, if you open the `Readme.txt` file in `Append` mode and add the text `Thank you for reading this document` to this file, Visual Basic won't delete or alter the text that is currently in the file but will add the new text to the end of the file.

- `Output` mode—In `Output` mode, Visual Basic will delete the data that is currently in the file. If the file does not exist, a brand-new file will be created. For example, if you open the `Readme.txt` file in `Output` mode and attempt to write some text to it, the previous text that was stored in this file will be removed. If you don't back up the file prior to writing the data, this mistake may be quite costly. You should open an existing file in `Output` mode only if you want to replace its entire contents with new data.

Here are some examples of when to open a file in `Append` mode or `Output` mode:

- To create a brand-new text file called `C:\VBAExcel2024_ByExample\ReadmeNew.txt`, open the file in `Output` mode, as follows:

```
Open "C:\VBAExcel2024_ByExample\ReadmeNew.txt" For Output As #1
```

- To add new text to the end of `C:\VBAExcel2024_ByExample\ReadmeMe.txt`, open the file in `Append` mode, as follows:

```
Open "C:\VBAExcel2024_ByExample\ReadmeNew.txt" For Append As #1
```

● To replace the contents of an existing file, `C:\VBAExcel2024_ByExample\ Winners.csv`, with a list of new winners, first, prepare a backup copy of the original file, then open the original file in `Output` mode:

```
FileCopy "C:\VBAExcel2024_ByExample\Winners.csv",
    "C:\VBAExcel2024_ByExample\Winners.old"
Open "C:\VBAExcel2024_ByExample\Winners.csv" For Output As #1
```

Can't Read and Write at the Same Time

You cannot perform read-write operations simultaneously on an open sequential file. The file must be opened separately for each operation. For instance, after data has been written to a file that has been opened for output, the file must be closed before being opened for input.

Advantages and Disadvantages of Sequential Files

Although sequential files are easy to create and use, and don't waste any space, they have several disadvantages. For example, you can't easily find one specific item in the file without having to read through a large portion of the file. Also, you must rewrite the entire file to change or delete an individual item in the file. As stated above, sequential files must be opened for reading and again for writing.

Using Write # and Print # Statements

Now that you know both methods (`Append` and `Output`) for opening a text file with the intention of writing to it, let's look at the `Write #` and `Print #` statements that will allow you to send data to the file.

When you read data from a sequential file with the `Input #` statement, you usually write data to this file with the `Write #` statement. This statement looks like this:

```
Write #filenumber, [outputlist]
```

`filenumber` specifies the number of the file you're working with. It is the only required argument of the `Write #` statement.

`outputlist` is the text you want to write to the file and it can be a single text string or a list of variables that contain data that you want to write. If you specify only the `filenumber` argument, Visual Basic will write a single empty line to the open file.

To illustrate how data is written to a file, let's prepare a text file with the first name, last name, birth date, and number of siblings for three people.

Hands-On 12.5   Using the Write # Statement to Write Data to a File

1. In the SeqFiles module, enter the DataEntry procedure as shown below:

```
Sub DataEntry()
    Dim lname As String
    Dim fname As String
    Dim birthdate As Date
    Dim sib As Integer

    Open "C:\VBAExcel2024_ByExample\Friends.txt" For Output As #1
    lname = "Smith"
    fname = "Gregory"
    birthdate = #1/2/1963#
    sib = 3
    Write #1, lname, fname, birthdate, sib

    lname = "Conlin"
    fname = "Janice"
    birthdate = #5/12/1948#
    sib = 1
    Write #1, lname, fname, birthdate, sib

    lname = „Kaufman"
    fname = „Steven"
    birthdate = #4/7/1986#
    sib = 0
    Write #1, lname, fname, birthdate, sib

    Close #1
End Sub
```

The above procedure creates a brand-new file named C:\VBAExcel2024_ByExample\Friends.txt, opens it for Output, and then writes to it three records. The data written to the file is stored in variables. Notice that the strings are delimited with double quotes ("") and the birthdate is surrounded by pound signs, #.

2. Run the DataEntry procedure.
3. Locate the Friends.txt file created by the DataEntry procedure and open it using Windows Notepad.
   The Friends.txt file opened in Notepad looks as follows:

```
"Smith","Gregory",#1963-01-02#,3
"Conlin","Janice",#1948-05-12#,1
"Kaufman","Steven",#1986-04-07#,0
```

The `Write #` statement in the `DataEntry` procedure automatically inserted commas between the individual data items in each record and placed the carriage return linefeed sequence (`Chr(13) & Chr(10)`) at the end of each line of text so that each new record starts in a new line. Each line of text shows one record: each record begins with the last name and ends with the number of siblings.

To show the data separated by columns, instead of commas, write the data with the `Print #` statement. For example, if you replace the `Write #` statement in the `DataEntry` procedure with the `Print #` statement, Visual Basic will write the data as follows:

```
Smith Gregory 1/2/63 3
Conlin Janice 5/12/48 1
Kaufman Steven 4/7/86 0
```

Although the `Print #` statement has the same syntax as the `Write #` statement, `Print #` writes data to the sequential file in a format ready for printing. The variables in the list may be separated with semicolons or spaces. To print out several spaces, you should use the `Spc(n)` instruction, where n is the number of spaces. Similarly, to enter a word in the fifth column, you should use the instruction `Tab(5)`. Let's look at some formatting examples:

- To add an empty line to a file, use the `Write #` statement with a comma: `Write #,`
- To enter the text `fruits` in the fifth column: `Write #1, Tab(5); "fruits"`
- To separate the words `fruits` and `vegetables` with five spaces: `Write #1, "fruits"; Spc(5); "vegetables"`

WORKING WITH RANDOM-ACCESS FILES

When a file contains structured data, open the file in random access mode. A file opened for random access allows you to:

- Read and write data at the same time.
- Quickly access a particular record.

In random-access files, all records are of equal length, and each record has the same number of fixed-size fields. The length of a record or field must be determined prior to writing data to the file. If the length of a string that is being written to a field is less than the specified size of the field, Visual Basic automatically enters spaces at the end of the string to fill in the entire size of the field. If the text being written is longer than the size of the field, the text that does not fit will be truncated.

Creating Random-Access Files—Writing Records

For random-access files, you need to structure your data using a user-defined type (UDT) (see Hands-On 12.6). You write/read records using the Put and Get statements, respectively.

To find out how to work with random-access files, let's create a small database for use in a foreign language study. This database will contain records made up of two fields: an English term and its foreign language equivalent.

⊚ **Hands-On 12.6   Creating a Random-Access Database with a User-Defined Data Type**

1. Insert a new module into the FileMan_IO (Chap12.xlsm) project and rename it RandomFiles.
2. Enter the following statements just below the Option Explicit statement at the top of the RandomFiles module:

```
' create a user-defined data type called Dictionary
Type Dictionary
  en As String * 16 ' English word up to 16 characters
  sp As String * 20 ' Spanish word up to 20 characters
End Type
```

In addition to the built-in data types introduced in Chapter 3, Visual Basic allows you to define a non-standard (user-defined) data type using a Type...End Type statement placed at the top of the module. The UDT can contain items of various data types (String, Integer, Date, and so on). When you work with files open for random access, you often create a user-defined variable because such a variable provides you with easy access to the individual fields of a record.

The UDT called Dictionary that you just defined contains two items declared as String with the specified size. The en item can accept up to 16 characters. The size of the second item (sp) cannot exceed 20 characters. By adding up the lengths of both items, you will get a record length of 36 characters (16 + 20).

Understanding the Type Statement

The `Type` command allows you to create a custom group of mixed variable types called a UDT. This statement is generally used with random-access files to store data as fields within records of a fixed size. Instead of declaring a separate variable for each field, cluster the fields into a user-defined variable using the `Type` statement. For example, define a record containing three fields in the following way:

```
Type MyRecord
    country As String * 20
    city As String * 14
    rank As Integer
End Type
```

Once the general type is defined, you must give a name to the variable that will be of that type:

```
Dim myInfo As MyRecord
```

Access the interior variables (`country`, `city`, `rank`) by using the following format:

```
Variable_name.Interior_variable_name
```

For example, to specify the city, enter:

```
MyInfo.city = "New York"
```

3. Enter the `EnglishToSpanish` procedure as shown below:

```
Sub EnglishToSpanish()
   Dim d As Dictionary
   Dim recNr As Long
   Dim choice As String
   Dim totalRec As Long

   recNr = 1
   ' open the file for random access
   Open "C:\VBAExcel2024_ByExample\Translate.txt" _
       For Random As #1 Len = Len(d)

   Do
     ' get the English word
     choice = InputBox("Enter an English word", "ENGLISH")
     d.en = choice
     ' exit the loop if canceled
     If choice = "" Then Exit Do
     choice = InputBox("Enter the Spanish equivalent of " _
         & d.en, "SPANISH EQUIVALENT  " & d.en)
     If choice = "" Then Exit Do
     d.sp = choice
```

```
    ' write to the record
    Put #1, recNr, d
    ' increase record counter
    recNr = recNr + 1
  'ask for words until Cancel
  Loop Until choice = ""

  totalRec = LOF(1) / Len(d)
  MsgBox "This file contains " & totalRec & " record(s)."
  ' close the file
  Close #1
End Sub
```

The EnglishToSpanish procedure begins with the declaration of four variables. The variable d is declared as a UDT called Dictionary. This type was declared earlier with the Type statement (see Step 2 above) at the top of the code module.

After the initial value is assigned to the variable RecNr, Visual Basic opens the Translate.txt file for random access as file number 1. The Len(d) instruction tells Visual Basic that the size of each record is 36 characters. (The variable d contains two elements: sp is 20 characters and en is 16 characters. Consequently, the total size of a record is 36 characters.) Next, Visual Basic executes the statements inside the Do...Until loop.

The first statement in the loop prompts you to enter an English word and assigns it to the variable choice. The value of this item is then passed to the first element of the user-defined variable d (d.en). When you cancel or stop entering data, Visual Basic exits the Do loop and executes the final statements that count and display the total number of records in the file. The last statement closes the file.

After entering an English word and clicking OK, you will be prompted to supply a foreign language equivalent. If you do not enter a word, Visual Basic will exit the loop and continue with the remaining statements. If the foreign language equivalent is supplied, Visual Basic will assign it to the variable choice and then pass it to the second element of the user-defined variable d (d.sp). Next, Visual Basic will write the entire record to the file using the following statement:

```
Put #1, recNr, d
```

After writing the first record, Visual Basic will increase the record counter by one and repeat the statements inside the loop. The EnglishToSpanish procedure allows you to enter any number of records into your dictionary.

When you quit supplying the words, the procedure uses the LOF and Len functions to calculate the total number of records in the file and displays a message. After that, Visual Basic closes the text file (Translate.txt).

4. Run the EnglishToSpanish procedure. When prompted, enter data as shown in Figure 12.3. For example, enter the word mother. When prompted for a Spanish equivalent of mother, enter madre. Click Cancel when done.

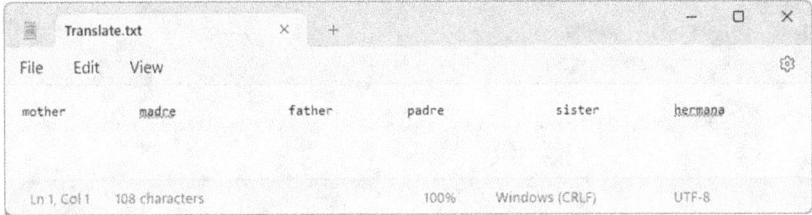

FIGURE 12.3. The contents of a random-access file opened in Notepad.

Reading Random-Access Files—Working with Records

Creating a random-access file is only the beginning. Next, we'll create the VocabularyDrill procedure to illustrate how to work with records in a file open for random access. Here, you will learn statements that allow you to quickly find the appropriate data in your file.

> ◉ **Hands-On 12.7 Reading Vocabulary Items Stored in the Translate.txt Random-Access file**

1. Below the EnglishToSpanish procedure in the RandomFiles module of the FileMan_IO (Chap12.xlsm) VBA project, enter the VocabularyDrill procedure as shown here:

```
Sub VocabularyDrill()
    Dim d As Dictionary
    Dim totalRec As Long
    Dim recNr As Long
    Dim randomNr As Long
    Dim question As String
    Dim answer As String

    ' open a random access file
    Open "C:\VBAExcel2024_ByExample\Translate.txt" _
        For Random As #1 Len = Len(d)

    ' print the total number of bytes in this file
    Debug.Print "There are " & LOF(1) & " bytes in this file."
```

```
' find and print the total number of records
recNr = LOF(1) / Len(d)
Debug.Print "Total number of records: " & recNr

Do
   ' get a random record number
   randomNr = Int(recNr * Rnd) + 1
   Debug.Print randomNr

   ' find the random record
   Seek #1, randomNr

   ' read the record
   Get #1, randomNr, d
   Debug.Print Trim(d.en); " "; Trim(d.sp)

   ' assign answer to a variable
   answer = InputBox("What's the Spanish equivalent?", d.en)

   ' finish if cancelled
   If answer = "" Then Close #1: Exit Sub
   Debug.Print answer
       ' check if the answer is correct
       If answer = Trim(d.sp) Then
           MsgBox "Congratulations!"
       Else
           MsgBox "Invalid Answer!!!"
       End If
' keep on asking questions until Cancel is pressed
   Loop While answer <> ""

   ' close file
   Close #1
End Sub
```

After declaring variables, the VocabularyDrill procedure opens a file for random access and tells Visual Basic the length of each record: Len = Len(d). Next, two statements print to the Immediate window the total number of bytes and number of records in the open file.

The number of bytes is returned by the LOF(1) statement. The number of records is computed by dividing the entire file (LOF) by the length of one record—Len(d). Next, Visual Basic executes the statements inside the loop until the Esc key is pressed or Cancel is clicked.

The first statement in the loop assigns the result of the `Rnd` function to the variable `randomNr`. The next statement writes this number to the Immediate window. The instruction,

```
Seek #1, randomNr
```

moves the cursor in the open file to the record number specified by the variable `randomNr`. The next instruction reads the contents of the found record. To read the data in a file open for random access, you must use the `Get` statement.

The instruction,

```
Get #1, randomNr, d
```

tells Visual Basic the record number (`randomNr`) to read and the variable (`d`) into which data is being read. The first record is at position 1, the second record is at position 2, and so on. Omitting a record number causes Visual Basic to read the next record. The values of both elements of the `Dictionary` UDT are then written to the Immediate window.

The `Trim(d.en)` and `Trim(d.sp)` functions print the values of the record being read without the leading and trailing spaces that the user may have entered.

Next, Visual Basic displays an input box with a prompt to supply the foreign language equivalent of the word shown in the input box. The word is assigned to the variable `answer`. If you press the Esc key instead of clicking OK, Visual Basic closes the file and ends the procedure. Otherwise, Visual Basic prints your answer to the Immediate window and notifies you whether your answer is correct. Just press the Esc key or click the Cancel button in the dialog box when you are ready to quit the vocabulary drill.

If you want to continue, click OK, and a new random number will be generated. The program will retrieve an English word and ask you for the Spanish equivalent.

2. Run the `VocabularyDrill` procedure. When prompted, type the Spanish equivalent of the English word shown in the title bar of the input box. Press Cancel to exit the vocabulary drill.

Opening Random-Access Files in Excel

When you double-click a random-access file, the file opens in Windows Notepad by default. This is because Windows associates the file type with Notepad, allowing you to view the file's contents as plain text. Notepad might not display the data in a structured or user-friendly format, however, especially when the file contains complex data. To bring this data into Excel and utilize its powerful data analysis tools, you should use File | Open in the Excel Application window. Hands-On 12.8 walks you through opening the `Translate.txt` file that was created earlier in Excel.

◉ Hands-On 12.8   Opening a Random-Access File in Excel

1. In the Microsoft Excel application window (`Chap12.xlsm`), click the File tab on the ribbon, then click Open.
2. In the Open dialog box, browse to the location of your random-access file (`c:\ VBAExcel2024_ByExample`).
3. Ensure that the file type dropdown is set to All Files (*.*) so you can see the file.
4. Select the Translate.txt file and click Open.
   Excel's Text Import Wizard - Step 1 of 3 window will launch, as shown in Figure 12.4.

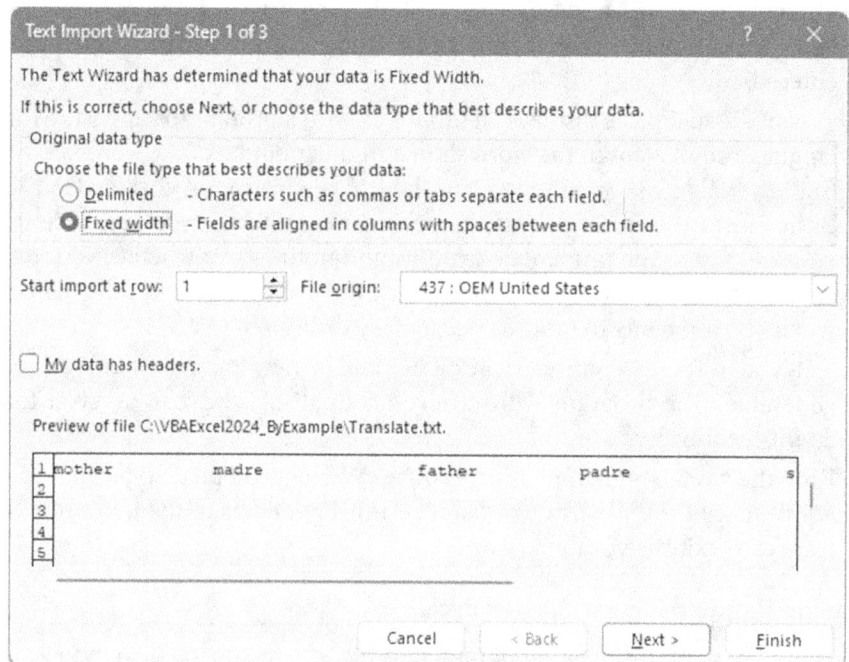

FIGURE 12.4.   Opening random-access files in Excel is handled by the Text Import Wizard. Notice that Excel correctly recognizes the original data type—the data in a random-access file is of a fixed width.

NOTE	*The Text Import Wizard offers three screens to help you import your data correctly:*
	• *Screen 1—Choose the data type (Delimited or Fixed Width) based on your file's structure. Use Delimited if data is separated by commas, tabs, etc., or Fixed Width if your data is in specific column positions.*
	• *Screen 2—Define the delimiters or column breaks. This step ensures that Excel recognizes the divisions between your data fields correctly.*
	• *Screen 3—Format each column as needed. You can specify the data type for each column, such as General, text, Data, etc.*

You can click the Next button to move through the Wizard's screens. If you're confident that the file data type is correct and you don't need to make any formatting changes, you can click Finish right away.

5. Click Finish to load your translation data file into Excel.

Figure 12.5 shows the results of this import. Notice that Excel imports the data from a random-access file, placing records in columns instead of rows. To get the layout we want (each record in a new row), you'll need to write a separate VBA code to reorganize the data after it's imported (see Hands-On 12.9).

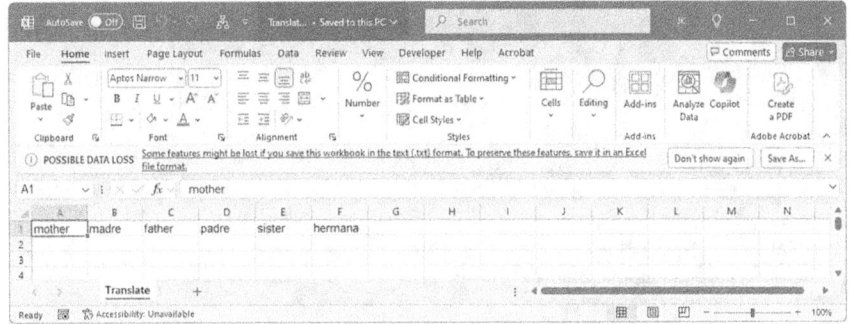

FIGURE 12.5. The random-access file data imported to Excel shows records in columns instead of rows.

6. Click the Save As button in the warning message and save the workbook file as `Translate.xlsx` in your `C:\VBAExcel2024_ByExample` folder.

Reformatting Data Imported from a Random-Access File in Excel

Let's write VBA code to transpose the imported data in the `Translate.xlsx` workbook so each record is placed in a new row.

⊙ Hands-On 12.9   Using VBA to Transpose Data in a Worksheet

1. Return to your `FileMan_IO (Chap12.xlsm)` project and insert a new module, then rename it `DataReorg`.

2. In the `DataReorg` module, enter the following `TransposeDataInTarget-Workbook` procedure:

```vba
Sub TransposeDataInTargetWorkbook()
    Dim sourceWs As Worksheet
    Dim lastCol As Long
    Dim lastRow As Long
    Dim recordCols As Long
    Dim i As Long, j As Long
    Dim targetWb As Workbook
    Dim targetWs As Worksheet
    Dim dataWs As Worksheet
    Dim wbName As String

    ' Name of the target workbook file
    wbName = "Translate.xlsx"

    ' Path to the target workbook file
    Dim wbPath As String
    wbPath = "C:\VBAExcel2024_ByExample\" & wbName

    ' Check if the target workbook is already open
    On Error Resume Next
    Set targetWb = Workbooks(wbName)
    On Error GoTo 0

    ' If the target workbook is not open, open it
    If targetWb Is Nothing Then
        Set targetWb = Workbooks.Open(wbPath)
    End If

    ' Set the data worksheet in the target workbook
    '(assuming data is in Translate sheet)
    Set dataWs = targetWb.Sheets("Translate")

    ' Add a new worksheet to the target workbook
    Set targetWs = targetWb.Sheets.Add
    targetWs.Name = "TransposedData"
```

```
    ' Find the last column with data in the data worksheet
    lastCol = dataWs.Cells(1, dataWs.Columns.Count).End(xlToLeft)
                                                    .Column

    ' Number of columns in each record
    recordCols = 2 ' Change this to the number of fields per
        record

    ' Loop through the data and place each record in a new row
    ' in the target worksheet
    For i = 1 To lastCol Step recordCols
        lastRow = targetWs.Cells(targetWs.Rows.Count, 1).
                    End(xlUp).Row + 1
        For j = 0 To recordCols - 1
            targetWs.Cells(lastRow, j + 1).Value = dataWs
                                .Cells(1, i + j).Value
        Next j
    Next i
    targetWs.Cells(1, 1).Value = "English Term"
    targetWs.Cells(1, 2).Value = "Spanish Equivalent"
    Columns("A:B").AutoFit

    MsgBox "Data transposed successfully in '" & wbName & "'!"

    ' Save and close the target workbook
    targetWb.Save
    targetWb.Close
End Sub
```

3. Run the `TransposeDataInTargetWorkbook` procedure.
4. When the procedure is completed your data layout should match Figure 12.6.

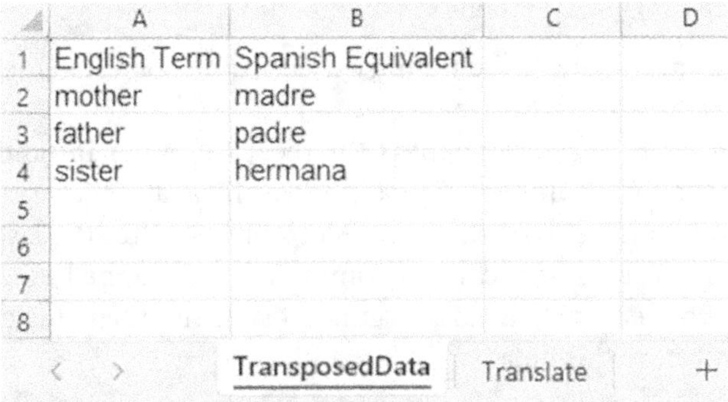

	A	B	C	D
1	English Term	Spanish Equivalent		
2	mother	madre		
3	father	padre		
4	sister	hermana		
5				
6				
7				
8				

‹ ›   **TransposedData**   Translate   +

FIGURE 12.6. The random-access file data imported to Excel has been reformatted with VBA to show records in rows instead of columns.

Advantages and Disadvantages of Random-Access Files

Unlike sequential files, data stored in random-access files can be accessed very quickly. Also, these files don't need to be closed before writing into them and reading from them, and they don't need to be read or written to in order.

Random-access files also have some disadvantages. For example, they often store the data inefficiently. Because they have fixed-length fields and records, the same number of bytes is used regardless of the number of characters being stored. So, if some fields are left blank or contain strings shorter than the declared field size, you may waste a lot of space.

WORKING WITH BINARY FILES

Unlike random-access files that store data in records of fixed length, binary files store records with variable lengths. For example, the first record may contain 10 bytes, the second record may have only 5 bytes, and the third record may have 15 bytes. This method of storing data saves a lot of disk space because Visual Basic doesn't need to add additional spaces to the stored string to ensure that all the fields are of the same length.

Just like random-access files, binary files can be opened for simultaneous read and write operations. As binary file records are of variable length, however, it is more difficult to manipulate these files. In order to retrieve the data correctly, you must store information about the size of each field and record.

To work with binary files, you will use the following four statements:

- The `Get` statement is used to read data. This statement has the following syntax:

```
Get [#]filenumber, [recnumber], varname
```

The `filenumber` argument is the number used in the `Open` statement to open a file.

The optional `recnumber` argument is the record number in random-access files, or the byte number in binary access files, at which reading begins. If you omit `recnumber`, the next record or byte after the last `Get` statement is read. You must include a comma for the skipped `recnumber` argument. The required `varname` argument specifies the name of the variable that will store this data.

- The `Put` statement allows you to enter new data into a binary file. This statement has the following syntax:

```
Put [#]filenumber, [recnumber], varname
```

The `filenumber` argument is the number used in the `Open` statement to open a file.

The optional `recnumber` argument is the record number in random-access files, or the byte number in binary access files, at which writing begins. If you omit `recnumber`, the next record or byte after the last `Put` statement is written. You must include a comma for the skipped `recnumber` argument. The required `varname` argument specifies the name of the variable containing data to be written to disk.

- The `Loc` statement returns the number of the last byte that was read. (In random-access files, the `Loc` statement returns the number of the record that was last read.)

- The `Seek` statement moves the cursor to the appropriate position inside the file.

To master the usage of the above statements, let's proceed to Hands-On 12.10.

⊙ Hands-On 12.10   Mastering the Get, Put, Loc and Seek Statements

1. Add a new module to the `FileMan_IO` (Chap12.xlsm) VBA project and name it `BinaryFiles`.
2. In the `BinaryFiles` module, enter the `CreateAndRead_BinaryFile` procedure located in the `BinaryData_Example.txt` file in the companion files.

   The `CreateAndRead_BinaryFile` procedure creates a binary file named `DataSample.dat`. The procedure begins with the definition of variables that will store `filePath`, `fileNum`, `firstName`, `lastName`, and `nameLength`. The user is prompted to enter their first and last names using `InputBox`. Then, the binary file is open for writing like this:

```
fileNum = FreeFile
Open filePath For Binary As #fileNum
```

The `#fileNum` is obtained using the `FreeFile` VBA function. When working with file input and output in VBA, each open file is assigned a unique file number. The `FreeFile` function helps you avoid conflicts by ensuring that you use a file number that is not currently in use by any other open files.

Once the file is open, the procedure writes the length of the first name followed by the first name itself. Then, the length of the last name is written followed by the last name itself:

```
' Write first name length and first name to the file
nameLength = Len(firstName)
Put #fileNum, , nameLength
Put #fileNum, , firstName
MsgBox "First name written to byte position: " & Loc(fileNum)

' Write last name length and last name to the file
nameLength = Len(lastName)
Put #fileNum, , nameLength
Put #fileNum, , lastName
MsgBox "Last name written to byte position: " & Loc(fileNum)
```

Next, we want to read the file contents. We must use the `Seek` statement to move the file pointer back to the beginning of the file:

```
Seek #fileNum, 1
```

To display the first name, we first read the length of the first name, create a string of that length, and read the first name, like this:

```
' Read and display the first name
  Get #fileNum, , nameLength
  firstName = Space(nameLength)
  Get #fileNum, , firstName
  MsgBox "Read from file - First Name: " & firstName & _
     " at byte position: " & Loc(fileNum)
```

Similarly, we read the last name, and then we close the file after reading:

```
Close #fileNum
```

3. Run the `CreateAndRead_BinaryFile` procedure.
4. Use Windows Notepad to open the `DataSample.dat` file created by the `CreateAndRead_BinaryFile` procedure and examine the file contents.

Advantages and Disadvantages of Binary Access Files

In comparison with sequential and random-access files, binary files are the smallest of all. Because they use variable-length records, they can conserve disk space. Like files open for random access, you can simultaneously read and write to a file open for binary access. One big disadvantage of binary access files is that you must know precisely how the data is stored in the file to retrieve or manipulate it correctly.

SUMMARY

This chapter has given you a working knowledge of writing to and retrieving data from three types of files: sequential, random-access, and binary. You learned how low-level file access allows for more control over file operations, enabling you to manipulate data at the byte level for greater precision and flexibility. By mastering these techniques, you should be able to handle a wide range of file operations, from simple text file manipulations to complex data processing, as required by your specific project.

Part III of this book will introduce you to automation tasks that will allow you to control other applications with Excel VBA.

Part III

INTERACTING WITH OTHER APPLICATIONS

We live in a connected world and often rely on many applications that perform various functions to meet our professional and personal needs. Integrating these applications can streamline workflows, enhance productivity, and provide a seamless user experience. Excel VBA is a powerful language tool that allows you to interact with other applications and automate complex tasks, which saves time and reduces manual effort.

In this part of the book, you will learn how to use your Excel VBA skills to open and manipulate various applications right from Excel. You'll work with other Microsoft 365 applications such as Word, Outlook, PowerPoint, and Access. Additionally, you'll learn how integrating PowerShell with Excel VBA allows you to harness PowerShell's capabilities directly from Excel, enabling you to perform system level tasks, automate workflows, and manipulate data efficiently.

Chapter **13** OFFICE *365*
AUTOMATION WITH
EXCEL *VBA*

One of the nicest things about the VBA language is that you can use it to launch and control other Microsoft 365 applications. For example, you can create, open, and format Word documents straight from your VBA procedure without ever leaving Excel or seeing the Word user interface. You can create PowerPoint presentations by populating slides with data from Excel, applying the required formatting, and even adding animations. In addition to document and presentation automation, Excel VBA can streamline communication tasks by accessing the Outlook object model, so you can automate the creation and sending of emails based on data within Excel. Furthermore, VBA can be used to connect Excel to Access databases to import/export data and automate database operations. In summary, Excel VBA provides the capability to manipulate objects, methods, and properties of other Office 365 applications.

This chapter will introduce various methods for launching Office applications and demonstrate how to transfer data between them seamlessly.

HOW APPLICATIONS COMMUNICATE WITH ONE ANOTHER

There are two standard ways in which applications can communicate with one another: through Application Programming Interfaces (APIs) and file-based data exchange:

- *APIs*: APIs allow different software applications to communicate by providing a set of functions and protocols. These functions enable one application to request services or data from another application. For instance, Office 365 applications such as Excel, Word, PowerPoint, Access, and Outlook have built-in APIs that can be accessed through VBA. By using APIs, developers can automate tasks, transfer data between applications, and integrate various software solutions seamlessly.

- *File-based data exchange*: This method involves exporting data from one application into a file format that another application can read and import. Common file formats include CSV (Comma-Separated Values), XML (Extensible Markup Language), and JSON (JavaScript Object Notation). For example, an Excel spreadsheet can be saved as a CSV file, which can then be imported into another application, such as a database or a different Office 365 application. File-based data exchange is a straightforward way to transfer information, especially when direct API integration is not available.

Understanding Office Automation

Office automation focuses on automating repetitive tasks and streamlining workflows within office environments. By leveraging the power of VBA and the object models of Office applications, businesses can achieve higher efficiency, accuracy, and consistency in their operations.

The applications that support automation are called *automation objects*. These objects are programmable entities within software applications that expose their functionality through an object model. This model consists of various objects, properties, methods, and events that developers can use to control and interact with the application programmatically. Automation objects can include documents, workbooks, slides, and email messages within Office applications such as Word, Excel, PowerPoint, and Outlook. The object model of each application provides the necessary tools to manipulate its contents.

Controlling one Office application from another requires an understanding of the object models and the use of binding techniques—specifically, early binding and late binding. You may recall these terms from Chapter 11, where we worked with the `FileSystemObject` object from the Microsoft Scripting Runtime library.

Now, let's examine early and late binding from the Office automation standpoint. Suppose you need to automate Word from Excel, meaning your Excel VBA application needs to create a new Word document with specific text and save it to disk. The example procedures in the following sections demonstrate how to automate Word with early and late binding.

Using Early Binding

To use early binding, you must set a reference to the relevant object library at design time (see Figure 13.1). This method allows you to take full advantage of many of the debugging tools that are available in the VBE window. For example, you can look up external objects, properties, and methods with the Object Browser. Visual Basic Auto Syntax Check, Auto List Members, and Auto Quick Info (all discussed in Chapter 2) can help you write your code faster and with fewer errors. In addition, early binding allows you to use built-in constants as arguments for methods and property settings. Because these constants are available in the type library at design time, you do not need to define them. The handy built-in syntax checking, IntelliSense features, and support for built-in constants aren't available with late binding. Although VBA procedures that use early binding execute faster, some very old Windows applications can only use late binding.

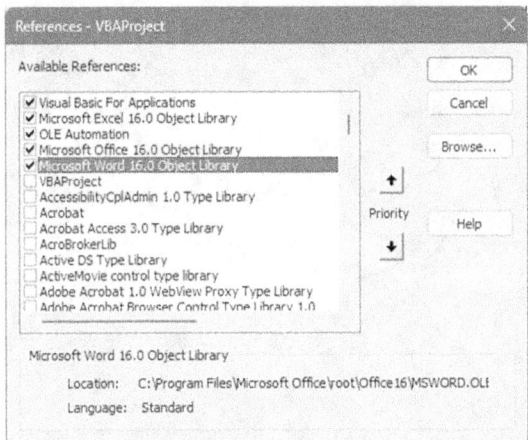

FIGURE 13.1.   Early binding requires setting a reference to the object library of the application you want to control with Excel. In this case, this is Microsoft Word 16.0 Object Library. This library contains objects, methods, and properties that you need to access from Excel VBA to create and manipulate Word documents.

The procedure shown below uses early binding to create a new Word document.

```
Sub WriteLetter_EarlyBinding()
  ' Set reference to Microsoft Word 16.0 Object Library
  ' Use Tools | References
  Dim wordAppl As Word.Application
  Set wordAppl = New Word.Application

  ' Create a new Word document
    Dim wordDoc As Word.Document
    Set wordDoc = wordAppl.Documents.Add

  ' Define other variables
    Dim strFilePath As String
    Dim flag As Boolean
    Dim strContent As String

    On Error GoTo ErrorHandler
  ' Initialize variables
    flag = True
    strFilePath = "C:\VBAExcel2024_ByExample\Invitation.docx"
    strContent = "We recently sent you an invitation & _
        "to answer two questions regarding your purchase " & _
        "experience at our new store. "

    ' Make Word application visible
    wordAppl.Visible = True

    ' Insert text into the document
    wordDoc.Content.Text = strContent

    ' Save the document
    wordDoc.SaveAs2 strFilePath
    wordAppl.Quit
ExitHere:
    If flag Then MsgBox "The Document " & strFilePath & _
        " was successfully created and saved.", _
        vbInformation, "Operation Completed"
    ' Cleanup
    Set wordDoc = Nothing
    Set wordAppl = Nothing
    Exit Sub
ErrorHandler:
  If Err.Number <> 0 Then
      MsgBox Err.Number & ":" & Err.Description
```

```
      flag = False
   End If
   Resume ExitHere
End Sub
```

In the `WriteLetter_EarlyBinding` procedure, we have declared a variable `wordAppl` of type `Word.Application`. This type is explicitly defined due to early binding. As you enter your code, the IntelliSense features will activate, providing you with the names of objects, properties, and methods from the Microsoft Word 16.0 Object Library. After declaring your variable that represents the Word application, you need to use the `New` keyword to create a new instance of the Word application and the `Set` keyword to assign that instance to the object variable.

Using the New Keyword

The `New` keyword tells Visual Basic to create a new instance of an object, return a reference to that instance, and assign the reference to the object variable being declared. For example, you can use the `New` keyword in the following way:

```
Dim objWord As Word.Application
Set objWord = New Word.Application
```

Object variables declared with the `New` keyword are always early bound. Using the `New` keyword is more efficient than using the `CreateObject` function, which is used with late binding, as discussed in the next section. Each time you use the `New` keyword, Visual Basic creates a new instance of the application. The `New` keyword can also be used to create a new instance of the object at the same time that you declare its object variable. For example:

```
Dim objWord As New Word.Application
```

Notice that when you declare the object variable with the `New` keyword in the `Dim` statement, you do not need to use the `Set` statement. This method of creating an object variable, however, is not recommended because you lose control over when the object variable is actually created. Using the `New` keyword in the declaration statement causes the object variable to be created even if it isn't used. Therefore, if you want control over when the object is created, always declare your object variables using the following syntax:

```
Dim objWord As Word.Application
Set objWord = New Word.Application
```

The `Set` statement can be placed further in your code where you need to use the object.

Once you have a reference to the `Word.Application` object, you should create an object variable to represent a Word document. This is done by declaring a variable of type `Word.Document` and using the `Set` keyword to associate this variable with a new or existing document. This allows the manipulation of the document's content, format, and other properties through the object variable.

In the `WriteLetter_EarlyBinding` procedure, we used the `Add` method of the Word `Documents` collection to create a new blank document. Next, we used the `Visible` property of the Word `Application` object to ensure that the Word application window is visible in the user interface. We added text to the document using the `Content` and `Text` properties.

The `Content` property is a member of the `Word.Document` object. It returns a `Range` object that represents the main body of the document, including all text, tables, and other elements within the document. Essentially, the `Content` property provides a way to access and manipulate the entire document's content.

The `Text` property is a member of the `Range` object. It allows you to get or set the text within the specified range. When used in conjunction with the `Content` property, it enables you to manipulate the text within the entire document's main body. If you use the `Content` and `Text` properties as we used them in the procedure example, the entire content of the document will be replaced with the specified text. If you want to append text instead of replacing it, you can modify the `Text` property by concatenating the existing content with the new text, like this:

```
wordDoc.Content.Text = wordDoc.Content.Text  & vbCrLf & strContent
```

The `vbCrLf` adds a newline character to separate the existing content from the new text. This approach ensures that new text is added to the document without removing the existing content.

Using Late Binding

Late binding does not require setting a reference to the object library at design time. Since you don't have a reference to the specific object library, you declare a generic object variable using the `Object` data type. Next, you use the `CreateObject` function to create an instance of the specified application and return a reference to it. You then assign this reference to your object variable:

```
' Create a new Word application instance
Dim wordAppl As Object
Set wordAppl = CreateObject("Word.Application")
```

Late binding is also known as *runtime binding*. Late binding simply means that Visual Basic doesn't associate your object variable with the automation object at design time but waits until you actually run the procedure. Because the declaration `As Object` is very general in nature, Visual Basic cannot determine at compile time that the object your variable refers to has the properties and methods your VBA procedure is using. The advantage of late binding is that all the automation objects know how to use it. Late binding provides more flexibility, as it does not depend on the version of the Office application installed on the user's system and avoids potential issues with missing references or version conflicts. The disadvantage is the lack of IntelliSense features and no support for built-in constants. Because Visual Basic does not know at design time the type library to which your object is referring, you must define constants in your code by looking up the values in the application's documentation.

In the `WriteLetter_LateBinding` procedure shown below, we use the late binding to open an existing Word document that was created with the `WriteLetter_EarlyBinding` procedure. We modify this document by appending the new text to the last paragraph, ensuring the new content is part of the same paragraph:

```
Sub WriteLetter_LateBinding()
    Dim strFilePath As String
    Dim strContent As String

    ' Create a new Word application instance
    Dim wordAppl As Object
    Set wordAppl = CreateObject("Word.Application")

    ' Initialize variables
    strFilePath = "C:\VBAExcel2024_ByExample\Invitation.docx"
    strContent = "Please take a few minutes to complete " & _
      "the review request."

    ' Open an existing Word document
    Dim wordDoc As Object
    Set wordDoc = wordAppl.Documents.Open(strFilePath)

    ' Make Word application visible
    wordAppl.Visible = True

    ' Access the last paragraph
    Dim wordRange As Object
    Set wordRange = wordDoc.Paragraphs(wordDoc.& _
        Paragraphs.Count).Range
```

```
' Move the end of the range to exclude the paragraph mark
' 1 represents wdCharacter
  wordRange.MoveEnd Unit:=1, Count:=-1

' Append text to the same paragraph
wordRange.Text = wordRange.Text & Chr(32) & strContent

' Save the document
wordDoc.SaveAs2 strFilePath
wordAppl.Quit

' Cleanup
Set wordRange = Nothing
Set wordDoc = Nothing
Set wordAppl = Nothing
End Sub
```

In the above example, a new `Word.Application` instance is created with the `CreateObject` function. After establishing a generic reference to a Word document object, we open an existing document with the `Open` method of the Word `Documents` collection. Before adding new text to the last paragraph, we set a reference to the range of the last paragraph and move the end of the range to exclude the paragraph mark. We then append the new text of the document, continuing from the previous paragraph. The `Chr(32)` represents a space character, which is needed to ensure there is a space between the previous sentence and the newly inserted one.

GetObject Versus CreateObject

As a rule, use the `CreateObject` function when there is no current instance of the object. If the instance of the object is already running, a new instance is created. If an application is already running and you want to connect to it, you can use `GetObject`. For example, if Word is already open, use the following code:

```
' Connect to an existing instance of Word
Dim worddAppl As Object
Set wordAppl = GetObject(, "Word.Application")
```

The `GetObject` function has two arguments. The first one is an optional `pathname` argument that specifies the full path and name of the file to open. If omitted (left as an empty string, or with just a comma), `GetObject` tries to connect to an existing instance of the application specified in the second argument. This argument is the class of the object you want to get. For instance, `Word.Application`, `PowerPoint.Application`, `Access.Application`, etc.

You can also use `GetObject` if you want to open a specific file and work with it directly. For example, the following procedure will open the `Invitation.docx` that we created and manipulated earlier:

```
Sub OpenSpecificWordDocument()
    ' Open an existing Word document
    Dim wordDoc As Object
    Set wordDoc = GetObject("C:\VBAExcel2024_ByExample
            \Invitation.docx")
    ' Get a reference to the Word application
    Dim wordAppl As Object
    Set wordAppl = wordDoc.Application

    ' Make Word visible
    wordAppl.Visible = True

    ' You can now manipulate the Word document as needed
    MsgBox "Opened the specified Word document."

    ' Clean up
    Set wordDoc = Nothing
    Set wordAppl = Nothing
End Sub
```

Figure 13.2 shows the contents of the Word document that was created and modified so far in this chapter's examples.

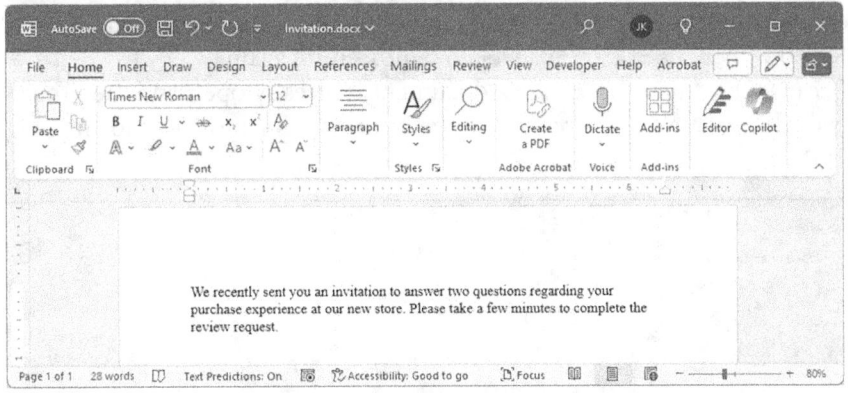

FIGURE 13.2.   A Word document created and modified using Office automation from Excel VBA.

> | **NOTE** | *Please note that the files for the "Hands-On" projects can be found in the companion files.* |
> |---|---|

To successfully work with other Office applications as outlined in this chapter, it is essential that Word, PowerPoint, Outlook, and Access are installed on your machine.

⊙ Hands-On 13.1   Working with Word from Excel

1. Open a new workbook and save it as `C:\VBAExcel2024_ByExample\Chap13.xlsm`.
2. Switch to the VBE window and insert a new module into `VBAProject (Chap13.xlsm)`.
3. Replace the `Module1` name with `WordAutomation`.
4. In the `WordAutomation` Code window, enter the example procedures that were demonstrated earlier in this chapter: `WriteLetter_EarlyBinding`, `WriteLetter_LateBinding`, and `OpenSpecificWordDocument`.
5. Set a reference to the Microsoft Word 16.0 Object Library using the References dialog box, as shown in Figure 13.1 earlier.
6. Run the `WriteLetter_EarlyBinding` procedure to create the `Invitation.docx` document.
7. Run the `WriteLetter_LateBinding` procedure to modify the `Invitation.docx` document.
8. Run the `OpenSpecificWordDocument` procedure to display the `Invitation.docx` document. The resulting document should match Figure 13.2.
9. Run the `OpenSpecificWordDocument` procedure again.
   Because the `Invitation.docx` document is already open and the procedure code does not include error handling, Excel displays the error shown in Figure 13.3.

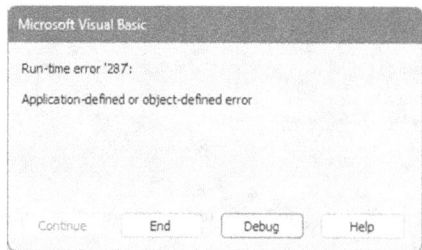

FIGURE 13.3. Run-time error 287 is an application-defined or object-defined error. This error typically occurs when there is an issue with the way an object or application is being accessed or manipulated.

10. Click the Debug button to dismiss the error.
    Excel enters break mode and highlights the statement it had trouble executing.
    To fix this error, we need to reset the project and write additional code that
    checks whether the document is already open.
11. Choose Run | Reset to exit break mode.
12. Add the following modified procedure in the WordAutomation module:

```
Sub OpenSpecificWordDocument_Modified()
    Dim wordAppl As Object
    Dim wordDoc As Object
    Dim strFilePath As String
    Dim docFound As Boolean

    strFilePath = "C:\VBAExcel2024_ByExample\Invitation.docx"
    docFound = False

    ' Attempt to get a running instance of Word
    On Error Resume Next
    Set wordAppl = GetObject(, "Word.Application")
    On Error GoTo 0

    ' If no running instance is found, create a new one
    If wordAppl Is Nothing Then
        Set wordAppl = CreateObject("Word.Application")
    End If

    ' Make Word visible
    wordAppl.Visible = True

    ' Check if the document is already open
    For Each wordDoc In wordAppl.Documents
        If wordDoc.FullName = strFilePath Then
            docFound = True
            Exit For
        End If
    Next wordDoc

    ' If the document is not already open, open it
    If Not docFound Then
        Set wordDoc = wordAppl.Documents.Open(strFilePath)
    Else
        Set wordDoc = wordAppl.Documents(strFilePath)
    End If

    ' Clean up
    Set wordDoc = Nothing
```

```
        Set wordAppl = Nothing
End Sub
```

In this modified version of `OpenSpecificWordDocument`, the `GetObject` is used to attempt to connect to a running instance of Word. If no running instance is found, `CreateObject` creates a new one. The code then checks whether the document is already open by iterating through the open documents in Word. If the document is not open, the document is opened, and if it is already open, a reference is set to the existing document.

13. To fully understand how this modified code works, be sure to run the `OpenSpecificWordDocument_Modified` procedure in step mode by pressing the F8 key. Run it a couple of times while the document is already open, and then again when it is closed.

14. Close the `Invitation.docx` document.

Understanding Linking and Embedding

Object Linking and Embedding (OLE) allows you to create compound documents. A *compound document* contains objects created by other applications. For example, if you embed a Word document in a Microsoft Excel worksheet, Excel only needs to know the name of the application that was used to create this object and the method of displaying the object on the screen.

Compound documents are created by either linking or embedding objects. When you use the manual method to embed an object, you first need to copy it in one application and then paste it into another. The main difference between a linked object and an embedded object is in the way the object is stored and updated:

- The embedded object becomes a part of the destination file. This may cause the file size to increase considerably. Because the embedded object is not connected with the original data, the information is static. When the data changes in the source file, the embedded object is not updated. To change the embedded data, you must double-click it. This will open the object for editing in the source program which must be installed on your computer.

- The linked object is updated automatically when the data in the source file changes. Because the destination document contains only information on how the object is linked with the source document, object linking doesn't increase the size of a destination file. The following formula is used to link an object in Microsoft Excel:

```
=Word.Document.12|'C:\VBAExcel2024_ByExample\Invitation.docx'!''''
```

15. When you enter this formula in the formula bar in an Excel worksheet, you should see the contents of the `Invitation.docx` document displayed in the selected cell.

The `InsertLetter` procedure shown in Hands-On 13.2 demonstrates how to programmatically embed a Word document in an Excel worksheet.

(◉) Hands-On 13.2   Embedding a Word Document in a Worksheet

1. Insert a new module into `VBAProject` (`Chap13.xlsm`) and rename it `OLE`.
2. In the `OLE` module Code window, enter the `InsertLetter` procedure as shown below:

```
Sub InsertLetter()
    Workbooks.Add
    ActiveSheet.Shapes.AddOLEObject _
    Filename:="C:\VBAExcel2024_ByExample\Invitation.docx"
End Sub
```

The `InsertLetter` procedure uses the `AddOLEObject` method. This method creates an `OLE` object and returns the `Shape` object that represents the new `OLE` object. Additional arguments that can be used with the `AddOLEObject` method are listed in the VBA online documentation.

3. Run the `InsertLetter` procedure.
The procedure opens a new workbook and embeds the indicated Word document in it (see Figure 13.4).
4. Make sure the object is selected, as shown in Figure 13.4. If the selection is elsewhere in the worksheet, click the object containing the embedded text to reselect it.

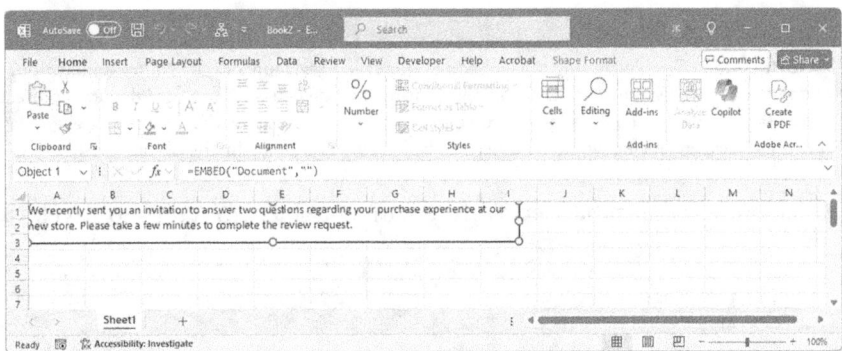

FIGURE 13.4.   A Word document embedded programmatically in an Excel worksheet.

If you are using Windows 11/Microsoft 365, when you embed an object in Excel and select it, the formula bar displays:

```
=EMBED("Document","")
```

If you are working on Windows 10, you might see:

```
=EMBED("Word.Document.12","")
```

The number following `Word.Document` denotes the current version of Word you are using. Version 12 indicates that we are bringing an object from Word 2007 or newer.

If your source program is a Word version prior to 2007, the formula bar displays:

```
=EMBED("Word.Document.8","")
```

5. Now, double-click the embedded object, and notice that the source application is launched right in Excel, as shown in Figure 13.5.
The embedded document can now be modified and formatted using the tools available in Word.

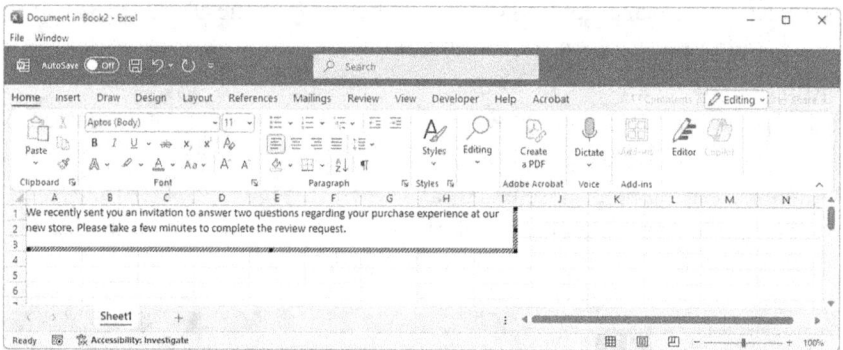

FIGURE 13.5. Double-clicking the embedded object in Excel launches the source application. It appears that Word is now part of the Excel window, and you can edit the Word document directly within the Excel interface. This integration allows you to seamlessly switch between Excel and Word functionalities, enhancing workflow efficiency by keeping everything in one place.

6. Click outside the embedded Word document object to return to Excel.

NOTE	*The* AddOLEObject *method of the* Shapes *object has a* Link *parameter that can be set to* True *to establish a link to a specified Word document. Unfortunately, using this parameter generates a "Run-time error '1004': Cannot insert object" message. In recent editions of Microsoft 365, the implemented security measures block linking of external content, including Word documents, to protect users from potential security risks. This practice is common and aims to prevent malicious content from being inadvertently accessed and executed.*

Working with Microsoft Word from Excel

Based on what we've discussed in this chapter, you should now be able to launch the Word application and create or modify a document directly from within Excel VBA using early and late binding methods.

The Word object model, similar to Excel's, is incredibly comprehensive. It contains hundreds of objects, methods, properties, and events that allow you to control virtually every aspect of the Word application. The range of objects includes documents, paragraphs, tables, bookmarks, and more, each with its own set of properties and methods.

After setting a reference to Microsoft Word 16.0 Object Library from your Excel VBA project, you can begin exploring and utilizing the Word object model. By accessing the Object Browser in the VBE screen (by pressing F2), you can search for and find the specific Word objects, methods, properties, and events you need for your project (see Figure 13.6).

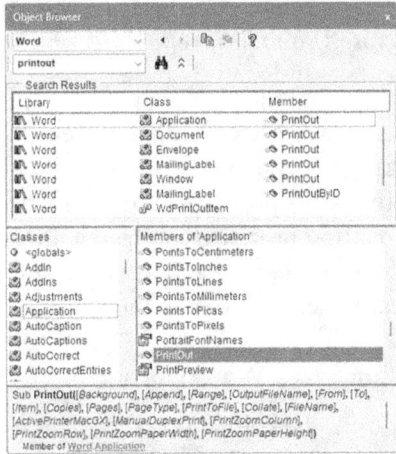

FIGURE 13.6. All of the Microsoft Word objects, properties, methods, and events can be accessed from a Microsoft Excel VBA project after adding a reference to Microsoft Word 16.0 Object Library (see Figure 13.1).

Let's write a VBA procedure that prints out the `Invitation.docx` Word document created in this chapter. We'll use late binding in this example. Make sure your default printer is turned on.

⊙ Hands-On 13.3 Printing a Word Document with VBA

1. In the `WordAutomation` module `VBAProject` (Chap13.xlsm), enter the `PrintWordDoc` procedure as shown below:

```
Sub PrintWordDoc()
   Dim objWord As Object
   Set objWord = CreateObject("Word.Application")

   With objWord
     .Visible = True
     .Documents.Open "C:\VBAExcel2024_ByExample\Invitation.docx"
     .Options.PrintBackground = False
     .ActiveDocument.PrintOut
     .Documents.Close
     .Quit
   End With

   Set objWord = Nothing
End Sub
```

2. Run the `PrintWordDoc` procedure.
   You should get a printout of your document. For more information on using the Word `PrintOut` method and its multiple parameters, check out the Microsoft official documentation at *https://docs.microsoft.com/en-us/office/vba/api/word. application.printout.*

The next Hands-On exercise prompts the user to select a range of cells, copy those cells to a new Word document, change the layout of the document to landscape, and format the table to ensure that it fits properly in the Word document. When the copy/paste and table formatting tasks are completed, the user is prompted to save the document with the desired name and location, and the Word application is closed and its objects are released from memory.

⊙ Hands-On 13.4 Copying an Excel Range to a Word Document

1. Open the `Instructors.xlsx` workbook from the companion files.
   This workbook contains data (see Figure 13.7) that will be copied to a new Word document during this exercise. If you prefer, you can open your own Excel workbook that contains some data you'd like to copy to Word.

2. In the `WordAutomation` module Code window, enter the `CopyWorksheetRangeTo_Word` procedure as shown below:

```vba
Sub CopyWorksheetRangeTo_Word()
    Dim wordAppl As Object
    Dim wordDoc As Object
    Dim wordTable As Object
    Dim selectedRange As Range
    Dim filePath As String

    ' Prompt user to select a range of cells
    On Error Resume Next
    Set selectedRange = Application.InputBox( _
        "Select the range of cells to copy to Word:", Type:=8)
    On Error GoTo 0

    If selectedRange Is Nothing Then
     MsgBox "No range selected. Exiting procedure.", vbExclamation
        Exit Sub
    End If

    ' Create a new Word application and document
    Set wordAppl = CreateObject("Word.Application")
    Set wordDoc = wordAppl.Documents.Add

    ' Change page layout to landscape
    wordDoc.PageSetup.Orientation = 1 ' wdOrientLandscape

    ' Set left and right margins to 0.3 inches
    With wordDoc.PageSetup
        .LeftMargin = wordAppl.InchesToPoints(0.3)
        .RightMargin = wordAppl.InchesToPoints(0.3)
    End With

    ' Copy selected range to the clipboard
    selectedRange.Copy

    ' Paste the range into the Word document as a table
    With wordDoc.Content
        .Paste
        Set wordTable = .Tables(.Tables.Count)
    End With

    ' Format the table to fit the content
    With wordTable
```

```
            .Rows.Alignment = 1 ' Align table to center
            .Borders.Enable = True
            .Range.Font.Size = 9
            .Columns.AutoFit
        End With

        ' Save the Word document
        filePath = Application.GetSaveAsFilename( _
            FileFilter:="Word Documents (*.docx), *.docx", _
            Title:="Save Word Document")
        If filePath <> "False" Then
            wordDoc.SaveAs2 filePath
          MsgBox "Word document saved as " & filePath, vbInformation
        Else
            MsgBox "Save operation cancelled.", vbExclamation
        End If

        ' Clean up
        wordAppl.Quit
        Set wordDoc = Nothing
        Set wordAppl = Nothing
    End Sub
```

3. Run the `CopyWorksheetRangeTo_Word` procedure.
   Excel will prompt you to select the range of cells to copy.

4. Switch to your worksheet containing the data and select the range of cells you'd like to copy (see Figure 13.7), then click OK.

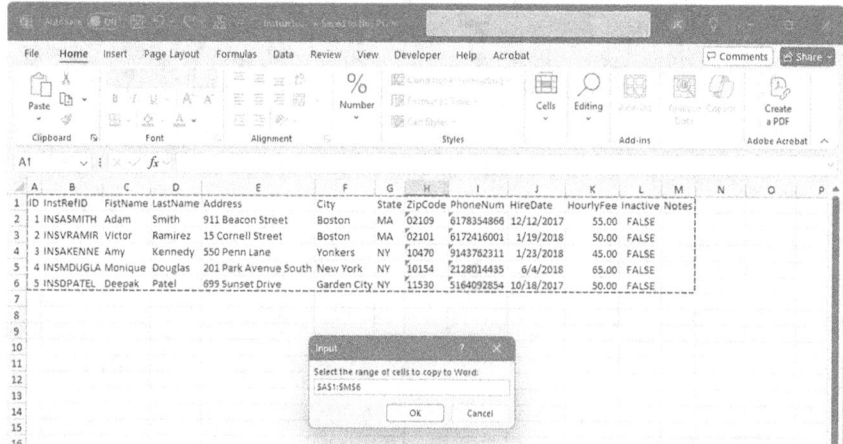

FIGURE 13.7. The CopyWorksheetRangeTo_Word procedure prompts the user to select a range of cells to copy to Word.

5. In the SaveAs dialog box that appears next, switch to your `C:\VBAExcel2024_ByExample` folder and enter `Active_Instructors` in the filename box, or specify your own filename, and click OK.

6. When the process is completed, open the newly created Word document to verify that the data was copied and formatted as intended (see Figure 13.8).

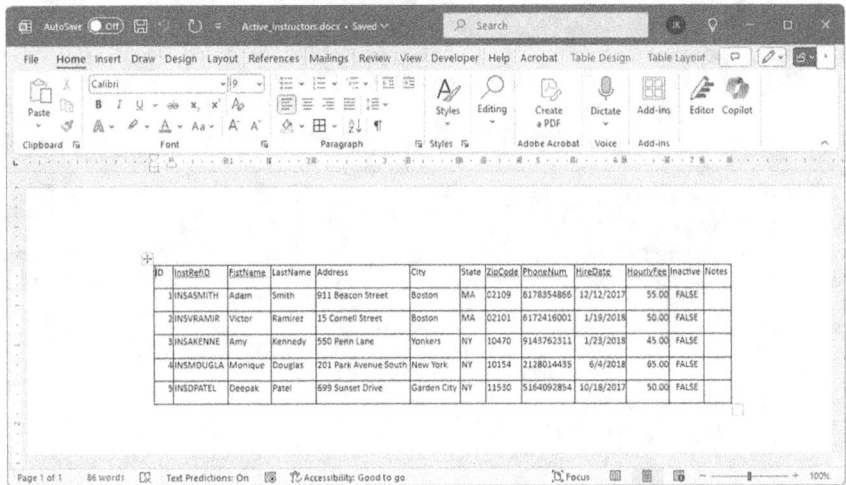

FIGURE 13.8. Excel data is shown here after it has been copied to a Word document.

Working with Microsoft Outlook from Excel

To access Outlook's object model directly from Excel, begin by establishing a reference to Microsoft Outlook 16.0 Object Library. The example procedure in the Hands-On 13.5 exercise will insert your Outlook contact information into an Excel spreadsheet.

(⊙) Hands-On 13.5   Bringing Outlook Contacts to Excel

1. In `VBAProject(Chap13.xlsm)` insert a new module and rename it `Outlook Automation`.

2. Use the References dialog box to set a reference to Microsoft Outlook 16.0 Object Library.

3. In the `OutlookAutomation` module Code window, enter the `GetContacts` procedure as shown below:

```
Sub GetContacts()
  Dim objOut As Outlook.Application
```

```
Dim objNspc As Namespace
Dim objItem As ContactItem
Dim r As Integer ' row index
Dim Headings As Variant
Dim i As Integer ' array element
Dim cell As Variant

r = 2
Set objOut = New Outlook.Application
Set objNspc = objOut.GetNamespace("MAPI")

Headings = Array("Full Name", "Street", "City", _
    "State", "Zip Code", "E-Mail")
Workbooks.Add
Sheets(1).Activate
For Each cell In Range("A1:F1")
    cell.FormulaR1C1 = Headings(i)
    i = i + 1
Next
For Each objItem In objNspc.GetDefaultFolder _
    (olFolderContacts).Items
    With ActiveSheet
        .Cells(r, 1).Value = objItem.FullName
        .Cells(r, 2).Value = objItem.BusinessAddress
        .Cells(r, 3).Value = objItem.BusinessAddressCity
        .Cells(r, 4).Value = objItem.BusinessAddressState
        .Cells(r, 5).Value = objItem.BusinessAddressPostalCode
        .Cells(r, 6).Value = objItem.Email1Address
    End With
    r = r + 1
Next objItem

Set objItem = Nothing
Set objNspc = Nothing
Set objOut = Nothing
MsgBox "Your contacts have been dumped to Excel."
End Sub
```

The GetContacts procedure starts by declaring an object variable called objOut to hold a reference to the Outlook application. This variable is defined by a specific object type (Outlook.Application); therefore, VBA will use early binding.

Notice that, in this procedure, we use the New keyword discussed earlier to create a new instance of an Outlook Application object, return a reference to that instance, and assign the reference to the objOut variable being declared.

In order to access contact items in Outlook, you also need to declare object variables to reference the Outlook `Namespace` and `Item` objects. The `Namespace` object represents the message store known as `MAPI` (Messaging Application Programming Interface). The `Namespace` object contains folders (`Contacts`, `Journal`, `Tasks`, etc.), which, in turn, contain items. An item is an instance of Outlook data, such as an email message or a contact.

After writing column headings to the worksheet using the `For Each...Next` loop, the procedure uses another `For Each...Next` loop to iterate through the `Items` collection in the `Contacts` folder. The `GetDefaultFolder` method returns an object variable for the `Contacts` folder. This method takes one argument, the constant representing the folder you want to access. After all the contact items are written on an Excel spreadsheet, the procedure releases all object variables by setting them to `Nothing`.

4. Run the `GetContacts` procedure and check out the new workbook file that was created by this procedure.

At times, you may need to send individual email messages to a list of people on your worksheet. The following procedure demonstrates how to process this kind of request from Excel using objects, properties, and methods provided by the Microsoft Outlook 16.0 Object Library.

(◉) Hands-On 13.6   Sending Bulk Emails from Excel

1. Prepare the worksheet shown in Figure 13.9. Enter the valid email addresses of your own contacts in column D. Type the full names of your contacts in the Employee Name column. This data can also be copied from the `Chap13_BulkMail.xlsx` workbook available in the companion files.

	A	B	C	D
1	Employee Name	Expense Type	Amount	Email
2	Margaret Hicks	Education	$140.00	Mhicks@test.com
3	Terry Bergman	Medical	$234.00	Tbergman@test.com
4	Michael DeCastro	Transportation	$100.00	Mdecastro@test.com
5	Tony Bennet	Parking	$ 30.00	Tbennet@test.com
6				

FIGURE 13.9.   Sample worksheet for bulk emailing demo.

2. In the `OutlookAutomation` module's Code window, enter the `SendBulkMail` procedure as shown below:

```
' This procedure requires a reference
' to the Microsoft Outlook 16.0 Object Library
Sub SendBulkMail(EmailCol, BeginRow, EndRow, SubjCol, _
    NameCol, AmountCol)
  Dim objOut As Outlook.Application
  Dim objMail As Outlook.MailItem
  Dim strEmail As String
  Dim strSubject As String
  Dim strBody As String
  Dim r As Integer

  On Error Resume Next

  Application.DisplayAlerts = False

  Set objOut = New Outlook.Application

  For r = BeginRow To EndRow
      Set objMail = objOut.CreateItem(olMailItem)
      strEmail = Cells(r, EmailCol)
      strSubject = Cells(r, SubjCol) & " reimbursement"

      strBody = "Dear " & Cells(r, NameCol).Text & ":" & _
          vbCrLf & vbCrLf
      strBody = strBody & "We have approved your request for " & _
          LCase(strSubject)
      strBody = strBody & " in the amount of " & Cells(r, _
          AmountCol).Text & "."
      strBody = strBody & vbCrLf & "Please allow 3 business " & _
          "days for this"
      strBody = strBody & " amount to appear on your "  & _
                      "bank statement."
      strBody = strBody & vbCrLf & vbCrLf & " Employee Services"

      With objMail
          .To = strEmail
          .Body = strBody
          .Subject = strSubject
          .Send
      End With
  Next
  Set objMail = Nothing
  Set objOut = Nothing
  Application.DisplayAlerts = True
End Sub
```

The above procedure uses the following parameters: `EmailCol`, `BeginRow`, `EndRow`, `SubjCol`, `NameCol`, and `AmountCol`.

`EmailCol` is the number of the column on the worksheet where the email address has been entered. In this example, it's the fourth column.

The `BeginRow` and `EndRow` parameters specify the first and last rows of your data range. In this example, the first row we want to process is 2 and the last row is 5.

`SubjCol` is the column number where the email subject is entered. In this example, it's the second column (Expense Type).

`NameCol` contains the employee's name and is the first column here.

`AmountCol` is the column number where the expense amount has been entered. In this example, it's the third column.

While the statement `Application.DisplayAlerts = False` will prevent Excel from displaying alert messages, this will not prevent Outlook's messages from appearing.

After setting a reference to the Outlook application, the procedure uses the `For...Next` loop to iterate through the worksheet data starting at row 2 and ending at row 5. Each time, in the loop, we set a reference to an Outlook `MailItem` and place the data we need for our email message in various variables. Once the procedure knows where the data is located in the worksheet, we set the required properties of Microsoft Outlook.

The `To` property returns or sets a semicolon-delimited string list of display names for the `To` recipients for the Outlook item. In this example, we use one recipient for each email we send. The `Body` property returns or sets a string representing the text message we want to send in the email.

The `Subject` property is used to specify the email subject. Finally, the `Send` method sends the email message. If you'd rather not send the email at this time, you can view it by replacing the `Send` method with the `Display` method:

```
With objMail
    .To = strEmail
    .Body = strBody
    .Subject = strSubject
    .Display
End With
```

3. Enter the following `Call_SendBulkMail` in the `OutlookAutomation` module:

```
Sub Call_SendBulkMail()
    SendBulkMail EmailCol:=4, _
```

```
          BeginRow:=2, _
          EndRow:=5, _
          SubjCol:=2, _
          NameCol:=1, _
          AmountCol:=3
End Sub
```

The above procedure calls the `SendBulkMail` procedure and passes it the parameters indicating the column number of the recipient's address (4), the beginning and ending rows of the data (2, 5), the column where the email subject is located (2), the column containing the employee's name (1), and the column number with the amount of reimbursement (3).

4. Run the `Call_SendBulkMail` procedure.

   Excel begins to execute the specified procedure. The first recipient listed in the worksheet should receive an email like the one shown in Figure 13.10.

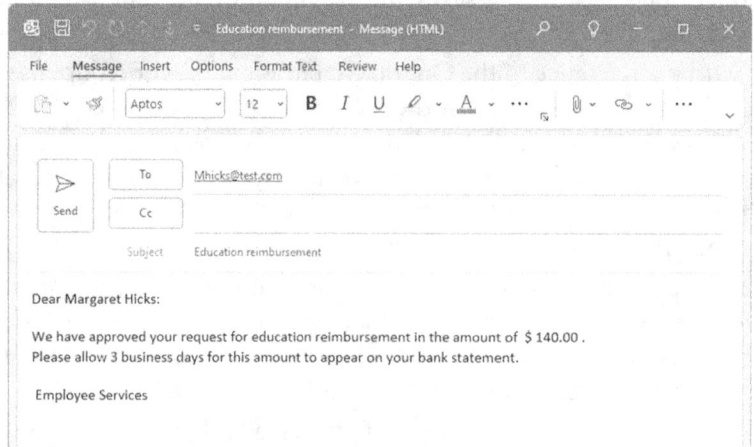

FIGURE 13.10. This sample Microsoft Outlook email message was put together based on the data in an Excel worksheet.

Working with Microsoft PowerPoint from Excel

If you ever need to create and edit PowerPoint presentations programmatically, you can do it directly from Excel VBA. The process is similar to automating Microsoft Word and Outlook through VBA, but with key differences in the specific PowerPoint objects and methods you'll need to use.

To enable the PowerPoint object library, open the VBE screen and choose Tools | References, then locate and select Microsoft PowerPoint 16.0 Object Library (see Figure 13.11) and click OK.

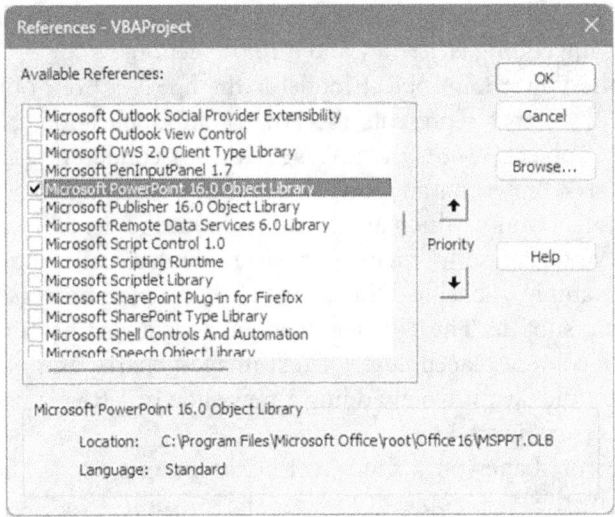

FIGURE 13.11.   Adding a reference to PowerPoint 16.0 Object Library.

While late binding can be used to program PowerPoint, opting for early binding offers the advantage of exploring the PowerPoint object library via the Object Browser, as depicted in Figure 13.12. Recall that early binding involves setting a

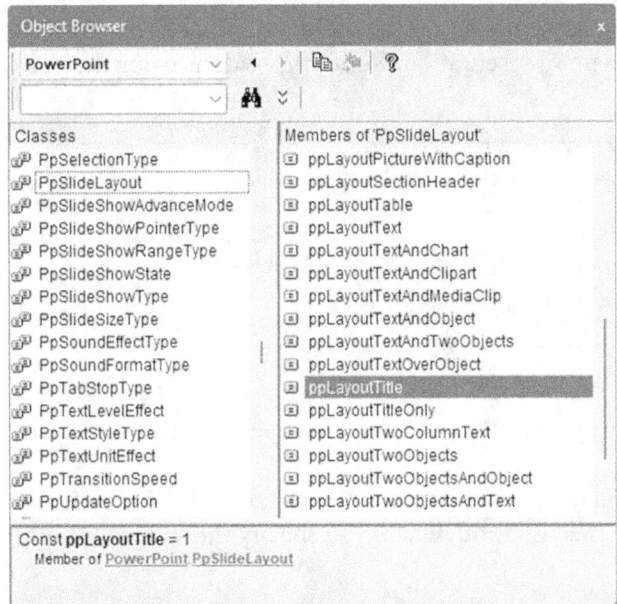

FIGURE 13.12.   Exploring the PowerPoint objects, properties, methods, and events. PowerPoint slides can be created using various auto layouts, as shown here.

reference to the object library whose objects you want to manipulate, enabling IntelliSense and improving coding efficiency and error detection.

The key object in the PowerPoint object model is the `Application` object (`PowerPoint.Application`) that represents the PowerPoint application. Next comes, the `Presentation` object (`PowerPoint.Presentation`), containing `Slide` objects. `PowerPoint.Slide` represents an individual slide in a presentation.

Slides can have various layouts, which are predefined arrangements of content placeholders that you can use to create a consistent and visually appealing presentation. For example, the Title slide layout includes placeholders for the presentation title and subtitle. The Title and Content slide layout has a title placeholder and a large content placeholder for text images, charts, and other content. You can specify the layout when adding a new slide in VBA by using the `ppLayout` constants (see Figure 13.12).

There may be different shapes on a slide (text boxes, images, etc.). These shapes are represented by the `Shape` object (`PowerPoint.Shape`).

When you create a PowerPoint presentation programmatically using VBA, you use the `Add` method to add a new presentation, and the `Add` method to add individual slides to that presentation.

Adding a New PowerPoint Presentation

The following code snippet demonstrates how you can initialize the PowerPoint application and create a new presentation. Notice this is done using the `Presentations.Add` method:

```
Dim pptApp As PowerPoint.Application
Dim pptPres As PowerPoint.Presentation

' Initialize PowerPoint
Set pptApp = New PowerPoint.Application
pptApp.Visible = True

' Create a new presentation
Set pptPres = pptApp.Presentations.Add
```

The statement `Set pptApp = New PowerPoint.Application` creates a new, empty presentation.

Once you have created a new presentation, you can add slides to it. This is done using the `Slides.Add` method, where you specify the slide index and the layout:

```
' Add a new slide with a specific layout (e.g., Title Slide)
```

```
Set pptSlide = pptPres.Slides.Add(1, ppLayoutTitle)
pptSlide.Shapes.Title.TextFrame.TextRange.Text = _
   "Bicycle Sales 2024"
pptSlide.Shapes.Placeholders(2).TextFrame.TextRange.Text = _
   "Regional Sales Comparison ."
```

Here, we've added a first slide to our presentation with a Title Slide layout (`ppLayoutTitle`). You can use various slide layouts to suit different types of content, as shown in Hands-On 13.7.

Hands-On 13.7   Creating a PowerPoint Presentation with Multiple Slides

1. Open the `SalesData_Bicycles.xlsx` workbook from the companion files. This workbook (see Figure 13.13) contains a `SalesData_2024` worksheet with bicycle sales for various regions. We will use this data to create a PowerPoint presentation.

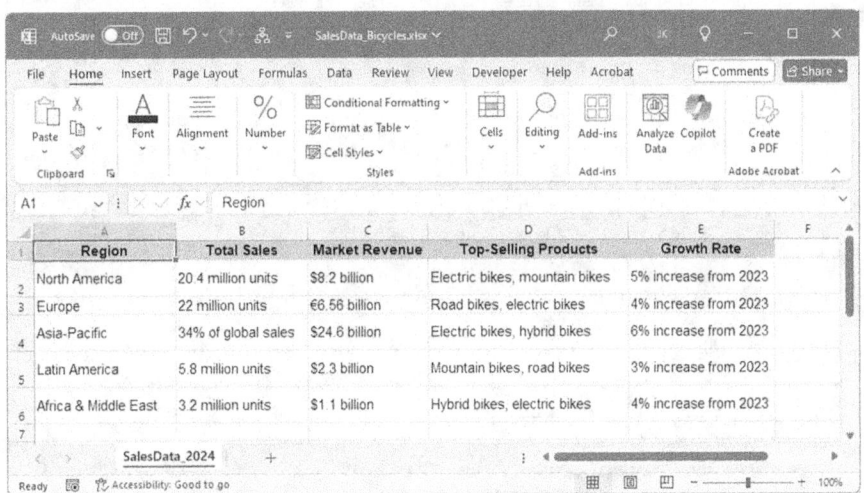

FIGURE 13.13. The sales data in this worksheet will be used to generate a PowerPoint presentation with automation.

2. In the VBE screen of `VBAProject(Chap13.xlsm)`, insert a new module and rename it `PowerPointAutomation`.
3. Write the following procedure in the `PowerPointAutomation` module:

```
Sub CreateBicycleSalesPresentation()
    Dim pptApp As PowerPoint.Application
    Dim pptPres As PowerPoint.Presentation
```

```
Dim pptSlide As PowerPoint.Slide
Dim ws As Worksheet
Dim lastRow As Long
Dim i As Long

' Reference the Excel worksheet
Set ws = Workbooks("SalesData_Bicycles.xlsx"). & _
        Sheets("SalesData_2024")
lastRow = ws.Cells(ws.Rows.Count, 1).End(xlUp).Row

' Initialize PowerPoint
Set pptApp = New PowerPoint.Application
pptApp.Visible = True

' Create a new presentation
Set pptPres = pptApp.Presentations.Add

' Apply a theme
pptPres.ApplyTemplate "C:\Program Files\Microsoft Office\" & _
    "Root\Document Themes 16\Facet.thmx"

' Add a title slide
Set pptSlide = pptPres.Slides.Add(1, ppLayoutTitle)
pptSlide.Shapes.Title.TextFrame.TextRange.Text = & _
    "Bicycle Sales 2024"
pptSlide.Shapes.Placeholders(2).TextFrame.TextRange.Text = _
    "Regional Sales Comparison"

' Loop through each row and create a slide with data
For i = 2 To lastRow
    Set pptSlide = pptPres.Slides.Add(i, ppLayoutText)
    pptSlide.Shapes.Title.TextFrame.TextRange.& _
    Text = ws.Cells(i, 1).Value

    Dim slideContent As String
    slideContent = "Total Sales: " & ws.Cells(i, 2).& _
                    Value & vbCrLf
                    "Market Revenue: " & ws.Cells(i, 3).& _
                    Value & vbCrLf & _
                    "Top-Selling Products: " & & _
                        ws.Cells(i, 4).Value & vbCrLf & _
                    "Growth Rate: " & ws.Cells(i, 5).Value

    pptSlide.Shapes.Placeholders(2).TextFrame.TextRange.& _
        Text = slideContent Next i
```

```
' Save the presentation to disk
Dim savePath As String
savePath = ThisWorkbook.Path & "\Bicycle_Sales_2024.pptx"
pptPres.SaveAs savePath

' Cleanup
Set pptSlide = Nothing
Set pptPres = Nothing
Set pptApp = Nothing
End Sub
```

The `CreateBicycleSalesPresentation` procedure creates a blank PowerPoint presentation and then loops through the data in the Excel sheet to create a title slide and a slide for each region, populating it with the corresponding data. To make the presentation visually appealing, a theme is applied to all slides. You can replace the theme with any theme file available in your PowerPoint installation. Adjust the theme path if necessary to match your system configuration.

4. Run the `CreateBicycleSalesPresentation` procedure.

   Running this procedure generates a PowerPoint presentation, as shown in Figure 13.14, and saves the file in your working directory as `Bicycle_Sales_2024.pptx`.

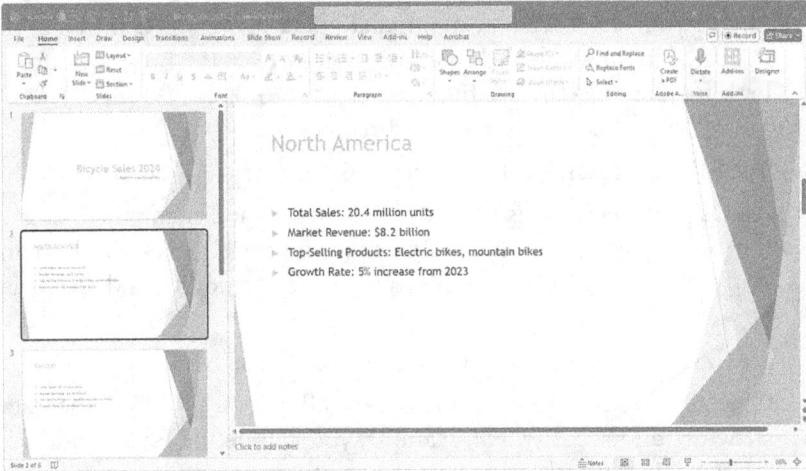

FIGURE 13.14.   This PowerPoint presentation was created from Excel data via automation.

When you create a PowerPoint presentation via automation and your Excel data changes, you can quickly rerun the procedure to ensure that your presentation is up to date with the latest data.

Working with PowerPoint via automation can be very rewarding if you often need to create presentations for various purposes from your Excel data. Automating PowerPoint tasks can save you a significant amount of time, ensure consistency, and enhance productivity.

WORKING WITH MICROSOFT ACCESS FROM EXCEL

In this section, we focus on using automation as well as Data Access Objects (DAOs) and ActiveX Data Objects (ADOs) to work with Microsoft Access data from Excel.

Using Object Libraries in Microsoft Access

A Microsoft Access database consists of various types of objects stored in different object libraries, which are shown in Figure 13.5 and described below:

- The Microsoft Access 16.0 Object Library (Access):

  This library provides objects that are used to display data and work with Microsoft Access.

- The Microsoft DAO 3.6 Object Library (DAO):

  DAOs that are provided by this library allow you to determine the structure of your database and manipulate data using VBA. If this library is not listed in the available library references, click the Browse button, then look for the `DAO360.dll` file in your `\Program Files (x86)\Common Files\Microsoft Shared\DAO` directory.

- The Microsoft ActiveX Data Objects 6.1 Library (ADODB):

  ADOs provided by this library let you access and manipulate data using the OLE database provider. ADOs make it possible to establish a connection with a data source in order to read, insert, modify, and delete data in an Access database.

- The Microsoft ADO Ext. 6.0 for DDL and Security Library (ADOX):

  Objects that are stored in this library allow you to define the database structure and security. For example, you can define tables, indexes, and relationships, as well as create and modify user and group accounts.

- The Microsoft Jet and Replication Objects 2.6 Library (JRO):

  Objects contained in this library are used in the replication of a database. Database replication is no longer supported in Access 2013–2024, so there is no need to set up a reference to this library for this section's Hands-On exercises.

- The Visual Basic for Applications object library (VBA):

  Objects contained in this library allow you to access your computer's file system, work with date and time functions, perform mathematical and financial computations, interact with users, convert data, and read text files. The reference to this library is automatically set when you install Excel.

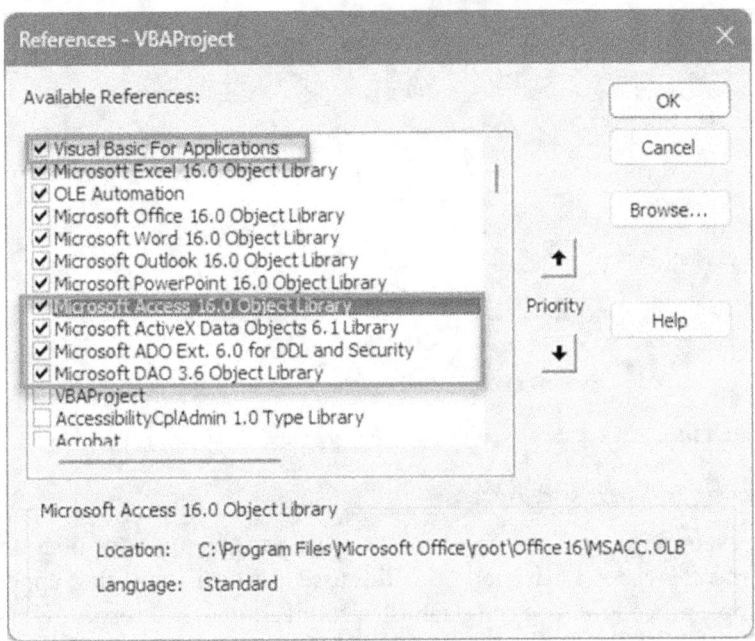

FIGURE 13.15.   To work with Microsoft Access from Excel, we need to access multiple object libraries.

To view objects of a specific library, choose View | Object Browser or press F2, then use the dropdown in the Object Browser to select the library name. For example, to see objects in the Microsoft Access 16.0 Object Library, choose Access. All library names for this chapter's automation projects are shown in Figure 13.16.

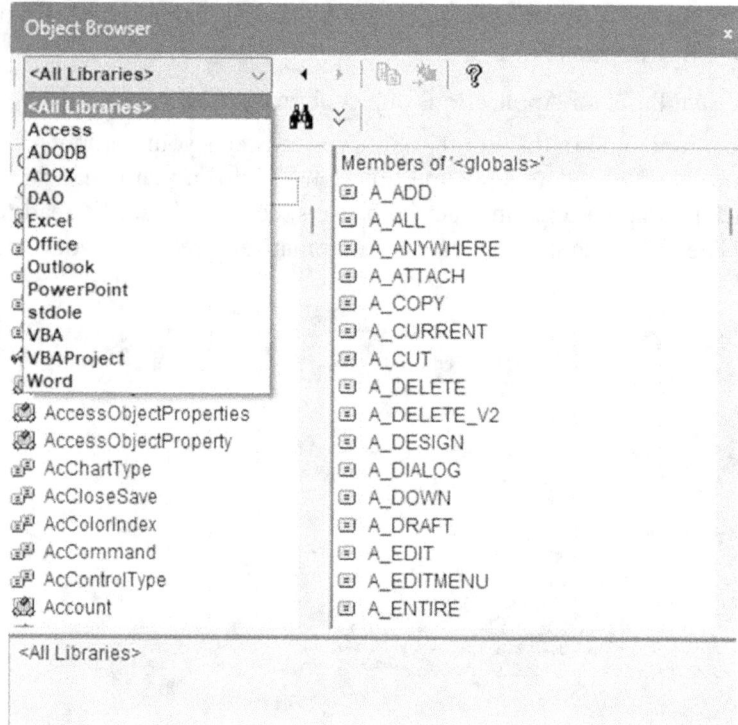

FIGURE 13.16.   Object libraries, as they appear in the Object Browser.

Connecting to Microsoft Access

The example procedures in this section use various methods of connecting to Microsoft Access. Each method is discussed in detail as it first appears in the example procedure. You can establish a connection to Microsoft Access by using one of the following three methods:

- Automation
- DAOs
- ADOs

Connecting to Microsoft Access Using Automation

When working with Microsoft Access from Excel (or another application) using automation, you must take the same steps as we used earlier in this chapter when we worked with automation of other Office applications.

- Early binding

  Early binding requires that you set a reference to the Microsoft Access 16.0 Object Library. The following snippet sets the connection to the Access application using early binding:

  ```
  ' You must set a reference to the Microsoft Access 16.0
  ' Object Library
  ' in the References dialog box
  Dim accApp As Access.Application
  Set accApp = New Access.Application
  ```

- Late binding

  Use the `CreateObject` function to instantiate the Access application:

  ```
  Dim accApp As Object
    ' Initialize Access
    Set accApp = CreateObject("Access.Application.16")
  ```

NOTE	*If you need to access a specific version of the Access application, include the version number in the argument of the* `CreateObject` *or* `GetObject` *function. For example,* `Access.Application.15` *for Access 2013,* `Access.Application.14` *for Access 2010, and* `Access.Application.12` *for Access 2007.*

Connecting to an Access Database Using the DAO Library

The DAO library is a powerful library for accessing and manipulating Access databases. To connect to an Access database using DAOs, follow these steps:

1. Set a reference to the Microsoft DAO 3.6 Object Library.
2. Create an instance of the Access application, as you did in the early binding example.
3. Use the `DBEngine.OpenDatabase` method to open the Access database:

   ```
   ' You must set a reference to the Microsoft Access DAO 3.6 &_
       Object Library
   ' in the References dialog box
   Dim accApp As Access.Application
   Set accApp = New Access.Application

   Dim db As DAO.Database
   Dim dbPath As String

   dbPath = "C:\VBAExcel2024_ByExample\"
   ```

```
Set db = accApp.DBEngine.OpenDatabase(dbPath &_
        "Northwind 2007.accdb")
```

See the complete VBA procedure in Hands-On 13.9 later in this chapter.

Connecting to an Access Database Using the ADO Library

The ADO library is another powerful library for accessing and manipulating Access databases.

To connect to an Access database using ADOs, follow these steps:

1. Set a reference to the Microsoft ActiveX Data Objects 6,1 Object Library.
2. Define the connection string that specifies the provider (`Microsoft.ACE.OLEDB.12.0`) and the path to the Access database.
3. Initialize ADOs. To do this, create instances of the `ADODB.Connection` and `ADODB.Recordset` objects.
4. Open a connection to the database using the `conn.Open` method.
5. Specify the SQL query to retrieve data from the Access database.
6. Open the recordset. To do this, use the `rs.Open` method to execute the query and retrieve data into the recordset.
7. Loop through the recordset and write field names and data to an Excel worksheet.

See the complete VBA procedure in Hands-On 13.10 later in this chapter.

Should You Connect to Access Using the DAO or ADO Library?

The choice between using the DAO and ADO library to connect to an Access database depends on several factors, including the specific requirements of your project, compatibility considerations, and your familiarity with each library. Here are some key points to help you decide when to use DAO and when to use ADO:

- Using DAO

  DAO is designed specifically for Microsoft Access databases, and it provides native support and optimized performance. All Access-specific features, such as complex queries, multi-value fields, and attachments, are supported by DAO. Therefore, if you need to use these features, DAO is the preferred choice. DAO is also more intuitive and simpler to use. If you need to maintain or extend an existing Access application that already uses DAOs, it makes sense to continue using the DAO library for consistency and compatibility.

- Using ADO

  The ADO library supports various data sources, including SQL Server, Oracle, and others, in addition to Access. If you need to connect to multiple types of databases, ADO offers greater flexibility. ADO can be used in environments that support ActiveX, such as Web applications and COM-based applications. It also supports disconnected recordsets, allowing you to retrieve data, disconnect from the data source, and work with data offline. This can be useful for scenarios where you need to manipulate data without maintaining an open connection to the database.

Opening a Microsoft Access Database and Retrieving Data

Once you establish a connection to the Access database, you will need to open the database to perform common automation tasks such as running Access queries that execute various Access SQL commands and retrieve results to Excel. You can add, update, or delete records in Access tables directly from Excel. You can also generate Access reports and export them as PDF or Excel.

In Hands-On 13.8, you will learn how to establish a connection to Microsoft Access using automation (early binding method) and how to use the `OpenCurrentDatabase` method of the Access `Application` object to retrieve the names of its tables into an Excel worksheet.

(◎) Hands-On 13.8   Retrieving Access Table Names with Automation

1. Copy the `Northwind 2007.accdb` database from the companion files to your `C:\VBAExcel2024_ByExample` folder. This is a Microsoft-designed sample database provided with earlier versions of the Microsoft Access product.
2. Insert a new module into `VBAProject   (Chap13.xlsm)` and rename it `AccessAutomation`.
3. In the `AccessAutomation` module, enter the following procedure:

```
Sub RetrieveAccessTableNames_EarlyBinding()
    Dim accApp As Access.Application
    Dim db As DAO.Database
    Dim tdf As DAO.TableDef
    Dim wb As Workbook
    Dim ws As Worksheet
    Dim i As Integer
    Dim strDb As String
    Dim strPath As String

    On Error Resume Next
    strPath = "C:\VBAExcel2024_ByExample\"
```

```
strDb = "Northwind 2007.accdb"


Set accApp = GetObject(, " Access.Application.16")
If accApp Is Nothing Then
   ' Get a reference to the Access Application object
   Set accApp = New Access.Application
End If

' Open the Northwind database
accApp.OpenCurrentDatabase strPath & strDb

' Set the database object
Set db = accApp.CurrentDb

' Create a new Excel workbook
 Set wb = Workbooks.Add
' Use the first sheet in the new workbook
Set ws = wb.Sheets(1)
ws.Cells(1, 1).Value = "Access Tables"
ws.Name = "Access Table Names"


' Loop through all table definitions and retrieve table names
i = 2 ' Start from the second row
For Each tdf In db.TableDefs
    ' Skip system tables (those that start with "MSys")
    If Left(tdf.Name, 4) <> "MSys" Then
        ws.Cells(i, 1).Value = tdf.Name
        i = i + 1
    End If
Next tdf
ws.Columns("A").AutoFit
ws.Cells(1, 1).Interior.Color = RGB(173, 216, 230)


' Cleanup
Set tdf = Nothing
Set db = Nothing
accApp.Quit

Set accApp = Nothing

' Save the new workbook to disk
Dim savePath As String
savePath = strPath & "List of Access Tables.xlsx"
```

```
        wb.SaveAs fileName:=savePath

End Sub
```

The `RetrieveAccessTableNames_EarlyBinding` procedure uses a current instance of Access if it is available. If Access isn't running, a runtime error will occur and the object variable will be set to `Nothing`. By placing the `On Error Resume Next` statement inside this procedure, you can trap this error. Therefore, if Access isn't running, a new instance of Access will be started. This particular example uses the `New` keyword to start a new instance of Access. The `OpenCurrentDatabase` method is used to open an existing Access database.

`strPath` and `strDb` variables are combined to form the full path to the database file on your computer. Once the Northwind 2007 database is opened with the `OpenCurrentDatabase` method, the next statement uses the `CurrentDb` property to return a reference to the database currently open in the Access application instance (`accApp`). This allows you to use the `db` variable to interact with the database, such as running queries, opening tables, or manipulating data. In this procedure, we access the DAO `TableDef` collection to retrieve the names of Access tables in the open database.

4. Run the `RetrieveAccessTableNames_EarlyBinding` procedure by stepping through its code with the F8 key.

After running this procedure, you should see the names of tables in the `Northwind 2007.accdb` database entered in the first column of a new Excel workbook (see Figure 13.17).

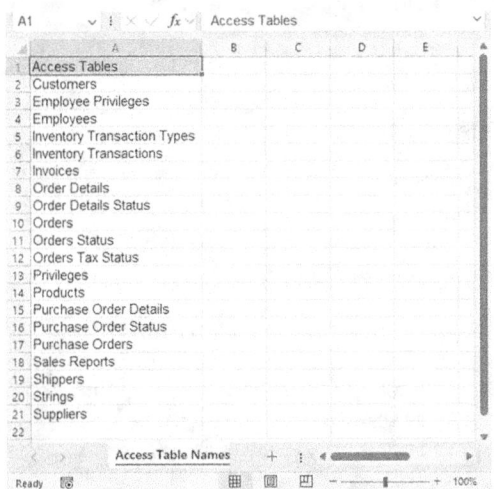

FIGURE 13.17.   Access table names were retrieved to Excel using Access automation.

Now let's write a procedure to connect to an Access database using DAO and retrieve data from the Orders table.

Hands-On 13.9   Retrieving Data from an Access Table Using DAO

1. Insert a new module into `VBAProject(Chap13.xlsm)` and rename it `AccessDAO`.
2. In the `AccessDAO` module, enter the following `RetrieveCustomerOrders` procedure:

```
Sub RetrieveCustomerOrders()
    Dim accApp As Access.Application
    Dim db As DAO.Database
    Dim rs As DAO.Recordset
    Dim wb As Workbook
    Dim ws As Worksheet
    Dim i As Integer
    Dim j As Integer
    Dim dbPath As String

    On Error Resume Next

    dbPath = "C:\VBAExcel2024_ByExample\Northwind 2007.accdb"

    ' Create a new Excel workbook
    Set wb = Workbooks.Add
    Set ws = wb.Sheets(1)
    ws.Name = "Customer Orders"


    ' Initialize Access
    Set accApp = New Access.Application

    ' Open the Northwind database
    accApp.OpenCurrentDatabase dbPath

    ' Set the database and recordset objects
    Set db = accApp.CurrentDb
    Set rs = db.OpenRecordset("SELECT * FROM Orders")

    ' Write the field names to the first row of the worksheet
    For j = 0 To rs.Fields.Count - 1
        ws.Cells(1, j + 1).Value = rs.Fields(j).Name
    Next j
```

```
' Loop through the records and write data to the worksheet
i = 2 ' Start from the second row
Do While Not rs.EOF
    For j = 0 To rs.Fields.Count - 1
        ws.Cells(i, j + 1).Value = rs.Fields(j).Value
    Next j
    rs.MoveNext
    i = i + 1
Loop

Columns.AutoFit

' Cleanup
rs.Close
Set rs = Nothing
Set db = Nothing
accApp.Quit
Set accApp = Nothing

' Save the new workbook to disk
Dim savePath As String
savePath = Left(dbPath, InStrRev(dbPath, "\")) & _
        "CustomerOrders.xlsx"
wb.SaveAs fileName:=savePath

' Inform user of completion
MsgBox "Customer orders are listed in the " & _
        ws.Name & " worksheet."
End Sub
```

After establishing a connection to the `Northwind 2007.accdb` database, you can use the DAO `Recordset` object to access data. `Recordset` objects are used to manipulate data at the record level. The `Recordset` object is made up of records (rows) and fields (columns). To obtain a set of records, you need to use the `OpenRecordset` method and specify information such as the source of records for the recordset. The source of records can be the name of a database table, a query, or the SQL statement that returns records, such as the one used in this example: `SELECT * FROM Orders`.

3. Run the `RetrieveCustomerOrders` procedure.

The resulting worksheet is depicted in Figure 13.18.

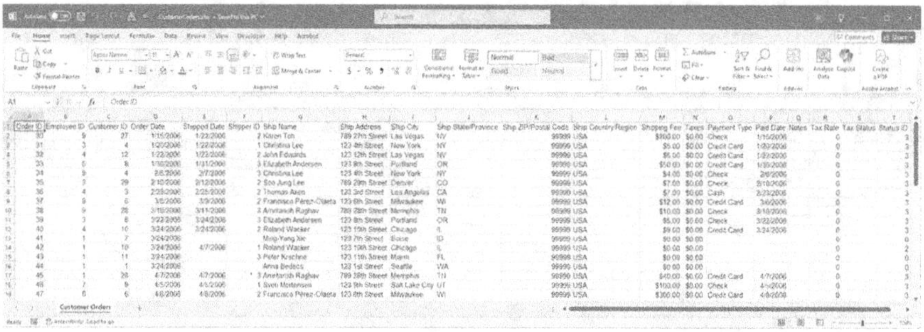

FIGURE 13.18. Partial output of data retrieved from the Access database using the DAO technique.

Now let's write a procedure to connect to an Access database using ADO and retrieve data from the Employees table.

Hands-On 13.10   Retrieving Data from an Access Table Using ADO

1. Insert a new module into VBAProject(Chap13.xlsm) and rename it AccessADO.

2. In the AccessADO module, enter the following RetrieveCustomerOrders procedure:

```
Sub RetrieveDataUsingADO()
    Dim conn As ADODB.Connection
    Dim rs As ADODB.Recordset
    Dim connString As String
    Dim query As String
    Dim wb As Workbook
    Dim ws As Worksheet
    Dim i As Long
    Dim j As Long

    ' Define the connection string
    connString = "Provider=Microsoft.ACE.OLEDB.12.0;" & _
      "Data Source=C:\VBAExcel2024_ByExample\" & _
      "Northwind 2007.accdb;Persist Security Info=False;"

    ' Define the query
    query = "SELECT * FROM Employees"

    ' Initialize the connection and recordset objects
```

```vba
    Set conn = New ADODB.Connection
    Set rs = New ADODB.Recordset

    ' Open the connection
    conn.Open connString

    ' Open the recordset
    rs.Open query, conn, adOpenStatic, adLockReadOnly

    ' Create a new Excel workbook
    Set wb = Workbooks.Add
    Set ws = wb.Sheets(1)
    ws.Name = "Employees Data"

    ' Write the field names to the first row of the worksheet
    For i = 0 To rs.Fields.Count - 1
        ws.Cells(1, i + 1).Value = rs.Fields(i).Name
    Next i

    ' Write the data to the worksheet
    j = 2 ' Start from the second row
    Do Until rs.EOF
        For i = 0 To rs.Fields.Count - 1
            ws.Cells(j, i + 1).Value = rs.Fields(i).Value
        Next i
        rs.MoveNext
        j = j + 1
    Loop
    Columns.AutoFit

    ' Cleanup
    rs.Close
    conn.Close
    Set rs = Nothing
    Set conn = Nothing

    ' Save the new workbook to disk
    Dim savePath As String
    savePath = "C:\VBAExcel2024_ByExample\EmployeesData.xlsx"
    wb.SaveAs fileName:=savePath

    ' Inform user of completion
    MsgBox "Employees data has been retrieved into " & ws.Name
End Sub
```

The `RetrieveDataUsingADO` procedure uses the ADO connection string to connect to an Access database. Note that the `Could not find installable ISAM` error typically occurs when there is an issue with the connection string or the necessary database drivers aren't properly installed. The connection string must be correctly formatted. If you encounter this error, check for any missing punctuation marks.

For Access 2007 and later, you should use the `Microsoft.ACE.OLEDB.12.0` provider. For older versions, use `Microsoft.Jet.OLEDB.4.0`. You must also ensure the correct drivers are installed on your system. For Access 2007 and later, you need the Microsoft Access Database Engine (ACE) installed. You can download it from the Microsoft Web site.

If you are running a 64-bit version of Office, you need the 64-bit version of the Access Database Engine; otherwise, the 32-bit version should be used. Verify that the file path to your Access database is correct and the database file exists at that location. By following these guidelines, you should be able to resolve the ISAM error and successfully connect to your Access database using ADO. For more detailed troubleshooting, you can refer to Microsoft's documentation.

3. Run the `RetrieveDataUsingADO` procedure.
   The resulting worksheet is depicted in Figure 13.19.

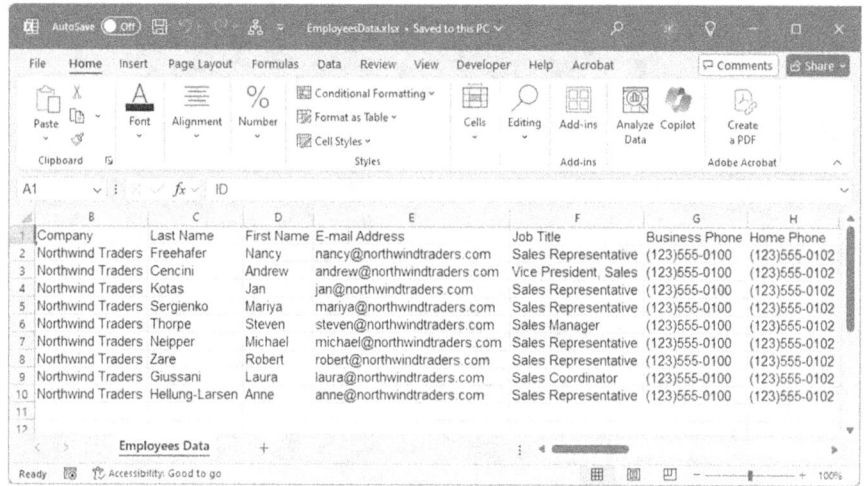

FIGURE 13.19. Partial output of data retrieved from the Access database using the ADO technique.

Using Various Data Retrieval Methods

In earlier Hands-On exercises, you retrieved data from Access by accessing the `Recordset` object. There are many other ways you can use to get external data into an Excel worksheet. In this section, you'll work with the `GetRows`, `CopyFromRecordset`, and `TransferSpreadsheet` methods.

Using the GetRows Method

The `GetRows` method allows you to retrieve multiple rows from a `Recordset` object and place them into an array. This can be useful when you want to manipulate or process data before inserting it into an Excel worksheet.

The `GetRows` method returns a two-dimensional array, where the first subscript is a number representing the field and the second subscript is the number representing the record. Record and field numbering begins with 0. The following Hands-On exercise demonstrates how to use the `GetRows` method in a VBA procedure. We will run the `Invoices` query in the Northwind database and return records to a worksheet.

⊙ Hands-On 13.11   Retrieving Access Data Using the GetRows Method

1. Copy the `NorthwindStarter.accdb` database from the companion files to your `C:\VBAExcel2024_ByExample` folder.
2. Insert a new module into `VBAProject(Chap13.xlsm)` and rename it `Access_GetRows`.
3. Make sure you have set a reference to the Microsoft DAO 3.6 Object Library in the References dialog box.
4. In the `Access_GetRows` module Code window, enter the `GetData_withGetRows` procedure, as shown below:

```
Sub GetData_withGetRows()
   Dim db As DAO.Database
   Dim qdf As DAO.QueryDef
   Dim rs As DAO.Recordset
   Dim arrData As Variant
   Dim i As Long
   Dim j As Long
   Dim strPath As String
   Dim a As Variant
   Dim countR As Long
   Dim strShtName As String

   strPath = "C:\VBAExcel2024_ByExample\NorthwindStarter.accdb"
```

```
    strShtName = "Invoice Data"

    Set db = DBEngine.OpenDatabase(strPath)
    Set qdf = db.QueryDefs("qryInvoice")
    Set rs = qdf.OpenRecordset

    rs.MoveLast
    countR = rs.RecordCount
    a = InputBox("This recordset contains " & _
        countR & " records." & vbCrLf _
        & "Enter number of records to return: ", _
        "Get Number of Records")

    If a = "" Or a = 0 Then Exit Sub
    If a > countR Then
        a = countR
        MsgBox "The number you entered is too large." & vbCrLf _
            & "All records will be returned."
    End If

    Workbooks.Add
    ActiveWorkbook.Worksheets(1).Name = strShtName
    rs.MoveFirst
    With Worksheets(strShtName).Range("A1")
      arrData = rs.GetRows(a)
      For i = 0 To UBound(arrData, 2)
          For j = 0 To UBound(arrData, 1)
              .Offset(i + 1, j) = arrData(j, i)
          Next j
      Next i
      For j = 0 To rs.Fields.Count - 1
          .Offset(0, j) = rs.Fields(j).Name
          .Offset(0, j).EntireColumn.AutoFit
      Next j
    End With
' cleanup
  Set rs = Nothing
  Set qdf = Nothing
  db.Close
End Sub
```

After opening the specified Access database with the OpenDatabase method of the DBEngine object, the GetData_withGetRows procedure runs the qryInvoice query. In the Microsoft DAO 3.6 Object Library, the QueryDefs object represents a Select or action query.

`Select` queries return data from one or more tables or queries, while action queries allow you to add, modify, or delete records. After executing the query, the procedure places the records returned by the query in the object variable of type `Recordset` using the `OpenRecordset` method. Next, the record count is retrieved using the `RecordCount` method and placed in the `countR` variable. Notice that to obtain the correct record count, the record pointer must first be moved to the last record in the recordset by using the `MoveLast` method.

The procedure then prompts the user to enter the number of records to return to the worksheet. You can cancel at this point by clicking the Cancel button in the input dialog box or you can type the number of records to retrieve. If you enter a number that is greater than the record count, the procedure will retrieve all the records.

Before retrieving records, you must move the record pointer to the first record by using the `MoveFirst` method. If you forget to do this, the record pointer will remain on the last record and only one record will be retrieved.

The procedure then goes on to activate the worksheet. The records are first returned to the `Variant` variable containing a two-dimensional array by using the `GetRows` method of the `Recordset` object. Next, the procedure loops through both dimensions of the array to place the records in the worksheet starting at cell A2. When this is done, another loop will fill in the first worksheet row with the names of fields and autofit each column so that the data is displayed correctly.

5. Run the `GetData_withGetRows` procedure. When prompted for the number of records, type `10` and click OK. Next, switch to the Microsoft Excel application window to view the results.

6. Close the Excel workbook created in this Hands-on exercise.

Using the CopyFromRecordset Method

The `CopyFromRecordset` method of the `Range` object allows you to copy data directly from a `Recordset` object to an Excel worksheet. This method is very efficient, making it ideal for quickly transferring data.

The `CopyFromRecordset` method can take up to three arguments: `Data`, `MaxRows`, and `MaxColumns`. Only the first argument, `Data`, is required. This argument can be the `Recordset` object. The optional arguments, `MaxRows` and `MaxColumns`, allow you to specify the number of records (`MaxRows`) and the number of fields (`MaxColumns`) that should be returned.

If you omit the `MaxRows` argument, all the returned records will be copied to the worksheet. If you omit the `MaxColumns` argument, all the fields will be retrieved.

The following procedure uses the ADOs and the `CopyFromRecordset` method to retrieve all the records from the `NorthwindStarter.accdb` database `Products` table.

Hands-On 13.12   Retrieving Access Data Using the opyFromRecordset Method

1. Insert a new module into `VBAProject(Chap13.xlsm)` and rename it `Access_CopyFromRecordset`.
2. Make sure you've copied the `NorthwindStarter.accdb` database from companion files to your working directory.
3. Make sure you have set a reference to the Microsoft ActiveX Data Objects 6.1 Library in the References dialog box.
4. In the `Access_CopyFromRecordset` module Code window, enter the `GetProducts` procedure, as below:

```
Sub GetProducts()
    Dim conn As New ADODB.Connection
    Dim rst As ADODB.Recordset
    Dim strPath As String
    Dim j As Long


    strPath = "C:\VBAExcel2024_ByExample\NorthwindStarter.accdb"

    conn.Open "Provider=Microsoft.ACE.OLEDB.12.0;" _
        & "Data Source=" & strPath & ";"
    conn.CursorLocation = adUseClient

    ' Create a Recordset from all the records
    ' in the Products table

    Set rst = conn.Execute(CommandText:="Products", _
        Options:=adCmdTable)

    rst.MoveFirst

    ' transfer the data to Excel
    ' get the names of fields first
    Worksheets.Add
```

```
    With ActiveSheet.Range("A1")
        '.CurrentRegion.Clear
        For j = 0 To rst.Fields.Count - 1
            .Offset(0, j) = rst.Fields(j).Name
        Next j
        .Offset(1, 0).CopyFromRecordset rst
        .CurrentRegion.Columns.AutoFit
    End With
    rst.Close
    conn.Close

    Set rst = Nothing
    Set conn = Nothing
End Sub
```

The `GetProducts` procedure copies all the records from the `Products` table into an Excel worksheet. If you want to copy fewer records, use the `MaxRows` argument as follows:

```
.Offset(1, 0).CopyFromRecordset rst, 5
```

The above statement tells Visual Basic to copy only five records. The `Offset` method causes the records to be entered into a worksheet, starting with the second row. To send all the records to the worksheet using the data from only two table fields, use the following statement:

```
.Offset(1, 0).CopyFromRecordset rst, , 2
```

The above statement tells Visual Basic to copy all the data from the first two columns. The comma between `rst` and the number 2 is a placeholder for the omitted `MaxRows` argument.

5. Run the `GetProducts` procedure and switch to the Excel application window to view the results.

Using the TransferSpreadsheet Method

`TransferSpreadsheet` is part of the Access object model and allows you to export data from Access to an Excel workbook or import data from Excel to Access. This method is especially convenient when working with large datasets or when you need to automate data transfers.

Using this method, you can also link the data in an Excel worksheet to the current Access database. With a linked worksheet, you can view and edit the worksheet data with Access while still allowing complete access to the data from your Excel application.

The `TransferSpreadsheet` method carries out the `TransferSpreadsheet` action in Visual Basic and has the following syntax:

```
DoCmd.TransferSpreadsheet [transfertype][, spreadsheettype],
    tablename, filename [, hasfieldnames][, range]
```

The `transfertype` argument can be one of the following constants: `acImport` (default setting), `acExport`, or `acLink`. These constants define whether data has to be imported, exported, or linked to the database. The `spreadsheettype` argument specifies the worksheet name and the version number. The `tablename` argument is a string expression that specifies the name of the Access table you want to import worksheet data into, export worksheet data from, or link worksheet data to. Instead of the table name, you may also specify the name of the `Select` query whose results you want to export to a worksheet.

The `filename` argument is a string expression that specifies the filename and path of the spreadsheet you want to import from, export to, or link to.

The `hasfieldnames` argument is a logical value of `True` (-1) or `False` (0). `True` indicates that the first worksheet row contains the field names. `False` denotes that the first row contains normal data. The default setting is `False` (no field names in the first row).

The `range` argument is a string expression that specifies the range of cells or the name of the range in the worksheet. This argument applies only to importing. If you omit the `range` argument, the entire spreadsheet will be imported. Leave this argument blank if you want to export, unless you need to specify the worksheet name.

The `ExportData` example procedure shown below exports data from the `Shippers` table in the `Northwind 2007.accdb` database to the `Shippers.xlsx` workbook using the `TransferSpreadsheet` method.

Hands-On 13.13   Retrieving Data with the TransferSpreadsheet Method

1. Insert a new module into `VBAProject(Chap13.xlsm)` and rename it `Access_TransferSpreadsheet`.
2. In the `Access_TransferSpreadsheet` module Code window, enter the `ExportData` procedure as shown below:

```
Sub ExportData()
  Dim objAccess As Access.Application
  Set objAccess = CreateObject("Access.Application")

  objAccess.OpenCurrentDatabase filePath:= _
```

```
     "C:\VBAExcel2024_ByExample\Northwind 2007.accdb"

  objAccess.DoCmd.TransferSpreadsheet _
      TransferType:=acExport, _
      SpreadsheetType:=acSpreadsheetTypeExcel12Xml, _
      TableName:="Shippers", _
      fileName:="C:\VBAExcel2024_ByExample\Shippers.xlsx", _
      HasFieldNames:=True, _
      Range:="Sheet1"

  objAccess.Quit
  Set objAccess = Nothing
End Sub
```

The `ExportData` procedure uses automation to establish a connection to Microsoft Access. The database is opened using the `OpenCurrentDatabase` method. The `TransferSpreadsheet` method of the `DoCmd` object is used to specify that the data from the `Shippers` table should be exported into an Excel workbook named `Shippers.xlsx` and placed in `Sheet1`. The first row of the worksheet is to be used by field headings. When data is retrieved, the Access application is closed and the object variable pointing to the Access application is destroyed. When using the `TransferSpreadsheet` method, it's important to ensure that the correct spreadsheet type is specified. For `.xlsx` files, you should use the `acSpreadsheetTypeExcel12Xml` constant instead of `acSpreadsheetTypeExcel12`, which is used for the older `.xls` file format.

3. Switch to the Excel application window and choose View | Macros | View Macros. In the Macros dialog box, select the `ExportData` procedure and click Run.

4. Open the `Shippers.xlsx` file created by the `ExportData` procedure to view the retrieved data.

Creating an Embedded Chart from Microsoft Access Data

With so many powerful formatting and data analysis tools available in Excel, data retrieved from Access can be easily transformed with VBA to provide meaningful insights and visual representations. For example, your VBA procedure can load data directly into a PivotTable so you can analyze relationships and trends. You can also create data tables and charts based on your Microsoft Access data.

Charts are created by using the `Add` method of the `Charts` collection. Let's spend some time now creating a procedure that fetches data from the Access database and creates an embedded chart.

⊙ Hands-On 13.14 Creating an Embedded Chart from Access Data

1. Insert a new module into `VBAProject(Chap13.xlsm)` and rename it `Access_ChartingData`.

2. In the `Access_ChartingData` module Code window, enter the `ChartData_withADO` procedure, as shown below:

```
Sub ChartData_withADO()
  Dim conn As New ADODB.Connection
  Dim rst As New ADODB.Recordset
  Dim mySheet As Worksheet
  Dim recArray As Variant
  Dim i As Long
  Dim j As Long
  Dim sSQL As String

  sSQL = "SELECT Products.ProductName AS [Product Name],"
  sSQL = sSQL & " Sum([Quantity]*[OrderDetails].[UnitPrice]) " & _
      "AS [Total Sales]"
  sSQL = sSQL & " FROM Products INNER JOIN (Orders INNER JOIN"
  sSQL = sSQL & " OrderDetails ON Orders.OrderID = " & _
      "OrderDetails.OrderID)"
  sSQL = sSQL & " ON Products.ProductID = OrderDetails.ProductID"
  sSQL = sSQL & " GROUP BY Products.ProductName, Orders.EmployeeID"
  sSQL = sSQL & " HAVING (((Orders.EmployeeID)=10))"
  sSQL = sSQL & " ORDER BY Products.ProductName;"

  ' Connect with the database
    conn.Open _
  "Provider=Microsoft.ACE.OLEDB.12.0;Data Source=" & _
      "C:\VBAExcel2024_ByExample\NorthwindStarter.accdb;"
  ' Open Recordset based on the SQL statement
    rst.Open sSQL, conn, adOpenForwardOnly, adLockReadOnly

  Workbooks.Add
  Set mySheet = Worksheets("Sheet1")
  With mySheet.Range("A1")
    recArray = rst.GetRows()
    For i = 0 To UBound(recArray, 2)
      For j = 0 To UBound(recArray, 1)
        .Offset(i + 1, j) = recArray(j, i)
      Next j
    Next i
    For j = 0 To rst.Fields.Count - 1
```

```
            .Offset(0, j) = rst.Fields(j).Name
            .Offset(0, j).EntireColumn.AutoFit
        Next j
    End With
    rst.Close
    conn.Close
    Set rst = Nothing
    Set conn = Nothing

    mySheet.Activate
    Charts.Add
    ActiveChart.ChartType = xl3DColumnClustered
    ActiveChart.SetSourceData _
        Source:=mySheet.Cells(1, 1).CurrentRegion, PlotBy:=xlRows
    ActiveChart.Location Where:=xlLocationAsObject, _
        Name:=mySheet.Name


    With ActiveChart
        .HasTitle = True
        .ChartTitle.Characters.Text = "Internet Sales by Product"
        .Axes(xlCategory).HasTitle = True
        .Axes(xlCategory).AxisTitle.Characters.Text = _
            "Popular Products"
        .Axes(xlValue).HasTitle = True
        .Axes(xlValue).AxisTitle. _
            Characters.Text = mySheet.Range("B1") & " ($)"
        .Axes(xlValue).AxisTitle.Orientation = xlUpward
        .Legend.Font.Size = 9
    End With

    With ActiveSheet.Shapes("Chart 1")
        .ScaleWidth 1.24, msoFalse, msoScaleFromTopLeft
        .ScaleHeight 1.5, msoFalse, msoScaleFromTopLeft
    End With
End Sub
```

The `ChartData_withADO` procedure retrieves data using the `GetRows` method that was discussed in an earlier section. Once the data is placed in an Excel worksheet, a chart is added with the `Add` method. The `SetSourceData` method of the `Chart` object sets the source data range for the chart like this:

```
ActiveChart.SetSourceData Source:=mySheet. _
Cells(1, 1).CurrentRegion, PlotBy:=xlRows
```

`Source` is the range that contains the source data that we have just placed on the worksheet beginning with cell A1. `PlotBy` will cause the embedded chart to plot data by rows.

Next, the `Location` method of the `Chart` object specifies where the chart should be placed. This method has two arguments: `Where` and `Name`. The `Where` argument is required. You can use one of the following constants for this argument: `xlLocationAsNewSheet`, `xlLocationAsObject`, or `xlLocationAutomatic`. The `Name` argument is required if `Where` is set to `xlLocationAsObject`. In this procedure, the `Location` method specifies that the chart should be embedded in the active worksheet:

```
ActiveChart.Location Where:=xlLocationAsObject, _
    Name:=mySheet.Name
```

Next, a group of statements formats the chart by setting various properties.

3. Run the `ChartData_withADO` procedure.
   The resulting chart is shown in Figure 13.20.

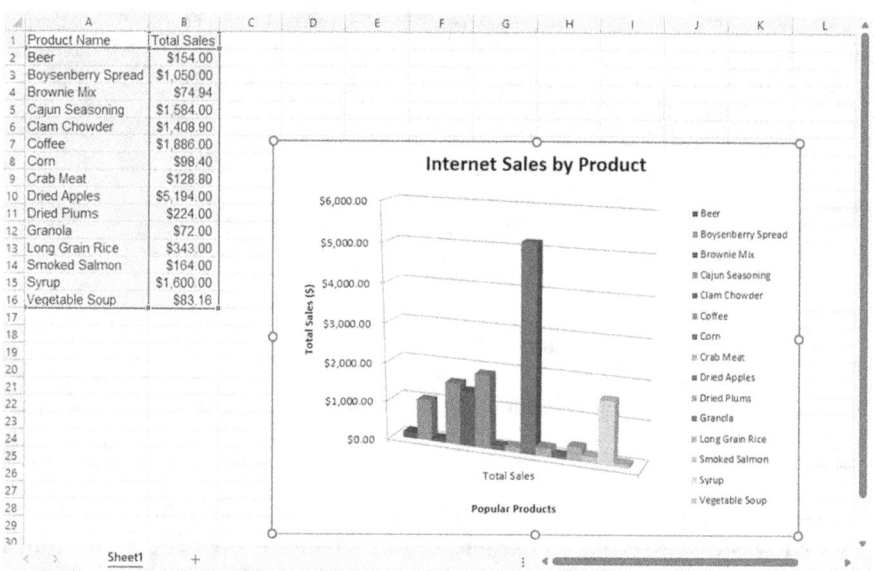

FIGURE 13.20.   You can create an embedded chart programmatically with VBA based on the data retrieved from a Microsoft Access table, a query, or a SQL statement.

LAUNCHING APPLICATIONS

If a program you want to launch is a Microsoft application, use the `Activa-teMicrosoftApp` method. This method is available from the Microsoft Excel `Application` object. For example, to launch PowerPoint from the Immediate window, all you need to do is type the following instruction and press Enter:

```
Application.ActivateMicrosoftApp xlMicrosoftPowerPoint
```

Notice that the `ActivateMicrosoftApp` method requires a constant to indicate which program to start. The above statement starts Microsoft PowerPoint if it is not already running. If the program is already open, this instruction does not open a new occurrence of the program; it simply activates the already running application. You can use the constants shown in Table 13.1 with the `Activate-MicrosoftApp` method.

TABLE 13.1.   ActivateMicrosoftApp method constants.

Application Name	Constant
Access	`xlMicrosoftAccess`
Mail	`xlMicrosoftMail`
PowerPoint	`xlMicrosoftPowerPoint`
Word	`xlMicrosoftWord`

Any application, including Office 365 applications, can be launched directly from a VBA procedure using the `Shell` function. For example, to launch Notepad, all you need is one statement between the keywords `Sub` and `End Sub`. Better yet, you can type the following statement in the Immediate window and press Enter to see the result immediately:

```
Shell "notepad.exe", vbMaximizedFocus
```

sHere, `notepad.exe` is the name of the program you want to start. This name should include the complete path (the drive and folder name) if you have any concerns that the program may not be found. Notice that the program name is in double quotes. The second argument of the `Shell` function is optional. This argument specifies the window style. In this example, Notepad will appear in a maximized window. If the window style is not specified, the program will be minimized with focus. Table 13.2 lists the window style constants and appearance options.

TABLE 13.2. Window styles used in the Shell function.

Window Style Constant	Value	Window Appearance
vbHide	0	The window is hidden and the focus is passed to the hidden window. Not applicable on Macintosh platforms.
vbNormalFocus	1	The window has focus and is restored to its original size and position.
vbMinimizedFocus	2	The window is displayed as an icon with focus.
vbMaximizedFocus	3	The window is maximized with focus.
vbNormalNoFocus	4	The window is restored to its most recent size and position. The currently active window remains active.
vbMinimizedNoFocus	6	The window is displayed as an icon. The currently active window remains active.

If the Shell function is successful in launching the specified executable file, it will return a number called a task ID, which uniquely identifies the application that has been launched. If the Shell function cannot start the specified program, Visual Basic generates an error. The Shell function works asynchronously. This means that Visual Basic starts the program specified by the Shell function and, immediately after launching, it returns to the procedure to continue with the execution of the remaining instructions without giving you a chance to work with the application. If you want to work with the program launched by the Shell function, do not enter any other statements in the procedure after the Shell function.

MOVING BETWEEN APPLICATIONS

Because users can work simultaneously with several applications in the Windows environment, your VBA procedure must also know how to switch between the open programs.

Suppose that, in addition to Microsoft Excel, you have two other applications open: Microsoft Word and Windows Explorer. To activate an already open program, use the AppActivate statement using the following syntax:

```
AppActivate title [, wait]
```

Only the title argument is required. This is the name of the application as it appears in the title bar of the active application window or its task ID number as returned by the Shell function.

The optional argument `wait` is a Boolean value (`True/False`) that specifies when Visual Basic activates the application. The value of `False` in this position immediately activates the specified application, even if the calling application does not have the focus. If you place `True` in the position of the `wait` argument, the calling application waits until it has the focus before it activates the specified application.

For example, here's how you can activate Microsoft Word:

```
AppActivate "Microsoft Word"
```

Notice that the name of the application is surrounded by double quotation marks.

You can also use the return value of the `Shell` function as the argument of the `AppActivate` statement:

```
' run Microsoft Word
Sub RunWord()
    Dim ReturnValue As Variant
    ReturnValue = Shell("C:\Program Files\Microsoft Office\" & _
      "root\office16\WINWORD.EXE /w", 1)
    ' activate Microsoft Word
    AppActivate ReturnValue
End Sub
```

In the `RunWord` procedure, the `/w` startup switch after `WINWORD.EXE` will start a new instance of Word with a blank document. `1` indicates that the application window has a focus and will be displayed in its original size and position (see Table 13.2 earlier).

The `AppActivate` statement is used for moving between applications and requires that the program is already running. This statement merely changes focus. The specified application becomes the active window. The `AppActivate` statement will not start an application running.

SUMMARY

In this chapter, we delved into the topic of automating various Microsoft Office products using Excel VBA. You learned how to use both early and late binding to access the object models of other applications. You expanded your knowledge of VBA statements by mastering the use of the `CreateObject` and `GetObject` functions and the `New` keyword for instantiating objects.

You automated Microsoft Word by creating a new document, adding content, applying formatting, and embedding it in an Excel worksheet. You explored how to retrieve contact addresses from Microsoft Outlook and place them in an Excel worksheet. You also sent bulk emails directly from Excel. Furthermore, you leveraged Excel data to create a PowerPoint presentation, dynamically adding and formatting slides. Additionally, various techniques for retrieving Microsoft Access data were covered, including creating an embedded chart in Excel. Lastly, you learned new VBA commands to start programs and switch between applications.

In the next chapter, you will learn how to use PowerShell from Excel to automate various administrative tasks.

Chapter **14** *USING POWERSHELL WITH EXCEL VBA*

In Chapter 13, you explored how to use VBA to control other Microsoft 365 applications directly from Excel. Building on that foundation, this chapter delves into PowerShell—a robust, task-oriented scripting language and command-line shell developed by Microsoft. By integrating PowerShell with Excel VBA, you can perform a wide range of system-level tasks on both local and remote Windows systems. If you've ever worked with DOS commands, you'll find some familiarity here, but PowerShell offers far more flexibility, functionality, and power.

GETTING STARTED WITH POWERSHELL

The good news is that PowerShell is already included with most modern versions of Windows and accessing it on your Windows machine is quite straightforward. You can launch it by searching for PowerShell in the Start menu or by using the built-in Run dialog box (press Windows key+R, type `powershell`, and hit Enter). Depending on your version of Windows, you may see different editions, such as Windows PowerShell (see Figure 14.1) or the newer PowerShell 7.5 (see Figure 14.2), which is cross-platform and more feature-rich.

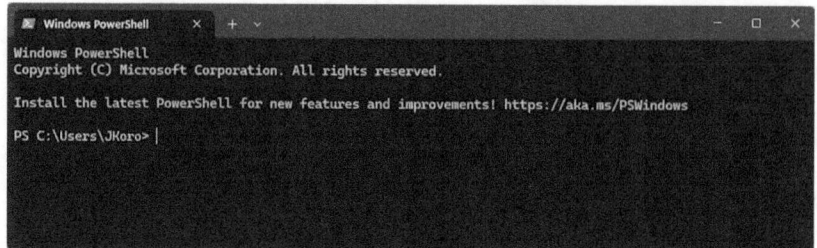

FIGURE 14.1.   To upgrade Windows PowerShell to the newest edition, use the provided link (*https://aka.ms/PSWindows*) for the installation guidelines.

FIGURE 14.2.   The current version of PowerShell at the time of writing. Follow the link provided in Figure 14.1 for the installation guidelines. You can deploy PowerShell using the MSI package or ZIP package.

PowerShell 7 installs to a new directory, allowing you to run both Windows PowerShell 5.1 and PowerShell. In Windows PowerShell, the PowerShell executable is named `powershell.exe`. In PowerShell 7, it is named `pwsh.exe`. The new name enables side-by-side execution of both versions.

If you're opening PowerShell for the first time, it might look intimidating at first glance, but it is extremely easy to use for both beginners and experienced users. Here, you can enter and run simple commands such as `Get-Date` to display the current date and time or explore your computer's files and folders using commands such as `Get-ChildItem` (see Figure 14.3).

PowerShell can save you time by automating repetitive tasks. You can get a full list of available commands by typing `Get-Command` and pressing Enter. For more detailed information on each command, type `Get-Help` followed by the command from the list you just obtained. For example, type `Get-Help Write-Output` and press Enter. At this point, you should see the syntax for the `Write-Output` command as well as some remarks. You might need to download and install `Help` files for the `Get-Help` command to access them on your computer, or use `Get-Help Write-Output -Online` to view the Help topic online.

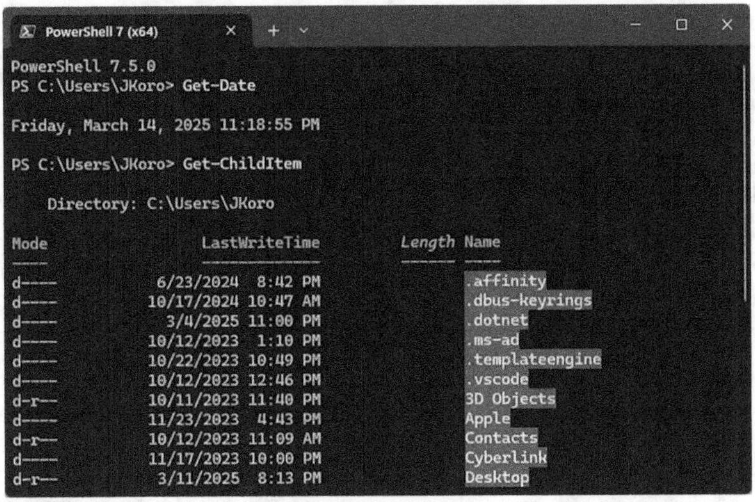

FIGURE 14.3. Entering commands in the PowerShell console window.

What Are Cmdlets and Pipelines in PowerShell?

Cmdlets (pronounced "command-lets") are lightweight, single-function commands used in PowerShell to perform specific tasks. They are designed to handle administrative and system-related tasks efficiently. They are part of the PowerShell environment and are run within its command-line interface. Cmdlets follow the same naming convention: a *Verb-Noun* structure. For example, `Get-Process` retrieves information about running processes on your system. `Add-Content` appends content, such as words or data, to a file. Cmdlets are designed to work well together. They can output data to other cmdlets using pipelines. In PowerShell, pipelines are a way of passing output from one cmdlet as input to another. The pipeline operator (|) is used to link cmdlets together. For example, if you have a folder containing multiple text files and you want to add the same line of text to all of them, you can use the following command:

```
Get-ChildItem -Path "C:\TextFiles_Examples" -Filter *.txt | Add-Content
-Value "This is a new line added to the file."
```

Let's break this down:

- `Get-ChildItem -Path "C:\TextFiles_Examples" -Filter *.txt` retrieves all `.txt` files in the specified folder (`C:\TextFiles_Examples`).
- The pipeline (|) takes the output (a collection of `.txt` files) and passes it to `Add-Content`.
- `Add-Content -Value "This is a new line added to the file."` appends the specified text to each `.txt` file retrieved by the first cmdlet.

This chain of commands allows you to achieve a complex result without manually intervening at each step. In this case, it eliminates the need to manually open and edit each file. You can further customize this example by filtering files based on their names, extensions, or other properties.

To exit the PowerShell console window, simply type `Exit` and press Enter. In the remaining sections of this chapter, we'll be working with PowerShell from Excel VBA.

Launching PowerShell from Excel VBA

You can launch PowerShell directly from VBA by using the `Shell` function. Let's look at a few examples. The procedure in Hands-On 14.1 opens the PowerShell console window, where you can begin to manually write various PowerShell commands.

NOTE	*Please note that the files for the "Hands-On" projects can be found in the companion files.*

⊙ Hands-On 14.1   Launching the PowerShell Console Window

1. Create a new Excel workbook named `C:\VBAExcel2024_ByExample\Chap14.xlsm`.
2. Switch to the VBE window, insert a new module, and enter the following `Open_PowerShell1` procedure:

```
Sub Open_PowerShell1()
    'Use the executable for your version of PowerShell

    'Shell "powershell.exe -noexit", vbNormalFocus
    Shell "pwsh.exe -noexit", vbNormalFocus
End Sub
```

3. Choose Run / RunSub/UserForm or press F5 to run the `Open_PowerShell` procedure.
   When you launch PowerShell using the `-noexit` switch, the PowerShell console window stays open so you can work in it. You can now enter any PowerShell command.
4. In the PowerShell console window, type the following command and press Enter:

```
Get-Process | Sort-Object CPU -Descending | Select-Object -First 5
```

You should get a list of the top 5 processes with the highest CPU usage:

- `Get-Process` retrieves a list of all running processes on your machine.
- `Sort-Object CPU -Descending` takes the output and sorts the processes by CPU usage in descending order.

* `Select-Object -First 5` then picks the top 5 processes with the highest CPU usage from the sorted list.

5. In the PowerShell console window, type `Exit` and press Enter.
Now let's write another procedure, in which you will issue a command to get your computer information.

6. In the VBE window, enter the following procedure in the same code module:

```
Sub GetComputerInfo_PowerShell()

    'Use the executable for your version of PowerShell

    'Shell "powershell.exe –noexit –command Get-ComputerInfo", _
        vbNormalFocus

    Shell "pwsh.exe –noexit –command Get-ComputerInfo", _
        vbNormalFocus

End Sub
```

Again, we kept the PowerShell console window open so we could see the results of the `Get-ComputerInfo` command, which provides detailed information about your computer's hardware and operating system.

7. Position the insertion point anywhere within the code of the `GetComputerInfo_PowerShell` procedure and choose Run / Run Sub/UserForm or press F5 to run it.
When you run `Get-ComputerInfo` without any parameters, it retrieves all available information, including operating system details, system hardware, network configuration, and BIOS information. If you're looking for specific details, you can filter the output by selecting particular properties. For example, the following command will fetch only the computer's name, Windows version, and operating system architecture:

```
Get-ComputerInfo | Select-Object –Property CsName, WindowsVersion,
OsArchitecture
```

8. In the PowerShell console window, type `Exit` and press Enter.
Oftentimes, when working with PowerShell in VBA, you will need to write longer commands, so it will be easier to store them in strings, as demonstrated in the next example procedure.

9. In the VBE window, in the same code module, enter the following procedure:

```
Sub CreateFolder_PowerShell()
    Dim sCmdPs As String
```

```
        Dim sFolderName As String

        sFolderName = "'C:\PowerShell_Test'"
        sCmdPs = "-noexit -command New-Item " & sFolderName
        sCmdPs = sCmdPs & "-ItemType Directory"

        Shell "powershell.exe " & sCmdPs, vbNormalFocus

        Application.Wait (Now + TimeValue("0:00:05"))
        SendKeys "exit~"
    End Sub
```

In the above procedure, we issue a `New-Item` command to create a folder on a computer and, after a five-second delay, we exit the PowerShell console window by using the `SendKeys` method.

10. Position the insertion point anywhere within the code of the `CreateFolder_PowerShell()` procedure and choose Run / Run Sub/UserForm or press F5 to run it.

Notice that when you run this procedure, Excel opens the PowerShell console window and passes it a command to create a new folder; then, the application waits for five seconds, and the PowerShell console window is automatically closed using the VBA `SendKeys` statement. This statement is covered in detail in the next section of this chapter.

Using the SendKeys Statement in Excel VBA

The `SendKeys` statement allows you to send a series of keystrokes to the active application window. You can send a key or a combination of keys and achieve the same result as if you worked directly in the active window using the keyboard. The `SendKeys` statement looks as follows:

```
SendKeys string [, wait]
```

The required argument, `string`, is the key or key combination that you want to send to the active application. For example, to exit the PowerShell console window (see Hands-On 14.1), we needed to send the string `"exit"` followed by the Enter key. The Enter key is represented by the ~ character; therefore, we sent `"exit~"`.

Let's look at other examples of usage, not related to PowerShell. To send a letter `f`, use the following instruction:

```
SendKeys "f"
```

To send the key combination Alt+f, use:

```
SendKeys "%f"
```

The percent sign (%) is the symbol used for the Alt key.

To send a combination of keys, such as Shift+Tab, use the following statement:

```
SendKeys "+{TAB}"
```

The plus sign (+) denotes the Shift key.

To send other keys and combinations of keys, see Table 14.1.

The `SendKeys` statement's second argument, `wait`, is optional. This is a logical value that is `True` or `False`. If `False` (default), Visual Basic returns to the procedure immediately upon sending the keystrokes. If `wait` is `True`, Visual Basic returns to the procedure only after the sent keystrokes have been executed. To send characters that aren't displayed when you press a key, use the codes in Table 14.1. Remember to enclose these codes in quotes. For example:

```
SendKeys "{BACKSPACE}"
```

TABLE 14.1.   Keycodes used with the SendKeys statement.

Key	Code
Backspace	{BACKSPACE}{BS}{BKSP}
Break	{BREAK}
Caps Lock	{CAPSLOCK}
Del or Delete	{DELETE}{DEL}
Down Arrow	{DOWN}
End	{END}
Enter	{ENTER} or ~
Esc	{ESC}
Help	{HELP}
Home	{HOME}
Ins or Insert	{INSERT}{INS}
Left Arrow	{LEFT}
Num Lock	{NUMLOCK}
Page Down	{PGDN}
Page Up	{PGUP}
Print Screen	{PRTSC}

(Contd.)

Key	Code
Right Arrow	{RIGHT}
Scroll Lock	{SCROLLLOCK}
Tab	{TAB}
Up Arrow	{UP}
F1	{F1}
F2	{F2}
F3	{F3}
F4	{F4}
F5	{F5}
F6	{F6}
F7	{F7}
F8	{F8}
F9	{F9}
F10	{F10}
F11	{F11}
F12	{F12}
F13	{F13}
F14	{F14}
F15	{F15}
F16	{F16}
Shift	+
Ctrl	^
Alt	%

Note that you can only send keystrokes to applications that were designed for the Microsoft Windows operating system. Some characters have a special meaning when used with the `SendKeys` statement. These keys are the plus sign (+), caret (^), tilde (~), and parentheses, (). To send these characters to another application, you must enclose them in braces, {}. To send braces, enter {{} and {}}.

The SendKeys Statement Is Case-Sensitive

When you send keystrokes with the `SendKeys` statement, bear in mind that you must distinguish between lowercase and uppercase characters. Therefore, to send the key combination Ctrl+d, you must use ^d, and to send Ctrl+Shift+d, you should use ^+d.

You can use the `SendKeys` statement to activate a ribbon tab. For example, to display the Insert Picture dialog in Excel, enter the following procedure in a VBE module, then run it via the Excel Macro dialog box (press Alt+F8 in the Excel application window):

```
Sub InsertPicture_UseKeys()

  SendKeys "%n+pi+d"

End Sub
```

To view the list of keys assigned to individual ribbon tabs, press the Alt key. You should see the ribbon mappings as shown in Figure 14.4. After pressing the key for the tab you'd like to activate, the ribbon will now display the access keys assigned to individual commands, as illustrated in Figure 14.5.

FIGURE 14.4. Each ribbon tab has an access key that can be used with the SendKeys statement to activate a particular tab. To activate the Insert tab, press Alt+N. In the SendKeys statement, you can achieve this by passing "%n" as the string parameter.

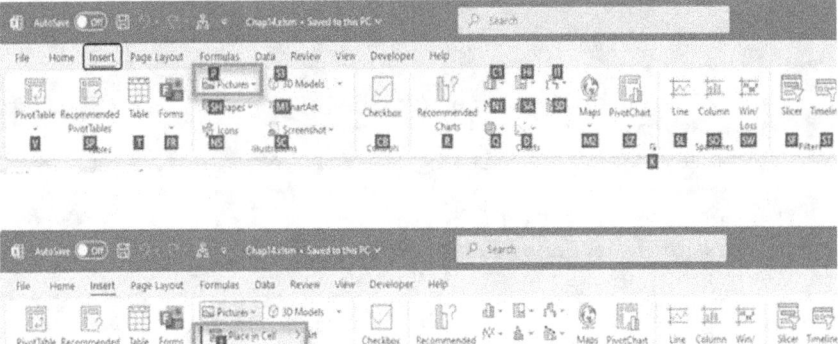

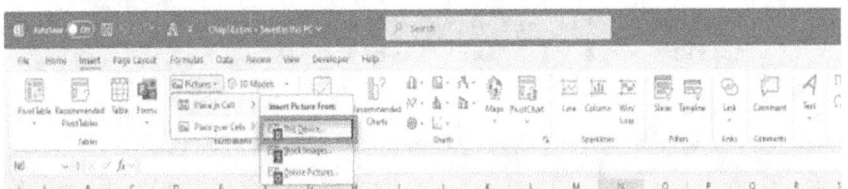

FIGURE 14.5. Each icon/command in the ribbon has an access key that can be used to activate a particular command using the SendKeys statement. Therefore, to display the Insert Picture dialog, append the additional keys to the previous string parameter: "+pi+d".

Passing Data Between Excel and PowerShell

You can pass data from Excel to PowerShell and vice versa to leverage Power-Shell's advanced capabilities in conjunction with Excel's data handling.

In the following Hands-On exercise, we will write a VBA procedure that outputs data from cells A1:A3 to a text file using a PowerShell script. This script will utilize the Shell function to invoke PowerShell commands and pass the data as arguments.

⊚ Hands-On 14.2   Passing Data from Excel to PowerShell

1. In the VBE screen of VBAProject (Chap14.xlsm), insert a new module and enter the following SendDataToPowerShell procedure:

```
Sub SendDataToPowerShell()
    Dim firstName As String
    Dim lastName As String
    Dim hireDate As String
    Dim psCommand As String
    Dim ws As Worksheet
    Dim strPath As String

    strPath = "C:\VBAExcel2024_ByExample\outputFromExcel.txt"

    Set ws = ThisWorkbook.Worksheets(1)

    ' Write some data to the worksheet
    ws.Range("A1:C1").Value = _
        Array("John", "Smith", "12/5/2024")

    ' Read data from Excel cells
    firstName = ws.Range("A1").Value
    lastName = ws.Range("B1").Value
    hireDate = ws.Range("C1").Value


    ' Build the PowerShell command to save the data to a file
    psCommand = "powershell.exe -Command ""& {" & _
        "$data = @{'First Name'='" & _
        firstName & "'; 'Last Name'='" & _
        lastName & "'; 'Hire Date'='" & _
        hireDate & "'}; " & _
        "$data | Out-File -FilePath '" & strPath & "' _
            -Encoding UTF8}"""
```

```
    ' Execute the PowerShell command
    shell psCommand, vbNormalFocus
End Sub
```

Here, `powershell.exe -Command` launches PowerShell from the command line. The `-Command` parameter specifies that the following script will be executed within PowerShell.

The ampersand (`&`) is used to run a script block (`{...}`). This ensures that all the commands inside the curly braces are executed as a single unit within PowerShell. `$data` is a hash table, which is a collection of key-value pairs in PowerShell. In our example, the keys are `First Name`, `Last Name`, and `Hire Date`. The corresponding values are pulled dynamically from the VBA variables: `firstName`, `lastName`, and `hireDate`.

The pipeline operator (`|`) sends the `$data` hash table to the `Out-File` cmdlet. `Out-File` is used to write the data to a file.

The `-FilePath` parameter specifies the location and name of the output file. In our procedure, the file path is determined by the `strPath` variable in VBA, making the destination dynamic. Make sure that the specified folder exists on your computer.

`-Encoding UTF8` ensures the files are saved with UTF-8 encoding, which supports a wide range of characters, including special or non-English characters.

2. Run the `SendDataToPowerShell` procedure.

   After running this procedure, you should see the output in the `outpFromExcel.txt` file as shown in Figure 14.6.

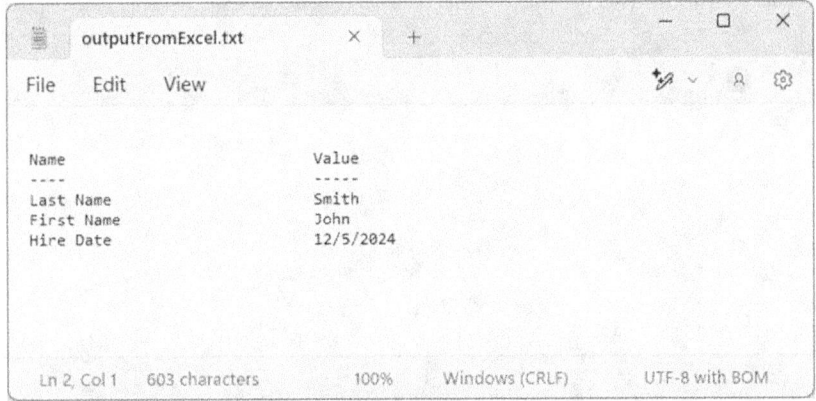

FIGURE 14.6.   Data exported from an Excel worksheet to a text file via a PowerShell script.

Now, let's use PowerShell to retrieve data to an Excel worksheet. How about retrieving the names, modification dates, extensions, and sizes of all the files in the `C:\VBAExcel2024_ByExample` folder?

Hands-On 14.3.   Retrieving Data to Excel Using PowerShell Commands

1. In the VBE screen of `VBAProject (Chap14.xlsm)`, insert a new module and enter the following `GetFolderFileDetails` procedure:

```
Sub GetFolderFileDetails()
    Dim folderPath As String
    Dim psCommand As String
    Dim shellOutput As String
    Dim objShell As Object
    Dim objExec As Object
    Dim outputLines As Variant
    Dim i As Long

    ' Change this to your desired folder
    folderPath = "C:\VBAExcel2024_ByExample"

    ' PowerShell command to retrieve file details
    psCommand = "powershell.exe -Command ""Get-ChildItem & _
        -Path '" folderPath & "' | " & _
        "Select-Object Name, LastWriteTime, Extension, & _
        Length | "
        "ForEach-Object {($_.Name + '|' + & _
        $_.LastWriteTime) + " "('|' + $_.Extension + '|' _
        + $_.Length)}"""

    ' Create a shell object to run the PowerShell command
    Set objShell = CreateObject("WScript.Shell")
    Set objExec = objShell.Exec(psCommand)

    ' Capture the PowerShell output
    shellOutput = objExec.StdOut.ReadAll

    ' Split the output into lines
    outputLines = Split(shellOutput, vbCrLf)

    Workbooks.Add

    ' Write headers to the worksheet
    With ActiveWorkbook.Sheets(1)
```

```
' Loop through the output lines and write to Excel
For i = LBound(outputLines) To UBound(outputLines)
    If Trim(outputLines(i)) <> "" Then
        Dim fileDetails As Variant
        fileDetails = Split(outputLines(i), "|")
        .Cells(i + 1, 1).Value = fileDetails(0)
        .Cells(i + 1, 2).Value = fileDetails(1)
        .Cells(i + 1, 3).Value = fileDetails(2)
        .Cells(i + 1, 4).Value = fileDetails(3)
    End If
Next i
End With

' Insert Headings
Rows("1:1").Select
Selection.Insert Shift:=xlDown, _
    CopyOrigin:=xlFormatFromLeftOrAbove
Range("A1:D1").Value = _
    Array("Name", "Date Modified", "Extension", "Size
(Bytes)")

' Format Headings and Auto Size Columns
Range("A1:D1").Font.Bold = True
Columns.AutoFit

' Cleanup
Set objShell = Nothing
Set objExec = Nothing
End Sub
```

Notice that the PowerShell command consists of several parts. `Get-ChildItem` retrieves files in the specified folder. `Select-Object` extracts the `Name`, `LastWriteTime`, and `Length` (file size in bytes) of each file. `ForEach-Object` concatenates the details with a delimiter (`|`) for parsing in VBA. A shell object (`Wscript.Shell`) runs the PowerShell command, and the output is read using the `StdOut.ReadAll` method. The output is split into lines, then parsed and written to the worksheet. At the end, we add and format headers for better readability.

2. Run the `GetFolderFileDetails` procedure.
   You should see the output as shown in Figure 14.7.

	A	B	C	D	E
1	Name	Date Modified	Extension	Size (Bytes)	
2	GetTransform	2/25/2025 18:45			
3	Images	2/19/2025 21:57			
4	WorksheetWithChart_files	3/6/2025 21:14			
5	Active_Instructors.docx	2/7/2025 19:36	.docx	15842	
6	AllMakesList.xlsx	3/9/2025 23:59	.xlsx	290558	
7	ArraysDemo_3D.xlsm	1/26/2025 17:56	.xlsm	22684	
8	AssetInfo.txt	3/3/2025 18:34	.txt	366	
9	Bicycle_Sales_2024.pptx	2/9/2025 17:22	.pptx	52845	
10	Chap01.xlsm	1/18/2025 2:41	.xlsm	33449	
11	Chap01.xlsx	1/18/2025 0:25	.xlsx	8453	
12	Chap01_Supplement.xlsm	1/18/2025 2:40	.xlsm	21019	
13	Chap02.xlsm	1/18/2025 20:04	.xlsm	17788	
14	Chap03.xlsm	1/22/2025 0:42	.xlsm	23990	
15	Chap04.xlsm	1/23/2025 2:04	.xlsm	27226	
16	Chap05.xlsm	1/23/2025 14:46	.xlsm	22644	
17	Chap06.xlsm	1/24/2025 14:47	.xlsm	24846	
18	Chap07.xlsm	1/26/2025 20:02	.xlsm	37232	
19	Chap08.xlsm	1/27/2025 15:31	.xlsm	32787	
20	Chap09.xlsm	1/27/2025 21:15	.xlsm	23701	

< > Sheet1 +

FIGURE 14.7. File details are retrieved to an Excel worksheet via a PowerShell script.

Automating Tasks with PowerShell Scripts

So far, we've utilized PowerShell commands directly embedded within our VBA procedures. This approach works well for relatively simple tasks where the required commands are straightforward and easily managed within a single VBA statement. When dealing with more complex operations, however—such as retrieving detailed system information, performing large-scale file operations, or executing multi-step automation processes—creating dedicated script files and running them separately can offer significant advantages.

A script file, often referred to as a PowerShell script, is a plain text file with a `.ps1` extension that contains a series of PowerShell commands. VBA can dynamically create a PowerShell script file by writing the necessary commands into a `.ps1` file. Using VBA's `Shell` function, you can run the script directly from Excel.

Figure 14.8 shows the VBA procedure that creates a PowerShell script file (`SystemInfoDemo.ps1`) designed to retrieve operating system information for a demo. It generates the script file in the specified folder. You can run this script file from the Immediate window by passing the script filename (including its path) to the `RunPowerShellScript` procedure. You can import the code of both

these procedures to your VBA project from the companion files—see the file
`PowerShellDemo.bas`.

FIGURE 14.8.   Creating and running a PowerShell script file from Excel VBA.

Importing CSV Data with PowerShell

The `Import-Csv` command in PowerShell is used to import data from a *CSV*
(*Comma Separated Values*) file and convert it into objects that you can manipulate within a PowerShell session. This cmdlet is very useful for processing

structured data and performing operations such as filtering, transforming, or exporting data.

In the following procedure, we open a CSV file (`Orders.csv`) in a Power-Shell session. You can copy `Orders.csv` from the companion files to your `C:\VBAExcel2024_ByExample` folder.

```
Sub ImportCSVWithPowerShell()
    Dim shell As Object
    Dim filePath As String

    filePath = "C:\VBAExcel2024_ByExample\Orders.csv"
    Set shell = CreateObject("WScript.Shell")

    Dim script As String
    script = "powershell -noexit -command ""Import-Csv -Path '" & _
            filePath & "' | Out-GridView"""
    shell.Run script
End Sub
```

The `Out-GridView` cmdlet in PowerShell is used to display data in an interactive, graphical grid view. It is particularly useful for inspecting, filtering, and selecting data in a table-like format, providing a visual alternative to viewing raw data in the console.

When you run the `ImportCSVWithPowerShell` procedure, a separate, resizable window (called Grid View) opens, displaying the data in a tabular format with rows and columns (see Figure 14.9).

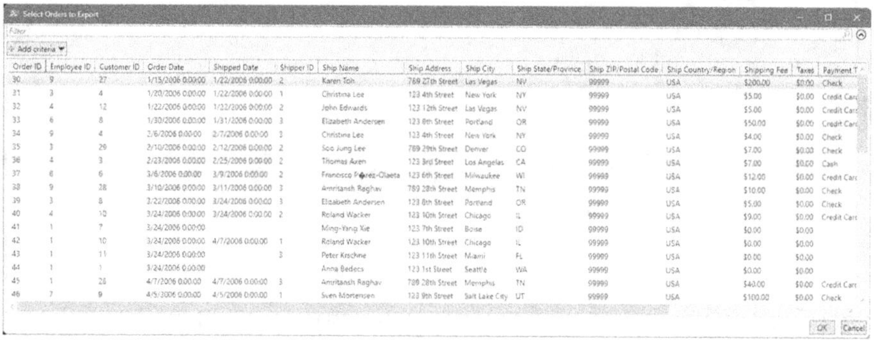

FIGURE 14.9. Contents of the Orders.csv file shown in a grid window.

The grid view allows you to scroll, sort, and resize columns to explore the data interactively. You can click the column header to sort data into ascending and descending order. At the top of the grid view, there is a Filter box. You can type search queries to dynamically filter rows based on their content. For example, entering `Chicago` will only display rows containing the word `Chicago`. You can also add additional criteria to filter the data. This is just the beginning. If you include the `-PassThru` parameter with `Out-GridView`, you can let users interactively select data and then send the selected data to Excel for further processing. The procedure in Hands-On 14.4 demonstrates this approach.

⊙ Hands-On 14.4   Importing CSV Data from PowerShell Grid View to Excel

1. In the VBE screen of `VBAProject (Chap14.xlsm)`, insert a new module and enter the following `ImportCSVWithPowerShell_ToExcel` procedure:

```vba
Sub ImportCSVWithPowerShell_ToExcel()
    Dim shell As Object
    Dim csvFilePath As String
    Dim excelFile As String
    Dim excelFilePath As String
    Dim script As String
    Dim selectedData As String
    Dim tempFile As String

    ' Define the paths for the CSV file and the Excel export file
    excelFile = "SelectedOrders.xlsx"

    ' CSV file to import
    csvFilePath = "C:\VBAExcel2024_ByExample\Orders.csv"

    ' File to save selected data
    excelFilePath = "C:\VBAExcel2024_ByExample\" & excelFile

    ' PowerShell script to retrieve the active Excel object
    script = "powershell -command ""$excel = " & _
        "[Runtime.InteropServices.Marshal]::GetActiveObject(" & _
        "'Excel.Application'); " & _
      "Write-Host 'Attached to the running Excel instance.'; " & _
        "$data = Import-Csv -Path '" & csvFilePath & _
        "' | Out-GridView -PassThru -Title 'Select Orders & _
          to Export'; "
        "if ($data -ne $null) { " & _
        "$workbook = $excel.Workbooks.Add(); " & _
        "$worksheet = $workbook.Sheets.Item(1); " & _
        "$columns = $data[0].PsObject.Properties.Name; " & _
        "for ($col = 0; $col -lt $columns.Count; $col++) {" & _
```

```
      "$worksheet.Cells.Item(1, $col + 1).Value = & _
      "$columns[$col]; } "
      "$row = 2; " & _
      "foreach ($item in $data) { " & _
      "for ($col = 0; $col -lt $columns.Count; $col++) {" & _
     "$worksheet.Cells.Item($row, $col + 1).Value = $item.(" & _
      "$columns[$col]); } $row++; } " & _
      "$workbook.SaveAs('" & excelFilePath & "', 51); }"""


    ' Initialize the shell object and run the PowerShell script
    Set shell = CreateObject("WScript.Shell")
    shell.Run script, vbNormalFocus, True

    ' Check if the specified Excel file was created
    If Dir(excelFile) <> "" Then
        Workbooks(excelFile).Activate
    Else
      MsgBox "You've either cancelled the CSV import" & vbCrLf _
      & " or selected 'No' when prompted to overwrite" & vbCrLf _
      & " the existing file.", vbExclamation
      GoTo ExitHere
    End If
    Columns.AutoFit
    Workbooks(excelFile).Save

ExitHere:
    Set shell = Nothing
```

The PowerShell command that is constructed in the `ImportCSVWithPowerShell_ToExcel` procedure connects to a running Excel instance with:

```
$excel = [Runtime.InteropServices.Marshal]::GetActiveObject('Excel.Application')
```

The script then uses the `Import-Csv` cmdlet to load the data from the specified CSV file.

The loaded data is displayed in a grid view (`Out-GridView`) with the option to filter and select rows (`-PassThru`). The `-PassThru` parameter ensures that the selected rows are returned for further processing.

The next part of the script handles user selections.

If rows are selected, PowerShell:

- Creates a new workbook in Excel.

- Writes the selected data into the workbook, including column headers.
- Saves it as a file in the specified location. The number 51 specifies the .xlsx file format.

The constructed script is then executed via VBA using WScriptShell. The vbNormalFocus ensures the PowerShell window is brought to the foreground, and True makes the VBA procedure wait until the PowerShell script completes before continuing. Finally, the VBA procedure activates the newly created workbook and autofits all columns for better readability, and the workbook changes are saved.

Let's execute this procedure to see its final output.

2. Run the ImportCSVWithPowerShell_ToExcel procedure.
   You should see the PowerShell console window displaying the text: Attached to the running Excel instance. The CSV data is then presented in an interactive grid view. You can now start filtering and selecting rows, as shown in Figure 14.10. We'll be filtering orders data by Ship City, where Ship City is Denver or Chicago.

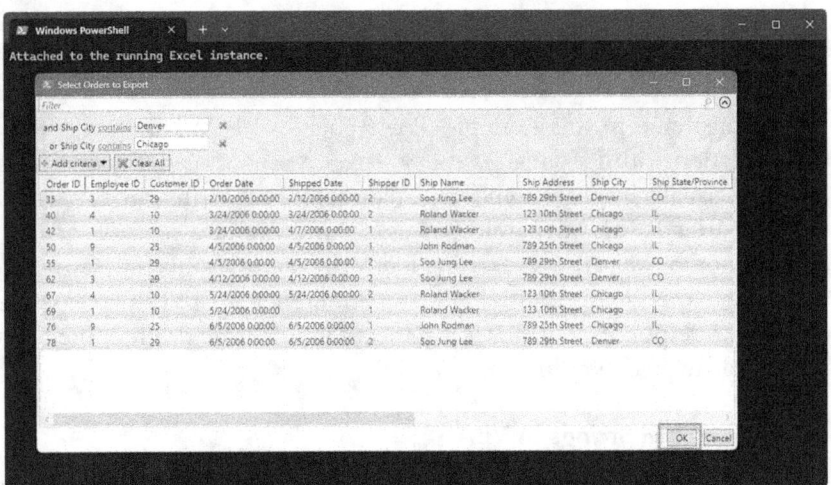

FIGURE 14.10.   CSV data displayed and filtered by two cities in an interactive grid view. The highlighted data will be imported to Excel after clicking OK.

3. In the grid view, click the Add Criteria button and click the checkbox next to Ship City, then click Add. This will add the Ship City criteria text box to the grid view. Add another Ship City criteria and enter the cities, as shown in Figure 14.10. Notice the grid is now filtered to what we've specified.

4. You must now manually select the rows. With the first row already highlighted, hold down the Shift key and click on the last row in the grid. To choose only certain rows, hold the Ctrl key.

5. With the rows selected, click the OK button to pass the selected rows to the script.
   The PowerShell console window and the grid view should now close, and the selected orders are now available in Excel, as shown in Figure 14.11.

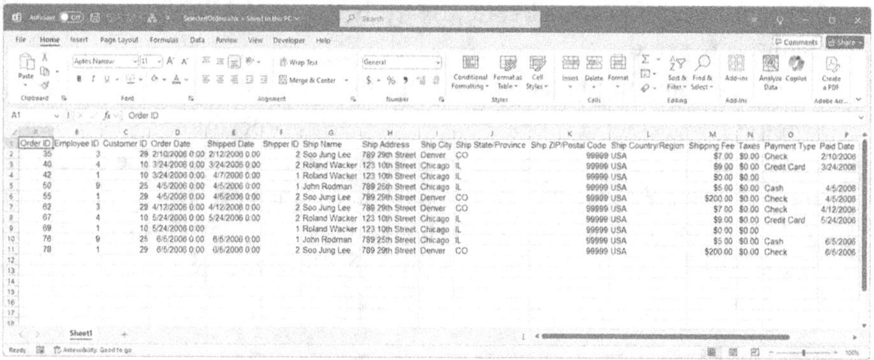

FIGURE 14.11. Selected orders data interactively imported to Excel using a PowerShell script from VBA.

As demonstrated in this example, PowerShell excels in handling interactive data selection with its `Out-GridView -PassThru` cmdlet. The user gets the ability to visually inspect and filter the data before exporting it to Excel, making it a user-friendly and dynamic process. If no rows are selected or the user chooses not to overwrite an existing file, VBA can exit gracefully by providing appropriate messages. Additionally, the PowerShell script can be extended to manage `No` responses, but this is a subject for a dedicated PowerShell study if you decide to go down this route.

Creating PowerShell Strings in VBA

PowerShell and VBA have distinct syntax rules. Adapting PowerShell commands into a format that VBA can handle, especially a multi-line string, requires careful attention to detail. Special characters, such as `vbCrLf` to represent line breaks and double quotation marks (`"`), within the PowerShell code must be escaped using `" "`. Embedding PowerShell scripts in VBA can make them harder to read, debug, and maintain due to the additional VBA syntax overhead. Debugging PowerShell scripts embedded in VBA string can be tedious as errors in the

PowerShell code can often be seen only during runtime. If the PowerShell script includes dynamic variables (e.g., file paths or user input), you need to correctly integrate these variables within the VBA string.

Calling a .ps1 Script from VBA

Instead of embedding the PowerShell script as a string, it is possible to call a .ps1 script file from VBA. This approach can simplify your VBA code and make it easier to maintain and debug the PowerShell script. Let's see how this is done. We will convert the PowerShell string we used in the previous Hands-On exercise to a fully-fledged .ps1 script and then call that script file from VBA.

⊙ Hands-On 14.5   Calling a .ps1 Script to Import Specified Orders

1. Open Notepad and enter the following PowerShell script (see the ImportCSV. ps1 file in the companion files for proper formatting):

```
# Define the CSV file path and Excel file path
$csvFilePath = "C:\VBAExcel2024_ByExample\Orders.csv"
$excelFilePath = "C:\VBAExcel2024_ByExample\
                    SelectedOrders.xlsx"

# Attach to the running instance of Excel
$excel = [Runtime.InteropServices.Marshal]::GetActiveObject
    ("Excel.Application")
Write-Host "Attached to the running Excel instance."

# Import data from the CSV and open Out-GridView for selection
$data = Import-Csv -Path $csvFilePath | Out-GridView -PassThru
    -Title "Select Orders to Export"
# Check if data was selected
if ($data -ne $null) {
    # Create a new workbook and get the first worksheet
    $workbook = $excel.Workbooks.Add()
    $worksheet = $workbook.Sheets.Item(1)

    # Add column headers
    $columns = $data[0].PsObject.Properties.Name
    for ($col = 0; $col -lt $columns.Count; $col++) {
        $worksheet.Cells.Item(1, $col + 1).
            Value = $columns[$col]
    }

    # Add data rows
    $row = 2
```

```
    foreach ($item in $data) {
        for ($col = 0; $col -lt $columns.Count; $col++) {
            $worksheet.Cells.Item($row, $col + 1).
                Value = $item.($columns[$col])
        }
        $row++
    }

    # Select all columns and autofit
    $worksheet.Columns.AutoFit()
    # Save the workbook as an Excel file
    # File format 51 corresponds to .xlsx format
    $workbook.SaveAs($excelFilePath, 51)
    $workbook.Close()
    Write-Host "Workbook saved to $excelFilePath"
} else {
    Write-Host "No data was selected. The process was
      canceled."
}
```

Note that in the PowerShell script, the character # denotes comments.

This script displays CSV data in an interactive grid, allowing you to select rows for import. The script activates an already running instance of Excel, creates a new workbook, adds headers, and writes the selected data row by row to the Excel file. The workbook is then formatted and saved in the .xlsx format.

2. Save the script file as ImportCSV.ps1 in your C:\VBAExcel2024_ByExample folder.

3. In the VBE window of VBAProject (Chap14.xlsm), insert a new module and enter the following procedure:

```
Sub InteractWithCSV_PowerShell()
    Dim WshShell As Object
    Dim PSFilePath As String
    Dim PSCommand As String

    ' Define the path to the PowerShell script
    PSFilePath = "C:\VBAExcel2024_ByExample\ImportCSV.ps1"
    ' Create a Shell object
    Set WshShell = CreateObject("WScript.Shell")

    ' Build the command to run PowerShell and execute the script
    PSCommand = "powershell.exe " & _
        "-ExecutionPolicy Bypass -File """ & PSFilePath & """"
```

```
    ' Run the command
    ' 1 = window shown, True = wait for completion
    WshShell.Run PSCommand, 1, True
    MsgBox "PowerShell script executed successfully.", _
        vbInformation, "Done"
End Sub
```

The `-ExecutionPolicy Bypass` parameter allows the script to run without being restricted by PowerShell's execution policy. This should be used cautiously and only with trusted scripts.

4. Run the `InteractWithCSV_PowerShell` procedure.
5. Select the rows you want from the Orders grid and click OK to export them to Excel.
   The workbook file is created and then closed. If the file already exists, you can save the data under a different filename. If you don't want the workbook to be closed after it's been created, simply delete the `$workbook.Close()` line from the `.ps1` script. As you can see, working with a standalone PowerShell script makes it easy to modify it. It also avoids the complexity of embedding long PowerShell commands in your procedures as VBA strings.

SUMMARY

In this chapter, you learned how to get started with PowerShell to unlock its potential for automating tasks directly from your VBA procedures. You were introduced to the basics of PowerShell, including its command-line interface and scripting capabilities. You explored essential cmdlets and how they work as building blocks for managing files and processes. You also utilized the pipeline to seamlessly pass data between commands. You constructed a complex string to call PowerShell from VBA and created a workbook with the data selected from a CSV file. You then learned how to perform the same operation by calling a standalone `.ps1` script. This chapter also covered how to send keystrokes to another application by using the `SendKeys` statement. There is a lot more to discover about PowerShell. You can start by reading the documentation at *https:// docs.microsoft.com/en-us/powershell/*.

In the next chapter, you will learn how event programming can help you build spreadsheet applications that respond to or limit user actions.

Part IV ENHANCING USER EXPERIENCE

Excel VBA can greatly enhance the efficiency, usability, and interactivity of your spreadsheets. Beyond the basic formatting techniques, you can utilize VBA to create custom interface elements that cater to your user's specific needs. By customizing the ribbon, designing custom user forms, and creating dialog boxes, you can make data entry and manipulation more intuitive and streamlined. Furthermore, Excel VBA's event-driven programming capabilities will allow you to respond dynamically to users' actions within the worksheet.

To provide a comprehensive understanding of these concepts, we will cover them in several chapters:

Chapter 15

Utilizing Event-Driven Programming

How can you validate data entered in a worksheet? How do you display a custom message before a workbook is opened or closed? To gain complete control over Microsoft Excel, you must learn how Excel VBA's event-driven programming can help you respond dynamically to users' actions within the spreadsheet. Learning how to program events will allow you to implement your own functionality in an Excel application.

An *event* is an action recognized by an object in Excel, and it can be triggered by a user, another program, or the system itself. For instance, when you right-click a worksheet cell, this action triggers an event that displays a context menu. You can, however, use VBA to react to these events and customize the behavior, such as disabling the context menu or displaying a custom menu based on specific conditions. Once you learn about events in Excel, you will find it easier to understand events that occur to objects in other Microsoft 365 applications such as Word, Outlook, PowerPoint, or Access, as they share similar event-driven principles.

The following Microsoft Excel objects can respond to events:

- Worksheet
- Chart sheet
- Query table
- Workbook
- Application

You can decide what should happen when a particular event occurs by writing an event procedure.

INTRODUCTION TO EVENT PROCEDURES

A special type of VBA procedure, known as an *event procedure*, is used to react to specific events within your Excel application. The event procedure contains VBA code that handles a particular event, which can range from simple, single-line commands to more complex routines.

Event procedures have specific names, which are created in the following format:

```
ObjectName_EventName()
```

The parameters that need to be sent to the procedure are placed in parentheses after the name of the event. The programmer cannot change the name of the event procedure, as it is predefined by Excel.

Before you can write an event procedure to react to an Excel event, you need to know:

- *The object that will trigger the event*: This could be a worksheet, workbook, or a specific range of cells. Understanding the object is essential for correctly handling the event.
- *The event that you want to handle*: Events can include actions such as opening a workbook, changing a cell value, clicking a button, or activating a sheet. Each of these actions corresponds to a different event procedure.
- *The parameters associated with the event*: Some events require additional information to be passed to the procedure. For example, the `Worksheet_Change` event needs to know which cell or range was changed. These parameters help you write more precise and efficient code.
- *The desired outcome of the event procedure*: Clearly defining what you want to achieve with the event procedure will guide you in writing the

appropriate code. This could involve validating data, updating the interface, or triggering other actions.

By understanding these components, you can effectively create event procedures to enhance the functionality and interactivity of your Excel applications.

THE PLACEMENT OF EVENT CODE

When working with events in Excel VBA, it's crucial to know where to place your event code. Some events are coded in standard modules, while others must be entered in class modules. For example, workbook, chart sheet, and worksheet events are typically placed within their respective object modules (e.g., `This-Workbook`, `Chart1`, and `Sheet1`). Creating event procedures for embedded charts, query tables, or the `Application` object, however, requires you to create a new object in a class module using the `With Events` keyword, write your event code in a class module, and then initialize the event in a standard module (see Hands-On 15.22 later in this chapter).

When you're in the VBE, you can select an object from the left drop-down list and then view all the available events for that object in the right drop-down list. This feature makes it easier to write event procedures since you can directly select the event you want to handle and the VBE will automatically generate the event procedure's skeleton code for you (see Figure 15.1).

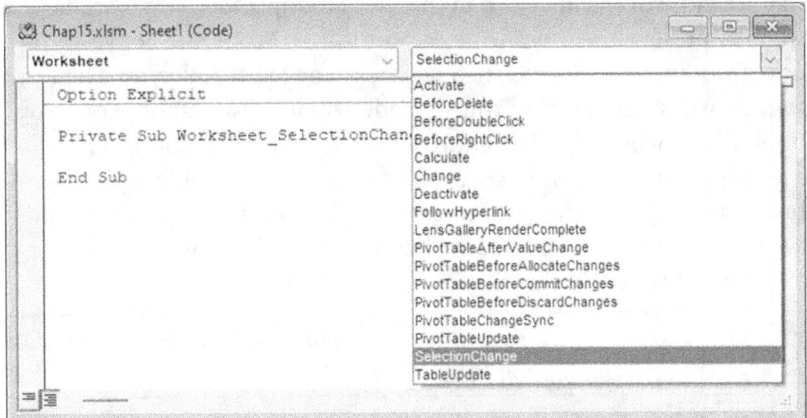

FIGURE 15.1.   You can find out the names of the events that are available for a selected object (e.g., Worksheet) in the VBE Code window's procedure drop-down list.

Additionally, you can use the Object Browser to find out event names for the specific Excel object you want to control (see Figure 15.2).

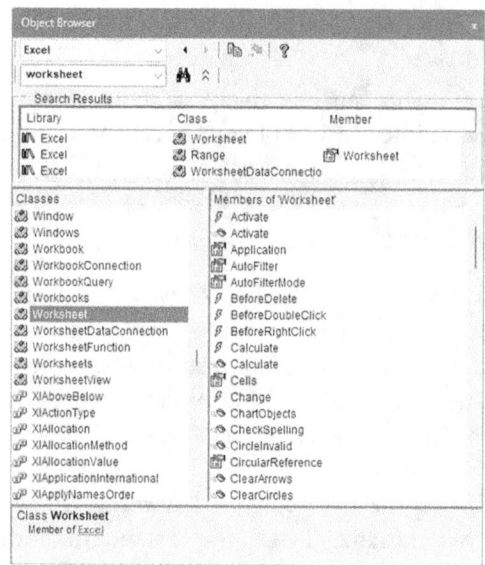

FIGURE 15.2.   Use the Object Browser to find various event names.

WRITING YOUR FIRST EVENT PROCEDURE

At times, you will want to trigger a specific operation when a user invokes an Excel command. For instance, when the user attempts to save a workbook, you may want to present them with the opportunity to copy the active worksheet to another workbook. This scenario can be managed through an event procedure, which will automatically run its code when a user attempts to save the workbook file in which the procedure is located. In VBA, you can handle this scenario using the `Workbook_BeforeSave` event. This event triggers before the workbook is saved, allowing you to execute your custom code. In Hands-On 15.1, we will write an event procedure that offers the user the option to copy the active worksheet to another workbook before saving it.

NOTE	*Please note that the files for the "Hands-On" projects can be found in the companion files.*

⊙ Hands-On 15.1   Writing an Event Procedure

1. Open a new workbook and save it as `C:\VBAExcel2024_ByExample\Chap15.xlsm`.

2. Change the name of `Sheet1` in the `Chap15.xlsm` workbook to `Test`.
3. Type anything in cell A1 and press Enter.
4. Switch to the VBE screen, and in Project Explorer, double-click This Workbook in the Microsoft Excel Objects folder under `VBAProject(Chap15.xlsm)`.
5. In the ThisWorkbook Code window, enter the following `Workbook_BeforeSave` event procedure:

```
Private Sub Workbook_BeforeSave(ByVal SaveAsUI As Boolean, _
    Cancel As Boolean)
    Dim Response As VbMsgBoxResult
    Dim newWkb As Workbook
    Dim strMsg As String

    strMsg = "Would you like to copy the active sheet" & vbCrLf & _
            " to another workbook before saving?"

    'Prompt the user with an option to copy the active sheet
    Response = MsgBox(strMsg, vbYesNo + vbQuestion, _
            "Copy Active Sheet")

    If Response = vbYes Then
        ' Create a new workbook
        Set newWkb = Workbooks.Add

        'Copy the active sheet to the new workbook
        ThisWorkbook.ActiveSheet.Copy Before:=newWkb.Sheets(1)

        MsgBox "Active sheet has been copied to a new workbook.", _
            vbInformation

    End If
End Sub
```

6. Switch to the Microsoft Excel application window and click Save.
   The `Workbook_BeforeSave` event procedure will be triggered at this time. Click Yes to copy the active sheet or No to cancel. If you select Yes, a new workbook is created, and the active sheet is copied to this new workbook.
7. Close the workbook file created by Excel without saving any changes. Do not close the `Chap15.xlsm` workbook.
8. Click the File tab and choose Save to save `Chap15.xlsm`.
   Notice that, again, you are prompted with the dialog box. Click No to the message. Notice that the workbook file is now being saved.

In the `Workbook_BeforeSave` event, the arguments `SaveAsUI` and `Cancel` provide additional control over the `Save` operation.

The `SaveAsUI` argument is a `Boolean` that indicates whether the Save As dialog box should be displayed. If `SaveAsUI` is `True`, the Save As dialog box will appear, allowing the user to specify a different filename, location, or file type. If `SaveAsUI` is `False`, the workbook will be saved without displaying the Save As dialog box. You can use this argument to control whether the Save As prompt should be shown or suppressed based on certain conditions.

The `Cancel` argument is also a `Boolean`, and it determines whether the `Save` operation should be canceled. If `Cancel` is set to `True`, the `Save` operation is aborted and the workbook is not saved. If `Cancel` is `False`, the `Save` operation proceeds as usual. This argument is useful for implementing custom logic that may prevent the user from saving the workbook under certain conditions, such as when specific validation checks fail.

The following example demonstrates how to use both `SaveAsUI` and `Cancel` arguments in the `Workbook_BeforeSave` event:

```
Private Sub Workbook_BeforeSave(ByVal SaveAsUI As Boolean, _
    Cancel As Boolean)
Dim Response As VbMsgBoxResult
Dim newWkb As Workbook
Dim strMsg As String

' Validate data before proceeding with the save operation
If ThisWorkbook.ActiveSheet.Range("A1").Value = "" Then

    MsgBox "Please enter a value in cell A1 before saving.", _
        vbExclamation
    Cancel = True ' Cancel the save operation
    Exit Sub
End If
strMsg = "Would you like to copy the active sheet" & vbCrLf & _
        " to another workbook before saving?"

' Prompt the user with an option to copy the active sheet
Response = MsgBox(strMsg, vbYesNo + vbQuestion, _
                    "Copy Active Sheet")
If Response = vbYes Then
    ' Create a new workbook and copy the active sheet to it

    Set newWkb = Workbooks.Add
    ThisWorkbook.ActiveSheet.Copy Before:=newWkb.Sheets(1)
```

```
    MsgBox "Active sheet has been copied to the new workbook.", _
        vbInformation
End If
' Check if SaveAsUI is True
' and ask for confirmation before showing the Save As dialog
If SaveAsUI Then
    Response = MsgBox("Do you want to proceed with Save As?", _
        vbYesNo + vbQuestion, "Save As Confirmation")
    If Response = vbNo Then
        Cancel = True ' Cancel the Save As operation
    End If
End If
End Sub
```

In this modified `Workbook_BeforeSave` event procedure, the data validation logic is placed at the beginning of the procedure. This ensures that the validation is performed before any other actions, and if the validation fails, the `Save` operation is canceled immediately using `Cancel = True` and `Exit Sub`.

The prompt asking the user whether they want to copy the active sheet is executed only after the validation check has passed. This ensures the `Copy` operation only occurs if the data is valid.

The confirmation for the Save As dialog code only executes when the File | Save As option is selected for saving the workbook.

ENABLING AND DISABLING EVENTS

You can use the `Application` object's `EnableEvents` property to enable or disable events. If you are writing a VBA procedure and don't want a particular event to occur, set the `EnableEvents` property to `False`.

To demonstrate how you can prevent a custom event procedure from running, we will write a procedure in a standard module that will save the workbook after making a few changes in the active sheet. We will continue working with the `Chap15.xlsm` file because it already contains the `Worksheet_Before-Save` event procedure we want to block in this demonstration.

⊙ Hands-On 15.2   Disabling a Custom Event Procedure

This Hands-On requires prior completion of Hands-On 15.1.

1. Choose Insert | Module to add a standard module to `VBAProject(Chap15.xlsm)` and rename it `StandardProcedures`.

2. In the StandardProcedures Code window, enter the following `EnterData` procedure:

```
Sub EnterData()
  With ActiveSheet.Range("A1:B1")
    .Font.Color = vbRed
    .Value = 15
  End With
  Application.EnableEvents = False
  ActiveWorkbook.Save
  Application.EnableEvents = True
End Sub
```

Notice that prior to calling the `Save` method of the `ActiveWorkbook` property, we have disabled events by setting the `EnableEvents` property of the `Application` object to `False`. This will prevent the `Workbook_BeforeSave` event procedure from running when Visual Basic encounters the statement to save the workbook. We don't want the user to be prompted to copy the worksheet while running the `EnterData` procedure. When Visual Basic has completed the saving process, we want the system to respond to the events as we programmed them, so we enable the events with the `Application.EnableEvents` statement set to `True`.

3. Switch to the `Chap15.xlsm` application window and choose View | Macros | View Macros. In the Macro dialog box, select EnterData and click Run.
Notice that when you run the `EnterData` procedure, you are not prompted to copy the worksheet before saving. This indicates that the code you entered in the Hands-On 15.1 `Workbook_BeforeSave` event procedure is not running.

4. Close the `Chap15.xlsm` workbook.

EVENT SEQUENCES

Events occur in response to specific actions. Events also occur in a predefined sequence. Table 15.1 demonstrates the sequence of events that occur while opening a new workbook, adding a new worksheet to a workbook, and closing the workbook.

TABLE 15.1. Event sequences.

Action	Object	Event Sequence

Opening a new workbook	`Workbook`	`NewWorkbook`
		↓
		`WindowDeactivate`
		↓
		`WorkbookDeactivate`
		↓
		`WorkbookActivate`
		↓
		`WindowActivate`

Inserting a new sheet into a workbook	`Workbook`	`WorkbookNewSheet`
		↓
		`SheetDeactivate`
		↓
		`SheetActivate`

Closing a workbook	`Workbook`	`WorkbookBeforeClose`
		↓
		`WindowDeactivate`
		↓
		`WorkbookDeactivate`
		↓
		`WorkbookActivate`
		↓
		`WindowActivate`

WORKSHEET EVENTS

A `Worksheet` object responds to such events as activating and deactivating a worksheet, calculating data in a worksheet, making a change to a worksheet, and double-clicking or right-clicking a worksheet. Table 15.2 lists some of the events the `Worksheet` object can respond to.

TABLE 15.2. Worksheet events (a partial listing).

Worksheet Event Name	Event Description
`Activate`	This event occurs upon activating a worksheet.
`Deactivate`	This event occurs when the user activates a different sheet.
`SelectionChange`	This event occurs when the user selects a worksheet cell.
`Change`	This event occurs when the user changes a cell formula.

(Contd.)

Worksheet Event Name	Event Description
Calculate	This event occurs when the user recalculates the worksheet.
BeforeDoubleClick	This event occurs when the user double-clicks a worksheet cell.
BeforeRightClick	This event occurs when the user right-clicks a worksheet cell.

Let's try out these events to understand how they work.

Worksheet_Activate()

This event occurs upon activating a worksheet.

(⊙) Hands-On 15.3   Writing the Worksheet_Activate() Event Procedure

1. Open a new workbook and save it as Chap15_WorksheetEvents.xlsm in your VBAExcel2024_ByExample folder.
2. Insert a new worksheet into the current workbook.
3. Switch to the VBE window and, in the Project Explorer, double-click Sheet2 under VBAProject (Chap15_WorksheetEvents.xlsm).
4. In the Sheet2 Code window, enter the code shown below:

```
Dim shtName As String

Private Sub Worksheet_Activate()
  shtName = ActiveSheet.Name
  Range("B2").Select
End Sub
```

The example procedure selects cell B2 each time the sheet is activated. Notice that the shtName variable is declared at the top of the module.

5. Switch to the Microsoft Excel application window and activate Sheet2.
6. Notice that when Sheet2 is activated, the selection is moved to cell B2. Excel also stores the sheet name in the shtName variable that was declared at the top of the module. We will need this value as we work with other event procedures in this section.

Worksheet_Deactivate()

This event occurs when the user activates a different sheet in a workbook.

(⊙) Hands-On 15.4   Writing the Worksheet_Deactivate() Event Procedure

This Hands-On exercise uses the Chap15_WorksheetEvents workbook created in Hands-On 15.3.

1. Switch to the VBE window. In the Sheet2 Code window, enter the `Worksheet_Deactivate` procedure, as shown below:

```
Private Sub Worksheet_Deactivate()
  MsgBox "You deactivated " & _
    shtName & "." & vbCrLf & _
    "You switched to " & _
    ActiveSheet.Name & "."
End Sub
```

The example procedure displays a message when `Sheet2` is deactivated.

2. Switch to the Microsoft Excel application window and click the Sheet2 tab. The `Worksheet_Activate` procedure that you created in Hands-On 15.3 will run first. Excel will select cell B2 and store the name of the worksheet in the `shtName` global variable declared at the top of the `Sheet2` code module.

3. Now, click any other sheet in the active workbook.
Notice that Excel displays the name of the worksheet that you deactivated and the name of the worksheet you switched to.

Worksheet_SelectionChange(ByVal Target As Range)

This event occurs when the user selects a worksheet cell.

⊙ Hands-On 15.5   Writing the Worksheet_SelectionChange() Event Procedure

1. In the current `Chap15_WorksheetEvents.xlsm` workbook, insert a new worksheet.

2. Switch to the VBE window and, in the Project Explorer window, double-click Sheet3 under `VBAProject (Chap15_WorksheetEvents.xlsm)`.

3. In the Sheet3 Code window, enter the `Worksheet_SelectionChange` procedure, as shown below:

```
Private Sub Worksheet_SelectionChange(ByVal Target As Range)
  Dim myRange As Range

  On Error Resume Next
  Set myRange = Intersect(Range("A1:A10"), Target)
  On Error GoTo 0
  If Not myRange Is Nothing Then
    MsgBox "Data entry or edits are not permitted " & _
      "in this range.", vbExclamation
  End If
End Sub
```

The example procedure will notify the user that data entry or edits are not permitted when they select a cell in the restricted range (myRange).

4. Switch to the Microsoft Excel application window and activate Sheet3. Click on any cell within the specified range, A1:A10.
Notice that Excel displays a message whenever you click a cell in the restricted area.

   To ensure that data entry or editing is indeed not permitted, you can combine the existing Worksheet_SelectionChange procedure with Worksheet_Change to fully prevent modifications. The Worksheet_SelectionChange event can alert users when they select restricted cells, and the Worksheet_Change event can revert any changes made to those cells.

5. Enter the following Worksheet_Change event procedure in the Sheet3 code module:

```
Private Sub Worksheet_Change(ByVal Target As Range)
    Dim RestrictedRange As Range
    Set RestrictedRange = Me.Range("A1:A10")

    If Not Intersect(Target, RestrictedRange) Is Nothing Then
        Application.EnableEvents = False
        MsgBox "Data entry or edits are not permitted " & _
            "in this range.", vbExclamation
        Application.Undo ' Revert the change
        Application.EnableEvents = True
    End If
End Sub
```

Worksheet_Change(ByVal Target As Range)

This event occurs when the user changes a cell formula.

⊙ Hands-On 15.6   Writing the Worksheet_Change() Event Procedure

1. In the VBE window, activate the Project Explorer window and double-click Sheet1 in the Microsoft Excel Objects folder of Chap15_WorksheetEvents. xlsm.

2. In the Sheet1 Code window, enter the Worksheet_Change event procedure, as shown below:

```
Private Sub Worksheet_Change(ByVal Target As Range)
  Application.EnableEvents = False
    Target = UCase(Target)
    Columns(Target.Column).AutoFit
```

```
    Application.EnableEvents = True
End Sub
```

The example procedure changes what you type in a cell to uppercase. The column where the target cell is located is then auto-sized.

3. Switch to the Microsoft Excel application window and activate `Sheet1`. Enter any text in any cell.
   Notice that as soon as you press the Enter key, Excel changes the text you typed to uppercase and auto-sizes the column.

Worksheet_Calculate()

This event occurs when the user recalculates the worksheet.

⊚ Hands-On 15.7   Writing the Worksheet_Calculate() Event Procedure

1. Add a new sheet to the `Chap15_WorksheetEvents` workbook.
2. In cell A2 of the newly added sheet, enter 1, and in cell B2, enter 2. Enter the following formula in cell C2: = `A2+B2`.
3. Switch to the VBE window, activate the Project Explorer window, and double-click the sheet you added in Step 1.
4. In the Code window, enter the code of the `Worksheet_Calculate` procedure, as shown below:

```
Private Sub Worksheet_Calculate()
   MsgBox "The worksheet was recalculated."
End Sub
```

5. Switch to the Microsoft Excel application window and modify the entry in cell B2 on the sheet you added in Step 1 by typing any number.
   Notice that after leaving Edit mode, the `Worksheet_Calculate` event procedure is triggered and you are presented with a custom message.
6. Switch to the Microsoft Excel application window and activate `Sheet3`. Click on any cell within the specified range, A1:A10. Try to add some data in the restricted range.
   You should be alerted that data entry or edits are not permitted (`Worksheet_SelectionChange`). Any modifications to the restricted range will be immediately undone (`Worksheet_Change`).

Worksheet_BeforeDoubleClick(ByVal Target As Range, Cancel As Boolean)

This event occurs when the user double-clicks a worksheet.

◉ Hands-On 15.8　Writing the Worksheet_BeforeDoubleClick()
　　　　　　　　　　Event Procedure

1. Enter any data in cell C9 on Sheet2 of the Chap15_WorksheetEvents workbook.
2. In the VBE window, activate the Project Explorer window and double-click Sheet2.
3. In the Sheet2 Code window, type the code of the procedure as shown below:

```
Private Sub Worksheet_BeforeDoubleClick(ByVal _
    Target As Range, Cancel As Boolean)
  If Target.Address = "$C$9" Then
    MsgBox "No double-clicking, please."
    Cancel = True
  Else
    MsgBox "You may edit this cell."
  End If
End Sub
```

The Worksheet_BeforeDoubleClick procedure disallows in-cell editing when cell C9 is double-clicked.

4. Switch to the Microsoft Excel application window and double-click cell C9 on Sheet2.

The Worksheet_BeforeDoubleClick event procedure cancels the built-in Excel behavior, and the user is not allowed to edit the data inside the cell. The user can get around this restriction, however, by clicking on the formula bar or pressing F2. When writing event procedures that restrict access to certain program features, you should write additional code that prevents any workaround.

Worksheet_BeforeRightClick(ByVal Target As Range, Cancel As Boolean)

This event occurs when the user right-clicks a worksheet cell.

◉ Hands-On 15.9　Writing the Worksheet_BeforeRightClick()
　　　　　　　　　　Event Procedure

1. In the VBE window, activate the Project Explorer window and double-click Sheet2 in the Microsoft Excel Objects folder.
2. In the Sheet2 Code window, enter the code of the Worksheet_BeforeRightClick procedure, as shown below:

```
Private Sub Worksheet_BeforeRightClick(ByVal _
    Target As Range, Cancel As Boolean)
```

```
With Application.CommandBars("Cell")
  .Reset
  If Target.Rows.Count > 1 Or _
      Target.Columns.Count > 1 Then
      With .Controls.Add(Type:=msoControlButton, _
          before:=1, temporary:=True)
        .Caption = "Print..."
        .OnAction = "PrintMe"
      End With
  End If
End With
End Sub
```

The example procedure adds a Print option to the cell context menu when the user selects more than one cell on the worksheet.

3. Insert a new module into the current project and enter the `PrintMe` procedure, as shown below:

```
Sub PrintMe()
  Application.Dialogs(xlDialogPrint).Show arg12:=1
End Sub
```

The `PrintMe` procedure is called by the `Worksheet_BeforeRightClick` event when the user selects the Print option from the context menu. Notice that the `Show` method of the `Dialogs` collection is followed by a named argument: `arg12:=1`. This argument will display the Print dialog box with the preselected option button, Selection, in the Print area. Excel dialog boxes are covered in the next chapter.

4. Switch to the Microsoft Excel application window and right-click on any single cell in `Sheet2`.
   Notice that the context menu appears with the default options.

5. Now, select at least two cells in the `Sheet2` worksheet and right-click the selected area.
   You should see the Print option as the first menu entry. Click the Print option and notice that, instead of the default, Print active sheet, the Print dialog displays Print Selection.

6. Close the Print dialog box.

7. Save and close the `Chap15_WorksheetEvents.xlsm` workbook file.

NOTE	*The* `Worksheet_BeforeRightClick` *event procedure relies on the* `CommandBar` *object to customize Excel's built-in context menu. Before Excel 2010, the* `CommandBar` *object was the only way to create, modify, or disable context menus. Excel 2024 continues to support the* `CommandBars` *collection for backward compatibility; however, it is recommended that you add your own customizations to context menus by using the RibbonX model, which is discussed in Chapter 18.*

WORKBOOK EVENTS

`Workbook` object events occur when the user performs such tasks as opening, activating, deactivating, printing, saving, and closing a workbook. To write code that responds to a particular workbook, you can:

- Double-click the ThisWorkbook object in the VBE Project Explorer.

  In the code window that appears, open the object drop-down list on the left-hand side and select the Workbook object.
- In the drop-down list on the right, select the event you want, such as Open. The selected event procedure stub will appear in the code window, as shown below:

```
Private Sub Workbook_Open()
   [place your event handling code here]
End Sub
```

Table 15.3 lists some of the events to which the `Workbook` object can respond.

TABLE 15.3. Workbook events (a partial listing).

Workbook Event Name	Event Description
`Activate`	This event occurs when the user activates the workbook. This event will not occur when the user activates the workbook by switching from another application.
`Deactivate`	This event occurs when the user activates a different workbook within Excel. This event will not occur when the user switches to a different application.
`Open`	This event occurs when the user opens a workbook.

Workbook Event Name	Event Description
BeforeSave	This event occurs before the workbook is saved. The SaveAsUI argument is read-only and refers to the Save As dialog box. If the workbook has not been saved, the value of SaveAsUI is True; otherwise, it is False.
BeforePrint	This event occurs before the workbook is printed and before the Print dialog appears. The example procedure places the full workbook's name in the document footer prior to printing if the user clicks Yes in the message box.
BeforeClose	This event occurs before the workbook is closed and before the user is asked to save changes.
NewSheet	This event occurs after the user creates a new sheet in a workbook.
WindowActivate	This event occurs when the user shifts the focus to any window showing the workbook.
WindowDeactivate	This event occurs when the user shifts the focus away from any window showing the workbook.
WindowResize	This event occurs when the user opens, resizes, maximizes, or minimizes any window showing the workbook.

Let's try out the above events to find out how they work.

Workbook_Activate()

This event occurs when the user activates the workbook. The Workbook_Activate event will not occur when the user activates the workbook by switching from another application.

⊙ Hands-On 15.10   Writing the Workbook_Activate() Event Procedure

1. Open a new workbook and save it as Chap15_WorkbookEvents.xlsm in your C:\VBAExcel2024__ByExample folder.
2. Switch to the VBE window and in the Project Explorer, double-click ThisWorkbook.
3. In the ThisWorkbook code window, type the Workbook_Activate procedure, as shown below:

```
Private Sub Workbook_Activate()
  MsgBox "This workbook contains " & _
    ThisWorkbook.Sheets.Count & " sheets."
End Sub
```

The example procedure displays the total number of worksheets when the user activates the workbook containing the Workbook_Activate event procedure.

4. Switch to the Microsoft Excel application window and open a new workbook.
5. Activate the `Chap15_WorkbookEvents.xlsm` workbook.
   Excel should display the total number of sheets in this workbook.

Workbook_Deactivate()

This event occurs when the user activates a different workbook within Excel. The `Workbook_Deactivate` event does not occur when the user switches to a different application.

> **Hands-On 15.11   Writing the Workbook_Deactivate() Event Procedure**

1. In the VBE window, activate the Project Explorer window and double-click ThisWorkbook.
2. In the ThisWorkbook code window, type the `Workbook_Deactivate` procedure, as shown below:

```
Private Sub Workbook_Deactivate()
   Dim cell As Range
   For Each cell In ActiveSheet.UsedRange
     If Not IsEmpty(cell) Then
         Debug.Print cell.Address & ":" & cell.Value
     End If
   Next
End Sub
```

The example procedure will print to the Immediate window the addresses and values of cells containing entries in the current workbook when the user activates a different workbook.

3. Switch to the Microsoft Excel application window and make some entries on the active sheet. Next, activate a different workbook.
   This action will trigger the `Workbook_Deactivate` event procedure.
4. Switch to the VBE screen and open the Immediate window to see what entries were reported.

Workbook_Open()

This event occurs when the user opens a workbook.

◉ Hands-On 15.12   Writing the Workbook_Open() Event Procedure

1. Double-click the ThisWorkbook object in the Microsoft Excel `Objects` folder under `VBAProject` (`Chap15_WorkbookEvents.xlsm`).
2. In the ThisWorkbook Code window, type the `Workbook_Open` procedure, as shown below:

```
Private Sub Workbook_Open()
  ActiveSheet.Range("A1").Value = Format(Now(), "mm/dd/yyyy")
  Columns("A").AutoFit
End Sub
```

The example procedure places the current date in cell A1 when the workbook is opened.

3. Save and close `Chap15_WorkbookEvents.xlsm` and then reopen it.
   When you open the workbook file again, the `Workbook_Open` event procedure will be triggered, and the current date will be placed in cell A1 on the active sheet.

Workbook_BeforeSave(ByVal SaveAsUI As Boolean, Cancel As Boolean)

This event occurs before the workbook is saved. The `SaveAsUI` argument is read-only and refers to the Save As dialog box. If the workbook has not been saved, the value of `SaveAsUI` is `True`; otherwise, it is `False`.

◉ Hands-On 15.13   Writing the Workbook_BeforeSave() Event Procedure

1. In the VBE screen, activate the Project Explorer window and double-click ThisWorkbook.
2. In the ThisWorkbook Code window, type the `Workbook_BeforeSave` procedure, as shown below:

```
Private Sub Workbook_BeforeSave(ByVal _
    SaveAsUI As Boolean, Cancel As Boolean)
  If SaveAsUI = True And _
    ThisWorkbook.Path = vbNullString Then
    MsgBox "This document has not yet " _
      & "been saved." & vbCrLf _
    & "The Save As dialog box will be displayed."
  ElseIf SaveAsUI = True Then
    MsgBox "You are not allowed to use " _
      & "the SaveAs option. "
    Cancel = True
```

```
    End If
End Sub
```

The example procedure displays the Save As dialog box if the workbook hasn't been saved before. The workbook's pathname will be a null string (vbNullString) if the file has not been saved before. The procedure will not let the user save the workbook under a different name—the SaveAs operation will be aborted by setting the Cancel argument to True. The user will need to choose the Save option to have the workbook saved.

3. Switch to the Microsoft Excel application window and activate any sheet in the Chap15_WorkbookEvents.xlsm workbook.

4. Make an entry in any cell of this workbook, click the File tab, and choose Save As | Excel Macro-Enabled Workbook.
   The Workbook_BeforeSave event procedure will be activated, and the ElseIf clause will be executed. Notice that you are not allowed to save the workbook by using the Save As option.

Workbook_BeforePrint(Cancel As Boolean)

This event occurs before the workbook is printed and before the Print dialog appears.

⊙ Hands-On 15.14 Writing the Workbook_BeforePrint() Event Procedure

1. In the VBE window, activate the Project Explorer window and double-click ThisWorkbook.

2. In the ThisWorkbook Code window, type the Workbook_BeforePrint event procedure, as shown below:

```
Private Sub Workbook_BeforePrint(Cancel As Boolean)
  Dim response As Integer
  response = MsgBox("Do you want to " & vbCrLf & _
      "print the workbook's full name in the footer?", _
          vbYesNo)
  If response = vbYes Then
    ActiveSheet.PageSetup.LeftFooter = _
        ThisWorkbook.FullName
  Else
      ActiveSheet.PageSetup.LeftFooter = ""
  End If
End Sub
```

The example procedure places the workbook's full name in the document footer prior to printing if the user clicks Yes in the message box.

3. Switch to the Microsoft Excel application window and activate any sheet in the Chap15_WorkbookEvents.xlsm workbook.
4. Enter anything you want in any worksheet cell.
5. Click File | Print and click the Print button.
   Excel will ask you whether you want to place the workbook's name and path in the footer.

Workbook_BeforeClose(Cancel As Boolean)

This event occurs before the workbook is closed and before the user is asked to save changes.

> **Hands-On 15.15   Writing the Workbook_BeforeClose() Event Procedure**

1. In the VBE window, activate the Project Explorer window and double-click ThisWorkbook.
2. In the ThisWorkbook Code window, type the Workbook_BeforeClose event procedure, as shown below:

```
Private Sub Workbook_BeforeClose(Cancel As Boolean)
   If MsgBox("Do you want to change " & vbCrLf _
      & " workbook properties before closing?", _
      vbYesNo) = vbYes Then
    Application.Dialogs(xlDialogProperties).Show
   End If
End Sub
```

The example procedure displays the Properties dialog box if the user responds Yes to the message box.

3. Switch to the Microsoft Excel application window and close the Chap15_WorkbookEvents.xlsm workbook.
   Upon closing, you should see a message box asking you to view the Properties dialog box prior to closing. After viewing or modifying the workbook properties, the procedure closes the workbook. If there are any changes that you have not yet saved, you are given the chance to save the workbook, cancel the changes, or abort the closing operation altogether.

Workbook_NewSheet(ByVal Sh As Object)

This event occurs after the user creates a new sheet in a workbook.

> ### Hands-On 15.16   Writing the Workbook_NewSheet()
> Event Procedure

1. Open a new workbook and save it as `Chap15_WorkbookEvents2.xlsm` in your `C:\VBAExcel2024__ByExample` folder.
2. Switch to the VBE window, and in the Project Explorer window, double-click ThisWorkbook.
3. In the ThisWorkbook Code window, type the `Workbook_NewSheet` event procedure, as shown below:

```
Private Sub Workbook_NewSheet(ByVal Sh As Object)
  If MsgBox("Do you want to place  " & vbCrLf _
      & "the new sheet at the beginning " & vbCrLf _
      & "of the workbook?", vbYesNo) = vbYes Then
        Sh.Move before:=ThisWorkbook.Sheets(1)
  Else
    Sh.Move After:=ThisWorkbook.Sheets( _
        ThisWorkbook.Sheets.Count)
    MsgBox Sh.Name & _
        " is now the last sheet in the workbook."
  End If
End Sub
```

The example procedure places the new sheet at the beginning of the workbook if the user responds Yes to the message box; otherwise, the new sheet is placed at the end of the workbook.

4. Switch to the Microsoft Excel application window and click the New Sheet button (+). Excel will ask where to place the new sheet.

Let's try out some of the events related to operations on workbook windows.

Workbook_WindowActivate(ByVal Wn As Window)

This event occurs when the user shifts the focus to any window showing the workbook.

> ### Hands-On 15.17   Writing the Workbook_WindowActivate()
> Event Procedure

1. In the VBE window, activate the Project Explorer window and double-click ThisWorkbook.

2. In the ThisWorkbook Code window, enter the `Workbook_WindowActivate` event procedure, as shown below:

```
Private Sub Workbook_WindowActivate(ByVal Wn As Window)
  Wn.GridlineColor = vbYellow
End Sub
```

The example procedure changes the color of the worksheet gridlines to yellow when the user activates the workbook containing the code of the `Workbook_WindowActivate` procedure.

3. Switch to the Microsoft Excel application window and open a new workbook.

4. Arrange Microsoft Excel workbooks vertically on the screen, by choosing View | Arrange All to open the Arrange Windows dialog. Select the Vertical option button and click OK. When you activate the worksheet of the workbook in which you entered the code of the `Workbook_WindowActivate` event procedure, the color of the gridlines should change to yellow.

Workbook_WindowDeactivate(ByVal Wn As Window)

This event occurs when the user shifts the focus away from any window showing the workbook.

Hands-On 15.18   Writing the Workbook_WindowDeactivate() Event Procedure

1. In the VBE window, activate the Project Explorer window and double-click ThisWorkbook.

2. In the ThisWorkbook Code window, enter the `Workbook_WindowDeactivate` procedure, as shown below:

```
Private Sub Workbook_WindowDeactivate(ByVal Wn As Window)
  MsgBox "You have just deactivated " & Wn.Caption
End Sub
```

The example procedure displays the name of the deactivated workbook when the user switches to another workbook from the workbook containing the code of the `Workbook_WindowDeactivate` procedure.

3. Switch to the Microsoft Excel application window and open a new workbook.

Excel displays the name of the deactivated workbook in a message box.

Workbook_WindowResize(ByVal Wn As Window)

This event occurs when the user opens, resizes, maximizes, or minimizes any window showing the workbook.

Hands-On 15.19   Writing the Workbook_WindowResize() Event Procedure

1. In the VBE window, activate the Project Explorer window and double-click ThisWorkbook.
2. In the ThisWorkbook Code window, enter the `Workbook_WindowResize` procedure, as shown below:

```
Private Sub Workbook_WindowResize(ByVal Wn As Window)
    If Wn.WindowState <> xlMaximized Then
        Wn.Left = 0
        Wn.Top = 0
    End If
End Sub
```

The example procedure moves the workbook window to the top left-hand corner of the screen when the user resizes it.

3. Activate the `Chap15_WorkbookEvents2.xlsm` workbook in the Excel application window.
4. Click the Restore Window button to the left of the Close button.
5. Move the Chap15_WorkbookEvents2.xlsm window to the middle of the screen by dragging its title bar.
6. Change the size of the active window by dragging the window borders in or out.

   As you complete the sizing operation, the workbook window should automatically jump to the top left-hand corner of the screen.
7. Click the Maximize button to restore the Chap15_WorkbookEvents2.xlsm workbook window to its full size.

An Excel workbook can respond to a number of other events. Some of them are listed in Table 15.4. For a complete list of the events and their usage, go to *https://docs.microsoft.com/en-us/office/vba/api/excel.workbook* and expand the Events topic in the left pane.

TABLE 15.4.   Additional workbook events.

Workbook Event Name	Event Description
SheetActivate	This event occurs when the user activates any sheet in the workbook. This event also occurs at the application level when any sheet in any open workbook is activated.

Workbook Event Name	Event Description
SheetDeactivate	This event occurs when the user activates a different sheet in a workbook.
SheetSelectionChange	This event occurs when the user changes the selection on a worksheet. This event happens for each sheet in a workbook.
SheetChange	This event occurs when the user changes a cell formula.
SheetCalculate	This event occurs when the user recalculates a worksheet.
SheetBeforeDoubleClick	This event occurs when the user double-clicks a cell on a worksheet.
SheetBeforeRightClick	This event occurs when the user right-clicks a cell on a worksheet.
NewChart	This event occurs when a new chart is created in the workbook.
AfterSave	This event occurs after the workbook is saved.
AddinInstall	This event occurs after the workbook is installed as an add-in.
AddinUninstall	This event occurs when the workbook is uninstalled as an add-in.
SheetFollowHyperlink	This event occurs when you click any hyperlink in Microsoft Excel.

PIVOTTABLE EVENTS

In Excel, PivotTable reports provide a powerful way of analyzing and comparing large amounts of information stored in a database. By rotating rows and columns of a PivotTable report, you can see different views of the source data or see details of the data that interests you the most.

When working with PivotTable reports programmatically, you can determine when a PivotTable report opened or closed the connection to its data source by using the PivotTableOpenConnection and PivotTableCloseConnection workbook events. You can determine when the PivotTable was updated using the SheetPivotTableUpdate event. Table 15.5 lists events related to PivotTable reports. Chapter 20, Programming PivotTables and PivotCharts, will get you started writing VBA code for creating and manipulating PivotTables and Pivot-Charts. You will find it easier to delve into the PivotTable event programming after working through Chapter 20.

TABLE 15.5. Workbook events related to PivotTable reports.

Workbook Event Name	Event Description
`PivotTableOpen-Connection`	This event occurs after a PivotTable report opens the connection to its data source. This event requires that you declare an object of type `Application` or `Workbook` using the `WithEvents` keyword in a class module (see examples of using this keyword further in this chapter).
`PivotTableClo-seConnection`	This event occurs after a PivotTable report closes the connection to its data source. This event requires that you declare an object of type `Application` or `Workbook` using the `WithEvents` keyword in a class module (see examples of using this keyword further in this chapter).
| `SheetPivotTa-bleUpdate`<br><br>The `SheetPivot-TableUpdate` event procedure takes the following two arguments:<br><br>`Sh`—The selected sheet<br>`Target`—The selected PivotTable report | This event occurs after the sheet of the PivotTable report has been updated. This event requires that you declare an object of type `Application` or `Workbook` using the `WithEvents` keyword in a class module (see examples of using this keyword at the end of this chapter). *Note:* The example event procedure shown below, along with other procedures related to the PivotTable reports, can be found in `Chap15_PivotReportEvents.xlsm` in the companion files:<br><br>```vba
Private Sub pivTbl_SheetPivotTableUpdate( _
    ByVal Sh As Object, _
    ByVal Target As PivotTable)
    MsgBox Target.Name & _
    " report has been updated." & vbCrLf _
    & "The PivotReport is located in cells " & _
    Target.DataBodyRange.Address
End Sub
``` |
| `SheetPivotTa-bleChangeSync`<br><br>This event takes the following two arguments:<br><br>`Sh`—The worksheet that contains the Pivot-Table<br><br>`Target`—The Pivot-Table that was changed | This event occurs after changes to a PivotTable. For example, after making changes to a PivotTable, you can write code to display a message:<br><br>```vba
Private Sub _
    Workbook_SheetPivotTableChangeSync( _
    ByVal Sh As Object, _
    ByVal Target As PivotTable)

    MsgBox "Thanks for working with " & _
        "PivotTable (" & Target.Name & _
        ") on " & Sh.Name & _
        " worksheet."
End Sub
``` |

Workbook Event Name	Event Description
SheetPivotTa- bleAfter ValueChange	This event occurs after a cell or range of cells (that contain formulas) inside a PivotTable are edited or recalculated. This event will not occur when a PivotTable is refreshed, sorted, filtered, or drilled down on.
SheetPivotTa- bleBefore DiscardChanges	This event occurs immediately before changes to a PivotTable are discarded. It is used with the PivotTable's OLAP (Online Analytical Processing) data source.
SheetPivotTa- bleBefore CommitChanges	This event occurs immediately before changes are committed against the OLAP data source for a PivotTable.
SheetPivotTa- bleBefore AllocateChanges	This event occurs immediately before changes are applied to the Pivot-Table's OLAP data source.

CHART EVENTS

As you know, Excel allows you to create charts that are embedded in a worksheet or located on a separate chart sheet. In this section, you will learn how to control chart events no matter where you've decided to place your chart. Before you try out selected chart events, perform the tasks in Hands-On 15.20.

Hands-On 15.20   Creating Charts for Trying Out Chart Events

1. Open a new Excel workbook and save it as `Chap15_ChartEvents.xlsm`.
2. Enter sample data, as shown in Figure 15.3.

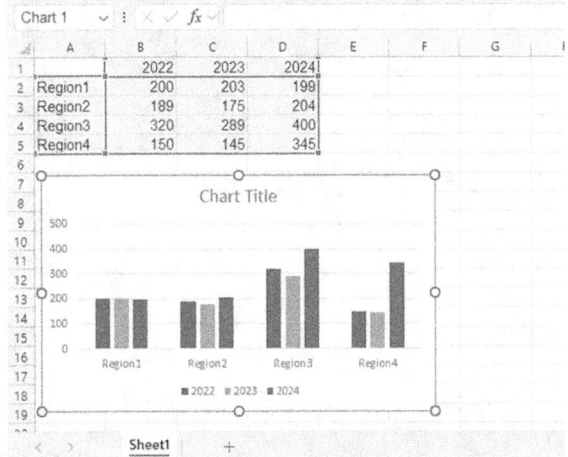

FIGURE 15.3. Column chart embedded in a worksheet.

3. Select cells A1:D5 and choose Insert | Charts | Insert Column Chart | 2-D | Clustered Column.
4. Resize the chart, as shown in Figure 15.3.
5. Using the same data, create a line chart on a separate chart sheet, as shown in Figure 15.5:

    i. To add a new chart sheet, right-click any sheet tab in the workbook and choose Insert.

    ii. In the Insert dialog box, select Chart and click OK.

    iii. On the Chart Design tab, click the Select Data button.

    iv. Excel will display the Select Data Source dialog box.

    v. Click the Sheet1 tab and select cells A1:D5.

    vi. Excel will fill in the Chart data range box in the dialog box, as shown in Figure 15.4.

    vii. Click OK to complete the chart.

    viii. Now, change the chart type to Line chart with Markers by choosing the Change Chart Type button in the Chart Design tab.

    ix. In the Change Chart Type dialog box, select Line in the left pane, and click the button representing Line chart with Markers.

    x. Click OK to close the dialog box.

    xi. Add a legend to the chart using the Add Chart Element button on the Chart Layout group of the Chart Design tab.

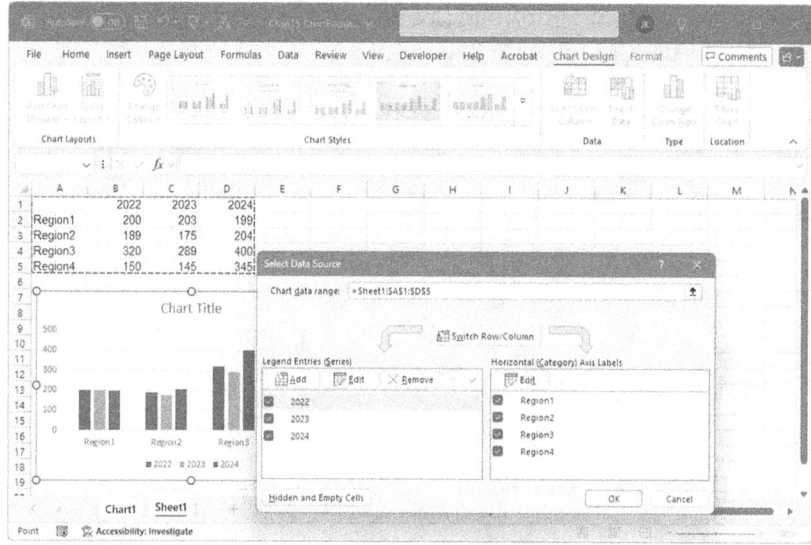

FIGURE 15.4. Creating a chart in a chart sheet.

6. Change the name of the chart sheet to `Sales Analysis Chart` (see Figure 15.5).

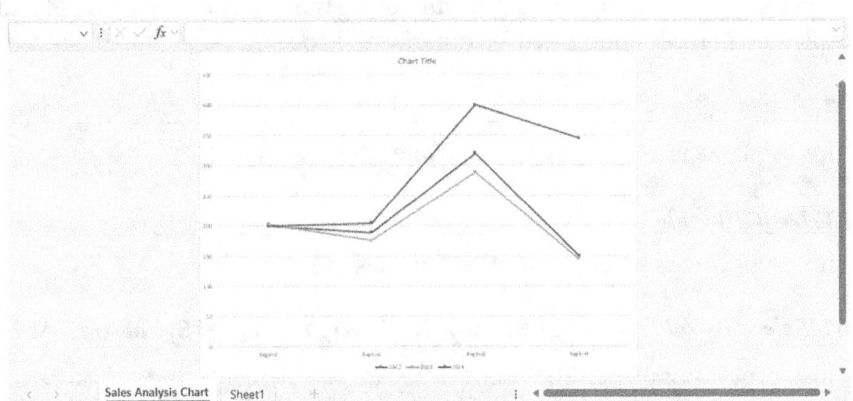

FIGURE 15.5. Line chart placed in a chart sheet.

Writing Event Procedures for a Chart Located on a Chart Sheet

Excel charts can respond to a number of events, as shown in Table 15.6.

TABLE 15.6. Chart events.

Chart Event Name	This event occurs when...
`Activate`	The user activates the chart sheet.
`Deactivate`	The user deactivates the chart sheet.
`Select`	The user selects a chart element.
`SeriesChange`	The user changes the value of a chart data point. The `Chart` object should be declared in the class module using the `With-Events` keyword.
`Calculate`	The user plots new or changed data on the chart.
`Resize`	The user changes the size of the chart. The `Chart` object should be declared in the class module using the `WithEvents` keyword.
`BeforeDoubleClick`	An embedded chart is double-clicked before the default double-click action.
`BeforeRightClick`	An embedded chart is right-clicked before the default right-click action.
`MouseDown`	A mouse button is pressed while the pointer is over a chart.
`MouseMove`	The position of a mouse pointer changes over a chart.
`MouseUp`	A mouse button is released while the pointer is over a chart.

We will begin by writing event procedures that control a chart placed on a separate chart sheet, as shown in Figure 15.5. Events for a chart embedded in a worksheet like the one in Figure 15.3 require using the `WithEvents` keyword and are explained in the section titled Writing Event Procedures for Embedded Charts.

Chart_Activate()

This event occurs when the user activates the chart sheet.

Chart_Deactivate()

This event occurs when the user deactivates the chart sheet.

Chart_Select(ByVal ElementID As Long, ByVal Arg1 As Long, ByVal Arg2 As Long)

This event occurs when the user selects a chart element. `ElementID` returns a constant representing the type of the selected chart element. Arguments `Arg1` and `Arg2` are used in relation to some chart elements. For example, the chart axis (`ElementID = 21`) can be specified as the main axis (`Arg1 = 0`) or a secondary axis (`Arg1 = 1`), while the axis type is specified by `Arg2`, which can be one of the following three values: `0` (Category axis), `1` (Value axis), and `3` (Series axis).

Chart_Calculate()

This event occurs when the user plots new or changed data on the chart.

Chart_BeforeRightClick()

This event occurs when the user right-clicks the chart.

Chart_MouseDown(ByVal Button As Long, ByVal Shift As Long, ByVal x As Long, ByVal y As Long)

This event occurs when a mouse button is pressed while the pointer is over a chart. The `Button` argument determines which mouse button was pressed (`MouseDown` event) or released (`MouseUp` event): `1` indicates the left button, `2` is for the right button, and `4` is the middle button. The `Shift` argument specifies the state of the Shift, Ctrl, and Alt keys: `1` means that Shift was selected, `2` means that Ctrl was selected, and `4` means that Alt was selected. The `x` and `y` arguments specify the mouse pointer coordinates.

(◉) Hands-On 15.21   Writing Event Procedures for a Chart Sheet

1. In the VBE window, activate the Project Explorer window and open the Microsoft Excel `Objects` folder under `VBAProject(Chap15_ChartEvents.xlsm)`.
2. Double-click the chart object `Sales Analysis Chart`.
3. In the Code window, enter the code of the following event procedures:

```vba
Private Sub Chart_Activate()
  MsgBox "You've activated the chart sheet."
End Sub


Private Sub Chart_Deactivate()
  MsgBox "It looks like you want to leave the " _
      & "chart sheet."
End Sub


Private Sub Chart_Select(ByVal ElementID As Long, _
    ByVal Arg1 As Long, ByVal Arg2 As Long)
  If Arg1 <> 0 And Arg2 <> 0 Then
    MsgBox ElementID & ", " & Arg1 & ", " & Arg2
  End If
  If ElementID = 4 Then
    MsgBox "You've selected the chart title."
  ElseIf ElementID = 24 Then
    MsgBox "You've selected the chart legend."
  ElseIf ElementID = 12 Then
    MsgBox "You've selected the legend key."
  ElseIf ElementID = 13 Then
    MsgBox "You've selected the legend entry."
  End If
End Sub


Private Sub Chart_Calculate()
  MsgBox "The data in your spreadsheet has " & vbCrLf _
      & "changed. Your chart has been updated."
End Sub


Private Sub Chart_BeforeRightClick(Cancel As Boolean)
  Cancel = True
End Sub


Private Sub Chart_MouseDown(ByVal Button As Long, _
    ByVal Shift As Long, ByVal x As Long, ByVal y As Long)
  If Button = 1 Then
```

```
    MsgBox "You pressed the left mouse button."
  ElseIf Button = 2 Then
    MsgBox "You pressed the right mouse button."
  Else
    MsgBox "You pressed the middle mouse button."
  End If
End Sub
```

4. Activate the chart sheet and perform the actions that will trigger the event procedures that you've written. For example, click the chart legend and notice that this action triggers two events: `Chart_MouseDown` and `Chart_Select`.

Writing Event Procedures for Embedded Charts

To capture events raised by a chart embedded in a worksheet, you must first create a new object in the class module using the keyword `WithEvents`.

The `WithEvents` keyword allows you to specify an object variable that will be used to respond to events triggered by an ActiveX object. This keyword can only be used in class modules in the declaration section.

In the following example procedure, we will learn how to use the `With-Events` keyword to capture the `Chart_Activate` event for the embedded chart shown in Figure 15.3.

Hands-On 15.22   Writing the Chart_Activate() Event Procedure for an Embedded Chart

1. Activate the VBE window. In the Project Explorer, select VBAProject (Chap15_ChartEvents.xlsm).
2. Choose Insert | Class Module.
   In the `Class Modules` folder, you will see a module named `Class1`.
3. In the Properties window, rename `Class1` to `clsChart`.
4. In the clsChart class module Code window, type the following declaration, just below the Option Explicit statement:

```
Public WithEvents xlChart As Excel.Chart
```

The above statement declares an object variable that will represent the events generated by the `Chart` object. The `Public` keyword will make the object variable `xlChart` available to all modules in the current VBA project. Declaring an object variable using the `WithEvents` keyword exposes all of the events defined for that particular object type.

After typing the above declaration, the `xlChart` object variable is added to the drop-down Object list in the upper-left corner of the Code window, and the events associated with this object variable appear in the Procedure drop-down list box in the upper-right corner of the Code window.

5. Open the Object drop-down list box and select the xlChart variable.

The Code window should now show the skeleton of the `xlChart_Activate` event procedure:

```
Private Sub xlChart_Activate()

End Sub
```

6. Add your VBA code to the event procedure. In this example, we will add a statement to display a message box. After adding this statement, your VBA procedure should look like the following:

```
Private Sub xlChart_Activate()
  MsgBox "You've activated a chart embedded in  " & _
      ActiveSheet.Name & "."
End Sub
```

After entering the code of the event procedure, you need to inform Visual Basic that you are planning on using it (see Step 7).

7. In the Project Explorer window, double-click the object named ThisWork-book, and in the first line of the ThisWorkbook Code window, enter the statement to create a new instance of the class named `clsChart`:

```
Dim myChart As New clsChart
```

This instruction declares an object variable named `myChart`. This variable will refer to the `xlChart` object located in the class module `clsChart`. The `New` keyword tells Visual Basic to create a new instance of the specified object.

Before you can use the `myChart` object variable, you must write a VBA procedure that initializes it (see Step 8).

8. Enter the following procedure in the ThisWorkbook Code window to initialize the object variable `myChart`:

```
Sub InitializeChart()

' you must run this procedure before event procedures
' written in clsChart class module can be triggered for
' the chart embedded in Sheet1

' connect the class module with the Excel chart object
  Set myChart.xlChart = _
      Worksheets("Sheet1").ChartObjects(1).Chart
End Sub
```

9. Position the insertion point anywhere within the `InitializeChart` procedure and press F5 to run this procedure.

After executing this procedure, the event procedures entered in the `clsChart` class module will be triggered in response to an event. Recall that, right now, the `clsChart` class module contains the `Chart_Activate` event procedure. Later on, in the `clsChart` class module, you may want to write additional event procedures to capture other events for your embedded chart.

10. Activate the Microsoft Excel application window and click the embedded chart in `Sheet1`.

At this time, the `xlChart_Activate` event procedure that you entered in Step 6 should be triggered.

Save and close the `Chap15_ChartEvents.xlsm` workbook file.

EVENTS RECOGNIZED BY THE APPLICATION OBJECT

If you want your event procedure to execute no matter which Excel workbook is currently active, you must create the event procedure for the `Application` object. Event procedures for the `Application` object have a global scope. This means that the procedure code will be executed in response to a certain event as long as the Microsoft Excel application remains open.

Some events for the `Application` object are listed in Table 15.7. For a complete list of events and their usage, go to *https://learn.microsoft.com/en-us/office/vba/api/excel.application.aftercalculate*.

Similar to an embedded chart, event procedures for the `Application` object require that you create a new object using the `WithEvents` keyword in a class module.

TABLE 15.7.   Partial listing of Application events.

Application Event Name	Event Description
AfterCalculate	This event occurs whenever all pending calculations and all of the resultant calculation activities have been completed and there are no outstanding queries.
NewWorkbook	This event occurs when the user creates a new workbook.
ProtectedViewWindowActivate	This event occurs when a Protected View window is activated.

Application Event Name	Event Description
ProtectedViewWindowBeforeClose	This event occurs immediately before a Protected View window or a workbook in a Protected View window opens.
ProtectedViewWindowBeforeEdit	This event occurs immediately before editing is enabled on the workbook in the specified Protected View window.
ProtectedViewWindowDeactivate	This event occurs when a Protected View window is deactivated.
ProtectedViewWindowOpen	This event occurs when a workbook is opened in Protected View.
ProtectedViewWindowResize	This event occurs when any Protected View window is resized.
WorkbookOpen	This event occurs when the user opens a workbook.
WorkbookActivate	This event occurs when the user shifts the focus to an open workbook.
WorkbookDeactivate	This event occurs when the user shifts the focus away from an open workbook.
WorkbookNewSheet	This event occurs when the user adds a new sheet to an open workbook.
WorkbookNewChart	This event occurs when a new chart is created in any open workbook. If multiple charts are inserted or pasted, the event will occur for each chart in the insertion order. If a chart object or chart sheet is moved from one location to another, the event will not occur; however, the event will occur if the chart is moved between a chart object and a chart sheet.
WorkbookBeforeSave	This event occurs before an open workbook is saved.
WorkbookBeforePrint	This event occurs before an open workbook is printed.
WorkbookBeforeClose	This event occurs before an open workbook is closed.
WorkbookAddInInstall	This event occurs when the user installs a workbook as an add-in.
WorkbookAddInUninstall	This event occurs when the user uninstalls a workbook as an add-in.
WorkbookAfterSave	This event occurs after the workbook is saved.

Application Event Name	Event Description
WorkbookRowsetComplete	This event occurs when the user either drills through the recordset or invokes the rowset action on an OLAP PivotTable.
SheetActivate	This event occurs when the user activates a sheet in an open workbook.
SheetDeactivate	This event occurs when the user deactivates a sheet in an open workbook.
SheetFollowHyperlink	This event occurs when the user clicks any hyperlink in Microsoft Excel.
SheetPivotTableAfterValueChanged	This event occurs after a cell or range of cells that contain formulas inside a PivotTable are edited or recalculated.
SheetPivotTableBeforeAllocateChanges	This event occurs before changes are applied to a PivotTable.
SheetPivotTableBeforeCommitChanges	This event occurs before changes are committed against the OLAP data source for a PivotTable (immediately before Excel executes a COMMIT transaction).
SheetPivotTableBeforeDiscardChanges	This event occurs before changes to a PivotTable are discarded.
SheetPivotTableUpdate	This event occurs after the sheet of the PivotTable report has been updated.
SheetSelectionChange	This event occurs when the user changes the selection on a sheet in an open workbook.
SheetChange	This event occurs when the user changes a cell formula in an open workbook.
SheetCalculate	This event occurs when the user recalculates a worksheet in an open workbook.
SheetBeforeDoubleClick	This event occurs when the user double-clicks a worksheet cell in an open workbook.
SheetBeforeRightClick	This event occurs when the user right-clicks a worksheet cell in an open workbook.
WindowActivate	This event occurs when the user shifts the focus to an open window.
WindowDeactivate	This event occurs when the user shifts the focus away from the open window.
WindowResize	This event occurs when the user resizes an open window.

Application Event Name	Event Description
`WorkbookPivotTableCloseConnection`	This event occurs after a PivotTable report connection has been closed.
`WorkbookPivotTableOpenConnection`	This event occurs after a PivotTable report connection has been opened.
`WorkbookAfterXmlExport`	This event occurs after Microsoft Excel saves or exports data from any open workbook to an XML data file.
`WorkbookAfterXmlImport`	This event occurs after an existing XML data connection is refreshed or new XML data is imported into any open Microsoft Excel workbook.
`WorkbookBeforeXmlExport`	This event occurs before Microsoft Excel saves or exports data from any open workbook to an XML data file.
`WorkbookBeforeXmlImport`	This event occurs before an existing XML data connection is refreshed or new XML data is imported into any open Microsoft Excel workbook.
`WorkbookSync`	This event occurs when the local copy of a workbook that is part of a document workspace is synchronized with the copy on the server. This event has been deprecated; it's used only for backward compatibility.

Let's create a couple of event procedures for the `Application` object.

Hands-On 15.23 Writing Event Procedures for the Application Object

1. Open a new workbook and save it as `Chap15_ApplicationEvents.xlsm` in `C:\VBAExcel2021__ByExample`.
2. Switch to the VBE window, and in the Project Explorer window, select VBAProject (Chap15_ApplicationEvents.xlsm).
3. Choose Insert | Class Module.
4. In the Properties window, change the class module name to `clsApplication`.
5. In the clsApplication Code window, type the following declaration statement:

```
Public WithEvents App As Application
```

This statement uses the `WithEvents` keyword to declare an `Application` object variable.

6. Below the declaration statement, enter the event procedures, as shown below:

```
Private Sub App_WorkbookOpen(ByVal Wb As Workbook)
   If Wb.FileFormat = xlCSV Then
      If MsgBox("Do you want to save this " & vbCrLf _
         & "file as an Excel workbook?", vbYesNo, _
         "Original file format: " _
         & "comma delimited file") = vbYes Then
         Wb.SaveAs FileFormat:=xlWorkbookNormal
      End If
   End If
End Sub


Private Sub App_WorkbookBeforeSave(ByVal _
      Wb As Workbook, ByVal SaveAsUI As Boolean, _
      Cancel As Boolean)
   If Wb.Path <> vbNullString Then
      ActiveWindow.Caption = Wb.FullName & _
         " [Last Saved: " & Time & "]"
   End If
End Sub


Private Sub App_WorkbookBeforePrint(ByVal _
      Wb As Workbook, Cancel As Boolean)
   Wb.PrintOut Copies:=2
End Sub


Private Sub App_WorkbookBeforeClose(ByVal _
      Wb As Workbook, Cancel As Boolean)
   Dim r As Integer
   Dim p As Variant

   Sheets.Add
   r = 1
   For Each p In Wb.BuiltinDocumentProperties
      On Error GoTo ErrorHandle
      Cells(r, 1).Value = p.Name & " = " & _
         ActiveWorkbook.BuiltinDocumentProperties _
         .Item(p.Name).Value
      r = r + 1
   Next
   Exit Sub
ErrorHandle:
      Cells(r, 1).Value = p.Name
      Resume Next
End Sub
```

```
Private Sub App_SheetSelectionChange(ByVal Sh _
    As Object, ByVal Target As Range)

  If Selection.Count > 1 Or _
      (Selection.Count < 2 And _
    IsEmpty(Target.Value)) Then
      Application.StatusBar = Target.Address
    Else
      Application.StatusBar = Target.Address & _
        "(" & Target.Value & ")"
  End If
End Sub

Private Sub App_WindowActivate(ByVal _
    Wb As Workbook, ByVal Wn As Window)
  Wn.DisplayFormulas = True
End Sub
```

7. After you've entered the code of the above event procedures in the class module, choose Insert | Module to insert a standard module into your current VBA project.

8. In the newly inserted standard module, create a new instance of the clsApplication class and connect the object located in the class module clsApplication with the object variable App, representing the Application object, as shown below:

```
Dim DoThis As New clsApplication

Public Sub InitializeAppEvents()
  Set DoThis.App = Application
End Sub
```

Recall that you declared the App object variable to point to the Application object in Step 5 above.

9. Now, place the mouse pointer within the InitializeAppEvents procedure and press F5 to run it.

10. As a result of running the InitializeAppEvents procedure, the App object in the class module will refer to the Excel application. From now on, when a specific event occurs, the code of the event procedures you've entered in the class module will be executed.

11. If you don't want to respond to events generated by the Application object, you can break the connection between the object and its object variable by entering (and then running) the following procedure in a standard module:

```
Public Sub CancelAppEvents()
  Set DoThis.App = Nothing
End Sub
```

When you set the object variable to `Nothing`, you release the memory and break the connection between the object variable and the object to which this variable refers. When you run the `CancelAppEvents` procedure, the code of the event procedures written in the class module will not be automatically executed when a specific event occurs.

Now, let's proceed to try triggering the application events you coded in the class module.

12. Activate `Chap15_ApplicationEvents` workbook. Click the File tab and choose New. Select Blank Workbook and click Create.

13. Click the File tab and choose Save As. Save the workbook opened in Step 10 as `TestBeforeSaveEvent.xlsx`.

14. Type anything in `Sheet1` of the `TestBeforeSaveEvent.xlsx` workbook and save this workbook.
Notice that Excel writes the full name of the workbook file and the time the workbook was last saved in the workbook's title bar, as coded in the `WorkbookBeforeSave` event procedure (see Step 6). Every time you save this workbook file, Excel will update the last saved time in the workbook's title bar.

15. Look at the code in the other event procedures you entered in Step 6 and perform actions that will trigger these events.

16. Close the `Chap15_ApplicationEvents.xlsm` file and other workbooks if they are currently open.

QUERY TABLE EVENTS

A query table is a table in an Excel worksheet that represents data returned from an external data source, such as a SQL Server database, a Microsoft Access database, a Web page, or a text file. Excel provides two events for the `QueryTable` object: `BeforeRefresh` and `AfterRefresh`. These events are triggered before or after the query table is refreshed. You can create a query table as a standalone object or as a list object whose data source is a query table. The list object is discussed in detail in Chapter 19, Using and Programming Excel Tables.

When you retrieve data from an external data source such as Access or SQL Server using the controls available on the Excel ribbon's Data tab, Excel creates a query table that is associated with a list object. The resulting table is easier to use thanks to some built-in data management features available on the ribbon.

The next Hands-On demonstrates how to create a query table associated with a list object and enable the `QueryTable` object's `BeforeRefresh` and `AfterRefresh` events. You must have Microsoft Access and a sample `Northwind 2007.accdb` database installed on your computer.

⊙ Hands-On 15.24   Writing Event Procedures for a Query Table

1. Open a new Microsoft Excel workbook and save it as `Chap15_QueryTableEvents.xlsm` in your `C:\VBAExcel2024_ByExample` folder.
2. Choose the Data tab. In the Get & Transform Data group, click Get Data and select From Database | From Microsoft Access Database.
3. In the Import Data dialog, select the `Northwind 2007.accdb` database and click Import. This file is included in the companion files.
4. In the Navigator screen, select the Inventory on Order table and click the Load button (see Figure 15.6).

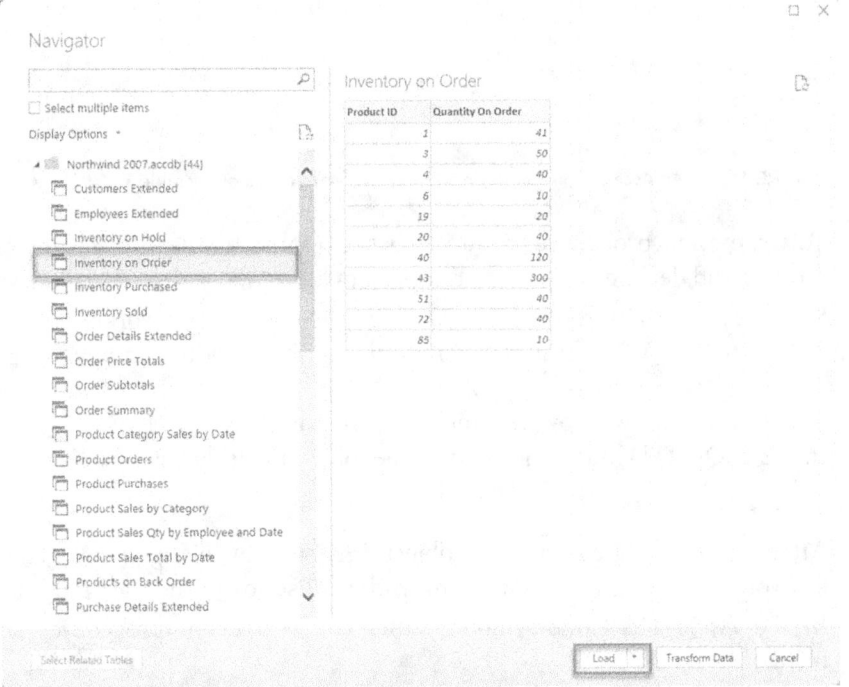

FIGURE 15.6.   After you choose to load data from a database, the Navigator screen appears. This screen allows you to browse and select the specific tables you want to import into Excel.

5. The Navigator screen is designed to make it easy to explore, preview, and select the data you want to import from a database into Excel. When you click the Load button in the Navigator screen, Excel loads data directly into a worksheet formatted as a table, as shown in Figure 15.7.

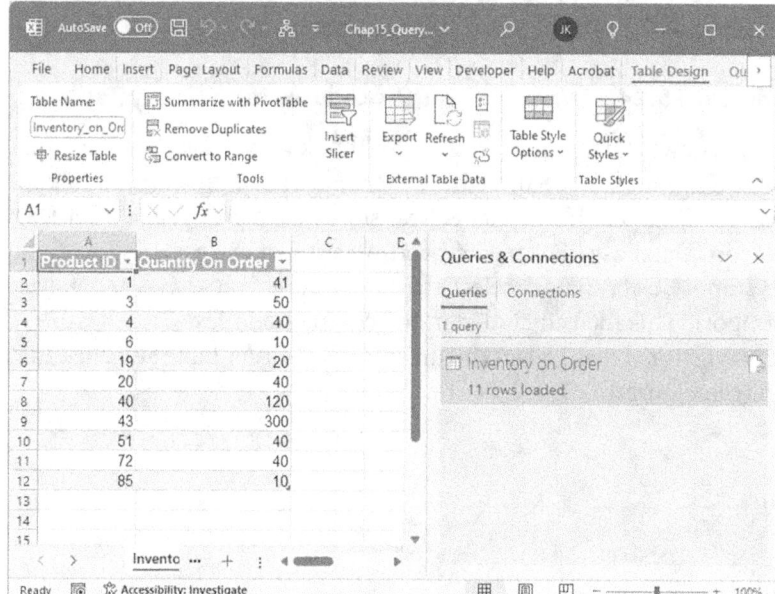

FIGURE 15.7.   An Access table loaded into Excel is formatted as a query table.

To write event procedures for a `QueryTable` object, you must create a class module and declare a `QueryTable` object by using the `WithEvents` keyword.

6. Save the changes in the `Chap15_QueryTableEvents.xlsm` workbook.

7. Switch to the VBE window and insert a class module into `VBAProject` (`Chap15_QueryTableEvents.xlsm`).

8. In the Properties window, rename the class module `clsQryTbl`.

9. In the clsQryTbl Code window, type the following declaration statement:

```
Public WithEvents qryTbl As QueryTable
```

After you've declared the new object (`qryTbl`) by using the `WithEvents` keyword, it appears in the Object drop-down list box in the class module.

10. In the clsQryTbl Code window, enter the two event procedures shown below:

```
Private Sub qryTbl_BeforeRefresh(Cancel As Boolean)
```

```
Dim Response As Integer

Response = MsgBox("Are you sure you " _
    & " want to refresh now?", vbYesNoCancel)
If Response = vbNo Then Cancel = True
End Sub

Private Sub qryTbl_AfterRefresh(ByVal Success As Boolean)
If Success Then
  MsgBox "The data has been refreshed."
Else
  MsgBox "The query failed."
End If
End Sub
```

The `BeforeRefresh` event of the `QueryTable` object occurs before the query table is refreshed. The `AfterRefresh` event occurs after a query is completed or canceled. The `Success` argument is `True` if the query was completed successfully.

Before you can trigger these event procedures, you must connect the object that you declared in the class module (`qryTbl`) to the specified `QueryTable` object. This is done in a standard module, as shown in Step 11.

11. Insert a module into `VBAProject (Chap15_QueryTableEvents.xlsm)` and rename it `QueryTableListObj`.

12. In the QueryTableListObj Code window, enter the declaration line and the procedure, as shown below:

```
Dim sampleQry As New clsQryTbl

Public Sub Auto_Open()
 ' connect the class module and its objects with the Query object

 Set sampleQry.qryTbl = ActiveSheet.ListObjects(1).QueryTable
End Sub
```

This procedure creates a new instance of the `QueryTable` class (`clsQryTbl`) and connects this instance with the first list object on the active worksheet.

NOTE	*A query table associated with a list object can only be accessed through the* ListObject.QueryTable *property. This query table is not a part of the* Worksheet.QueryTables *collection. To find out whether a query table exists on a worksheet, be sure to check both the* QueryTables *and* ListObjects *collections. This can be done easily by entering the following statements in the Immediate window:*
	
	`?ActiveSheet.ListObjects.Count`
	`?ActiveSheet.QueryTables.Count`

13. Run the Auto_Open procedure.

    After you run this initialization procedure, the object that you declared in the class module points to the specified QueryTable object.

NOTE	*The next time you want to work with the* QueryTable *object in this workbook file, you won't need to run the* Auto_Open *procedure. This procedure will run automatically upon opening the workbook file.*

14. Switch to the Microsoft Excel application window.
15. In the worksheet where you placed the Inventory on Order table from the Microsoft Access database, choose Data | Refresh All. Excel will now trigger the qryTbl_BeforeRefresh event procedure, and you should see the custom message box. If you click Yes, the data in the worksheet will be refreshed with the existing data in the database. Excel will then trigger the qryTbl_AfterRefresh event procedure, and another custom message will be displayed.
16. Save and close the Chap15_QueryTableEvents.xlsm workbook file.

OTHER EXCEL EVENTS

There are two events in Excel that are not associated with a specific object: the OnTime and OnKey events. These events are accessed using the methods of the Application object: OnTime and OnKey.

OnTime Method

The OnTime event uses the OnTime method of the Application object to trigger an event at a specific time. The syntax is:

```
Application.OnTime(EarliestTime, Procedure, LatestTime, Schedule)
```

The `Procedure` parameter is the name of the VBA procedure to run. The `EarliestTime` parameter is the time you would like the procedure to run. Use the `TimeValue` function to specify the time, as shown in the examples below. `LatestTime` is an optional parameter that allows you to specify the latest time the procedure can be run. Again, you can use the `TimeValue` function to specify a time for this parameter. The `Schedule` parameter allows you to clear a previously set `OnTime` event. Set this parameter to `False` to cancel the event. The default value for `Schedule` is `True`.

For example, you can have Excel run the specified procedure at 4:00 p.m., as shown below:

```
Application.OnTime TimeValue("4:00PM"), "YourProcedureName"
```

To cancel the above event, run the following code:

```
Application.OnTime TimeValue("4:00PM"),
    "YourProcedureName", , False
```

To schedule the procedure five minutes after the current time, use the following code:

```
Application.OnTime Now + TimeValue("00:05:00"),
    "YourProcedureName"
```

The `Now` function returns the current time. Therefore, to schedule the procedure to occur in the future (a certain amount from the current time), you need to set the value of the `EarliestTime` parameter to:

```
Now + TimeValue(time)
```

To trigger your procedure on July 4, 2022, at 12:01 a.m., type the following statement on one line in the Immediate window and press Enter:

```
Application.OnTime DateSerial(2022, 7, 4) +
        TimeValue("00:00:01"), "YourProcedureName"
```

OnKey Method

You can use the `Application` object's `OnKey` method to trigger a procedure whenever a particular key or key combination is pressed. The syntax of the `OnKey` method is as follows:

```
Application.OnKey(Key, Procedure)
```

Here, `Key` is a string indicating the key to be pressed, and `Procedure` is the name of the procedure you want to execute. If you pass an empty string, `""`, as the second parameter for the `OnKey` method, Excel will ignore the keystroke.

The `Key` parameter can specify a single key or any key combined with Shift, Alt, and/or Ctrl. For a letter or any other character key, use that character. To specify a key combination, use the plus sign (+) for Shift, percent sign (%) for Alt, and caret (^) for Ctrl in front of the keycode.

For example, to run your procedure when you press Ctrl+a, you would write the following statement:

```
Application.OnKey "^a", "YourProcedureName"
```

Special keys are entered using curly braces: `{Enter}`, `{Down}`, `{Esc}`, `{Home}`, `{Backspace}`, `{F1}`, or `{Right}`. See the list of keycodes in Table 13.4 (in Chapter 13). For example, to run the procedure named `NewFolder` when the user presses Alt+F10, use the following code:

```
Application.OnKey "%{F10}", "NewFolder"
```

To cancel an `OnKey` event and return the key to its normal function, call the `OnKey` method without the `Procedure` parameter:

```
Application.OnKey "%{F10}"
```

The above code will return the key combination Alt+F10 to its default function in Excel, which is to display the Selection and Visibility pane on the right side of the Excel screen.

While using the `OnKey` method is a quick way to assign a shortcut to execute a VBA procedure or macro, a better way is to use the Options button in the Macro dialog to assign a Ctrl+key combination to a procedure.

When using the `OnKey` events, keep in mind that reassigning frequently used Excel shortcuts (such as Ctrl+P for Print) to perform other customized processes may make you an unpopular developer among your users.

SUMMARY

In this chapter, you gained hands-on experience with events and event-driven programming in Excel. These are invaluable skills, whether you are planning to create spreadsheet applications for others to use or simply automating your daily worksheet tasks. Excel provides many events to which you can respond. By writing event procedures, you can change the way objects respond to events. Your event procedures can be as simple as a single line of code displaying a custom message or more complex, with code that includes decision-making statements and other programming structures. When a certain event occurs, Visual

Basic will simply run your custom event procedure instead of responding in the standard way.

You've learned that some event procedures are written in a standard module (workbook, sheet, or standalone chart sheet) while others (embedded chart, application, and query table) require that you create a new object by using the WithEvents keyword in a class module. You've also learned that you can enable or disable events using the EnableEvents property.

In the final section of this chapter, you worked with two Application object methods to execute procedures at a specific time or in response to the user pressing a key or a key combination.

The next chapter will take you through the process of designing and implementing custom forms and dialog boxes.

16 WORKING WITH DIALOG BOXES AND USER FORMS

arlier in this book, you learned how to use the built-in `InputBox` function to collect single items of data from the user during the execution of your VBA procedure. What if your procedure requires more data at runtime, though? The user may want to supply all the data at once or make appropriate selections from a list of items. If your procedure must collect data, you can:

- Use the collection of built-in dialog boxes.
- Create a custom form.

This chapter introduces you to interacting with Excel's built-in dialog boxes from your VBA procedures and designing your own custom forms from scratch.

EXCEL DIALOG BOXES

Before you start creating your own forms, you should spend some time learning how to take advantage of the dialog boxes that are built into Excel and are therefore ready for you to use. I'm not talking about your ability to manually select appropriate options, but how to call these dialog boxes from your own VBA procedures.

Microsoft Excel has a special collection of built-in dialog boxes that are represented by constants beginning with `xlDialog`, such as `xlDialogClear`, `xlDialogFont`, `xlDialogDefineName`, and `xlDialogOptionsView`. These built-in dialog boxes, some of which are listed in Table 16.1, are Microsoft Excel objects that belong to the built-in collection of dialog boxes. Each `Dialog` object represents a built-in dialog box.

TABLE 16.1.   Frequently used built-in dialog boxes.

Dialog Box Name	Constant
New	xlDialogNew
Open	xlDialogOpen
Save As	xlDialogSaveAs
Page Setup	xlDialogPageSetup
Print	xlDialogPrint
Fonts	xlDialogFont

To display a dialog box, use the `Show` method in the following format:

`Application.Dialogs(constant).Show`

For example, the following statement displays the Fonts dialog box:

`Application.Dialogs(xlDialogFont).Show`

Figure 16.1 shows a list of constants identifying Excel built-in dialog boxes, which is available in the Object Browser window after selecting the Excel library and searching for `xlDialog`.

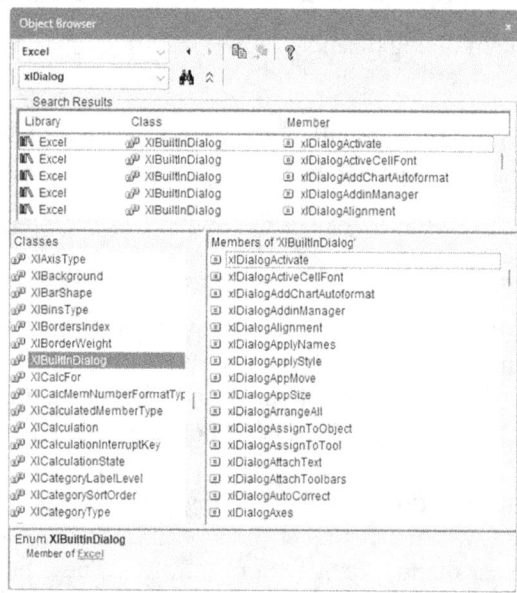

FIGURE 16.1.   Constants prefixed with xlDialog identify Excel's built-in dialog boxes.

Let's practice displaying some of the Excel dialog boxes straight from the Immediate window.

NOTE	*Please note that the files for the "Hands-On" projects can be found in the companion files.*

⊙ Hands-On 16.1   Using Excel Dialog Boxes from the Immediate Window

1. Open a new workbook and save it as `C:\VBAExcel2024_ByExample\Chap16.xlsm`.
2. Switch to the VBE window and choose View | Immediate window.
3. In the Immediate window, type the following statement and press Enter:

```
Application.Dialogs(xlDialogFont).Show
```

The above instruction displays the Fonts dialog box, where you can select an appropriate option and Excel will format the selected cell, range, or the entire sheet.

   Although you can't modify the appearance or behavior of a built-in dialog box, you can decide which initial setting the built-in dialog box will display when you show it from your VBA procedure. If you don't change the initial settings, VBA will display the dialog box with its default settings.

4. Press Cancel to exit the Fonts dialog box.
5. In the Immediate window, type the following statement and press Enter:

```
Application.Dialogs(xlDialogFontProperties).Show
```

The above instruction displays the Format Cells dialog box with the Font tab active.

6. Press Cancel to exit the Format Cells dialog box.
7. In the Immediate window, type the following statement and press Enter:

```
Application.Dialogs(xlDialogDefineName).Show
```

The above statement displays the Define Name dialog box, where you can define a name for a cell or a range of cells.

8. Press Close to exit the Define Name dialog box.
9. In the Immediate window, type the following statement and press Enter:

```
Application.Dialogs(xlDialogOptionsView).Show
```

The above instruction opens the Excel Options dialog box with the Advanced options displayed, as shown in Figure 16.2.

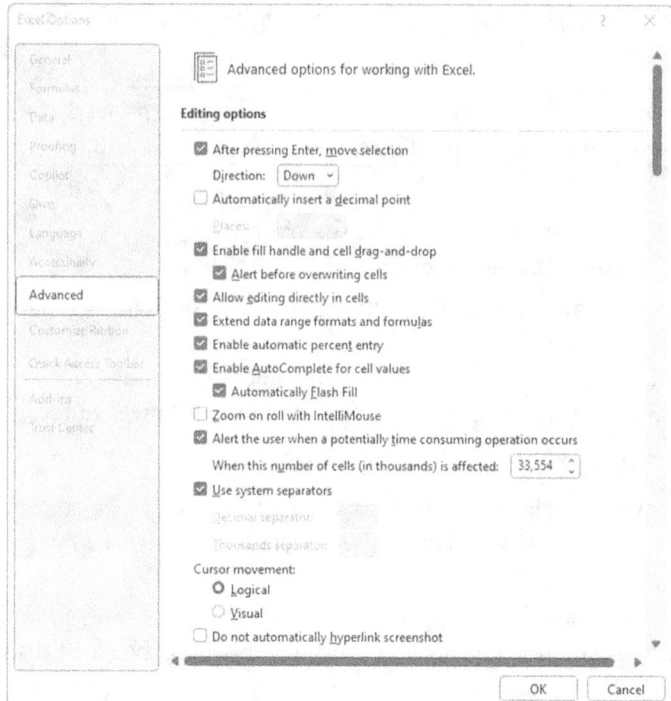

FIGURE 16.2. The advanced settings available in the Excel Options dialog box are identified by the xlDialogOptionsView constant.

10. Press Cancel to exit the Excel Options dialog box.
11. Type the following statement in the Immediate window and press Enter:

```
Application.Dialogs(xlDialogClear).Show
```

12. Excel shows the Clear dialog box with four option buttons: All, Formats, Contents, and Comments. Normally, the Contents option button is selected when Excel displays this dialog box. What if you wanted to invoke this dialog with a different option selected as the default, though? To do this, you can include a list of arguments. Arguments are entered after the Show method. For example, to display the Clear dialog box with the first option button (All) selected, you would enter the following statement:

```
Application.Dialogs(xlDialogClear).Show 1
```

Excel often numbers the available options. Therefore, All is 1, Formats is 2, Contents is 3, Comments is 4, and Hyperlinks is 5.

13. Press Cancel to close the Clear dialog box.

About Dialog Box Arguments

The built-in dialog box argument lists are available at *https://docs.microsoft.com/en-us/ office/vba/excel/Concepts/Controls-DialogBoxes-Forms/built-in-dialog-box-argument-lists* (see Figure 16.3).

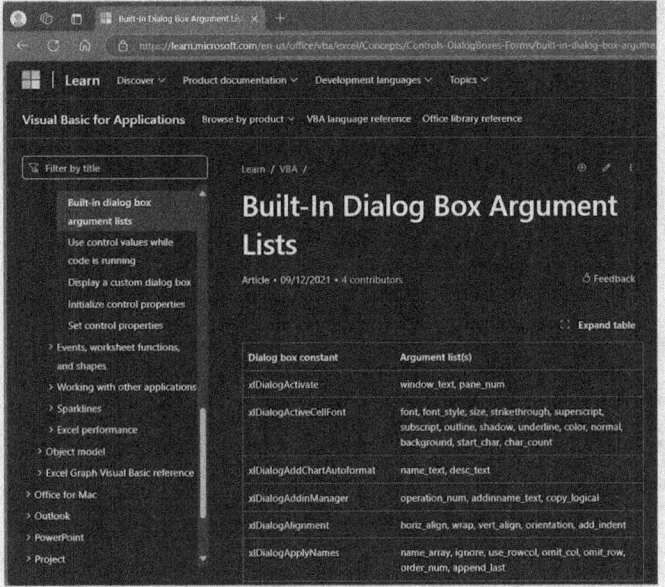

FIGURE 16.3. Microsoft Excel built-in dialog box arguments list.

14. To display the Fonts dialog box in which the Arial 14-point font is already selected, type the following instruction in the Immediate window and press Enter:

```
Application.Dialogs(xlDialogFont).Show "Arial", 14
```

15. Press Cancel to close the Fonts dialog box.
16. To specify only the font size, enter a comma in the position of the first argument:

```
Application.Dialogs(xlDialogFont).Show , 8
```

17. Press Cancel to close the Fonts dialog box.
18. Type the following instruction in the Immediate window and press Enter:

```
Application.Dialogs(xlDialogDefineName).Show "John", "=$A$1"
```

19. The above statement displays the Define Name dialog box, enters "John" in the Names in workbook text box, and places the reference to cell A1 in the Refers to box.

The Show method returns True if you click OK and False if you cancel.

20. Press Close to close the Define Name dialog box.

FILE OPEN AND FILE SAVE AS DIALOG BOXES

FileDialog is a very powerful dialog object. This object allows you to display the File Open and File Save As dialog boxes from your VBA procedures. Because the FileDialog object is a part of the Microsoft Office 16.0 Object Library, it is available to all Microsoft 365 applications. You can also use two methods of the Application object (GetOpenFilename and GetSaveAsFilename) to display the File Open and File Save As dialog boxes without actually opening or saving any files (this is discussed later in this chapter).

Let's practice using the FileDialog object from the Immediate window.

⊙ Hands-On 16.2   Using the FileDialog Object from the Immediate Window

1. To display the File Open dialog box, type the following statement in the Immediate window and press Enter:

```
Application.FileDialog(msoFileDialogOpen).Show
```

2. Press Cancel to close the File Open dialog box.
3. To display the File Save As dialog box, type the following statement and press Enter:

```
Application.FileDialog(msoFileDialogSaveAs).Show
```

4. Press Cancel to close the File Save As dialog box.
5. In addition to the File Open and File Save As dialog boxes, the FileDialog object is capable of displaying a dialog box with a list of files or a list of files and folders. Let's take a quick look at these dialog boxes.
Type the following statement in the Immediate window and press Enter to display the Browse dialog box.

```
Application.FileDialog(msoFileDialogFilePicker).Show
```

6. Press Cancel in the dialog box to return to the Immediate window.
7. Type the following statement in the Immediate window and press Enter:

```
Application.FileDialog(msoFileDialogFolderPicker).Show
```

Excel displays a dialog box with a list of directories.

8. Press Cancel in the dialog box to return to the Immediate window.

The constants that the `FileDialog` object uses are listed in Table 16.2. The `mso` prefix denotes that the constant is a part of the Microsoft Office object model.

TABLE 16.2. The FileDialog object's constants.

Constant Name	Value
msoFileDialogOpen	1
msoFileDialogSaveAs	2
msoFileDialogFilePicker	3
msoFileDialogFolderPicker	4

Filtering Files

When you choose File | Open and click Browse, Excel displays the Open dialog box listing Excel files. You can control the types of files that are displayed in this window via the drop-down box located to the right of the File name drop-down box, or you can do this programmatically by using the `Filters` property. If the filter you need is not listed in the Open dialog, you can add it to the Filters list. Filters are stored in the `FileDialogFilters` collection for the `FileDialog` object.

In the following Hands-On, you will create a simple procedure that returns the list of default file filters to an Excel worksheet.

> **Hands-On 16.3   Writing a List of Default File Filters to an Excel Worksheet**

1. In the Project Explorer window, select VBAProject (Chap16.xlsm).
2. In the Properties window, rename the project `VBA_Dialogs`.
3. Insert a new module into the `VBA_Dialogs` project and rename it `DialogBoxes`.
4. In the DialogBoxes Code window, enter the `ListFilters` procedure, as shown below:

```
Sub ListFilters()
    Dim fdf As FileDialogFilters
    Dim fltr As FileDialogFilter
    Dim c As Integer

    Set fdf = Application.FileDialog(msoFileDialogOpen).Filters

    Workbooks.Add
```

484 Excel 2024/Microsoft 365 Programming By Example

```
    Cells(1, 1).Select
    Selection.Formula = "List of Default filters"
    With fdf
      c = .Count
      For Each fltr In fdf
        Selection.Offset(1, 0).Formula = fltr.Description & _
            ": " & fltr.Extensions
        Selection.Offset(1, 0).Select
      Next
    MsgBox c & " filters were written to a worksheet."
    End With
    Columns("A").AutoFit

End Sub
```

The above procedure declares two object variables. The `fdf` variable returns a reference to the `FileDialogFilters` collection of the `FileDialog` object, and the `fltr` variable stores a reference to the `FileDialogFilter` object.

The `Count` property of the `FileDialogFilters` collection returns the total number of filters.

Next, the procedure iterates through the `FileDialogFilters` collection and retrieves the description and extension of each defined filter.

5. Run the ListFilters procedure.

When the procedure is completed, you should see a list of preset filters in the worksheet of a new workbook.

By using the `Add` method of the `FileDialogFilters` collection, you can easily add your own filter to the default filters. The following modified `ListFilters_Modified` procedure demonstrates how to add a filter to filter out temporary files, `*.tmp`.

The last statement in this procedure will open the File Open dialog box so that you can check for yourself that the custom filter Temporary files (*.tmp) has indeed been added to the list of filters in the drop-down list:

```
Sub ListFilters_Modified()
    Dim fdf As FileDialogFilters
    Dim fltr As FileDialogFilter
    Dim c As Integer

    Set fdf = Application.FileDialog(msoFileDialogOpen).Filters

    Workbooks.Add
    Cells(1, 1).Select
    Selection.Formula = "List of Default filters"
```

```
  With fdf
    c = .Count
    For Each fltr In fdf
      Selection.Offset(1, 0).Formula = fltr.Description & _
          ": " & fltr.Extensions
      Selection.Offset(1, 0).Select
    Next
    MsgBox c & " filters were written to a worksheet."
    .Add "Temporary Files", "*.tmp", 1
    c = .Count
    MsgBox "There are now " & c & " filters." & vbCrLf _
        & "Check for yourself."
    Application.FileDialog(msoFileDialogOpen).Show
  End With
  Columns("A").AutoFit
End Sub
```

You can remove all the preset filters using the `Clear` method of the `FileDi-alogFilters` collection. For example, you could modify the `ListFilters_Modified` procedure to clear the built-in filters prior to adding the custom filter—Temporary files (*.tmp).

Selecting Files

When you select a file in the Open File dialog box, the selected filename and path are placed in the `FileDialogSelectedItems` collection. Use the `Select-edItems` property to return the `FileDialogSelectedItems` collection.

By setting the `AllowMultiSelect` property of the `FileDialog` object to `True`, a user can select one or more files by holding down the Shift or Control keys while clicking filenames.

The following procedure demonstrates how to use the above-mentioned properties. This procedure will open a new workbook and insert a list box control. The user will be allowed to select more than one file. The selected files will then be loaded into the list box control, and the first filename will be highlighted.

⊚ Hands-On 16.4   Loading Files into a Worksheet List Box Control

1. In the DialogBoxes module Code window, enter the `ListSelectedFiles` procedure, as shown below:

```
Sub ListSelectedFiles()
  Dim fd As FileDialog
  Dim myFile As Variant
```

```
Dim lbox As Object

Application.FileDialog(msoFileDialogOpen).Filters.Clear
Set fd = Application.FileDialog(msoFileDialogOpen)
With fd
    .AllowMultiSelect = True
    If .Show Then
        Workbooks.Add
    Set lbox = Worksheets(1).Shapes. _
        AddFormControl(xlListBox, _
        Left:=20, Top:=60, Height:=40, Width:=300)
    lbox.ControlFormat.MultiSelect = xlNone
    For Each myFile In .SelectedItems
        lbox.ControlFormat.AddItem myFile
    Next
    Range("B4").Formula = _
        "You've selected the following " & _
        lbox.ControlFormat.ListCount & " files:"
    lbox.ControlFormat.ListIndex = 1
    End If
  End With
End Sub
```

The above procedure uses the following statement to clear the list of filters in the File Open dialog box to ensure that only the preset filters are listed:

```
Application.FileDialog(msoFileDialogOpen).Filters.Clear
```

Next, the reference to the `FileDialog` object is stored in the object variable `fd`:

```
Set fd = Application.FileDialog(msoFileDialogOpen)
```

Prior to displaying the File Open dialog box, we set the `AllowMultiSelect` property to `True` so that users can select more than one file. Next, the `Show` method is used to display the File Open dialog box. This method does not open the files selected by the user. When the Open button is clicked, the names of the files are retrieved from the `SelectedItems` collection via the `SelectedItems` property and placed in a list box on a worksheet.

2. Run the `ListSelectedFiles` procedure. When the File Open dialog box appears on the screen, switch to the `VBAExcel2024_ByExample` folder, select a couple of files (hold down the Shift or Ctrl key to choose contiguous or nonadjacent files), and then click Open.

   Note that the selected files will not be opened. The procedure simply loads the names of the files you selected in a list box control that has been added to a worksheet (see Figure 16.4).

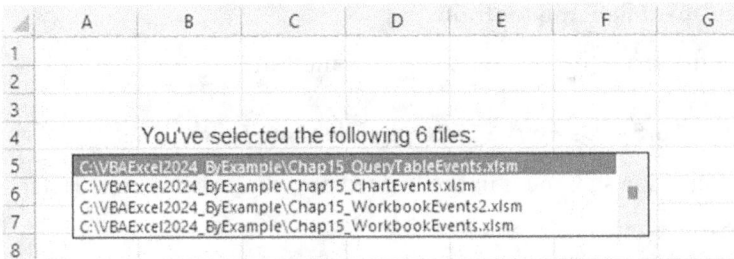

FIGURE 16.4.   User-selected files are loaded into a list box control placed in a worksheet by the ListSelectedFiles procedure.

If you'd like to immediately carry out the `File Open` operation when the user clicks the Open button, you must use the `Execute` method of the `FileDialog` object. The `OpenRightAway` procedure shown below demonstrates how to open the user-selected files right away:

```
Sub OpenRightAway()
  Dim fd As FileDialog
  Dim myFile As Variant

  Set fd = Application.FileDialog(msoFileDialogOpen)
  With fd
    .AllowMultiSelect = True
    If .Show Then
      For Each myFile In .SelectedItems
        .Execute
      Next
    End If
  End With
End Sub
```

THE GETOPENFILENAME AND GETSAVEASFILENAME METHODS

For many years now, Excel has offered its programmers two handy VBA methods for displaying the File Open and File Save As dialog boxes: `GetOpenFilename` and `GetSaveAsFilename`. These methods are available only in Excel and can still be used in Excel if backward compatibility is required. The `GetOpenFilename` method displays the Open dialog box, where you can select the name of a file to open. The `GetSaveAsFilename` method shows the Save As dialog box.

Let's try out these methods from the Immediate window.

Using the GetOpenFilename Method

Let's open a file using the `GetOpenFilename` method.

(⦿) Hands-On 16.5   Using the GetOpenFilename Method

1. Type the following statement in the Immediate window and press Enter:

```
Application.GetOpenFilename
```

The above statement displays the Open dialog box where you can select a file. The `GetOpenFilename` method gets a filename from the user without opening the specified file. This method has four optional arguments. The most often used are the first and third arguments, shown in Table 16.3.

TABLE 16.3.   Arguments of the GetOpenFilename method.

Argument Name	Description
`FileFilter`	This argument determines what appears in the dialog box's Save as type field. For example, the filter `excel files (*.xlsx), .xlsx` displays the following text in the Save As drop-down list of files: excel files. The first part of the filter, `excel files (.xlsx)`, determines the text to be displayed. The second part, `.xlsx`, specifies which files are displayed. The filter parts are separated by a comma.
`title`	This is the title of the dialog box. If omitted, the dialog box will appear with the default title, Open.

2. Click Cancel to close the dialog box opened in Step 1.

3. To see how arguments are used with the `GetOpenFilename` method, enter the following statement in the Immediate window (be sure to enter it on one line) and press Enter:

```
Application.GetOpenFilename("excel macro-enabled files(*.xlsm),
    *.xlsm"),,"Highlight the File"
```

Notice that the Open dialog box now has the text "Highlight the File" in the title bar. Also, the Files of type drop-down list box is filtered to display only the specified file type.

4. Click Cancel to close the dialog box that you opened in Step 3.

The `GetOpenFilename` method returns the name of the selected or specified file. This name can be used later by your VBA procedure to open the file. Let's see how this is done.

5. In the Immediate window, type the following statement and press Enter:

```
yourFile = Application.GetOpenFilename
```

This statement displays the Open dialog box. The file you select while this dialog box is open will be stored in the `yourFile` variable.

6. Select an Excel file and click Open.
Notice that Excel did not open the selected file. All it did was store its name in the `yourFile` variable. Let's check this out.

7. In the Immediate window, type the following statement and press Enter:

```
?yourFile
```

Excel prints the name of the selected file in the Immediate window. Now that you have a filename, you can write a statement to actually open this file (see the next step).

8. In the Immediate window, type the following statement and press Enter:

```
Workbooks.Open Filename:=yourFile
```

Notice that the file you picked is now opened in Excel.

9. Close the file you opened in Step 8.

> **NOTE** *The* `GetOpenFilename` *method returns* `False` *if you cancel the dialog box by pressing the Esc key or clicking Cancel.*

Using the GetSaveAsFilename Method

Having mastered the `GetOpenFilename` method, let's examine a similar method that allows you to save a file. We will continue to work in the Immediate window.

(•) Hands-On 16.6   Using the GetSaveAsFilename Method

1. Open a new workbook and switch to the VBE window.
2. In the Immediate window, type the following statement and press Enter:

```
yourFile = Application.GetSaveAsFilename
```

The above statement displays the Save As dialog box. The suggested filename is automatically entered in the File name box at the bottom of the dialog box.

The `GetSaveAsFilename` method is convenient for obtaining the name of the file the workbook should be saved as. The filename that the user enters in the File name box will be stored in the `yourFile` variable.

3. Type `Test1.xlsx` in the File name box and click Save.
When you click the Save button, the `GetSaveAsFilename` method will store the filename and its path in the `yourFile` variable.

4. Check out the value of the `yourFile` variable in the Immediate window by entering the following statement and pressing Enter:

```
?yourFile
```

To actually save the file, you have to enter a different statement, as demonstrated in the next step.

5. In the Immediate window, type the following statement and press Enter:

```
ActiveWorkbook.SaveAs yourFile
```

Now, the workbook file opened in Step 1 has been saved as `Test1.xlsx`.

6. To close the `Test1.xlsx` file, type the following statement in the Immediate window and press Enter:

```
Workbooks("Test1.xlsx").Close
```

NOTE	*Because the file we are working with does not contain any VBA code, we have saved it with the* `.xlsx` *extension instead of using the macro-enabled file format (*`.xlsm`*).*
	
	When using the `GetSaveAsFilename` *method, you can specify the filename, file filter, and custom title for the dialog box:*
	
	`yourFile = Application.GetSaveAsFilename("Test1.`
	`xlsx", "Excel files(*.xlsx), *.xlsx",,"Name of`
	`your file")`

CREATING CUSTOM FORMS

Although ready to use and convenient, the built-in dialog boxes will not meet all of your VBA application's requirements. Apart from displaying a dialog box on the screen and specifying its initial settings, you can't control the dialog box's appearance. You can't decide which buttons to add, which ones to remove, and which ones to move around. Also, you can't change the size of a built-in dialog box. Therefore, if you're looking to provide a custom interface, you need to create a user form.

A user form is like a custom dialog box. You can add various controls to the form, set properties for these controls, and write VBA procedures that respond to form and control events.

Forms are separate objects that you add to a VBA project by choosing Insert | UserForm from the VBE screen. Forms can be shared across applications. For example, you can reuse the form you designed in Microsoft Excel, Microsoft Word, or another application that uses VBE.

To create a custom form, follow these steps:

1. Press Alt+F11 or select Developer | Visual Basic to switch to the VBE window.
2. Choose Insert | UserForm.

A new folder called `Forms` appears in the Project Explorer window. This folder contains a blank `UserForm` object. The work area automatically displays the form and the toolbox with the necessary tools for adding controls (see Figure 16.5).

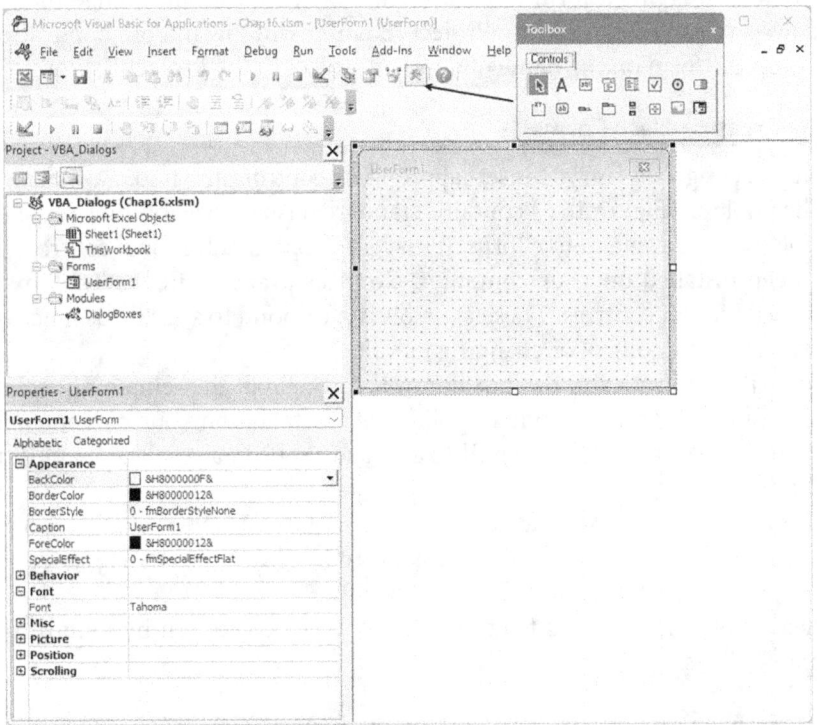

FIGURE 16.5.   A new form can be added to the open VBA project by selecting UserForm from the Insert menu. The Toolbox lists different types of controls you can add to your form. The Properties window displays a number of properties you can set. The properties can be accessed using the Alphabetic or Categorized tabs.

The Properties window (see Figure 16.5) displays seven properties that you can set: Appearance, Behavior, Font, Misc, Picture, Position, and Scrolling.

To list form properties by category, click the Categorized tab in the Properties window. To find out information about a specific property, click the property name and press F1. The online help will be launched with the property description topic.

After adding a new form to your VBA project, you should assign a unique name to it by setting the Name property. You can also set the form's title by using the Caption property.

All VBA applications that use the VBE share features for creating custom forms. You can share your forms by exporting and importing form files or by dragging a form object to another project. To import or export a form file, choose File | Import File or File | Export File.

Before you export a form file, be sure to select it in the Project Explorer window. Before dragging a form to a different VBA application, arrange the VBE windows so that you can see the Project Explorer window in both applications, then drop the form on the name of another project in the Project Explorer.

Tools for Creating User Forms

When you design a form, you insert appropriate controls into it to make it useful.

The Toolbox (Figure 16.5) contains standard Visual Basic buttons for all the controls that you can add to a form. It may also contain additional controls that have been installed on your computer. Controls available in the Toolbox are known as ActiveX controls. These controls can respond to specific user actions, such as clicking a control or changing its value.

You will work with the Toolbox controls throughout the remaining sections of this chapter. If you have other applications installed on your computer that contain ActiveX controls that you'd like to use, you can also place them in the Toolbox.

Let's take a few minutes and add a TreeView ActiveX control to the Toolbox. This control is one of several controls included in Microsoft Windows Common Controls 6.0 located in the MSCOMCT1.OCX file.

Please note that the files for the "Hands-On" projects can be found in the companion files.

⊙ Hands-On 16.7   Adding an ActiveX TreeView Control to the Toolbox

1. Switch to the VBE window and select VBA_Dialogs(Chap16.xlsm) in the Project Explorer window.

2. Choose Insert | UserForm to add a new form to the selected project.
   A default user form named UserForm1 appears with the accompanying Toolbox.

3. Right-click the Controls tab in the Toolbox and choose New Page from the context menu.
   A New Page tab appears in the Toolbox.

4. Right-click the New Page tab in the Toolbox and choose Rename. If this option is not available, make sure you are right-clicking the New Page tab.

5. In the Caption box, type `Extra Controls` as the new name.

6. In the Control Tip text box, type the following description: `Additional ActiveX Controls`.

7. Click OK to return to the Toolbox.

8. Right-click anywhere within the new page area and choose Additional Controls from the context menu. If this option is not available, make sure you are right-clicking the page area in the Toolbox and not the Extra Controls tab itself.

9. When the Additional Controls dialog box appears, click the checkbox next to Microsoft TreeView Control, version 6.0, or any other control from the list of choices (see Figure 16.6).

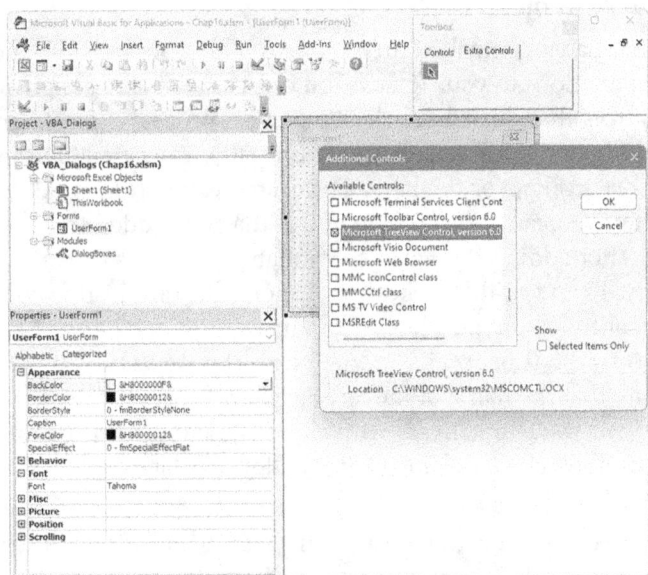

FIGURE 16.6. You can add to the Toolbox additional ActiveX controls that are installed on your computer.

10. Click OK to close the Additional Controls dialog box.
The selected control now appears on the Extra Controls page in the Toolbox (see Figure 16.7).

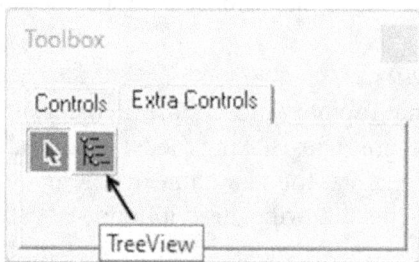

FIGURE 16.7. The TreeView control has been added to the custom Extra Controls tab in the Toolbox.

Default Toolbox Controls

The standard Visual Basic controls are described below. You will use many of these controls in this chapter's Hands-On projects:

- *Select Objects*: Select Objects is the only item in the Toolbox that doesn't draw a control. Use it to resize or move a control that has already been drawn on a form.

- *Label*: The Label control is often used to add captions, titles, headings, and explanations to your forms. You can use a label to assign a title to those controls that don't have the **Caption** property (for example, text boxes, list boxes, scrollbars, and spin buttons). You can define an accelerator (shortcut) key for the label. For example, by pressing Alt and a specified letter, you can activate the control that was added to the form immediately after adding the Label control and setting its **Accelerator** property. To add a title or a keyboard shortcut to an existing control, add a Label control and type a letter from its caption in its **Accelerator** property in the Properties window. Next, choose View | Tab Order, and make sure that the name of the label appears before the name of the control that you want to activate with the assigned keyboard shortcut. You will learn how to use the Tab Order dialog box later in this chapter (see Figure 16.10).

- *TextBox*: Text boxes are the most popular form controls because they can be used to either display or request data from the user. You can enter text, numbers, cell references, or formulas in them. By changing the setting of the **MultiLine** property, you can enter more than one line of text in a text box. The text lines can automatically wrap when you set the **WordWrap** property. If you set the **EnterKeyBehavior** property to **True** when the **MultiLine** property is also set to **True**, you'll be able to start a new line in the text box by pressing Enter.

Another property, EnterFieldBehavior, determines whether the text is selected when the user selects the text field. Setting this property to 0 (fmEnterFieldBehaviorSelectAll) will select the text within the field. Setting this property to 1 (fmEnterFieldBehaviorRecallSelect) will only select the text that the user selected the last time they activated this field.

If you want to limit the number of characters the user can enter in a text box, you can do this by specifying the exact number of characters in the MaxLength property.

- *ComboBox*: The combo box is a control that combines a text box with a list box. This control is often used to save space on the form. When you click the down arrow located to the right of the combo box, the box will drop open to reveal a number of items from which to choose. You may enter a new value if you set the **MatchRequired** property to **False**. The **ListRows** property determines how many items will appear when the user drops down the list. The **Style** property determines the type of combo box. To select an item from the list, use **0** (**fmStyleDrop-DownCombo**). Set the Style property to **2** (**fmStyleDropDownList**) to limit the user's selection to the items available in the combo box.

- *ListBox*: Instead of prompting the user to enter a specific value in a text box, sometimes it's better to present a list of available choices from which to select. The list box reduces the possibility of data entry errors. List box entries can be typed in a worksheet, or they can be loaded directly from a VBA procedure using the **AddItem** method. The **Row-Source** property indicates the source of data displayed in the list box. For example, the reference **$A$1: $B$8** will display in the list box the contents of the specified range of cells.

The list box can display one or more columns when you set the ColumnCount property. Another property, ColumnHeads, can be set to True to display the column titles in the list box.

The user may select more than one item in the list box if the MultiSelect property is set to True.

- ☑ *CheckBox*: Checkboxes are used for turning specific options on and off. Unlike option buttons, you can select one or more checkboxes. If the checkbox is selected, its **Value** property is set to **True**; if the checkbox is not selected, its **Value** property is set to **False**.

- ◉ *OptionButton*: An option button lets you select one of several options. Option buttons usually appear in groups of two or more buttons surrounded by a frame control. Only one option button can be selected. When you select a new option button, the previously selected option button is automatically deselected. To activate or deactivate an option button, set its **Value** property to **True** or **False**. **True** means that the option is activated; **False** indicates that the option is deactivated.

- ☐ *ToggleButton*: A toggle button looks like a command button and works similarly to an option button. When you click a toggle button, the button stays pressed. The next click on the button returns it to the normal (unpressed) state. The pressed toggle button has its **Value** property set to **True**. The unpressed toggle button has its **Value** property set to **False**.

- ⌗ *Frame*: Frames allow you to visually organize and logically group various controls placed on the form. Later in this chapter, you will find an example of the **Info Survey** form that uses four frames. One of them organizes Hardware and Software option buttons into one logical group; the second frame groups the checkboxes related to gender; the third one organizes computer types, and the fourth one groups checkboxes listing various operating systems (see Figure 16.8).

- ⌗ *CommandButton*: A command button carries out a command when it is clicked. In this chapter, you will learn how to execute VBA procedures from command buttons.

- ⌗ *TabStrip*: Although the **TabStrip** and **MultiPage** controls look almost alike, each has a different function. **TabStrip** lets you use the same controls for displaying multiple sets of the same data. Suppose that the form shows students' exams. Each student has to pass an exam in the same subjects. Each subject can be placed on a separate page (tab), and each tab will contain the same controls to collect data, such as the grade received and the date of the exam. When you activate any subject tab, you will see the same controls. Only the data in these controls will change.

- ⌗ *MultiPage*: The **MultiPage** control displays a series of tabs at the top of the form. Each tab acts as a separate page. Using the **MultiPage** control, you can design forms that contain two or more pages. You can place a different set of controls on each form page to make the data more readable. It's much easier to click a form tab than move around in a long form using scrollbars.

By default, each `MultiPage` control appears on your form with two pages. New pages can be added by using the shortcut menu or the `Add` method from within a VBA procedure.

• *Scrollbar*: This control allows you to place horizontal and vertical scrollbars on your form. Although normally used to navigate windows, scrollbars can be used on your form to enter values in a predefined range. The current value of the scrollbar is set or returned by the **Value** property.

The scrollbar's `Max` property lets you set its maximum value. The `Min` property determines the minimum value. The `LargeChange` property determines by what value the `Value` property should change when the user clicks inside the scrollbar. When programming the behavior of the scrollbar, don't forget to set the `SmallChange` property, which determines how the `Value` property changes when you click one of the scroll arrows.

• *SpinButton*: The spin button works similarly to a scrollbar. You can click an arrow to increment or decrement a value. The spin button is often used with a text box. The user can then type the exact value in the text box or select a value by using the arrows. The technique of using the spin button with a text box is discussed later in this chapter.

• *Image*: The image control lets you display a graphical image on a form. This control supports the following file formats: ***.bmp, *.cur, *gif, *.ico, *.jpg**, and ***.wmf**. Like other controls in the Toolbox, the image control has a number of properties that you can set. For example, you can control the appearance of the picture with the **PictureSizeMode** property. This property has the following three settings:

▪ 0 (fmPictureSizeModeClip) crops the part of a picture that does not fit within the picture frame.

▪ 1 (fmPictureSizeModeStretch) stretches the picture horizontally or vertically until it fills the entire frame area.

▪ 3 (fmPictureSizeModeZoom) enlarges the picture without distorting its proportions.

- 🖼 *RefEdit*: This control is specific to forms created in Microsoft Excel, as it allows you to select a cell or a range of cells in a worksheet and pass it to your VBA procedure. You can see how this control works by taking a look at some of the built-in dialog boxes in Excel. For example, the Consolidate dialog box accessed from the Data tab's Data Tools group has a `RefEdit` control labeled Reference that lets you specify the range of data that you want to consolidate. To temporarily hide the dialog box while selecting a range of cells, click the button on the right of the RefEdit control.

Placing Controls on a Form

When you create a custom form, you place various controls on the form using the Toolbox. The type of control you select depends on the type of data the control will have to store and the functionality of your form. The Toolbox can be moved around on the screen. You can also change its size or close it when all controls are already on the form and all you want to do is work with their properties. The Toolbox display can be toggled on and off by choosing View | Toolbox.

To add a new control to a form, first, click the control image in the Toolbox and then click the form or draw a frame. Clicking on a form (without drawing a frame) will place a control in its default size. The standard settings of each control can be looked up in the Properties window. For example, the standard text box size is 18 x 72 points (see the Height and Width properties of the text box). After placing a control on a form, the Select Object button (represented by the arrow) becomes the active control in the Toolbox. When you double-click a control in the Toolbox, you can draw as many instances of that control as you want. For example, to quickly place three text boxes on your form, double-click the text box control in the Toolbox and then click three times on the form.

Setting Grid Options

When you drag a control on a form, Visual Basic adjusts the control so that it aligns with the form's grid. You can set the grid to your liking by using the Options dialog box.

To access grid options, perform the following:

1. Choose Tools | Options.
2. Click the General tab in the Options dialog box.

The Form Grid Settings area lets you turn off the grid, adjust the grid size, and decide whether you want the controls aligned to the grid.

SAMPLE APPLICATION: INFO SURVEY

As you already know, the best way to understand a complex feature is to apply it in a real-life project. In this section, you will create a custom form for a coworker who requested that you streamline the tedious process of entering survey data into a worksheet. While working with this form (see Figure 16.8), you will experiment with many controls and their properties. Also, you will learn how to transfer data from your custom form to a worksheet.

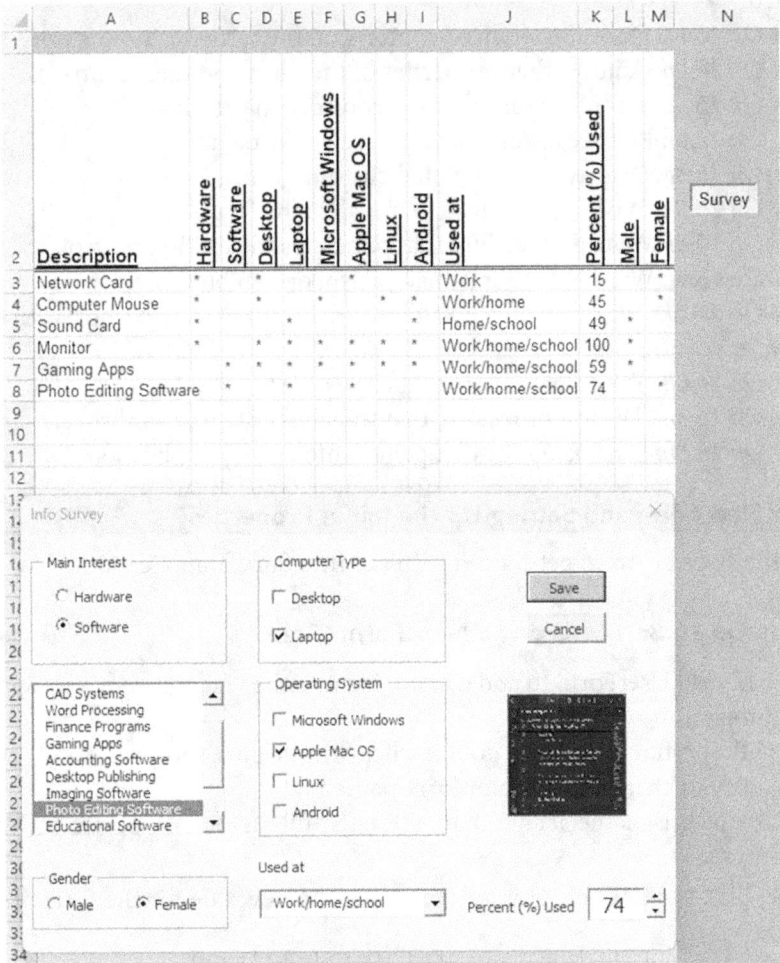

Description	Hardware	Software	Desktop	Laptop	Microsoft Windows	Apple Mac OS	Linux	Android	Used at	Percent (%) Used	Male	Female
Network Card	*		*			*			Work	15		*
Computer Mouse	*		*	*			*	*	Work/home	45		
Sound Card	*			*					Home/school	49	*	
Monitor	*		*	*	*	*	*	*	Work/home/school	100	*	
Gaming Apps		*	*	*	*			*	Work/home/school	59		
Photo Editing Software	*		*			*			Work/home/school	74		*

FIGURE 16.8.   The Info Survey custom form is used to enter data by making appropriate selections from various controls placed on the form. Each time the form is used, the user's selections are written to the worksheet.

Setting Up the Custom Form

Before we get into programming, you need to perform several tasks, which are broken down into the different Hands-On exercises listed below:

1. *Hands-On 16.8a*: Insert a new form into your VBA project and set up this form's initial properties, such as the Name and Caption properties that will allow you to identify the form.
2. *Hands-On 16.8b*: Adjust the size of the form so that all controls required by the application can be easily placed on the form and the form does not look crowded.
3. *Hands-On 16.8c*: Place the required controls on the form.
4. *Hands-On 16.8d*: Adjust other properties of the form and their controls.
5. *Hands-On 16.8e*: Set the tab order of the controls on the form.
6. *Hands-On 16.8f*: Prepare a worksheet to receive the data.
7. *Hands-On 16.8g*: Display the completed custom form.
8. *Hands-On 16.8h*: Write a procedure to initialize the form.
9. *Hands-On 16.8i*: Write a procedure that populates the list box control.
10. *Hands-On 16.8j*: Write procedures that control option buttons.
11. *Hands-On 16.8k*: Write a procedure that synchronizes the text box with the spin button control.
12. *Hands-On 16.8l*: Write a procedure that closes the form.
13. *Hands-On 16.8m*: Write a procedure that transfers data to a worksheet.
14. *Hands-On 16.8n*: Start using the completed Info Survey application.

Inserting a New Form and Setting Up the Initial Properties

Follow the steps below to get started with the Info Survey application.

(◉) **Hands-On 16.8a   Inserting a New Form (Step 1)**

1. Choose Insert | UserForm to add a blank form to the VBA_Dialogs (Chap16. xlsm) project.
2. In the Properties window, double-click the Name property and type InfoSurvey to change the default form name.
   We will use this name later on to refer to this UserForm object in VBA procedures.
3. Double-click the Caption property and type the new title for the form: Info Survey.
   The name Info Survey will appear in your form's title bar.
4. Double-click the BackColor property, click the Palette tab, and select a color for the form background.

Resizing the Form

When a default form is too large or too small to fit all the controls properly, you can change its size by using the mouse or by setting the form properties in the Properties window.

To resize the form with the mouse, click on an empty part of the form, and notice that several selection handles appear around the form. Place the mouse pointer over any selection handle located in the middle of a side, drag the handle to the position you want, and then release the mouse button. You can also place the mouse pointer over the selection handle located in the undocked corner and drag the handle to the position you want. Release the mouse button.

> **NOTE** *Each new form has a default size of 180 x 240. The form's dimensions are in points. One point equals 1/72 inch.*

To resize the form using the Properties window, you will need to enter new values for the form's Height and Width. Click on the form's title bar. In the Properties window, double-click the Height property and enter a new value. Do the same for the Width property if you need to adjust the form's width as well. To avoid extra work, figure out how much space you really need and resize the form before adding the desired controls.

After setting the initial properties for our custom Info Survey form, let's adjust the size of the form so that all the controls that we need to place on this form will fit nicely.

(◉) Hands-On 16.8b   Adjusting the Size of the Form (Step 2)

1. Click the form's title bar (where the words "Info Survey" appear).
2. In the Properties window, double-click the Height property and enter the value 276.
3. In the Properties window, double-click the Width property and enter the value 408.

Adding Buttons, Checkboxes, and Other Controls to a Form

Now, we are ready to proceed with placing the required controls on the Info Survey form. We will model this form after Figure 16.8.

The UserForm toolbar (see Figure 16.9) contains a number of useful shortcuts for working with forms, such as making controls the same size, centering a control horizontally or vertically, aligning control edges, and grouping and ungrouping controls. To display this toolbar, choose View | Toolbars | UserForm.

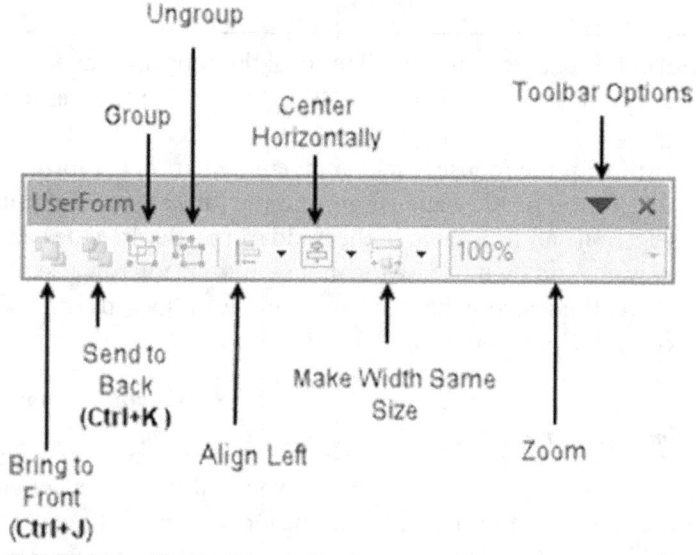

FIGURE 16.9. The UserForm toolbar.

(◉) Hands-On 16.8c Adding Buttons, Checkboxes, and Other Controls to a Form (Step 3)

1. Click the Frame control in the Toolbox.
   The mouse pointer will change to a cross accompanied by the symbol of the selected control.
2. Point to the upper left-hand side of the form, then click and drag the mouse to draw a small rectangle.
   When you release the mouse button, you will see a small rectangle titled Frame1. When the frame is selected, various selection handles will appear on its sides, and the Properties window's title bar will display Properties-Frame1.
3. In the Properties window, double-click the Caption property and replace the selected default caption, Frame1, with `Main Interest`.
4. Click the OptionButton control in the Toolbox. Next, click inside the Main Interest frame that you've just added to your form. Click and drag the mouse to the right until you see a rectangle with the default label OptionButton1.
5. In the Properties window, change the option button's Caption property to `Hardware`.
6. Use the method presented in Step 4 to add another option button to the Main Interest frame. Change the Caption property of this option button to `Software`.

The option buttons are used whenever the user must select one choice from a group of mutually exclusive choices. If the user can select more than one choice, checkboxes are used.

7. Click the ListBox control in the Toolbox.

   The mouse pointer will change to a cross accompanied by the symbol of the selected control.

8. Click below the Main Interest frame and drag the mouse down and to the right to draw a list box.

   When you release the mouse button, you will see a white rectangle. Figure 16.8 shows the list box populated with hardware entries. This list will be added within a VBA procedure when we get into form programming.

9. Insert a frame below the list box. Change the frame's Caption property to Gender. Add two option buttons inside this frame and change the first button's Caption property to Male and the second one to Female (see Figure 16.8).

10. Click the Frame control in the Toolbox and draw a rectangle to the right of the frame labeled Main Interest.

11. Change the Caption property of the new frame to Computer Type.

12. Click the CheckBox control in the Toolbox and click inside the empty frame that you have just added. The CheckBox1 control should appear inside the frame.

13. Change the Caption property of the CheckBox1 control to Desktop.

14. Place one more checkbox inside the frame labeled Computer Type. Use the Caption property to assign the following title to this checkbox: Laptop. The result should match Figure 16.8.

    Unlike option buttons, which are mutually exclusive, checkboxes allow the user to activate one or more options simultaneously. The checkbox can be checked, unchecked, or unavailable at any time. An unavailable checkbox has its label grayed out and is therefore inactive (cannot be selected). A checked box has an x in front of its caption. A checkbox that has the focus is indicated by a dotted line around the caption.

    Use option buttons when only one option can be selected at a given time. Use checkboxes to have the user select any number of options that apply.

15. Create the frame for Operating System with four checkboxes, as depicted in Figure 16.8.

16. Click the Label control in the Toolbox.

17. Click the empty space below the frame labeled Operating System. The Label1 control should appear.

18. Change the Caption property of Label1 to Used at.

19. Click the ComboBox control in the Toolbox.

20. Click the empty space below the Used at label and drag the mouse to draw a rectangle, then release the mouse button.

The combo box displays a list of available choices only after you click the down arrow placed at the right of this control. The combo box is sometimes referred to as a drop-down list and is used to save valuable space on the screen. Although the user can only see one element of the list at a given time, the current selection can be quickly changed by clicking on the arrow button.

21. Click the Label control in the Toolbox.

22. Click on the empty part of the form just below the Used at combo box. A label control will appear. Change the Caption property for this label to `Percent (%) Used`.

23. Click the TextBox control in the Toolbox.

24. Click to the right of the Percent (%) Used label control to place a default size text box.

25. Click the SpinButton control in the Toolbox, and then click to the right side of the text box control. A default-sized spin button will appear.

The spin button has two arrows that are used to increment or decrement a value in a given range. The maximum value is determined by the setting of the Max property, and the minimum value is set with the Min property.

The spin button has the same properties as the scrollbar, with two differences. The spin button does not have a scroll box, and it lacks the LargeChange property. A text box is usually placed next to the spin button. This allows the user to enter a value directly into the text box or use the spin buttons to determine the value. If the spin button must work with the text box, your VBA procedure must ensure that the value of the text box and the spin button are synchronized. In this example, you will use the spin button to indicate the percentage of interest that the user has in the selected hardware or software product.

26. Double-click the CommandButton control in the Toolbox. By doing so, you can create more than one control using the selected tool.

27. Click in the top right-hand corner of the form. This will cause CommandButton1 to appear.

28. Click below CommandButton1. CommandButton2 will appear.

29. Change the Caption property of CommandButton1 to `Save` and CommandButton2 to `Cancel`.

Most custom forms have two command buttons that enable the user to accept the data entered on the form or dismiss the form. In this example, the Save button will transfer the data entered on the form to a worksheet. The user will be able to click the Cancel button to exit the form. To make the buttons respond to user actions, you will write appropriate VBA procedures later in this chapter.

30. Click the Image control in the Toolbox. Click the mouse below the Cancel button and drag the mouse to draw a rectangle. Release the mouse button. The result is shown in Figure 16.8.

The form will display a different picture depending on whether the Hardware or Software option button is selected.

31. Spend a few minutes now copying the `hardware.jpg` and `software. jpg` images from the companion files to your `C:\VBAExcel2024_ByExample` folder.

32. While the Image control is selected, select the Picture property in the Properties window and click the button with three dots next to this property. This will invoke the Load Picture dialog. Select the `hardware.jpg` file and click Open. The Properties window should now list (Bitmap) for the Picture property setting and the selected picture should appear in the image control. Change the PictureSizeMode property to `1 -fmPictureSizeModeStretch`.

33. Click the title bar or click on any empty area of the form to select it. Press F5 or choose Run | Run Sub/UserForm to display the form as the user will see it.

Visual Basic switches to the active sheet in the Microsoft Excel window and displays the custom form you designed.

If you forget to select the form, the Macro dialog box will appear. Close the dialog box and try again.

34. Click the Close button (x) in the top right-hand corner of the form to close the form and return to the VBE window.

Note that the Save and Cancel buttons placed on the form aren't functional yet. They require VBA procedures to make them work. After you've added controls to the form, use the mouse or the Format menu commands to adjust the alignment and spacing of the controls.

Copying and Moving Form Controls

The Info Survey form design is complete. Now, you should feel comfortable designing any form you want. When working with controls, it is worthwhile to learn some shortcuts. For example, here's how you can quickly copy and move controls:

- To copy a control, click the Select Objects tool in the Toolbox and select the control (a selected control will have handles at its sides), hold down the Ctrl key, position the mouse pointer inside the control, and press the left mouse button. Drag the pointer to the position you want and release the mouse button. Then, change the control's Caption property.

- To select an entire group of controls, click the Select Objects tool in the Toolbox and start drawing a rectangle around the group of controls that you want to move together. When you release the mouse button, all the controls will be selected. (You can also select more than one control by holding down the Ctrl key while clicking each of the controls you want to select.)

- To move the selected group of controls to another position on the form, click within the selected area and drag the mouse to the desired position.

Changing Control Names and Properties

After you have placed controls on your form (but before you begin to write procedures to control the form), you should assign your own names to the controls. Although Visual Basic automatically assigns a default name to each control (`OptionButton1`, `OptionButton2`, and so on), these names are difficult to distinguish in a procedure that may reference objects of the same class that have almost identical names. Assigning meaningful names to the controls placed on your form makes VBA procedures referencing these controls much more readable.

Before you change the Name property, make sure that the title bar of the Properties window displays the correct type of control.

For example, to assign a new name to the frame control, click the frame control on the form. When the Properties window displays Properties-Frame1, double-click the Name property and type the new name in place of the highlighted default name. Do not confuse the name of the control with the control's title (caption). For example, on the Info Survey form, the default name of the frame control is Frame1, but the title of this control is Main Interest. The control's title can be changed by setting the Caption property. While the control's caption allows the user to identify the purpose of the control and may suggest the type of data expected, it is the name of the control that will be used in the code of your VBA procedures to make things happen.

Let's go back to our form to make adjustments to the controls' properties.

⊙ Hands-On 16.8d   Naming Form Controls (Step 4)

1. Assign names to the controls placed on the Info Survey form as shown below. To assign a new name to a control, perform these steps:
   a. Click the appropriate control on the form.
   b. Double-click the Name property in the Properties window.
   c. Type the corresponding name in the Name property column:

Option button Hardware	`optHard`
Option button Software	`optSoft`
List box control under Main Interest	`lboxSystems`
Option button Male	`optMale`
Option button Female	`optFemale`
Checkbox Desktop	`chkDesktop`
Checkbox Laptop	`chkLaptop`
Checkbox Microsoft Window	`chkWindows`
Checkbox Apple Mac OS	`chkMac`
Checkbox Linux	`chkLinux`
Checkbox Android	`chkAndroid`
Combo box below Used at	`cboxWhereUsed`
Text box next to Percent (%) Used	`txtPercent`
Spin button next to txtPercent	`spPercent`
First command button	`cmdSave`
Second command button	`cmdCancel`
Image	`picImage`

The controls that you placed on the Info Survey form are objects. Each of these objects has its own properties and methods. You've just changed the Name property for all the objects that will be referenced later from within VBA procedures. The control properties can be set during the design phase of your custom form or at runtime (that is, when your VBA procedure is executed).

Let's set other properties for selected controls.

2. Change the object properties as shown below.
   To set a property, click a control on the form, locate the desired property in the Properties window, and type the new value in the space to the right of the property name.

For example, to set the ControlTipText property of the lboxSystems control, click the list box control on the Info Survey form and locate the ControlTipText property in the Properties window. In the right-hand column of the Properties window, type the text you want to display when the user positions the mouse pointer over the list box control—in this case, `Select only one item.`

Object Name	Property	Change to:
lboxSystems	ControlTipText	Select only one item.
spPercent	Max	100
spPercent	Min	0
cmdSave	Accelerator	O
cmdCancel	Accelerator	C
picImage	PictureSizeMode	0-fmPictureSizeModeClip

The Accelerator property indicates which letter in the object name can be used to activate the control with the keyboard shortcut combination. The specified letter will appear underlined in the object's caption (title). For example, after displaying the form, you will be able to quickly select OK by pressing Alt+O.

The remaining properties of the Info Survey form objects will be set directly from VBA procedures.

Setting the Tab Order

The user can move around a form by using the mouse or the Tab key. Because many users prefer to navigate through the form using the keyboard, it is important to determine the order in which each control on the form is activated. Follow these steps to set the tab order in the Info Survey form.

Hands-On 16.8e   Setting the Tab Order in a Form (Step 5)

1. In the Forms folder in the Project Explorer window, double-click the Info Survey form.
2. Choose View | Tab Order.
   The Tab Order dialog box appears. This box displays the names of all the controls on the Info Survey form in the order that they were added. The right side of the dialog box has buttons that allow you to move the selected control up or down. To move a control, click its name and click the Move Up or Move Down button until the control appears in the position you want.
3. Rearrange the order of controls for the Info Survey form as shown in Figure 16.10.

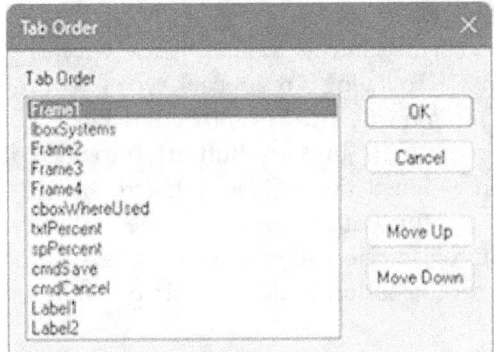

FIGURE 16.10. The Tab Order dialog box lets you organize the controls on the form in the order in which you would like to access them.

4. Click OK to close the Tab Order dialog box.
5. Activate the Info Survey user form and tab through the controls. Press the Tab key to move forward. Press Shift+Tab to move backward.
6. Close the Info Survey form.
   If you'd like to change the order in which the controls are activated, reopen the Tab Order dialog box and make the appropriate changes.

Preparing a Worksheet to Store Custom Form Data

When you create a custom form for your users, you should provide an easy interface for launching your form and reporting the data. Follow the steps below to get your worksheet ready.

Hands-On 16.8f   Preparing a Worksheet to Store Custom Form Data (Step 6)

1. Activate the Microsoft Excel window and add a new sheet to the `Chap16.xlsm` workbook. Rename it `Info Survey`.
2. In the Info Survey worksheet, enter the column headings as shown in Figure 16.8 earlier in this chapter.
3. Select row 1 through column N and change the background of all cells to your favorite color (use the Fill Color button in the Font section of the Home tab). You may also want to change the background color of column N, as shown in Figure 16.8.
   The easiest way to launch a custom form from a worksheet is by clicking a button. The remaining steps walk you through the process of adding the Survey button to your Info Survey worksheet.

4. Choose Developer | Controls | Insert.
5. Click the Button control on the Form Controls toolbar. Click in cell N2 to place a button. When the Assign Macro dialog box appears, type DoSurvey in the Macro name box and click OK. You will write this procedure later.
6. When you return to the worksheet, the button (Button1, provided it is your first button) to which you assigned the DoSurvey macro should still be selected. Type the new name for this button: Survey. If the button is not selected, use the right mouse button to select it. Choose Edit Text from the shortcut menu and type Survey for the button's new name. To exit Edit mode, click outside the button.
7. Save the changes you've made to Chap16.xlsm.

Displaying a Custom Form

Each UserForm object has a Show method that allows you to display the form to the user. In the example below, you will prepare the DoSurvey procedure. Recall that, in the previous section, you assigned the DoSurvey procedure to the Survey button placed in the Info Survey worksheet.

(◉) Hands-On 16.8g   Displaying a Custom Form (Step 7)

1. In the VBE window, select the VBA_Dialogs (Chap16.xlsm) VBA project in the Project Explorer window and choose Insert | Module.
2. In the Properties window, change the new module's name to ShowSurvey.
3. Enter the following procedure to display the custom form:

```
Sub DoSurvey()
   InfoSurvey.Show
End Sub
```

Notice that the Show method is preceded by the name of the form object as it appears in the Forms folder (InfoSurvey).

4. Save the changes made to the Chap16.xlsm workbook.
5. Switch to the Microsoft Excel window and click the Survey button. The Info Survey form appears.

NOTE	*If an error message appears after you click the Survey button, you have not assigned the required macro to this button as instructed in Step 6 (Hands-On 16.8f). To correct this problem, click OK on the message, right-click the Survey button, and choose Assign Macro from the shortcut menu. Click the DoSurvey macro name in the list box and click OK. Now, click the Survey button to display the form.*

6. Close the Info Survey form by clicking the Close button (x) in the top right-hand corner of the form.

Before we can utilize this form, we need to program some events. We've already covered events in Chapter 15. In this chapter, we will only look at the events related to creating and using `UserForm`.

Understanding Form and Control Events

In addition to having properties and methods, each form and control has a pre-defined set of events. As you recall from the previous chapter, an *event* is some type of action, such as clicking a mouse button, pressing a key, selecting an item from a list, or changing a list of items available in a list box. Events can be triggered by the user or the system.

To specify how a form or control should respond to events, you write event procedures. When you design a custom form, you should anticipate and program events that can occur at runtime (while the form is being used). A form itself can respond to more than 20 separate events, including `Click`, `DblClick`, `Activate`, `Initialize`, and `Resize`. The most popular event is the `Click` event. Every time a command button is clicked, it triggers the appropriate event procedure to respond to the `Click` event for that button. Similarly, the `Change` event for a text box is triggered whenever the content of the text box is modified. This event can be used to perform real-time data validation or calculations, ensuring that the data entered by the user meets specific criteria. Moreover, Excel VBA also offers events that are tied to the lifecycle of the `UserForm` itself. For example, the `Initialize` event occurs when a `UserForm` is created and is typically used to set initial values for controls or load data from an external source. The `Terminate` event is triggered when the `UserForm` is closed and can be used to clean up resources or save state information.

Each form you create contains a form module for storing VBA event procedures. To access the form module to write an event procedure or to find out the events recognized by a specific control, you can:

- Double-click a control.
- Right-click the control and choose View Code from the shortcut menu.
- Click the View Code button in the Project Explorer window.
- Double-click any unused area of the `UserForm`.

When you execute any of the above actions, a Code window will open for the form, as shown in Figure 16.11. The procedure box lists all the procedures that are available for the object that is currently selected in the Object box. In Figure 16.11, the `Initialize` event procedure for the `UserForm` is shown.

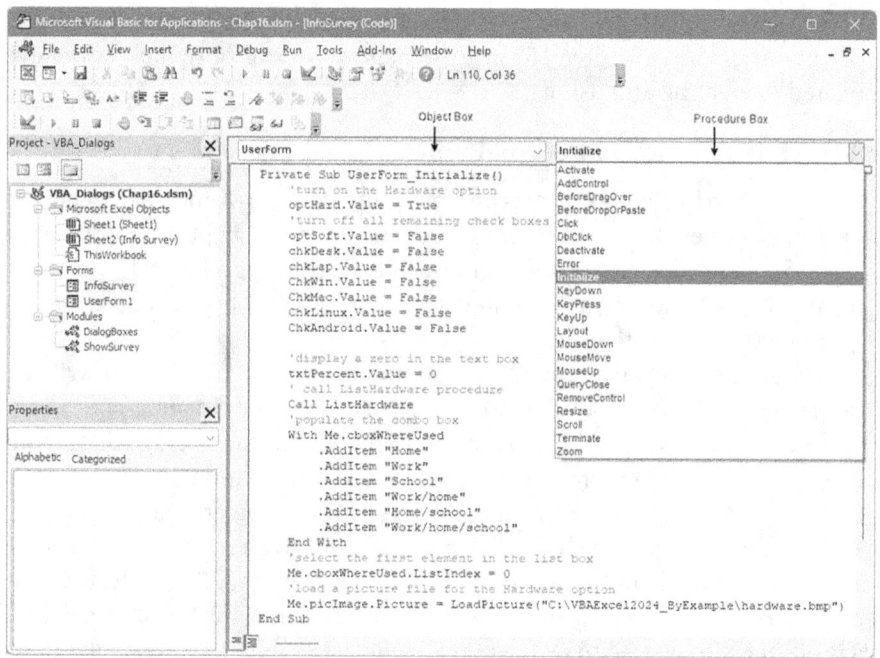

FIGURE 16.11. The Procedure box above the Code window lists the available event procedures for the UserForm.

If you need to write an event procedure for a control placed on the form, locate the name of your control in the Object box and the Procedure box will list all available events for the selected control.

Writing VBA Procedures to Respond to Form and Control Events

Before you can accomplish specific tasks with a custom form, you must write VBA procedures to handle these tasks. As mentioned earlier, each form created in the VBE has a module for storing procedures used by that form. Before displaying a custom form, you may want to set initial values for various form controls. This is usually accomplished by writing an `Initialize` event procedure. The `Initialize` event occurs when the `UserForm` is loaded but before it's shown on the screen.

Writing a Procedure to Initialize the Form

Suppose that you want the Info Survey form to appear with the following initial settings:

- The Hardware button is selected in the Main Interest frame.
- The list box below contains the items that correspond to the selected Hardware option button.
- None of the Computer Type checkboxes are selected.
- None of the Operating System checkboxes are selected.
- The combo box below the Used at label displays the first available item, and the user cannot add a new item to the combo box.
- The text box next to the spin button displays the initial value of zero (0).
- The image control displays a related picture when the Hardware option button is selected.

Hands-On 16.8h   Writing a Procedure to Initialize the Form (Step 8)

1. In the Project Explorer window, double-click the InfoSurvey form.
2. Double-click the background of the form to open the Code window for the active form.

   When you double-click the form or control, the Code window opens to the form or control's `Click` event. In the procedure definition, Visual Basic automatically adds the keyword `Private` before the `Sub` keyword. Private procedures can be called only from the current form module. In other words, a procedure that is in another module of the current project cannot call this particular (`Private`) procedure.
3. Click the down arrow in the Procedure box on the right and select the Initialize event. Visual Basic displays the `InfoSurveyUserForm_Initialize` procedure in the Code window:

```
Private Sub UserForm_Initialize()

End Sub
```

4. Type the form's initial settings between the `Private Sub` and `EndSub` keywords. The complete `UserForm_Initialize` procedure is shown below:

```
Private Sub UserForm_Initialize()
    'turn on the Hardware option
    optHard.Value = True
```

```
'turn off all remaining check boxes
optSoft.Value = False
chkDesk.Value = False
chkLap.Value = False
ChkWin.Value = False
ChkMac.Value = False
ChkLinux.Value = False
ChkAndroid.Value = False

'display a zero in the text box
txtPercent.Value = 0
' call ListHardware procedure
Call ListHardware
'populate the combo box
With Me.cboxWhereUsed
    .AddItem "Home"
    .AddItem "Work"
    .AddItem "School"
    .AddItem "Work/home"
    .AddItem "Home/school"
    .AddItem "Work/home/school"
End With
'select the first element in the list box
Me.cboxWhereUsed.ListIndex = 0
'load a picture file for the Hardware option
Me.picImage.Picture = LoadPicture _
        ("C:\VBAExcel2024_ByExample\hardware.jpg")
End Sub
```

To simplify the event procedure code, you can use the `Me` keyword instead of the actual form name. For example, instead of using the statement,

```
InfoSurvey.cboxWhereUsed.ListIndex = 0
```

you can use the following statement:

```
Me.cboxWhereUsed.ListIndex = 0
```

This technique is especially useful when the form name is long. Notice also that the first element of the list box has the index number zero (`0`). Therefore, if you'd like to select the second item in the list, you must set the `ListIndex` property to `1`.

The `UserForm_Initialize` procedure calls the `ListHardware` procedure to populate its list box control with the hardware items. The code of this procedure is shown in Step 5 below.

Notice that the `UserForm_Initialize` procedure ends with loading a picture into the image control. Make sure that the specified graphics file can be located in the indicated folder. If you don't have this file, enter the complete path of a valid picture file that you want to display.

5. Double-click the ShowSurvey module in the Project Explorer window and enter the `ListHardware` procedure in the Code window, as shown below:

```
Sub ListHardware()
    With InfoSurvey.lboxSystems
        .AddItem "Optical Drive"
        .AddItem "Printer"
        .AddItem "Network Card"
        .AddItem "Sound Card"
        .AddItem "Graphics Card"
        .AddItem "Monitor"
        .AddItem "Computer Mouse"
        .AddItem "Computer Keyboard"
        .AddItem "External Drive"
        .AddItem "Scanner"
    End With
End Sub
```

Now that you have the `UserForm_Initialize` and `ListHardware` procedures ready, you should run the form to see how it displays with the initial settings.

6. Launch the form by clicking the Survey button in the Info Survey worksheet.

When the form is displayed, the user can select appropriate options or dismiss the form. If the user clicks the Software option button, the list box below should display different items. At the same time, the image control should load a different picture. The next section explains how you can program these events.

Writing a Procedure to Populate the List Box Control

In the preceding section, you prepared the ListHardware procedure to populate the lboxSystems list box with the hardware items. You can use the same method to load the software items into the list box.

(◉) **Hands-On 16.8i Populating the List Box Control (Step 9)**

Activate the `ShowSurvey` module and enter the code of the `ListSoftware` procedure, as shown below:

```
Sub ListSoftware()
    With InfoSurvey.lboxSystems
        .AddItem "Spreadsheets"
```

```
      .AddItem "Databases"
      .AddItem "CAD Systems"
      .AddItem "Word Processing"
      .AddItem "Finance Programs"
      .AddItem "Gaming Apps"
      .AddItem "Accounting Software"
      .AddItem "Desktop Publishing"
      .AddItem "Imaging Software"
      .AddItem "Photo Editing Software"
      .AddItem "Educational Software"
      .AddItem "Entertainment Software"
      .AddItem "Health and Fitness Applications"
      .AddItem "Antivirus Software"
    End With
  End Sub
```

Writing a Procedure to Control Option Buttons

When you click the Software button in the Info Survey form, the hardware items from the list box should be replaced with the software items, and vice versa. Let's write procedures that will control the Hardware and Software buttons located in the Main Interest frame.

⊙ Hands-On 16.8j   Controlling Option Buttons (Step 10)

1. Activate the `InfoSurvey` form and double-click the Software option button located in the Main Interest frame.
2. When the Code window appears with the `optSoft_Click` procedure skeleton, highlight the code and press Delete.
3. Click the down arrow in the Procedure box and select the Change event procedure.
4. Visual Basic will automatically enter the beginning and end of the `optSoft_Change` procedure for you.
5. Enter the code of the `optSoft_Change` procedure, as shown below:

```
Private Sub optSoft_Change()
  Me.lboxSystems.Clear
  Call ListSoftware
  Me.lboxSystems.ListIndex = 0
  Me.picImage.Picture = _
    LoadPicture("C:\VBAExcel2024_ByExample\software.jpg")
End Sub
```

The `optSoft_Change` procedure begins with a statement that uses the `Clear` method to remove the current list of items from the lboxSystems list box. The next statement calls for the ListSoftware procedure to populate the list box with software items. In other words, when the user clicks the Software button, the procedure removes the hardware items from the list box and adds software items. If you don't clear the list box prior to adding new items, the new items will be appended to the current list. The statement Me.lboxSystems.ListIndex = 0 selects the first item in the list.

The final statement in this procedure loads a picture file to the Image control. Be sure to replace the reference to this file with the complete path to a valid picture file on your computer. Because the user may want to reselect the Hardware button after selecting the Software button, you must create a similar Change event procedure for the optHard option button.

6. Enter the following optHard_Change procedure, just below the optSoft_Change procedure:

```
Private Sub optHard_Change()
    Me.lboxSystems.Clear
    Call ListHardware
    Me.lboxSystems.ListIndex = 0
    Me.picImage.Picture = _
    LoadPicture("C:\VBAExcel2024_ByExample\hardware.jpg")
End Sub
```

7. Launch the form by clicking the Survey button in the Info Survey worksheet and check the results.
8. Close the form by clicking the Close button in the form's upper-right corner.

Writing Procedures to Synchronize the Text Box with the Spin Button

The Info Survey form has a text box in front of the spin button control. To indicate the percentage of time that the selected Hardware or Software item is used, the user can type a value in a text box or use the spin button. The initial value of the text box is set to zero (0). Suppose the user entered 10 in the text box and now wants to increase this value to 15 by using the spin button. To enable this action, the text box and the spin button have to be synchronized. Each of these objects requires a separate Change event procedure.

⊙ **Hands-On 16.8k   Synchronizing the Text Box with the Spin Button (Step 11)**

1. Right-click the spin button and choose View Code from the shortcut menu.

2. Enter the `spPercent_Change` procedure as shown below:

```
Private Sub spPercent_Change()
   txtPercent.Value = spPercent.Value
End Sub
```

Using the spin buttons will cause the text box value to go up or down.

3. Working in the same Code window, enter the following `txtPercent_Change` procedure:

```
Private Sub txtPercent_Change()
   Dim entry As String

   On Error Resume Next

   entry = Me.txtPercent.Value
     If entry > 100 Then
         entry = 0
         Me.txtPercent.Value = entry
     End If
   spPercent.Value = txtPercent.Value
End Sub
```

The `txtPercent_Change` procedure ensures that only values from 0 to 100 can be entered into the text box. The procedure uses the `On Error Resume Next` statement to ignore data entry errors. If the user enters a non-numeric value in the text box (or a number greater than 100), Visual Basic will reset the text box value to zero (0). Each time a spin button is pressed, a text box value is incremented or decremented by one.

Writing a Procedure That Closes the User Form

After displaying the form, you may want to cancel it by pressing the Esc key or clicking the Cancel button. To remove the form from the screen, let's prepare a simple procedure that uses the `Hide` method.

⊙ **Hands-On 16.8l   Writing a Procedure That Closes the Form (Step 12)**

Double-click the Cancel button and enter the `cmdCancel_Click` procedure as shown below:

```
Private Sub cmdCancel_Click()
   Me.Hide
End Sub
```

The `Hide` method hides the object but does not remove it from memory. This way, your VBA procedure can use the form's objects and properties behind the scenes when the form isn't visible to the user. Use the `Unload` method to remove the form both from the screen and from memory resources:

```
Unload Me
```

When the form is unloaded, all memory associated with it is reclaimed. The user can't interact with the form, and the form's objects can't be accessed by the VBA procedure until the form is placed in memory again by using the `Load` statement.

Transferring Form Data to the Worksheet

After clicking the Save button, the form's selections should be written to the worksheet. You can quit using the form at any time by clicking the Cancel button. Let's write a procedure that will copy the form's data to the worksheet when the Save button is clicked.

Hands-On 16.1m   Transferring Form Data to the Worksheet (Step 13)

1. In the VBE window, double-click the InfoSurvey form in the Project Explorer.
2. Double-click the Save button on the Info Survey form and enter the `cmdSave_Click` procedure shown below:

```
Private Sub cmdSave_Click()
    Dim r As Integer
    Me.Hide

    r = Application.CountA(Range("A:A"))
    Range("A1").Offset(r + 1, 0) = Me.lboxSystems.Value

    If Me.optHard.Value = True Then
        Range("A1").Offset(r + 1, 1) = "*"
    End If
    If Me.optSoft.Value = True Then
        Range("A1").Offset(r + 1, 2) = "*"
    End If
    If Me.chkDesk.Value = True Then
        Range("A1").Offset(r + 1, 3) = "*"
    End If
    If Me.chkLap.Value = True Then
        Range("A1").Offset(r + 1, 4) = "*"
```

```
    End If
    If Me.ChkWin.Value = True Then
        Range("A1").Offset(r + 1, 5) = "*"
    End If
        If Me.ChkMac.Value = True Then
        Range("A1").Offset(r + 1, 6) = "*"
    End If
    If Me.ChkLinux.Value = True Then
        Range("A1").Offset(r + 1, 7) = "*"
    End If
    If Me.ChkAndroid.Value = True Then
        Range("A1").Offset(r + 1, 8) = "*"
    End If

    Range("A1").Offset(r + 1, 9) = Me.cboxWhereUsed.Value
    Range("A1").Offset(r + 1, 10) = Me.txtPercent.Value

    If Me.optMale.Value = True Then
        Range("A1").Offset(r + 1, 11) = "*"
    End If
    If Me.optFemale.Value = True Then
        Range("A1").Offset(r + 1, 12) = "*"
    End If
    Unload Me
End Sub
```

The `cmdSave_Click` procedure begins by hiding the user form. The statement

```
r = Application.CountA(Range("A:A"))
```

uses the Visual Basic CountA function to count the number of cells that contain data in column A. The result of the function is assigned to the variable r. The next statement

```
Range("A1").Offset(r + 1, 0) = Me.lboxSystems.Value
```

enters the selected list box item in a cell located one row below the last used cell in column A (r + 1).

Next, there are several conditional statements. The first one tells Visual Basic to place an asterisk in the appropriate cell in column B if the Hardware option button is selected.

Column B is one column to the right of column A; hence there's a 1 in the position of the second argument of the Offset method. The second If statement enters the asterisk in column C if the user selected the Software option button.

Similar instructions record the actual checkbox values. In columns F, G, H, and I, the operating system will be reported. In column J, the procedure will enter the item selected in the Used at combo box. Column K will show the value entered in the Percent (%) Used text box, and columns L and M will identify the gender of the person who submitted the survey.

Using the Info Survey Application

Your application is now ready for the final test. Take off your programming hat and enjoy the result of your work from the user's standpoint. As you work with the form, think of any improvements you would like to make to enhance your user experience.

⊙ Hands-On 16.8n   Using the Info Survey Application (Step 14)

1. Switch to the Microsoft Excel Info Survey worksheet and click the Survey button.
2. When the form appears, select the appropriate options, and click Save.
3. Activate the form several times, each time selecting different options.
4. Save the changes made to the `Chap16.xlsm` workbook.

UserForm: Modal Versus Modeless

By default, `UserForm` is *modal*, which means that the user cannot interact with the parent application while the form is visible. The `Info Survey` application that you created in this chapter behaves exactly like that. Each time you click the Survey button on the worksheet, the form pops up and you are not allowed to interact with any other Excel screen until the form is dismissed. Sometimes, however, you may want to provide access to other parts of the application while the form is visible. For example, if you are creating a custom search form, users may be required to perform specific operations in Excel outside of your form interface. The *modeless* form will allow you to do just that. Making `UserForm` modeless is quite simple. Simply pass the `vbModeless` constant to the `Show` method, like this:

```
Sub DoSurvey()
  InfoSurvey.Show vbModeless
End Sub
```

Run the modified `DoSurvey` procedure to observe the modeless form behavior. Notice that each time you click the Save button, the form's data is written on the worksheet, and the form is not dismissed. You have full control over the worksheet; you can even delete rows of data and go back to the form to create new entries.

The `vbModal` constant passed to the `Show` method will make the form modal; however, you can omit it as this is the default.

SUMMARY

You began this chapter by learning how to use VBA statements to display various built-in dialog boxes. You practiced selecting files by using the `FileDialog` object. You learned about older methods of displaying File Open and File Save As dialog boxes that you may often encounter in VBA procedures written in older versions of Excel. Additionally, this chapter has shown you how you can create and program custom user forms. Let's quickly summarize what you've learned in this chapter's main project:

- For custom VBA applications that require user input, you placed the desired controls on a custom UserForm. You made sure the user could move around the form in a logical order by setting the tab order.
- For the form to respond to user actions, you wrote VBA procedures in a form module. You set the initial values of controls by using the Properties window and writing the `UserForm_Initialize` event procedure.
- To ensure that the data collected via the custom form is properly reported in Excel, you wrote VBA procedures that transferred the form's data to a worksheet.

In the next chapter, you will learn how to format Excel worksheets with VBA.

Chapter 17

FORMATTING WORKSHEETS FOR DISPLAY AND PRINTING

Microsoft Excel offers a fairly comprehensive selection of worksheet formatting features. By applying different fonts, colors, borders, and patterns, or using conditional formatting and built-in styles, you can easily transform any raw and unfriendly worksheet data into a visually appealing and easy-to-understand document. Even if you don't care about cell appearance and your only desire is to provide a no-frills data dump, chances are that before you share your worksheet with others, you will spend ample time formatting cell values.

To better highlight important information, you can apply conditional formatting with a number of visual features such as data bars, color scales, icon sets, and sparklines. You can produce consistent-looking worksheets by using document themes and styles. When you select a theme, Excel automatically makes changes to text, charts, drawing objects, and graphics to reflect the theme you selected. By using various shapes and SmartArt graphics, you can bring different artistic effects to your worksheets.

This chapter assumes that you are already a master formatter and what really interests you is how the basic and advanced formatting features can be applied to your worksheets programmatically to prepare them for display and printing. So, let's get to work.

PERFORMING BASIC FORMATTING TASKS WITH VBA

This section focuses on cell value formatting, which controls the relationship between the values that you enter in a worksheet cell and the cell's format. Cell *value formatting* should always be attempted prior to cell *appearance formatting*.

Always begin your formatting tasks by checking that Excel correctly interprets the values you entered or copied over from an external data source, such as a text file, an SQL server, or a Microsoft Access database.

When you enter data into a worksheet yourself, you are more likely to stop right away when Excel incorrectly interprets the data. When the data comes from an external source, it is much harder to pinpoint the cell formatting problems unless you run custom VBA procedures that check for specific problems and automatically fix them when found. The meaning of your data largely depends on how Excel interprets your cell entries; therefore, to avoid confusing the end user, you must take the cell formatting control into your own hands.

Formatting Numbers

Depending on how you have formatted the cell, the number that you actually see in a worksheet can differ from the underlying value stored by Excel. As you know, each new Excel worksheet will have its default cell format set to the built-in number format named General. In this format, Excel removes the leading and trailing zeros. For example, if you entered `08.40`, Excel displays 8.4. In VBA, you can write the following statement to have Excel retain both zeros:

```
ActiveCell.NumberFormat = "00.00"
```

The `NumberFormat` property of the `CellFormat` object is used to set or return the format for a specific cell or cell range. The format code is the string displayed in the Format Cells dialog box (see Figure 17.1) or your custom string.

If the number has decimal places, Excel will only display as many decimal places as it can fit in the current column width. For example, if you enter `9.34512344443` in a cell that is formatted with the General format, Excel displays 9.345123 but keeps the full value you entered. When you widen the column, it will adjust the number of displayed digits. While determining the display of your number, Excel will also determine whether the last displayed digit needs to be rounded.

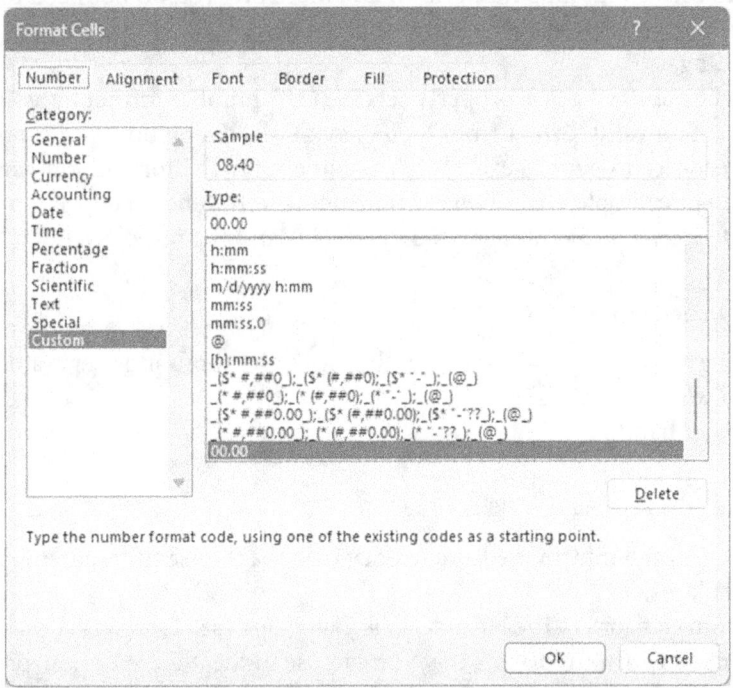

FIGURE 17.1. Format codes for your VBA procedures can be looked up in the Format Cells dialog box (choose Home | Format | Format Cells, or press Alt+H+O+E). In the Custom category, you will find a list of ready-to-use number formats and an option to create a new format or edit the existing format code to suit your particular needs, such as 00.00, as shown here.

You can use the `NumberFormat` property to determine the format that Excel applied to cells. For example, to find out the format used in the third column of your worksheet, you can enter the following statement in the Immediate window:

```
?Columns(3).NumberFormat
```

If all cells in the specified column have the same formatting applied, Excel displays the name of the format, such as General, or the format code that has been applied, such as $#,##0.00.

If different formatting is found in different cells of the specified column, Excel prints out Null in the Immediate window. It is recommended that you apply the same number formatting for the entire column. Whether you are doing this programmatically or manually via the Excel user interface, the formatting can be applied before or after you enter the numbers. Excel will only apply number formatting to cells containing numeric values; therefore, you do not need to be concerned if the first cell in the selected column contains text that defines the column heading.

Because the `NumberFormat` property sets the cell's number format by assigning a string with a valid format from the Format Cells dialog box, you can use the macro recorder to get the exact VBA statement for the format you would like to apply. For example, the following statements were generated by the macro recorder to format the values entered in cells D1 and E1, respectively:

```
Range("D1").Select
Selection.NumberFormat = "#,##0"
```

The preceding code displays a large number with the thousands separator (a comma) and with no decimal places.

On the other hand,

```
Range("E1").Select
Selection.NumberFormat = "$#,##0.00"
```

displays a large number formatted as currency with the thousands separator and two decimal places.

The Custom category in the Format Cells dialog box (see Figure 17.1 earlier) lists many built-in custom formats that you can use in the `NumberFormat` property to control how values are displayed in cells. You can also create your own custom number format strings by using the formatting codes shown in Table 17.1.

TABLE 17.1.   Number formatting codes.

Code	Description
0	Digit placeholder. Use it to force a zero. For example, to display `.5` as 0.50, use the following VBA statement: `Selection.NumberFormat = "0.00"` or enter `0.00` in the Type box in the Format Cells dialog box.
#	Digit placeholder. Use it to indicate the position where the number can be placed. For example, the code `#,###` will display the number 2345 as 2,345.
.	Decimal placeholder. In the United States, a period is used as the decimal separator. In Germany, it is a comma.

Code	Description

,	Thousands separator (comma). In the United States, one thousand two hundred and five is displayed as 1,205. In other countries, the thousands separator can be a period (e.g., Germany) or a space (e.g., Sweden).
	In the United States, placing a single comma after the number format indicates that you want to display numbers in thousands. To make it clear to the user that the number is in thousands, you may want to place the letter K after the comma:
	
	`Selection.NumberFormat = "#, ##0, K"`
	
	Use two commas at the end to display the number in millions:
	
	`Selection.NumberFormat = "#,##0.0,, "`
	
	To indicate that the number is in millions, add a backslash followed by M:
	
	`Selection.NumberFormat = "#,##0.0,,\M"`
	
	Or surround the letter M with double quotes:
	
	`Selection.NumberFormat = "#,##0.0,, ""M"""`
	
	This will cause the number 23093456 to appear as 23.1M.

/	Forward slash character. Used for formatting a number as a fraction. For example, to format 1.25 as 1¼, use the following statement: `Selection.NumberFormat = "# ?/?"`

_	The underscore character is used for aligning formatting codes. For example, to ensure that positive and negative numbers are aligned as shown below, apply the format as follows:
	
	`Selection.NumberFormat = "#,##0.00_);(#,##0.00)"`
	
	`    234.23`
	`   (234.12)`

*	The asterisk in a number format allows you to fill in the cell with the character that follows the asterisk. For example, the following VBA statement produces the output shown below:
	
	`Selection.NumberFormat = "*_0000"`
	
	`_____1045`
	`_____23455`

Number formats have four parts separated by semicolons. The first part is applied to positive numbers, the second to negative numbers, the third to zero, and the fourth to text. For example, take a look at the following VBA statement:

```
Range("A1:A4").NumberFormat = "#,##0;[red](#,#0);""zero"";@"
```

This statement tells Excel to format positive numbers with the thousands separator, display negative numbers in red and in parentheses, display the text "zero" whenever 0 is entered, and format any text entered in the cell as text. For example, if you make the following entries in cells A1:A4:

```
2870
-3456
0
Test text
```

and apply the above format to these cells, you will see them formatted as shown below:

```
2,870
(3,456)
Zero
Test text
```

You can hide the content of any cell by using the following VBA statement:

```
Selection.NumberFormat = ";;;"
```

When the number format is set to three semicolons, Excel hides the display of the cell entry on the worksheet and in printouts. You can only see the actual value or text stored in the cell by taking a look at the Formula bar.

By using the number format codes, it is also possible to apply conditional formats with one or two conditions. Consider the following VBA procedure:

```
Sub FormatUsedRange()
  ActiveSheet.UsedRange.Select
  Selection.SpecialCells(xlCellTypeConstants, 1).Select
  Selection.NumberFormat = "[<150][Red];[>250][Green];[Yellow]"
End Sub
```

This procedure tells Excel to select all the values in the UsedRange on the active sheet and display them as follows: values less than 150 in red, values over 250 in green, and all the other values in the range from 150 to 250 in yellow.

Excel supports eight popular colors: [white], [black], [blue], [cyan], [green], [magenta], [red], and [yellow], as well as 56 colors from the predefined color palette that you can access by indicating a number between 1 and 56, such as [Color 11] or [Color 32]. Be sure to use closed brackets to enclose conditions and colors.

In addition to the NumberFormat property, Excel VBA has a Format function that you can use to apply a specific format to a variable. For example, take a look at the following procedure that formats a number prior to entering it in a worksheet cell:

```
Sub FormatVariable()
  Dim myResult, frmResult
  myResult = "1435.60"
  frmResult = Format(myResult, "Currency")
  Debug.Print frmResult
  ActiveSheet.Range("G1").FormulaR1C1 = frmResult
End Sub
```

When you run the `FormatVariable` procedure, cell G1 in the active worksheet will contain the entry $1,435.60. The `Format` function specifies the expression to format (in this case, the expression is the name of the variable that stores a specified value) and the number format to apply to the expression. You can use one of the predefined number formats such as General, Currency, Standard, Percent, and Fixed, or you can specify a custom format using the formatting codes from Table 17.1.

For example, the following statement will apply number formatting to the value stored in the `myResult` variable and assign the result to the `frmResult` variable:

```
frmResult = Format(myResult, "#.##0.00")
```

For more information about the `Format` function and the complete list of formatting codes, refer to the online help.

To check whether the cell value is a number, use the `IsNumber` function, as shown below:

```
MsgBox Application.WorksheetFunction.IsNumber(ActiveCell.Value)
```

If the active cell contains a number, Excel returns `True`; otherwise, it returns `False`.

To use an Excel function in VBA, you must prefix it with `Application.WorksheetFunction`. The `WorksheetFunction` property returns the `WorksheetFunction` object that contains functions that can be called from VBA.

Formatting Text

To format a cell as a text string, use the following VBA statement:

```
Selection.NumberFormat = "@"
```

To find out whether a cell value is a text string, use the following statement:

```
MsgBox Application.WorksheetFunction.IsText(ActiveCell.Value)
```

Use the UCase function to convert a cell entry to uppercase:

```
Range("K3").value = UCase(ActiveCell.Value)
```

Use the LCase function to convert a cell entry to lowercase if the cell is not a formula:

```
If Not Range(ActiveWindow.Selection.Address).HasFormula Then
   ActiveCell.Value = LCase(ActiveCell.Value)
End If
```

Use the Proper function to capitalize the first letter of each word in a text string:

```
ActiveCell.Value = Application.WorksheetFunction.
    Proper(ActiveCell.Value)
```

Use the Replace function to replace a specified character within text. For example, the following statement replaces a space with an underscore (_) in the active cell:

```
ActiveCell.Value = Replace(ActiveCell.Value, " ", "_")
```

To ensure that the text entries don't have leading or trailing spaces, use the following VBA functions:

- LTrim—Removes the leading spaces
- RTrim—Removes the trailing spaces
- Trim—Removes both the leading and trailing spaces

For example, the following statement written in the Immediate window will remove the trailing spaces from the text found in the active cell:

```
ActiveCell.value = RTrim(ActiveCell.value)
```

Use the Font property to format the text displayed in a cell.

For example, the following statement changes the font of the selected range to Verdana:

```
Selection.Font.Name = "Verdana"
```

You can also format parts of the text in a cell by using the Characters collection.

For example, to display the first character of the text entry in red, use the following statement:

```
ActiveCell.Characters(1,1).Font.ColorIndex = 3
```

Formatting Dates

Microsoft Excel stores dates as serial numbers. In the Windows operating system, the serial number 1 represents January 1, 1900. If you enter the number 1 in a worksheet cell and then format this cell as Short Date using the Number Format drop-down in the Number section of the ribbon's Home tab, Excel will display the date formatted as 1/1/1900 and will store the value of 1 (you can check this out by looking at the General category in the Format Number dialog box).

By storing dates as serial numbers, Excel can easily perform date calculations.

To apply a date format to a particular cell or range of cells using VBA, use the NumberFormat property of the Range object, like this:

```
Range("A1").NumberFormat = "mm/dd/yyyy"
```

Table 17.2 lists the formatting codes for dates and times.

TABLE 17.2.   Date and time formatting codes.

Code	Description
D	Day of the month. Single-digit number for days from 1 to 9.
Dd	Day of the month (two-digit). Leading zeros appear for days from 1 to 9.
ddd	A three-letter day of the week abbreviation (Mon, Tue, Wed, Thu, Fri, Sat, and Sun).
m	Month number from 1 to 12. Zeros are not used for single-digit month numbers.
mm	Two-digit month number.
mmm	Three-letter month name abbreviation (e.g., Jan, Jun, Sep).
yy	Two-digit year number (e.g., 13).
yyyy	Four-digit year number (e.g., 2016).
h	The hour from 0 to 23 (no leading zeros).
hh	The hour from 0 to 23 (with leading zeros).
:m	The minute from 0 to 59 (no leading zeros).
:mm	The minute from 0 to 59 (with leading zeros)
:s :s.0 :s.00	The second from 0 to 59 (no leading zeros). To add tenths of a second, follow this with a period and a zero (.0), and to add hundredths of a second, follow this code with a period and two zeros (.00).
:ss :ss.0 :ss.00	The second from 0 to 59 (with leading zeros). To add tenths of a second, follow this code with a period and a zero (.0), and to add hundredths of a second, follow this code with a period and two zeros (.00).

(Contd.)

Code	Description
AM/PM am/pm A/P a/p	Use for a 12-hour clock, with AM or PM. Use for a 12-hour clock, with am or pm. Use for a 12-hour clock, with A or P. Use for a 12-hour clock, with a or p.
[ ]	Bracket the time component (hour, minute, second) to prevent Excel from rolling over hours, minutes, or seconds when they hit the 24-hour mark (hours become days) or the 60 mark (minutes become hours, and seconds become minutes). For example, to display time as 25 hours, 59 minutes, and 12 seconds, use the following format code: [hh]:[mm]:ss.

The following VBA procedure applies a date format to the Inspection Date Formatted column, as shown in Figure 17.2.

```
Sub FormatDateFields()
    Dim ws As Worksheet
    Set ws = ThisWorkbook.Worksheets("Sheet1")

    Dim lastRow As Long
    lastRow = ws.Cells(ws.Rows.Count, "A").End(xlUp).Row

    Dim i As Long
    For i = 2 To lastRow 'Assuming row 1 contains headers
        ws.Cells(i, 4).NumberFormat = "m/dd/yyyy h:mm:ss AM/PM"
    Next i
End Sub
```

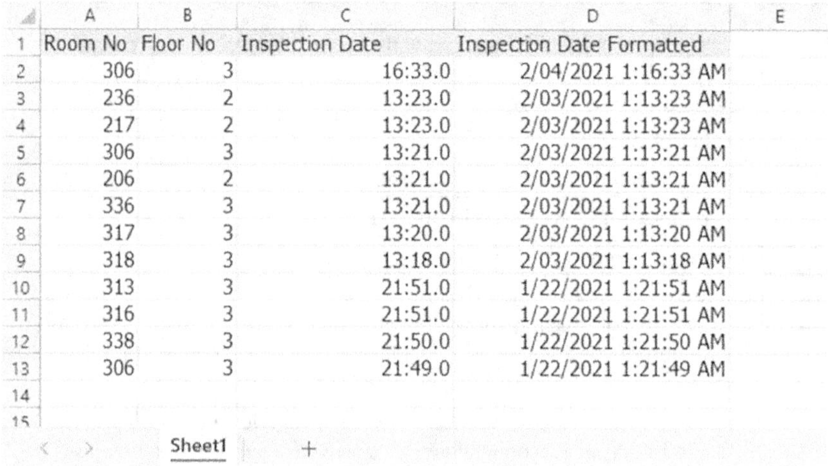

	A	B	C	D	E
1	Room No	Floor No	Inspection Date	Inspection Date Formatted	
2	306	3	16:33.0	2/04/2021 1:16:33 AM	
3	236	2	13:23.0	2/03/2021 1:13:23 AM	
4	217	2	13:23.0	2/03/2021 1:13:23 AM	
5	306	3	13:21.0	2/03/2021 1:13:21 AM	
6	206	2	13:21.0	2/03/2021 1:13:21 AM	
7	336	3	13:21.0	2/03/2021 1:13:21 AM	
8	317	3	13:20.0	2/03/2021 1:13:20 AM	
9	318	3	13:18.0	2/03/2021 1:13:18 AM	
10	313	3	21:51.0	1/22/2021 1:21:51 AM	
11	316	3	21:51.0	1/22/2021 1:21:51 AM	
12	338	3	21:50.0	1/22/2021 1:21:50 AM	
13	306	3	21:49.0	1/22/2021 1:21:49 AM	
14					
15					

Sheet1 +

FIGURE 17.2. Initially, columns C and D contained the same unformatted values. After running the FormatDateFields procedure, the values of column D are now formatted as dates.

Formatting Columns and Rows

To speed up your worksheet formatting tasks, you can apply formatting to entire rows and columns instead of single cells. The best way to find the required VBA statement is by using the built-in macro recorder. Keep in mind that Excel will record more code than is necessary for your specific task and you'll need to clean it up before copying it to your VBA procedure.

For example, here's the recorded code for setting the horizontal alignment of data in row 7:

```
Rows("7:7").Select
  With Selection
      .HorizontalAlignment = xlRight
      .VerticalAlignment = xlBottom
      .WrapText = False
      .Orientation = 0
      .AddIndent = False
      .IndentLevel = 0
      .ShrinkToFit = False
      .ReadingOrder = xlContext
      .MergeCells = False
  End With
```

To set the horizontal alignment for row 7, you can write a single VBA statement like this:

```
Rows(7).HorizontalAlignment = xlRight
```

You can use the macro recorder to help you find out the names of properties that should be used to turn a specific formatting feature on or off. Once you know the property and the required setting, you can write your own short statement to get the job done. Table 17.3 shows examples of VBA statements that can be used to format columns and rows.

TABLE 17.3.   Formatting columns and rows—example VBA statements.

Formatting Columns and Rows	VBA Statement
To format column D as a date using the `NumberFormat` property:	`Columns("D").NumberFormat = "mm/dd/yyyy"`
To format column G as currency:	`Columns("G").NumberFormat = "$###,##0.00"`
To format column G as currency using the `Style` property:	`Columns("G").Style = "Currency"`

(Contd.)

Formatting Columns and Rows	VBA Statement
To set column width or row height:	`Columns(2).ColumnWidth = 21.5` `Rows(2).RowHeight = 55.55`
To auto-fit column width or row height:	`Columns(2).AutoFit` `Rows(2).Autofit`
To apply a bold font to the first row:	`Rows(1).Font.Bold = True`
To right-align data in row 1:	`Rows(1).HorizontalAlignment = xlRight`
To center data in column B:	`Columns("B").HorizontalAlignment = xlCenter`
To set the background of the column where the active cell is located to yellow:	`Columns(ActiveCell.Column).interior.color = vbYellow`
To check the width of a column, use the `ColumnWidth` method:	`MsgBox Columns(ActiveCell.Column).ColumnWidth`
To auto-fit all rows and columns:	`ActiveSheet.Cells.EntireRow.AutoFit` `ActiveSheet.Cells.EntireColumn.AutoFit`

Formatting Headers and Footers

Headers and footers are made of three sections each: `LeftHeader`, `CenterHeader`, and `RightHeader`, and `LeftFooter`, `CenterFooter`, and `RightFooter`. Using the special formatting codes shown in Table 17.4, you can customize your worksheet's header or footer according to your needs.

TABLE 17.4.   Header and footer formatting codes.

Format Code	Description
`&D`	Prints the current date
`&T`	Prints the current time
`&F`	Prints the name of the workbook
`&A`	Prints the name of the sheet tab
`&P`	Prints the page number
`&P+number`	Prints the page number plus the specified number
`&P-number`	Prints the page number minus the specified number
`&N`	Prints the total number of pages in the workbook
`&Z`	Prints the workbook's path
`&G`	Inserts an image
`&&`	Prints a single ampersand

Format Code	Description
&nn	Prints the characters that follow in the specified font size in points
&color	Prints the characters in the specified color using the hexadecimal color value
&"fontname"	Prints the characters that follow in the specified font
&L	Left-aligns the characters that follow
&C	Centers the characters that follow
&R	Right-aligns the characters that follow
&B	Turns bold printing on or off
&I	Turns italic printing on or off
&U	Turns underline printing on or off
&E	Turns double-underline printing on or off
&S	Turns strikethrough printing on or off
&X	Turns superscript printing on or off
&Y	Turns subscript printing on or off

Table 17.5 shows example VBA statements demonstrating how to apply custom formatting to a header or footer:

TABLE 17.5. Header and footer formatting—example VBA statements.

Formatting Headers and Footers	VBA Statement (enter on one line)
To create a two-line header with bold text in the first line and italic text in the second line:	`ActiveSheet.PageSetup.LeftHeader = "&BYour Company Name" & Chr(13) & "&IYour Company Department"`
To place the workbook creation date in the footer:	`ActiveSheet.PageSetup.RightFooter = "Created on: " & ActiveWorkbook.BuiltinDocument Properties ("Creation Date")`
To place a cell's contents in the header:	`ActiveSheet.PageSetup.CenterHeader = ActiveSheet.Cells(2,2).value`
To insert the filename and path in the footer:	`ActiveSheet.PageSetup.CenterFooter = ActiveWorkbook.FullName`
To place text in the header using the Arial Narrow font and italic formatting, with the last word in bold, italic, and red:	`ActiveSheet.PageSetup.CenterHeader = "&""ArialNarrow""&IYour text goes here &I&B&KFF0000now."`
To remove the formatting and text entries from the center header:	`ActiveSheet.PageSetup.CenterHeader = ""`

If you need to use a different date format in your header or footer than the date format shown in the Regional settings of the Windows Control panel, use the `Format` function like this:

```
ActiveSheet.PageSetup.RightFooter = Format(Date, "mm-dd-yyyy")
```

The above statement inserts in the right footer a current system date returned by the `Date` function. The date is formatted as a two-digit month number followed by a dash, a two-digit day number followed by a dash, and a four-digit year number (see Table 17.2 for the explanation of date and time formatting codes).

Formatting Cell Appearance

Adding cosmetic touches to your worksheets such as fonts, color, borders, shading, and alignment should be undertaken after applying the required formatting to numbers, dates, and times. Formatting cell appearance makes your worksheet easier to read and interpret. By applying borders to cells, and with the clever use of font colors, background shading, and patterns, you can draw the reader's attention to particularly important information. Use the `Font` object to change the font format in VBA. You can easily apply multiple format properties using the `With...End With` statement block shown below:

```
Sub ApplyCellFormat()
  With ActiveSheet.Range("A1").Font
    .Name = "Tahoma"
    .FontStyle = "italic"
    .Size = 14
    .Underline = xlUnderlineStyleDouble
    .ColorIndex = 3
  End With
End Sub
```

The `ColorIndex` property refers to the 56 colors that are available in the color palette. Number 3 represents red. The following procedure prints a color palette to the active sheet of a new workbook:

```
Sub ColorLoop()
  Dim r As Integer
  Dim c As Integer
  Dim k As Integer

  Workbooks.Add

  k = 0
```

```
   For r = 1 To 8
     For c = 1 To 7
         Cells(r, c).Select
         k = k + 1
         ActiveCell.Value = k
         With Selection.Interior
             .ColorIndex = k
             .Pattern = xlSolid
         End With
     Next c
   Next r
End Sub
```

To change the background color of a single cell or a range of cells, use one of the following VBA statements:

```
Selection.Interior.Color = vbBlue
Selection.Interior.ColorIndex = 5
```

To change the font color, use the following VBA statement:

```
Selection.Font.Color = vbMagenta
```

The font can be made bold, italic, underlined, or a combination of the three using the following With...End With block statement:

```
With Selection.Font
    .Italic = True
    .Bold = True
    .Underline = xlUnderlineStyleSingle
End With
```

To apply borders to your cells and ranges in VBA, use the following statement examples:

```
Selection.BorderAround Weight:=xlMedium, ColorIndex:=3
Selection.BorderAround Weight:=xlThin, Color:=vbBlack
```

The BorderAround method of the Range object places a border around all edges of the selected cells. The following xlBorderWeightEnumeration constants can be used to specify the thickness weight of the border: xlHairline, xlMedium, xlThick, and xlDash. When specifying border color, you can use either ColorIndex or Color but not both. Instead of specifying the thickness of the border, you may want to use LineStyle, as in the following example:

```
Selection.BorderAround LineStyle:=xlDashDotDot, Color:=vbBlack
```

You can use any of the following `xlLineStyle` enumeration constants:

- `xlContinuous`
- `xlDash`
- `xlDashDot`
- `xlDashDotDot`
- `xlDot`
- `xlDouble`
- `xlLineStyleNone`
- `xlSlantDashDot`

Use `xlLineStyleNone` to clear the border. VBA has a `Borders` collection that contains the four borders of a `Range` or `Style` object. To set just a bottom border for cells A1:C1, use the following statement:

```
ActiveSheet.Range("A1:C1").Borders(xlEdgeBottom).Weight = xlThick
```

You may specify any of the following border types:

- `xlDiagonalDown`
- `xlDiagonalUp`
- `xlEdgeBottom`
- `xlEdgeLeft`
- `xlEdgeRight`
- `xlEdgeTop`
- `xlEdgeHorizontal`
- `xlEdgeVertical`

You can change the appearance of cells by specifying the horizontal or vertical alignment:

```
Selection.HorizontalAlignment = xlCenter
Selection.VerticalAlignment = xlTop
```

The value of the `HorizontalAlignment` property can be one of the following constants:

- `xlCenter`
- `xlDistributed`
- `xlJustify`
- `xlLeft`
- `xlRight`

The `VerticalAlignment` property can be one of the following constants:

- `xlBottom`
- `xlCenter`
- `xlDistributed`
- `xlJustify`
- `xlTop`

Removing Formatting from Cells and Ranges

To remove cell formatting, use the `ClearFormats` method of the `Range` object. This method restores the formatting to the original General format without removing the cell's content. To remove the content of the cell, use the `Clear-Contents` method.

PERFORMING ADVANCED FORMATTING TASKS WITH VBA

Let's focus on how you can use VBA with the enhanced formatting features that are available in the Styles group on the ribbon's Home tab and in the Themes group of the Page Layout tab.

We will work with the `FormatConditions` collection of the `Range` object and explore the conditional formatting tools: data bars, color scales, and icon sets. Then, we'll look at how sparklines can be used to enhance your worksheets. We'll also take a look at document themes that can be applied to a workbook and see how they affect another formatting feature—styles.

Conditional Formatting Using VBA

To help you automatically highlight important parts of your worksheet, Excel provides a feature known as *conditional formatting*. This feature allows you to set a condition (a formatting rule) and specify the type of formatting that should be applied to cells and ranges when the condition is met. For example, you can use conditional formatting to apply different background colors, fonts, or borders to a cell based on its value.

Conditional formatting provides users with various types of common rules and formatting tools such as data bars, color scales, and icon sets, which make it easy to highlight certain worksheet data. For example, you can highlight the top or bottom 10% of values, locate duplicate or unique values, or indicate values above or below the average. You can specify an unlimited number of conditional formats and refer to ranges in other worksheets when using conditional formats.

The Conditional Formatting Rules Manager shown in Figure 17.3 simplifies the creation, modification, and removal of conditional rules. All conditional formatting features that are available in the Excel application window can be accessed via VBA.

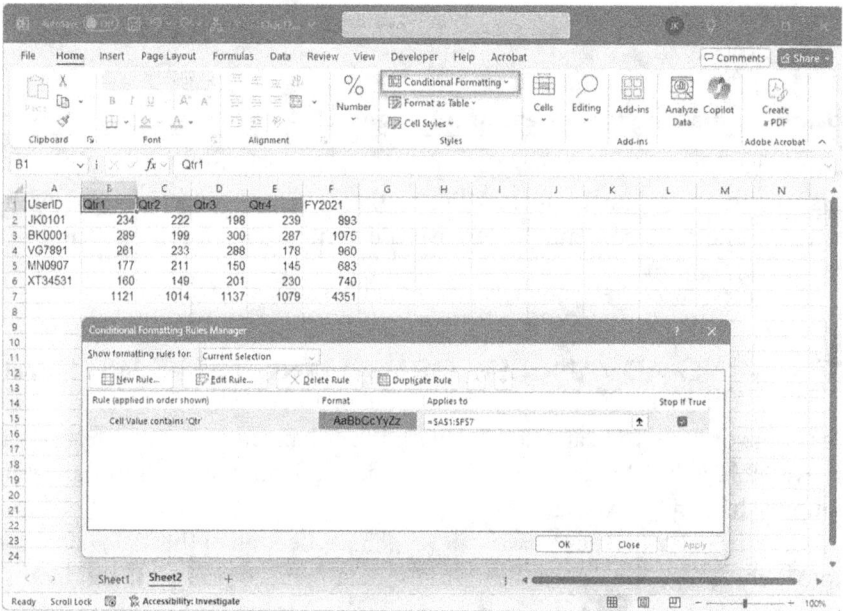

FIGURE 17.3. To activate Conditional Formatting Rules Manager, choose Home | Conditional Formatting | Manage Rules.

To create a new conditional formatting rule, click the New Rule button in the Conditional Formatting Rules Manager dialog box. You will see the list of built-in rules that you can select from. In VBA, use the `Add` method of the `Format-Conditions` collection to create a new rule.

For example, to format cells containing `Qtr` in the text string, enter the following procedure in a standard module and then run it:

```
Sub FormatQtrText()
  With ActiveSheet.UsedRange
      .FormatConditions.Delete
      .FormatConditions.Add Type:=xlTextString, String:="Qtr", _
      TextOperator:=xlContains
      .FormatConditions(1).Interior.Color = RGB(123, 130, 0)
  End With
End Sub
```

Notice that before creating and applying a new conditional format to a range of cells, it's a good idea to delete the existing format condition from the selection using the `Delete` method. The `Add` method that is used to add a new condition requires at least the `Type` argument that specifies whether the conditional format is based on a cell value or an expression.

Use the `xlFormatConditionType` enumeration constants listed in Table 17.6 to set the condition type. For example, to format cells that contain dates, use the `xlTimePeriod` constant in the `Type` argument and specify the `DateOperator` using one of the following constants: `xlToday`, `xlYesterday`, `xlTomorrow`, `xlLastWeek`, `xlThisWeek`, `xlNextWeek`, `xlLast7Days`, `xlLastMonth`, `xlThisMonth`, or `xlNextMonth`:

```
Selection.FormatConditions.Add Type:=xlTimePeriod,
    DateOperator:=xlLast7Days
```

TABLE 17.6. Conditional format Type settings (xlFormatConditionType enumeration).

Constant	Description

xlAboveAverageCondition	Above/below average condition:
	`With Selection`
	`    .FormatConditions.Delete`
	`    .FormatConditions.AddAboveAverage`
	`    .FormatConditions(1).AboveBelow =`
	`        xlAboveAverage`
	`    .FormatConditions(1).Font.Bold = True`
	`End With`

xlBlanksCondition	Format cells that contain blanks:
	`With Selection`
	`    .FormatConditions.Add _`
	`        Type:=xlBlanksCondition`
	`End With`

xlCellValue	Format a cell value:
	`With Selection`
	`    .FormatConditions.Add Type:=xlCellValue,`
	`        Operator:=xlLess`
	`        Formula1:="=2000"`
	`    .FormatConditions(1).NumberFormat = "#, ##0"`
	`End With`
	You can use the following constants in the `Operator` argument: `xlBetween`, `xlEqual`, `xlGreater`, `xlGreaterEqual`, `xlLess`, `xlLessEqual`, `xlNotBetween`, or `xlNotEqual`.
	To specify the numeric value for the operator, use the `Formula1` argument. The `xlBetween` and `xlNotBetween` operators require that you also specify a second value in `Formula2`.

(Contd.)

Constant	Description

xlColorScale	Format the color scale:
	
	``
	If Selection.FormatConditions(1).Type = 3 Then
	    MsgBox "Formatted with " & _
	        "ColorScale conditional format."
	End If
	``

xlDataBar	Format the data bar:
	
	``
	If Selection.FormatConditions(1).Type = 4 Then
	    MsgBox "Formatted with " & _
	        "DataBar conditional format."
	End If
	``

xlErrorsCondition	Format cells that contain errors: ``` Selection.FormatConditions.Add _ Type:=xlErrorsCondition ```

xlExpression	Expression to specify a custom formula that identifies the cells that the conditional format applies to. For example, the following procedure changes the background color of alternate rows in the used range:
	
	``
	Sub HighlightAltRows()
	  With ActiveSheet.UsedRange
	    .FormatConditions.Add Type:=xlExpression, _
	        Formula1:="=MOD(ROW(),2)=0"
	    .FormatConditions(1).Interior.ColorIndex = 6
	  End With
	End Sub
	``
	To highlight every third row, use this formula:
	`= MOD(ROW(),3)= 0`

xlIconSet	Format the icon set:
	
	``
	If Selection.FormatConditions(1).Type = 6 Then
	    MsgBox "Formatted with " & _
	        "IconSet conditional format."
	End If
	``

xlNoBlanksCondition	Format cells that do not contain blanks:
	
	``
	Sub HighlightNonEmptyCells()
	  Range("A1:B12").Select
	  Selection.FormatConditions.Add _
	      Type:=xlNoBlanksCondition
	    With Selection.FormatConditions(1).Interior
	    .ThemeColor = xlThemeColorAccent4
	    .TintAndShade = 0.399945066682943
	    End With
	End Sub
	``

Constant	Description

`xlNoErrorsCondition`	Format cells that do not contain errors:
	```Sub HighlightCellsWithNoErrors()```
	``` Range("F1:F7").Select```
	``` Selection.FormatConditions.Add _```
	``` Type:=xlNoErrorsCondition```
	``` With Selection.FormatConditions(1).Interior```
	``` .ThemeColor = xlThemeColorAccent4```
	``` .TintAndShade = 0.399945066682943```
	``` End With```
	```End Sub```
	Before running the above procedure, enter any number in cell F1, and enter zero (0) in cell F2. In cell F3, enter the following formula: `=F1/F2`. Because there is no division by zero, Excel will display the following error code: #DIV/0! When you run the procedure, all cells in the selected range except for cell F2 will be shaded with the specified color.

`xlTextString`	Format cells that contain text:
	```With ActiveSheet.UsedRange```
	``` .FormatConditions.Add Type:=xlTextString, _```
	``` String:="es", TextOperator:=xlContains```
	``` .FormatConditions(1).Font.Bold = True```
	```End With```
	Other text operators you can use: `xlBeginsWith`, `xlDoesNot-Contain`, and `xlEndsWith`.

`xlTimePeriod`	Format cells that contain dates:
	```With ActiveSheet.UsedRange```
	``` .FormatConditions.Add Type:=xlTimePeriod, _```
	``` DateOperator:=xlLastMonth```
	``` .FormatConditions(1).Interior.ColorIndex = 6```
	```End With```
	Other date operators you can use: `xlToday`, `xlYesterday`, `xlTomorrow`, `xlLastWeekn`, `xlThisWeek`, `xlNextWeek`, `xlLast7Days`, `xlThisMonth`, and `xlNextMonth`.

`xlTop10`	Format the 10 top values:
	```With Selection```
	``` .FormatConditions.AddTop10```
	``` .FormatConditions(1).TopBottom = xlTop10Top```
	``` .FormatConditions(1).Value = 5```
	``` .FormatConditions(1).Percent = False```
	``` .FormatConditions(1).Interior.Color = _```
	``` RGB(255,0,0)```
	```End With```

(Contd.)

Constant	Description

xlUniqueValue	Format unique values:
	```With Selection```
	``` .FormatConditions.AddUniqueValues```
	``` .FormatConditions(1).DupeUnique = xlUnique```
	``` Formula1:="=200"```
	```End With```
	
	By replacing the xlUnique constant with xlDuplicate, you can select duplicate values.

Conditional Formatting Rule Precedence

You can apply multiple conditional formats to a cell. For example, you can apply a conditional format to make the cell bold, and then another one to make a red border around the cell. Because these two formats do not conflict with one another, they can both be applied to the same cell. If you create another format that tells Excel to apply a blue border to the cell, however, this rule will not be applied because it conflicts with the previous rule that told Excel to apply the red border.

In order to control multiple conditions applied to a range of cells, Excel uses *rule precedence*. When rules conflict with one another, Excel applies the rule that is higher in precedence. Rules are evaluated in order of precedence by how they are listed in the Conditional Formatting Rules Manager dialog box. In VBA, this is controlled by the Priority property of the FormatConditions object. For example, to assign a second priority to the first rule, use the following statement:

```
Range("B2:B17").FormatConditions(1).Priority = 2
```

You can make the rule the lowest priority with the following statement:

```
Range("B2:B17").FormatConditions(1).SetLastPriority
```

In cases where the same format is applied both manually and via conditional formatting to a range of cells, the conditional formatting rule takes precedence over the manual format. Formats applied manually are not considered when determining conditional formatting rule precedence and do not appear in the Conditional Formatting Rules Manager dialog box.

Deleting Rules with VBA

You can use the following statement to delete all rules applied to a specific range of cells:

```
Range("B2:B17").FormatConditions.Delete
```

To delete a particular rule, refer to its index number before calling the `Delete` method of the `FormatConditions` collection:

```
Range("B2:B17").FormatConditions(2).Delete
```

Using Data Bars

The data visualization tool known as the *data bar* allows users to easily see how data values relate to each other. Data bars can be added via conditional formatting using the New Formatting Rule dialog box (choose Home | Conditional Formatting | Data Bars | More Rules) or by a VBA procedure.

Excel draws data bars proportionally according to their values. Thanks to this feature, data bars can be used to compare values. When you select the lowest value in the Type dropdown, the data bar will not be drawn. When you select a maximum value, Excel will draw a bar that covers the entire cell.

You can easily format the bar appearance with additional formatting options, such as solid fills and borders. By using this feature, you can see which cell has the highest value. Keep in mind, however, that applying a solid fill to a data bar may make some portions of the text harder to read, especially when using darker colors.

In VBA, you can create a data bar formatting rule by using the `AddDatabar` or `Add` method of the `FormatConditions` collection, as shown below:

```
Sub FormatWithDataBars()
    With Range("B2:E6").FormatConditions
    .AddDatabar
    .Add Type:=xlDatabar, _
        Operator:=xlGreaterEqual, Formula1:="200"
    End With
End Sub
```

This procedure will place a blue bar in the worksheet cells, as illustrated in Figure 17.4.

Notice that the length of the data bar corresponds to the cell's value. If you change or recalculate the worksheet data, the data bar is automatically reapplied to the specified range. Instead of using the lowest and highest values to specify the shortest and longest bars, you can specify that the bar be based on numbers, percentages, formulas, or percentiles. For example, you can use the following statements to change the color, type, and threshold parameters of the data bar:

```
set mBar = Selection.FormatConditions.AddDatabar
mBar.MinPoint.Modify NewType:=xlConditionValuePercentile, _
    NewValue:=20
```

```
mBar.MaxPoint.Modify NewType:=xlConditionValuePercentile, _
    NewValue:=80
mBar.BarColor.ColorIndex = 7
```

In the previous statements, the `MinPoint` and `MaxPoint` properties of the `DataBar` object are used to set the values of the shortest and longest bars of a range of data, and the `BarColor` property is used to modify the color of the bars in the data bar conditional format.

	A	B	C	D	E	F
1	UserID	Qtr1	Qtr2	Qtr3	Qtr4	FY2021
2	JK0101	234	222	198	239	893
3	BK0001	289	199	300	287	1075
4	VG7891	261	233	288	178	960
5	MN0907	177	211	150	145	683
6	XT34531	160	149	201	230	740
7		1121	1014	1137	1079	4351
8						

FIGURE 17.4. A worksheet shown with the new data bar formatting.

Using Color Scales

You can create special visual effects in your worksheet by selecting a range of values and applying a color scale. Color scales use cell shading to help you understand variation in your data. When you apply a color scale conditional format via the user interface (Home | Conditional Formatting | Color Scales | More Rules) or from your VBA procedure, Excel uses the lowest, highest, and midpoint values in the range to determine the color gradients. You can apply a two-color or a three-color scale to your data.

To create a color scale conditional formatting rule in VBA, use the `AddColorScale` or `Add` method of the `FormatConditions` collection:

```
set cScale = Selection.FormatConditions.AddColorScale _
    (ColorScaleType:=2)
```

The previous statement creates a two-color `ColorScale` object in the selected worksheet cells.

To change the minimum threshold to green and the maximum threshold to blue, use the following statements:

```
cScale.ColorScaleCriteria(1).FormatColor.Color = RGB(0, 255, 0)
cScale.ColorScaleCriteria(2).FormatColor.Color = RGB(0, 0, 255)
```

For darker color scales, it makes sense to change the font color to white:

```
Selection.Font.ColorIndex = 2
```

As with a data bar, you can change the type of threshold value for a color scale to a number, percent, formula, or percentile. To create striking visual effects, try applying both data bar and color scale conditional formatting to the same range of data.

Using Icon Sets

Icon sets are visualization features that place icons in cells, making your data more comprehensive and visually appealing. They enable users to easily see the relationship between data values and recognize trends. To view the available icon choices, choose Home | Conditional Formatting | Icon Sets.

Each icon in an icon set represents a range of values. For example, in the three-icon set, `3 Symbols (Circled)`, shown in Figure 17.5, Excel uses the check mark symbol in a green circle for values that are greater than or equal to 67%, an exclamation point in an orange circle for values that are less than 67% and greater than or equal to 33%, and an X symbol in a red circle for values that are less than 33%.

Icon sets have become a very effective highlighting tool, thanks to the ability to apply icons only to specific cells instead of the entire range of cells. You can hide the icon for cells that meet the specified criteria by selecting No Cell Icon from the Icon dropdown.

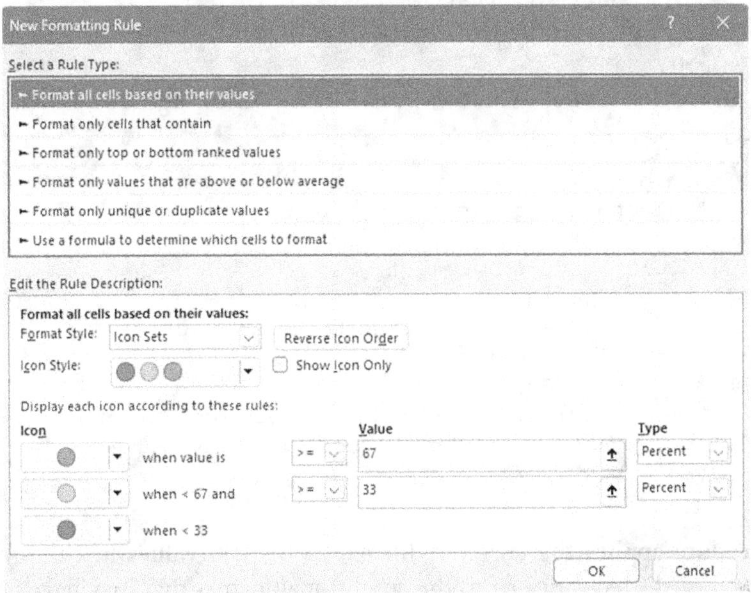

FIGURE 17.5. You can display each icon according to the rules set in the New Formatting Rule dialog box.

The Icon Style dropdown also allows you to select four- and five-icon sets. These sets display each icon according to which quartile or quintile the value falls into. You may change the default threshold value and its type (number, percent, formula, or percentile) for each icon in an icon set by editing the formatting rule or with a VBA procedure, as demonstrated in Hands-On 17.1.

NOTE	*Icons can be made larger or smaller by increasing or decreasing the font size.*

In VBA, the IconSet object in the IconSets collection represents a single set of icons. To create a conditional formatting rule that uses icon sets, use the IconSetCondition object. You can add criteria for an icon set conditional formatting rule with the IconCriteria collection.

NOTE	*Please note that the files for the "Hands-On" projects can be found in the companion files.*

(◉) Hands-On 17.1   Using Icon Sets with VBA

1. Copy the Chap17.xlsm workbook from the companion files to your VBAExcel2024_ByExample folder.
2. Open Chap17.xlsm and select Sheet3.
3. Switch to the VBE screen and insert a new module into VBAProject (Chap17.xlsm).
4. In the module's Code window, enter the following IconSetRules procedure:

```
Sub IconSetRules()
  Dim iSC As IconSetCondition

  Sheets(3).Columns("C:C").Select
  With Selection
    .SpecialCells(xlCellTypeConstants, 23).Select
    .FormatConditions.Delete
    .NumberFormat = "$#,##0.00"
    Set iSC = Selection.FormatConditions.AddIconSetCondition
    iSC.IconSet = ActiveWorkbook.IconSets(xl3Symbols)
  End With
End Sub
```

This procedure applies the currency format to the cell values in column C and clears the selected range from the conditional format that may have been applied earlier. Next, the AddIconSetCondition method is used to create an icon set conditional format for the selected range of cells.

In Step 5, when you run the procedure in step mode by pressing F8, you will notice colored circles being applied to the cells. The next statement changes the default icon set to x13Symbols, as shown in Figure 17.6.

5. Place the insertion point anywhere inside the code of the IconSetRules procedure and press F8 after each statement to execute the code in step mode.

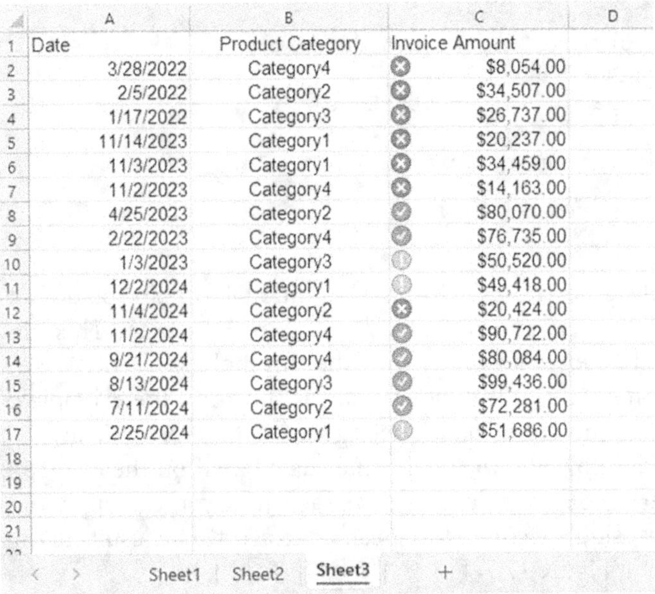

	A	B	C	D
1	Date	Product Category	Invoice Amount	
2	3/28/2022	Category4	⊗ $8,054.00	
3	2/5/2022	Category2	⊗ $34,507.00	
4	1/17/2022	Category3	⊗ $26,737.00	
5	11/14/2023	Category1	⊗ $20,237.00	
6	11/3/2023	Category1	⊗ $34,459.00	
7	11/2/2023	Category4	⊗ $14,163.00	
8	4/25/2023	Category2	✓ $80,070.00	
9	2/22/2023	Category4	✓ $76,735.00	
10	1/3/2023	Category3	◐ $50,520.00	
11	12/2/2024	Category1	◐ $49,418.00	
12	11/4/2024	Category2	⊗ $20,424.00	
13	11/2/2024	Category4	✓ $90,722.00	
14	9/21/2024	Category4	✓ $80,084.00	
15	8/13/2024	Category3	✓ $99,436.00	
16	7/11/2024	Category2	✓ $72,281.00	
17	2/25/2024	Category1	◐ $51,686.00	
18				
19				
20				
21				

Sheet1 Sheet2 **Sheet3** +

FIGURE 17.6. A column of data with an icon set used in the conditional format.

As mentioned earlier, you can modify the icon set conditions using the dialog box or with VBA. Suppose that instead of using the default percentage distribution with the threshold of >=67, >=33, and <33, you want to use the following criteria: >=80000, >=50000, and <50000.

Let's take a look at the revised procedure that modifies the formatting rule and applies a filter by cell icon criteria:

```
Sub IconSetRules_Revised()
  Dim iSC As IconSetCondition

  Sheets(3).Columns("C:C").Select
  Selection.SpecialCells(xlCellTypeConstants, 23).Select
  With Selection
    .FormatConditions.Delete
    .AutoFilter
    .NumberFormat = "$#,##0.00"
    Set iSC = Selection.FormatConditions.AddIconSetCondition
```

```
iSC.IconSet = ActiveWorkbook.IconSets(xl3Symbols)
    With iSC.IconCriteria(2)
        .Type = xlConditionValueNumber
        .Value = 50000
        .Operator = xlGreaterEqual
    End With

    With iSC.IconCriteria(3)
        .Type = xlConditionValueNumber
        .Value = 80000
        .Operator = xlGreaterEqual
    End With
    .AutoFilter Field:=1, Criteria1:=iSC.IconSet.Item(3),
Operator:=xlFilterIcon
  End With
End Sub
```

When changing the criteria for the icon set conditional format, you do not need to specify the type, value, and operator for `IconCriteria(1)`. This property is read-only. Excel determines on its own the threshold value of `IconCriteria(1)`, and if you try to set it in your code, as shown in the example procedure available in the online help, you will get a runtime error. The `Sort` and `Filter` commands allow you to sort or filter data based on the cell icon. The `IconSetRules_Revised` procedure demonstrates how you can apply the filter programmatically using an icon in the specified icon set. The results of this procedure are shown in Figure 17.7.

	A	B	C	D	E	F
1	Date	Product Category	Invoice Amount			
8	4/25/2023	Category2	✅ $80,070.00			
13	11/2/2024	Category4	✅ $90,722.00	Applied Filter		
14	9/21/2024	Category4	✅ $80,084.00			
15	8/13/2024	Category3	✅ $99,436.00			
18						
19						

FIGURE 17.7.   After running the IconSetRules_Revised procedure, a filter is applied to the Invoice Amount column to display only the cells with invoice values >=80000.

Suppose that you only want to highlight cells with values less than 50000. The following procedure creates a formatting rule that produces the output shown in Figure 17.8:

```
Sub IconSetHideIcons()
    Dim iSC As IconSetCondition

    ThisWorkbook.Worksheets(3).Activate
    ActiveSheet.AutoFilter.ShowAllData
    ActiveSheet.Columns("C:C").Select
```

```
Selection.SpecialCells(xlCellTypeConstants, 23).Select
With Selection

    .FormatConditions.Delete
    .NumberFormat = "$#,##0.00"
    Set iSC = Selection.FormatConditions.AddIconSetCondition
    iSC.IconSet = ActiveWorkbook.IconSets(xl3Symbols)
    .FormatConditions(1).IconCriteria(1)._
        Icon = xlIconRedCrossSymbol

    With iSC.IconCriteria(2)
        .Type = xlConditionValueNumber
        .Value = 50000
        .Operator = xlGreaterEqual
        .Icon = xlIconNoCellIcon
    End With

    With iSC.IconCriteria(3)
        .Type = xlConditionValueNumber
        .Value = 80000
        .Operator = xlGreaterEqual
        .Icon = xlIconNoCellIcon
    End With
    End With
End Sub
```

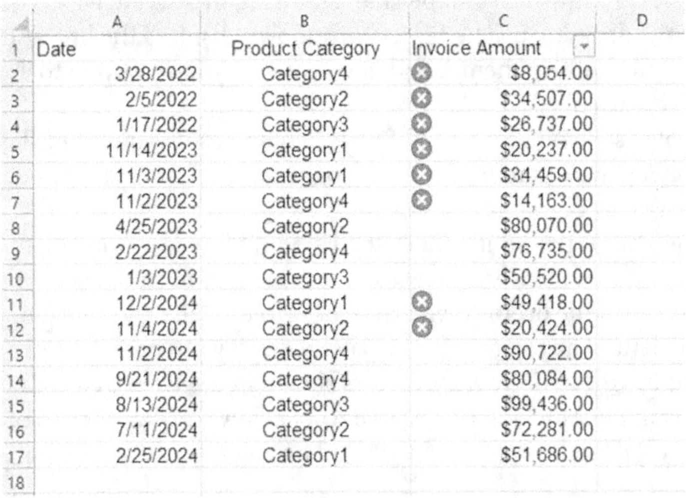

⊿	A	B	C	D
1	Date	Product Category	Invoice Amount ▼	
2	3/28/2022	Category4	⊗ $8,054.00	
3	2/5/2022	Category2	⊗ $34,507.00	
4	1/17/2022	Category3	⊗ $26,737.00	
5	11/14/2023	Category1	⊗ $20,237.00	
6	11/3/2023	Category1	⊗ $34,459.00	
7	11/2/2023	Category4	⊗ $14,163.00	
8	4/25/2023	Category2	$80,070.00	
9	2/22/2023	Category4	$76,735.00	
10	1/3/2023	Category3	$50,520.00	
11	12/2/2024	Category1	⊗ $49,418.00	
12	11/4/2024	Category2	⊗ $20,424.00	
13	11/2/2024	Category4	$90,722.00	
14	9/21/2024	Category4	$80,084.00	
15	8/13/2024	Category3	$99,436.00	
16	7/11/2024	Category2	$72,281.00	
17	2/25/2024	Category1	$51,686.00	
18				

FIGURE 17.8.   In this scenario, we use the icon set to draw attention to cells with invoice amounts less than 50000. Notice that by not applying icons to other cells, you can easily highlight the problem areas.

Formatting with Themes

If you need to change the look of the entire workbook, you ought to spend some time familiarizing yourself with document themes. A *document theme* consists of a predefined set of fonts, colors, and effects, such as lines and fills, that you can apply to the workbook and also share between other Office documents.

It is important to note that a theme applies to the entire workbook, not just the active worksheet. There are numerous document themes available in the user interface (choose Page Layout | Themes). You can also create custom themes by mixing and matching different theme elements using the three drop-down controls found in the Themes group on the Page Layout tab (Colors, Fonts, and Effects). Additional themes can be downloaded from Microsoft 365 online.

When you change the theme, the font and color pickers on the Home tab and other galleries such as Cell Styles or Table Styles are automatically updated to reflect the new theme. Therefore, if you are looking to apply a cell background with a particular color and that color is not listed in the color picker, apply a predefined theme that contains that color, or create your own custom color by choosing the More Colors option in the Font Color drop-down menu.

On the Page Layout tab, the Colors dropdown displays the color groups for each theme and gives you an option to create new theme colors. The Fonts dropdown shows a list of fonts for the theme. Theme fonts contain a heading font and a body text font, which can be changed using the Create New Theme Fonts option in the dropdown. The Effects dropdown displays the line and fill effects for each of the built-in themes and does not give you an option to create your own set of theme effects.

The theme color scheme consists of 12 base colors, as illustrated in Figure 17.9. When applying a color to a cell, the color is selected from the Fill Color dropdown, as shown in Figure 17.10.

On the Home Tab in the Font area of the ribbon, there is a convenient Fill Color button that makes it easy to view the available theme colors (see Figure 17.10). When you click the drop-down arrow next to the Fill Color button, you will see a color palette. The top row in the palette displays 10 base colors in the current color theme (the two hyperlink colors shown in the Create New Theme Colors dialog box are not included). The five rows below show variations of the base color. The color can be lighter or darker. The color name is shown in the tooltip.

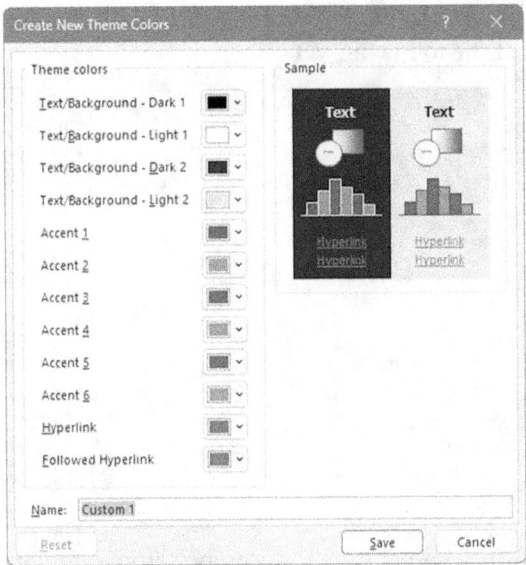

FIGURE 17.9.   The theme colors consist of four text/ background colors, six accent colors, and two hyperlink colors. To display this dialog box, choose Page Layout | Colors | Customize Colors.

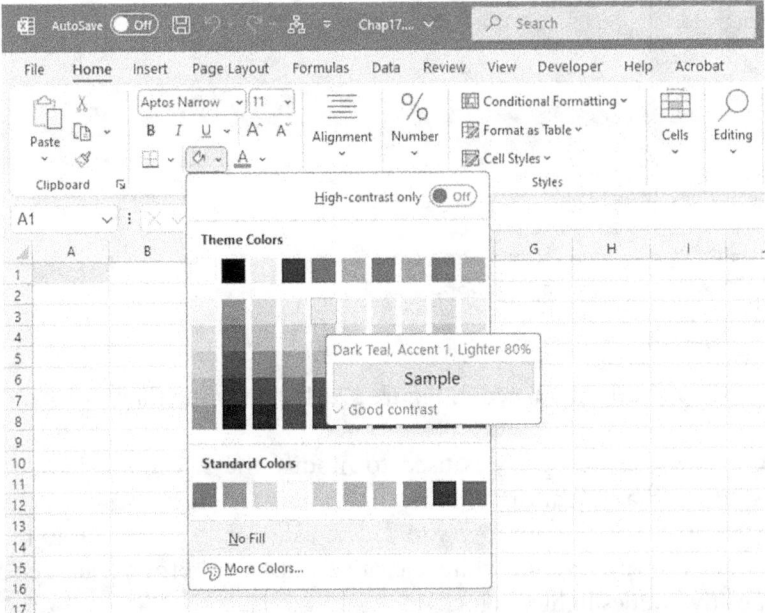

FIGURE 17.10.   The Fill Color control is used to color the background of selected cells.

If you record a macro while applying the Blue, Accent 1, Lighter 40% color to the cell background, you will get the following VBA code:

```
Sub Macro1()
'
' Macro1 Macro
'


'
    Range("F4").Select
    With Selection.Interior
        .Pattern = xlSolid
        .PatternColorIndex = xlAutomatic
        .ThemeColor = xlThemeColorAccent1
        .TintAndShade = 0.399975585192419
        .PatternTintAndShade = 0
    End With
End Sub
```

The pattern properties refer to the cell patterns that can be set via the Format Cells dialog box (Home | Format | Format Cells | Fill). These properties can be ignored if all that's required is setting the cell background color. The above code can be modified as follows:

```
Sub Macro1_Modified()
'
' Macro1 Macro
'


'
    Range("F4").Select
    With Selection.Interior
      .ThemeColor = xlThemeColorAccent1
      .TintAndShade = 0.399975585192419
    End With
End Sub
```

The `ThemeColor` properties listed in Figure 17.11 specify the theme color to be used.

The `TintAndShade` property is used to modify the selected color. A tint (lightness) and a shade (darkness) is a value from –1 to 1. If you want a pure color, set the `TintAndShade` property to 0. The value of –1 will result in black, and the value of 1 will produce white. Negative values will produce darker colors, and positive values lighter colors.

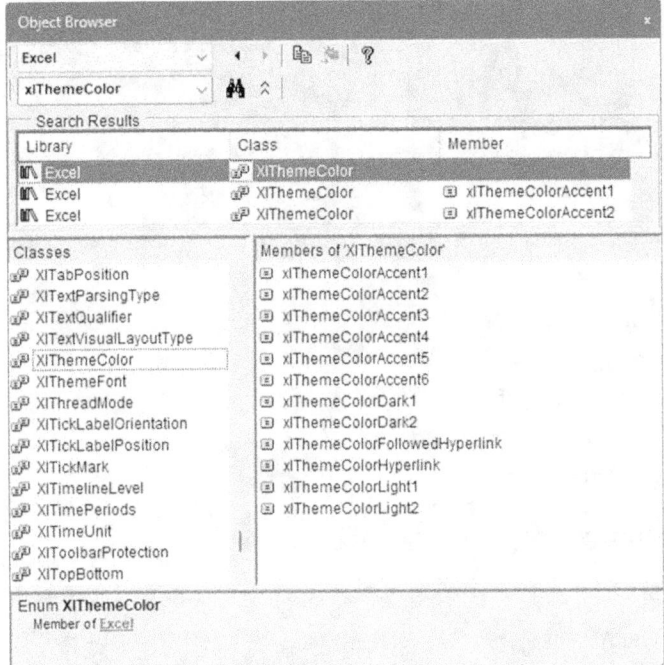

FIGURE 17.11.   Theme color constants and values as shown in the Object Browser.

The `TintAndShade` value of `0.399975585192419` means a 40% tint (or 40% lighter than the base color). If you change this number to `-0.399975585192419`, you will get a 40% darker color.

The following procedure loops through the colors in themes 4 through 10 and writes the color index and color variations to the range of cells shown in Figure 17.12:

```
Sub Themes4Thru10()
  Dim tintshade As Variant
  Dim heading As Variant
  Dim cell As Range
  Dim themeC As Integer
  Dim r As Integer
  Dim c As Integer
  Dim i As Integer


  heading = Array("ThemeColorIndex", "Neutral", "Lighter 80%", _
      "Lighter 60%", "Lighter 40%", "Darker 25%", "Darker 50%")
  tintshade = Array(0, 0.8, 0.6, 0.4, -0.25, -0.5)
```

```
  i = 0

  Workbooks.Add

  For Each cell In Range("A1:G1")
    cell.Formula = heading(i)
    i = i + 1
  Next

  For r = 2 To 8
    themeC = r + 2
      For c = 1 To 7
      If c = 1 Then
        Cells(r, c).Formula = themeC
      Else
        With Cells(r, c)
            With .Interior
              .ThemeColor = themeC
              .TintAndShade = tintshade(c - 2)
            End With
        End With
      End If
    Next c
  Next r
  ActiveSheet.Columns("A:G").AutoFit
End Sub
```

	A	B	C	D	E	F	G
1	ThemeColorIndex	Neutral	Lighter 80%	Lighter 60%	Lighter 40%	Darker 25%	Darker 50%
2		4						
3		5						
4		6						
5		7						
6		8						
7		9						
8		10						
9								
10								

FIGURE 17.12. This worksheet was generated by a VBA procedure. When you apply a different document theme, the colors will be replaced by those from the new theme.

The following procedure applies the current theme colors to a range of cells in an active worksheet:

```
Sub GetThemeColors()
  Dim tColorScheme As ThemeColorScheme
  Dim colorArray(10) As Variant
  Dim i As Long
  Dim r As Long

  Workbooks.Add
```

```
Set tColorScheme = ActiveWorkbook.Theme.ThemeColorScheme
For i = 1 To 10
  colorArray(i) = tColorScheme.Colors(i).RGB
  ActiveSheet.Cells(i, 1).Value = colorArray(i)
Next i
i = 0
For r = 1 To 10
  ActiveSheet.Cells(r, 2).Interior.Color = colorArray(i + 1)
  i = i + 1
Next r
End Sub
```

In the `GetThemeColors` procedure, the `ThemeColorScheme` object represents the color scheme of a Microsoft Office theme. In the first `For Next` loop, the `Colors` method and the `RGB` property are used to return a specific color. The color value is then stored in the `colorArray` array variable and entered in the specified row of the first worksheet column. The second `For Next` loop applies background color to cells based on the color values stored in the `colorArray` variable.

The following procedure does more color work in the active sheet—this time, using the `Interior.ThemeColor` property:

```
Sub ApplyThemeColors()
  Dim i As Integer

  For i = 1 To 10
    ActiveSheet.Cells(i, 3).Interior.ThemeColor = i
    ActiveSheet.Cells(i, 4).Value = i
  Next i
End Sub
```

In this procedure, we use the `Interior.ThemeColor` property to set the background color of cells in the third worksheet column using colors available in the current color scheme. The color scheme value is then written in the corresponding cell in the fourth column. The resulting worksheet (after running the `GetThemeColors` and `ApplyThemeColors` procedures) is shown in Figure 17.13.

Each new workbook is created with a default theme named Office. The theme information is stored in a separate theme file with the extension .thmx. When you change the theme in the workbook, the workbook's theme file is automatically updated with the new settings.

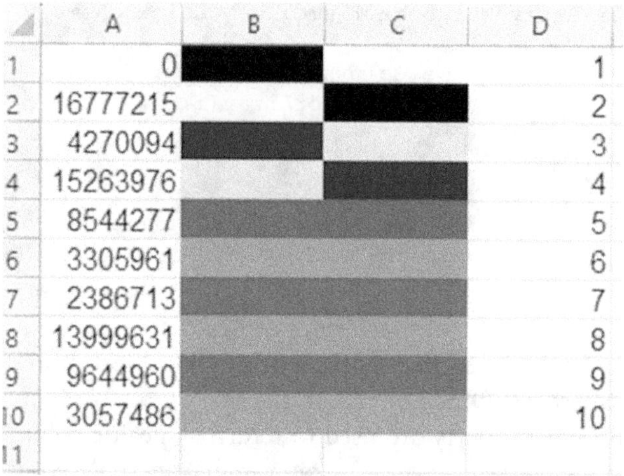

	A	B	C	D
1	0			1
2	16777215			2
3	4270094			3
4	15263976			4
5	8544277			5
6	3305961			6
7	2386713			7
8	13999631			8
9	9644960			9
10	3057486			10
11				

FIGURE 17.13. The background color of the cells was applied with VBA. When you select a different color theme, the background color of these cells (C1:C10) will automatically adjust.

When creating a custom theme with a new set of fonts, colors, or effects, save the theme in a file so that it can be used in any Office document or shared with other users. You will find your custom theme file in the following location:

```
\Users\<user name>\AppData\Roaming\Microsoft\Templates _
    \Document Themes
```

By default, the `Application Data (AppData)` folder is hidden. To access this folder, you need to modify the folder and search options in Windows Explorer.

For example, let's create a test theme named `MyTheme.thmx`. Choose Page Layout | Themes | Save Current Theme. Change the name to `MyTheme.thmx` and press Save.

Now that you've created a custom theme, you can apply it programmatically to the workbook using the `ApplyTheme` method of the `Workbook` object. Try it out now by entering the following statement on one line in the Immediate window (revise the path as necessary to match the location of the theme file on your computer):

```
ActiveWorkbook.ApplyTheme "C:\Users\<username>\AppData\ _
    Roaming\Microsoft\Templates\Document Themes\MyTheme.thmx"
```

After you apply a custom theme to a workbook, the theme name should appear in the Custom group of the Themes control, as shown in Figure 17.14.

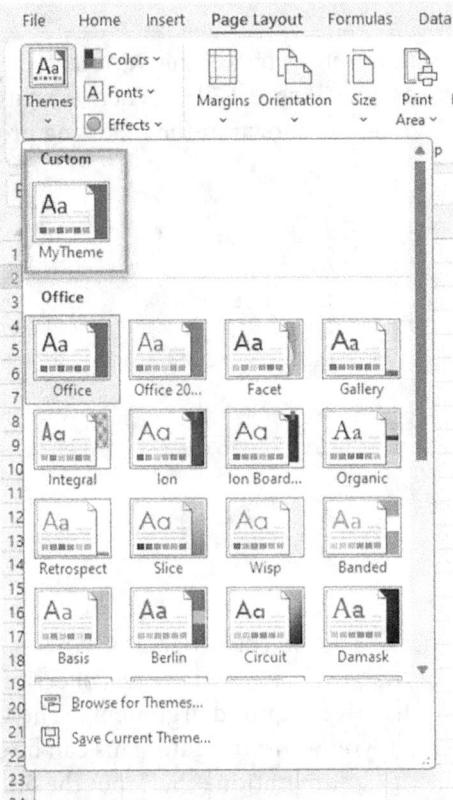

FIGURE 17.14. After applying a custom theme to a workbook, the theme name appears in the Custom group of the Themes control.

To programmatically load a color theme or font theme from a file, the following VBA statements can be used:

```
ActiveWorkbook.Theme.ThemeColorScheme.Load ("C:\Program Files\ _
    Microsoft Office\root\Document Themes 16\Theme Colors\ _
    Paper.xml")
```

```
ActiveWorkbook.Theme.ThemeFontScheme.Load " C:\Program Files\ _
    Microsoft Office\root\Document Themes 16\Theme Fonts\ _
    Calibri.xml")
```

To customize some theme components, you need to know how to work with document parts in the Office Open XML file format. You will find information on how to open, read, and modify data in Office XML files in Chapter 26.

Formatting with Shapes

You can make your worksheets more interesting by adding various types of shapes, such as the cylinder object shown in Figure 17.15. When formatting shapes, you can use document theme colors, as shown in the following procedure:

```
Sub AddCanShape()
    Dim oShape As Shape
    Worksheets(2).Activate
    ThisWorkbook.Worksheets(2).Cells.ClearFormats

    Set oShape = ActiveSheet.Shapes.AddShape(msoShapeCan, 54, _
        0, 54, 110)
    With oShape
        .Fill.ForeColor.ObjectThemeColor = msoThemeColorAccent4
        .Fill.Transparency = 0.5
        .Line.Visible = msoFalse
    End With
    Set oShape = Nothing
End Sub
```

In the `AddCanShape` procedure, we declare an object variable of type `Shape` and then use the `AddShape` method of the `ActiveSheet Shapes` collection to add a new `Shape` object. This method has five required arguments. The first one specifies the type of `Shape` object that you want to create. This can be one of the constants in the `msoAutoShapeType` enumeration (check out the online help). Excel offers a large number of shapes. The next two arguments tell Excel how far the object should be placed from the left and top corners of the worksheet. The last two arguments specify the width and height of the shape (in points). To specify the theme color of the `Shape` object, set the `Object-ThemeColor` property of the `ColorFormat` object to the required theme. To return the `ColorFormat` object, you must use the `ForeColor` property of the `FillFormat` object. The `FillFormat` object is returned by the `Fill` property of the `Shape` object:

```
oShape.Fill.ForeColor.ObjectThemeColor = msoThemeColorAccent4
```

Next, set the degree of transparency to make sure that the shape does not obstruct the data. The last line removes the border from the shape. The previous procedure places a `Shape` object over the data in column B.

	A	B	C	D	E	F
1	UserID	Qtr1	Qtr2	Qtr3	Qtr4	FY2021
2	JK0101	234	222	198	239	893
3	BK0001	289	199	300	287	1075
4	VG7891	261	233	288	178	960
5	MN0907	177	211	150	145	683
6	XT34531	160	149	201	230	740
7		1121	1014	1137	1079	4351
8						
9						

FIGURE 17.15. A Shape object placed on the worksheet uses the theme color scheme.

The following procedure can be run to programmatically remove shapes from the worksheet:

```
Sub RemoveShapes()
    Dim oShape As Shape
    Dim strShapeName As String
    With ActiveSheet
        For Each oShape In .Shapes
            strShapeName = oShape.Name
            oShape.Delete
            Debug.Print "The Shape Object named " _
            & strShapeName & " was deleted."
        Next oShape
    End With
End Sub
```

Working with Shapes in VBA

While you can look up the properties and methods of the Shape object in the online help, don't forget Excel's most useful programming tool—the macro recorder. Use Shape object recording to get a quick start in writing VBA code that inserts, positions, formats, and deletes shapes.

Formatting with Sparklines

Sparklines are tiny charts that can be inserted into a single cell to highlight important data trends and increase readers' comprehension. There are three types of sparklines in Excel: Line, Column, and Win/Loss. They can be inserted via the corresponding button in the Sparklines group on the ribbon's Insert tab (Figure 17.16).

To manually insert a sparkline graphic, click in the cell where you want the sparkline to appear and choose Insert. In the Sparklines group, click the type of sparkline you want to insert. At this point, Excel will pop up the Create Sparklines dialog box (Figure 17.17), where you can choose or enter the data range you want to include as the data source for the sparklines and the location range where you want them to be placed.

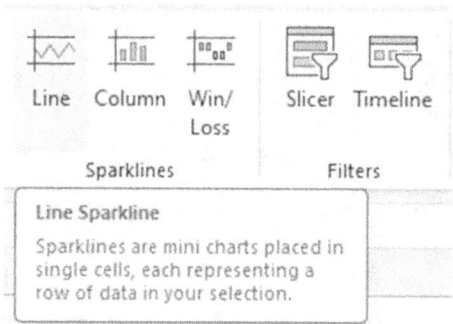

FIGURE 17.16. The Sparklines group on the Insert tab offers three buttons for inserting tiny charts into a cell.

	A	B	C	D	E	F	G	H
1	UserID	Qtr1	Qtr2	Qtr3	Qtr4	FY2021		
2	JK0101	234	222	198	239	893		
3	BK0001	289	199	300	287	1075		
4	VG7891	261	233	288	178	960		
5	MN0907	177	211	150	145	683		
6	XT34531	160	149	201	230	740		
7		1121	1014	1137	1079	4351		
8								

Create Sparklines   ?   ✕

Choose the data that you want

Data Range: B2:E6

Choose where you want the sparklines to be placed

Location Range: $G$2:$G$6

OK   Cancel

Sheet1   **Sheet2**   Sheet3   Sheet4   +

FIGURE 17.17. Selecting the location and data range for sparklines.

In Figure 17.17, we'd like to compare the performance of users during each quarter of fiscal year 2021. After making selections in the Create Sparklines dialog box, Excel activates the Sparkline Tools–Design context tab, where you can format the sparkline by showing points and markers, changing the sparkline or marker colors, switching between the types of sparklines, and more. Figure

17.18 displays the result of inserting and applying data point formatting to the sparklines.

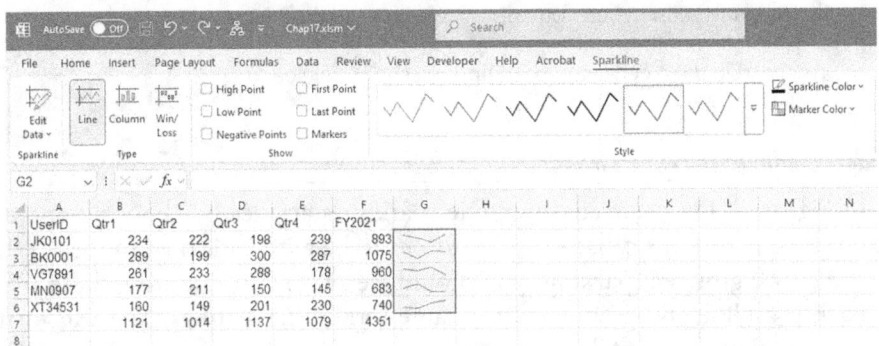

FIGURE 17.18. Sparklines in column G make it easy to spot the trends in users' performance.

Handling Hidden Data and Empty Cells by Sparklines

If you hide rows or columns, the hidden data will not appear in the sparkline. You can specify how empty cells should be handled in the Hidden and Empty Cells Settings dialog box (choose Sparkline | Edit Data | Hidden and Empty Cells).

Sparklines are dynamic. They will automatically adjust whenever the data in cells they are based on changes. Sparklines can be copied, cut, and pasted just like formulas. For bigger or wider sparklines, simply increase the width of the row or column. To get rid of the sparklines, use the Clear button in the ribbon's Sparkline tab, and select Clear Selected Sparklines or Clear Selected Sparkline Groups. Because a sparkline is a part of a cell's background, you can display text or formulas in cells containing sparklines, as shown in Figure 17.19.

	G2			f_x	=AVERAGE(B2:E2)	

	A	B	C	D	E	F	G
1	UserID	Qtr1	Qtr2	Qtr3	Qtr4	FY2021	
2	JK0101	234	222	198	239	893	223.25
3	BK0001	289	199	300	287	1075	268.75
4	VG7891	261	233	288	178	960	240
5	MN0907	177	211	150	145	683	170.75
6	XT34531	160	149	201	230	740	185
7		1121	1014	1137	1079	4351	
8							

FIGURE 17.19. Sparklines can share a cell with text or formulas.

> ### Sparklines and Backward Compatibility
>
> When you open a workbook containing sparklines in older versions of Excel (2007/2003), you will see blank cells. When you edit a file with sparklines in Excel 2007, the sparklines will not be visible, but they will appear again if the file is loaded in more recent versions of Excel, provided that the cells with sparklines were not deleted.

Understanding Sparkline Groups

In Figure 17.18, we created multiple sparklines at once by choosing one data range (B2:E6). This caused the sparklines to be automatically grouped. When you click a grouped sparkline, you should see a thin blue line around the group. Each sparkline group contains the same formatting settings. You can format each sparkline separately by breaking the group. Simply select the sparkline you want to format differently and choose Sparkline | Ungroup. The selected cell will be ungrouped from the group. Now, you can format that sparkline as desired without affecting other sparklines. For example, in Figure 17.20, the sparkline in row 4 was ungrouped and the type of sparkline was then changed to the Column type.

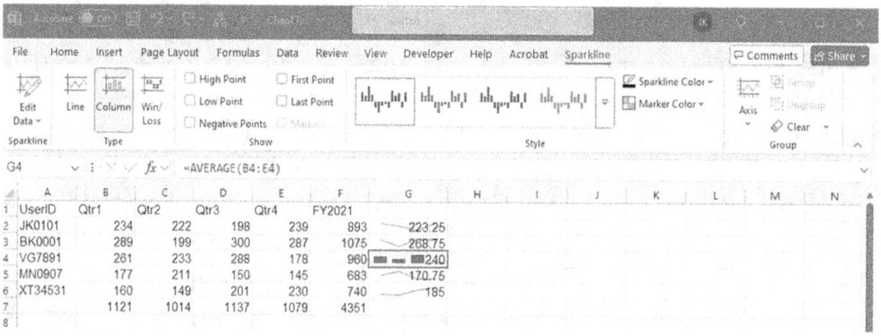

FIGURE 17.20. Ungrouping sparklines allows you to format them independently of other sparklines.

To break the entire sparkline group, select all the cells containing sparklines and click Ungroup.

Programming Sparklines with VBA

The Excel object model contains objects, properties, and methods that allow developers to use VBA to create and modify sparklines in their worksheets. To get an idea of how sparklines fit into the object model, take a look at Figure 17.21.

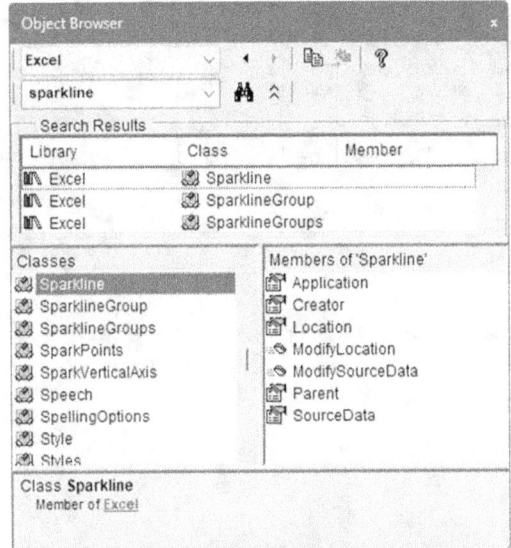

FIGURE 17.21.   Choose the Object Browser from the VBE View menu to quickly locate objects, methods, and properties that can be used to program sparklines with VBA.

Each sparkline is represented by a `Sparkline` object, which is a member of a `SparklineGroup` object. A `SparklineGroup` object can contain one or more `Sparkline` objects. You can have multiple `SparklineGroup` objects (collections of sparkline groups) in a worksheet. The `Range` object's `SparklineGroup` property returns a `SparklineGroups` object that represents an existing group of sparklines from the specified range.

Hands-On 17.2 demonstrates how to use VBA to read the information about sparklines contained in a worksheet.

(◉) Hands-On 17.2   Retrieving Information About Sparklines

1. In the `Chap17.xlsm` workbook, switch to the VBE screen and insert a new module.
2. In the module's Code window, enter the following `GetSparklineInfo` procedure:

```
Sub GetSparklineInfo()
    Dim spGrp As SparklineGroup
    Dim spCount As Long
    Dim i As Long

    spCount = Cells.SparklineGroups.count
    If spCount <> 0 Then
```

```
        For i = 1 To spCount
            Set spGrp = Cells.SparklineGroups(i)
            Debug.Print "Sparkline Group:" & i
            Select Case spGrp.Type
                Case 1
                    Debug.Print "Type:Line"
                Case 2
                    Debug.Print "Type:Column"
                Case 3
                    Debug.Print "Type:Win/Loss"
            End Select
            Debug.Print "Location:" & spGrp.Location.Address
            Debug.Print "Data Source: " & spGrp.SourceData
        Next i
    Else
        MsgBox "There are no sparklines in the active sheet."
    End If
End Sub
```

3. Activate the worksheet that contains sparklines and press Alt+F8 to display the Macro dialog box, then select and run the GetSparklineInfo macro.

4. Assuming you ran the procedure while the sheet displayed in Figure 17.20 was active, the following information should be retrieved into the Immediate window:

```
Sparkline Group:1
Type:Column
Location:$G$4
Data Source: B4:E4
Sparkline Group:2
Type:Line
Location:$G$2,$G$3,$G$5,$G$6
Data Source: B2:E2,B3:E3,B5:E5,B6:E6
```

In the next Hands-On, you will create a sparkline Win/Loss report from scratch using VBA. The Win/Loss sparkline bar chart type is used to display profit versus loss or positive versus negative comparisons.

(◉) Hands-On 17.3   Using VBA to Create Sparklines

1. In the Chap17.xlsm workbook, switch to the VBE screen and add the following two procedures just below the GetSparklineInfo procedure that you created in the previous Hands-On exercise:

```
Sub CreateSparklineReport()
  Dim spGrp As SparklineGroup
```

```
   Dim sht As Worksheet
   Dim cell As Range
   Dim spLocation As Range

   Workbooks.Add
   Set sht = ActiveSheet

   EnterData sht, 3, "Month", "Sales Quota", "Sales $", "Difference"
   EnterData sht, 4, "January", "234000", "250000", "=C4-B4"
   EnterData sht, 5, "February", "211000", "180000", "=C5-B5"
   EnterData sht, 6, "March", "304000", "370000", "=C6-B6"
   Range("B4:D6").Style = "Currency"

   Columns("A:D").AutoFit

   Range("A1").Value = "Win/Loss"
   Set spLocation = sht.Range("B1")
   Set spGrp = spLocation.SparklineGroups _
       .Add(xlSparkColumnStacked100, "D4:D6")
   spGrp.SeriesColor.ThemeColor = 2
   spLocation.SparklineGroups.Item(1) _
       .Axes.Horizontal.Axis.Visible = True
End Sub

Sub EnterData(sht As Worksheet, rowNum As Integer, _
   ParamArray myValues() As Variant)

   Dim j As Integer
   Dim count As Integer

   count = UBound(myValues()) + 1
   j = 1
   For j = j To count
       sht.Range(Cells(rowNum, 1), Cells(rowNum, count)) = myValues()
   Next
End Sub
```

2. Run the `CreateSparklineReport` procedure.
3. After running the procedure, you should see the Win/Loss report in a new workbook, as depicted in Figure 17.22. Notice that the Win/Loss sparkline in cell B1 compares sales data during the first three months of the year with the sales quota for each month. The source range for the sparkline is located in column D, which contains the difference between monthly sales and the monthly quota. A quick glance at cell B1 reveals that the sales quota was not met in February.

	A	B	C	D	E
1	Win/Loss	▬▬▬ ▬▬▬			
2					
3	Month	Sales Quota	Sales $	Difference	
4	January	$234,000.00	$250,000.00	$ 16,000.00	
5	February	$211,000.00	$180,000.00	$(31,000.00)	
6	March	$304,000.00	$370,000.00	$ 66,000.00	
7					

FIGURE 17.22.   This worksheet, including its Win/Loss sparkline in cell B1, was created programmatically in Hands-On 17.3.

Formatting with Styles

Most people use the Format Painter tool on the ribbon's Home tab (the paint-brush icon) to quickly copy formatting to other cells of the same worksheet or from one worksheet to another. When you create complex worksheets with different types of formatting, however, it is a good idea to save all your formatting settings in a file so you can reuse them whenever you need them. This can be done via the Styles feature. Cell styles can contain format options such as Number, Alignment, Font, Border, Fill, and Protection. If you change the style after you have applied it to your worksheet, Excel will automatically update the cells that have been formatted using that style.

Styles are easier to find and apply, thanks to the existence of galleries such as those shown in Figure 17.23. To apply a style to a cell, simply select the cells you want to format with the style and click on the appropriate style in the gallery

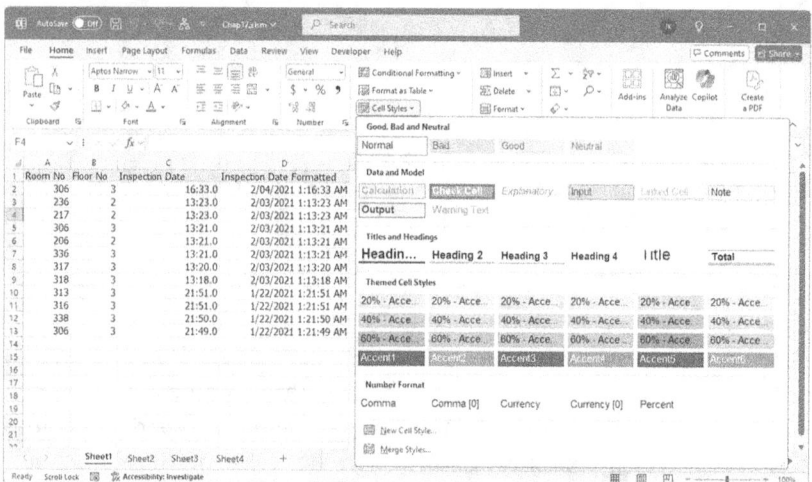

FIGURE 17.23.   You can see a preview of how the style will look before you apply the style. If the built-in style does not suit your needs, you can create your own custom style.

(available by clicking Cell Styles). Styles are based on the current theme. You can also apply them to Excel tables, PivotTables, charts, and shapes. Excel offers a large number of built-in styles. You can modify, duplicate, or delete the existing styles and add your own—simply right-click the style in the gallery and select the option you need.

To find out the number of styles in the active workbook, use the following statement:

```
MsgBox "Number of styles=" & ActiveWorkbook.Styles.Count
```

Excel tells us that there are 47 styles defined for a 2024 workbook. Use the `Styles` collection and the `Style` object to control the styles in a workbook.

To get a list of style names, let's iterate through the `Styles` collection:

```
Sub GetStyleNames()
  Dim i As Integer

  For i = 1 To ActiveWorkbook.Styles.Count
    Debug.Print "Style " & i & ":" & _
      ActiveWorkbook.Styles(i).Name
  Next i
End Sub
```

The above procedure prints the names of all workbook styles to the Immediate window. The style names are listed alphabetically.

To add a style, use the `Add` method, as shown in the following example procedure:

```
Sub AddAStyle()
  Dim newStyleName As String
  Dim curStyle As Variant
  Dim i As Integer

  newStyleName = "SimpleFormat"
  i = 0

  For Each curStyle In ActiveWorkbook.Styles
      i = i + 1
      If curStyle.Name = newStyleName Then
          MsgBox "This style " & "(" & newStyleName & _
              ") already exists. " & Chr(13) & _
              "It's the " & i & " style in the Styles _
                  collection."
          Exit Sub
      End If
  Next
```

```
With ActiveWorkbook.Styles.Add(newStyleName)
    .Font.Name = "Arial Narrow"
    .Font.Size = "12"
    .Borders.LineStyle = xlThin
    .NumberFormat = "$#,##0_);[Red]($#,##0)"
    .IncludeAlignment = False
End With
End Sub
```

The AddStyle procedure adds a specified style to the workbook's Styles collection provided that the style name is unique. The procedure begins by checking whether the style name has already been defined. If the workbook has a style with the specified name, the procedure ends after displaying a message to the user. If the style name does not exist, then the procedure creates the style with the specified formatting. Notice that if you do not wish to include a specific formatting feature in the style, you can set the following properties to False: IncludeAlignment, IncludeFont, IncludeBorder, IncludeNumber, IncludePatterns, and IncludeProtection.

For example, a setting of False omits the HorizontalAlignment, VerticalAlignment, WrapText, and Orientation properties in the style. The default setting for these properties is True.

The custom style is added to the Styles collection.

To find out the index number of the newly added style, simply rerun this procedure.

To programmatically apply your custom style to a selected range, run the following code in the Immediate window:

```
Selection.Style = "SimpleFormat"
```

To check out the settings the specific style includes, select the formatted range of cells and choose Home | Cell Styles to display the gallery of styles. Right-click the name of the selected style and choose Modify. Excel displays the dialog box shown in Figure 17.24.

The following code removes formatting applied to the selected range:

```
Selection.ClearFormats
```

The previous statement returns selection formats to the original state but does not remove the style from the Styles collection. To delete a style from a workbook, use the following statement:

```
ActiveWorkbook.Styles("SimpleFormat").Delete
```

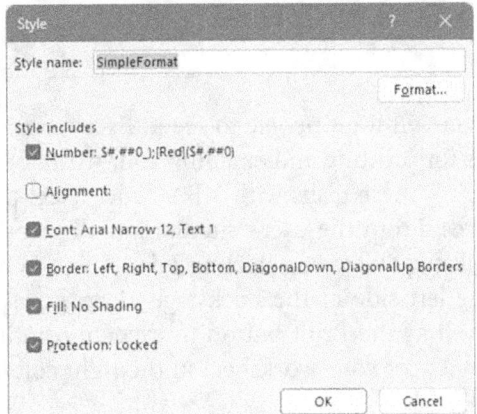

FIGURE 17.24. This dialog box displays the formatting settings of the SimpleFormat style that was created earlier by a VBA procedure.

If you have already applied formats to a cell, you can create a new style based on the active cell:

```
Sub AddSelectionStyle()
  Dim newStyleName As String

  newStyleName = "InvoiceAmount"
  ActiveWorkbook.Styles.Add Name:=newStyleName, _
      BasedOn:=ActiveCell
End Sub
```

By default, Excel creates a new style based on the Normal style. If you have already applied formatting to a specific cell and would like to save the settings in a style, however, use the optional BasedOn argument of the Styles collection Add method to specify a cell on which to base the new style.

The custom styles you create can be reused in other workbooks. To do this, you must copy the style information from one workbook to another. In VBA, this can be done by using the Merge method of the Workbook object Styles collection:

```
ActiveWorkbook.Styles.Merge "YourWorkbook.xlsx"
```

Assuming that you have defined some cool styles in the YourWorkbook.xlsx workbook, the above statement copies the styles found in the specified workbook to the active workbook.

FORMATTING WORKSHEETS FOR PRINTING AND EMAILING

After you set up your worksheet, you will want people to see it. Excel provides easy-to-use commands and buttons for printing and emailing your workbooks. In addition, programmers can control these tasks with VBA code. Excel provides easy access to all print features. From the user's standpoint, all printing options can be accessed in the Backstage view by selecting File | Print. When the Print command is selected, the left side of the Backstage view shows all available options for printing, as well as the Print button to execute printing. You automatically see the print preview of your worksheet in the right column of that window.

In the following sections, you will work with printing features as a developer. You will learn about methods of accessing and setting printing options and displaying print previews using VBA statements. These statements will allow you also to display the Print dialog box and Print Preview window as they were present in earlier versions of Excel. You will use these statements to automatically set printers and printing options in your VBA programs.

Controlling the Page Setup

You can control the look of your printed worksheet pages via the Page Layout tab on the ribbon. The Page Layout tab, shown in Figure 17.25, is divided into groups that include settings related to page setup (margins, orientation, and size), scaling, and sheet options. You may programmatically access these settings via the Page Setup dialog box, using the properties of the `PageSetup` object. To display the Page Setup dialog box, type the following statement in the Immediate window and press Enter:

```
Application.Dialogs(xlDialogPageSetup).Show
```

The previous statement uses the `Show` method of the `Dialogs` object to display the built-in Page Setup dialog box. You can include a list of arguments after the `Show` method. To set initial values in the Page Setup dialog box, use the arguments in Table 17.7.

FIGURE 17.25. The Page Layout tab allows you to specify the page margins, orientation, paper size, and scaling and sheet options, along with other settings.

TABLE 17.7. Show method arguments for the Dialogs object.

Argument Number	Argument Name
Arg1	Head
Arg2	Foot
Arg3	Left
Arg4	Right
Arg5	Top
Arg6	Bot
Arg7	Hdng
Arg8	Grid
Arg9	h_cntr
Arg10	v_cntr
Arg11	Orient
Arg12	paper_size
Arg13	Scale
Arg14	pg_num
Arg15	pg_order
Arg16	bw_cells
Arg17	Quality
Arg18	head_margin
Arg19	foot_margin
Arg20	Notes
Arg21	Draft

If you don't specify the initial settings, the Page Setup dialog box appears with its default settings. Let's focus on using these arguments. If you wish to display the Page Setup dialog box with the page orientation set to landscape, use the following statement:

```
Application.Dialogs(xlDialogPageSetup).Show Arg11:=2
```

Excel uses 1 for portrait and 2 for landscape orientation.

The following statement displays the Page Setup dialog box in which the Center on page Horizontally setting is selected on the Margins tab (Arg9:=1), and the Page tab has the Orientation option set to Portrait (Arg11:=1):

```
Application.Dialogs(xlDialogPageSetup).Show Arg9:=1, Arg11:=1
```

You can also set the initial values in the Page Setup dialog box by using the PageSetup object with its appropriate properties. For example, to set the page orientation as landscape, type the following statements on one line in the Immediate window and press Enter:

```
ActiveSheet.PageSetup.Orientation = 2 :Application. _
    Dialogs(xlDialogPageSetup).Show
```

These two statements are executed one after another. The colon indicates the end of the first statement and the beginning of another. This is a handy shortcut that can be used in the Immediate window to run a block of code.

The following sections describe various page settings (and the corresponding properties of the `PageSetup` object) that you may want to specify before printing your worksheets.

Controlling the Settings on the Page Layout Tab

The settings on the Page Layout tab (see Figure 17.25 in the previous section) are grouped into five main areas: Themes, Page Setup, Scale to Fit, Sheet Options, and Arrange. The Orientation settings in the Page Setup group indicate whether the page will be printed in portrait or landscape view (`Orientation` property). The Size setting lets you select one of the common paper sizes, such as Letter, Legal, Executive, A4, and so on (`PaperSize` property). The options in the Scale to Fit group enable you to adjust the printout according to your needs. You can reduce or enlarge the worksheet by using the Scale setting in the Scale to Fit group of the Page Layout tab. Excel can automatically scale a printout to fit a specified number of pages with the Width and Height settings.

Here are some VBA statements that you can use to access and set various options on the Page Layout tab:

- Set `Sheet1` to be printed in landscape orientation:

  ```
  Worksheets("Sheet1").PageSetup.Orientation = xlLandscape
  ```

- Scale `Sheet1` for printing by 200%:

  ```
  Worksheets("Sheet1").PageSetup.Zoom = 200
  ```

- Scale the worksheet so it prints exactly one page tall and wide:

  ```
  With Worksheets("Sheet1").PageSetup
    .FitToPagesTall = 1
    .FitToPagesWide = 1
  End With
  ```

- Set the paper size to Legal for `Sheet1`.

  ```
  Worksheets("Sheet1").PageSetup.PaperSize = xlPaperLegal
  ```

- Return the current setting for the horizontal and vertical print quality:

```
Debug.Print "Horizontal Print Quality = " &
     Worksheets("Sheet1").PageSetup.PrintQuality(1)
Debug.Print "Vertical Print Quality = " &
     Worksheets("Sheet1").PageSetup.PrintQuality(2)
```

Controlling the Settings on the Margins Tab

The settings available on the Margins tab of the Page Setup dialog box shown in Figure 17.26 allow you to specify the width of the top, bottom, left, and right margins (TopMargin, BottomMargin, LeftMargin, and RightMargin properties). The Header and Footer settings allow you to determine how far you'd like the header or footer to be printed from the top or bottom of the page (HeaderMargin and FooterMargin properties). The print area can be centered on the page horizontally and vertically (CenterHorizontally and CenterVertically properties).

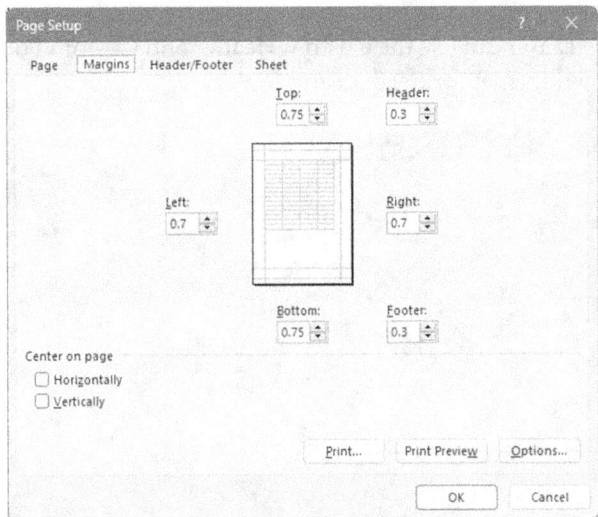

FIGURE 17.26. The settings on the Margins tab of the Page Setup dialog box determine the margins around the print area and the manner in which the print area should be centered on the printed page.

Let's look at some VBA statements below:

- Set all page margins (left, right, top, and bottom) to 1.5 inches:

```
With Worksheets("Sheet1").PageSetup
     .LeftMargin = Application.InchesToPoints(1.5)
     .RightMargin = Application.InchesToPoints(1.5)
     .TopMargin = Application.InchesToPoints(1.5)
```

```
        .BottomMargin = Application.InchesToPoints(1.5)
End With
```

- Set header and footer margins to 0.5 inch:

```
With Worksheets("Sheet1").PageSetup
    .HeaderMargin = Application.InchesToPoints(0.5)
    .FooterMargin = Application.InchesToPoints(0.5)
End With
```

- Center Sheet1 horizontally when it's printed:

```
With Worksheets("Sheet1").PageSetup
    .CenterHorizontally = True
    .CenterVertically = False
End With
```

Controlling the Settings on the Header/Footer Tab

Figure 17.27 shows the settings on the Header/Footer tab of the Page Setup dialog box. These settings allow you to add built-in or custom headers and footers to your printed worksheets. You can use the Custom Header and Custom Footer buttons to design your own format for headers and footers.

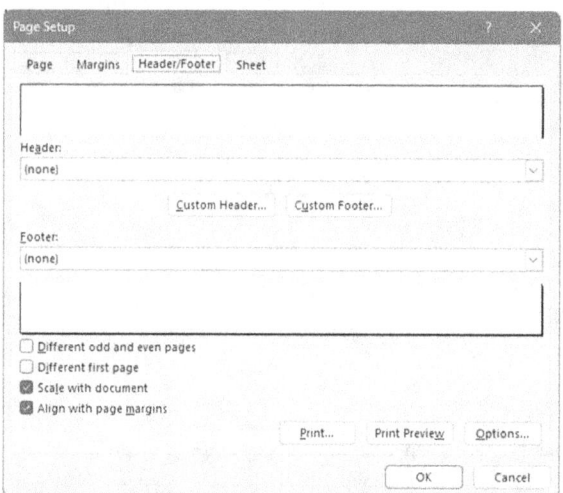

FIGURE 17.27. The Header/Footer tab of the Page Setup dialog box allows you to select one of the built-in headers or footers or create your own custom header and footer formats.

The PageSetup object has the following properties for setting up and controlling the creation of headers and footers: RightHeader, LeftHeader, Right-Footer, LeftFooter, CenterHeader, CenterFooter, RightHeaderPicture,

`RightFooterPicture`, `LeftHeaderPicture`, `LeftFooterPicture`, `Center-HeaderPicture`, and `CenterFooterPicture`.

The following settings are available on the Header/Footer tab:

- Different odd and even pages—Use the `PageSetup.OddAndEvenPages-HeaderFooter` property. This property returns `True` if the specified `PageSetup` object has different headers and footers for odd-numbered and even-numbered pages.

- Different first page—Use the `PageSetup.DifferentFirstPageHeaderFooter` property. This property returns `True` if a different header or footer is used on the first page.

- Scale with document—Use the `PageSetup.ScaleWithDocHeaderFooter` property. This property returns `True` if the header and footer should use the same font size and scale as the worksheet.

- Align with page margins—Use the `PageSetup.AlignMarginsHeaderFooter` property. This property returns `True` for Excel to align the header and the footer with the margins set in the page setup options.

Special formatting codes can be used in the header and footer text, as shown in Table 17.8.

TABLE 17.8. Formatting codes for headers and footers.

Format Code	Description
&L	Left-aligns the characters that follow.
&C	Centers the characters that follow.
&R	Right-aligns the characters that follow.
&E	Turns double-underline printing on or off.
&X	Turns superscript printing on or off.
&Y	Turns subscript printing on or off.
&B	Turns bold printing on or off.
&I	Turns italic printing on or off.
&U	Turns underline printing on or off.
&S	Turns strikethrough printing on or off.
&D	Prints the current date.
&T	Prints the current time.
&F	Prints the name of the document.
&A	Prints the name of the workbook tab.
&P	Prints the page number.

(Contd.)

Format Code	Description
&P+number	Prints the page number plus the specified number.
&P-number	Prints the page number minus the specified number.
&&	Prints a single ampersand.
& "fontname"	Prints the characters that follow in the specified font. Be sure to include the double quotation marks.
&nn	Prints the characters that follow in the specified font size. Use a two-digit number to specify a size in points.
&N	Prints the total number of pages in the document.
&G	Enables the image to show up in the header or footer.

Let's review VBA statements that use the above codes:

- Print the full path of the workbook in the upper-right corner of every page when Sheet1 is printed:

```
Worksheets("Sheet1").PageSetup.RightHeader = _
    ActiveWorkbook.FullName
```

- Print the date, page number, and total number of pages on the left at the bottom of each page when Sheet1 is printed:

```
Worksheets("Sheet1").PageSetup.LeftFooter = _
    "&D Page &P of &N"
```

- Display a watermark in the center section of the header on Sheet1:

```
Sub AddWatermarkToSheet1()
    Dim ws As Worksheet
    Dim picPath As String
    Dim header As String

    ' Change this to the path of your picture
    picPath = "C:\VBAExcel2024_ByExample\hardware.jpg"

    ' Define header with watermark picture
    header = "&C&G"

    ' Set the worksheet to Sheet1
    Set ws = ThisWorkbook.Sheets("Sheet1")

    ' Set the header to include the picture
    ws.PageSetup.CenterHeader = header
    ' Add the watermark picture to the header
    ws.PageSetup.CenterHeaderPicture.Filename = picPath
    ' Adjust picture settings
```

```
With ws.PageSetup.CenterHeaderPicture
.Height = 100 ' Adjust height as needed
.Width = 100 ' Adjust width as needed
.Brightness = 0.85 ' Adjust brightness as needed
.ColorType = msoPictureWatermark
.LockAspectRatio = msoTrue ' Maintain aspect ratio
End With
'Adjust margins
ws.PageSetup.TopMargin = Application.InchesToPoints(1.75)
ws.PageSetup.HeaderMargin = Application.
                                    InchesToPoints(0.5)


End Sub
```

The `AddWatermarkToSheet1` procedure will produce the output shown in Figure 17.28.

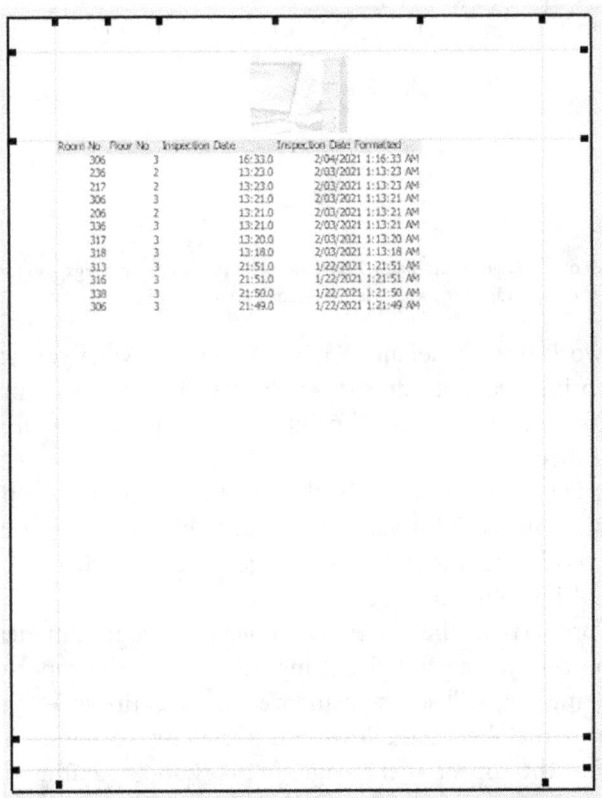

FIGURE 17.28. After running the AddWatermarkToSheet1 VBA procedure, a watermark picture is added to the center of a header in an Excel worksheet.

Controlling the Settings on the Sheet Tab

The settings available on the Sheet tab of the Page Setup dialog box (shown in Figure 17.29) determine the types of data to be included in the printout and the order in which Excel should proceed to print data ranges if the printout will span multiple pages.

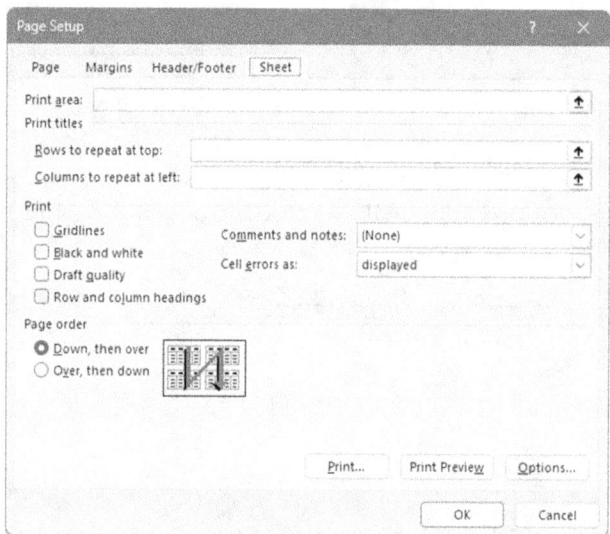

FIGURE 17.29. The Sheet tab of the Page Setup dialog box allows you to specify headings and ranges of data to appear on your printout and adjust the appearance of each page.

Excel prints the entire worksheet by default. You can print only what you actually want by defining a print area. You can specify the range address to print in the Print area setting. If you do not specify a print area, Excel will print all the data in the current worksheet.

If you specify the range of cells to print in the Print area setting and then choose the Selection option in the Print dialog box, Excel will print the current selection of cells in the worksheet instead of the range of cells specified in the Print area setting of the Print Setup dialog box.

Use the `PrintArea` property of the `PageSetup` object to programmatically return or set the range of cells to be printed. Setting the `PrintArea` property to `False` or to an empty string (`""`) will set the print area to the entire sheet.

The Print titles area on the Sheet tab allows you to specify workbook rows that should be printed at the top of every page, or workbook columns that should be printed on the left side of every page. These settings are especially useful for printing very large worksheets.

By default, Excel prints your row and column titles only on the first page, making it very difficult to understand data on subsequent pages. To fix this problem, you can tell Excel to print the specified row and column headings on every page. Specify the rows that contain the cells to be repeated at the top of each page in the Rows to repeat at top setting (`PrintTitleRows` property), and specify the columns that contain cells to be repeated on the left side of each page in the Columns to repeat at left setting (`PrintTitleColumns` property). You should specify both of these settings for extremely large worksheets. To turn off the title rows or title columns, you may want to set the corresponding property (`PrintTitleRows` or `PrintTitleColumns`) to `False` or to an empty string (`""`).

The Print settings control the look of your printed pages. To print the worksheet with gridlines, check the Gridlines box (`PrintGridlines` property). To print colors as shades of gray, select the Black and white box (`BlackandWhite` property). Draft-quality printing will be faster since Excel does not print gridlines and suppresses some graphics in this mode.

To show row and column headings on the printed pages, check the Row and column headings box (`PrintHeadings` property). Excel will identify the rows with numbers and worksheet columns with letters or numbers, depending on the style setting in the Excel Options dialog box (choose File | Excel Options | Formulas and see the R1C1 Reference style box).

If your worksheet contains comments, you can indicate the position on the printed page where you would like to have them printed by choosing an option from the Comments and notes drop-down box (`PrintComments` property).

If the worksheet contains errors, you can suppress the display of error values when printing a worksheet by making a selection from the Cell errors as drop-down box (`PrintErrors` property). When using the `PrintErrors` property, specify how you would like errors to be displayed with one of the following constants:

- `xlPrintErrorsBlank`
- `xlPrintErrorsDash`
- `xlPrintErrorsDisplayed`
- `xlPrintErrorsNA`

The settings in the Page order area of the Sheet tab allow you to specify how Excel should print and number pages when printing large spreadsheets. The default printing order is from top to bottom. You may request this order to be changed to left to right, which is a convenient way to print wide tables. Use the `OrderProperty` of the `PageSetup` object to set or return the print order. The page order can be one of the following constants: xlDownThenOver or xlOver-ThenDown.

Here are some examples of VBA statements:

- Set the print area as cells A2:D10 on Sheet1:

```
Worksheets("Sheet1").PageSetup.PrintArea = "$A$2:$D$10"
```

- Specify row 1 as the title row and columns A and B as title columns:

```
ActiveSheet.PageSetup.PrintTitleRows = _
    ActiveSheet.Rows(1).Address
ActiveSheet.PageSetup.PrintTitleColumns = _
    ActiveSheet.Columns("A:B").Address
```

- Print gridlines and column headings on Sheet1:

```
With Worksheets("Sheet1").PageSetup
   .PrintHeadings = True
   .PrintGridlines = True
End With
```

- Number and print the worksheet starting from the first page to the pages to the right, and then move down and continue printing across the sheet:

```
Worksheets("Sheet1").PageSetup.Order = xlOverThenDown
```

Retrieving Current Values from the Page Setup Dialog Box

Now that you are familiar with the many settings available in the Page Setup dialog box and know the names of the corresponding properties that can be used in VBA to write code that sets up your worksheets for printing, it's time for a Hands-On exercise.

Hands-On 17.4   Printing Page Setup Settings to the Immediate Window

1. In the Chap17.xlsm workbook VBE screen, insert a new module, and enter the ShowPageSettings procedure, as shown below:

```
Sub ShowPageSettings()
    ThisWorkbook.Sheets("Sheet2").Activate
    With ActiveSheet.PageSetup
        Debug.Print "Orientation=" & .Orientation
        Debug.Print "Paper Size = " & .PaperSize
        Debug.Print "Print Gridlines = " & .PrintGridlines
        Debug.Print "Print Area = " & .PrintArea
    End With
End Sub
```

2. Run the `ShowPageSettings` procedure.

The results of the procedure are printed to the Immediate window. Because we have not changed any settings in the Page Setup dialog box, the values you see after the equal signs are the default values.

3. Modify the `ShowPageSettings` procedure as follows:

```
Sub ShowPageSettings_Modified()
    ThisWorkbook.Sheets("Sheet2").Activate
    With ActiveSheet.PageSetup
        Debug.Print "Orientation=" & .Orientation
        Debug.Print "Paper Size = " & .PaperSize
        .PrintGridlines = True
        Debug.Print "Print Gridlines = " & .PrintGridlines
        Cells(1, 1).Select
        .PrintArea = ActiveCell.CurrentRegion.Address
        Debug.Print "Print Area = " & .PrintArea
        .CenterHeader = Chr(10) & "Quarterly Performance _
            Comparison"
    End With
    Application.Dialogs(xlDialogPrintPreview).Show
End Sub
```

4. Run the `ShowPageSettings_Modified` procedure.

Now, in addition to writing selected settings to the Immediate window, the test worksheet is formatted and displayed in the Print Preview window.

When printing worksheets that contain a large number of rows, it is a good idea to separately set the print titles and print area so that each page is printed with the column titles. The following procedure demonstrates this particular scenario. Notice how the `CurrentRegion` property of the `Range` collection is used together with the `Offset` and `Resize` properties to resize the print area so that it does not include the header row (row 1). The procedure sets the header row using the `PrintTitleRows` property of the `PageSetup` object:

```
Sub FormatSheet()
    Dim curReg As Range

    ThisWorkbook.Sheets("Sheet3").Activate
    Cells(1, 1).Select
    Set curReg = ActiveCell.CurrentRegion

    With ActiveSheet.PageSetup
        .PrintTitleRows = "$1:$1"
        Cells(1, 1).Select
```

```
        .PrintArea = curReg.Offset(1, 0).Resize _
            (curReg.Rows.count - 1, curReg.Columns.count)._
                Address
        Debug.Print "Print Area = " & .PrintArea
        .CenterHeader = Chr(10) & "Bonus Information Sheet"
        .PrintGridlines = True
    End With
    Application.Dialogs(xlDialogPrintPreview).Show
End Sub
```

PREVIEWING A WORKSHEET

As you can see in Figure 17.30, the Page Layout view (View | Page Layout) makes it easy to see how the worksheet will print.

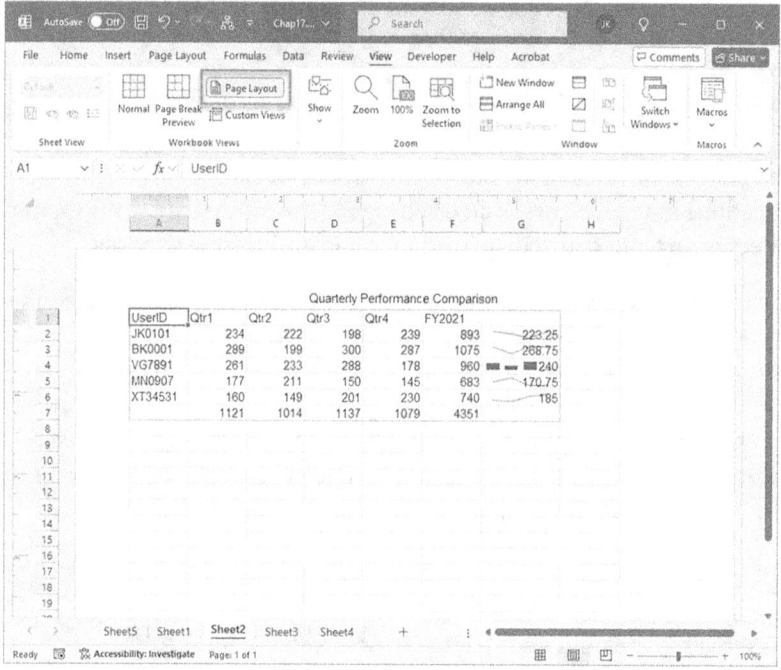

FIGURE 17.30. To view the worksheet as it would look when printed, choose View | Page Layout.

To add a header, simply click the top area of the worksheet and type your text, or click the appropriate buttons in the Design tab of the Header & Footer Tools on the ribbon. For example, to add today's date, click the Current Date button.

To add the footer, click the Go to Footer button in the Navigation group of the Header & Footer Tools Design tab.

Use the following VBA statement to activate the Page Layout view:

```
ActiveWindow.View =xlPageLayoutView
```

You can also display your worksheet in the Print Preview window by typing either one of the following statements in the Immediate window or in your VBA procedure:

```
Application.Dialogs(xlDialogPrintPreview).Show
```

or

```
Worksheets("Sheet2").PrintPreview
```

The above statements will not work if the worksheet has no data.

As you can see in Figure 17.31, there are buttons at the top of the Print Preview window that allow you to move between individual pages of your printout, make adjustments to the page setup and margins, get a closer look at the data or any part of the printout (Zoom button), and print your worksheet.

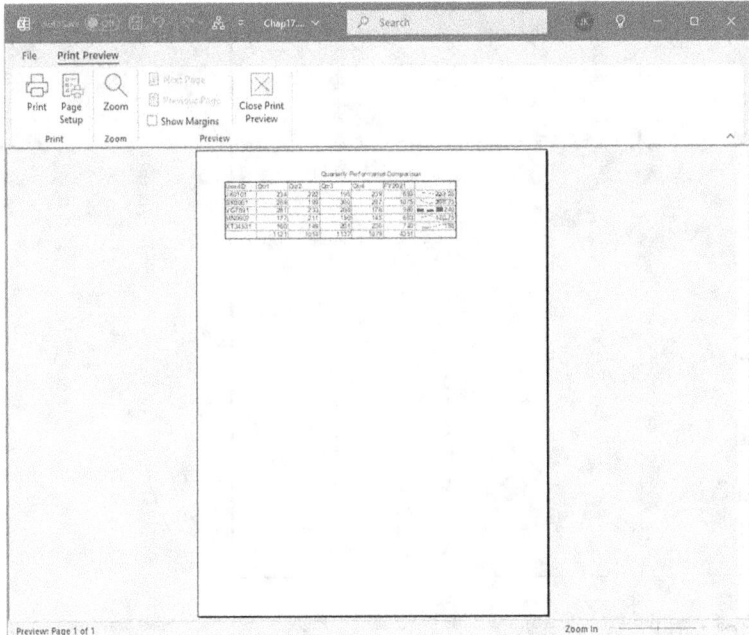

FIGURE 17.31. This Print Preview window displays a scaled-down version of the worksheet pages. Notice that this is an entirely different window from the one opened via the File | Print command or the Print Preview and Print button, which is available in the Quick Access Toolbar.

Sometimes you may want to prevent users from modifying the page setup or printing from the Print Preview window. This can be accomplished with VBA.

You can disable the Show Margins and Page Setup buttons in the Print Preview window in one of the following ways:

```
Application.Dialogs(xlDialogPrintPreview).Show False
```

or

```
Worksheets("Sheet2").PrintPreview EnableChanges:=False
```

or

```
Worksheets("Sheet2").PrintPreview False
```

Figure 17.32 shows the Print Preview window with the Show Margins option and Page Setup button disabled.

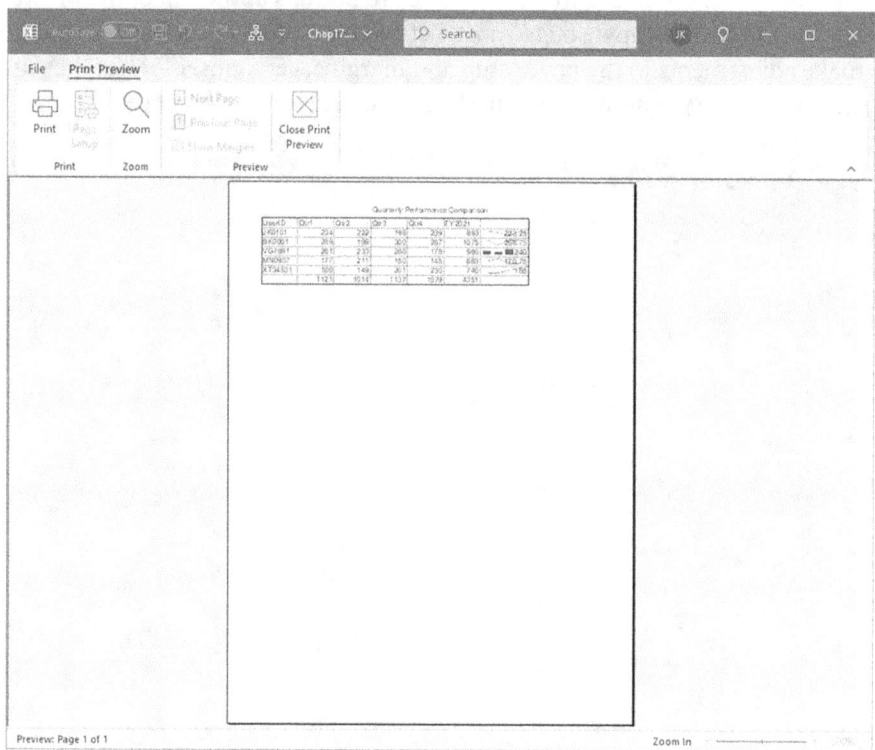

FIGURE 17.32.   The Print Preview window with the disabled Page Setup and Show Margins options.

CHANGING THE ACTIVE PRINTER

Before printing, you may want to display a list of printers for the users to select from or force a print job to go to a specific printer. The Printer Setup dialog box is shown in Figure 17.33. This dialog box can be displayed with the following statement:

```
Application.Dialogs(xlDialogPrinterSetup).Show
```

To find out the name of the active printer, use the `ActivePrinter` property of the `Application` object:

```
MsgBox Application.ActivePrinter
```

To change the active printer, use the following statement, replacing the printer's name and port with your own:

```
Application.ActivePrinter = "Brother DCP-7065DN Printer on Ne08:"
```

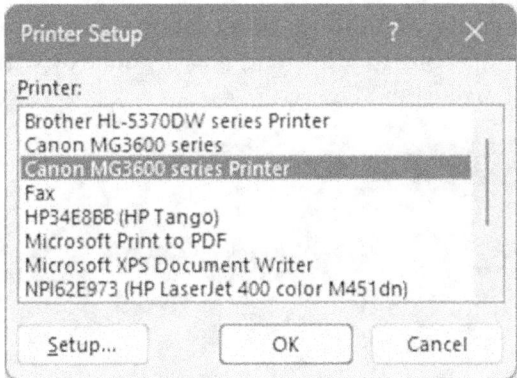

FIGURE 17.33.   The Printer Setup dialog box will display a list of printers available on your computer/ network.

You can tell Excel to set your default printer on opening a specific workbook by writing a simple `Auto_Open` macro.

Hands-On 17.5   Setting a Default Printer When Opening a Specific Workbook

1. In the `Chap17.xlsm` workbook, switch to the VBE screen and insert a new module.

2. In the module Code window, enter the `Auto_Open` procedure as shown below, replacing the printer's name with the name of your own printer:

```
Sub Auto_Open()
  Application.ActivePrinter = "Brother DCP-7065DN _
    Printer on Ne08:"
  MsgBox Application.ActivePrinter
End Sub
```

The printer must be connected for this code to execute.

3. Save the `Chap17.xlsm` workbook and close it. Do not exit Excel.
4. Reopen the `Chap17.xlsm` workbook.
   Excel will run the `Auto_Open` macro and display the active printer's name in the message box.

PRINTING A WORKSHEET WITH VBA

Before printing a worksheet, you may want to set print options such as print ranges, collation, or the number of copies to print. This is easily done by setting appropriate options in the Print dialog box (Figure 17.34).

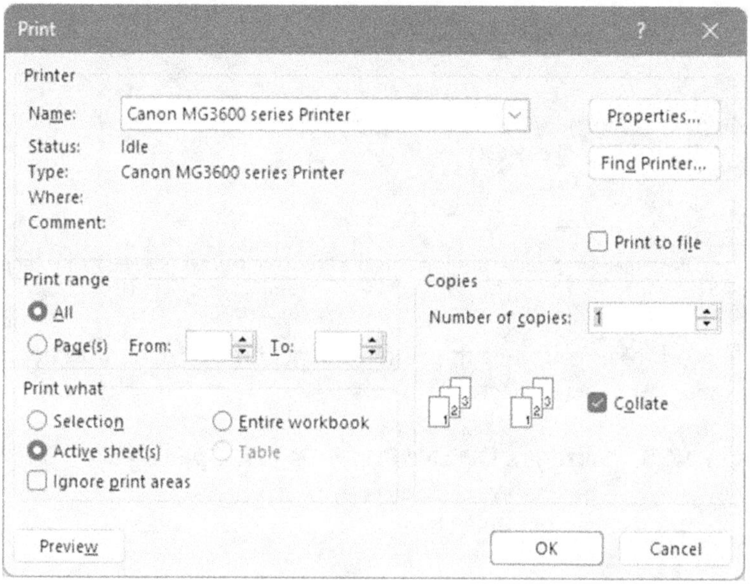

FIGURE 17.34. This Print dialog box allows you to specify various print options.

To display the Print dialog box programmatically, use the following statement:

```
Application.Dialogs(xlDialogPrint).Show
```

You can include a list of arguments after the `Show` method to set initial values in the Print dialog box (see Table 17.9).

TABLE 17.9.   Show method arguments for the Print dialog.

Argument Number	Argument Description
Arg1	range_num
Arg2	From
Arg3	To
Arg4	Copies
Arg5	Draft
Arg6	Preview
Arg7	print_what
Arg8	Color
Arg9	Feed
Arg10	Quality
Arg11	y_resolution
Arg12	Selection
Arg13	printer_text
Arg14	print_to_file
Arg15	Collate

For example, the following statement will print pages 1 to 2 of the active worksheet (assuming that the worksheet consists of two or more pages):

```
Application.Dialogs(xlDialogPrint).Show Arg1:=2, Arg2:=1, Arg3:=2
```

The first argument specifies the Page(s) option button in the Print range area of the Print dialog box. To select the All option button in the Print range area, set `Arg1` to 1.

The second and third arguments specify the pages you want to print (the beginning page number and the last page number to print should be specified).

To send your worksheet directly to the printer (without going through the Print dialog box), use the following statement:

```
ActiveSheet.PrintOut
```

The `PrintOut` method can take the arguments listed in Table 17.10.

TABLE 17.10.   PrintOut method arguments.

Argument Name	Argument Description
`From`	The number of the first page to print. If omitted, printing will start from the first page.
`To`	The number of the last page to print. If omitted, printing will end with the last page.
`Copies`	The number of copies to print. If omitted, one copy will be printed.
`Preview`	If set to `True`, Excel will display the Print Preview window before printing. If omitted or set to `False`, printing will begin immediately.
`ActivePrinter`	Sets the name of the active printer.
`PrintToFile`	If `True`, the worksheet is printed to a file. This is convenient when you want to print a worksheet on an offsite printer such as a PostScript printer. You should supply the filename in the `PrToFileName` argument.
`Collate`	Set this argument to `True` to collate multiple copies.
`PrToFileName`	Specifies the name of the file you want to print to if the `PrintToFile` argument is set to `True`.

DISABLING PRINTING AND PRINT PREVIEWING

At times, you may not want to allow the printing or print previewing of the worksheet. You can remove these features by customizing the ribbon's File tab. To disable the Print Preview window, write the `Workbook_BeforePrint` event procedure (see Figure 17.35 in the next section).

USING PRINTING EVENTS

Before the workbook is printed (and before the Print dialog box appears), Excel triggers the `Workbook_BeforePrint` event. You can use this event to perform certain formatting or calculating tasks prior to printing or to cancel printing and print previewing entirely when these features are requested. The code for the `Workbook_BeforePrint` event procedure must be placed in the ThisWorkbook Code window (Figure 17.35).

The ThisWorkbook Code window can be accessed by double-clicking the appropriate workbook name in the Project Explorer window of the VBE screen and double-clicking the ThisWorkbook object. Next, at the top of the ThisWorkbook Code window, select Workbook from the Object drop-down list on the left. The Procedure drop-down list on the right will display the names of

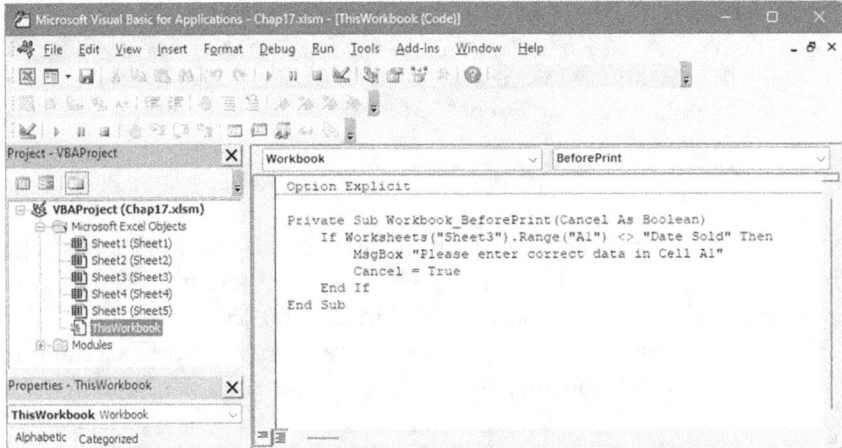

FIGURE 17.35. Writing the Workbook_BeforePrint event procedure in the ThisWorkbook Code window.

the events that the `Workbook` object can respond to. Select the BeforePrint event name; Excel will place the skeleton of this procedure in the Code window. Type your VBA code between the `Sub` and `End Sub` lines. The next time you print, Excel will run your code first and then proceed to print the worksheet.

The `Workbook_BeforePrint` event code is triggered whether you have requested printing via Excel's built-in tools or have written your own VBA procedure to control printing.

The following tasks can be performed via the VBA code placed in the `Workbook_BeforePrint` event procedure:

- Disabling printing and print previewing:

```
Private Sub Workbook_BeforePrint(Cancel As Boolean)
  If Weekday(Date, vbSunday) = 7 Then Cancel = True
End Sub
```

When you set the `Cancel` argument to `True`, the worksheet isn't printed when the procedure ends. The above procedure disallows printing on Saturdays (the seventh day of the week). The `Weekday` function specifies that `Sunday` is the first day of the week.

- Placing the full workbook's name in the page footer.

- Changing worksheet formatting prior to printing.

- Validating data upon printing (see the procedure code in Figure 17.35).

- Calculating all worksheets in the active workbook:

```
Private Sub Workbook_BeforePrint(Cancel As Boolean)
    Dim sh as Variant
    For Each sh in Worksheets
      sh.Calculate
    Next
End Sub
```

If you need to perform certain formatting tasks for all your workbooks prior to printing, you must create the `WorkbookBeforePrint` event procedure for the `Application` object, which was discussed in Chapter 15.

The following Hands-On exercise demonstrates how to have Excel automatically print in the footer the full path and filename for all existing and new workbooks.

Hands-On 17.6   Automatically Adding a Footer to Each Workbook

You will begin this Hands-On exercise by creating the `Personal.xlsb` file. Macros and VBA procedures stored in this file are available each time you work with Excel. Personal Macro Workbook (`Personal.xlsb`) is stored in the `XLStart` folder, usually located in the `\[YourUsername]\AppData\Roaming\Microsoft\Excel` folder.

If the `Personal.xlsb` workbook does not exist, Excel creates it when you record a macro and select the option to store it in the Personal Macro Workbook.

1. In the Excel Application window, choose Developer | Record Macro.
2. In the Record Macro dialog box, choose Personal Macro Workbook in the Store macro in drop-down list.
   The `Personal` macro workbook loads automatically in the background each time you start Excel.
3. Click OK to start recording.
   In this Hands-On, you will not record anything.
4. Click the Stop Recording button to stop the macro recorder.
5. Switch to the VBE screen. In the Project Explorer window, select VBAProject (Personal.xlsb).
6. With the project selected, using the Properties window, rename VBAProject `Personal`.
7. Double-click the Personal (Personal.xlsb) workbook.
8. In the Modules folder, right-click Module1 and choose Remove Module1. Click No when asked to export the module.

9. Choose Insert | Class Module.
   Excel inserts a module named `Class1` in the `Class Modules` folder in the `Personal` (`Personal.xlsb`) workbook.

10. In the Properties window, rename `Class1` as `clsFooter`.

11. Enter the following declaration line and event procedure code in the clsFooter Code window:

```
Public WithEvents objApp As Application

Private Sub objApp_WorkbookBeforePrint(ByVal Wb As Workbook, _
    Cancel As Boolean)
  With Wb.ActiveSheet
    .PageSetup.RightFooter = Wb.FullName
  End With
End Sub
```

Recall from Chapter 15 that the `WithEvents` keyword is used in a class module to declare an object variable that points to the `Application` object. In this procedure, `objApp` is the variable name for the `Application` object. The `Public` statement before the `WithEvents` keyword allows the `objApp` variable to be accessed by all modules in the VBA project.

The above procedure simply tells Excel to place the full path and filename in the right footer of the workbook's active sheet.

After writing the event procedure code in the class module, we also must write some code in the `ThisWorkbook` class module.

12. In the Project Explorer window, double-click the ThisWorkbook object located in the `Microsoft Excel Objects` folder under the `Personal` (`Personal.xlsb`) project.

13. Type the following declaration and event code in the ThisWorkbook Code window:

```
Dim clsFullPath As New clsFooter

Private Sub Workbook_Open()
  Set clsFullPath.objApp = Application
End Sub
```

The first line above declares a variable named `clsFullPath`, which points to the object (`objApp`) in the `clsFooter` class module. The `New` keyword indicates that a new instance of the object should be created the first time the object is referenced. We do this in the `Workbook_Open` event procedure by using the `Set` keyword. This statement connects the object located in the `clsFooter` class module with the object variable `objApp` representing the `Application` object.

The code placed in the `Workbook_Open` event procedure is run whenever a workbook is opened. Therefore, when a workbook (an existing one or a new one) is opened, Excel will know that it must listen to the `Application` events; in particular, it must track events for the `objApp` object and execute the code of the `WorkbookBeforePrint` event procedure when a request for print or print preview is made through the Excel user interface or via the VBA code that is placed inside a custom printing procedure.

Before Excel can perform the programmed tasks, you must save the changes to the `Personal.xlsb` file and exit Excel.

14. Choose Debug | Compile Personal to ensure that Excel will be able to execute the VBA code you've added to the `Personal.xlsb` workbook. If Excel finds any errors, it will highlight the statement that you need to examine. Make the appropriate corrections and repeat the Debug | Compile Personal command. When there are no errors in the `Personal.xlsb` project, the Compile Personal command on the Debug menu is grayed out.

15. Close the `Chap17.xlsm` workbook file and any other workbooks that you may have opened.

16. Exit Microsoft Excel. When Excel asks whether you'd like to save changes to the `Personal.xlsb` file, click Yes.

17. Restart Microsoft Excel. Open a new workbook and type anything in any cell on any sheet of this new workbook, then save the file as `TestFooter.xlsx`.

18. Click File | Print, select your printer, and click the Print button. When you click the Print button, Excel will execute the `WorkbookBeforePrint` event procedure, and your printout should include the file path in the right footer.

19. Close the `TestFooter.xlsx` workbook.

20. Open any existing workbook that did not have footers set up. Choose File | Print. When you print out the file, the hardcopy now includes a complete filename in the right footer.

21. Close the workbook you have opened.

SENDING EMAILS FROM EXCEL

You can share your Excel workbooks with others by emailing them. To send emails from Excel, you need one of the following programs:

- Microsoft Outlook

- Microsoft Live Mail
- Microsoft Exchange Client
- Any MAPI-compatible email program (MAPI stands for Messaging Application Programming Interface)

Excel workbooks can be sent as attachments in PDF/XPS format or as faxes, or you can send a link to the file stored in a shared location. When you send an email with a workbook attachment, the file is larger but the recipient can open and edit the workbook in Excel.

To share a file via email, choose File | Share. You should be presented with the Share window, as shown in Figure 17.36.

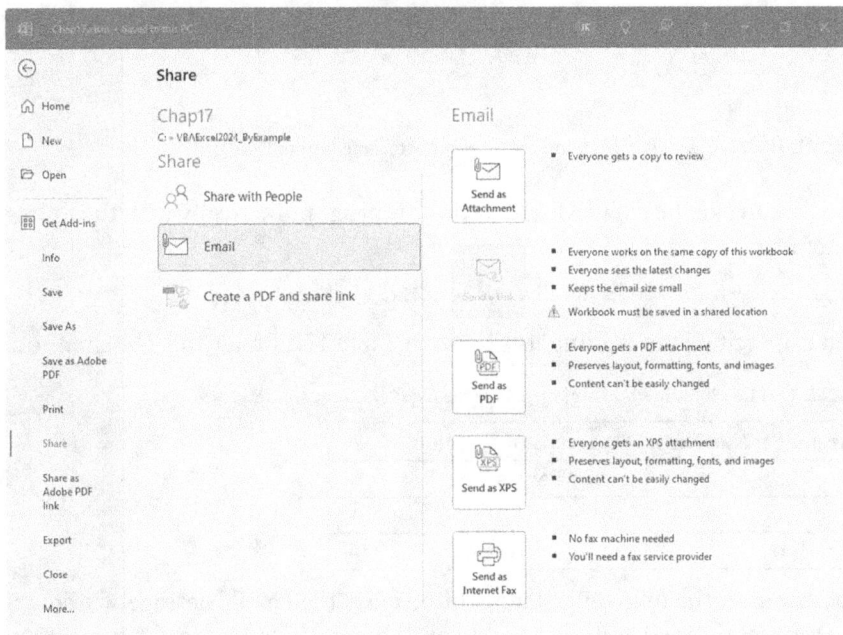

FIGURE 17.36. Share options allow you to send emails from Excel.

When you select to attach a copy of the Excel workbook, Excel displays an email Message window, as shown in Figure 17.37.

| **NOTE** | *You may be notified that you need to create a Microsoft Outlook profile. Follow the instructions in the message to add a profile to Outlook.* |

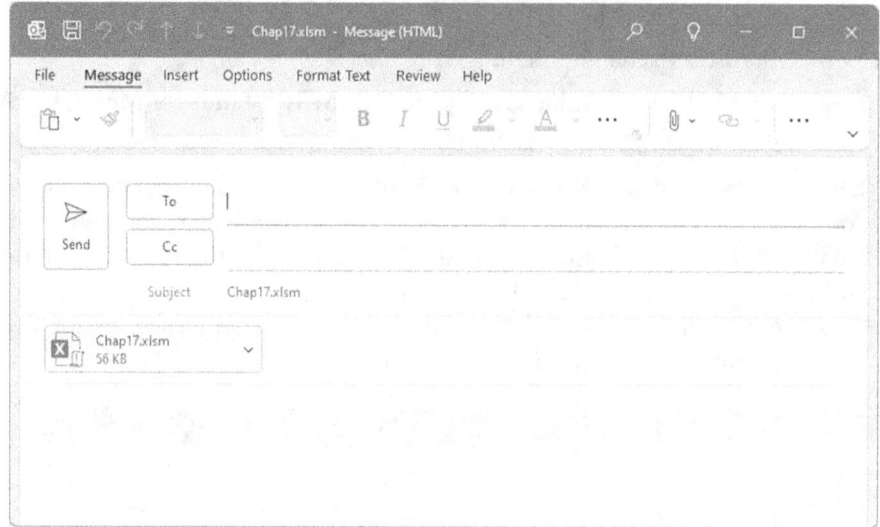

FIGURE 17.37. Sending a workbook as an email attachment from Excel.

You can invoke the email Message window programmatically with the following statement:

```
Application.Dialogs(xlDialogSendMail).Show
```

You can include the arguments shown in Table 17.11 after the `Show` method.

TABLE 17.11. Show method arguments for the email Message window.

Argument Number	Argument Description
Arg1	Recipients
Arg2	Subject
Arg3	return_receipt

For example, the following statement displays the email Message window with the recipient's email address filled in and the specified text in the subject line:

```
Application.Dialogs(xlDialogSendMail).Show
    Arg1:="SendToName@SendToProvider.com",
    Arg2:="New workbook file"
```

To check out the above statement, type the statement on one line in the Immediate window and press Enter.

Sending Emails Using the SendMail Method

Before you begin sending emails from your VBA procedures, it's a good idea to determine what email system is installed on your computer. You can do this with the `MailSystem` property of the `Application` object. This is a read-only property that uses the `xlMAPI`, `xlPowerTalk`, and `xlNoMailSystem` constants to determine the installed mail system.

MAPI is used for interfacing with email systems. The PowerTalk mail system was used on Macintosh systems in the early 1990s. Unfortunately, there is no constant for the current AppleMail, as it doesn't have a direct VBA object model such as MAPI for Outlook.

The following `Discover_EmailSystem` procedure demonstrates how to use the `MailSystem` property.

```
Sub Discover_EmailSystem()
  Select Case Application.MailSystem
    Case xlMAPI
      MsgBox "You have Microsoft Mail installed."
    Case xlNoMailSystem
      MsgBox "No mail system installed on this computer."
    Case xlPowerTalk
      MsgBox "Your mail system is PowerTalk"
  End Select
End Sub
```

The easiest way to send an email from Excel is by using the `SendMail` method of the `Application` object. This method allows you to specify the email address of the recipient, the subject of your email, and whether you'd like a return receipt. Let's create an email and send it to ourselves.

⊙ Hands-On 17.7   Using the SendMail Method to Send an Email

This Hands-On requires that you have a Microsoft Outlook account set up on your computer.

1. In the VBE window of the `Chap17.xlsm` workbook, insert a new module and enter the following `SendMailNow` procedure:

```
Sub SendMailNow()
  Dim strEAddress As String

  On Error GoTo ErrorHandler

  strEAddress = InputBox("Enter e-mail address", _
      "Recipient's E-mail Address ")
```

```
If IsNull(Application.MailSession) Then
  Application.MailLogon
End If

ActiveWorkbook.SendMail Recipients:=strEAddress, _
    Subject:="Test Mail"

Application.MailLogoff
Exit Sub

ErrorHandler:
  MsgBox "Some error occurred while sending e-mail."
End Sub
```

If Microsoft Mail isn't already running, you must use the `MailSession` property of the `Application` object to establish a mail session in Excel before sending emails. The `MailSession` property returns the MAPI mail session number as a hexadecimal string or `Null` if the mail session hasn't been established yet. The `MailSession` property isn't used on PowerTalk mail systems. To establish a mail session, use the `MailLogon` method of the `Application` object. To close a MAPI email session established by Microsoft Excel, use the `MailLogoff` method.

2. Run the `SendMailNow` procedure to email the active workbook. Type your email address when prompted and click OK.

3. When you see the message shown in Figure 17.38, click the Allow button to allow you to send the email.

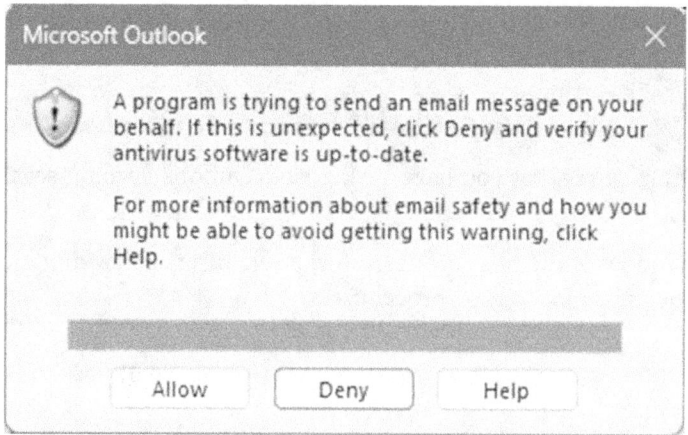

FIGURE 17.38. You will get a warning message when you try to send an email from Excel.

4. Open your email program and check the email received.

When the recipient receives an email with an attached workbook, they will need Excel to open the file.

Sending Emails Using the MsoEnvelope Object

You can send emails directly from Microsoft Excel and other Microsoft 365 applications via the `MsoEnvelope` object, which is included with the Microsoft Office 16.0 Object Library.

To return an `MsoEnvelope` object, use the `MailEnvelope` property of the `Worksheet` object. You also need to set up a reference to the `MailItem` object in the Microsoft Outlook 16.0 Object Library to access its properties and methods that format the email message.

The following procedure demonstrates sending an email from Excel using the `MsoEnvelope` object. Instead of attaching the entire workbook, we will embed the data from `Sheet2` of the current `Chap17.xlsm` workbook.

⊙ Hands-On 17.8   Sending an Email Using the MsoEnvelope Object

1. In the VBE screen of the `Chap17.xlsm` workbook, set up the reference to the Microsoft Outlook 16.0 and Microsoft Office 16.0 object libraries using the Tools | References dialog box.
2. Insert a new module in the current workbook and enter the following procedure:

```
Sub SendMsoMail(ByVal strRecipient As String)
' use MailEnvelope property of the Worksheet
' to return the msoEnvelope object

    ActiveWorkbook.EnvelopeVisible = True

    With ActiveSheet.MailEnvelope

        .Introduction = "Please see the list of " & _
                    "User IDs with their quarterly figures."
        With .Item
          ' Make sure the e-mail format is HTML
          .BodyFormat = olFormatHTML
          ' Add the recipient name
          .Recipients.Add strRecipient
          ' Add the subject
          .Subject = "Quarterly Figures"
          ' Send Mail
          .Send
```

```
        End With
    End With
End Sub
```

3. Run the `SendMsoMail` procedure by typing the following statement in the Immediate window (be sure to replace the email address with your own):

```
SendMsoMail "YourName@YourProvider.com"
```

When you press Enter, Excel calls the `SendMsoMail` procedure, passing the recipient's email address to it. The embedded worksheet is shown in Outlook in Figure 17.39.

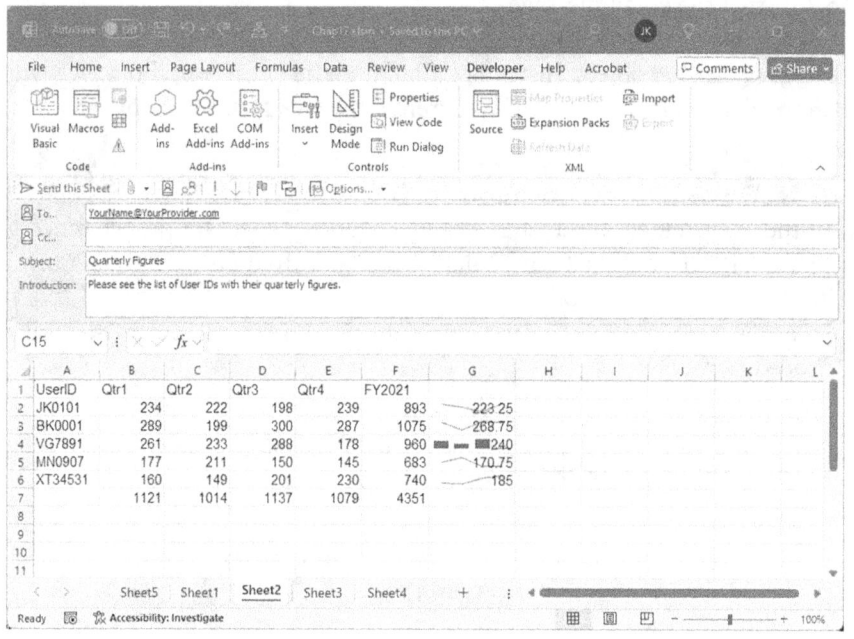

FIGURE 17.39. An email with an embedded worksheet generated by the VBA procedure in Hands-On 17.8.

SUMMARY

In this chapter, you learned how to use VBA to apply basic formatting features to your worksheets to make your data easier to read and interpret. You also explored advanced formatting features such as conditional formatting and the utilization of tools such as data bars, icon sets, color scales, shapes, and sparklines, as well as themes and cell styles. Additionally, you learned how to programmatically set page and print options, set up printers, and use printing events to perform formatting tasks prior to printing. Finally, you practiced various methods of sending your workbooks through electronic mail as attachments or embedded as the body of a message.

The next chapter focuses on the customization of Excel's ribbon interface and context menus.

18 CONTEXT MENU PROGRAMMING AND RIBBON CUSTOMIZATIONS

Users have come to expect easy ways to select commands and options in Microsoft 365 applications. Therefore, after you have written VBA procedures that provide solutions to specific worksheet automation dilemmas, you should spend additional time adding features that will make your application quick and easy to use. The most desired features of the User Interface (UI) in Excel are customizations of the context menus and the ribbon. While it is easy for users to get quick access to a specific command by placing it in the Quick Access Toolbar, your custom application's tools will need to appear either on the ribbon or in a context menu. This chapter shows you how to work with the context menus and the ribbon interface programmatically.

WORKING WITH CONTEXT MENUS

A *context menu* (also referred to as a *shortcut menu*) appears when you right-click on an object in the Microsoft Excel application window. You can customize built-in context menus by using the `CommandBar` object or by applying ribbon customizations, as demonstrated later in this chapter. This section focuses on using the `CommandBar` object's properties and method to create, modify, or disable context menus, depending on your application's needs.

Each object in the `CommandBars` collection is called `CommandBar`. The term `CommandBar` is used to refer to a context menu only. This object comes with a special `Type` property that can be used to return the specific type of the command bar (see Table 18.1).

TABLE 18.1.   Types of CommandBar objects in the CommandBars collection.

Type of Object	Index	Constant
Toolbar	0	`msoBarTypeNormal`
Menu Bar	1	`msoBarTypeMenuBar`
Context/Shortcut Menu	2	`msoBarTypePopup`

In versions of Excel prior to 2007, the `CommandBar` object was used to programmatically work with menu bars and toolbars. Since the introduction of the ribbon interface, the `CommandBar` object can only be used with context menus.

Modifying a Built-In Context Menu

Microsoft Excel offers numerous context menus with different sets of frequently used menu options. Let's write a VBA procedure that prints the names of the context menus to the Immediate window.

> | NOTE | *Please note that the files for the "Hands-On" projects can be found in the companion files.* |
> |---|---|

⊙ Hands-On 18.1   Enumerating Context Menus

1. Create a new workbook and save it as `C:\VBAExcel2024_ByExample\Chap18.xlsm`.
2. Switch to the VBE screen and insert a new module into `VBAProject (Chap18.xlsm)`.
3. Use the Properties window to rename the module `ContextMenus`.
4. In the ContextMenus Code window, enter the `ContextMenus` procedure as shown below:

```
Sub ContextMenus()
  Dim myBar As CommandBar
  Dim counter As Integer

  For Each myBar In CommandBars
    If myBar.Type = msoBarTypePopup Then
      counter = counter + 1
      Debug.Print counter & ": " & myBar.Name
```

```
    End If
  Next
End Sub
```

Notice the use of the msoBarTypePopup constant to identify the context menu in the collection of CommandBars.

5. Run the ContextMenus procedure.
   The result of this procedure is a list of context menus, which you can view in the Immediate window.

Now that you know the exact names of Excel's context menus (see Hands-On 18.1), you can easily add other frequently used commands to any of these menus. For instance, let's add the Insert Picture command to the context menu activated when you right-click a worksheet cell.

Hands-On 18.2   Adding a New Item to a Context Menu

1. In the ContextMenus Code window that you created in the previous Hands-On, enter the following procedures:

```
Sub AddToCellMenu()
  With Application.CommandBars("Cell")
    .Reset
    .Controls.Add(Type:=msoControlButton, _
        Before:=2).Caption = "Insert Picture..."
    .Controls("Insert Picture...").OnAction = "InsertPicture"
  End With
End Sub

Sub InsertPicture()
  CommandBars.ExecuteMso ("PictureInsertFromFile")
End Sub
```

The Reset method of the CommandBar object used in the AddToCellMenu procedure prevents placing the same option in the context menu again when you run the procedure more than once. To add a built-in or custom control to a context menu, use the Add method with the following syntax:

CommandBar.Controls.Add(Type, Id, Parameter, Before, Temporary)

CommandBar is the object to which you want to add a control. Type is a constant that determines the type of custom control you want to add. You may select one of the following types:

TABLE 18.2.   CommandBar object control types.

Control Type Constant	Constant Value
msoControlButton	1
msoControlPopup	10
msoControlEdit	2
msoControlDropDown	3
msoControlComboBox	4

Id is an integer that specifies the number of the built-in control you want to add.

Parameter is used to send information to a Visual Basic procedure or to store information about the control.

The Before argument is the index number of the control before which the new control will be added. If omitted, Visual Basic adds control at the end of the specified command bar.

The Temporary argument is a logical value (True or False) that determines when the control will be deleted. When you set this argument to True, the control will be automatically deleted when the Excel application is closed.

CommandBar controls have a number of properties that help you specify the appearance and functionality of a control. For example, the Caption property specifies the text displayed for the control. In the above procedure, you will see the Insert Picture… entry in the cell context menu. Note that it is customary to add an ellipsis (…) at the end of the menu option's text to indicate that the option will trigger a dialog box in which the user will need to make more selections.

The OnAction property specifies the name of a VBA procedure that will execute when the menu option is selected. In this example, upon selecting the Insert Picture… option, the InsertPicture procedure will be called. This procedure uses the ExecuteMso method of the CommandBar object to execute the ribbon's PictureInsertFromFile command.

2. Run the AddToCellMenu procedure.
3. Switch to the Microsoft Excel application window, right-click any cell in a worksheet, and select the Insert Picture… command (see Figure 18.1).
   Excel displays the Insert Picture dialog box from which you can insert a picture from a file. Note that the same dialog box is displayed when you click the Pictures button on the ribbon's Insert tab.

   Custom menu items added to Excel context menus are available in all open workbooks. It does not matter which workbook was used to add a custom

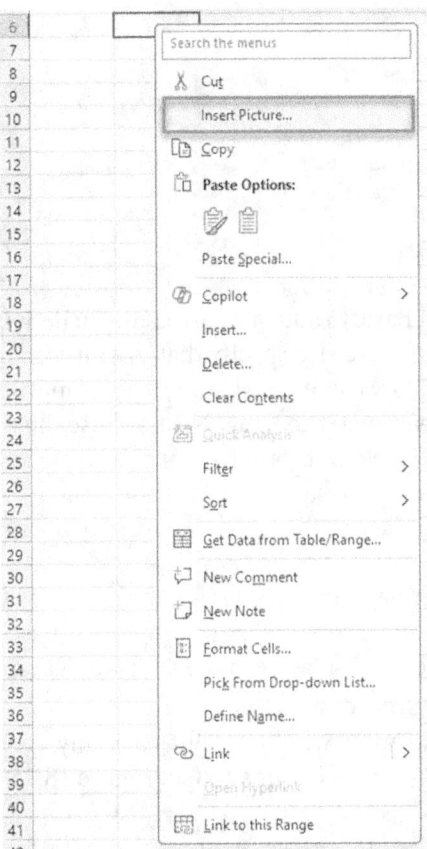

FIGURE 18.1. The built-in cell context menu displays a new item, Insert Picture..., which was added by a VBA procedure.

item. For this reason, it's a good idea to ensure that the custom menu item is removed when the workbook is closed. (See the next section titled Removing a Custom Item from a Context Menu.)

Notice in Figure 18.1 that some options in the context menu are preceded by a small graphic image. Let's write another version of the AddToCellMenu procedure to include an image next to the Insert Picture... command.

4. Enter the following procedure in the ContextMenus Code window:

```
Sub AddToCellMenu_Modified()
    Dim ct As CommandBarButton
    With Application.CommandBars("Cell")
        .Reset
        Set ct = .Controls.Add(Type:=msoControlButton, _
            Before:=11, Temporary:=True)
```

```
End With
With ct
    .Caption = "Insert Picture..."
    .OnAction = "InsertPicture"
    .Picture = Application.CommandBars. _
        GetImageMso("PictureInsertFromFile", 16, 16)
    .Style = msoButtonIconAndCaption

    End With
End Sub
```

In this procedure code, we tell Visual Basic to add our custom menu item in the 11th position on the cell context menu. We also specify that this custom menu option is removed automatically when we exit Excel. This is accomplished by setting the value of the `Temporary` parameter to `True`. Next, we use the `With... End With` statement block to set a couple of properties for the newly created control object (`ct`). In addition to setting two standard properties (`Caption` and `OnAction`), we assign the `imageMso` image to the `Picture` property of our new control. To return the image, you must use the `CommandBars.GetImageMso` method and specify the name of the image and its size (width and height). The size of the image is specified as 16 x 16 pixels. The `Style` property is used here to specify that the control button should display both the icon and its caption.

5. Run the `AddToCellMenu_Modified` procedure.
6. Switch to the Microsoft Excel application window, right-click any cell in a worksheet, and look for the Insert Picture... command (see Figure 18.2).

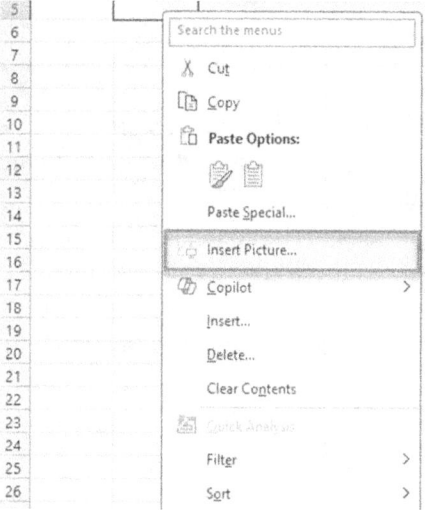

FIGURE 18.2. A custom Insert Picture... menu item is now identified by an icon and positioned just below the built-in Paste Special... command.

Notice that built-in context menu commands have a special hotkey indicated by the underlined letter. To invoke a menu option, you simply press the underlined letter after opening the menu. Let's add a hotkey to our custom menu option.

7. Modify the `Caption` property in the above procedure like this:

```
.Caption = "Insert Pict&ure..."
```

The `&` symbol in front of the letter u indicates that the lowercase u will serve as the hotkey. Remember that hotkeys are unique; you cannot use a letter that is already used by another menu item.

8. Run the `AddToCellMenu_Modified` procedure.

9. Switch to the Microsoft Excel application window, right-click any cell in a worksheet, and then press the lowercase u. You should see the Insert Picture dialog box.

Removing a Custom Item from a Context Menu

When you modify context menus, your customizations will not go away when you close the workbook. Restarting Excel will remove your custom changes to the context menu only if you set the value of the `Temporary` parameter to `True` when adding your custom menu item.

To ensure that the custom item is removed from the menu, consider writing a `Delete` procedure similar to the one shown below:

```
Sub DeleteInsertPicture()
    Dim c As CommandBarControl
    On Error Resume Next
    Set c = CommandBars("Cell").Controls("Insert Pict&ure...")
    c.Delete
End Sub
```

For automatic cleanup, call the previous procedure from the `Workbook_Before-Close` event procedure, like this:

```
Private Sub Workbook_BeforeClose(Cancel As Boolean)
    Call DeleteInsertPicture
End Sub
```

The previous event procedure must be entered in the `ThisWorkbook` code module. The `Workbook_BeforeClose` procedure will be executed just before the workbook is closed.

To ensure that your custom menu option is in place when you open the workbook, call the procedure that adds a custom menu item from the Work-book_Open event procedure entered in the ThisWorkbook Code window:

```
Private Sub Workbook_Open()
   Call AddToCellMenu_Modified
End Sub
```

Disabling and Hiding Items on a Context Menu

To disallow using a particular context menu item, you may want to disable it or hide it.

When a context menu item is disabled, its caption appears dimmed. When a menu item is hidden, it simply does not appear on the menu.

To disable a menu item, set the Enabled property of the control to False. For example, the following statement will disable the Insert Picture... command that you added earlier to the Cell context menu:

```
Application.CommandBars("Cell").Controls("Insert Pict&ure..")._
   Enabled = False
```

To enable a disabled menu item, simply set the Enabled property of the desired control to True.

To hide a menu item, set the Visible property of the control to False:

```
Application.CommandBars("Cell").Controls("Insert Pict&ure..."). _
   Visible = False
```

To unhide the hidden menu item, set the Visible property of the control to True:

```
Application.CommandBars("Cell").Controls("Insert Pict&ure..."). _
Visible = True
```

A good place to use the previous commands is in the Worksheet_Activate and Worksheet_Deactivate event procedures. For example, to disable the specific context menu item only when Sheet1 is active, write the following event procedures in the Sheet1 code module:

```
Private Sub Worksheet_Activate()
   Application.CommandBars("Cell").Controls("Sort"). _
     Enabled = False
End Sub
```

```
Private Sub Worksheet_Deactivate()
  Application.CommandBars("Cell").Controls("Sort"). _
    Enabled = True
End Sub
```

When writing code to control Excel context menus with the `CommandBar` object's properties and methods, you may find out that certain VBA statements will work in some but not all circumstances. Unless you need to write an application for Excel 2007 or earlier, you should move toward programming the ribbon interface, which allows you to control commands in the context menus (see the section titled Modifying Context Menus Using Ribbon Customizations later in this chapter).

Action Item 18.1

Excel has a `SpeakCells` command that will read the contents of a cell, a range of cells, or a worksheet. On your own, add the SpeakCells command to the Cells context menu. The solution can be found in the `Chap18_SpeakCells.txt` file.

Adding a Context Menu to a Command Button

When you design custom forms, you may want to add context menus to various controls placed on the form. The following set of VBA procedures demonstrates how right-clicking a command button can offer users a choice of options to select from.

⊙ Hands-On 18.3   Using Context Menus on User Forms

1. In the ContextMenus Code window, enter the `Create_ContextMenu` procedure as shown below:

```
Sub Create_ContextMenu()
  Dim sm As Object

  Set sm = Application.CommandBars.Add("MyComputer", msoBarPopup)
  With sm
    .Controls.Add(Type:=msoControlButton). _
        Caption = "Operating System"
    With .Controls("Operating System")
        .FaceId = 1954
        .OnAction = "OpSystem"
    End With
    .Controls.Add(Type:=msoControlButton).Caption _
        = "Active Printer"
    With .Controls("Active Printer")
        .FaceId = 4
```

```
            .OnAction = "ActivePrinter"
        End With
     .Controls.Add(Type:=msoControlButton).Caption _
        = "Active Workbook"
     With .Controls("Active Workbook")
            .FaceId = 247
            .OnAction = "ActiveWorkbook"
        End With
     .Controls.Add(Type:=msoControlButton).Caption _
        = "Active Sheet"
     With .Controls("Active Sheet")
            .FaceId = 18
            .OnAction = "ActiveSheet"
        End With
     End With
   End With
End Sub
```

The `Create_ContextMenu` procedure creates a custom context menu named `MyComputer` and adds four commands to it. Notice that each command is assigned an icon. When you select a command from this context menu, one of the procedures shown in Step 2 will run.

2. In the ContextMenus Code window, enter the following procedures that are called by the `Create_ContextMenu` procedure:

```
Sub OpSystem()
    MsgBox Application.OperatingSystem, , "Operating System"
End Sub

Sub ActivePrinter()
    MsgBox Application.ActivePrinter
End Sub

Sub ActiveWorkbook()
    MsgBox Application.ActiveWorkbook.Name
End Sub

Sub ActiveSheet()
    MsgBox Application.ActiveSheet.Name
End Sub
```

3. Run the `Create_ContextMenu` procedure.

To test the custom context menu you just created, use the `ShowPopup` method, as shown in Step 4.

4. Press Ctrl+G to activate the Immediate window and type the following statement, pressing Enter when done:

```
CommandBars("MyComputer").ShowPopup 0, 0
```

In the above example, the `MyComputer` context menu that was added by running the `Create_ContextMenu` procedure will appear at the top left-hand corner of the screen. It may appear on another monitor if you're using multiple monitors.

The `ShowPopup` method for the `CommandBar` object accepts two optional arguments (x, y) that determine the location of the context menu on the screen.

Let's make our context menu friendlier by attaching it to a command button placed on a user form.

5. In the VBE screen, choose Insert | UserForm to add a new form to the current VBA project.

Using the `CommandButton` control in the Toolbox, place a button anywhere on the empty user form. Use the Properties window to change the `Caption` property of the command button to `System Information`. You may need to resize the button on the form to fit this text.

6. Switch to the Code window for the form by clicking the View Code button in the Project Explorer window or double-clicking the form background.

7. Enter the following procedure in the UserForm1 Code window:

```
Private Sub CommandButton1_MouseDown(ByVal Button As Integer, _
    ByVal Shift As Integer, ByVal X As Single, ByVal Y As Single)
    If Button = 2 Then
        Call Show_ShortMenu
    Else
        MsgBox "You must right-click this button."
    End If
End Sub
```

This procedure calls the `Show_ShortMenu` procedure (see Step 9) when the user right-clicks the command button placed on the form. Visual Basic has two event procedures that are executed in response to a mouse click. When you click a mouse button, Visual Basic executes the `MouseDown` event procedure. When you release the mouse button, the `MouseUp` event occurs. The `MouseDown` and `MouseUp` event procedures require the following arguments:

- The `object` argument specifies the object. In this example, it's the name of the command button placed on the form.

- The Button argument is the integer value that specifies which mouse button was pressed. Use 1 for the left mouse button, 2 for the right mouse button, and 3 for the middle mouse button.
- The Shift argument determines whether the user was pressing the Shift, Ctrl, or Alt key when the event occurred. Use 1 for the Shift key, 2 for the Ctrl key, 3 for Shift and Ctrl keys, 4 for the Alt key, 5 for Alt and Shift keys, 6 for Alt and Ctrl keys, and 7 for Alt, Shift, and Ctrl keys.

8. In the ContextMenus Code window, enter the code of the Show_ShortMenu procedure:

```
Sub Show_ShortMenu()
  Dim shortMenu As Object

  Set shortMenu = Application.CommandBars("MyComputer")
  With shortMenu
     .ShowPopup
  End With
End Sub
```

9. In the Project Explorer window, double-click UserForm1 and press F5 to run the form.
10. Right-click the System Information button and select one of the options from the context menu.

Notice that the ShowPopup method used in this procedure does not include the optional arguments that determine the location of the context menu on the screen. Therefore, the menu appears wherever the mouse is clicked (see Figure 18.3).

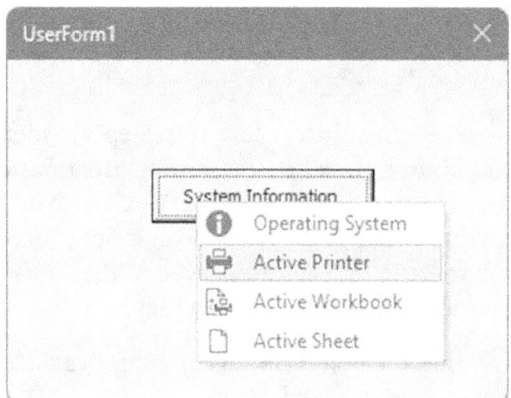

FIGURE 18.3. A custom context menu appears when you right-click an object.

11. To delete the context menu named `MyComputer`, enter and then run the following `Delete_ShortMenu` procedure in the ContextMenus Code window:

```
Sub Delete_ShortMenu()
    Application.CommandBars("MyComputer").Delete
End Sub
```

Finding a FaceID Value of an Image

When modifying context menus, you will most likely want to include an image next to the displayed text, as we did in Hands-On 18.3 (see Figure 18.3).

The `CommandBars` collection has hundreds of images that you can use. Each command bar control button has a `FaceID` value that determines the look of a control, but how do you know which ID belongs to which control button? The `FaceID` property returns or sets the ID number of the icon on the control button's face. In most cases, the icon ID number (`FaceID`) is the same as the control's `ID` property. The icon image can be copied to the Windows clipboard using the `CopyFace` method.

The `Images` procedure demonstrated below iterates through the `Command-Bars` collection and writes to a new workbook a list of control buttons that have a `FaceID` number. If you'd like to see this procedure in action, enter the code shown below in the ContextMenus Code window in the `Chap18.xlsm` workbook and then run it:

```
Sub Images()
    Dim i As Integer
    Dim j As Integer
    Dim total As Integer
    Dim buttonId As Integer
    Dim buttonName As String
    Dim myControl As CommandBarControl
    Dim bar As CommandBar

    On Error GoTo ErrorHandler
    Workbooks.Add
    Range("A1").Select
    With ActiveCell
      .Value = "Image"
      .Offset(0, 1) = "Index"
      .Offset(0, 2) = "Name"
      .Offset(0, 3) = "FaceID"
      .Offset(0, 4) = "CommandBar Name (Index)"
    End With
```

```
For j = 1 To Application.CommandBars.Count
    Set bar = CommandBars(j)
    total = bar.Controls.Count

    With bar
      For i = 1 To total
          buttonName = .Controls(i).Caption
          buttonId = .Controls(i).ID

          Set myControl = CommandBars._
              FindControl(ID:=buttonId)
          myControl.CopyFace ' error could occur here
          ActiveCell.Offset(1, 0).Select
          Sheets(1).Paste

          With ActiveCell
              .Offset(0, 1).Value = buttonId
              .Offset(0, 2).Value = buttonName
              .Offset(0, 3).Value = myControl.FaceId
              .Offset(0, 4).Value = bar.Name & " _
                  (" & j & ")"
          End With
StartNext:
        Next i
      End With
    Next j

    Columns("A:E").EntireColumn.AutoFit
    Exit Sub
ErrorHandler:
    Resume StartNext
End Sub
```

Because you cannot copy the image of an icon that is currently disabled, Visual Basic encounters an error when it attempts to copy the button's face to the clipboard. The procedure traps this error with the `On Error GoTo ErrorHandler` statement. This way, when Visual Basic encounters the error, it will jump to the `ErrorHandler` label and execute the instructions below this label. This will ensure that the problem control button is skipped and the procedure can continue without interruption. A partial result of this procedure is shown in Figure 18.4.

	Image	Index	Name		FaceID	CommandBar Name (Index)
1		Index	Name		FaceID	CommandBar Name (Index)
2		1031	&WordArt...		1031	WordArt (3)
3		2094	Edit Te&xt...		2094	WordArt (3)
4		1606	&WordArt Gallery		1606	WordArt (3)
5		962	Format &Object...		962	WordArt (3)
6		1063	&WordArt Same Letter Heights		1063	WordArt (3)
7						
8						
9		1064	&More Contrast		1064	Picture (4)
10		1065	&Less Contrast		1065	Picture (4)
11		1066	&More Brightness		1066	Picture (4)
12		1067	&Less Brightness		1067	Picture (4)
13		732	&Crop		732	Picture (4)
14		199	Rotate &Left 90°		199	Picture (4)
15		6382	&Compress Pictures...		6382	Picture (4)
16		962	Format &Object...		962	Picture (4)
17		2827	&Set Transparent Color		2827	Picture (4)
18						
19						
20		5916	Ex&pand		5916	Drawing Canvas (5)
21		7067	S&cale Drawing		7067	Drawing Canvas (5)
22		6933	I&nsert Shape		6933	Diagram (6)
23		6010	Move Shape &Backward		6010	Diagram (6)
24						
25		6099	&Reverse Diagram		6099	Diagram (6)
26		6068	&AutoFormat		6068	Diagram (6)
27		7502	&Eraser		7502	Ink Drawing and Writing (7)

Sheet1 +

FIGURE 18.4. A partial list of icon images and their corresponding FaceID values generated by a VBA procedure.

A QUICK OVERVIEW OF THE RIBBON INTERFACE

The Excel ribbon contains the title bar, the Quick Access Toolbar, and several tabs. Each tab on the ribbon provides access to features and commands related to a particular task. For instance, you can use the Insert tab to quickly insert tables, illustrations, charts, links, or text (see Figure 18.5). Related commands within a tab are organized into groups. This type of organization makes it easy to locate a particular command.

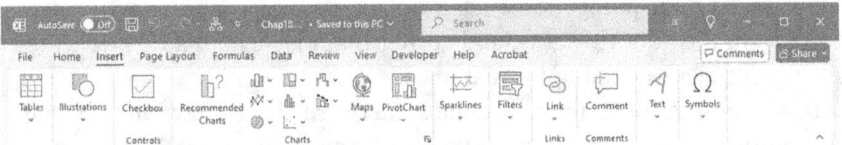

FIGURE 18.5. The rectangular area at the top of the Microsoft Excel window is called the ribbon. Each tab on the ribbon contains groups of related commands.

Various program commands are displayed as large or small buttons. A large button denotes a frequently used command, while a small button shows a specific feature of the main command that you may want to work with. Some large and small command buttons include drop-down lists of other specialized commands. For instance, the small More Functions button dropdown in the Function Library group on the Formulas tab contains additional types of functions you can insert: Statistical, Engineering, Cube, Information, Compatibility, and Web.

Some controls that you find on the ribbon do not display commands. Instead, they provide a visual clue of the output you might expect when a specific option is selected. These types of controls are known as *galleries*. The gallery control is often used to present various formatting options, such as the margin settings shown in Figure 18.6.

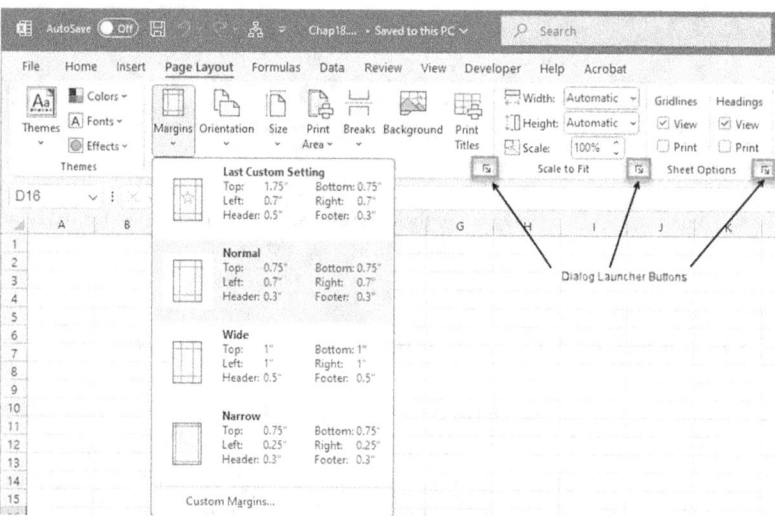

FIGURE 18.6.   The margin layouts are displayed in a gallery control.

As mentioned earlier, the commands on the ribbon tabs are organized into groups for easy browsing. Some tab groups have dialog box launchers (see Figure 18.6) in the bottom right-hand corner that display a dialog box where you can set several advanced options at once.

In addition to the main ribbon tabs, there are also *contextual* tabs that contain commands that apply to what you are doing. When a particular object is selected, the ribbon displays a contextual tab that provides commands for working with that object. For instance, when you select an image in a worksheet, the ribbon displays a contextual tab called Picture Format, as shown in Figure 18.7.

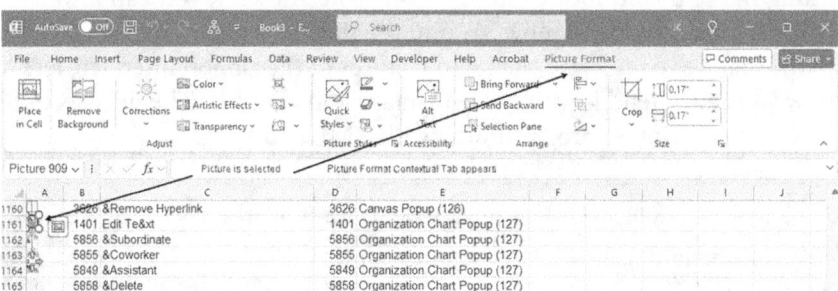

FIGURE 18.7. Contextual tabs will appear when you work with a particular object, such as a picture, PivotTable, or chart.

Clicking on the Picture Format tab brings up commands for dealing with a `Picture` object. The contextual tab disappears when you cancel the selection of the object. In other words, if you select a different cell in a worksheet, the Picture Format tab will be removed.

The tooltips of the controls display the name of the command, the control keyboard context (where available), and a description of what the command does (see Figure 18.8).

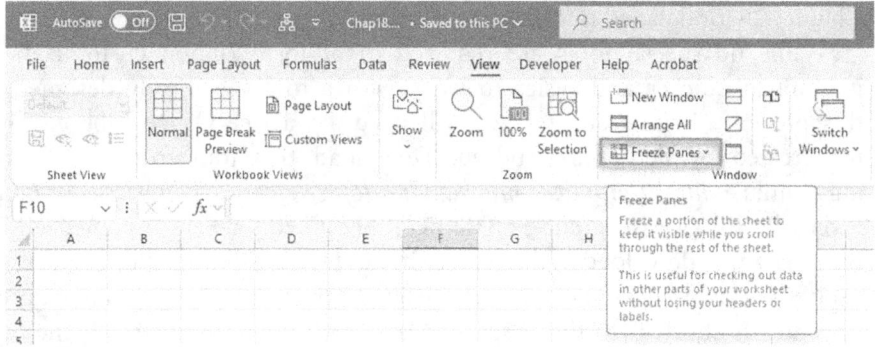

FIGURE 18.8. The enhanced tooltips, known as Super ToolTips, provide more information about the selected command.

All ribbon commands and the Quick Access Toolbar can be easily accessed via the keyboard. Simply press the Alt key on the keyboard to display small boxes with key tips. Every command has its own access key. For example, to access the File tab, press Alt and then F. Within the menus, you will see other key tips for every command.

To view key tips for the commands on a particular tab, first, select the access key for that tab. To remove the key tips, press the Alt key again. When you are working in command mode (after pressing the Alt key), you can also use the Tab key and arrow keys to move around the ribbon.

Now that we've reviewed the main features of the ribbon interface, let's look at how you can extend it with your own tabs and controls. The next section introduces you to ribbon programming with XML and VBA.

Ribbon Customizations via the User Interface

In addition to customizing the Quick Access Toolbar (QAT), you can create custom ribbon tabs and groups by choosing File | Options | Customize Ribbon. You can also rename and change the order of the built-in tabs and groups.

RIBBON PROGRAMMING WITH VBA AND XML

The components of the ribbon UI can be manipulated programmatically using Extensible Markup Language (XML) or other programming languages. All Microsoft 365 applications that use the ribbon interface rely on the programming model known as *Ribbon Extensibility*, or *RibbonX*.

This section introduces you to customizing the UI by using XML (refer to Chapter 25 for detailed information on using XML with Excel).

While no special tools are required to perform ribbon customizations, it is much quicker and easier to work with a tool specifically designed for this purpose. In this chapter, we will be using a free Office RibbonX Editor created by Fernando Andreu. Download this tool from the following link: *https://github.com/fernandreu/office-ribbonx-editor/releases/latest*. From the list of assets, click on OfficeRibbonXEditor-SelfContained-NET6-Binaries.zip to download it, then unzip it. Find OfficeRibbonXEditor.exe in the list of files, right-click on it, and use the Send command from the context menu to send the shortcut to the desktop. Note that under Windows 11, the Send command is available after choosing Show More Options.

If, for any reason, you cannot install the recommended tool, you can still customize the Microsoft 365 ribbon by creating a customization file in any text editor and saving it with the name `customUI.xml`. For details, see the following article that describes adding a custom tab, a custom group, and a button using a Word document as an example: *https://docs.microsoft.com/en-us/office/vba/library-reference/concepts/customize-the-office-fluent-ribbon-by-using-an-open-xml-formats-file*.

You can find out the names of the ribbon controls by downloading the Office Fluent User Interface Control Identifiers from the following links:

http://www.microsoft.com/en-us/download/details.aspx?id=50745 and

https://github.com/OfficeDev/office-fluent-ui-command-identifiers/blob/main/Microsoft%20365/Current%20Channel/excelcontrols.xlsx.

NOTE	*The identifier names can also be accessed in the QAT customization dialog box. Simply hover over the control that interests you and look at the screen tip of the control.* *To get the latest control identifiers for Office 365 applications, please visit https://github.com/OfficeDev/office-fluent-ui-command-identifiers.*

Creating the Ribbon Customization XML Markup

To make custom changes to the ribbon UI, you need to prepare an XML markup file that specifies all your customizations. The contents of the XML markup file that we will use in Hands-On 18.4 are shown in Figure 18.9, and the resulting output appears in Figure 18.10.

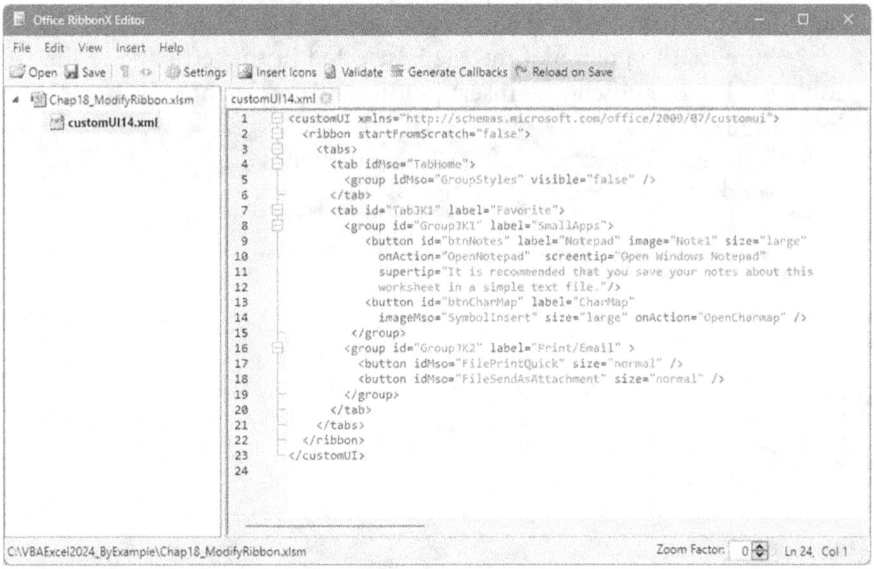

FIGURE 18.9. This XML file defines a new tab with two groups for the existing Excel 2024 ribbon. This file produces the output shown in Figure 18.10.

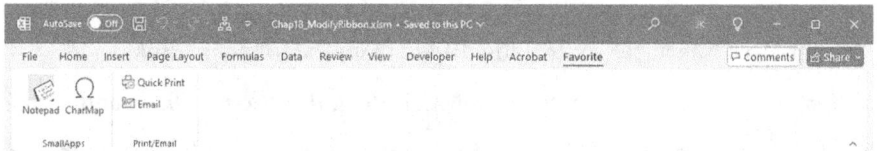

FIGURE 18.10. The custom Favorite tab is based on the custom XML markup file shown in Figure 18.9.

Hands-On 18.4 Creating an XML Document with Ribbon Customizations

This Hands-On and all the remaining Hands-On exercises in this chapter rely on the Office RibbonX Editor. See the instructions on how to get and install this free tool in the previous section.

1. Launch Microsoft Excel and create a new workbook. Save this workbook as `Chap18_ModifyRibbon.xlsm` in your `VBAExcel2024_ByExample` folder. Be sure to save the workbook as `Excel-Macro Enabled workbook (*.xlsm)`.
2. Close the workbook and exit Excel.
3. Launch the Office RibbonX Editor using the shortcut you created on the desktop.
4. In the Office RibbonX Editor, choose File | Open.
5. Select the C:\VBAExcel2024_ByExample\Chap18_ModifyRibbon.xlsm workbook file you created in Step 1 above and click Open.
6. Make sure that the name of the file is selected in the left pane of the Office RibbonX Editor and choose Insert | Office 2010+ Custom UI Part, as shown in Figure 18.11.

   This creates a `customUI14.xml` file in the workbook.

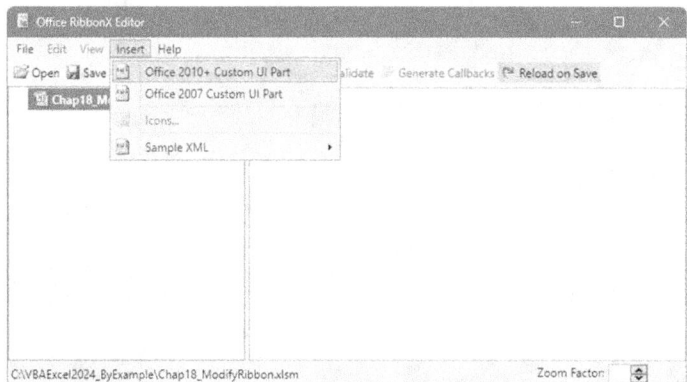

FIGURE 18.11. Use the Office RibbonX Editor to insert an Office 2010+ Custom UI Part into an Excel 2024 workbook you want to customize. This option works for ribbon customizations in Excel 2010–2024. To customize the ribbon in Excel 2007, you would choose the Office 2007 Custom UI Part instead.

7. Double-click the customUI14.xml file in the left pane to open it for editing.

8. In the right pane, enter the XML ribbon customization markup as shown below (see Figure 18.9 earlier). If you prefer, you can copy and paste the code from the companion files. Look for the file named `CustomUI14_ver01.txt`:

```
<customUI xmlns="http://schemas.microsoft.com/office/2009/07/
     customui">
  <ribbon startFromScratch="false">
    <tabs>
      <tab idMso="TabHome">
        <group idMso="GroupStyles" visible="false" />
      </tab>
      <tab id="TabJK1" label="Favorite">
        <group id="GroupJK1" label="SmallApps">
          <button id="btnNotes" label="Notepad" _
             image="Note1" size="large"
            onAction="OpenNotepad"  screentip="Open Windows _
              Notepad"
            supertip="It is recommended that you save your _
              notes about this
            worksheet in a simple text file."/>
          <button id="btnCharMap" label="CharMap"
            imageMso="SymbolInsert" size="large" _
              onAction="OpenCharmap" />
        </group>
        <group id="GroupJK2" label="Print/Email" >
          <button idMso="FilePrintQuick" size="normal" />
          <button idMso="FileSendAsAttachment" size="normal" />
        </group>
      </tab>
    </tabs>
  </ribbon>
</customUI>
```

XML is case-sensitive, so make sure you enter the statements exactly as shown above.

9. Click the Validate button ( ) on the Office RibbonX Editor toolbar to verify that your XML does not contain errors. You should see the message Custom UI XML is well formed. If there are errors, you must correct them to ensure that the XML is well formed.

At this point, you should have a well-formed `customUI14.xml` document containing ribbon customizations. Let's go over the XML document content.

As you will learn in Chapter 25, every XML document consists of a number of elements, called *nodes*. In any XML document, there must be a root node or a top-level element. In the ribbon customization file, the root tag is `<customUI>`. The root's purpose is to specify the Office RibbonX XML namespace:

```
<customUI xmlns="http://schemas.microsoft.com/office/2009/07/_
    customui">
```

Namespaces are used to identify elements in XML documents and avoid name collisions when elements with the same name are combined in the same document. If you were to customize the Office 2007 ribbon, you would use the following namespace instead:

```
<customUI xmlns="http://schemas.microsoft.com/office/2006/01/_
    customui">
```

The `xmlns` attribute of the `<customUI>` tag holds the name of the default namespace to be used in the ribbon customization. Notice that the root element encloses all other elements of this XML document: ribbon, tabs, tab, group, and button.

Each element consists of a beginning and ending tag. For example, `<customUI>` is the name of the beginning tag and `</customUI>` is the ending tag. The actual ribbon definition is contained within the `<ribbon>` tag:

```
<ribbon startFromScratch="false">
  [Include xml tags to specify the required ribbon customization]
</ribbon>
```

The `startFromScratch` attribute of the `<ribbon>` tag defines whether you want to replace the built-in ribbon with your own (`true`) or add a new tab to the existing ribbon (`false`).

Hiding the Elements of the Excel UI

Setting `startFromScratch="true"` in the `<ribbon>` tag will hide the default ribbon, including all built-in tabs, groups, and commands. Additionally, this setting will also hide the contents of the QAT. This provides a clean slate, allowing you to build a completely customized ribbon from the ground up, tailored to your specific needs and preferences.

To create a new tab in the ribbon, use the `<tabs>` tag. Each tab element is defined with the `<tab>` tag. The `label` attribute of the tab element specifies the name of your custom tab. The name in the `id` attribute is used to identify your custom tab:

```
<tabs>
<tab id="TabJK1" label="Favorite">
```

Ribbon tabs contain controls organized in groups. You can define a group for the controls on your tab with the `<group>` tag. The example XML markup file defines the following two groups for the Favorite tab:

```
<group id="GroupJK1" label="SmallApps">
<group id="GroupJK2" label="Print/Email">
```

Like the tab node, the group nodes of the XML document also contain the `id` and `label` attributes. Placing controls in groups is easy. The group labeled `SmallApps` has two custom button controls, identified by the `<button>` elements.

The group labeled `Print/Email` also contains two buttons; however, unlike the `SmallApps` group, the buttons placed here are built-in Office system controls rather than custom controls.

You can quickly determine this by looking at the `id` attribute for the control. Any attribute that ends with `Mso` refers to a built-in Office item:

```
<button idMso="FilePrintQuick" size="normal" />
```

Buttons placed on the ribbon can be large or small. You can define the size of the button with the `size` attribute set to `large` or `normal`. Buttons can have additional attributes:

```
<button id="btnNotes" label="Notepad" image="Note1" _
    size="large"
  onAction="OpenNotepad" screentip="Open Windows Notepad" _
    supertip="
  It is recommended that you save your notes about this _
    worksheet in a
    simple text file." />
<button id="btnCharMap" label="CharMap" imageMso="SymbolInsert"
  size="large" onAction="OpenCharmap" />
```

The `screentip` and `supertip` attributes allow you to specify the short and longer text that should appear when the mouse pointer is positioned over the button.

The `imageMso` attribute denotes the name of the existing Office icon. You can use images provided by any Office application. To provide your own image, use the `image` attribute, as shown in this Hands-On, or use the `getImage` attribute in the XML markup (see more information in the section titled Creating a Gallery Control later in this chapter).

The controls that you specify in the XML markup perform their designated actions via `callback` procedures. For example, the `onAction` attribute of a button control contains the name of the `callback` procedure that is executed when the button is clicked. When that procedure is completed, it calls back the ribbon to provide the status or modify the ribbon. You will write the callback procedures for the `onAction` attribute in Hands-On 18.5.

Buttons borrowed from the Office system do not require the `onAction` attribute. When clicked, these buttons will perform their default built-in action.

Before finishing off the XML ribbon customization document, always make sure that you have included all the end tags:

```
    </tab>
   </tabs>
  </ribbon>
</customUI>
```

Because our first ribbon customization calls upon a custom image, let's add it to the file.

10. Copy the `Images` folder from the companion files to your `VBAExcel2024_ByExample` folder.

11. In the Office RibbonX Editor, select customUI14.xml in the left pane and choose Insert | Icons…, or click the Insert Icons button in the toolbar.

12. Change the file filter to show all files in the `C:\VBAExcel2024_ByExample\Images` folder, then select Note.gif and click Open.
A note image should appear just below the `customUI14.xml` file in the left pane.

13. In the left pane of the Office RibbonX Editor window, right-click the image and choose Change ID. Rename the image to `Note1` to match the name of the `image` attribute in the XML markup:

```
<button id="btnNotes" label="Notepad" image="Note1" _
   size="large"
   onAction="OpenNotepad" screentip="Open Windows Notepad"
   supertip="It is recommended that you save your notes _
     about this
   worksheet in a simple text file."/>
```

14. Click the Save button or press Ctrl+S to save the current document.

15. Choose File | Exit to exit Office RibbonX Editor.

The first part of the ribbon customization is now completed. In the next part, you will load the workbook file into Excel and view the custom tab you have just created. You will also write callback procedures that perform specific actions.

Loading Ribbon Customizations

Hands-On 18.5 walks you through the remaining steps that are necessary for integrating ribbon customizations into your workbook.

 Hands-On 18.5   Adding VBA Code for Use by the Ribbon Customizations

1. Open the `C:\VBAExcel2024_ByExample\Chap18_ModifyRibbon.xlsm` workbook.
2. Notice the Favorite tab at the end of the ribbon (see Figure 18.10 earlier).
3. Switch to the VBE window and activate `VBAProject(Chap18_ModifyRibbon.xlsm)` in the Project Explorer window. Next, choose Insert | Module to add a new module to the selected project.
4. In the module's Code window, enter the following procedures:

```
Public Sub OpenNotepad(ctl As IRibbonControl)
  Shell "Notepad.exe", vbNormalFocus
End Sub

Public Sub OpenCharmap(ctl As IRibbonControl)
  Shell "Charmap.exe", vbNormalFocus
End Sub
```

`OpenNotepad` and `OpenCharmap` are the names of the callback procedures that were specified in the `onAction` attribute of the button (see Hands-On 18.4). As mentioned earlier, a callback procedure executes some action and then notifies the ribbon that the task has been completed.

The `onAction` callback is handled by a VBA procedure. The callback includes the `IRibbonControl` parameter, which is the control that was clicked. This control is passed to your VBA code by the ribbon:

```
Sub OpenNotepad(ctl as IRibbonControl)

Sub OpenCharmap(ctl as IRibbonControl)
```

For VBA to recognize this parameter, you must make sure that the References dialog box (Tools | References) has a reference to the Microsoft Office 16.0 Object Library.

The `OpenNotepad` and `OpenCharmap` procedures tell Excel to use the `Shell` function to open the Notepad or Charmap applications. Notice that the program's executable filename appears in double quotation marks. The second argument of the `Shell` function is optional. This argument specifies the window style, that is, how the program will appear once it is launched. The `vbNormalFocus` constant will open the application in a normal-size window with focus. If the window style is not specified, the program will be minimized with focus (`vbMinimizedFocus`).

The IRibbonControl Properties

You can view the properties of the `IRibbonControl` object (`Context`, `Id`, and `Tag`) in the Object Browser. The `Context` property returns the active window that contains the ribbon interface, in this case, Microsoft Excel. The `Id` property contains the ID of the control that was clicked. The `Tag` property can be used to store additional information with the control. To use this property, you need to add a tag attribute to the ribbon customization XML document. By using the `Tag` property, you can write a more generic procedure to handle the callbacks.

---

5. Switch to the Excel application window, activate the Favorite tab, and test the Notepad and Charmap buttons. These buttons should invoke the built-in Windows applications.
6. Save and close the `Chap18_ModifyRibbon.xlsm` workbook.
7. Keep the Excel application window open. Proceed to Step 6 to make sure that Excel is set up to display the RibbonX errors.
8. Click File | Options. In the Excel Options dialog box, click the Advanced tab and scroll down to the General section. Make sure that the Show Add-in User Interface Errors checkbox is selected and click OK.

   When you enable Show Add-in User Interface Errors, Excel will display errors in your ribbon customization when you load a workbook that contains errors in the custom RibbonX code. This is very helpful in the process of debugging. If you want to successfully use your customized ribbon interface, you must make sure that Excel does not find any errors when loading your workbook.
9. Exit Microsoft Excel.

Errors on Loading Ribbon Customizations

If Excel detects errors in the ribbon customization markup, it will display an error message. For example, if a matching opening or closing tag is missing, or an attribute name is typed in uppercase instead of lowercase, you will receive a message indicating the line and column number, as well as the attribute's name where the issue is found (see Figure 18.12). To resolve this, open the workbook file in the Office RibbonX Editor, identify and correct the error, and then attempt to open the workbook in Excel again. Error messages will persist until the entire file is debugged.

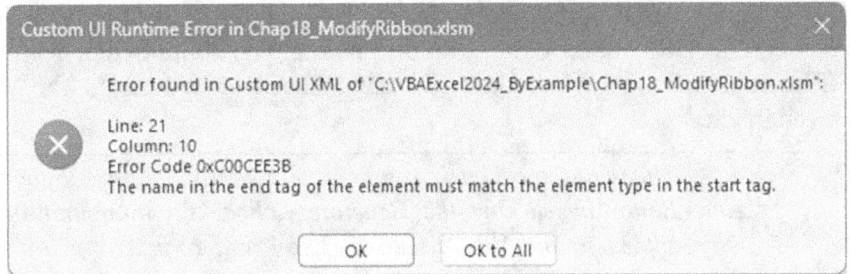

FIGURE 18.12. Excel displays an error message when it finds an error in the Custom UI XML code of the workbook file you are attempting to open.

Certain issues within the file might prompt Excel to display an error message regarding unreadable content, asking whether you'd like to recover the workbook's contents. By clicking Yes, Excel will attempt to repair the file and inform you of the outcome. Sometimes, Excel will make the necessary corrections automatically, but other times, you may need to identify and resolve the issue yourself.

Updating the Chapter18_ModifyRibbon.xlsm Workbook

As we proceed through this chapter, we'll be implementing new customizations to our custom ribbon tab. With each addition of new functionality, you'll need to update the XML markup code in the `Chap18_ModifyRibbon.xlsm` file using the Office RibbonX Editor. Afterward, open the file in Excel and add the necessary callbacks. The following steps will guide you through these revisions:

1. Open the `Chap18_ModifyRibbon.xlsm` workbook in the Office RibbonX Editor (make sure the file is not open in Excel).
2. Update the XML markup in the `CustomUI14.xml` part by entering or pasting the new XML markup segments. The new XML markup code is provided in text files included in the companion files.
3. Use the Validate icon in the Office RibbonX Editor's toolbar to make sure that the revised XML is well formed.
4. Save the modified file and exit the Office RibbonX Editor.
5. Open the `Chap18_ModifyRibbon.xlsm` file in Excel to view the updated ribbon customizations. You may encounter errors due to the missing callbacks.
6. Continue with the explanations in the sections that follow and enter any necessary callback procedures as instructed in the VBA module. Be sure to save the changes in the `Chap18_ModifyRibbon.xlsm` workbook.

7. Follow these steps as you proceed through the rest of the chapter. Whenever you see a revised ribbon customization markup, you should enter it in the Office RibbonX Editor and then add the callback procedures in the VBE screen in Excel.

NOTE	*The required callbacks can be generated in the Office RibbonX Editor by clicking the Generate Callbacks button and then copied to the VBE screen in Excel and completed with the required VBA code.*

Using Images in Ribbon Customizations

So far in this chapter, you have learned how to use built-in and custom images in your ribbon customizations. To reuse an Office icon, you must use the `imageMso` attribute of a control. To call your own BMP, GIF, and JPEG image files, you should use the `image` attribute.

Images can be added to the workbook file by using the Office RibbonX Editor (see Hands-On 18.4). You can also load the images at runtime when you open a workbook. For example, use the `loadImage` callback procedure in the `loadImage` attribute for the `customUI` element:

```
<customUI xmlns=http://schemas.microsoft.com/office/2009/07/
    customui loadImage="OnLoadImage">
```

The `loadImage` attribute specifies the following `OnLoadImage` callback procedure, which you must enter in a VBA code module of your `Chap18_ModifyRibbon.xlsm` workbook

```
Public Sub OnLoadImage(imgName As String, ByRef image)
    Dim strImgFileName As String
    strImgFileName = "C:\VBAExcel2024_ByExample\Images\" &
imgName
    Set image = LoadPicture(strImgFileName)
End Sub
```

You can load a picture from a file using the `LoadPicture` function. This function is a member of the `stdole.StdFunctions` library. The library file, which is called `stdole2.tlb`, is installed on your computer and is available for your VBA procedures without setting additional references. The `LoadPicture` function returns an object of type `IPictureDisp` that represents the image. You can view objects, methods, and properties available in the `stdole` library by using the Object Browser in the VBE window.

The following new button control in the `CustomUI14.xml` document has an image control that specifies the name of the image file:

```
<button id="btnCalc" label="Calculator"
  image="DownArrow.gif" onAction="OpenCalculator" />
```

The Calculator button uses the following callback procedure that must be entered in a VBA code module (`Chap18_ModifyRibbon.xlsm`):

```
Public Sub OpenCalculator(ctl As IRibbonControl)
    Shell "Calc.exe", vbNormalFocus
End Sub
```

Action Item 18.2

The example XML markup that adds the Calculator button to the Favorite tab is located in the `CustomUI14_ver02.txt` file in the companion files. After modifying the XML markup (following the instructions in the previous section), load the workbook file into Excel and add the `OnLoadImage` and `OpenCalculator` procedures to the VBAProject (Chap18_ModifyRibbon.xlsm) Code window. If you encounter errors while opening the Favorite tab, they may be caused by the missing callbacks.

About Tabs, Groups, and Controls

Built-in tabs and groups can be made invisible by setting the visible property of the `<tab>` or `<group>` elements to `false`. A built-in tab can contain a custom group. Built-in groups can also be added to other built-in or custom tabs. Some ribbon tabs, called contextual tabs, appear only when certain objects are in focus. For example, inserting a table will bring up the Table Tools contextual tab, which contains table-related options. You can add your custom groups to the built-in contextual tabs using the `<tabSet>` element within the `<contextualTabs>` element, like this:

```
<contextualTabs>
 <tabSet idMso="TabSetTableTools">
  <tab idMso="TabTableToolsDesign">
     <group id="CustomTools">
        <button id="btnID1"/>
     </group>
  </tab>
 </tabSet>
</contextualTabs>
```

Using Various Controls in Ribbon Customizations

By now, you should know how to go about creating the XML markup for your ribbon customizations and applying the custom ribbon to a workbook. Let's look at other types of controls you can show in the ribbon.

Creating Toggle Buttons

A *toggle button* is a button that alternates between two states. Many formatting features such as Bold, Italic, or Format Painter are implemented as toggle buttons. When you click a toggle button, the button will stay down until you click it again. To create a toggle button, use the `<toggleButton>` XML tag, as shown below:

```
<group id="GroupJK3" label="Various Controls">
<toggleButton id="tglR1C1" label="Reference Style" _
        size="normal" getPressed="onGetPressed" _
        onAction="SwitchRefStyle" />
</group>
```

Action Item 18.3
You will find the ribbon customization discussed in this section in the `Cus-tomUI14_ver03.txt` file in the companion files. Add this customization to the existing XML markup below the previous ending `</group>` tag. Don't forget to validate and save the file in the Office RibbonX Editor before opening it in Excel. Read the remaining text in this section to learn about the callbacks that need to be added to complete the implementation of the toggle button.

You can add a built-in image to the toggle button with the `imageMso` attribute or use a custom image, as discussed earlier in this chapter. To find out whether or not the toggle button is pressed, include the `getPressed` attribute in your XML markup. The `getPressed` callback procedure provides two arguments—the control that was clicked and the pressed state of the toggle button:

```
Sub onGetPressed(control As IRibbonControl, ByRef pressed)
  If control.ID = "tglR1C1" Then
    pressed = False
  End If
End Sub
```

Enter this callback procedure in a VBA code module of the `Chap18_Modify-Ribbon.xlsm` workbook to ensure that the specified toggle button is not pressed when the ribbon is loaded.

To perform an action when the toggle button is clicked, set the `onAction` attribute to the name of your custom callback procedure. This callback also provides two arguments: the control that was clicked and the state of the toggle button.

Add the following code to the VBA code module of the `Chap18_ModifyRibbon.xlsm` workbook:

```
Sub SwitchRefStyle(control As IRibbonControl, _
    pressed As Boolean)
    If pressed Then
        Application.ReferenceStyle = xlR1C1
    Else
        Application.ReferenceStyle = xlA1
    End If
End Sub
```

If the toggle button is pressed, the value of the `pressed` argument will be `True`; otherwise, it will be `False`. Figure 18.13 in the next section shows a custom toggle button named Reference Style. When you click this button, the worksheet headings will change to display letters or numbers. For more information on using R1C1 style references instead of A1 style, see the online help.

Creating Split Buttons, Menus, and Submenus

A *split button* is a combination of a button or toggle button and a menu. Clicking the button performs one default action, and clicking the drop-down arrow opens a menu with a list of related options to select from.

To create the split button, use the `<splitButton>` tag. Within this tag, you must define a `<button>` or `<toggleButton>` control and the `<menu>` control, as shown in the following XML markup:

```
<splitButton id="btnSplit1" size="large">
    <button id="btnGoTo" label="Navigate To..." imageMso="GoTo" />
  <menu id="mnuGoTo" label="Spreadsheet Navigation" itemSize="normal">
        <menuSeparator id="mnuDiv1" title="Formulas and Constants" />
        <button id="btnFormulas" label="Select Formulas"
            onAction="GoToSpecial" />
        <button id="btnNumbers" label="Select Numbers Only"
            onAction="GoToSpecial" />
        <button id="btnText" label="Select Text Only"
            onAction="GoToSpecial" />
        <menuSeparator id="mnuDiv2" title="Special Cells" />
```

```
        <button id="btnBlanks" label="Select blank cells"
          onAction="GoToSpecial" />
        <button id="btnLast" label="Select last cell"
          onAction="GoToSpecial" />
    </menu>
</splitButton>
```

Action Item 18.4

You will find the ribbon customization discussed in this section in the `Cus-tomUI14_ver04.txt` file in the companion files. Add this customization to the existing XML markup below the toggle button markup.

You can specify the size of the items in the menu with the `itemSize` attribute. The `<menuSeparator>` tag can be used inside the menu node to break the menu into sections. Each menu segment can then be titled using the `title` attribute, as shown in the previous example.

You can add the `onAction` attribute to each menu button to specify the call-back procedure or macro to execute when the menu item is clicked. The above XML markup uses the following callback procedure that should be entered in a VBA code module of the `Chap18_ModifyRibbon.xlsm` workbook:

```
Sub GoToSpecial(control As IRibbonControl)
    On Error Resume Next
    Range("A1").Select

    If control.ID = "btnFormulas" Then
        Selection.SpecialCells(xlCellTypeFormulas, 23).Select
    ElseIf control.ID = "btnNumbers" Then
```

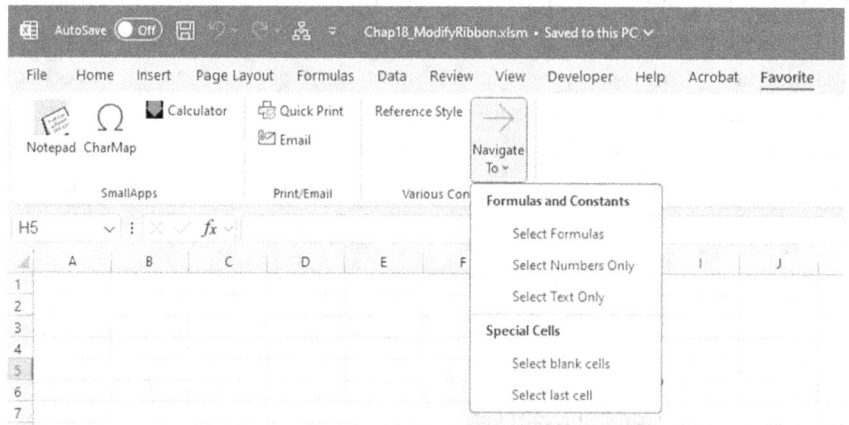

FIGURE 18.13. A toggle button (Reference Style) and a custom split button control (Navigate To) with a menu. See Chap18_ModifyRibbon_Fig13.xlsm in the companion files.

```
            Selection.SpecialCells(xlCellTypeConstants, 1).Select
        ElseIf control.ID = "btnText" Then
            Selection.SpecialCells(xlCellTypeConstants, 2).Select
        ElseIf control.ID = "btnBlanks" Then
            Selection.SpecialCells(xlCellTypeBlanks).Select
        ElseIf control.ID = "btnLast" Then
            Selection.SpecialCells(xlCellTypeLastCell).Select
        End If
End Sub
```

In addition to button controls, the ribbon's menus can contain toggle buttons, checkboxes, gallery controls, split buttons, and other menus.

Creating Checkboxes

The checkbox control is used to show the state—either `true` (on) or `false` (off). It can be included inside a menu control or used as a separate control on the ribbon.

To create a checkbox, use the `<checkBox>` tag, as shown in the following XML:

```
<separator id="OtherControlsDiv1" />
    <labelControl id="TitleForBox1" label="Show or Hide Screen
    Elements" />
    <box id="boxLayout1">
    <checkBox id="chkGridlines" label="Gridlines" visible="true"
        getPressed="onGetPressed" onAction="DoSomething" />
        <checkBox id="chkFormulaBar" label="Formula Bar" _
            visible="true"
            getPressed="onGetPressed" onAction="DoSomething" />
    </box>
```

Action Item 18.5

You will find the ribbon customization discussed in this section in the `CustomUI14_ver05.txt` file in the companion files. Add this customization to the existing XML markup below the ending `</splitButton>` tag.

In the above XML markup, the `<separator>` tag will produce the vertical bar that visually separates controls within the same ribbon group (see Figure 18.14). The `<labelControl>` tag can be used to display static text anywhere in the ribbon. In this example, we use it to place a header over a set of controls.

To control the layout of various controls (to display them horizontally instead of vertically), use the `<box>` tag. You can define whether a checkbox should be visible or hidden by setting the `visible` attribute to `true` or `false`. To disable a checkbox, set the `enabled` attribute to `false`; this will cause the checkbox to appear grayed out.

To get the checked state for a checkbox, add a callback procedure in the VBA Code window of the Chap18_ModifyRibbon.xlsm workbook. You can modify the onGetPressed procedure that we used earlier with the toggle button:

```
Sub onGetPressed(control As IRibbonControl, ByRef pressed)
    If control.ID = "tglR1C1" Then
        pressed = False
    End If

    If control.ID = "chkGridlines" And _
        ActiveWindow.DisplayGridlines = True Then
        pressed = True
    ElseIf control.ID = "chkGridlines" And _
        ActiveWindow.DisplayGridlines = False Then
        pressed = False
    End If

    If control.ID = "chkFormulaBar" And _
        Application.DisplayFormulaBar = True Then
        pressed = True
    ElseIf control.ID = "chkFormulaBar" And _
        Application.DisplayFormulaBar = False Then
        pressed = False
    End If
End Sub
```

The action of the checkbox control is handled by the callback procedure in the onAction attribute. To make this checkbox example work, enter the following procedure in a VBA code module of the Chap18_ModifyRibbon.xlsm workbook:

```
Sub DoSomething(ctl As IRibbonControl, pressed As Boolean)
    If ctl.ID = "chkGridlines" And pressed Then
        ActiveWindow.DisplayGridlines = True
    ElseIf ctl.ID = "chkGridlines" And Not pressed Then
        ActiveWindow.DisplayGridlines = False
    ElseIf ctl.ID = "chkFormulaBar" And pressed Then
        Application.DisplayFormulaBar = True
    ElseIf ctl.ID = "chkFormulaBar" And Not pressed Then
        Application.DisplayFormulaBar = False
    End If
End Sub
```

Like other controls, labels for checkboxes can contain static text in the label attribute, as shown in the above XML, or they can be assigned dynamically using the callback procedure in the getLabel attribute.

> | NOTE | *Callback procedures don't need to be named the same as the attribute they are used with. Also, you may change the callback's argument names as desired.* |
> |------|---|

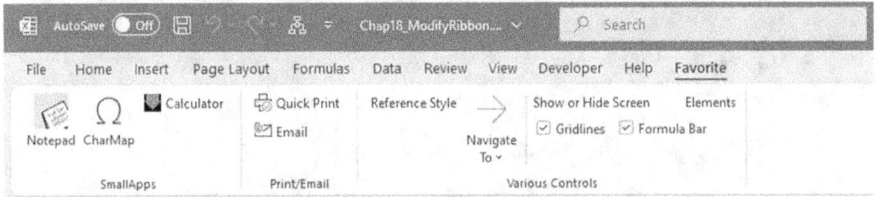

FIGURE 18.14.   The checkbox controls (Gridlines and Formula Bar) are laid out horizontally. See Chap18_ModifyRibbon_Fig14.xlsm in the companion files.

Creating Edit Boxes

Use the `<editBox>` tag to provide an area on the ribbon where users can type text or numbers:

```
<editBox id="txtFullName" label="First and Last Name:"
    sizeString="AAAAAAAAAAAAAAA" maxLength="25"
    onChange="onFullNameChange" />
```

Action Item 18.6
You will find the ribbon customization discussed in this section in the `CustomUI14_ver06.txt` file in the companion files. Add this customization to the existing XML markup below the ending `</box>` tag.

The `sizeString` attribute specifies the width of the edit box. Set it to a string that will give you the width you want. The `maxLength` attribute allows you to limit the number of characters and/or digits that can be typed in the edit box. If the text entered exceeds the specified number of characters (25 in this case), Excel automatically displays a balloon message on the ribbon: "The entry may contain no more than 25 characters." When the entry is updated in an edit box control, the callback procedure specified in the `onChange` attribute is called:

```
Public Sub onFullNameChange(ctl As IRibbonControl, text As _
    String)
  If text <> "" Then
    MsgBox "You've entered '" & text & _
    "' in the edit box."
  End If
End Sub
```

Enter the `onFullNameChange` callback procedure in the VBA code module of the `Chap18_ModifyRibbon.xlsm` workbook. When you enter some text in the edit box, the procedure will display a message box. The completed edit box control is shown in Figure 18.15.

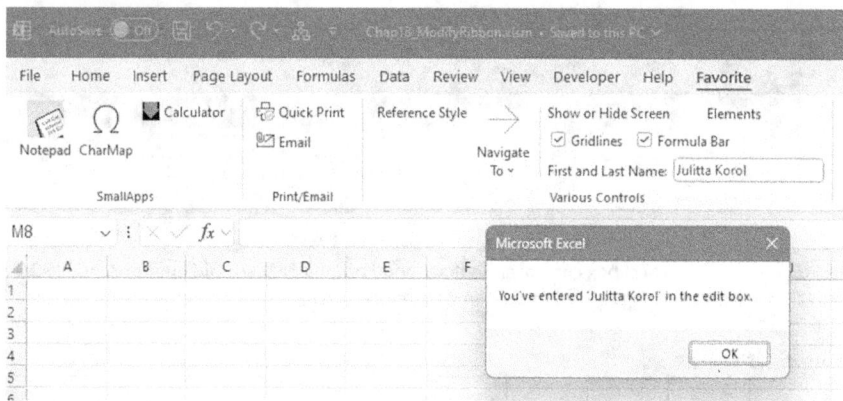

FIGURE 18.15.   An edit box control allows data entry directly on the ribbon. See Chap18_ModifyRibbon_Fig15.xlsm in the companion files.

Creating Combo Boxes and Dropdowns

There are three types of drop-down controls that can be placed on the ribbon: combo box, dropdown, and gallery. These controls can be dynamically populated at runtime by writing callbacks for their getItemCount, getItemID, getItemLabel, getItemImage, getItemScreentip, or getItemSupertip attributes. The combo box and drop-down controls can also be made static by defining their drop-down content using the <item> tag, as shown below:

```
<separator id="OtherControlsDiv2" />
  <comboBox id="cboDepartment" label="Departments"
        supertip="Select  Department" onChange="onChangeDept">
      <item id="Marketing" label="Marketing" />
      <item id="Sales" label="Sales" />
      <item id="Personnel" label="Personnel" />
      <item id="ResearchAndDevelopment" label="Research and
        Development" />
  </comboBox>
```

Action Item 18.7
You will find the ribbon customization discussed in this section in the CustomUI14_ver07.txt file in the companion files. Add this customization to the existing XML markup below the editBox XML markup.

To separate the combo box control from other controls in the same ribbon group, the previous example uses the `<separator>` tag. Notice that each `<item>` tag specifies a new drop-down row.

NOTE	*A combo box is a combination of a drop-down list and a single-line edit box, allowing the user to either type a value directly into the control or choose from the list of predefined options. Use the* `sizeString` *attribute to define the width of the edit box.*

The combo box control does not have the `onAction` attribute. It uses the `onChange` attribute, which specifies the callback to execute when the item selection changes:

```
Public Sub onChangeDept(ctl As IRibbonControl, text As String)
    MsgBox "You selected " & text & " department."
End Sub
```

Notice that the `onChange` callback provides only the text of the selected item; it does not give you access to the selected index. If you need the index of the selection, use the drop-down control instead, as shown below:

```
<dropDown id="drpBoro" label="City Borough"
      supertip="Select City Borough"
      onAction="onActionBoro">
      <item id="M" label="Manhattan" />
      <item id="B" label="Brooklyn" />
      <item id="Q" label="Queens" />
      <item id="I" label="Staten Island" />
      <item id="X" label="Bronx" />
</dropDown>
```

Action Item 18.8
You will find the `dropDown` ribbon customization discussed in this section in the `CustomUI14_ver08.txt` file in the companion files. Add this customization to the existing XML markup below `editBox`.

The `onAction` callback of the drop-down control will give you both the selected item's ID and its index:

```
Public Sub onActionBoro(ctl As IRibbonControl, _
      ByRef SelectedID As String, _
      ByRef SelectedIndex As Integer)
         MsgBox "Index=" & SelectedIndex & " ID=" & SelectedID
End Sub
```

Be sure to enter the above callback procedures in the VBA code module of the `Chap18_ModifyRibbon.xlsm` workbook. The combo box and a drop-down control are shown in Figures 18.16 and 18.17.

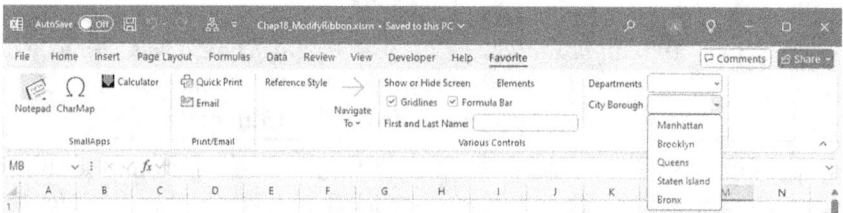

FIGURE 18.16.   Two combo boxes, when opened, display a list of departments and boroughs. See Chap18_ModifyRibbon_Fig16.xlsm in the companion files.

Creating a Gallery Control

A gallery control is a drop-down control that can display a grid of images with or without a label. Built-in galleries cannot be customized but you can build your own using the `<gallery>` tag.

The following XML markup dynamically populates a custom gallery control at runtime:

```
<gallery id="glHolidays" label="Holidays" columns="3" rows="4"
    getImage="onGetImage" getItemCount="onGetItemCount"
    getItemLabel="onGetItemLabel" getItemImage="onGetItemImage"
    getItemID="onGetItemID" onAction="onSelectedItem" />
```

Action Item 18.9
You will find the `gallery` ribbon customization discussed in this section in the `CustomUI14_ver09.txt` file in the companion files. Add this customization to the existing XML markup below the `dropDown` markup.

In the above XML markup, the gallery control will perform the action specified in the `onSelectedItem` callback procedure. Notice that the gallery control has many attributes that contain static text or define callbacks. We will discuss them later. Right now, let's focus on the image-loading process. The gallery control uses the `getImage` attribute with the `OnGetImage` callback procedure. This procedure entered in the VBA code module of the `Chap18_ModifyRibbon.xlsm` workbook tells Excel to load the appropriate image to the ribbon:

```
Public Sub onGetImage(ctl As IRibbonControl, ByRef image)
    Select Case ctl.ID
      Case "glHolidays"
        Set image = LoadPicture( _
```

```
        "C:\VBAExcel2024_ByExample\Images\Square0.gif")
    End Select
End Sub
```

The decision as to which image should be loaded is based on the ID of the control in the `Select Case` statement. The gallery control also uses the `OnGetItemImage` callback procedure (defined in the `getItemImage` attribute) to load custom images for its drop-down selection list (see Figure 18.17). Use the `columns` and `rows` attributes to specify the number of columns and rows in the gallery when it is opened. If you need to define the height and width of images in the gallery, use the `itemHeight` and `itemWidth` attributes (not used in this example due to the simplicity of the utilized images).

The `getItemCount` and `getItemLabel` attributes contain callback procedures that provide information to the ribbon on how many items should appear in the drop-down list and the names of those items. The `getItemImage` attribute contains a callback procedure that specifies the images to be displayed next to each gallery item. The `getItemID` attribute specifies the `onGetItemID` callback procedure that will provide a unique ID for each of the gallery items.

Now, let's go over other VBA callbacks that are used by the gallery control. All the VBA procedures in this section must be added to the VBA module for the previous XML markup to work:

```
Public Sub onGetItemCount(ctl As IRibbonControl, ByRef count)
    count = 12
End Sub
```

In the previous procedure, we use the `count` parameter to return to the ribbon the number of items we want to have in the gallery control.

The following procedure will label each of the gallery items:

```
Public Sub onGetItemLabel(ctl As IRibbonControl, _
    index As Integer, ByRef label)
    label = MonthName(index + 1)
End Sub
```

The VBA `MonthName` function is used to retrieve the name of the month based on the value of the index. The initial value of the index is zero (0). Therefore, `index + 1` will return `January`. To display an abbreviated form of the month's name (`Jan`, `Feb`, etc.), specify `True` as the second parameter to this function:

```
label = MonthName(index + 1, True)
```

If you are using a localized version of Microsoft Office (French, Spanish, etc.), the `MonthName` function will return the name of the month in the specified interface language.

The next callback procedure shows how to load images for each gallery item:

```
Public Sub onGetItemImage(ctl As IRibbonControl, _
    index As Integer, ByRef image)
    Dim imgPath As String

    imgPath = "C:\VBAExcel2024_ByExample\Images\square"
    Set image = LoadPicture(imgPath & index + 1 & ".gif")
End Sub
```

Each item in the gallery must have a unique ID, so the onGetItemID callback uses the MonthName function to use the month name as the ID:

```
Public Sub onGetItemID(ctl As IRibbonControl, _
    index As Integer, ByRef id)
    id = MonthName(index + 1)
End Sub
```

The last procedure you need to write for the Holidays gallery control should define the actions to be performed when an item in the gallery is clicked. This is done via the following onSelectedItem callback, which was specified in the onAction attribute of the XML markup:

```
Public Sub onSelectedItem(ctl As IRibbonControl, _
    selectedId As String, selectedIndex As Integer)
    Select Case selectedIndex
        Case 6
            MsgBox "Holiday 1: Independence Day, July 4th", _
                vbInformation + vbOKOnly, _
                selectedId & " Holidays"
        Case 11
            MsgBox "Holiday 1: Christmas Day, December 25th", _
                vbInformation + vbOKOnly, _
                selectedId & " Holidays"
        Case Else
            MsgBox "Please program holidays for " _
                & selectedId & ".", vbInformation + vbOKOnly, _
                " Under Construction"
    End Select
End Sub
```

In the previous callback procedure, the selectedId parameter returns the name that was assigned to the label, while the selectedIndex parameter is the position of the item in the list. The first item in the list (January) is indexed with zero (0), the second one with 1, and so forth. In the previous procedure, we just coded two holidays: one for the month of July (selectedIndex=6) and one for December (`selectedIndex=11`). The `Case Else` clause in the `Select Case` statement provides a message when other months are selected.

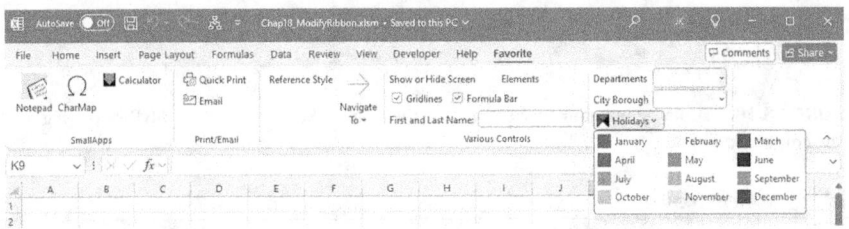

FIGURE 18.17. Customized ribbon with gallery control. See Chap18_ModifyRibbon_Fig17.xlsm in the companion files.

Creating a Dialog Box Launcher

On some ribbon tabs, you can see a small dialog box launcher button in the bottom-right corner of a group. You can use this button to open a special form that allows the user to set up many options at once, or you can display a form that contains specific information. To add a custom dialog box launcher button to the ribbon, use the `<dialogBoxLauncher>` tag, as shown below:

```
<dialogBoxLauncher>
    <button id="Launch1" screentip="Show Auto Correct Dialog"
    onAction="onActionLaunch" />
</dialogBoxLauncher>
```

Action Item 18.10

You will find the `dialogBoxLauncher` ribbon customization in the `CustomUI14_ver10.txt` file in the companion files. Add this customization to the existing XML markup below the `gallery` markup.

The dialog box launcher control must contain a button. The `onAction` attribute for the button contains the callback procedure that will execute when the button is clicked:

```
Public Sub onActionLaunch(ctl As IRibbonControl)
    Application.Dialogs(xlDialogAutoCorrect).Show
End Sub
```

The dialog box launcher control must appear as the last element within the containing group element.

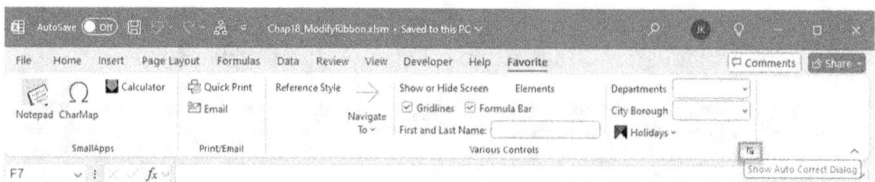

FIGURE 18.18. A dialog box launcher control on the ribbon. See Chap18_ModifyRibbon_Fig18.xlsm in the companion files.

Disabling a Control

You can disable a built-in or custom ribbon control by using the `enabled` or `getEnabled` attribute. The following XML markup uses the `enabled` attribute to disable our custom checkbox control that we created earlier:

```
<checkBox id="chkGridlines" label="Gridlines" visible="true"
    getPressed="onGetPressed" enabled="false"
onAction="DoSomething" />
```

You can use the `getEnabled` attribute to disable a control based on some conditions. For example, the following XML markup uses the `getEnabled` attribute to disable the custom checkbox control named Formula Bar:

```
<checkBox id="chkFormulaBar" label="Formula Bar" visible="true"
    getPressed="onGetPressed" getEnabled="onGetEnabled"
    onAction="DoSomething" />
```

Action Item 18.11

You will find the previous ribbon customizations for checkboxes in the `CustomUI14_ver11.txt` file in the companion files. Modify the existing `checkBox` markup for `chkFormulaBar` by adding the markup for the `getEnabled` attribute.

The checkbox customization requires the following variable declaration at the top of the VBA `Module1`:

```
'Add this just below the Option Explicit statement
'at the top of the Module1 code window
Public blnEnabled As Boolean
```

Next, add the following callback below other procedures in the module:

```
Public Sub onGetEnabled(ctl As IRibbonControl, ByRef returnedVal)

    returnedVal = blnEnabled

End Sub
```

NOTE	*Important Note: In addition to the previous procedure, you will need to implement the procedures and the markup as explained in the Refreshing the Ribbon section later in this chapter.*

Add two new sheets to the `Chap18_ModifyRibbon.xlsm` workbook and enter the following event procedures in the Sheet3 Code window. Notice that you will need to declare `objRibbon` at the top of `Module1`:

```
' enter the declaration at the top of
' the Module1 code window
Public objRibbon As IRibbonUI
```

In the Sheet3 Code window, enter the following two procedures:

```
Private Sub Worksheet_Activate()
    blnEnabled = False
    objRibbon.Invalidate
    MsgBox "Formula bar checkmark is disabled in this sheet only."
End Sub

Private Sub Worksheet_Deactivate()
    blnEnabled = True
    objRibbon.Invalidate
End Sub
```

You can use a callback procedure to display a "not authorized" message when a Formulas | Name Manager Ribbon control is selected. The following XML code shows how to disable the built-in Name Manager button:

```
<!-- Built-in commands section -->
<commands>
    <command idMso="NameManager" onAction="DisableNameManager" />
</commands>
```

Action Item 18.12

You will find the `NameManager` command ribbon customization in the `CustomUI14_ver12.txt` file in the companion files. Add this XML code just above the `<ribbon startFromScratch="false">` markup.

To make your XML code more readable, you can include comments between the `<!--` and `-->` characters. The `<command>` tag can be used to refer to any built-in command. This tag must appear in the `<commands>` section of the XML code. To see how this works, simply add the above code fragment to the XML code shown in the previous section just before the line:

```
<ribbon startFromScratch="false">
```

The `onAction` attribute contains the following callback procedure, which will display a message when the Name Manager button is clicked:

```
Sub DisableNameManager(ctl As IRibbonControl, ByRef
cancelDefault)
    MsgBox "You are not authorized to use this function."
    cancelDefault = True
End Sub
```

You can add more code to the above procedure if you need to cancel the control's default behavior only when certain conditions have been satisfied.

Repurposing a Built-In Control

It is possible to change the purpose of a built-in ribbon button. For example, when the user clicks the Picture button on the Insert tab when `Sheet1` is active, you could display a Copy Picture dialog box instead of the default Insert Picture dialog box, like this:

```
<command idMso="InsertCellPictureFromFile" onAction= _
    "ShowCopyPictureDialog"/>
```

Action Item 18.13

You will find this ribbon customization in the `CustomUI14_ver13.txt` file in the companion files. Add the shown XML markup to the command section of the `customUI14.xml` document in `Chap18_ModifyRibbon.xlsm`. In the Excel `Module1` of the `Chap18_ModifyRibbon.xlsm` workbook, add the following `ShowCopyPictureDialog` callback procedure:

```
Public Sub ShowCopyPictureDialog(ctl As IRibbonControl, _
    ByRef cancelDefault)
  If ActiveSheet.Name = "Sheet1" Then
    ' display the CopyPicture dialog box instead
    Application.Dialogs(xlDialogCopyPicture).Show
  Else
    cancelDefault = False
  End If
End Sub
```

Only simple controls that perform an action when clicked can be repurposed. You cannot repurpose advanced controls such as combo boxes, dropdowns, or galleries.

Refreshing the Ribbon

So far in this chapter, you've seen how to use callback procedures to specify the values of control attributes at runtime, but how should you handle the case of updating your custom ribbon or the controls placed in the ribbon based on what the user is doing in your application? By using the `InvalidateControl` method of the `IRibbonUI` object, you can change the attribute values at any time.

To use the `IRibbonUI` object, start by adding the `onLoad` attribute to the `customUI` element in your ribbon customization XML:

```
<customUI xmlns="http://schemas.microsoft.com/office/2009/07/ _
    customui" loadImage="OnLoadImage" onLoad="RefreshMe">
```

Action Item 18.14
You will find this ribbon customization in the `CustomUI14_ver14.txt` file in the companion files.

NOTE	*When you open the* `Chap18_ModifyRibbon.xlsm` *workbook with the previous customization in Excel, you will get errors because of the missing procedures. Click OK to dismiss the error message, switch to the VBE window, and enter the code as explained further in this section.*

In the preceding XML markup, the `onLoad` attribute points to the callback procedure that will give you a copy of the ribbon, which you can refresh anytime you want. In this example, the `onLoad` callback procedure is named `RefreshMe`. Let's say that, upon data entry, you want the text of the edit box to appear in uppercase. Implementing the `onLoad` callback requires the `Public` module-level variable of type `IRibbonUI` declared earlier at the top of the VBE code module of the `Chap18_ModifyRibbon.xlsm` workbook:

```
Public objRibbon As IRibbonUI
```

To keep track of the state of the edit box control, declare a `Private` module-level variable `strUserTxt` as follows:

```
Private strUserTxt As String
```

Next, enter the callback procedure in `Module1` of the `Chap18_ModifyRibbon.xlsm` workbook. This will store a copy of the ribbon in the `objRibbon` variable:

```
'callback for the onLoad attribute of customUI
Public Sub RefreshMe(ribbon As IRibbonUI)
```

```
        Set objRibbon = ribbon
End Sub
```

When the ribbon loads, you will have a copy of the IRibbonUI object saved for later use.

Now, let's take a look at the XML markup used in this scenario:

```
<editBox id="txtFullName" label="First and Last Name:"
    sizeString="AAAAAAAAAAAAAAAAAA" maxLength="25"
    getText="getEditBoxText" onChange="onFullNameChangeToUcase" />
```

Action Item 18.15

You will find the editBox ribbon customization in the CustomUI14_ver15.txt file in the companion files.

Note that the edit box control was introduced earlier in this chapter (see Figure 18.15). You need to modify the original XML markup for the edit box by adding the getText attribute, which points to the following callback:

```
Public Sub getEditBoxText(control As IRibbonControl, ByRef text)
    text = UCase(strUserTxt)
End Sub
```

The above callback getEditBoxText uses the VBA built-in UCase function to change the text entered in the edit box to uppercase letters. When text is updated, the procedure in the onChange attribute is called (be sure to change the procedure name in your original XML markup):

```
Public Sub onFullNameChangeToUcase(ByVal control _
        As IRibbonControl, text As String)
    If text <> "" Then
        strUserTxt = text
        objRibbon.InvalidateControl "txtFullName"
    End If
End Sub
```

The above callback begins by checking the value of the text parameter provided by the ribbon. If this parameter contains a value other than an empty string (""), the text the user entered is stored in the strUserTxt variable. Before a change can occur in the ribbon control, you need to mark the control as invalid. This is done by calling the InvalidateControl method of the IRibbonUI object that we have stored in the objRibbon variable:

```
objRibbon.InvalidateControl "txtFullName"
```

The above statement makes the txtFullName control refresh itself the next time it is displayed. When the control is invalidated, it will automatically call its callback functions. The `onFullNameChangeToUcase` callback procedure in the `onChange` attribute will execute, causing the text entered in the `txtFullName` edit box control to appear in uppercase letters.

NOTE	*The* `IRibbonUI` *object has only two methods:* `InvalidateControl` *and* `Invalidate`. *Use the* `InvalidateControl` *method to refresh an individual control. Use the* `Invalidate` *method to refresh all controls in the ribbon.*

NOTE	*If you find that some controls on the* `Favorite` *tab don't behave as programmed, make sure that the top of the VBA module contains the following three module-level variables and the* `RefreshMe` *procedure:*
	
	`Public objRibbon As IRibbonUI`
	`Private strUserTxt As String`
	`Public blnEnabled As Boolean`
	
	`'callback for the onLoad attribute of customUI`
	`Public Sub RefreshMe(ribbon As IRibbonUI)`
	`    Set objRibbon = ribbon`
	`End Sub`
	
	Reload the `Chap18_ModifyRibbon.xlsm` *workbook in Excel and check whether the problem was resolved. All the revisions we have implemented so far in this workbook are saved in* `Chap18_ModifyRibbon1.xslm` *in the companion files. You may want to check your work against this file.*

The CommandBar Object and the Ribbon

You can make your custom ribbon button match any built-in button by using the `CommandBar` object. This object has been extended with several `Get` methods that expose the state information for the built-in controls: `GetEnabledMso`, `GetImageMso`, `GetLabelMso`, `GetPressedMso`, `GetScreentipMso`, `GetSupertipMso`, and `GetVisibleMso`. Use these methods in your callbacks to check the built-in control's properties. For example, the following statement will return `False` if the ribbon's built-in Cut button is currently disabled (grayed out) or `True` if it is enabled (ready to use):

```
MsgBox Application.CommandBars.GetEnabledMso("Cut")
```

Notice that the `GetEnabledMso` method requires the name of the built-in control. To see the result of the above statement, try it out from the Immediate window.

The `GetImageMso` method is very useful if you'd like to reuse any of the built-in button images in your own controls. This method allows you to get the bitmap for any `imageMso` tag. For example, to retrieve the bitmap associated with the Cut button on the ribbon, enter the following statement in the Immediate window:

```
MsgBox Application.CommandBars.GetImageMso("Cut", 16, 16)
```

The `GetImageMso` method uses three arguments: the name of the built-in control and the width and height of the bitmap image in pixels. Because this method returns the `IPictureDisp` object, it is very easy to place the retrieved bitmap onto your own custom ribbon control by writing a simple VBA callback for your control's `getImage` attribute.

In addition to the methods that provide information about the properties of the built-in controls, the `CommandBar` object also includes a handy `ExecuteMso` method that can be used to trigger the built-in control's default action. This method is quite useful when you want to perform a click operation for the user from within a VBA procedure or want to conditionally run a built-in feature.

Let's take a look at the example implementation of the `GetImageMso` and `ExecuteMso` methods. Here's the XML definition for a custom ribbon button:

```
<button id="btnWordWizard" label="Use Thesaurus" size="normal"
    getImage="onGetBitmap" onAction="DoDefaultPlus" />
```

Action Item 18.16
You will find the above ribbon customization in the `CustomUI14_ver16.txt` file in the companion files. Add this XML code to the custom ribbon definition you've worked with in this chapter. Add it after the Reference Style toggle button.

Now, let's look at the VBA implementation. Let's assume that you want the button to use the same image as the built-in button labeled ResearchPane. When the button is clicked, you'd like to display the built-in Research pane set to Thesaurus only when a certain condition is true. Let's add the following code to your VBA module:

```
Sub onGetBitmap(ctl As IRibbonControl, ByRef image)
    Set image = Application.CommandBars.
GetImageMso("ResearchPane", 16, 16)
End Sub
```

When the ribbon is loaded, the `onGetBitmap` callback automatically retrieves the image bitmap from the ResearchPane button's `imageMso` attribute and assigns it to the `getImage` attribute of your button. When your button is clicked and the active cell contains a text entry, the Thesaurus opens up in the Research pane; if the active cell is empty or it contains a number, the user will see a message box instead:

```
Sub DoDefaultPlus(ctl As IRibbonControl)
    If Not IsNumeric(ActiveCell.Value) Then
        Application.CommandBars.ExecuteMso "Thesaurus"
    Else
        MsgBox "To use Thesaurus, select a cell " & _
        "containing text.", vbOKOnly + vbInformation, _
                    "Action Required"
    End If
End Sub
```

Be sure to enter the above procedures in the VBA code module of the `Chap18_ModifyRibbon.xlsm` workbook. Figure 18.19 shows the Thesaurus button in the Various Controls group of the Favorite tab.

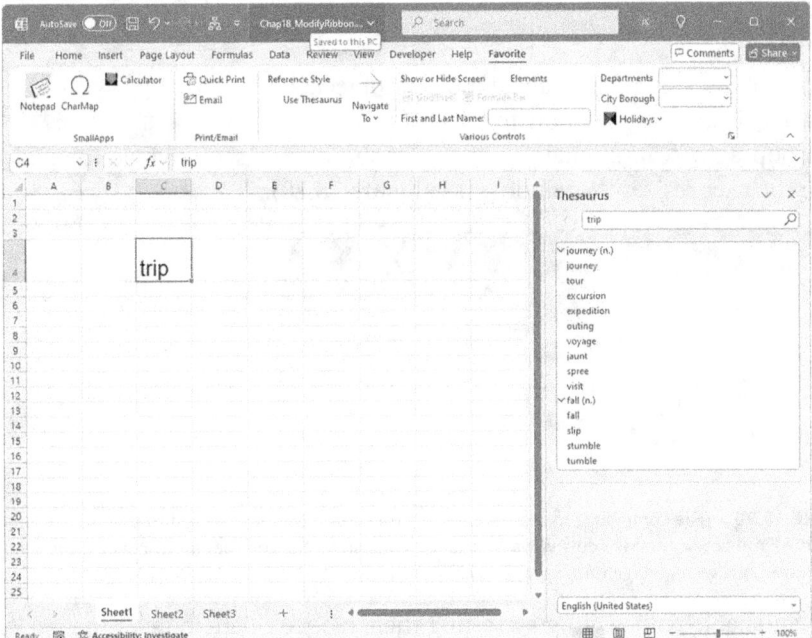

FIGURE 18.19. A custom button can conditionally trigger a built-in control's action. In this example, we are looking up the word trip using the Thesaurus button in the custom Favorite tab. See Chap18_ModifyRibbon_Fig19.xlsm in the companion files.

Tab Activation and Group Auto-Scaling

Tab activation makes it possible to activate a specific tab in response to some event.

To activate a custom tab on the Excel ribbon, use the `ActivateTab` method of the `IRibbonUI` object, passing to it the ID of the custom string. For example, to activate the Favorite tab you created in this chapter, use the following statement:

```
objRibbon.ActivateTab "TabJK1"
```

Recall that `objRibbon` is the module-level `Public` variable we declared earlier for accessing the `IRibbonUI` object. To activate a built-in tab, use the `ActivateTabMso` method.

For example, the following statement activates the Data tab:

```
objRibbon.ActivateTabMso "TabData"
```

Finally, there is also a special `ActivateTabQ` method used to activate a tab shared between multiple add-ins. In addition to the `tabID`, this method requires that you specify the namespace of the add-in. The syntax is shown below:

```
expression.ActivateTabQ(tabID As String, namespace as String)
```

where `expression` returns an `IRibbonUI` object.

Tab activation applies only to tabs that are visible.

Group auto-scaling enables custom ribbon groups to change their layout when the user resizes the window (see Figure 18.20).

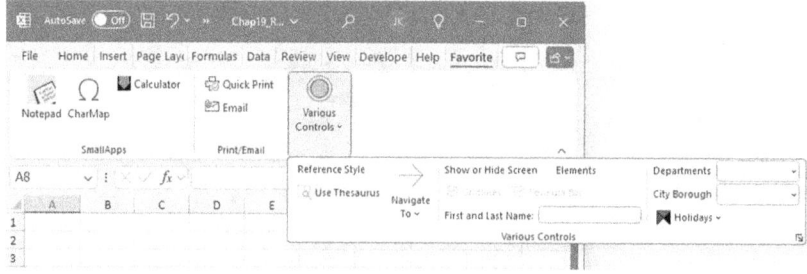

FIGURE 18.20. The commands in the Various Controls group are compressed to a single button when the Excel application window is made smaller. To change the icon that appears when the group is compressed, assign an image to the group itself.

You can enable auto-scaling by setting the `autoScale` attribute of the `<group>` tab to `true`, as in the following:

```
<group id="GroupJK3" label="Various Controls" autoScale="true">
```

Action Item 18.17

You will find the above ribbon customization in the `CustomUI14_ver17.txt` file in the companion files. Notice that the value of the `autoScale` attribute is entered in lowercase. Auto-scaling is set on a per-group basis.

The completed workbook file with all ribbon customizations that have been added up to this point can be found in the `Chap18_ModifyRibbon2.xlsm` workbook in the companion files.

CUSTOMIZING THE BACKSTAGE VIEW

The File tab provides an entry point to the Office UI known as the Backstage view. This view is specifically designed for working with workbooks. It contains commands known as *fast commands* that provide quick access to common functionality such as saving, opening, or closing workbooks. Here, you also find the Exit command for exiting Microsoft Excel and the Options command for customizing numerous Excel features. In addition to fast commands, the navigation bar on the left-hand side of the Backstage view includes several tabs that group related tasks. For example, clicking the Print tab in the navigation bar displays all the information related to the installed printers and allows you to easily access and change many of the print settings. A large area in the Print Backstage view is used for the presentation of the workbook's Print Preview.

The Info tab organizes tasks related to workbook permissions, versions, file sharing, and numerous other workbook properties.

As an Excel developer already familiar with ribbon UI customization, you will feel very comfortable customizing the Backstage view. Like the ribbon, the Backstage view uses XML markup, which you can add to the workbook file by using the Office RibbonX Editor.

The Backstage view is a perfect place to include custom solutions that present summaries of business processes or workflows (see the sidebar with links to Microsoft documents that will walk you through the process of customizing the Office Backstage view).

In this section, you'll try a couple of simple things in the Backstage view to get your feet wet so that you can later move on to more advanced customizations with the downloads recommended in the sidebar.

Backstage View Development

For an advanced introduction to the Backstage view, download the following Microsoft papers (note that the 2010 versions of these docs are still applicable to the current 2024 version):

❑ *Customizing the Office 2010 Backstage View for Developers*:
   http://msdn.microsoft.com/en-us/library/ee815851(printer).aspx

❑ *Dynamically Changing the Visibility of Groups and Controls in the Office 2010 Backstage View*:

http://msdn.microsoft.com/en-us/library/ff645396(printer).aspx

---

The Backstage view XML markup must be entered between the `<backstage></backstage>` elements within the `<customui></customui>` tags and below any ribbon customization markup.

The following XML markup adds a custom button named Synchronize and a custom tab named Endless Possibilities to the Backstage view:

```
<backstage>
    <button id="btnSync" label="Synchronize" _
       imageMso="SyncNow" isDefinitive="true"
     insertBeforeMso="FileClose" onAction=_
       "onActionCopyToArchive" />
    <tab id="mySpecialTab" label="Endless Possibilities" _
       insertAfterMso="TabRecent">
     <firstColumn>
        <group id="grp01" label="Home Group" helperText="_
        This is group 1 help text">
            <topItems>
                <button id="myButton1" label="My button" />
            </topItems>
        </group>
        <group id="gr02" label="Cheat Sheet">
            <topItems>
                <button id="myButton2" label="Cheat Ideas" />
            </topItems>
            <bottomItems>
              <layoutContainer id="set1" _
                 layoutChildren="horizontal" >
                 <editBox id="item1" label="Cheat Item 1"  />
                 <editBox id="item2" label="Cheat Item 2"  />
              </layoutContainer>
            </bottomItems>
```

```
        </group>
    </firstColumn>
    <secondColumn>
      <group id="grpHyperlinks" label="Frequently Accessed _
          Websites" visible="true">
        <primaryItem>
         <button id="top1" label="Primary Button" _
          imageMso="HyperlinkProperties" />
        </primaryItem>
        <topItems>
          <button id="msft" label="Microsoft" onAction= _
            "onActionExecHyperlink" />
          <layoutContainer id="set2" layoutChildren=_
            "vertical" >
          <hyperlink id="YouTube" label=_
              "http://www.YouTube.com"
              onAction="onActionExecHyperlink" />
          <hyperlink id="amazon" label=_
              "http://www.amazon.com"
              onAction="onActionExecHyperlink" />
          <hyperlink id="DeGru" label=_
              "http://www. DeGruyterBrill.com"
              onAction="onActionExecHyperlink" />
          </layoutContainer>
        </topItems>
      </group>
    </secondColumn>
  </tab>
</backstage>
```

Action Item 18.18

You will find the above Backstage customization in the `CustomUI14_ver18.txt`
file in the companion files. The resulting Backstage customization is shown in
Figure 18.21.

In the previous example XML markup, the `<button>` element is used to
incorporate a custom command labeled Synchronize into the Backstage view
navigation bar:

```
<button id="btnSync" label="Synchronize" _
     imageMso="SyncNow" isDefinitive="true"
   insertBeforeMso="FileClose" _
     onAction="onActionCopyToArchive" />
```

The `<button>` element contains the `isDefinitive` attribute. When this attribute is set to `true`, clicking the button will trigger the callback procedure defined

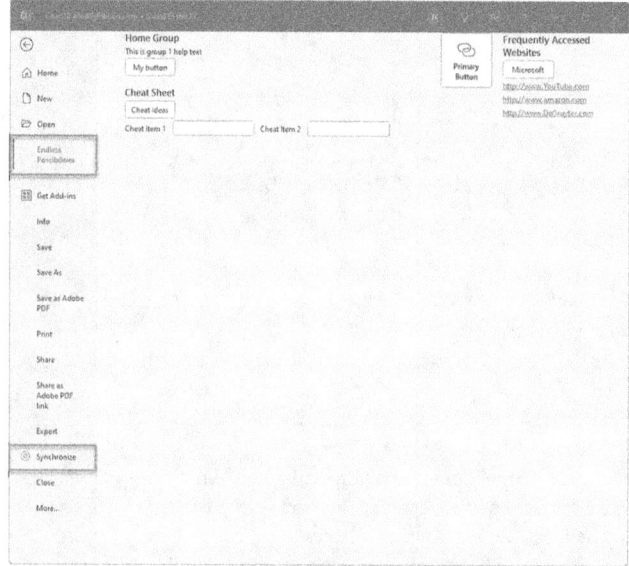

FIGURE 18.21.   The Backstage view is highly customizable. The Synchronize button and the Endless Possibilities tab were created by adding XML markup to the ribbon customization file.

in the `onAction` attribute and then automatically close the Backstage view and return to the worksheet.

The `onAction` callback for the custom Synchronize button is shown below. Notice that the callback calls the `CopyToArchive` procedure. This procedure allows you to make a copy of the current workbook file in a folder of your choice. Be sure to enter the procedure code in the VBA code module of the `Chap18_ModifyRibbon.xlsm` workbook:

```
Sub onActionCopyToArchive(ctl As IRibbonControl)
    Archive
End Sub

Sub Archive()
    Dim folderName As String
    Dim MyDrive As String
    Dim BackupName As String

    Application.DisplayAlerts = False

    On Error GoTo ErrorHandler

    folderName = ActiveWorkbook.Path
```

```
    If folderName = "" Then
        MsgBox "You can't copy this file. " & Chr(13) _
            & "This file has not been saved.", _
        vbInformation, "File Archive"
    Else
        With ActiveWorkbook
            If Not .Saved Then .Save
            MyDrive = InputBox("Enter the Pathname:" & _
                Chr(13) & "(for example: D:\, " & _
                    "E:\MyFolder\, etc.)", _
                    "Archive Location?", "D:\")
            If MyDrive <> "" Then
                If Right(MyDrive, 1) <> "\" Then
                    MyDrive = MyDrive & "\"
                End If
                BackupName = MyDrive & .Name
                .SaveCopyAs Filename:=BackupName
                MsgBox .Name & " was copied to: " _
                    & MyDrive, , "End of Archiving"
            End If
        End With
    End If
    GoTo ProcEnd
ErrorHandler:
    MsgBox "Visual Basic cannot find the " & _
        "specified path (" & MyDrive & ")" & Chr(13) & _
        "for the archive. Please try again.", _
        vbInformation + vbOKOnly, "Disk Drive or " & _
        "Folder does not exist"
ProcEnd:
    Application.DisplayAlerts = True
End Sub
```

The Backstage view XML markup also adds to the Backstage view navigation bar a custom tab labeled Endless Possibilities. Each <tab> element can have one or more columns. Our example contains two columns. Each tab can contain multiple <group> elements. Here, we have two groups in the first column and one group in the second column.

The Backstage group can contain different types of controls. You can group the controls into three types of sections:

- <primary item>

    This element is used to specify the most important item in the group. The primary item control can be a button or a menu with buttons, toggle buttons, checkboxes, or another menu.

- `<topItems>`

   This element defines controls that will appear at the top of the group.

- `<bottomItems>`

   This element defines the controls that will appear at the bottom of the group.

The layout of controls in the Backstage view is defined using the `<layoutContainer>` element. This element's `layoutChildren` attribute can define the layout of controls as horizontal or vertical. The second column of our example XML markup uses the following callback procedure for the button labeled Microsoft and the three hyperlinks. Enter this procedure in the VBA code module of the `Chap18_ModifyRibbon.xlsm` workbook:

```
Sub onActionExecHyperlink(ctl As IRibbonControl)
    Select Case ctl.id
        Case "YouTube"
            ThisWorkbook.FollowHyperlink Address:= _
                "http://www.YouTube.com", _
                NewWindow:=True
        Case "amazon"
            ThisWorkbook.FollowHyperlink Address:= _
                "http://www.amazon.com", _
                NewWindow:=True
        Case "DeGru"
            ThisWorkbook.FollowHyperlink Address:= _
                "http://www.DeGruyterBrill.com", _
                NewWindow:=True
        Case "msft"
            ThisWorkbook.FollowHyperlink Address:= _
                "http://www.Microsoft.com", _
                NewWindow:=True
        Case Else
            MsgBox "You clicked control id " & ctl.id & _
                " that has not been programmed!"
    End Select
End Sub
```

Hiding Backstage Buttons and Tabs

The following XML will hide the Save button in the Backstage view navigation bar:

```
<button idMso="FileSave" visible="false" />
```

The Backstage view uses the following button IDs: `FileSave`, `FileSaveAs`, `FileOpen`, `FileClose`, `ApplicationOptionsDialog`, and `FileExit`. To hide the Info tab in the Backstage view, use the following markup:

```
<tab idMso="TabInfo" visible="false" />
```

The Backstage view tab IDs are as follows: `TabInfo`, `TabRecent`, `TabNew`, `TabPrint`, `TabShare`, and `TabHelp`.

Things to Remember While Customizing the Backstage View

❑ The maximum number of allowed tabs is 255.

❑ You cannot reorder built-in tabs.

❑ You can add your custom tab before or after the built-in tab.

❑ You cannot modify the column layout of any built-in tab.

❑ You cannot reorder built-in groups; however, you can specify the order of groups you create.

CUSTOMIZING THE QUICK ACCESS TOOLBAR (QAT)

The QAT that appears just above the File tab gives application users quick access to the tools they use most frequently. These tools can be easily added to the toolbar by selecting More Commands from the Customize Quick Access Toolbar drop-down menu.

The QAT can only be customized in the start-from-scratch mode by setting the `startFromScratch` attribute to `true` in the ribbon XML customization file:

```
<ribbon startFromScratch="true">
```

After loading a workbook that contains this setting, Excel hides all built-in tabs. You must add your own custom tabs as demonstrated earlier in this chapter. QAT modifications are specified using the `<qat>` element. Within this element, use the `<sharedControls>` element to include controls that are shared by all open workbooks, and the `<documentControls>` element to specify the controls that should appear in the QAT when the workbook has the focus.

The following XML markup creates the custom QAT shown in Figure 18.22:

```
<customUI xmlns="http://schemas.microsoft.com/office/2009/07/ _
    customui" >
<ribbon startFromScratch="true">
  <qat>
    <sharedControls>
      <button idMso="FilePrintQuick" />
    </sharedControls>
    <documentControls>
      <button id="btnCalc2" label="Calculator"
      imageMso="SadFace" onAction="OpenCalculator" />
    </documentControls>
  </qat>
</ribbon>
</customUI>
```

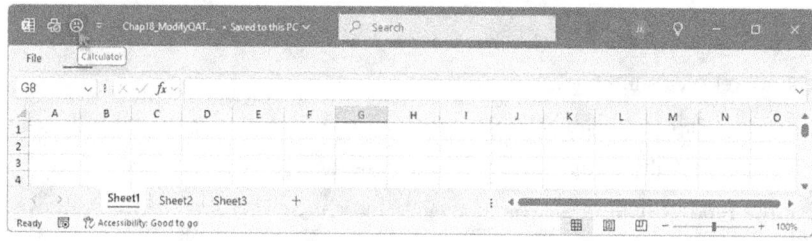

FIGURE 18.22. Customized Quick Access Toolbar.

The button labeled Calculator that is represented by the sad face image calls the OpenCalcuator procedure as shown below:

```
Public Sub OpenCalculator(ctl As IRibbonControl)
  Shell "Calc.exe", vbNormalFocus
End Sub
```

Action Item 18.19
You will find the XML markup code for the QAT in the `CustomUI_QAT.txt` file located in the companion files. The complete example, including the `OpenCalculator` callback procedure, is available in the `Chap18_ModifyQAT.xlsm` file in the companion files.

MODIFYING CONTEXT MENUS USING RIBBON CUSTOMIZATIONS

Ribbon extensibility refers to the ability to customize and extend the ribbon interface in Microsoft Office applications. With ribbon extensibility, you can use XML and VBA to modify the ribbon's appearance and functionality. The same XML markup and callbacks used to customize the ribbon UI can also be applied to modify context menus. This feature allows you to add both built-in and custom controls to menus and submenus, as well as hide controls in built-in menus.

Moreover, when creating custom submenus, you can dynamically populate them with controls using the `dynamicMenu` control. The following Hands-On exercise will help you get acquainted with context menu extensibility.

⊙ Hands-On 18.6   Customizing a Context Menu

1. Create a new workbook and save it in `C:\VBAExcel2024_ByExample` as `Chap18_ContextMenu.xlsm`.
2. Close the `Chap18_ContextMenu.xlsm` workbook and exit Excel.
3. Launch the Office RibbonX Editor, the standalone tool that you installed and worked with earlier in this chapter.
4. In the Office RibbonX Editor, open the `Chap18_ContextMenu.xlsm` workbook.
5. Choose Insert | Office 2010+ Custom UI Part.
   This creates a `CustomUI14.xml` file in the workbook.
6. Double-click customUI14.xml and, in the Editor's right pane, enter the context menu XML markup, as shown in Figure 18.23. This code is available in `customUI14_ContextMenu.txt` in the companion files.

FIGURE 18.23.   XML markup for customizing context menus.

7. Click the Validate button on the Office RibbonX Editor toolbar to verify that your XML does not contain errors. You should see the message "Custom UI

XML is well formed." If there are errors, you must correct them to ensure that the XML is well formed.

8. Save the file and close the Office RibbonX Editor.
9. Open the `Chap18_ContextMenu.xlsm` workbook in Excel, switch to the VBE window, and insert a module.
10. In the Module1 Code window, enter the callback procedures discussed below.
11. Switch to the Excel application window and right-click on any cell to view the custom commands added to the worksheet cell menu. Test each newly added command to ensure that it behaves as expected.

The context menu customization markup appears between the `<contextMenus>` `</contextMenus>` tags (see Figure 18.23). This XML markup adds three new button controls and a menu control to the context menu that appears when you click on any worksheet cell. These controls are shown in Figure 18.24.

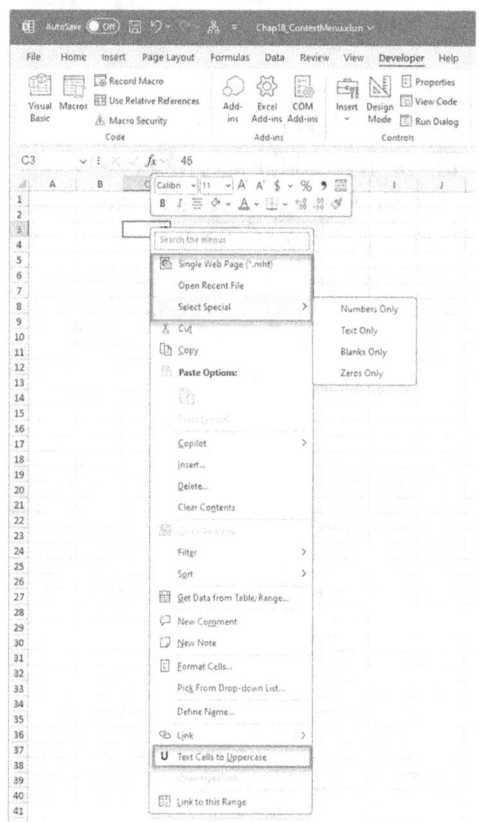

FIGURE 18.24. Standard worksheet cell context menu and the same context menu after applying the customization shown in Figure 18.23.

The first control added by the previous markup is a built-in Excel command with `idMso` set to `FileSaveAsWebPage`. This command appears at the top of the context menu and is labeled Single Web Page (*.mht). Recall from earlier sections of this chapter that built-in commands use the `idMso` attribute while the custom commands use the `id` attribute. When clicked, this command will execute Excel's built-in action that will allow you to save the worksheet as a Web page.

The second command in the previous markup adds a custom button labeled Open Recent File. When clicked, this command will run the following `onActionBuiltInCmd` callback procedure:

```
Sub onActionBuiltInCmd(ctl As IRibbonControl)
    CommandBars.ExecuteMso "FileOpenRecentFile"
End Sub
```

The `onActionBuiltInCmd` callback procedure uses the `ExecuteMso` method of the `CommandBars` object to run the built-in action assigned in Excel to the `FileOpenRecentFile` command.

The third button in the XML markup adds the custom control labeled Text Cells to Uppercase and designates the letter U as the keyboard accelerator. When clicked, this option will convert any text cell found within the selection of cells to uppercase letters by calling the following procedure:

```
Sub onActionUppercase(ctl As IRibbonControl)
    Dim cell As Variant
    For Each cell In Selection
      If WorksheetFunction.IsText(cell) Then
          cell.Value = UCase(cell.Value)
      End If
    Next
End Sub
```

The last command in the XML markup is a custom menu command labeled Select Special. When clicked, this option displays a menu of options. When you select a menu option, the following callback procedure is executed:

```
Sub onActionSelSpec(ctl As IRibbonControl)
  Select Case ctl.ID
      Case "text"
          Selection.SpecialCells(xlCellTypeConstants, 2).Select
      Case "num"
          Selection.SpecialCells(xlCellTypeConstants, 1).Select
      Case "blank"
          Selection.SpecialCells(xlCellTypeBlanks).Select
      Case "zero"
```

```
        Dim cell As Variant
        Dim myRange As Range
        Dim foundFirst As Boolean

        foundFirst = True

        Selection.SpecialCells(xlCellTypeConstants, 1).Select
            For Each cell In Selection
                If cell.Value = 0 Then
                    If foundFirst Then
                        Set myRange = cell
                        foundFirst = False
                    End If
                    Set myRange = Application. _
                        Union(myRange, cell)
                End If
            Next
        myRange.Select
    Case Else
        MsgBox "Missing Case statement for _
                control id=" & ctl.ID, _
            vbOKOnly + vbExclamation, "Check your VBA Procedure"
    End Select
End Sub
```

SUMMARY

In this chapter, you learned how to use VBA to work with built-in context menus and customize the ribbon interface as well as the Backstage view, using a combination of XML and VBA. While working with context menus, you explored various properties and methods of the CommandBar object. Next, you learned how to use the Office RibbonX Editor to create XML ribbon customization markup. You familiarized yourself with various controls that can be added to the ribbon and wrote VBA callback procedures to set your controls' attributes at runtime. Additionally, you learned how to modify the Backstage view and the Quick Access Toolbar. Finally, you mastered manipulating context menus using XML and VBA callbacks.

The knowledge and experience you gained in this chapter can be applied to make similar customizations in all of the Microsoft 365 applications that use the ribbon interface.

In the next chapter, we will focus on using and programming Excel tables.

Part V

EXCEL TOOLS FOR DATA ANALYSIS

Microsoft Excel offers powerful tools for organizing and presenting information from various sources. In this part of the book, you will learn how to work with Excel tables and how to analyze data from multiple perspectives using PivotTables and PivotCharts. In addition, you will use the Get & Transform Data feature to clean, load, and shape your data.

Chapter **19** *USING AND PROGRAMMING EXCEL TABLES*

O ver the years, people have used Excel spreadsheets to store and extract data from databases. Currently, a Microsoft Excel worksheet allows users to store as many as 1,048,576 rows by 16,384 columns. Furthermore, the data can be easily sorted, filtered, summarized, and validated. If you need to create any kind of table and store it in a spreadsheet, this chapter's tour of Excel table management will be helpful. We will look at the user interface for the table ranges and learn how to access and work with the table feature using VBA.

UNDERSTANDING EXCEL TABLES

Tables are groups of cells that store related data and are managed separately from data in other cells on the worksheet. Tables aren't a new feature in Excel 2024; however, they were known as *lists* when they were first introduced in Excel 2003. You can have many tables in a worksheet but a table cannot overlap another table. Each table is treated as a single entity and can be sorted, filtered, or shared. Tables can be easily recognized in worksheets as Excel automatically enables the filtering in the header row for each column, which you can see in Figure 19.1. You can use this feature to sort data in ascending or descending order or to create a custom view of your data.

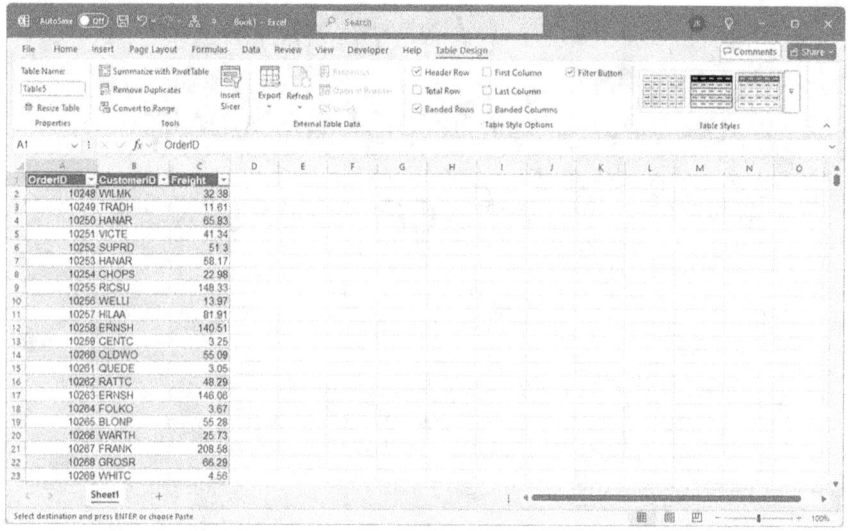

FIGURE 19.1.   A table in an Excel worksheet.

When you create a table from a cell range and don't specify column headers, Excel automatically adds column headers (Column1, Column2, etc.) to the range.

To subtotal data in the table, take the following steps:

1. Convert an Excel table into a standard worksheet range by clicking the Table Design tab and selecting Convert to Range in the Tools group. Click Yes in the dialog box when asked to convert the table to a normal range.

2. Sort the data according to your needs.
   Select any cell in the column you want to sort by and click Sort & Filter in the Editing group on the Home tab. Select the desired option from the Sort & Filter menu.

3. To add a subtotal, click anywhere within the data range, and then click Subtotal in the Outline group on the Data tab.

4. In the Subtotal dialog box (see Figure 19.2), make the appropriate selections:

   a. In the At each change in drop-down box, choose the column by which you want to subtotal.

   b. In the Use function dropdown, select a function that is appropriate for the type of summary you want to produce.

   c. In the Add subtotal to dropdown, check the appropriate column.

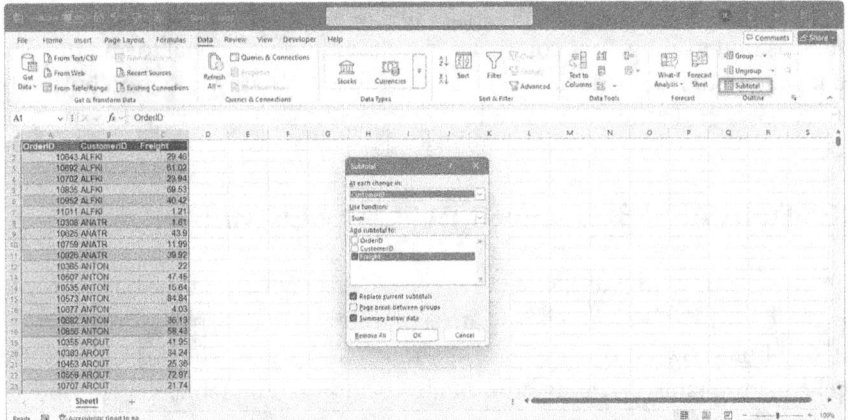

FIGURE 19.2.   Use the Subtotal dialog box to subtotal Excel tables. Note that the data is sorted by Customer ID and we are asking Excel to add the subtotal to the Freight column.

5. Click OK to finish adding the subtotals.

Excel subtotals the cost of Freight for each customer, as shown in Figure 19.3.

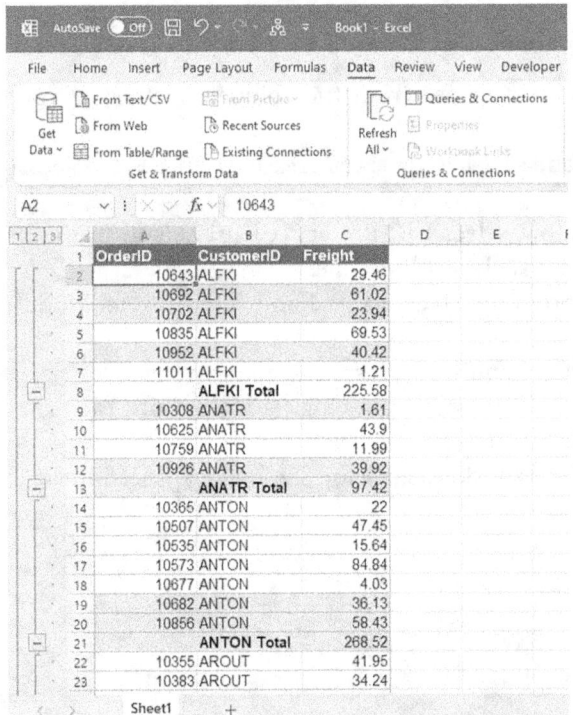

FIGURE 19.3.   Subtotals added to the table shown in Figure 19.2.

Cells that are part of a table can be formatted using standard formatting options such as underlining, font color, pattern, shading, conditional formatting, and so on. Data in the table may also be validated via the Data Validation button in the Data Tools group on the Data tab.

CREATING A TABLE USING BUILT-IN COMMANDS

To create a table in Excel, select a range of cells containing the data you want to include in the table and choose Insert | Table. Excel displays the Create Table dialog box shown in Figure 19.4, where you can accept the current selection of cells for the table or change the range of data for the table. To see how this is accomplished, we will begin by writing a VBA procedure that gets data from the Microsoft Access Northwind database (`Northwind.mdb`).

NOTE	*Please note that the files for the Hands-On projects can be found in the companion files.*

⊙ Hands-On 19.1 Obtaining Data from a Microsoft Access Database

1. Copy the `Northwind.mdb` database from the companion files to your `C:\VBAExcel2024_ByExample` folder.
2. Open a new workbook and save it as `Chap19.xlsm` in your `C:\VBAExcel2024_ByExample` folder.
3. Switch to the VBE window, select VBAProject (Chap19.xlsm) in the Project Explorer window, and choose Insert | Module.
4. Use the Properties window to change the Name property of Module1 to `Tables`.
5. Choose Tools | References, and in the list of available references, select the checkbox next to Microsoft ActiveX Data Objects Library 6.1, then click OK to exit the References dialog box.
6. In the Tables module Code window, enter the `GetOrders` procedure as shown below:

```
Sub GetOrders()
    Dim conn As New ADODB.Connection
    Dim rst As ADODB.Recordset
    Dim strPath As String
    Dim wks As Worksheet
    Dim j As Integer
```

```
strPath = "C:\VBAExcel2024_ByExample\Northwind.mdb"
Worksheets.Add
Set wks = ThisWorkbook.ActiveSheet

conn.Open "Provider=Microsoft.ACE.OLEDB.12.0;" _
          & "Data Source=" & strPath & ";"

' Create a Recordset from data in the Categories table
Set rst = conn.Execute(CommandText:="Select OrderID," & _
          "CustomerID, Freight from Orders", _
          Options:=adCmdText)
rst.MoveFirst

' transfer the data to Excel
' get the names of fields first
With wks.Range("A1")
    .CurrentRegion.Clear
    For j = 0 To rst.Fields.Count - 1
        .Offset(0, j) = rst.Fields(j).Name
    Next j
    .Offset(1, 0).CopyFromRecordset rst
    .CurrentRegion.Columns.AutoFit
    .Cells(1, 1).Select
End With
rst.Close
conn.Close

Set rst = Nothing
Set conn = Nothing
End Sub
```

7. Switch to the Microsoft Excel application window and select Sheet1.

8. Press Alt+F8 to display the Macro dialog box. Highlight the GetOrders procedure and click Run.

   The data is retrieved from the Orders table and placed in a new sheet.

9. Choose Insert | Tables | Table.

   Microsoft Excel displays the Create Table dialog box and highlights the range of cells identified as a table, as shown in Figure 19.4. You may change the range by making your own range selection in the worksheet.

10. Click OK to exit the Create Table dialog box.

    Your table is now ready to use or share with others (see Figure 19.1 earlier in this chapter).

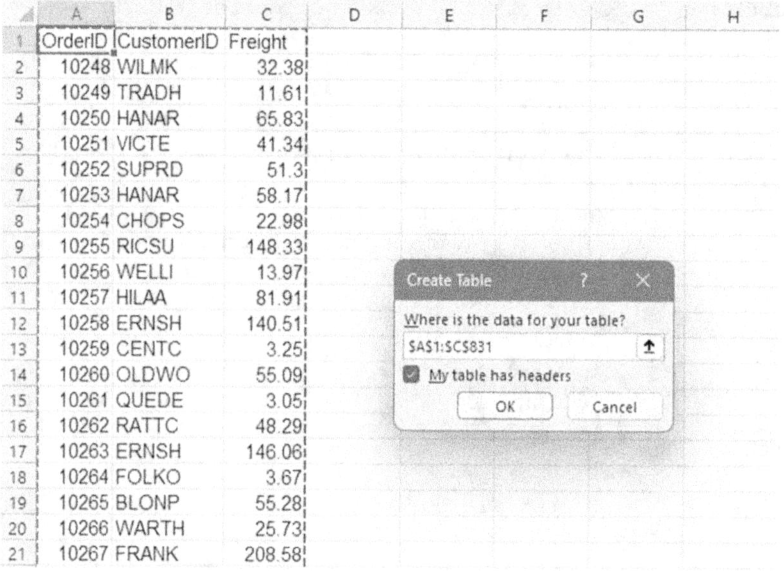

FIGURE 19.4. Converting a range of cells into an Excel table.

CREATING A TABLE USING VBA

To programmatically create an Excel table, use the ListObject object, which represents a list object in a worksheet. The ListObject object is a member of the ListObjects collection. This collection contains all the list objects on the worksheet. As mentioned earlier, you can have one or more tables in a single worksheet.

In the previous section, you learned how to use VBA to retrieve data from Access and how to manually convert it into an Excel table using the built-in ribbon commands. In this section, we will modify the GetOrders procedure you created in Hands-On 19.1 so that it automatically creates a table for us out of the Access data.

Hands-On 19.2   Creating a Table Using VBA

1. In the Tables module Code window, modify the GetOrders procedure as follows:

   - Add the following declaration to the procedure declaration section:
     ```
     Dim rng As Range
     ```

● Type the following statements just before the End Sub keywords:

```
'create a table in Excel

Set rng = wks.Range(Range("A1").CurrentRegion.Address)
wks.ListObjects.Add xlSrcRange, rng
```

The first statement that follows the comment will set the object variable (rng) to point to the range of cells that we want to convert into a table. The second statement uses the Add method of the ListObjects collection to create a table out of a specified range of cells. The xlSrcRange constant specifies that the source of the table is an Excel range, while the rng object variable indicates a Range object representing the data source.

The revised GetOrders procedure is shown below:

```
Sub GetOrders_Modified()
    Dim conn As New ADODB.Connection
    Dim rst As ADODB.Recordset
    Dim strPath As String
    Dim wks As Worksheet
    Dim j As Integer
    Dim rng As Range

    strPath = "C:\VBAExcel2024_ByExample\Northwind.mdb"
    Worksheets.Add
    Set wks = ThisWorkbook.ActiveSheet

    conn.Open "Provider=Microsoft.ACE.OLEDB.12.0;" _
            & "Data Source=" & strPath & ";"

    ' Create a Recordset from data in the Categories table

    Set rst = conn.Execute(CommandText:="Select OrderID," & _
            "CustomerID, Freight from Orders", _
            Options:=adCmdText)

    rst.MoveFirst

    ' transfer the data to Excel
    ' get the names of fields first
    With wks.Range("A1")
        .CurrentRegion.Clear
        For j = 0 To rst.Fields.Count - 1
            .Offset(0, j) = rst.Fields(j).Name
        Next j
        .Offset(1, 0).CopyFromRecordset rst
        .CurrentRegion.Columns.AutoFit
```

```
        .Cells(1, 1).Select
End With
rst.Close
conn.Close

Set rst = Nothing
Set conn = Nothing

'create a table in Excel

Set rng = wks.Range(Range("A1").CurrentRegion.Address)
wks.ListObjects.Add xlSrcRange, rng

End Sub
```

2. Switch to the Microsoft Excel application window and press Alt+F8 to display the Macro dialog box. Highlight the GetOrders_Modified procedure and click Run.

   The data is retrieved from the Orders table and placed on a new sheet as an Excel table.

   When you use the `Add` method of the `ListObjects` collection to create an Excel table, you may specify the arguments as shown in Table 19.1.

TABLE 19.1.   Arguments used with the Add method of the ListObjects collection.

Argument Name	Description
SourceType (optional)	Indicates the type of data for the list. You can use one of the following source types: • External data (`xlSrcExternal`) • Excel range (`xlSrcRange`) • XML data (`xlSrcXML`) If omitted, `SourceType` will default to `xlSrcRange`.
Source (optional when `SourceType` = `xlSrcRange` but required when `SourceType` = `xlSrcExternal`)	This argument can be an array of string values specifying a connection to the source: • Use 0 to indicate the SharePoint URL. • Use 1 to indicate the name of the list. • Use 2 to indicate the `ViewGUID` (identifies the view for a list on the SharePoint site). It can also be a `Range` object representing the data source. If this argument is omitted, `Source` is the range returned by the list range detection code.

Argument Name	Description
LinkSource (optional)	Indicates whether an external data source is to be linked to the ListObject object. The SourceType argument must be set to xl-SrcExternal.
HasHeaders (optional)	Indicates whether the data to be used for the list has column labels. You can use one of the following constants for this argument: xlGuess, xlNo, or xlYes. If Source does not have column headings, Excel automatically generates headers as Column1, Column2, etc.
Destination (required when SourceType = xlSrcExternal but ignored when SourceType = xlSrcRange)	Indicates the top-left corner of the new list object. Use a Range object with a single-cell reference. You cannot reference more than one cell. If the Destination range is not empty, new columns will be added to fit the new list (existing data will not be overwritten).

Notice that the arguments of the ListObject object's Add method are optional. If you omit the arguments, Excel will use its own logic to identify the range of cells for the table and will determine whether the table contains column headings. Contiguous cells containing data are always assumed to be a part of a table. If the first row of the identified data range contains text, Excel assumes that this is a header row.

Understanding Column Headings in the Table

When creating a table, Excel automatically adds column headings to the table. Depending on the type of data found in the first row of the data range, the first row may be designated as column headings or a new row may need to be inserted, causing other rows of data to shift down. Because you will not know exactly what Excel will do in a particular situation given a particular set of data, it is a good idea to supply the value for the HasHeaders argument in your VBA code, as shown earlier in Table 19.1.

Let's look at how we can control the location of the column headings in the Excel table.

⦿ Hands-On 19.3   Adding Headings to a Table

1. Insert a new sheet in the Chap19.xlsm workbook and type the sample data shown in Figure 19.5.

◢	A	B	C
1			
2	Evan DeCastro	89	
3	Mathew Lambert	95	
4	Barbara O'Connor	91	
5	Lucy Moreno	83	
6			

FIGURE 19.5. Data in a worksheet prior to conversion into a table.

2. Switch to the VBE screen and enter the following procedure in the `Tables` module:

```
Sub List_Headers()
    Dim rng As Range
    Dim wks As Worksheet

    Set wks = ActiveSheet
    Set rng = wks.Range("A2:B5")

    wks.ListObjects.Add SourceType:=xlSrcRange, Source:=rng, _
        XlListObjectHasHeaders:=xlNo
End Sub
```

Because the data in Figure 19.6 does not have column headings, we have specified `xlNo` for the `HasHeaders` argument. When Excel executes this procedure, it will add default headers (Column1, Column2) in row 2 and will shift the range down one row, as illustrated in Figure 19.6.

3. Position the insertion point anywhere within the `List_Headers` procedure and press F5 to run it.

4. Switch to the Microsoft Excel application window to view the formatted table.

◢	A	B	C
1			
2	**Column1** ▾	**Column2** ▾	
3	Evan DeCastro	89	
4	Mathew Lambert	95	
5	Barbara O'Connor	91	
6	Lucy Moreno	83	
7			
8			

FIGURE 19.6. A range of data after conversion to an Excel table. Notice that Excel has added default column headings in row 2 and shifted the data range one row down.

Sometimes you may not want Excel to shift data down when your table does not include column headings. To prevent this, it is recommended that you specify for your table a range of data in which the first row is blank. For example, to prevent Excel from shifting the data down one row, specify `A1:B5` as the range and use `xlYes` for the `HasHeaders` parameter. Before trying this out, let's convert the Excel table we have just created back to a normal range.

5. Select any cell within the table on the worksheet and choose Table Design | Convert to Range. Click Yes when Excel displays a confirmation message. Notice that after Excel creates a normal range out of a table, the default column headings are preserved.

6. Delete the headings row from this worksheet and save your changes. Switch back to the VBE screen and, in the `Tables` module, enter the following `List_Headers_Modified` procedure:

```
Sub List_Headers_Modified()
    Dim rng As Range
    Dim wks As Worksheet

    Set wks = ActiveSheet
    Set rng = wks.Range("A1:B5")

    wks.ListObjects.Add SourceType:=xlSrcRange, _
        Source:=rng, XlListObjectHasHeaders:=xlYes
End Sub
```

7. Run the `List_Headers_Modfied` procedure. The results are shown in Figure 19.7.

◢	A	B	C
1	Column1　▾	Column2　▾	
2	Evan DeCastro	89	
3	Mathew Lambert	95	
4	Barbara O'Connor	91	
5	Lucy Moreno	83	
6			

FIGURE 19.7. A range of data after conversion to an Excel table. Notice that Excel has added default column headings in row 1, which was empty when we specified the range of data for the table.

Multiple Tables in a Worksheet

You have seen, in the previous section, how Excel shifts the cells down when the data range specified for your table does not have column headings. As you may have more than one table in a worksheet, this behavior may cause problems when another table is placed right below the first table. Also, when you add new rows to the table, the table expands, so a conflict may occur if another table is placed in the rows below. Therefore, it is a good idea to avoid placing any data in the rows below a table.

When you have more than one table in a worksheet and want to manipulate these tables programmatically, it's a good practice to assign names to your tables so you can easily refer to them in your code. While you can always refer to a table by using its index number, names are more meaningful and easier to understand. By default, Excel assigns the names Table1, Table2, Table3, etc. to the tables in a worksheet. To name a table or retrieve the name of an existing table, use the `Name` property of the `ListObject` object. For example, the following statement entered in the Immediate window returns the name of the table in the active sheet:

```
?ActiveSheet.ListObjects(1).Name
```

To rename Table1 (your table number may be different depending on how many tables you've already added to the workbook), you can simply type the following statement in the Immediate window and press Enter:

```
ActiveSheet.ListObjects(1).Name = "Student Scores"
```

Now, you can refer to the first table in the active sheet as Student Scores.

The `DefineTableName` procedure shown below uses the `ListObjects` property to get a reference to the first table on the active sheet. Next, the `Name` property is used to assign a name to the referenced table:

```
Sub DefineTableName()
    Dim wks As Worksheet
    Dim lst As ListObject

    Set wks = ActiveWorkbook.Worksheets(ActiveSheet.Name)
    Set lst = wks.ListObjects(1)
    lst.Name = "1st Qtr. 2024 Student Scores"
End Sub
```

WORKING WITH THE EXCEL LISTOBJECT

The `ListObject` object represents a table on a worksheet. You can manipulate the table via the properties and methods of the `ListColumns` and `ListRows` collections:

- The `ListColumns` collection contains all the `ListColumn` objects in the specified `ListObject` object. Each `ListColumn` object is a column in the table.

• The `ListRows` collection contains all the `ListRow` objects in the specified `ListObject` object. Each `ListRow` object is a row in the table.

You can perform various operations on Excel tables using properties and methods of the `ListObject` object, as shown in Tables 19.2 and 19.3.

TABLE 19.2. Properties of the ListObject object.

Property Name	Description

`Active`	Indicates whether a list in a worksheet is currently active. Returns `True` or `False`. For example:
	
	`IsTblActive = ActiveSheet.ListObjects(1).active`
	`Debug.Print IsTblActive`
	
	There is no `Activate` method for the `ListObject`. To activate a table, you must activate a cell range within a table. For example:
	
	`ActiveSheet.ListObjects(1).Range.Activate`
	
	The above statement selects the entire range for the list.

`DataBodyRange`	Returns a `Range` object that represents the range of cells without the header row in a table. For example:
	
	`ActiveSheet.ListObjects(1).DataBodyRange.Select`
	
	or
	`dataRng = ActiveSheet.ListObjects(1). _ Data-`
	`BodyRange.Address`
	`Debug.Print dataRng`

`HeaderRowRange`	Returns a `Range` object that represents the range of the header row for a table. For example, use the following statement to select the header row in the table: `ActiveSheet.ListObjects(1).HeaderRowRange.Select`
`InsertRowRange`	In Excel 2003, this property returns a `Range` object representing the insert row. This property is not supported in Excel 2007/2024: `ActiveSheet.ListObjects(1).InsertRowRange.Activate`

`ListColumns`	Returns a `ListColumns` collection that represents all the columns in a `ListObject` object. For example, the following procedure deletes the last column from the table:
	
	`Sub DeleteLastCol()`
	`    Dim myList As ListObject`
	`    Dim lastCol As Integer`
	
	`    Set myList = ActiveSheet.ListObjects(1)`
	`    lastCol = myList.ListColumns.Count`
	`    myList.ListColumns(lastCol).Delete`
	`End Sub`

Property Name	Description
ListRows	Returns a ListRows object that represents all the rows of data in the ListObject object. For example, the following procedure prints to the Immediate window the total number of rows in the table: ```\nSub CountListRows()\n Dim objRows As ListRows\n Set objRows = ActiveSheet.ListObjects(1).ListRows\n Debug.Print objRows.Count\nEnd Sub\n```
Name	Returns or sets the name of the ListObject object. For example, use the following statement to assign a name to the first table in the active worksheet: ```\nActiveSheet.ListObjects(1).Name = "Student Scores"\n```
QueryTable	Returns the QueryTable object that provides a link for the ListObject object to the SharePoint site server.
Range	Returns a Range object that represents the range to which the specified list object applies. For example, the following statement prints to the Immediate window the range address of the entire list: ```\nDebug.Print ActiveSheet.ListObjects(1).Range.Address\n```
SharePointURL	Returns a string representing the URL of the SharePoint list. Use it to find the address of the shared list after it has been published: ```\nlistURL = ActiveSheet.ListObjects(1).SharePointURL\nDebug.Print listURL\n```
ShowAutoFilter	Indicates whether the AutoFilter will be displayed in the header row (True or False). Use the following statement to turn off the AutoFilter mode for a given table: ```\nActiveSheet.ListObjects(1).ShowAutoFilter = False\n```
ShowTotals	Indicates whether the Total row is visible (True) or hidden (False). The following statement turns on the display of the Total row: ```\nActiveSheet.ListObjects(1).ShowTotals = True\n```
SourceType	Returns one of the XlListObjectSourceType constants indicating the current source of the table (xlSrcRange, xlSrcExternal, or xlSrcXML). Please see Table 19.1.
TotalsRowRange	Returns a range representing the Total row for the specified ListObject object: ```\nDebug.Print ActiveSheet.ListObjects(1). _\n TotalsRowRange.Address\n```
XmlMap	Returns an XmlMap object that represents the schema map used for the specified table. See Chapter 25 for more information.

TABLE 19.3. Methods of the ListObject object.

Method Name	Description
Delete	Deletes the `ListObject` object and clears the cell data from the worksheet. If the list is linked to a SharePoint site, deleting it does not remove data on the server that is running Windows SharePoint Services. Any uncommitted changes not sent to the SharePoint list are lost when the list is deleted in Excel.
Publish	Publishes the `ListObject` object to a server that is running Microsoft Windows SharePoint Services. It returns a string indicating the URL of the published list on the SharePoint site. The `Publish` method requires two arguments: ● `Target`—This is a three-element string array that specifies the address of the SharePoint server (element 0), the name of the list (element 1), and an optional description of the list (element 2). ● `LinkSource`—A Boolean value (`True` or `False`). If the `ListObject` object is not currently linked to a list on a SharePoint site: ● `LinkSource = True` creates a new list on the specified SharePoint site. ● `LinkSource = False` leaves the list object unlinked. If the `ListObject` object is currently linked to a SharePoint site: ● `LinkSource = True` replaces the existing link—only one link to the list is allowed on the SharePoint site. ● `LinkSource = False` keeps the `ListObject` object linked to the current SharePoint site.
Refresh	Can be used only with tables that are linked to a SharePoint site. It retrieves the current data and schema for the table from the SharePoint server.
Resize	Allows a `ListObject` object to be resized over a new range. You must provide the range address as the argument to the `Resize` method. Assuming that the current table range is A1:B6, we can specify the new range for the table as: `ActiveSheet.ListObjects(1).Resize Range("A1:B3")`
Unlink	Removes the link to a SharePoint Services site from a list. For example: `ActiveSheet.ListObjects(1).Unlink`
Unlist	Converts an Excel table to a regular range of data. For example, the table on the active sheet is turned into a normal range like this: `ActiveSheet.ListObjects(1).Unlist`

Method Name	Description

UpdateChanges	Updates the list on a Microsoft Windows SharePoint Services site with the changes made to a table in the worksheet. You can specify how list/table conflicts should be resolved by using one of the xlListConflict resolution constants:
	• xlListConflictDialog (default)
	• xlListConflictRetryAllConflicts
	• xlListConflictDiscardAllConflicts
	• xlListConflictError
	For example:
	`ActiveSheet.ListObjects(1).UpdateChanges _`
	`    xlListConflictDialog`

The following Hands-On demonstrates how to use selected properties from Table 19.2.

Hands-On 19.4   Defining Table Names

1. Enter the following procedure in the `Tables` module of VBAProject (Chap19. xlsm):

```
Sub DefineTableName2()
    Dim wks As Worksheet
    Dim lst As ListObject
    Dim col As ListColumn
    Dim c As Variant

    Set wks = ActiveWorkbook.Worksheets(ActiveSheet.Name)

    Set lst = wks.ListObjects(1)
    With lst
        .Name = "1st Qtr. 2024 Student Scores"
        .ListColumns(1).Name = "Student Name"
        .ListColumns(2).Name = "Score"
        Set col = .ListColumns.Add
        col.Name = "Previous Score"
       Debug.Print "Header Address = " & .HeaderRowRange.Address
       Debug.Print "Data Range = " & .Range.Address
       Debug.Print "Data Body Range = " & .DataBodyRange.Address

        For Each c In wks.Range(.HeaderRowRange.Address)
            Debug.Print c
        Next
    End With
End Sub
```

2. Activate the sheet containing the data shown in Figure 19.7.
3. Press Alt+F8 to display the Macro dialog box. Highlight the DefineTableName2 procedure and click Run.
4. Check the Immediate window for the following procedure results:

```
Header Address = $A$1:$C$1
Data Range = $A$1:$C$5
Data Body Range = $A$2:$C$5
Student Name
Score
Previous Score
```

FILTERING DATA IN EXCEL TABLES USING AUTOFILTER

The AutoFilter drop-down box has a great search capability (see Figure 19.8), making it easy to find data in large tables.

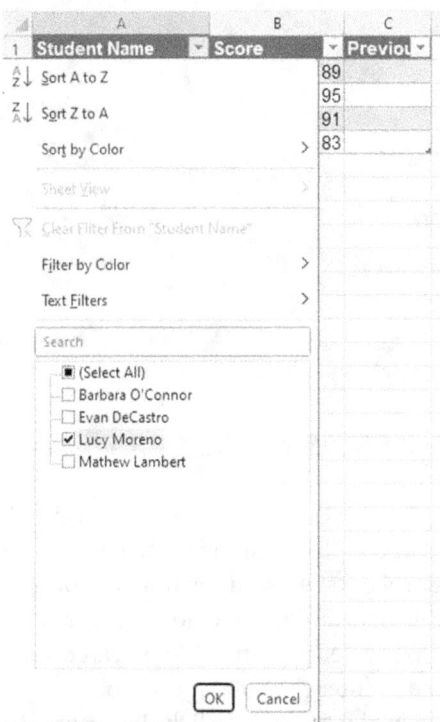

FIGURE 19.8.   You can quickly navigate to specific data in the table by entering the required search criteria in the AutoFilter drop-down search box.

To find out how to use the AutoFilter feature programmatically, all you have to do is turn on the macro recorder and work with the AutoFilter dropdown. The following macro was recorded while searching for the record of Barbara O'Connor. To remove the filter, simply comment out the first statement in the macro code below, and uncomment the second statement:

```
Sub Macro1()
'
' Macro1 Macro
'
  ActiveSheet.ListObjects("1st Qtr. 2024 Student Scores"). _
      Range.AutoFilter Field:=1, Criteria1:="Barbara O'Connor"
  ' ActiveSheet.ListObjects("1st Qtr. 2024 Student Scores"). _
      Range.AutoFilter Field:=1
End Sub
```

To make the recorded macro more dynamic, modify it to pass a variable to the AutoFilter criteria like this:

```
Sub Macro2()
  '
  ' Macro1 Macro
  '
  Dim strInput As String

  strInput = InputBox("Enter the search string:", "Find What")

  ActiveSheet.ListObjects("1st Qtr. 2024 Student Scores"). _
  Range.AutoFilter Field:=1, Criteria1:="=*" & strInput & "*"

End Sub
```

FILTERING DATA IN EXCEL TABLES USING SLICERS

Slicers can be used on any Excel table. This makes filtering a table as simple as it can be. Because slicers are floating controls, they can be placed anywhere on the worksheet. The look and feel of your worksheets can be much improved by turning off the filter buttons on table headers and using colorful slicers instead to filter the data. With slicers used as your filtering controls, you can even hide columns you don't need and still be able to filter the data in the table.

To use slicers, simply select any cell in the table and click the Insert Slicers button on the Design tab of the Table Tools (Figure 19.9).

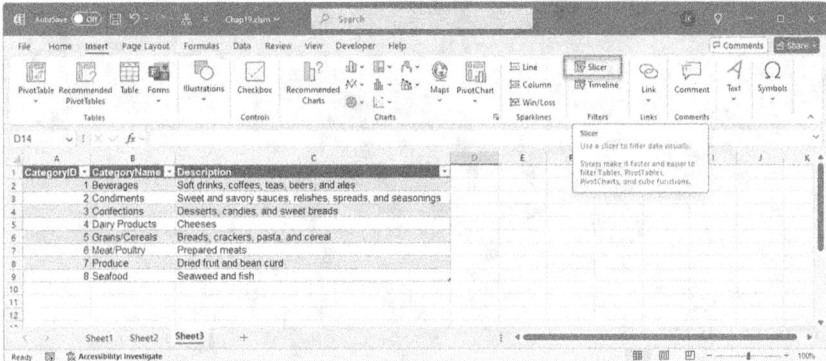

FIGURE 19.9.   Inserting a slicer on a worksheet containing a table.

Excel displays the Insert Slicers dialog (Figure 19.10), where you can choose the columns for which you want to create slicers. Each slicer is used to filter one column.

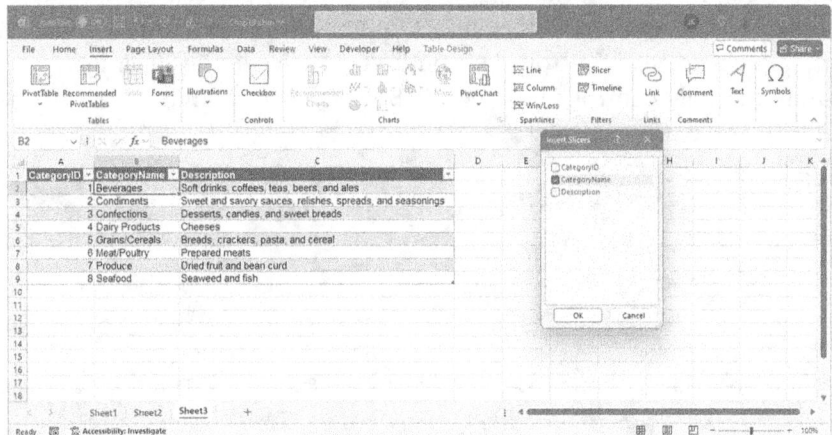

FIGURE 19.10.   Choosing columns for slicers.

When you make selections in the Insert Slicers dialog and click OK, the slicers are created and dropped on the worksheet (Figure 19.11). You can drag them anywhere you want.

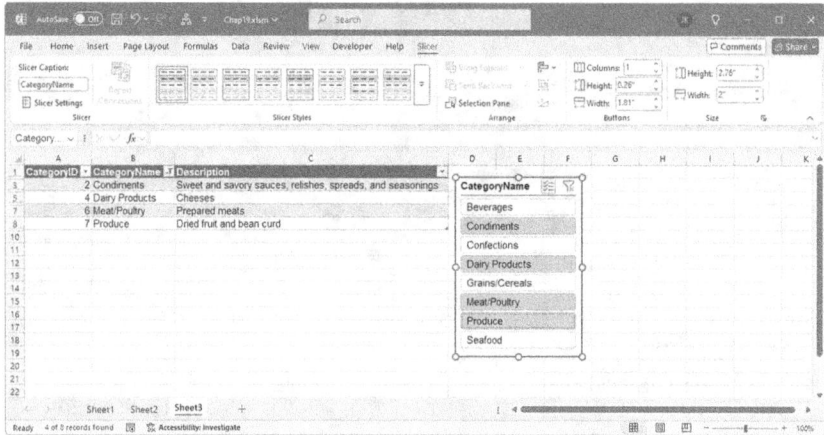

FIGURE 19.11.   CategoryName slicers are used to filter an Excel table.

To clear the filtering from the table, click the Clear Filter icon at the top right of the slicer. This will return the table to the full view. You can change the Slicer header by typing new text in the Slicer Caption box in the Slicer tab. Slicer formatting styles are also controlled via buttons available in the Slicer tab.

You can easily control slicers with VBA. Here's an example macro created with the Excel macro recorder:

```
Sub Macro3()
'
' Macro3 Macro
'
'

    ActiveSheet.Shapes.Range(Array("CategoryName")).Select
    ActiveSheet.Shapes("CategoryName").IncrementLeft 0.75
    ActiveSheet.Shapes("CategoryName").IncrementTop 41.25
    With ActiveWorkbook.SlicerCaches("Slicer_CategoryName")
        .SlicerItems("Dairy Products").Selected = True
        .SlicerItems("Beverages").Selected = False
        .SlicerItems("Condiments").Selected = False
        .SlicerItems("Confections").Selected = False
        .SlicerItems("Grains/Cereals").Selected = False
        .SlicerItems("Meat/Poultry").Selected = False
        .SlicerItems("Produce").Selected = False
        .SlicerItems("Seafood").Selected = False
    End With
    ActiveWorkbook.SlicerCaches("Slicer_CategoryName")._
        ClearManualFilter
```

```
With ActiveWorkbook.SlicerCaches("Slicer_CategoryName")
      .SlicerItems("Condiments").Selected = True
      .SlicerItems("Beverages").Selected = False
      .SlicerItems("Confections").Selected = False
      .SlicerItems("Dairy Products").Selected = False
      .SlicerItems("Grains/Cereals").Selected = False
      .SlicerItems("Meat/Poultry").Selected = False
      .SlicerItems("Produce").Selected = False
      .SlicerItems("Seafood").Selected = False
   End With
   With ActiveWorkbook.SlicerCaches("Slicer_CategoryName")
      .SlicerItems("Condiments").Selected = True
      .SlicerItems("Dairy Products").Selected = True
      .SlicerItems("Beverages").Selected = False
      .SlicerItems("Confections").Selected = False
      .SlicerItems("Grains/Cereals").Selected = False
      .SlicerItems("Meat/Poultry").Selected = False
      .SlicerItems("Produce").Selected = False
      .SlicerItems("Seafood").Selected = False
   End With
   ActiveSheet.Shapes("CategoryName").IncrementTop -23.25
End Sub
```

Creating slicers using VBA is covered in Chapter 20—see the Hands-On exercises.

DELETING WORKSHEET TABLES

You can delete an Excel table using one of the following methods:

- *User interface:*
  - Select the table on the worksheet and choose Home | Cells | Delete | Delete Cells.
  - If you don't need the sheet with the table, delete the entire worksheet.

- *VBA code:*
  - Use the `Delete` method to delete the worksheet table and its data.
  - Use the `Unlist` method to convert the worksheet table to a normal data range.

■ Use the `Unlink` method to remove the link between the worksheet table and the list on the SharePoint site. An unlinked list cannot be re-linked. The SharePoint lists can only be deleted on the SharePoint site or by using the Lists Web service provided by SharePoint Services.

SUMMARY

This chapter introduced you to Excel tables. You learned how to retrieve information from a Microsoft Access database, convert it into a table, and enjoy database-like functionality in the spreadsheet. You also learned how tables are exposed through Excel's object model and manipulated via VBA.

In the next chapter, you will learn how to program two Microsoft Excel objects that are used for data analysis: the PivotTable and PivotChart.

20 PROGRAMMING PIVOTTABLES AND PIVOTCHARTS

PivotTables serve millions of Microsoft 365 application users as powerful tools for organizing and presenting information from various sources. Using PivotTables and PivotCharts, you can analyze your data from multiple perspectives. PivotTables make it possible to drag headings around a table to rearrange them so that your data is displayed dynamically in any way you want. Similar to PivotTables, PivotCharts are interactive and allow you to view data in different ways by changing the position or detail of the PivotChart fields. Both PivotTables and PivotCharts allow you to focus on understanding your data rather than on organizing it.

CREATING A PIVOTTABLE REPORT

Before you can create a PivotTable, you need to prepare the data. You can get the data from one of the following sources:

- A range on an Excel worksheet (you can type in your data or paste it from other sources).
- An external data source (you can connect to a Microsoft Access file or a SQL Server database and get data directly).

Figure 20.1 displays the data that was dumped into a Microsoft Excel worksheet from a SQL Server database. The downloadable workbook file is named `EquipmentList.xlsx`. This file contains over 1,400 rows of data, which would be difficult to summarize if it weren't for the built-in Excel PivotTable feature.

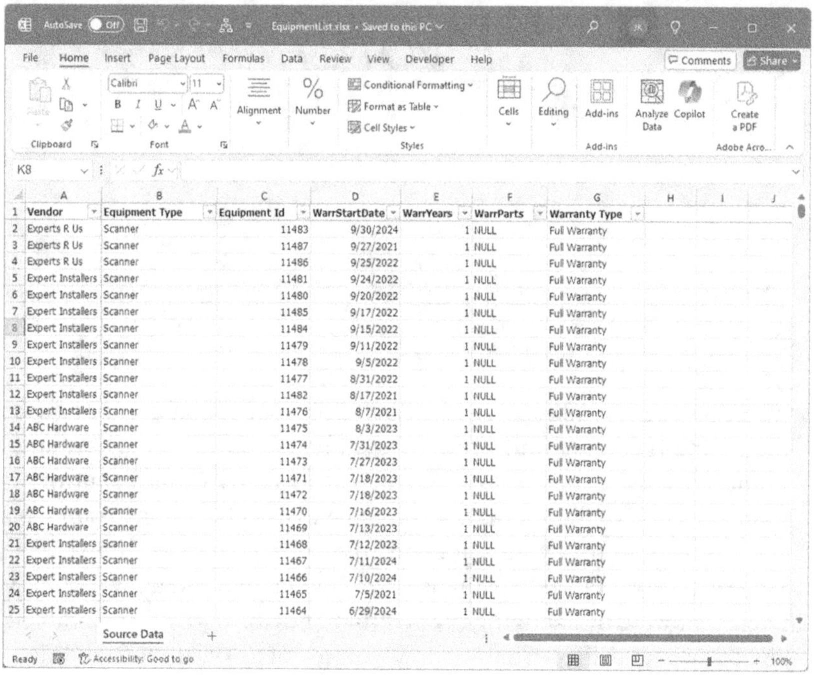

FIGURE 20.1.   Source data for the PivotTable.

Let's start working with PivotTables by using built-in ribbon commands.

NOTE	*Please note that the files for the Hands-On projects can be found in the companion files.*

Hands-On 20.1   Creating a PivotTable

1. Copy the `EquipmentList.xlsx` workbook from the companion files to your `VBAExcel2024_ByExample` folder and then open it in Microsoft Excel.
   Select any cell anywhere in the data range. For example, select cell A2 in the Source Data worksheet.
2. Choose Insert | Tables | PivotTable.

The Create PivotTable dialog box appears, as shown in Figure 20.2.

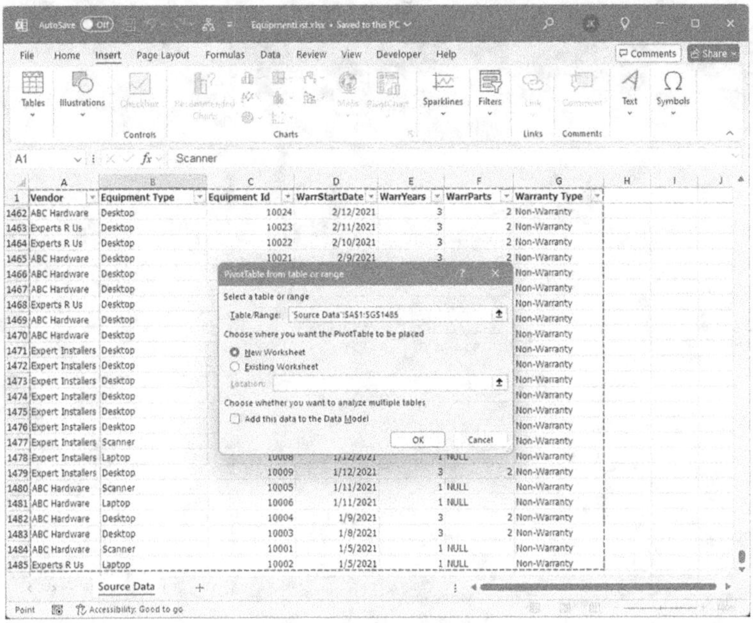

FIGURE 20.2. Create PivotTable dialog box.

Notice three sections in the Create PivotTable dialog box. In the top section, you must choose the data source for your report. This can be a table or range within a Microsoft Excel worksheet. The middle section specifies the location where you want the PivotTable to be placed. You can choose between placing it in a new worksheet or in the existing worksheet. The bottom section allows you to create a data model based on multiple tables. This topic is discussed in the last section of this chapter.

3. Make sure the Select a table or range and New Worksheet option buttons are specified.
   Ensure that the range displayed in the Table/Range box includes all the data on which you want to report. The range will appear automatically if the active cell is within the data range. If the currently selected cell is outside of the data range, you will need to make your own selection.

4. Click OK.

A blank PivotTable report is inserted in a new sheet and the PivotTable Field List pane is displayed, as shown in Figure 20.3.

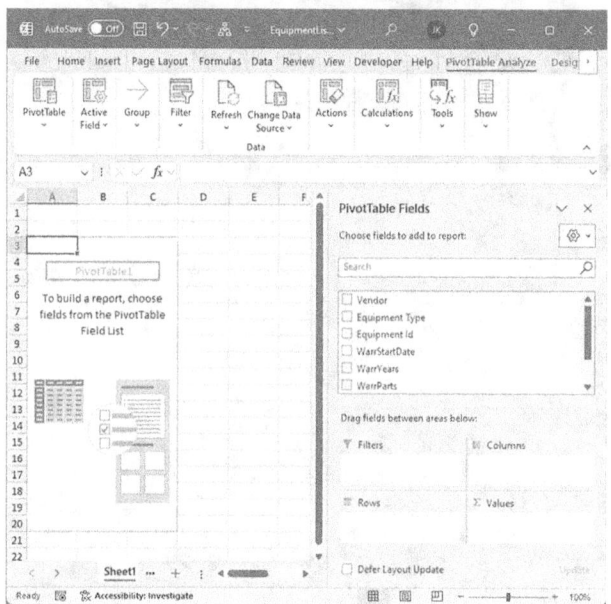

FIGURE 20.3. The PivotTable report waits for you to make field selections.

Notice that all the fields are listed in the PivotTable Field List pane to the right of the worksheet. Each field has a checkbox so you can easily indicate which fields to include in the report. The report is built as you make field selections. For example, when you check the box next to Vendor, you will notice that the Vendor field is automatically added to the Row Labels box at the bottom of the PivotTable Field List pane, and the PivotTable updates accordingly to show your selection.

By default, text fields are placed in the Rows list and numeric fields appear in the Values list, as shown in Figure 20.4. You can specify the type of calculation you want to use to summarize data by clicking on the down arrow next to the field name in the Values area and selecting Value Field Settings. The default for the Values area is the Sum function. Figure 20.4 displays the PivotTable using the Count function. You can easily adjust the position of the fields by dragging them between areas in the PivotTable Field List pane.

Figure 20.5 shows the views made available in the PivotTable Field List pane by clicking on the button at the top of the pane.

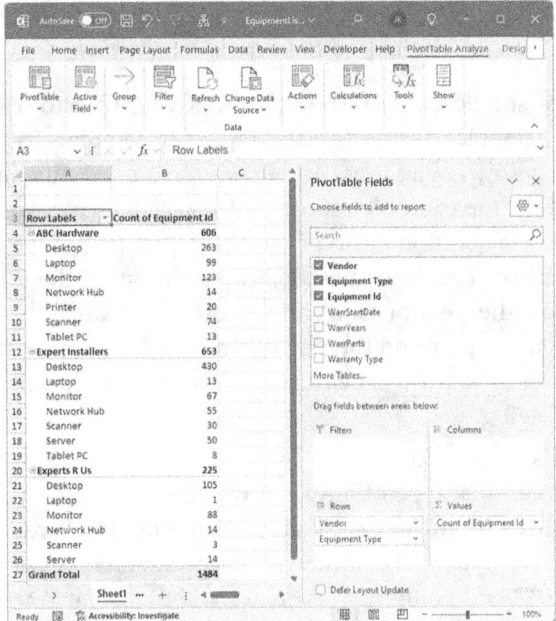

FIGURE 20.4.   Adding fields to the PivotTable report.

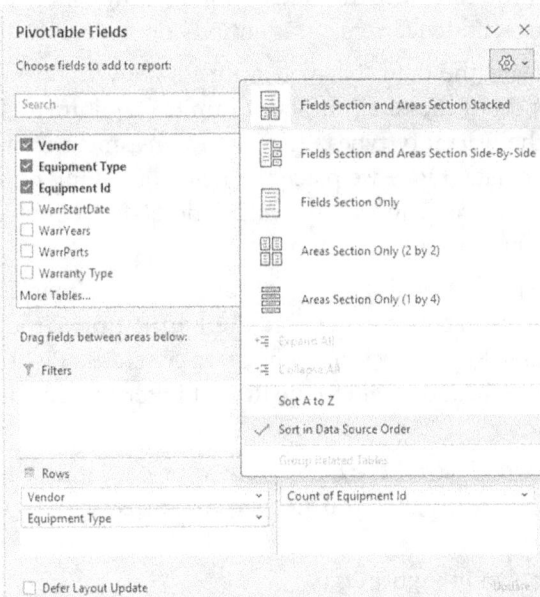

FIGURE 20.5.   You can easily change the layout of the sections that are available in the PivotTable Field List pane on the right.

At the bottom of the PivotTable Field List pane, there are four areas where you can place the fields:

- The Row Labels area should contain the fields that you want to display your data "by." For example, to produce the report by vendor, drag the Vendor field onto the Row Labels area. The Row Labels area can contain more than one field. In the example report that you will create, we also want to see the report by equipment type, so the Equipment Type field will be placed in the Row Labels area as well. By positioning the Equipment Type field below the Vendor field in the Row Labels area, the data will be grouped first by vendor and then by equipment type within those vendors. Fields listed in the Row Labels area can be moved into the desired position by dragging.

- The Column Labels area should contain fields that answer the question of "what." For example, what type of information do you want to display for each of the fields in the Row Labels area? Our example PivotTable will report on the warranty type. Because we want to see all types of warranties for each vendor and equipment type, we will place the Warranty Type field in the Column Labels area. To view your data from a different perspective, however, you can place the fields from the Row Labels area in the Column Labels area, and vice versa. It is entirely up to you.

- The Values area displays the data that you want to analyze. In our example, we want to find out the total number of units (Equipment Type) covered by each of the warranty types. The Values area must contain a field that has numeric data. Once we place the field containing numeric data in the Values area, we can choose what calculation (sum, count, average, and so on) we want to perform on the data.

- The Report Filter area is optional. Filter fields add a third dimension to your data analysis. Later in this chapter (see the Formatting, Grouping, and Sorting a PivotTable Report section), when you generate a PivotTable programmatically, you will add a field to the Report Filter area so you can experiment with the data.

Note that you do not have to place all the fields from your data source in the PivotTable. Place only those fields that you need; you can easily add other fields later.

5. Make your selections as shown in Figure 20.6.

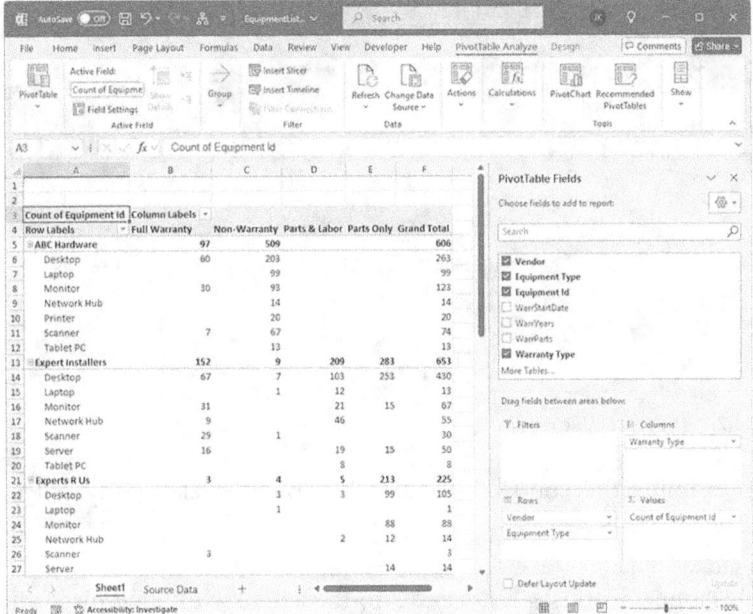

FIGURE 20.6. The completed PivotTable report.

While the PivotTable is selected, the PivotTable Field List pane is visible on the right-hand side so that you can easily modify the PivotTable by adding or removing fields. You can temporarily remove the PivotTable Field List pane by clicking outside the PivotTable selection in the worksheet. For example, if you click any cell in rows 1 or 2, the pane will disappear. Click again anywhere in the area containing the pivot data and the pane will reappear.

PivotTables are used for data analysis and presentation only. This means that you are not permitted to enter data directly into a PivotTable. You must make any changes or additions to the data in the underlying source data and then use the Refresh button on the PivotTable Analyze tab (see Figure 20.6) to update the PivotTable. If you add new rows to the source data, you must use the Change Data Source button in the Data group to expand the data range.

6. To see the data from a different point of view, reposition the selected fields in the PivotTable Field List, as shown in Figure 20.7.

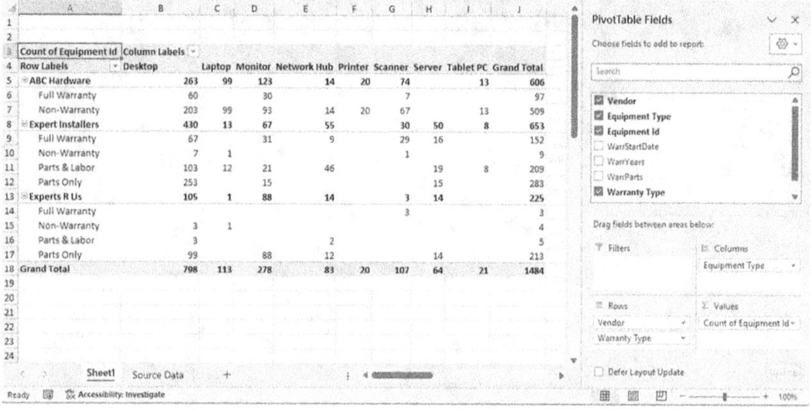

FIGURE 20.7. The PivotTable report—another view of the source data.

You can examine the contributing data by double-clicking any cell in the Grand Total column.

7. Double-click cell J13.

Excel adds a new worksheet, `Detail1`, to the active workbook displaying all the records that contributed to the selected total value (Figure 20.8).

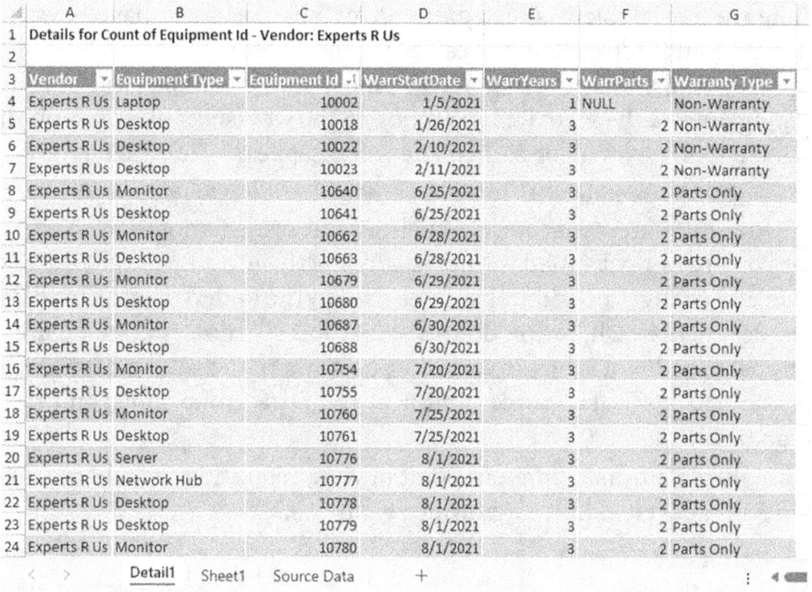

	A	B	C	D	E	F	G
1	Details for Count of Equipment Id - Vendor: Experts R Us						
2							
3	Vendor	Equipment Type	Equipment Id	WarrStartDate	WarrYears	WarrParts	Warranty Type
4	Experts R Us	Laptop	10002	1/5/2021	1	NULL	Non-Warranty
5	Experts R Us	Desktop	10018	1/26/2021	3	2	Non-Warranty
6	Experts R Us	Desktop	10022	2/10/2021	3	2	Non-Warranty
7	Experts R Us	Desktop	10023	2/11/2021	3	2	Non-Warranty
8	Experts R Us	Monitor	10640	6/25/2021	3	2	Parts Only
9	Experts R Us	Desktop	10641	6/25/2021	3	2	Parts Only
10	Experts R Us	Monitor	10662	6/28/2021	3	2	Parts Only
11	Experts R Us	Desktop	10663	6/28/2021	3	2	Parts Only
12	Experts R Us	Monitor	10679	6/29/2021	3	2	Parts Only
13	Experts R Us	Desktop	10680	6/29/2021	3	2	Parts Only
14	Experts R Us	Monitor	10687	6/30/2021	3	2	Parts Only
15	Experts R Us	Desktop	10688	6/30/2021	3	2	Parts Only
16	Experts R Us	Monitor	10754	7/20/2021	3	2	Parts Only
17	Experts R Us	Desktop	10755	7/20/2021	3	2	Parts Only
18	Experts R Us	Monitor	10760	7/25/2021	3	2	Parts Only
19	Experts R Us	Desktop	10761	7/25/2021	3	2	Parts Only
20	Experts R Us	Server	10776	8/1/2021	3	2	Parts Only
21	Experts R Us	Network Hub	10777	8/1/2021	3	2	Parts Only
22	Experts R Us	Desktop	10778	8/1/2021	3	2	Parts Only
23	Experts R Us	Desktop	10779	8/1/2021	3	2	Parts Only
24	Experts R Us	Monitor	10780	8/1/2021	3	2	Parts Only

Detail1   Sheet1   Source Data   +

FIGURE 20.8. You can view the breakdown of the data by double-clicking on a data field in the PivotTable.

8. Save the changes in the `EquipmentList.xlsx` workbook. Do not close this file as it will be used in the next Hands-On.

Drilling down on the data is a nice feature, except for the fact that if you do a lot of double-clicking, you will end up with many additional and, most likely, unwanted worksheets in your workbook. You may want to delete the drill-down worksheet after examining the detailed data. You can do this manually or you can perform the cleanup programmatically by writing VBA event procedures, as described in the next section.

REMOVING PIVOTTABLE DETAIL WORKSHEETS WITH VBA

In the previous Hands-On, Excel added a new worksheet named `Detail1` to the `EquipmentList.xlsx` workbook with the detailed data included in the total figure that was selected in the PivotTable report. The following example demonstrates two event procedures that allow you to specify whether you want to keep the detailed worksheet or delete it automatically after you've examined the data. This exercise requires completion of the steps outlined in Hands-On 20.1.

> **Hands-On 20.2   Writing VBA Procedures to Remove a PivotTable Detail Worksheet**

1. In the `EquipmentList.xlsx` workbook, rename the sheet with your PivotTable `PivotReport`.

2. Save the workbook in the Excel macro-enabled format as `EquipmentListPivot.xlsm`.

3. Press Alt+F11 to switch to the VBE screen.

4. In the Project Explorer window, double-click the ThisWorkbook object in the Microsoft Excel object folder under `VBAProject (EquipmentListPivot.xlsm)`.

5. In the EquipmentListPivot.xlsm – ThisWorkbook Code window, enter the global variable declaration and two event procedures, as shown below:

```
' Global variables
Dim flag As Boolean          ' Boolean variable to indicate whether
                             ' to delete a drill-down worksheet
Dim strPivSheet As String    ' String to hold the name of the sheet
                             ' containing the PivotTable
Dim strDrillSheet As String  ' String to hold the name of the
                             ' drill-down sheet
```

```
Dim strPivSource As String   ' String to hold the name of the
                             ' worksheet with the PivotTable
                             ' source data

Private Sub Workbook_SheetActivate(ByVal Sh As Object)
    If strPivSheet = "" Then Exit Sub
    If Sh.Name <> strPivSheet Then
    If InStr(1, strPivSource, Sh.Name) = 0 Then
       If MsgBox("Do you want to Delete " & Sh.Name & _
           " from the workbook" & vbCrLf _
           & "upon returning to PivotTable report?", _
           vbYesNo + vbQuestion, _
           "Sheet: Delete or Keep") = vbYes Then
           flag = True
           strDrillSheet = Sh.Name
        Else
          flag = False
          Exit Sub
        End If
     End If
    End If
    If ActiveSheet.Name = strPivSheet And flag = True Then
       Application.DisplayAlerts = False
       Worksheets(strDrillSheet).Delete
       Application.DisplayAlerts = True
       flag = False
    End If
End Sub

Private Sub Workbook_SheetBeforeDoubleClick(ByVal Sh As Object, _
    ByVal Target As Range, Cancel As Boolean)
    With ActiveSheet
      If .PivotTables.Count > 0 Then
          strPivSource = ActiveSheet.PivotTables(1).SourceData
          If ActiveCell.PivotField.Name <> "" And _
              IsEmpty(Target) Then
                MsgBox "Selected cell has no data " & _
                        "- cannot drill down."
              Cancel = True
              Exit Sub
          End If
          strPivSheet = ActiveSheet.Name
        End If
    End With
End Sub
```

When the user returns to the worksheet with the PivotTable report, the `Workbook_SheetActivate` event procedure will prompt them to confirm whether the drill-down worksheet should be deleted. If the user selects Yes in the message box, the Boolean variable `flag` will be set to `True`. To prevent Excel's default confirmation message when deleting a worksheet, the procedure temporarily disables application messages, allowing for a seamless deletion. Remember to re-enable the alerts after the deletion.

The `Workbook_SheetBeforeDoubleClick` event procedure will disable the drill-down function if the user double-clicks on an empty PivotTable cell. If the cell is not empty, the name of the worksheet containing the PivotTable will be assigned to the global variable `strPivSheet`. Additionally, to avoid deleting the worksheet with the PivotTable source data, the `SourceData` property of the `PivotTables` collection is used to store the name of the source data worksheet and the underlying data range in the global variable `strPivSource`.

To find out exactly how these two event procedures work together, use some of the debugging skills that you acquired in Chapter 8.

6. Press Ctrl+S to save the changes you made in the VBE window.
7. Return to the Microsoft Excel application window and manually delete the `Detail1` worksheet. Next, select the PivotReport worksheet and double-click cell J9.
   Excel will execute the code inside the `Workbook_SheetBeforeDoubleClick` event procedure and proceed to execute the code inside the `Workbook_SheetActivate` procedure. Because cell J9 is not empty, Excel will ask you whether you want to delete the drill-down worksheet upon returning to the `PivotTable` worksheet.
8. Click the Yes button in the message box.
   Nothing happens at this point. Excel has simply set the flag to delete this drill-down worksheet when you are done viewing it.
9. Click the PivotReport worksheet tab.
   At this point, Excel deletes the drill-down worksheet and activates the `PivotReport` worksheet.
10. Double-click cell B14 in the `PivotReport` worksheet. Excel displays the message that the drill-down is not allowed because there is no data in this cell. Recall that this message was coded inside the `Workbook_SheetBeforeDoubleClick` event procedure.
11. Click OK to dismiss the message box.
12. Save the `EquipmentListPivot.xlsm` workbook and then close it.

CREATING A PIVOTTABLE REPORT WITH VBA

Although the PivotTable creation process has undergone many improvements in Excel, some users may still find creating PivotTable reports confusing. For those users, you may want to generate PivotTables via VBA code. With VBA, you can also make many formatting changes to the existing PivotTables. This section demonstrates how to work with PivotTables programmatically. We will start by creating the PivotTable report shown earlier in Figure 20.6 using the data source presented in Figure 20.1.

(◉) **Hands-On 20.3   Creating a PivotTable Report with VBA**

1. Open the EquipmentList.xlsx workbook.
2. Right-click the Source Data sheet tab and choose the Move or Copy option from the context menu.
3. In the Move or Copy dialog box, choose the (new book) entry from the To book drop-down list. Indicate that you want to make a copy of the selected sheet by clicking the checkbox next to Create a copy label. Click OK to proceed with the copy operation.
   Excel will create a new workbook with one sheet named Source Data. This sheet has been copied from the EquipmentList.xlsx file.
4. Save this new workbook in the Excel macro-enabled format as C:\VBAExcel2024_ByExample\Chap20.xlsm.
5. Insert three new sheets into the Chap20.xlsm file and save the changes made to the workbook.
6. Close the EquipmentList.xlsx workbook. Leave the Chap20.xlsm file open.
7. Press Alt+F11 to switch to the VBE screen.
8. In the Project Explorer window, highlight VBAProject (Chap20.xlsm) and choose Insert | Module.
9. In the Module1 Code window, enter the CreateNewPivotReport procedure as shown below:

```
Sub CreateNewPivotReport()
    Dim wksData As Worksheet
    Dim rngData As Range
    Dim wksDest As Worksheet
    Dim pvtTable As PivotTable

    ' Set up object variables
    Set wksData = ThisWorkbook.Worksheets("Source Data")
    Set rngData = wksData.UsedRange
    Set wksDest = ThisWorkbook.Worksheets("Sheet2")
```

```vba
' Create a skeleton of a PivotTable

Set pvtTable = wksData.PivotTableWizard(SourceType:= _
    xlDatabase, SourceData:=rngData, TableDestination:= _
    wksDest.Range("B5"))

' Close the PivotTable Field List that appears automatically
ActiveWorkbook.ShowPivotTableFieldList = False

' Add fields to the PivotTable
With pvtTable
    .PivotFields("Vendor").Orientation = xlRowField
    .PivotFields("Equipment Type").Orientation = xlRowField
    .PivotFields("Warranty Type").Orientation = xlColumnField
    With .PivotFields("Equipment Id")
        .Orientation = xlDataField
        .Function = xlCount
    End With
    .PivotFields("Equipment Id").Orientation = xlPageField
End With

' Autofit columns so all headings are visible
wksDest.UsedRange.Columns.AutoFit
End Sub
```

The `CreateNewPivotReport` procedure uses the `PivotTableWizard` method of a `Worksheet` object to create a new PivotTable report. This method accepts several optional arguments to specify the data source type, its location, and the placement of the PivotTable report. It's advisable to use these arguments as demonstrated in our example code.

Using the `xlDatabase` constant in the `SourceType` argument allows you to create a PivotTable from various data sources. Our code indicates that the data comes from an Excel range. If you want to create a PivotTable report from another PivotTable, use `xlPivotTable` for this argument. To pull data from an external database, specify `xlExternal` as the `SourceType`.

In the example procedure, the `SourceData` argument references the used range on the worksheet containing the source data, and the `TableDestination` argument refers to cell B5 on `Sheet2` of the current workbook, marking the upper-left corner of the report.

The code assumes that `Sheet2` exists in the workbook. If it doesn't, you can add it via the VBA code before setting the reference. When you call the `PivotTableWizard` method, it creates a blank PivotTable report with all the fields from the data source hidden. To display the fields, add them to the appropriate areas of the PivotTable Field List pane (Row Labels, Column Labels, Values, and Report Filter).

Although the PivotTable Field List pane appears automatically on the right-hand side of the worksheet during manual creation, it's unnecessary to display this list when creating a PivotTable programmatically. Set the `ShowPivotTableFieldList` property to hide the pane.

To display each field in the PivotTable report, set the `Orientation` property of the `PivotField` object using the constants `xlRowField`, `xlColumnField`, `xlDataField`, and `xlPageField`. For the Total Units field in the Values area, set the `Function` property of the `PivotField` object to `xlSum`.

10. When you are creating a PivotTable report via code, you may need to check whether a PivotTable already exists in the destination worksheet. You can place the following code just below the code that sets up object variables (see the previous `CreateNewPivotReport` procedure):

```
' Check if PivotTable already exists
 If wksDest.PivotTables.Count > 0 Then
     MsgBox "Worksheet " & wksDest.Name & _
         " already contains a pivot table."
     Exit Sub
 End If
```

11. Run the `CreateNewPivotReport` procedure.
When you switch to the Microsoft Excel application window, Sheet2 should contain the PivotTable report shown in Figure 20.9.

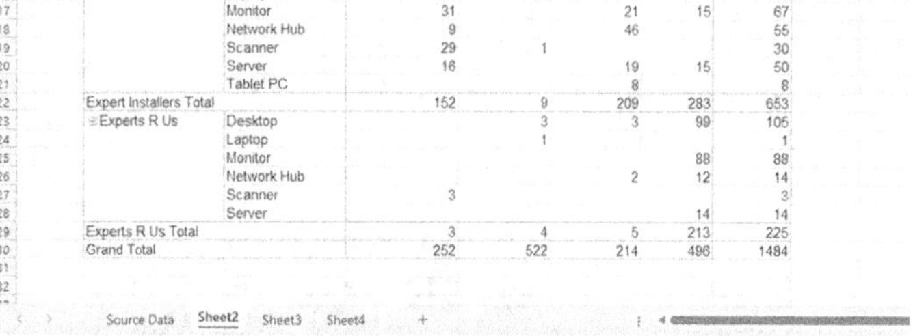

Equipment Id	(All)					
Count of Equipment Id		Warranty Type				
Vendor	Equipment Type	Full Warranty	Non-Warranty	Parts & Labor	Parts Only	Grand Total
ABC Hardware	Desktop	60	203			263
	Laptop		99			99
	Monitor	30	93			123
	Network Hub		14			14
	Printer		20			20
	Scanner	7	67			74
	Tablet PC		13			13
ABC Hardware Total		97	509			606
Expert Installers	Desktop	67	7	103	253	430
	Laptop		1	12		13
	Monitor	31		21	15	67
	Network Hub	9		46		55
	Scanner	29	1			30
	Server	16		19	15	50
	Tablet PC			8		8
Expert Installers Total		152	9	209	283	653
Experts R Us	Desktop		3	3	99	105
	Laptop		1			1
	Monitor				88	88
	Network Hub			2	12	14
	Scanner	3				3
	Server				14	14
Experts R Us Total		3	4	5	213	225
Grand Total		252	522	214	496	1484

FIGURE 20.9. A PivotTable report created with VBA code.

CREATING A PIVOTTABLE REPORT FROM AN ACCESS DATABASE

You can use the same `PivotTableWizard` method of the `Worksheet` object (demonstrated in Hands-On 20.3) to create a PivotTable report from an external data source. Let's start by creating a PivotTable report from a Microsoft Access sample database.

Hands-On 20.4   Creating a PivotTable Report from Access with VBA

1. Add a new module to `VBAProject` (Chap20.xlsm) and enter the `PivotTable_External` procedure as shown below:

```
Sub PivotTable_External()
    Dim strConn As String
    Dim strQuery_1 As String
    Dim strQuery_2 As String
    Dim myArray As Variant
    Dim destRange As Range
    Dim strPivot As String

    strConn = "Driver={Microsoft Access Driver" & _
        "(*.mdb, *.accdb)};" & _
        "DBQ=C:\VBAExcel2024_ByExample\Northwind.mdb;"

    strQuery_1 = "SELECT Customers.CustomerID, " & _
        "Customers.CompanyName," & _
        "Orders.OrderDate, Products.ProductName, Sum([Order " & _
        "Details].[UnitPrice]*[Quantity]*(1-[Discount])) " & _
            "AS Total " & _
        "FROM Products INNER JOIN ((Customers INNER JOIN " & _
        "Orders ON Customers.CustomerID = "

    strQuery_2 = "Orders.CustomerID) INNER JOIN " & _
        "[Order Details] ON Orders.OrderID = " & _
        "[Order Details].OrderID) ON Products.ProductID = " & _
        "[Order Details].ProductID " & _
        "GROUP BY Customers.CustomerID, " & _
        "Customers.CompanyName, Orders.OrderDate, " & _
        "Products.ProductName;"

    myArray = Array(strConn, strQuery_1, strQuery_2)
    Worksheets.Add

    Set destRange = ActiveSheet.Range("B5")
    strPivot = "PivotFromAccess"
```

```
    ActiveSheet.PivotTableWizard _
     SourceType:=xlExternal, _
     SourceData:=myArray, _
     TableDestination:=destRange, _
     TableName:=strPivot, _
     SaveData:=False, _
     BackgroundQuery:=False

    ' Close the PivotTable Field List that appears automatically
    ActiveWorkbook.ShowPivotTableFieldList = False

    ' Add fields to the PivotTable
    With ActiveSheet.PivotTables(strPivot)
    .PivotFields("ProductName").Orientation = xlRowField
    .PivotFields("CompanyName").Orientation = xlRowField
    With .PivotFields("Total")
        .Orientation = xlDataField
        .Function = xlSum
        .NumberFormat = "$#,##0.00"
    End With
    .PivotFields("CustomerID").Orientation = xlPageField
    .PivotFields("OrderDate").Orientation = xlPageField
    End With
    ' Autofit columns so all headings are visible
    ActiveSheet.UsedRange.Columns.AutoFit
End Sub
```

When using the `PivotTableWizard` method of the `Worksheet` object to create a PivotTable report from an external data source, you need to specify several arguments, which are listed in Table 20.1.

TABLE 20.1. Required arguments used with the PivotTableWizard method.

Name	Description
SourceType	Use the `xlExternal` constant to indicate that the data for the Pivot-Table comes from an external data source.
SourceData	Specify an array containing two or more elements. The first element of the array must be a connection string to the database. The second argument is the SQL statement for querying an external database. If the SQL statement is longer than 255 characters, break up the statement into several strings and pass each string as a separate element of the array.
	In the example procedure above, the SQL statement necessary for obtaining the required data from an external database is longer than 255 characters; therefore, the SQL string is broken into two strings:

Name	Description
	`strQuery_1` and `strQuery_2`. Next, the connection string and the SQL statement are placed in an array like this: `myArray = Array(strConn, strQuery_1, strQuery_2)` `myArray` is then used as the `SourceData` argument of the `Pivot-TableWizard` method.
`TableDestination`	Specify a worksheet range where the PivotTable should be placed.
`TableName`	Specify the name of the PivotTable that you want to create.

In addition to the above arguments, the `PivotTable_External` example procedure uses the optional `SaveData` and `BackgroundQuery` arguments. By setting the `SaveData` argument to `False`, the PivotTable will not be saved. This setting allows you to save space on the disk. When the `BackgroundQuery` argument is set to `False`, Visual Basic will refrain from executing other operations in Excel in the background until the query is complete.

After creating the PivotTable, the procedure specifies where the fields returned by the SQL statement should be placed in the PivotTable report.

2. Run the `PivotTable_External` procedure to generate the PivotTable. The resulting PivotTable report is illustrated in Figure 20.10.

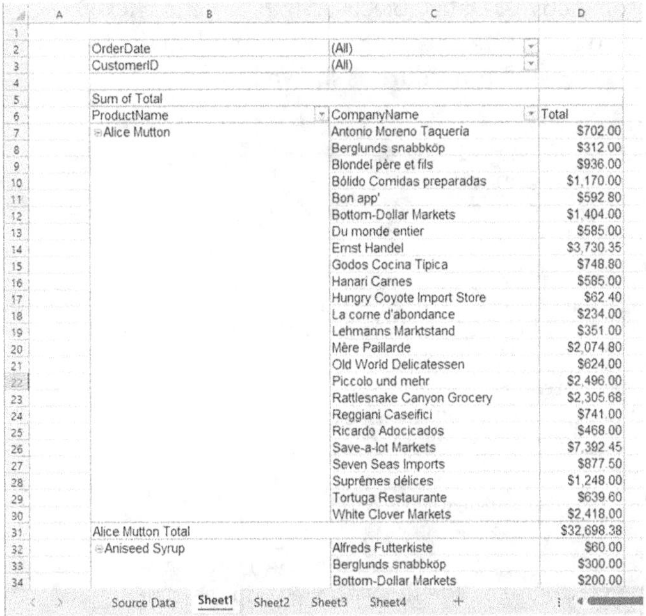

	A	B	C	D
1				
2		OrderDate	(All)	
3		CustomerID	(All)	
4				
5		Sum of Total		
6		ProductName	CompanyName	Total
7		Alice Mutton	Antonio Moreno Taquería	$702.00
8			Berglunds snabbköp	$312.00
9			Blondel père et fils	$936.00
10			Bólido Comidas preparadas	$1,170.00
11			Bon app'	$592.80
12			Bottom-Dollar Markets	$1,404.00
13			Du monde entier	$585.00
14			Ernst Handel	$3,730.35
15			Godos Cocina Típica	$748.80
16			Hanari Carnes	$585.00
17			Hungry Coyote Import Store	$62.40
18			La corne d'abondance	$234.00
19			Lehmanns Marktstand	$351.00
20			Mère Paillarde	$2,074.80
21			Old World Delicatessen	$624.00
22			Piccolo und mehr	$2,496.00
23			Rattlesnake Canyon Grocery	$2,305.68
24			Reggiani Caseifici	$741.00
25			Ricardo Adocicados	$468.00
26			Save-a-lot Markets	$7,392.45
27			Seven Seas Imports	$877.50
28			Suprêmes délices	$1,248.00
29			Tortuga Restaurante	$639.60
30			White Clover Markets	$2,418.00
31		Alice Mutton Total		$32,698.38
32		Aniseed Syrup	Alfreds Futterkiste	$60.00
33			Berglunds snabbköp	$300.00
34			Bottom-Dollar Markets	$200.00

Source Data Sheet1 Sheet2 Sheet3 Sheet4

FIGURE 20.10. A PivotTable report can be created programmatically from an external data source such as a Microsoft Access database.

USING THE CREATEPIVOTTABLE METHOD OF THE PIVOTCACHE OBJECT

When you use the macro recorder to generate VBA code for creating a PivotTable programmatically, Excel uses the Add method of the PivotCaches collection to create a new PivotCache object. A PivotCache object represents the data behind a PivotTable. It is an area in memory where data is stored and accessed as required from a data source. Use PivotCache when you need to generate multiple PivotTables from the same data source. By using PivotCache, you can gain a high level of control over your external data source. The PivotCache object can also be used to change and refresh data stored in the cache.

The example procedure in Hands-On 20.5 connects to the Microsoft Access database (Northwind 2007.accdb) using the Microsoft.ACE.OLEDB.12.0 provider. Recall that to use this type of connection, you must set up a reference to Microsoft ActiveX Data Objects (ADO) in the References dialog box.

Hands-On 20.5   Creating a PivotTable Report Using the PivotCache Object

1. In the VBE screen, choose Tools | References. In the Available References list box, select Microsoft ActiveX Data Objects 6.1 Library and click OK.
2. Add a new module to VBAProject (Chap20.xlsm) and, in the Code window, enter the Pivot_External_2 procedure as shown below:

```
Sub Pivot_External_2()
    Dim objPivotCache As PivotCache
    Dim conn As New ADODB.Connection
    Dim rst As New ADODB.Recordset
    Dim dbPath As String
    Dim strSQL As String

    dbPath = "C:\VBAExcel2024_ByExample\Northwind 2007.accdb"

    conn.Open "Provider=Microsoft.ACE.OLEDB.12.0;" _
            & "Data Source=" & dbPath & _
            "; Persist Security Info=False;"

    strSQL = "SELECT Products.[Product Name], " & _
            "Orders.[Order Date], " & _
            "Sum([Unit Price]*[Quantity]) AS Amount " & _
            "FROM Orders INNER JOIN (Products INNER JOIN " & _
            "[Order Details] ON Products.ID = " & _
            "[Order Details].[Product ID]) ON " & _
```

```
            "Orders.[Order ID] = [Order Details].[Order ID] " & _
            "GROUP BY Products.[Product Name], " & _
             "Orders.[Order Date], Products.[Product Name]" & _
             "ORDER BY Sum([Unit Price]*[Quantity]) DESC , " & _
             "Products.[Product Name];"

Set rst = conn.Execute(strSQL)

' Create a PivotTable cache and report
Set objPivotCache = ActiveWorkbook.PivotCaches.Add( _
    SourceType:=xlExternal)
Set objPivotCache.Recordset = rst

Worksheets.Add
With objPivotCache
    .CreatePivotTable TableDestination:=Range("B6"), _
        TableName:="Invoices"
End With

' Add fields to the PivotTable
With ActiveSheet.PivotTables("Invoices")
    .SmallGrid = False
    With .PivotFields("Product Name")
        .Orientation = xlRowField
        .Position = 1
    End With
    With .PivotFields("Order Date")
    .Orientation = xlRowField
        .Position = 2
        .Name = "Date"
    End With
    With .PivotFields("Amount")
        .Orientation = xlDataField
        .Position = 1
        .NumberFormat = "$#,##0.00"
    End With
End With

' Autofit columns so all headings are visible
ActiveSheet.UsedRange.Columns.AutoFit

' Clean up
rst.Close
conn.Close
Set rst = Nothing
Set conn = Nothing
```

```
    ' Obtain information about PivotCache
    With ActiveSheet.PivotTables("Invoices").PivotCache
        Debug.Print "Information about the PivotCache"
        Debug.Print "Number of Records: " & .RecordCount
        Debug.Print "Data was last refreshed on: " & .RefreshDate
        Debug.Print "Data was last refreshed by: " & .RefreshName
        Debug.Print "Memory used by PivotCache: " & _
            .MemoryUsed " (bytes)"
    End With
End Sub
```

After establishing a connection with a database and executing the SQL statement to obtain the data, the `Pivot_External_2` procedure creates a `PivotCache` using the following line of code:

```
Set objPivotCache = ActiveWorkbook.PivotCaches.Add( _
    SourceType:=xlExternal)
```

The code then places the data from the external data source in the `PivotCache` by assigning a `Recordset` object to the `PivotCache` object, like this:

```
Set objPivotCache.Recordset = rst
```

Next, the `CreatePivotTable` method of the `PivotCache` object is used to create an empty PivotTable:

```
    With objPivotCache
    .CreatePivotTable TableDestination:=Range("B6"), _
        TableName:="Invoices"
    End With
```

Once the skeleton of the PivotTable is created, we add appropriate fields to the PivotTable. The last few lines of the example procedure demonstrate how to find out information about the `PivotCache`.

To force the `PivotCache` to refresh automatically when a workbook containing the PivotTable is opened, set the `RefreshOnFileOpen` property to `True`. To do this, you may want to add the following statement at the end of the `Pivot_External_2` procedure:

```
    ActiveSheet.PivotTables("Invoices").PivotCache. _
    RefreshOnFileOpen = True
```

3. Run the `Pivot_External_2` procedure to generate the PivotTable. The resulting PivotTable report is illustrated in Figure 20.11.

	A	B	C	D
1				
2				
3				
4				
5				
6		Sum of Amount		
7		Product Name ▾	Date ▾	Total
8		⊟Northwind Traders Almonds	5/24/2006	$200.00
9		Northwind Traders Almonds Total		$200.00
10		⊟Northwind Traders Beer	1/15/2006	$1,400.00
11			4/5/2006	$1,218.00
12			4/8/2006	$4,200.00
13		Northwind Traders Beer Total		$6,818.00
14		⊟Northwind Traders Boysenberry Spread	3/24/2006	$250.00
15			6/5/2006	$2,250.00
16		Northwind Traders Boysenberry Spread Total		$2,500.00
17		⊟Northwind Traders Cajun Seasoning	3/24/2006	$220.00
18			6/5/2006	$660.00
19		Northwind Traders Cajun Seasoning Total		$880.00
20		⊟Northwind Traders Chai	1/22/2006	$270.00
21			3/24/2006	$450.00
22		Northwind Traders Chai Total		$720.00
23		⊟Northwind Traders Chocolate	2/10/2006	$127.50
24			3/22/2006	$1,275.00
25			4/3/2006	$127.50
26			6/5/2006	$510.00
27			6/8/2006	$510.00
28		Northwind Traders Chocolate Total		$2,550.00
29		⊟Northwind Traders Chocolate Biscuits Mix	1/30/2006	$276.00
30			2/6/2006	$184.00
31			3/24/2006	$92.00
32			4/5/2006	$230.00
33		Northwind Traders Chocolate Biscuits Mix Total		$782.00
34		⊟Northwind Traders Clam Chowder	2/23/2006	$1,930.00
35			4/5/2006	$289.50

Source Data   **Sheet5**   Sheet1   Sheet2   Sheet3   Sheet4   +

FIGURE 20.11.   A PivotTable report created using the CreatePivotTable method of the PivotCache object.

FORMATTING, GROUPING, AND SORTING A PIVOTTABLE REPORT

You can modify the display and format of a PivotTable programmatically by using a number of different properties of the `PivotTable` object. For example, you may want to reposition the fields within the PivotTable report, sort the data by a specific field, or group your data by years, quarters, months, and so on. The example procedure in the next Hands-On exercise reformats the PivotTable report shown in Figure 20.10 to look like the one shown in Figure 20.12.

⊙ Hands-On 20.6   Formatting a PivotTable Report

1. Add a new module to VBAProject (Chap20.xlsm) and enter the FormatPivotTable procedure as shown below:

```vba
Sub FormatPivotTable()
    Dim pvtTable As PivotTable
    Dim strPiv As String

    If ActiveSheet.PivotTables.Count > 0 Then
        strPiv = ActiveSheet.PivotTables(1).Name
            Set pvtTable = ActiveSheet.PivotTables(strPiv)
    Else
        Exit Sub
    End If

    With pvtTable
        .PivotFields("OrderDate").Orientation = xlColumns
        .PivotFields("CompanyName").Orientation = xlHidden
        ' use this statement to group OrderDate by year
        .PivotFields("OrderDate").DataRange.Cells(1).Group _
            Start:=True, End:=True, _
            periods:=Array(False, False, False, False, False, _
            False, True)

        .TableRange1.AutoFormat Format:=xlRangeAutoFormatColor2
        .PivotFields("ProductName").DataRange.Select

        ' sort the Product Name field in descending order based
        ' on the Sum of Total
        .PivotFields("ProductName").AutoSort xlDescending, _
            "Sum of Total """
        Selection.IndentLevel = 2
        With Selection.Font
            .Name = "Times New Roman"
            .FontStyle = "Bold"
            .Size = 10
        End With
        With Selection.Borders(xlInsideHorizontal)
            .LineStyle = xlContinuous
            .Weight = xlThin
            .ColorIndex = xlAutomatic
        End With
    End With
End Sub
```

By studying the code of the `FormatPivotTable` procedure, you can easily conclude that:

- To change the layout of a PivotTable, set the `Orientation` property of the required field to a different constant. The previous example code moves the OrderDate field from the Report Filter area to the Row Labels area of the PivotTable.

- To display a PivotTable without a particular field, set the `Orientation` property of the required field to `xlHidden`.

- To group the `OrderDate` field by year, use the `Group` method of the `Range` object. For example, the code uses the following statement to group the data in the `OrderDate` field by year:

```
PivotFields("OrderDate").DataRange.Cells(1). _
    Group Start:=True,End:=True, periods:=Array(False, _
    False, False, False, False, False, True)
```

- The `Start` and `End` arguments specify the start and end dates to be included in the grouping. By setting these arguments to `True`, all dates are included. The `periods` argument is a 7-element array of `Boolean` values that specify which periods you want to group by. Each `False` or `True` value represents a different time period. The order of the `periods` in the array is as follows: seconds (1), minutes (2), hours (3), days (4), months (5), quarters (6), and years (7). `True` means the data will be grouped by that period. `False` means the data will not be grouped by that period.

- You can apply automatic formatting to the entire PivotTable report by using the `AutoFormat` property of the `Range` object. The `TableRange1` property returns a `Range` object that represents the range containing the entire PivotTable report without the page fields:

```
.TableRange1.AutoFormat Format:=xlRangeAutoFormatColor2
```

- You can select the data items in a particular field by using the `DataRange` property and the `Select` method, like this:

```
.PivotFields("ProductName").DataRange.Select
```

- You can sort a particular field in descending or ascending order. The example procedure uses the following statement to sort the `ProductName` field in descending order based on the `Sum` of `Total`:

```
.PivotFields("ProductName").AutoSort xlDescending, _
    "Sum of Total"
```

- You can change the text indentation, font name, size, and style, as well as the borders of the selected range, as demonstrated in the last statements of the example procedure shown above.

NOTE	*The following statement will ungroup the dates:*
	`ActiveSheet.PivotTables(1).`
	`PivotFields("OrderDate").LabelRange.Ungroup`

2. Switch to the Microsoft Excel application window and activate the sheet containing the PivotTable report shown in Figure 20.10 earlier in this chapter.
3. Press Alt+F8 to open the Macro dialog box. Highlight the FormatPivotTable procedure and click Run.
   The resulting reformatted PivotTable report is illustrated in Figure 20.12.

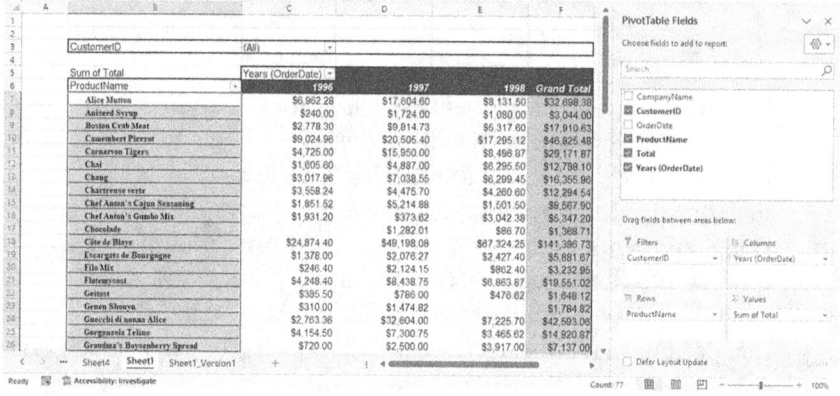

FIGURE 20.12. The PivotTable report shown in Figure 20.10 was reformatted to view data from a different perspective.

HIDING ITEMS IN A PIVOTTABLE

In the previous example procedure, you grouped the data in the PivotTable by year based on the `OrderDate` field. To hide some of the grouped data, you can set the `Visible` property of the `PivotItem` object to `False`. For instance, the following procedure demonstrates how to hide the 1996 column of data in the PivotTable report depicted in Figure 20.12:

```
Sub Hide1996YearColumn()
    Dim pt As PivotTable
    Dim pf As PivotField
    Dim pi As PivotItem
```

```
    Set pt = ActiveSheet.PivotTables(1)
    Set pf = pt.PivotFields("Years (OrderDate)")

    ' Loop through the PivotItems
    ' and hide the one labeled "1996"
    For Each pi In pf.PivotItems
        If pi.Name = "1996" Then
            pi.Visible = False
            Exit For
        End If
    Next pi
End Sub
```

ADDING CALCULATED FIELDS TO A PIVOTTABLE

You can customize a PivotTable report by defining calculated fields and items. Using the contents of other numeric fields in a PivotTable, you can create a calculated field that performs the required calculation. For example, let's create a procedure with two calculated fields named Change: 2010/2009 and Change: 2009/2008 to calculate the difference in number of products sold from year to year.

⊚ Hands-On 20.7 Creating a PivotTable Report with Calculated Fields

1. Save and close the Chap20.xlsm workbook.
2. Copy the Chap20b.xlsm workbook from the companion files to your C:\ VBAExcel2024_ByExample folder.
3. Open the Chap20b.xlsm workbook. This workbook contains the data shown in Figure 20.13.

⊿	A	B	C	D	E
1	Product	2022	2023	2024	
2	Prod1	904	614	694	
3	Prod2	456	139	755	
4	Prod3	1522	1009	1002	
5					

FIGURE 20.13. Sample data for the PivotTable report.

4. Switch to the VBE screen and highlight VBAProject(Chap20b.xlsm) in the Project Explorer.

5. Choose Insert | Module to add a new module and enter the `PivotWithCalcFields` procedure as shown below:

```
Sub PivotWithCalcFields()
    ActiveWorkbook.PivotCaches.Add( _
        SourceType:=xlDatabase, _
        SourceData:="Sheet1!R1C1:R4C4").CreatePivotTable _
        TableDestination:="'[Chap20b.xlsm]Sheet1'!R4C7", _
        TableName:="Piv1", _
        DefaultVersion:=xlPivotTableVersion10

    With ActiveSheet.PivotTables("Piv1").PivotFields("Product")
        .Orientation = xlRowField
        .Position = 1
    End With

    ActiveSheet.PivotTables("Piv1").AddDataField _
        ActiveSheet.PivotTables("Piv1").PivotFields("2024"), _
        "Sum of 2024", xlSum
    ActiveSheet.PivotTables("Piv1").AddDataField _
        ActiveSheet.PivotTables("Piv1").PivotFields("2023"), _
        "Sum of 2023", xlSum
    ActiveSheet.PivotTables("Piv1").AddDataField _
        ActiveSheet.PivotTables("Piv1").PivotFields("2022"), _
        "Sum of 2022", xlSum
    ActiveSheet.PivotTables("Piv1").CalculatedFields.Add _
        "Change: 2024/2023", "='2024' - '2023'", True
    ActiveSheet.PivotTables("Piv1").CalculatedFields.Add _
        "Change: 2023/2022", "='2023' - '2022'", True
    ActiveSheet.PivotTables("Piv1"). _
        PivotFields("Change: 2024/2023"). _
        Orientation = xlDataField
    ActiveSheet.PivotTables("Piv1"). _
        PivotFields("Change: 2023/2022"). _
        Orientation = xlDataField

End Sub
```

Notice that calculated fields are defined by using the `Add` method of the `CalculatedFields` object and supplying the name for the new field and a formula:

```
ActiveSheet.PivotTables("Piv1").CalculatedFields.Add _
        "Change: 2024/2023", "='2024' - '2023'", True
ActiveSheet.PivotTables("Piv1").CalculatedFields.Add _
        "Change: 2023/2022", "='2023' - '2022'", True
```

The third (optional) argument set to `True` indicates that the strings in field names will be interpreted as having been formatted in standard U.S. English instead of using local settings. The default setting is `False`.

A calculated field uses a formula that refers to other pivot fields that contain numeric data. This can be a simple formula, such as addition (+), subtraction (–), multiplication (*), or division (/), or an Excel function. In the procedure example above, we created the two calculated fields.

Figure 20.14 shows a new pivot field named Data, which Excel creates when you use multiple pivot fields in the Values area. The labels for the multiple pivot fields in the Data area can be displayed going down the rows or across the columns. You can specify the orientation of the labels by setting the Orientation property of the Data field to xlRowField or xlColumnField. Once you define a calculated field, the field is added to the PivotTable Field List and maintained in the PivotTable cache.

NOTE	*You can add a calculated field manually. Click Fields, Items, & Sets in the Calculations group of the PivotTable Analyze tab. Then, click Calculated Field to open the Insert Calculated Field dialog box.*

6. Run the `PivotWithCalcFields` procedure.
The resulting PivotTable report is shown in Figure 20.14.

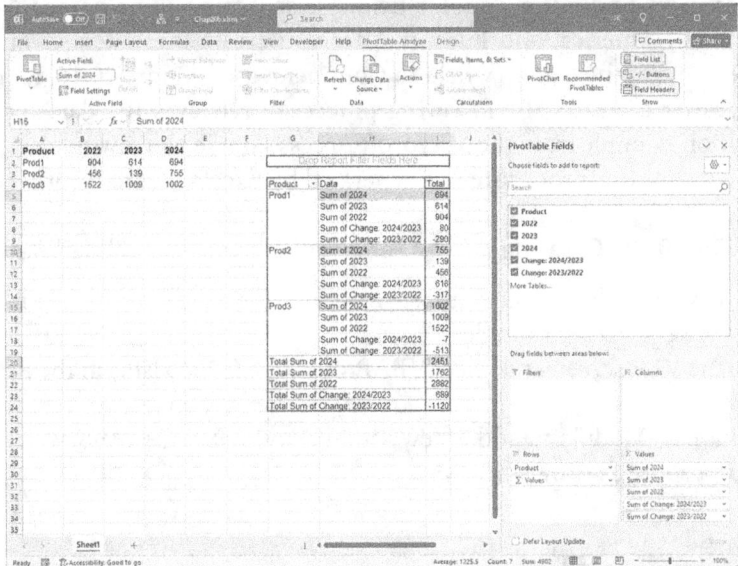

FIGURE 20.14. You can add additional calculations to a PivotTable by defining additional fields.

By adding the following statement at the end of the `PivotWithCalcFields` procedure, deleting columns G:I in Sheet1, and rerunning the procedure, the PivotTable depicted in Figure 20.14 will look like the one shown in Figure 20.15:

```
ActiveSheet.PivotTables("Piv1"). _
    PivotFields("Data").Orientation = xlColumnField
```

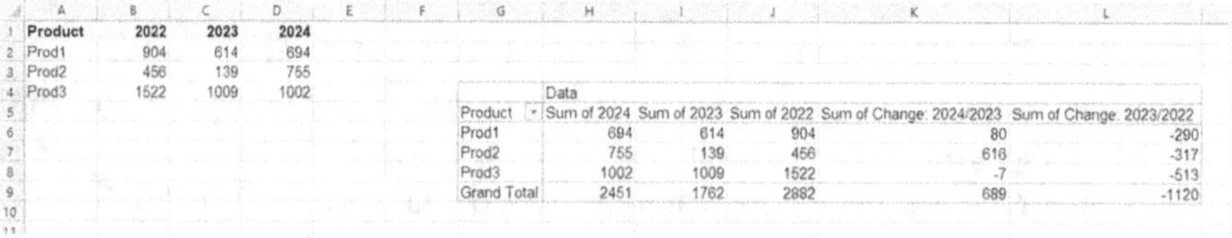

Product	2022	2023	2024
Prod1	904	614	694
Prod2	456	139	755
Prod3	1522	1009	1002

Data					
Product	Sum of 2024	Sum of 2023	Sum of 2022	Sum of Change: 2024/2023	Sum of Change: 2023/2022
Prod1	694	614	904	80	-290
Prod2	755	139	456	616	-317
Prod3	1002	1009	1522	-7	-513
Grand Total	2451	1762	2882	689	-1120

FIGURE 20.15. Changing the orientation of the PivotTable data.

7. Save and close the `Chap20b.xlsm` workbook.

You must not confuse a calculated item with a calculated field. A calculated item is a custom item you define in a PivotTable field to perform calculations using the contents of other fields and items in the PivotTable.

Suppose you created a report showing the total product sales for each of your salespeople by country. Then, you want to look at the data differently and show the sales made by each salesperson on three continents. You will need three new (calculated) items under the Country field. These items will be named North America, South America, and Europe. After you create these items, you can change the name of the Country field to Continent, as in Figure 20.16, to make your data easier to read. The following procedure retrieves the data for this demonstration example from the Microsoft Access sample Northwind database. The VBA code for this procedure was generated by Excel's macro recorder.

◉ Hands-On 20.8 Creating a PivotTable Report with Calculated Items

1. Open a new workbook file and save it as `Chap20c.xlsm` in your `C:\VBAExcel2024_ByExample` folder.
2. Switch to the VBE screen and select VBAProject(Chap20c.xlsm) in the Project Explorer.
3. Choose Insert | Module to add a new module and enter the `PivotWithCalcItems` procedure as shown below:

```
Sub PivotWithCalcItems()
    Dim strConn As String
```

```
Dim strSQL As String
Dim myArray As Variant
Dim destRng As Range
Dim strPivot As String

strConn = "Driver={Microsoft Access Driver & _
        (*.mdb, *.accdb)};" _
        "DBQ=C:\VBAExcel2024_ByExample\" & _
        "Northwind.mdb;"

strSQL = "SELECT Invoices.Customers.CompanyName, " & _
        "Invoices.Country, Invoices.Salesperson, " & _
        "Invoices.ProductName, Invoices.ExtendedPrice " & _
        "FROM Invoices ORDER BY Invoices.Country"

myArray = Array(strConn, strSQL)
Worksheets.Add

Set destRng = ActiveSheet.Range("B5")
strPivot = "PivotTable1"

ActiveSheet.PivotTableWizard _
    SourceType:=xlExternal, _
    SourceData:=myArray, _
    TableDestination:=destRng, _
    TableName:=strPivot, _
    SaveData:=False, _
    BackgroundQuery:=False

With ActiveSheet.PivotTables(strPivot).& _
     PivotFields("CompanyName")
    .Orientation = xlPageField
    .Position = 1
End With

With ActiveSheet.PivotTables(strPivot).PivotFields("Country")
    .Orientation = xlRowField
    .Position = 1
End With

ActiveSheet.PivotTables(strPivot).AddDataField _
ActiveSheet.PivotTables(strPivot).PivotFields _
    ("ExtendedPrice"), "Sum of ExtendedPrice", xlSum
```

```
With ActiveSheet.PivotTables(strPivot)& _
    .PivotFields("Salesperson")
    .Orientation = xlRowField
    .Position = 1
End With

With ActiveSheet& _
    .PivotTables(strPivot).PivotFields("Salesperson")
    .Orientation = xlPageField
    .Position = 1
End With

With ActiveSheet& _
    .PivotTables(strPivot).PivotFields("Salesperson")
    .Orientation = xlColumnField
    .Position = 1
End With

ActiveSheet.PivotTables(strPivot).PivotFields("Country"). _
    CalculatedItems.Add "North America", "=USA+Canada", True
ActiveSheet.PivotTables(strPivot).PivotFields("Country"). _
    CalculatedItems.Add "South America", _
    "=Argentina+Brazil+Venezuela ", True
ActiveSheet.PivotTables(strPivot).PivotFields("Country"). _
    CalculatedItems("North America").StandardFormula = _
    "=USA+Canada+Mexico"
ActiveSheet.PivotTables(strPivot).PivotFields("Country"). _
    CalculatedItems.Add "Europe", _
    "=Austria+Belgium+Denmark+Finland+" & _
    "France+Germany+Ireland+Italy+Norway+Poland+" & _
    "Portugal+Spain+Sweden+Switzerland+UK", True

With ActiveSheet.PivotTables(strPivot).PivotFields("Country")
    .PivotItems("Argentina").Visible = False
    .PivotItems("Austria").Visible = False
    .PivotItems("Belgium").Visible = False
    .PivotItems("Brazil").Visible = False
    .PivotItems("Canada").Visible = False
    .PivotItems("Denmark").Visible = False
    .PivotItems("Finland").Visible = False
    .PivotItems("France").Visible = False
    .PivotItems("Germany").Visible = False
    .PivotItems("Ireland").Visible = False
    .PivotItems("Italy").Visible = False
    .PivotItems("Mexico").Visible = False
    .PivotItems("Norway").Visible = False
```

```
            .PivotItems("Poland").Visible = False
            .PivotItems("Portugal").Visible = False
            .PivotItems("Spain").Visible = False
            .PivotItems("Sweden").Visible = False
            .PivotItems("Switzerland").Visible = False
            .PivotItems("UK").Visible = False
            .PivotItems("USA").Visible = False
            .PivotItems("Venezuela").Visible = False
    End With

    ActiveSheet.PivotTables(strPivot). _
            PivotFields("Country").Caption = "Continent"

    With ActiveSheet.PivotTables(strPivot). _
      PivotFields("Sum of ExtendedPrice"). _
        NumberFormat = "$#,##0.00"
    End With

    With ActiveSheet.PivotTables(strPivot)& _
        .PivotFields("ProductName")
        .Orientation = xlRowField
        .Position = 2
    End With

    ActiveSheet.PivotTables(strPivot). _
        PivotFields("ProductName").Orientation = xlHidden
End Sub
```

A calculated item uses a formula that refers to other items in the specified PivotTable field. For example, a PivotTable that contains a Country field listing a number of different country items (Austria, UK, Brazil, Argentina, etc.) could have a calculated item named South America defined as the sum of countries located on the South American continent.

All of the calculated items in the specified PivotTable are members of the CalculatedItems collection. Calculated items are defined by using the Add method of the CalculatedItems object and supplying two arguments—the name for the new item and a formula—as shown below:

```
ActiveSheet.PivotTables(strPivot).PivotFields("Country"). _
    CalculatedItems.Add "South America", _
    "=Argentina+Brazil+Venezuela", True
```

The third (optional) argument set to True indicates that any strings in field names will be interpreted as having been formatted in standard U.S. English instead of using local settings. The default setting is False.

4. Run the `PivotWithCalcItems` procedure.
   The resulting PivotTable report is shown in Figure 20.16.

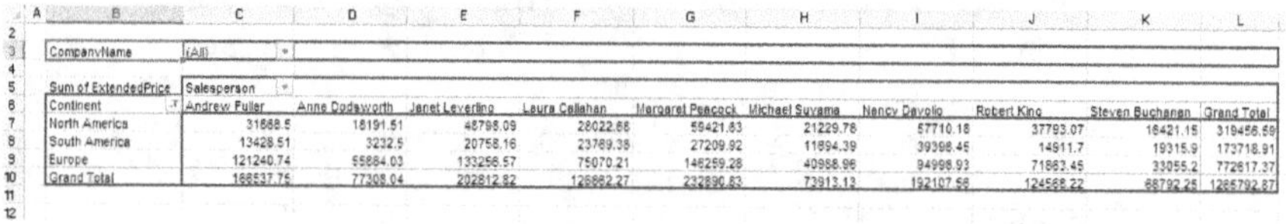

CompanyName	(All)									
Sum of ExtendedPrice	Salesperson									
Continent	Andrew Fuller	Anne Dodsworth	Janet Leverling	Laura Callahan	Margaret Peacock	Michael Suyama	Nancy Davolio	Robert King	Steven Buchanan	Grand Total
North America	31688.5	18191.51	48798.09	28022.66	59421.63	21229.78	57710.18	37793.07	16421.15	319456.59
South America	13428.51	3232.5	20758.16	23769.38	27209.92	11894.39	39398.45	14911.7	19315.9	173718.91
Europe	121240.74	55884.03	133256.57	75070.21	146259.28	40988.96	94998.93	71863.45	33055.2	772617.37
Grand Total	166537.75	77308.04	202812.82	126862.27	232890.83	73913.13	192107.56	124568.22	68792.25	1265792.87

FIGURE 20.16. By defining new items in a PivotTable report, you can present information summaries according to specific needs. Here, the Country field has been renamed Continent to present information summarized by continent. North America, South America, and Europe are calculated items in this PivotTable report.

To find out whether `PivotField` or `PivotItem` is a calculated item, use the `IsCalculated` property of the `PivotField` or `PivotItem` object. The procedure below prints a list of fields and items in the PivotTable to the Immediate window (see Figure 20.17), indicating whether the field or item is calculated. Additionally, the names of all calculated items and their formulas are written in `Sheet1` of the current workbook:

```
Sub ListCalcFieldsItems()
    Dim pivTable As PivotTable
    Dim fld As PivotField    ' field enumerator
    Dim itm As PivotItem    ' item enumerator
    Dim r As Integer    ' row number

    Set pivTable = Worksheets(2).PivotTables(1)

    On Error Resume Next

    ' print to the Immediate window the names of fields
    ' and calculated items
    For Each fld In pivTable.PivotFields
      If fld.IsCalculated Then
        Debug.Print fld.Name & ":" & _
        fld.Name & vbTab & "-->Calculated field"
      Else
        Debug.Print fld.Name
      End If
      For Each itm In pivTable. _
        PivotFields(fld.Name).CalculatedItems
          Debug.Print fld.Name & ":" & _
            itm.Name & vbTab & "--> Calculated item"
          ' enter information about Calculated items
```

```
    ' in a worksheet
    r = r + 1
    With Worksheets("Sheet1")
        .Cells(r, 1).Value = itm.Name
        .Cells(r, 2).Value = Chr(39) & itm.Formula
    End With
  Next
Next
End Sub
```

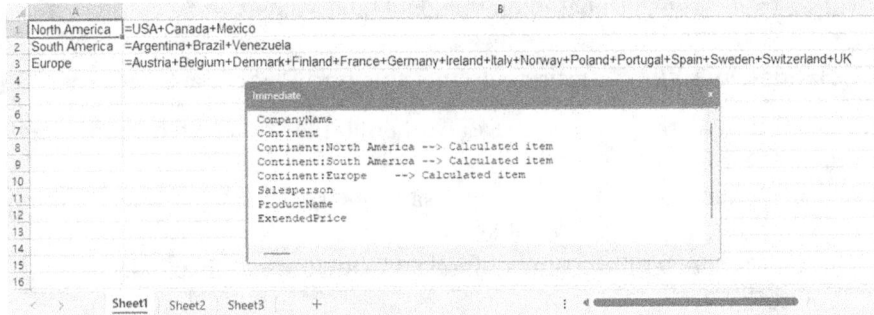

FIGURE 20.17. The output of the ListCalcFieldsItems procedure displays the names and formulas of calculated items within a worksheet. It also shows all PivotTable fields in the Immediate window, indicating whether the field is a calculated item.

CREATING A PIVOTCHART REPORT USING VBA

A PivotChart represents the data in a PivotTable report. Using VBA code, you can create a PivotChart based on an existing PivotTable report, and you can change the layout and data displayed in a PivotChart just as easily as you can reformat a PivotTable report.

A PivotChart report is linked to a PivotTable report. This means that when you rearrange the data in a PivotTable report, the PivotChart report displays the same view of the data, and vice versa. The default chart type for a PivotChart is a stacked column chart. This type of chart is useful for comparing the contribution of each value to a total across categories. You can generate any type of PivotChart report except for XY (Scatter), Stock, or Bubble.

You can create a PivotChart manually by choosing Insert | PivotChart. To create a PivotChart report programmatically, use the SetDataSource method of the PivotChart object and specify a reference to the PivotTable range. The

PivotTable object has the following two properties that return ranges representing part or all of the PivotTable report:

- TableRange1—Returns a range representing the PivotTable report without page fields
- TableRange2—Returns a range representing the entire PivotTable report

The procedures in Hands-On 20.9 generate a PivotTable report from the Microsoft Access sample Northwind database. Another procedure in this Hands-On will set up a PivotChart based on the PivotTable's data.

Hands-On 20.9   Creating a PivotTable and PivotChart Reports

1. Save and close the Chap20c.xlsm workbook file you created in the previous Hands-On.
2. Create a new workbook file and save it as Chap20d.xlsm in your C:\ VBAExcel2024_ByExample folder.
3. Switch to the VBE screen and select VBAProject(Chap20d.xlsm) in the Project Explorer.
4. Choose Insert | Module and enter the GeneratePivotReport procedure as shown below:

```
Sub GeneratePivotReport()
    Dim strConn As String
    Dim strSQL As String
    Dim myArray As Variant
    Dim destRng As Range
    Dim strPivot As String

    strConn = "Driver={Microsoft Access Driver & _
        (*.mdb, *.accdb)};"
        "DBQ=C:\VBAExcel2024_ByExample\Northwind.mdb;"

    strSQL = "SELECT Invoices.Customers.CompanyName, " & _
        "Invoices.Country, Invoices.Salesperson, " & _
        "Invoices.ProductName, Invoices.ExtendedPrice " & _
        "FROM Invoices ORDER BY Invoices.Country"

    myArray = Array(strConn, strSQL)
    Worksheets.Add

    Set destRng = ActiveSheet.Range("B5")
    strPivot = "PivotTable1"
```

```
    ActiveSheet.PivotTableWizard _
        SourceType:=xlExternal, _
        SourceData:=myArray, _
        TableDestination:=destRng, _
        TableName:=strPivot, _
        SaveData:=False, _
        BackgroundQuery:=False

    With ActiveSheet.PivotTables(strPivot)& _
        .PivotFields("ProductName")
        .Orientation = xlPageField
        .Position = 1
    End With


  With ActiveSheet.PivotTables(strPivot).PivotFields("Country")
        .Orientation = xlRowField
        .Position = 1
    End With

    With ActiveSheet.PivotTables(strPivot)& _
        .PivotFields("Salesperson")
        .Orientation = xlColumnField
        .Position = 1
    End With

    ActiveSheet.PivotTables(strPivot).AddDataField _
    ActiveSheet.PivotTables(strPivot)& _
    .PivotFields("ExtendedPrice"), _
    "Sum of ExtendedPrice", xlSum

    With ActiveSheet.PivotTables(strPivot). _
  PivotFields("Sum of ExtendedPrice").NumberFormat = "$#,##0.00"
    End With
End Sub
```

5. Run the `GeneratePivotReport` procedure.
Excel adds a new worksheet with a PivotTable to the current workbook, as shown in Figure 20.18.

Country	Andrew Fuller	Anne Dodsworth	Janet Levering	Laura Callahan	Margaret Peacock	Michael Suyama	Nancy Davolio	Robert King	Steven Buchanan	Grand Total
Argentina	477	944.5	319.2	2750.5	1329.4	76	688.7	1535.8		8119.1
Austria	16603.08	8967.6	23941.35	10979.59	17959.66	6728.93	17087.27	25745.15		128003.63
Belgium	2666.5	2808.37	295.38		13597.2	1209	732.6	4841.5	7674.3	33824.85
Brazil	9985.03	1910	9192.59	11118.58	17770.57	8444.87	29459.37	6200.2	14707.97	108789.18
Canada	9034.5	986.8	12196.73	1278.4	4826.05	3412.83	8801.41	9719.96		50196.28
Denmark	2345.7		1684.27	1814.35	17291.81	736	6874	2114.85		32661.01
Finland	5876.93	1590.56	957.86	4131.8	2117.7	270	1826.9	642	1833.2	19259.05
France	9434.28	3828.73	15471.13	5356	24340.8	4470.38	12487.44	2985.9	2543.85	80918.31
Germany	53627.17	15763.53	45976.81	26497.97	36836.67	7953.53	24611.38	7608.93	8046.54	220916.53
Ireland	10604.98	7403.9	16615.04	1313.82	1366.4	7658	2519	2598.76		49979.9
Italy	5422.05	563.8		88	2078.86	3686	55.2	1138.44	1025	15770.15
Mexico	2190.65		3076.9	988.8	6706.1		5147.48	4223.06	1249.1	23582.07
Norway	622.35		2684.4				1728.4	700		5735.15
Poland				666	1019.1		808	858.86		3531.96
Portugal	1411	57.8	987.5	2893.4	4454.58	206		285.12	1274.72	12683.36
Spain	977.5	224	3375.25	206	7729.89		1241	1861.1	2368.46	17983.2
Sweden	8036.7	4879.75	11520.4	9303.52	1826.3	3240.62	7491.18	6395.44	1801.23	54495.14
Switzerland		2949.24	5049.08	496.1	5282.08	3431.8	4135.5	9790.24	556.62	31892.64
UK	3411.4	6836.55	4808.1	9319.8	8751.11	4527.5	10946.73	5469.43	5101.68	58971.3
USA	20643.35	17224.71	33564.48	25755.46	47889.48	17816.95	43761.31	23850.45	15172.05	245676.24
Venezuela	2966.48	378	11246.37	9900.3	8109.95	3173.52	9252.38	7176.7	4807.93	56810.63
Grand Total	166537.75	77308.04	202812.82	126862.27	232890.03	73913.13	192107.56	124568.22	68792.25	1265792.9

ProductName (All)
Sum of ExtendedPrice | Salesperson

FIGURE 20.18. This PivotTable report will be used to graph data in the PivotChart report.

6. In the same code module where you entered the `GeneratePivotReport` procedure, enter the code of the `CreatePivotChart` procedure, as shown below:

```
Sub CreatePivotChart()
    Dim shp As Shape
    Dim rngSource As Range
    Dim pvtTable As PivotTable
    Dim r As Integer

    Set pvtTable = Worksheets(ActiveSheet.Name).PivotTables(1)

    ' set the current page for the PivotTable report to the
    ' page named "Tofu"
    pvtTable.PivotFields("ProductName").CurrentPage = "Tofu"

    Set rngSource = pvtTable.TableRange2
    Set shp = ActiveSheet.Shapes.AddChart

    shp.Chart.SetSourceData Source:=rngSource
    shp.Chart.SetElement (msoElementChartTitleAboveChart)
    shp.Chart.ChartTitle.Caption = _
        pvtTable.PivotFields("ProductName").CurrentPage

    r = ActiveSheet.UsedRange.Rows.Count + 3

    With Range("B" & r & ":E" & r + 15)
        shp.Width = .Width
        shp.Height = .Height
        shp.Left = .Left
```

```
        shp.Top = .Top
    End With
End Sub
```

The `CreatePivotChart` procedure changes the current page for the PivotTable report to display information about the product named Tofu. The `AddChart` method of the `Shapes` collection is used to create a `Chart` object.

The `SetSourceData` method of the `Chart` object is then used to specify the `PivotTable` range as the chart's data source. It's always a good idea to add a chart title, so the next two lines of code make sure that the title is positioned above the chart area and its text is set to the current product name in the PivotTable.

To ensure that the chart appears just below the PivotTable report, we calculate the active worksheet's used range and add to it three rows. The `Top`, `Left`, `Width`, and `Height` properties are used to position the chart over the specified range.

7. Run the `CreatePivotChart` procedure.

The resulting PivotChart report is shown in Figure 20.19.

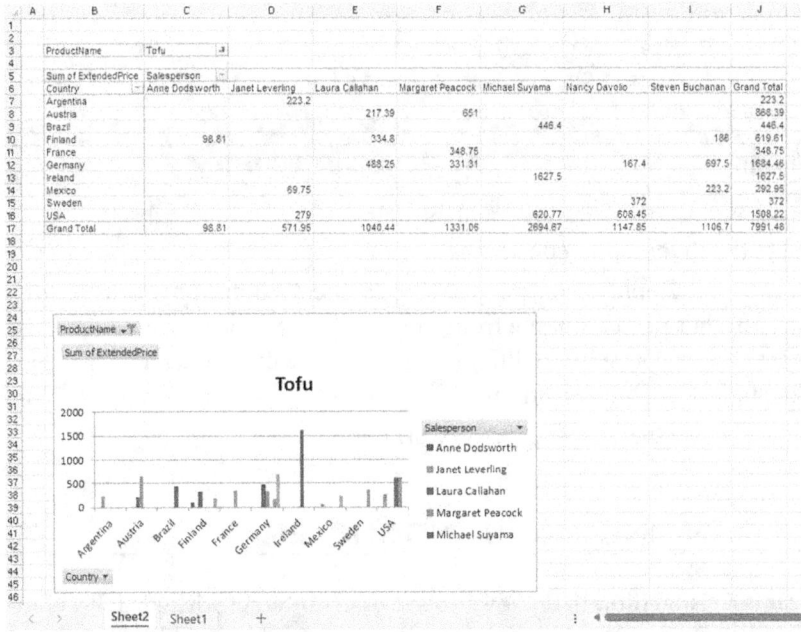

FIGURE 20.19. The PivotChart report is generated from the PivotTable report data embedded in the same worksheet.

To ensure that the chart title changes when you select a different product in the Product Name field of the PivotTable report, you must create the `Worksheet_PivotTableUpdate` event procedure. If your sheet with the PivotChart has a different number, use that sheet reference instead.

8. In the Project Explorer window, double-click the sheet number containing the PivotChart under VBAProject (Chap20d.xlsm) and enter the following event procedure in the Code window:

```
Private Sub Worksheet_PivotTableUpdate(ByVal Target As
PivotTable)
    Dim strPivotPage As String
    Dim r As Integer

    strPivotPage = Target.PivotFields("ProductName"). _
        CurrentPage.Value

    If ActiveSheet.ChartObjects.Count > 0 Then
        ActiveSheet.ChartObjects(1).Activate
        ActiveChart.ChartTitle.Text = strPivotPage

        r = ActiveSheet.UsedRange.Rows.Count + 3

        With Range("B" & r)
            ActiveSheet.ChartObjects(1).Top = .Top
    End With
    End If
End Sub
```

The above event procedure will be triggered automatically each time you update the PivotTable report.

9. In the Excel Application window in the sheet containing your PivotChart, select another product name from the PivotTable ProductName field.
   Notice that as you select a different product, the chart data and the chart title adjust to reflect your selection.

10. Save and close the `Chap20d.xlsm` workbook.

UNDERSTANDING AND USING SLICERS

A *slicer* is a visual filter that allows you to easily interact with data. Previously, in Chapter 19, we discussed how slicers are used with Excel tables. Slicers can also be applied to PivotTable and PivotChart reports. When you select values from the slicers, Excel automatically adjusts the PivotTable or PivotChart. Before you can start working with slicers, you need to create a PivotTable or PivotChart associated with a data source. Slicers are based on the row labels in your Pivot-Table. Depending on the analysis you want to perform, you can use one or more slicers.

Creating Slicers Manually

In the next Hands-On, you will learn how slicers are applied to PivotTables using the Insert Slicer button available on the ribbon's PivotTable Tools tab. We will reuse the `EquipmentList.xlsx` workbook that you worked with in Hands-On 20.1. This file already contains a PivotTable that's ready to go and is associated with the data stored in the Source Data sheet.

Hands-On 20.10   Creating a Slicer Using the Ribbon

1. Open the `EquipmentList.xlsx` workbook file and save it as `EquipmentListSlicers.xlsm` in the `VBAExcel2024_ByExample` folder. Be sure to save the file in the macro-enabled format.
2. Select Row Labels in the PivotTable report located on the PivotReport sheet and click the PivotTable Analyze | Insert Slicer button in the ribbon's Filter group, as shown in Figure 20.20.

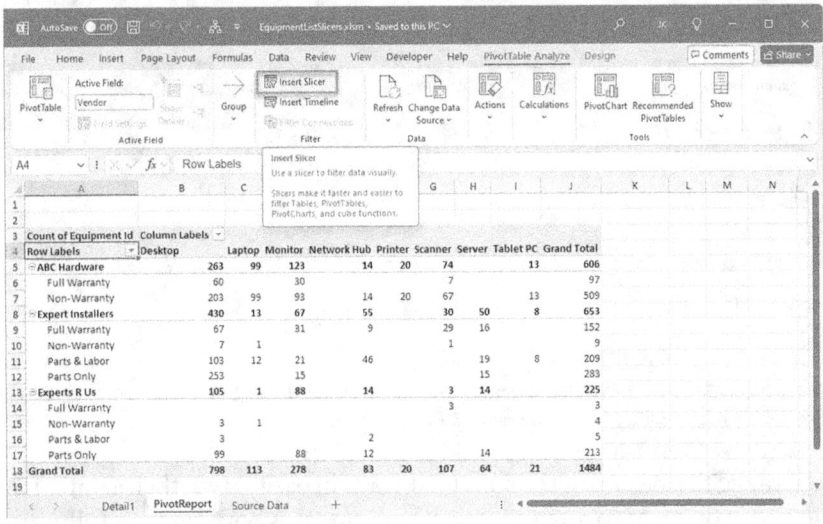

FIGURE 20.20.   Inserting a slicer using the Insert Slicer button.

The Insert Slicers dialog box appears with the list of all the fields that are available in the PivotTable.

3. Select the labels as shown in Figure 20.21 (Vendor, Equipment Type, and Warranty Type) and click OK.

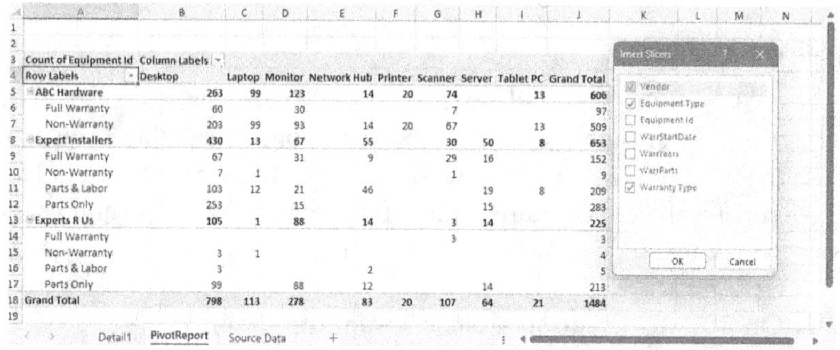

FIGURE 20.21. Selecting labels for slicers.

Excel creates a slicer for each label you selected (Figure 20.22).

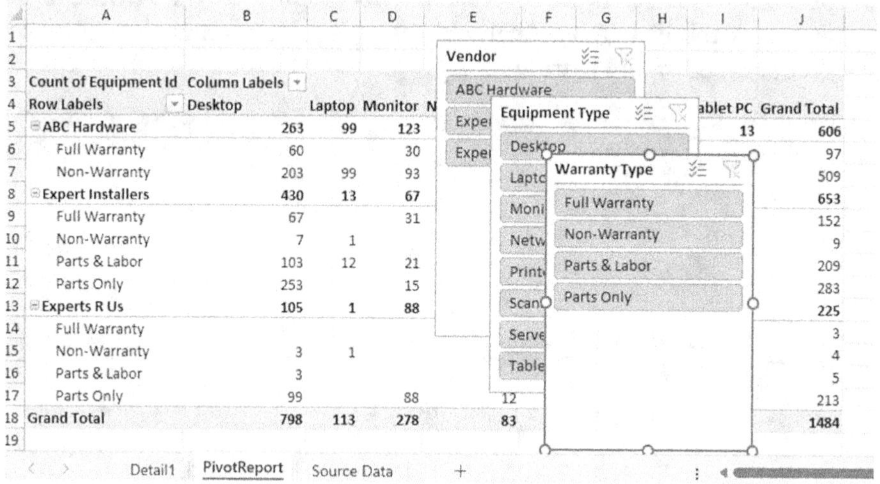

FIGURE 20.22. Each field label selected in the Insert Slicers dialog box (Figure 20.21) gets one slicer with a relevant list of values found in the data source.

4. Rearrange the slicers on the screen so you can easily view their content. Suppose you want to analyze which equipment is covered by a full warranty. Let's find out how slicers work together to give you immediate feedback on the data.

5. In the Warranty Type slicer, select Full Warranty. Notice that not only does the PivotTable adjust automatically but the other slicers automatically disable values that don't meet the selected criteria. In this case, the Equipment Type slicer tells us that we no longer have laptops, printers, and tablet PCs covered by a full warranty (see Figure 20.23).

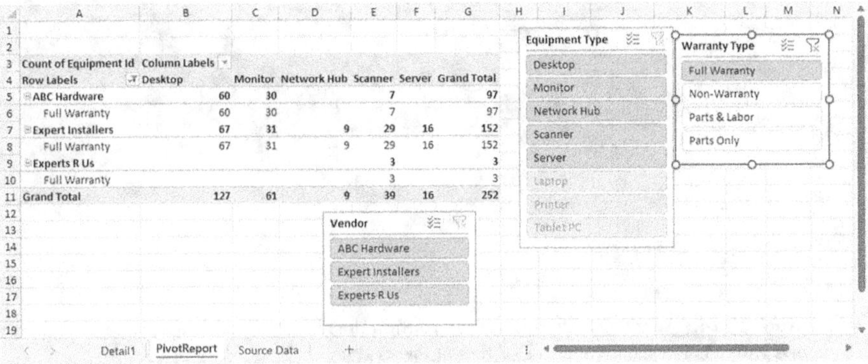

FIGURE 20.23. Using slicers to filter and analyze data.

You can continue drilling down on your data by clicking on different values shown in the slicers or remove the filters by clicking the filter icon in the top-right corner of each slicer. You can select non-consecutive items in the slicer by holding down the Ctrl key or select a series of sequential items by holding down the Shift key.

6. You can also explore different options that are available in the Slicer tab on the ribbon. Because slicers are `Shape` objects, it's easy to move, resize, or delete them. You can change the look of slicers by applying different styles to each one. The items in the slicer can be laid out in one or more columns. Simply select the slicer you want to change and look for the Columns box in the Buttons group of the Slicer tab.

7. You can control one or more PivotTables using the same set of slicers. You can also manage which PivotTables the slicer is connected to by clicking the Report Connections button in the Slicer tab (Figure 20.24). This button itself doesn't display the current filters used; instead, it helps manage the connections between slicers and PivotTables. If you want to see the current filters applied, you would need to look at the slicer itself or the filtered PivotTable.

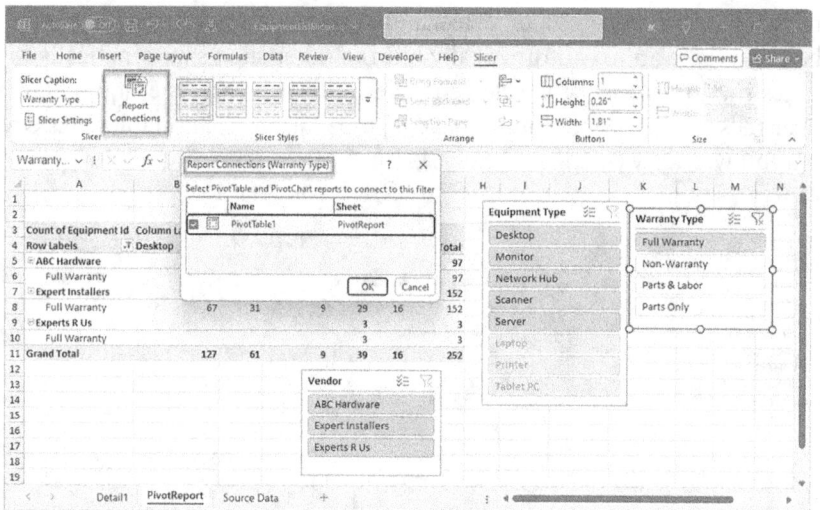

FIGURE 20.24. Working with the Slicer options on the ribbon. The Report Connections button allows you to link or unlink slicers to different PivotTables.

8. Save the changes you made in the `EquipmentListSlicers` workbook and keep the file open for the next Hands-On.

Working with Slicers Using VBA

In the previous Hands-On, you manually added three slicers to visually filter the PivotTable report. Each of these added slicers is represented by a `Slicer` object. Each `Slicer` object belongs to a workbook's `Slicers` object collection. Recall that before you were able to add slicers to your worksheet, you had to have an existing PivotTable report associated with a data source.

In VBA, before you can add a slicer, you must first define a slicer cache at the workbook level. Use the `Add` method of the `SlicerCaches` object collection to define a slicer cache. To do this, specify the name of the PivotTable from which the slicer will be created and the name of the column header of the field the slicer will be based on, as in the following code snippet:

```
Dim oSlicerCache As SlicerCache
Set oSlicerCache = ActiveWorkbook.SlicerCaches _
    .Add(Source:=ActiveSheet. _
    PivotTables(1), SourceField:="WarrYears")
```

A slicer cache can have multiple slicers. Once you have defined a slicer cache, you are ready to add the slicer. Use the `Add` method of the `Slicers` object like this:

```
Dim oSlicer as Slicer
```

```
Set oSlicer = oSlicerCache.Slicers.Add( _
SlicerDestination:=ActiveSheet, _
    Name:="Warranty Years", Caption:="Warranty Years", _
    Top:=14.6551181102362, Left:=481.034409448819)
```

The first argument of the `Slicers.Add` method specifies the sheet where the slicer should be placed. All the other arguments listed in the above statement are optional. The `Name` argument specifies the name of the slicer. If omitted, Excel will automatically generate a name for the slicer. The name must be unique across all slicers within a workbook. The `Caption` argument is the name that appears in the header of the slicer. The `Top` argument (in points) specifies the initial vertical position of the slicer relative to the upper-left corner of cell A1 on the worksheet.

The `Left` argument (in points) specifies the initial horizontal position of the slicer relative to the upper-left corner of cell A1 on the worksheet. You can also specify the initial width and height of the slicer by including the optional `Width` and `Height` arguments.

To best position your slicer on the worksheet, record a macro to get the required settings for your statement. By default, slicers are created with one column; however, you can easily change the number of columns like this:

```
oSlicer.NumberOfColumns = 3
oSlicer.Height = 50
```

When creating a slicer, you may want to specify which button in the slicer should be activated. To do this, you'll need to access the `SlicerItem` object.

To access the `SlicerItems` collection that represents all the items in a slicer for a PivotTable, use the `SlicerItems` property of the `SlicerCache` object that is associated with the `Slicer` object. For example, the following code ensures that only items with the value of 3 are selected:

```
Dim oItem As SlicerItem

With ActiveWorkbook.SlicerCaches("Slicer_WarrYears")
    For Each oItem In .SlicerItems
        If oItem.Value = "3" Then
            oItem.Selected = True
        Else
            oItem.Selected = False
        End If
    Next
End With
```

To remove the filter from the `Slicer` object, use the following code:

```
ActiveWorkbook.SlicerCaches("Slicer_WarrYears").ClearManualFilter
```

Creating Slicers Using VBA

Let's use VBA to add a fourth slicer to the PivotReport worksheet in the open **EquipmentListSlicers** workbook.

Hands-On 20.11 Creating a Slicer Using VBA

1. Switch to the VBE window and insert a new module into VBAProject (EquipmentListSlicers.xlsm).
2. In the Code window, enter the AddSlicer procedure as shown in the following:

```
Sub AddSlicer()
    Dim oSlicerCache As SlicerCache
    Dim oSlicer As Slicer
    Dim oItem As SlicerItem

    Set oSlicerCache = ActiveWorkbook.SlicerCaches.Add( _
        Source:=ActiveSheet.PivotTables(1), _
        SourceField:="WarrYears")

    Set oSlicer = oSlicerCache.Slicers.Add( _
        SlicerDestination:=ActiveSheet, _
        Name:="Warranty Years", Caption:="Warranty Years", _
        Top:=14.6551181102362, Left:=481.034409448819)

    oSlicer.NumberOfColumns = 3
    oSlicer.Height = 50

    With ActiveWorkbook.SlicerCaches("Slicer_WarrYears")
        For Each oItem In .SlicerItems
            If oItem.Value = "3" Then
                oItem.Selected = True
            Else
                oItem.Selected = False
            End If
        Next
    End With
End Sub
```

3. Run the AddSlicer procedure and then switch to the Excel application window to view the result (see Figure 20.25).

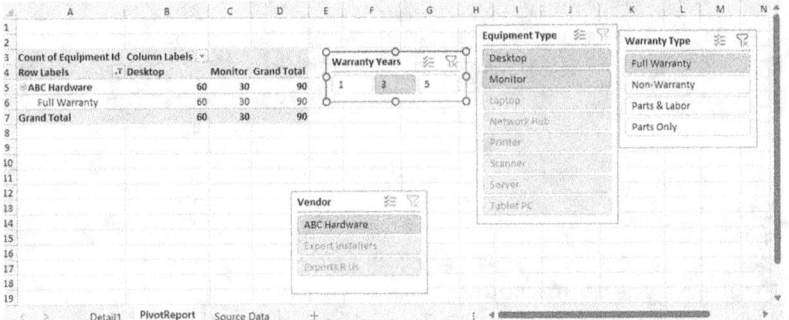

FIGURE 20.25. The Warranty Years slicer was added to this worksheet via a VBA procedure.

Retrieving Information About Slicers

Hands-On 20.12 demonstrates a procedure that retrieves information about slicers.

⊚ Hands-On 20.12 Retrieving Information About Slicers

1. In the VBE window, insert a new module into VBAProject (EquipmentList-Slicers.xlsm).

2. In the Code window, enter the ListSlicers procedure as shown below:

```
Sub ListSlicers()
    Dim oSlicerCache As SlicerCache
    Dim oSlicerCaches As SlicerCaches
    Dim oSlicer As Slicer
    Dim cnt As Integer

    Set oSlicerCaches = ActiveWorkbook.SlicerCaches
    cnt = oSlicerCaches.Count

    If cnt > 0 Then
        For Each oSlicerCache In oSlicerCaches
            Debug.Print "Slicer Cache Index|Name:" & _
                oSlicerCache.Index & "|" & oSlicerCache.Name
            Debug.Print "Source Type: " & oSlicerCache.SourceType
            For Each oSlicer In oSlicerCache.Slicers
                Debug.Print vbTab & "Name:" & oSlicer.Name
                Debug.Print vbTab & "Caption:" & oSlicer.Caption
                Debug.Print vbTab & "Cols:" & & _
                    oSlicer.NumberOfColumns
                Debug.Print vbTab & "Col Width:" & & _
                    oSlicer.ColumnWidth
                Debug.Print vbTab & "Height:" & oSlicer.Height
```

```
                    Debug.Print vbTab & "Top:" & oSlicer.Top
                    Debug.Print vbTab & "Left:" & oSlicer.Left
                    Debug.Print vbTab & "Style:" & oSlicer.Style
                    Debug.Print vbTab & "Cache level:" & _
                        oSlicer.SlicerCache.CrossFilterType
                Next
            Next
        End If
    End Sub
```

3. Run the `ListSlicers` procedure.
Check the procedure output in the Immediate window.

Deleting Slicers

It's quite simple to delete a slicer you no longer need. For example, to remove the Warranty Years slicer that was added by the `AddSlicer` procedure in Hands-On 20.11, you would run the following code (try this out from the Immediate window):

```
ActiveWorkbook.SlicerCaches("Slicer_WarrYears").Delete
```

You can also delete the slicer by accessing the `Shapes` collection like this:

```
ActiveSheet.Shapes.Range(Array("Warranty Years")).Delete
```

Moving Slicers

You can move slicers to a different sheet using VBA. The `MoveSlicers` procedure shown below removes three slicers from the `PivotReport` sheet and places them on another sheet of the current worksheet. Next, the procedure renames the newly inserted sheet `Slicers` and arranges the workbook sheets so that it is possible to view the changes in the PivotTable when slicing the data from the `Slicers` sheet:

```
Sub MoveSlicers()
    ActiveSheet.Shapes.Range(Array("Vendor", _
        "Equipment Type", "Warranty Type")).Select
    Selection.Cut
    Sheets.Add
    Range("b3").Select
    With ActiveSheet
        .Name = "Slicers"
        .Paste
    End With

    'arrange windows
    With ActiveWindow
```

```
        .DisplayGridlines = False
        .DisplayHeadings = False
        .NewWindow
    End With
    Sheets("PivotReport").Select
    ActiveWorkbook.Windows.Arrange ArrangeStyle:=xlVertical
End Sub
```

DATA MODEL FUNCTIONALITY AND PIVOTTABLES

An Excel feature called Data Model makes it possible to work with disparate data sources simultaneously in the same PivotTable and PivotChart reports. With the Data Model built directly into Excel, you can manage various data connections, import millions of rows from multiple data sources, and create relationships between multiple tables. When you use the built-in Data Model, not only is the data processed very fast but it is also highly compressed, so you don't need to worry about handling a large-sized workbook file.

Hands-On 20.13 walks you through the steps required to create a Data Model. In this example, you will relate three tables from the Northwind 2007 database (`Products`, `Orders`, and `Order Details`) to analyze product sales by city.

> ### (◉) Hands-On 20.13   Creating a Data Model and Exposing Its Data Through a PivotTable

1. Open a new workbook and save it as `Chap20_DataModel.xlsm` in the `C:\VBAExcel2024_ByExample` folder.
2. Choose Insert | PivotTable | From External Data Source.
   The Create PivotTable dialog box appears, as shown in Figure 20.26.

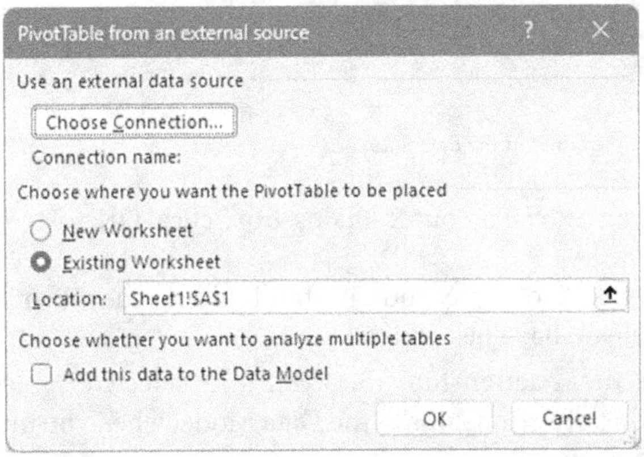

FIGURE 20.26.   Using the external data source for a PivotTable.

3. Click the Choose Connection… button.
4. In the Existing Connections dialog box, click the `Browse for More…` button.
5. In the Select Data Source dialog box, choose C:\VBAExcel2024_ByExample\ Northwind 2007.accdb from the File name dropdown and click Open.
   Excel displays the Select Table dialog box listing all the tables that are available in the chosen database.
6. Click the Enable selection of multiple tables checkbox. Place a check next to the Order Details, Orders, and Products tables and click OK (see Figure 20.27). When you click OK, you are returned to the PivotTable from the external data source dialog box. Notice that Excel has placed the connection name Northwind 2007 below the Choose Connection… button, and the setting Add this data to the Data Model is now checked (Figure 20.28).

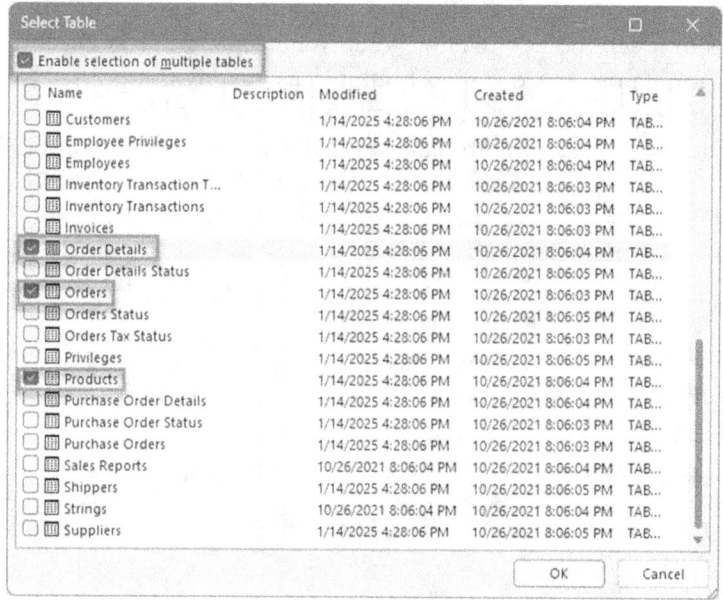

FIGURE 20.27. Selecting multiple tables in the data source.

7. In the PivotTable from an external source dialog box, click OK to create a PivotTable.
   Now, you should see a blank PivotTable. You can start building your PivotTable based on the Data Model you have just created.
   If Excel can detect the relationships between the selected tables, it automatically recreates these relationships in the Data Model when you import all the tables in a single operation.

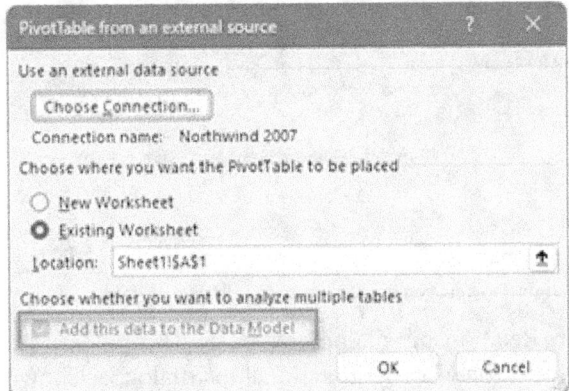

FIGURE 20.28. When you choose multiple tables or views in the Select Table dialog box (Figure 20.27), Excel adds the selected data to the Data Model. If you already have a data source connection with the same name, Excel adds a number to the end of the name, making it unique.

If Excel fails to determine how your tables are related, you will need to explicitly define table relationships before Excel can handle the data in the Data Model. This is done via the Manage Relationships dialog box available by selecting Relationships on the ribbon's PivotTable Analyze tab. Let's create relationships between tables in the Data Model.

8. Click the Relationships button on the ribbon's PivotTable Analyze tab.

9. In the Manage Relationships dialog, click the New button.

10. In the Create Relationship dialog, make selections as shown in Figure 20.29 to define the relationship between the Order Details and Products tables.

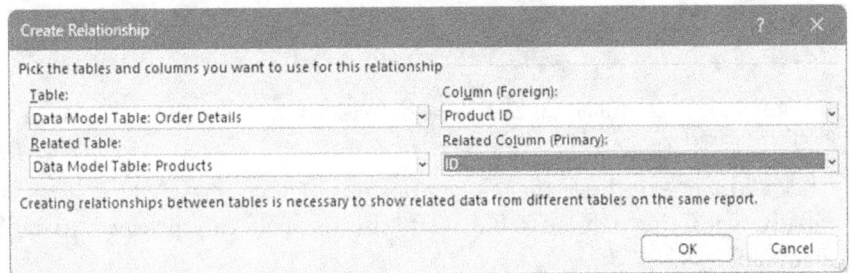

FIGURE 20.29. Defining the relationship between the Order Details and Products tables.

11. Click OK to exit the Create Relationship dialog box.
Excel adds the relationship to the Manage Relationships dialog.

12. Click the New button again to create another relationship.

13. In the Create Relationship dialog, make selections as shown in Figure 20.30 to define the relationship between the Orders and Order Details tables.

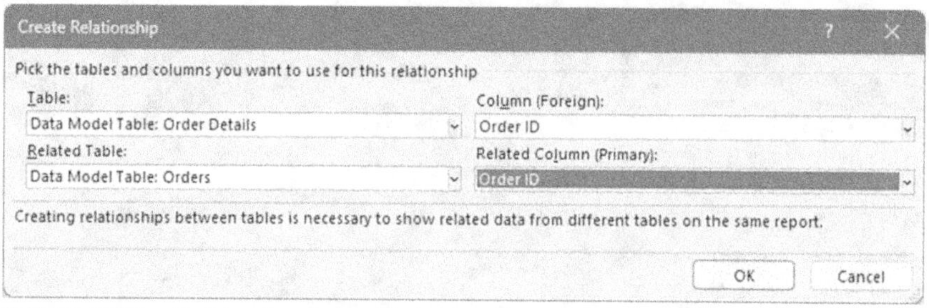

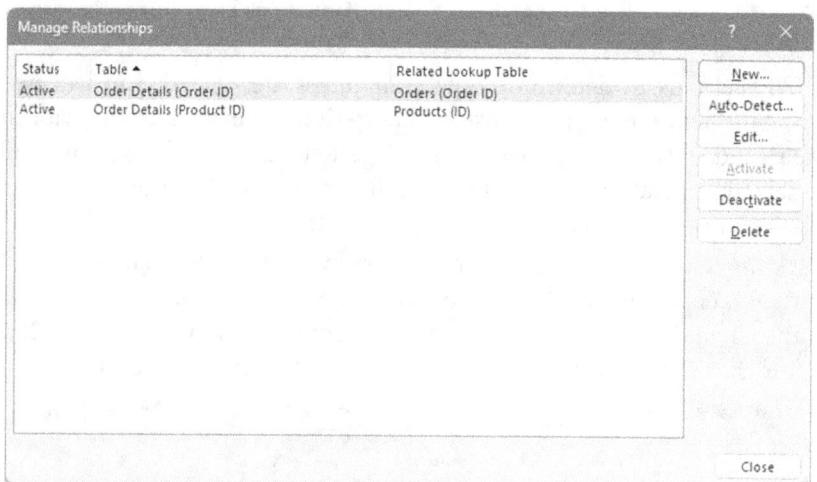

FIGURE 20.30. Defining the relationship between the Orders and Order Details tables.

14. Click OK to exit the Create Relationship dialog box.

Excel adds the relationship to the Manage Relationships dialog, as shown in Figure 20.31.

FIGURE 20.31. The relationships between tables in the Data Model are listed in the Manage Relationships dialog.

15. Click Close to exit the Manage Relationships dialog.

Finally, with the defined table relationships, you can proceed to build a PivotTable.

Let's choose fields from the PivotTable Fields list.

16. In the PivotTable Fields panel, expand the Products table and drag Product Name to the Rows area.

17. Expand the Orders table and drag Ship City to the Columns area.

18. Expand the Order Details table and drag Quantity to the Values area.

19. Make the formatting adjustments to your liking by using the buttons on the Design tab.

The completed PivotTable is shown in Figure 20.32.

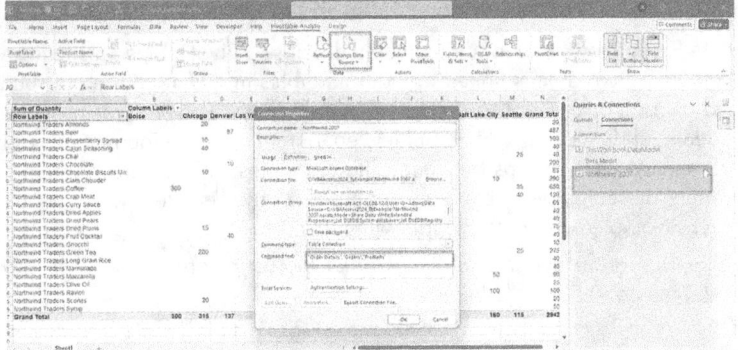

FIGURE 20.32. This PivotTable summarizes product sales by city.

> **NOTE**
>
> *At any time, you can quickly add additional database tables to the Data Model by choosing PivotTable Analyze | Change Data Source | Connection Properties. Activate the Properties dialog box by right-clicking on the Northwind 2007 database name in the Queries & Connections pane and choosing Properties. In the Connection Properties dialog box, modify the Command text section on the Definition tab (see Figure 20.33).*
>
>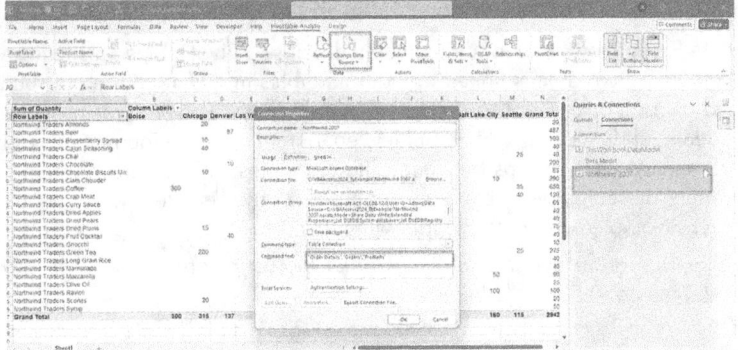
>
> **FIGURE 20.33.** Use the Connection Properties dialog box to fine-tune the settings of your data connections related to PivotTables and the Data Model in Excel. The Command text section allows you to customize the data retrieval process by writing complex queries. This can include filtering, joining tables, grouping, and other SQL operations to shape the data as needed.

Deferring PivotTable Layout Updates

If you take a close look at Figure 20.32, at the very bottom of the PivotTable Fields pane, you will see the Defer Layout Update checkbox. By placing a check in this box, you can prevent Excel from making real-time updates to your PivotTable as you work with its fields. Only click the Update button when you are ready to update your PivotTable layout. Note that this button is disabled if the Defer Layout Update checkbox is not checked. When you are done building your PivotTable, remove the check from the Defer Layout Update setting. You can also control this setting using VBA. Simply set the `ManualUpdate` property of the `PivotTable` object to `True` or `False`.

For example, the following `DeferLayoutUpdate` procedure temporarily turns off automatic updates to a PivotTable while making changes and then re-enables updates once the changes are complete. This helps to make the update process more efficient by preventing intermediate refreshes:

```
Sub DeferLayoutUpdate()
    With Sheet1.PivotTables(1)
        .ManualUpdate = True
        With .CubeFields("[Products].[Category]")
            .Orientation = xlRowField
            .Position = 1
        End With
        .ManualUpdate = False
    End With
End Sub
```

Adding Calculated Fields to the Tables in the Data Model

While Excel allows you to add calculated fields to PivotTables, this feature is not available when your PivotTable is linked to multiple tables in a Data Model. In such cases, you will need to use Power Pivot to create calculated fields. *Power Pivot* provides advanced data modeling and calculation capabilities, enabling you to define more complex calculations and manage relationships between different tables. This ensures that your PivotTable can leverage the combined data from multiple sources efficiently. For more information on Power Pivot, please refer to the following link:

https://support.microsoft.com/en-us/office/power-pivot-overview-and-learning-f9001958-7901-4caa-ad80-028a6d2432ed.

PROGRAMMATIC ACCESS TO THE DATA MODEL

In addition to the existing VBA Object Model (OM), Excel has introduced a Data Model OM to enhance the way users interact with complex data structures programmatically. This OM is designed to work with the Power Pivot Data Model, providing advanced functionality for data manipulation and analysis.

Data Model OM is a set of objects, properties, methods, and events that allow users to programmatically load, manage, and manipulate data in Excel's Data Model.

The following VBA procedure demonstrates how to use the `Model` property of the `Workbook` object to get some information about the Data Model you created in the previous section:

```
Sub GetDataModel_Info()
    Dim wkb As Workbook
    Dim tbl As Variant

    Set wkb = ActiveWorkbook

    Debug.Print "Model Name: " & wkb.Model.Name
    Debug.Print "Relationships: " & _
       wkb.Model.ModelRelationships.Count
    Debug.Print "Number of Tables: " & wkb.Model.ModelTables.Count
    Debug.Print "--TABLE NAMES--"
    For Each tbl In wkb.Model.ModelTables
        Debug.Print tbl.Name
    Next
End Sub
```

The `GetDataModel_Info` procedure displays information about the Data Model, as shown in Figure 20.34.

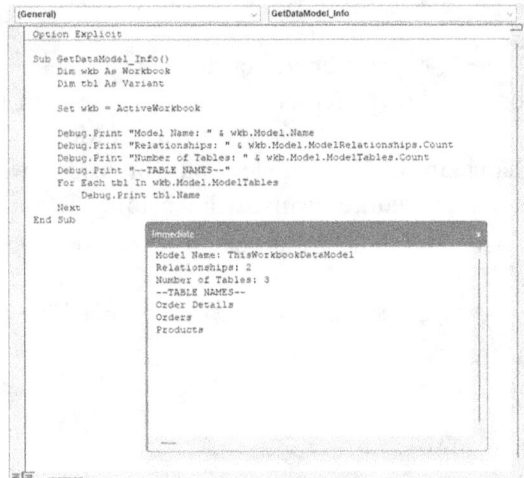

FIGURE 20.34.   Using the Model property of the Workbook object allows you to display information about the Data Model in Excel. This includes details such as the name of the model, the number of defined relationships, table names used in the Data Model, and other relevant information.

In Hands-On 20.14, you will work with the `ModelChanges` object, which contains information about which changes were made to the Data Model when the `Workbook_ModelChange` event occurred. While various changes can be made to the Data Model, this exercise focuses on detecting whether any new tables were added to the existing Data Model.

⦿ Hands-On 20.14   Detecting Changes to the Data Model.

You must complete Hands-On 20.13 prior to working with this Hands-On.

1. In the VBE window, select VBAProject(Chap20_DataModel.xlsm) and choose Insert | Module.
2. In the Module code window, enter the following procedure:

```
Sub DataModel_TableChanges()
    Dim strCmdTxt_1 As String
    Dim strCmdTxt_2 As String

    strCmdTxt_1 = """Order Details"",""Orders"",""Products"""
    strCmdTxt_2 = strCmdTxt_1 & "",""Customers"",""Employees"""

    With ActiveWorkbook.Connections("Northwind 2007") _
        .OLEDBConnection
        .CommandText = strCmdTxt_2
        .Refresh
    End With
End Sub
```

This procedure modifies the `CommandText` property of the `OLEDBConnection` object to include two additional tables (`Customers` and `Employees`) in the Data Model. The `Refresh` method tells Excel to update the Data Model with the new data. To detect the change in the Data Model, you will need to write the `Workbook_ModelChange` event procedure, as instructed below.

3. In the Project Explorer, double-click the ThisWorkbook object of `VBAProject (Chap20_DataModel.xlsm)`.
4. In the ThisWorkbook Code window, enter the following event procedure:

```
Private Sub Workbook_ModelChange( _
    ByVal Changes As ModelChanges)

    Dim colTblNames As ModelTableNames
    Dim tblCount As Long
    Dim i As Integer

    Set colTblNames = Changes.TablesAdded
    tblCount = colTblNames.Count
```

```
    If tblCount > 0 Then
        Debug.Print tblCount & " tables were added."
    Else
        Debug.Print "There are no new tables in the data model."
    End If
    For i = 1 To tblCount
        Debug.Print colTblNames.Item(i)
    Next i
End Sub
```

The `Workbook_ModelChange` event procedure is triggered when Excel detects that changes were made to the Data Model. The `Changes` variable represents the `ModelChanges` object and denotes the type of change that was made. Changes include:

- Adding, changing, and deleting columns (`ColumnsAdded`, `Column-sChanged`, and `ColumnsDeleted` properties)

- Adding, changing, deleting, renaming, and refreshing (recalculating) tables (`TablesAdded`, `TablesChanged`, `TablesDeleted`, `TableName-sChanged`, and `TablesModified` properties)

- Changing one or more relationships in the model (`RelationshipChange` property)

- Adding measures (`MeasuresAdded` property)

- Making an unknown change (`UnknownChange` property)

When tables are added to the model as part of the model operation, we use the `Changes.TablesAdded` property to find out the names of the tables that were added. This property returns a `ModelTableNames` object (Excel) collection of table names as strings representing all tables that were added to the model as part of a model operation:

```
Set colTblNames = Changes.TablesAdded
```

You can get the number of objects in the `colTblNames` collection using the `Count` property of the `ModelTableNames` object:

```
tblCount = colTblNames.Count
```

The following `For...Next` loop retrieves the names of tables that were added to the Data Model:

```
For i = 1 To tblCount
    Debug.Print colTblNames.Item(i)
Next i
```

5. Run the `DataModel_TableChanges` procedure.

This procedure will automatically trigger the `Workbook_ModelChange` procedure you wrote in the previous step. You should see the following output in the Immediate window when the procedure completes:

```
2 tables were added.
Customers
Employees
```

After running this procedure, the Customers and Employees tables should appear in the PivotTable Fields pane when you click the All tab (see Figure 20.35).

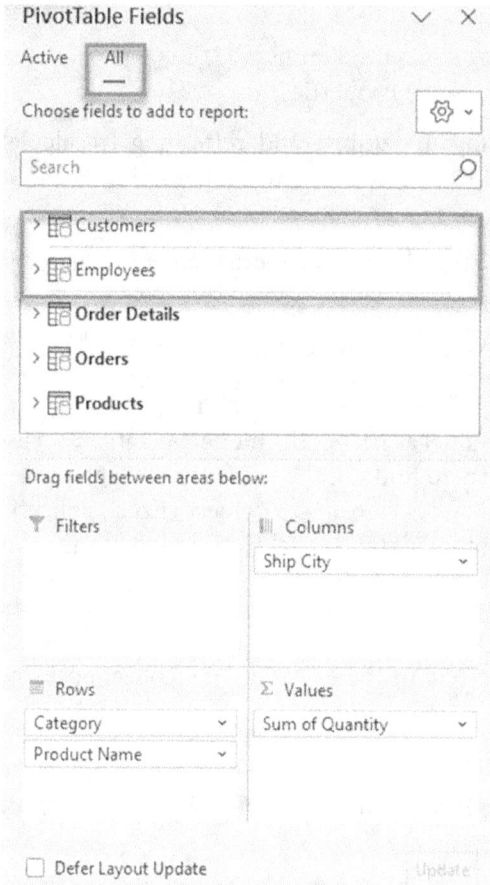

FIGURE 20.35. The PivotTable Fields pane listing additional tables.

6. Modify the `CommandText` property in the `DataModel_TableChanges` procedure by inserting the following line of code:

```
.CommandText = strCmdTxt_1
```

7. Run the `DataModel_TableChanges` procedure again.
   You have just removed the `Customers` and `Employees` tables from the Data Model. You should see only the original three tables in the PivotTable Fields pane and the following text in the Immediate window:

```
There are no new tables in the data model.
```

SUMMARY

In this chapter, you worked with two powerful Microsoft Excel objects that are used for data analysis: `PivotTable` and `PivotChart`. You learned how to use VBA to manipulate these two objects to quickly produce reports that allow you to easily examine large amounts of data pulled from an Excel worksheet range or from an external data source such as a Microsoft Access database. Additionally, this chapter demonstrated the use of the slicer feature, which makes it possible to visually filter PivotTable data.

In the last two sections of this chapter, you learned how to use the Excel Data Model to load data from multiple tables, create relationships between tables, expose the data through a PivotTable, and use several new VBA objects and properties to obtain information about the Data Model and specific changes that were made to it.

The next chapter features the data gathering, shaping, and modeling capabilities available in Excel.

Chapter 21 GETTING AND TRANSFORMING DATA

If you frequently handle data and need to automate processes such as transforming, cleaning, and loading, Excel has a robust feature that can greatly assist you. You can access this feature under the Get & Transform Data section of the ribbon's Data tab (see Figure 21.1).

With the Power Query technology integrated into Excel, you can create advanced queries that simplify the process of importing data into Excel from both external and internal sources. This tool also enables you to combine and transform data with ease.

This chapter provides a quick hands-on introduction to the data analysis and transformation process with Power Query.

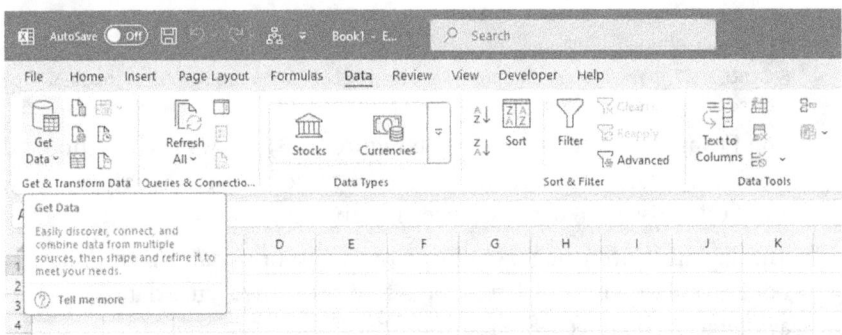

FIGURE 21.1. The Get Data button in the ribbon's Get & Transform Data group is used for loading, shaping, and refining data.

USING THE GET DATA BUTTON

When you click the Get Data button, Excel lists the types of data sources you can use for creating a query (see Figure 21.2):

NOTE	*Not all data sources shown in Figures 21.2–21.5 are available in all versions of Excel. The Hands-On exercises in this chapter rely on sources that are available with the Microsoft 365 license.*

- The From File category (see Figure 21.2) allows you to import data from a file such as Microsoft Excel Workbook, Text/CSV, XML, JSON, and PDF. In addition, you can import data into a single consolidated file from a folder containing multiple files of the same type. The latter is very convenient when you need to create a report based on data spread over numerous files.

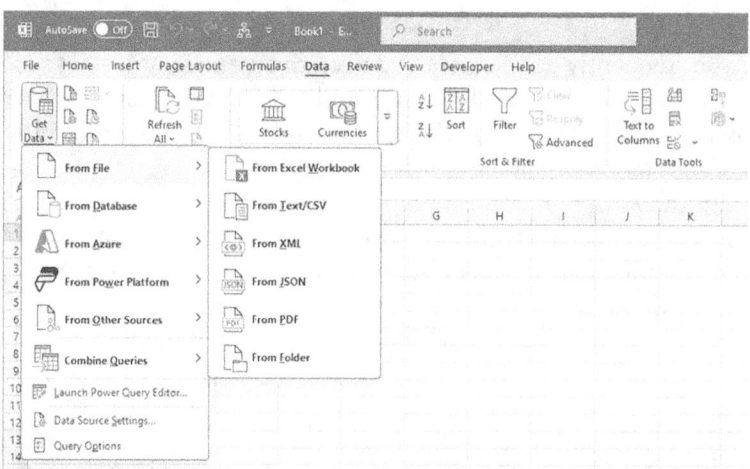

FIGURE 21.2.   You can get data from various types of files including files of the same type located in a particular folder.

- The From Database category (see Figure 21.3) allows you to import data from a database such as SQL Server, Microsoft Access, Analysis Services, and SQL Server Analysis Services Database.

- The From Azure category allows you to import data from Azure Data Lake Storage Gen2 and Azure Data Explorer. The Azure data source is not available if you have a Standard license.

- The From Power Platform category has an option to import data from Dataflows and Dataverse.

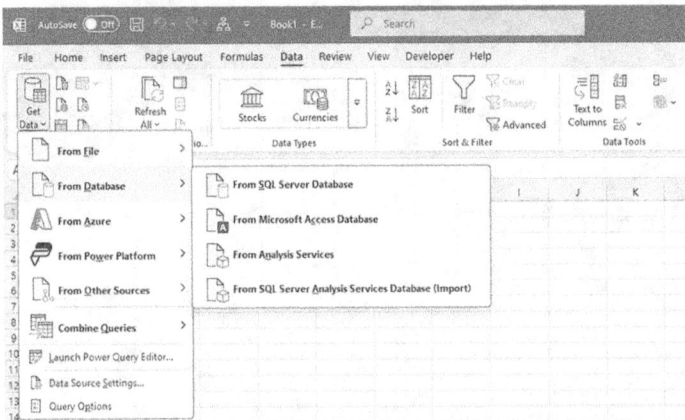

FIGURE 21.3.   You can access data from various types of databases depending on your type of Excel license.

- The From Other Sources category (see Figure 21.4) lists all the other sources that can provide data for your query. These include Excel Table/Range, Web, OData Feed, ODBC, and OLEDB. You can even start with a clean slate by choosing Blank Query.

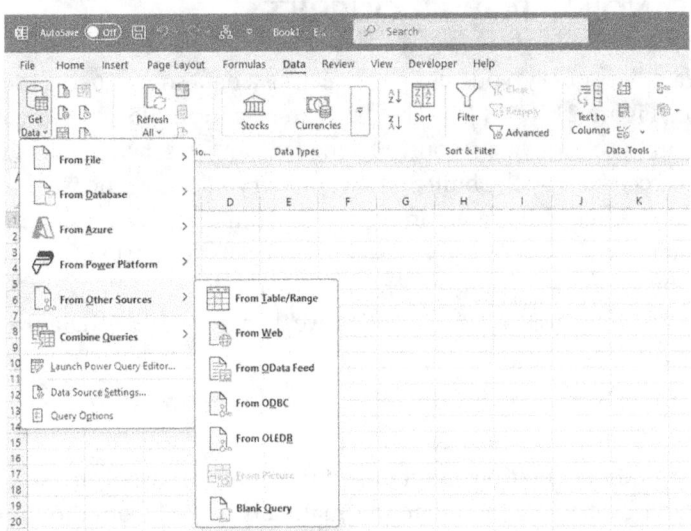

FIGURE 21.4.   A query can be based on various other data sources depending on the type of Excel license you have purchased.

- The Combine Queries category (see Figure 21.5) has options for merging and appending queries, which allow you to create complex queries.

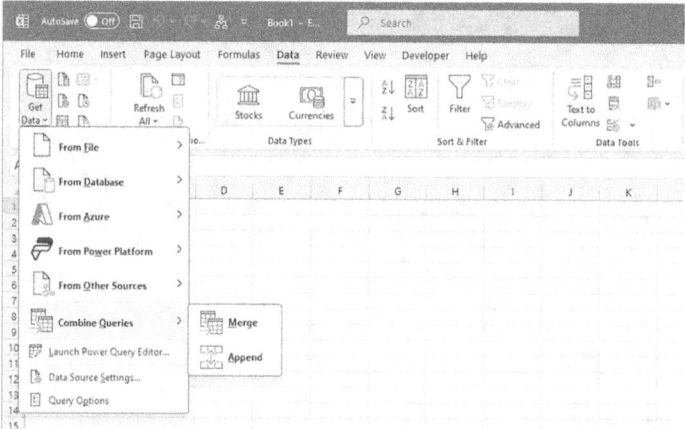

FIGURE 21.5. Queries can be combined by merging or appending.

At the bottom of the Get Data dropdown (see Figure 21.5), you will find options that will allow you to launch Power Query Editor, manage data source settings, and view query options.

UNDERSTANDING POWER QUERIES

To create a power query, you begin by selecting a data source using the Get Data button (see Figure 21.2 earlier). The data can be a local file on your computer, a file stored in the cloud, or a resource fetched via a Web service. Your data source selection is recorded by the built-in Power Query Editor as the first step in your power query. After making a connection to your data, you will be put in the Power Query Editor where you can shape your data to meet your needs.

To get hands-on experience with the Power Query user interface, let's proceed with Custom Project 21.1, in which we'll bring together data from the following sources:

- An Excel workbook (.xlsx file) containing two worksheets
- A .csv file

These files hold information about post offices in the tri-state area (NY, NJ, and CT). The source data for each state was manually copied from the United States Postal Service Web resource at

https://about.usps.com/who/profile/history/postmaster-finder/

The goal of this project is to combine and clean the data so that you can produce a summary of active and discontinued post offices by state. Although this project can be executed with a simple copy-and-paste operation, using Power Query offers a more powerful solution by automating recurring data tasks. With Power Query, the process involves creating a series of steps, so you'll never have to repeat the same actions if the data changes. Simply refresh and the data will be reprocessed automatically, saving you time and effort.

NOTE	*Please note that the files for the "Hands-On" projects can be found in the companion files.*

⊙ Custom Project 21.1   Creating a Query from Multiple Sources

1. Copy the `GetTransform` folder from the companion files to your `VBAExcel2024_ByExample` folder.
2. Launch Excel and choose New | Blank workbook.
3. Save the empty workbook as `PQ_NJ_NY.xlsx`.

   Step 1: Getting Data from an Excel Workbook
4. Choose Get Data | From File | From Excel Workbook.
5. Select the PostOffice_NY_NJ.xlsx workbook file from the GetTransform folder and click Import.
6. In the Navigator window, click the checkbox next to Select Multiple Items and select the checkboxes for NY and NJ, as shown in Figure 21.6.

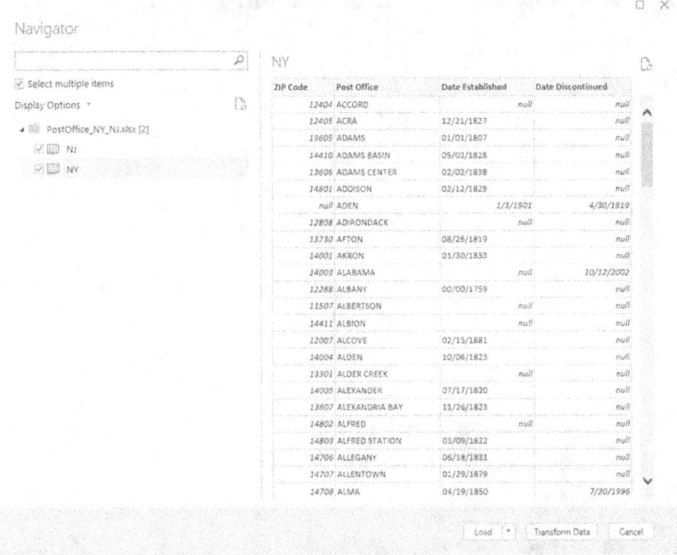

FIGURE 21.6.   The Navigator window displays data that is available in the requested data source.

Notice that the right pane displays the contents of the selected item.

7. Click the drop-down arrow next to Load and choose LoadTo.
8. In the Import Data window, click the radio button next to Table, choose New worksheet, and check the Add this data to the Data Model box (Figure 21.7).

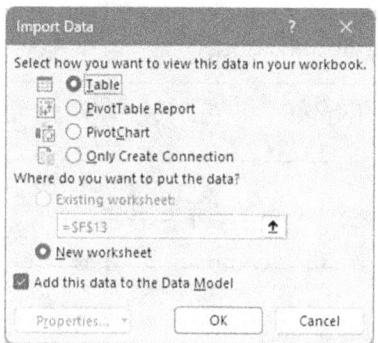

FIGURE 21.7.   In the Import Data dialog, you can specify how you want to view the data in your workbook and where it should be loaded (an existing worksheet or a new worksheet).

9. Click the OK button in the Import Data dialog.
   Excel creates two tables based on the two worksheets that you selected. The right pane displays the Workbook Queries pane with a list of queries (Figure 21.8).

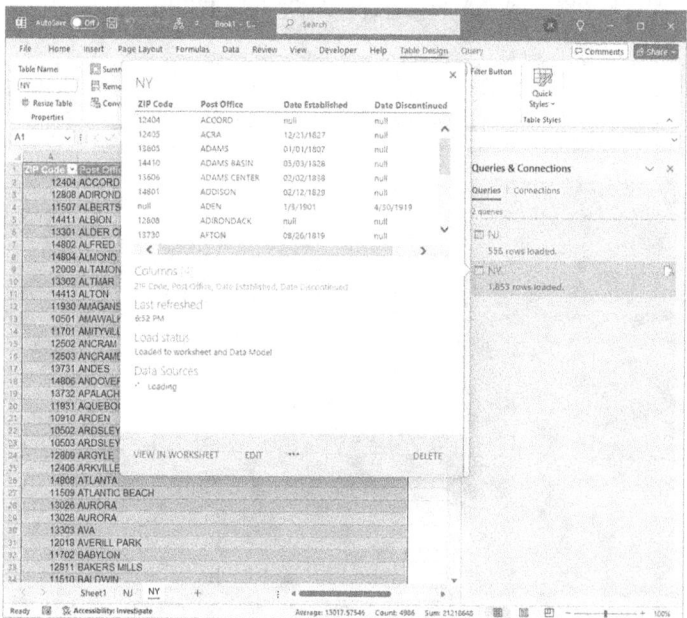

FIGURE 21.8.   The workbook with two queries.

10. To see the information about each query, point or click on the query name to view the preview screen (see Figure 21.8).

In the footer of the preview screen, you can find options that allow you to view the data in a worksheet, edit the query, or delete the query from the workbook. Notice that in the Queries & Connections pane, Excel displays the number of rows that were successfully loaded. If errors are encountered upon loading, the error count will be shown as a hyperlink so you can get more information about the error.

Looking at the initial data load in Figure 21.8, notice that the data has been converted to an Excel table for each worksheet found in the source file. The ribbon shows the Table Design tab. At this point, you could reformat your table or create a PivotTable to help in data analysis, but we have more important tasks to perform. Currently, there is no way to quickly identify the data as belonging to NY or NJ other than looking at the query names. We need the State column within each table.

Step 2: Adding, Renaming, and Moving Columns

11. In the Workbook Queries pane, right-click the NJ query and choose Edit.

The NJ - Power Query Editor window appears, showing you the four transformation steps in the APPLIED STEPS area of the Query Settings pane. Excel automatically created these steps when you loaded the data (Figure 21.9).

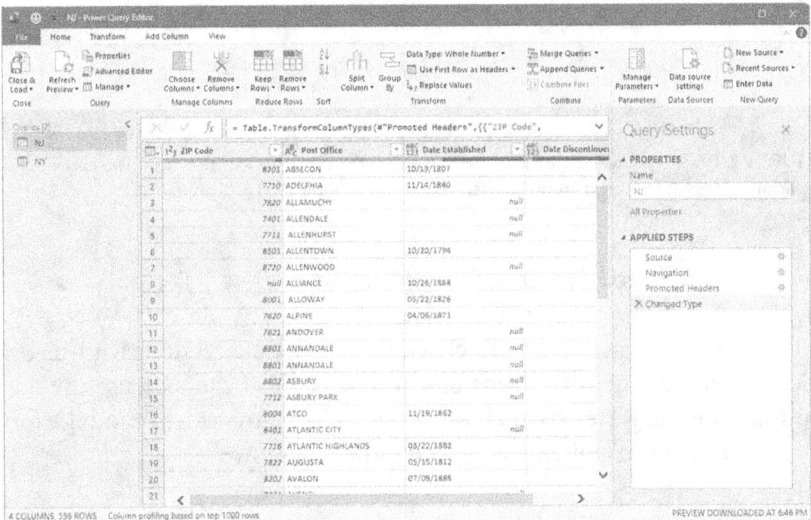

FIGURE 21.9. The NJ - Power Query Editor window displays the applied data transformation steps to the NJ table.

A step can be deleted by clicking the button (x) in front of the step name. You can use the Undo feature (Ctrl+Z) to reverse the deletion if it's the most recent action you've taken. If you've already performed other actions since deleting the step, you can manually recreate the step by following the same procedure you originally used. Some steps will show a gear icon next to them (see Figure 21.9). These steps can be edited using the same dialog that was used to create it. Simply click this icon or right-click it and choose Edit settings. The gear icon will disappear if you make an invalid change in the formula for the step. Figure 21.10 shows the dialog used to edit the Source step.

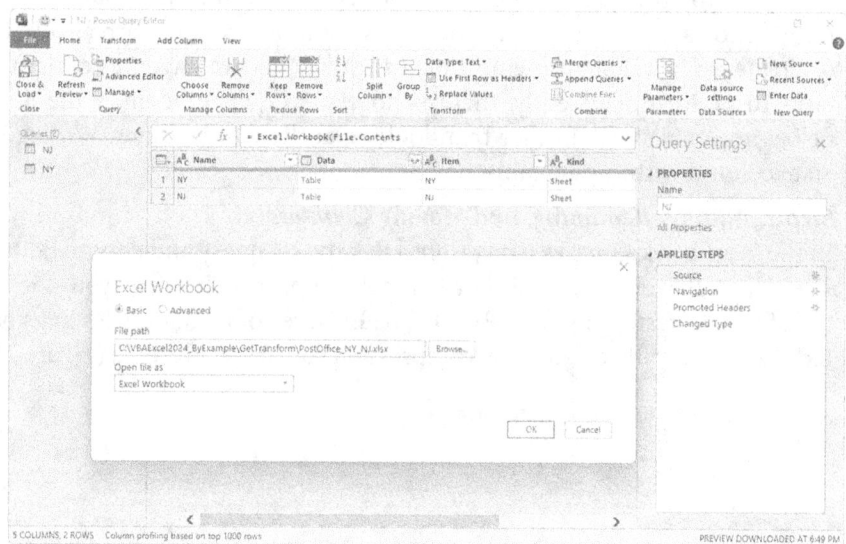

FIGURE 21.10.   The Excel dialog provides a way to edit the source of the data. You can access it by clicking the gear icon next to the Source step in the APPLIED STEPS area of the Query Settings pane.

12. Click the Promoted Headers step and choose View | Formula Bar (Figure 21.11).

You can use the formula bar to check or edit the expression that was used to perform the selected data transformation step. You can also use the formula bar to create a new step by clicking on the fx icon. You can expand the formula bar to see its entire contents by clicking the dropdown.

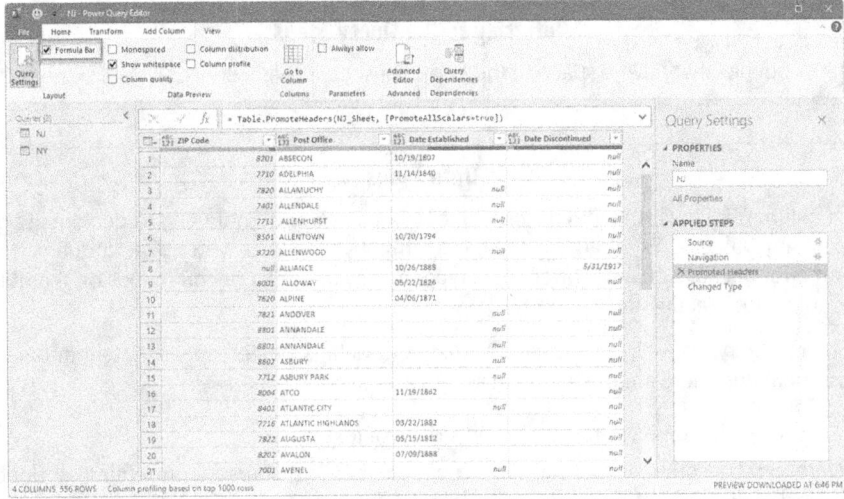

FIGURE 21.11. Examining the formula for the selected step.

13. Select the Changed Type step and notice that the formula references the previous step (Promoted Headers). See Figure 21.12.

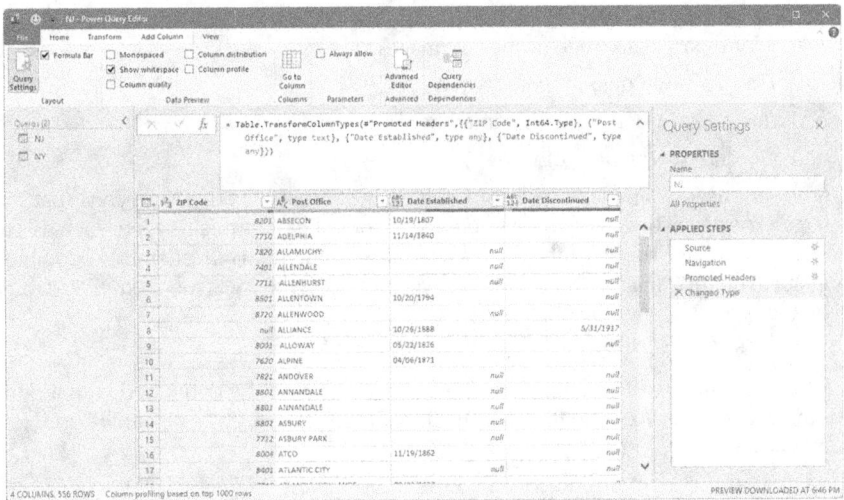

FIGURE 21.12. The expanded formula bar.

The Changed Type step automatically applies a data type to each of the columns.

Referring to the Query Steps

Steps or queries that have a space in the name must be referenced as #`"query name"`.

Data Types

Each column in Power Query has a specific data type. You do not need to declare the data type of any value as Excel automatically determines this when you create a query. If you get an error because of an incorrect data type, however, there are commands on the ribbon available for correcting this.

In Power Query, data types help ensure that your data is treated appropriately throughout transformations. Here are the main data types available:

❑ *Text*: For alphanumeric characters and strings of text
❑ *Number*: For numerical data, including decimal numbers (representing a double-precision floating-point number), fixed decimal numbers (with four decimal places), and whole numbers (integers)
❑ *Date/Time*: For date and time, including `date` values, `time` values, the combined `date` and `time` values, `date` and `time` values with timezone information, and `duration` values
❑ *Boolean*: For `True/False` values
❑ *Binary*: For binary data, such as files or images
❑ *Table*: For structured data in table format
❑ *Record*: For data structures with named fields
❑ *List*: For ordered sequences of values

Power Query uses a language called *M*, which is strongly typed. This means that the M language expects each value and function argument to be of a specific type, such as `Text`, `Number`, `Date`, and so on. If the values you pass to functions or attempt to combine in expressions do not match the expected type, errors will be generated. For example, combining text with a number such as:

```
"This is example " & 1
```

using the & operator will throw an error because text cannot be combined with a number without a conversion. To avoid the error, the function `Number.ToText` can be used to cast a number to text:

```
"This is example " & Number.ToText(1)
```

---

14. To add a new column to the current table, choose Add Column | Custom Column.

Be sure to click out of the formula bar if you find that the Add Column button is disabled.

15. Enter the data as shown in Figure 21.13 and click OK.

The Add Custom Column dialog specifies that you want to add a new column named State and fill it with the string NJ.

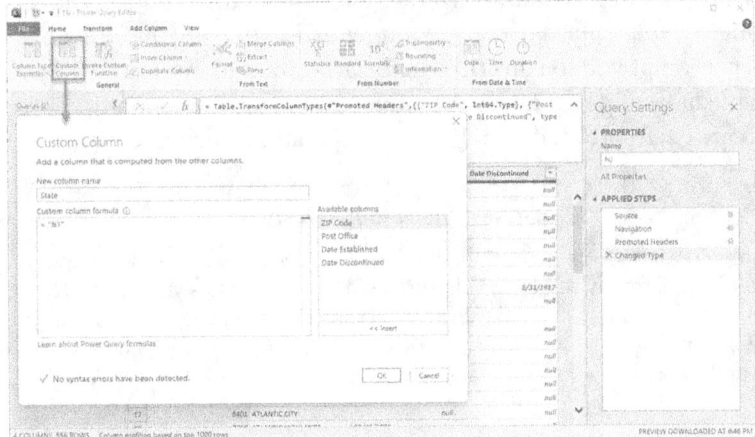

FIGURE 21.13.   Adding a new custom column.

Excel has now added the State column to our table, and we have a way to identify each zip code with the state of New Jersey (Figure 21.14).

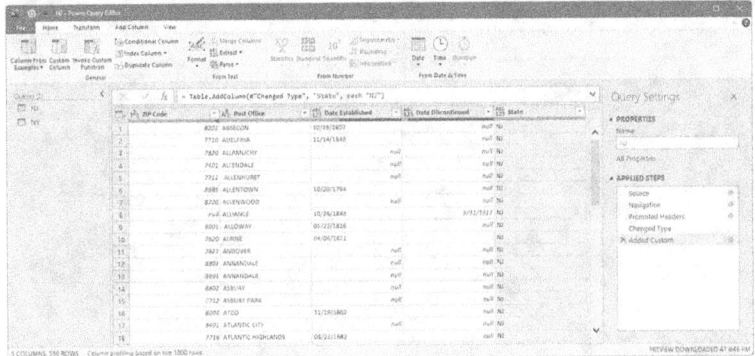

FIGURE 21.14.   The State column appears as a new transformation named Added Custom in the APPLIED STEPS area of the Query Settings pane.

> **NOTE**
> *Figure 21.14 shows four-digit zip codes in the ZIP Code column, which is incorrect. You can easily fix this by clicking the ZIP Code column heading to select the entire column, choosing Transform | Format | Add Prefix, and entering zero (0) in the dialog box. The Added Prefix step will appear in the APPLIED STEPS area of the Query Settings pane after you've done that.*

16. Right-click the Added Custom step in the APPLIED STEPS area and choose Rename. Type the new name for this step, `Added State Column`, and press Enter.

17. Right-click the State column in the data area and choose Move | To Beginning. If you followed the recommended action in the previous note, Excel may prompt you to confirm whether you want to insert a step. Inserting an intermediate step can affect subsequent steps and potentially cause your query to break. To proceed with the move operation, click OK.

Adding New Steps to the Query

When performing various operations in the Power Query Editor window, always pay attention to the step currently selected. This ensures that any changes you make are applied to the correct part of the query and helps prevent unintended modifications or errors in your data processing workflow. Keeping track of your steps will help you maintain the integrity of your queries and streamline your data transformation process.

The State column should now appear before the ZIP Code column. The APPLIED STEPS area contains a new step named Reordered Columns (see Figure 21.15).

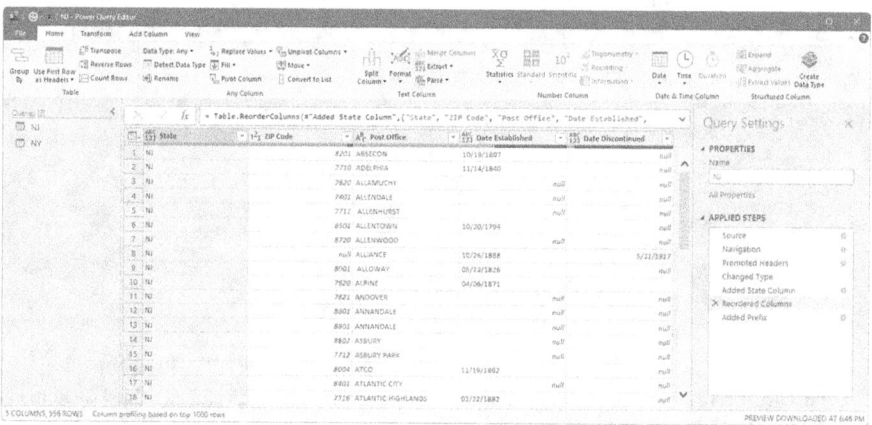

FIGURE 21.15.   The NJ - Power Query Editor window shows three additional steps that we've added in this project: Added State Column, Reordered Columns, and Added Prefix.

Let's perform the same steps in the NY query.

18. In the Queries Navigation pane on the very left of the Power Query Editor, select the NY query to open it in the Power Query Editor (see Figure 21.16).

> | | |
> |---|---|
> | **NOTE** | *The Navigation pane can be expanded or minimized by clicking on the small arrow icon on the pane's header. When expanded, the Navigation pane allows you to see a list of available queries and their structures, making it easier to manage and navigate between them. Minimizing the Navigation pane provides more screen real estate for working on your data transformations and viewing the Data Preview pane.* |

Notice that the NY query contains the initial steps that Excel created for us when we loaded the data in the Power Query Editor (Figure 21.16).

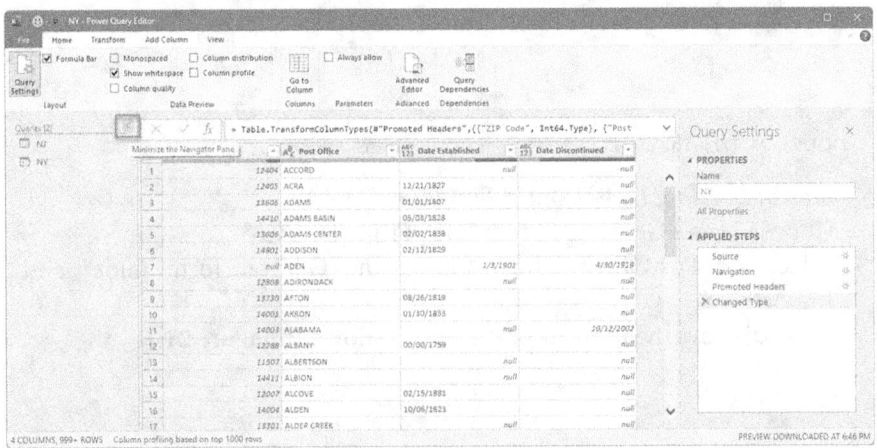

FIGURE 21.16.   Use the Query Navigation pane on the left to load the NY query into the Power Query Editor.

19. On your own, add the State column to the NY query, rename the custom query step, and move the State column to the beginning of the table.

20. Choose File | Close & Load to save the changes to the queries. This also closes the Power Query Editor window and loads the data in the default location you selected when you loaded data from the data source.

    Excel updates the tables to include the State column (Figure 21.17).

21. Save the Excel workbook with your `NY` and `NJ` queries as `PQ_NJ_NY.xlsx` in your `VBAExcel2024_ByExample` folder.

    When you save the file, Excel also saves the queries you created. Make sure to save often while working with Power Query as Excel may crash suddenly and you might lose your work.

    Let's not forget that we still must pull the Connecticut data from another data source—a CSV file.

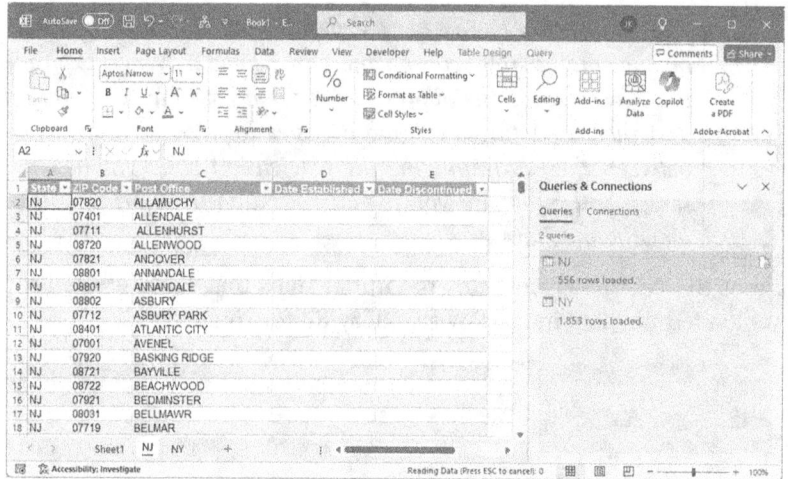

FIGURE 21.17.   The Excel table now contains the State column.

Step 3: Loading Data from a Text File

22. Choose Data | Get Data | From File | From Text/CSV.

23. Select the PostOffice_CT.csv file in the GetTransform folder and click Import.

Excel displays the preview of the file, as shown in Figure 21.18.

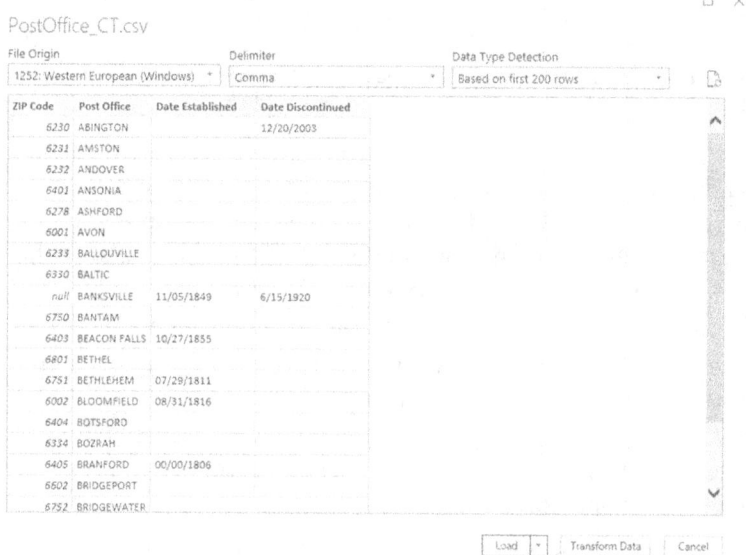

FIGURE 21.18.   Excel displays the preview screen when loading data from Text/CSV files.

24. Click the drop-down arrow next to Load and choose Load to.

25. In the Import Data dialog, choose Table, New worksheet, and click Add this data to the Data Model, then click OK.

Excel loads the data into a new worksheet of the existing PQ_NJ_NY.xlsx workbook and adds the new power query into the Queries & Connections pane (Figure 21.19).

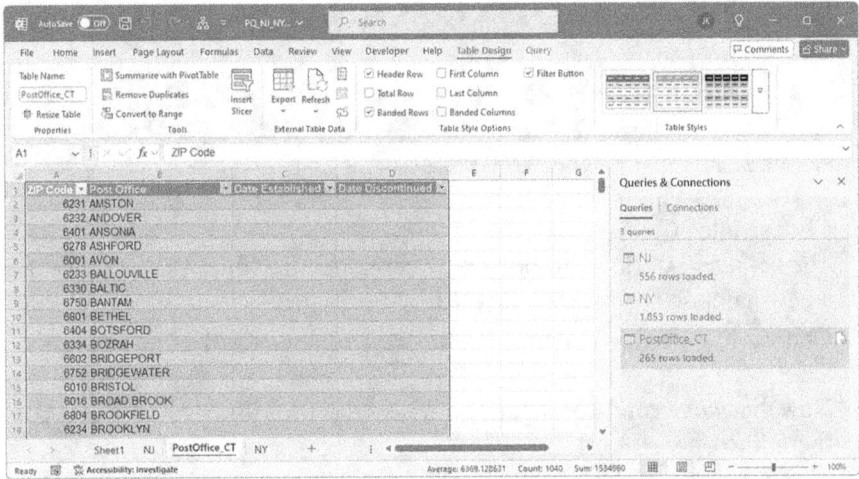

FIGURE 21.19. Excel loads and adds the new query to the Queries & Connections pane.

26. Highlight the PostOffice_CT query and click the Edit button in the preview window.

Excel loads the file into the PostOffice_CT - Power Query Editor window. It looks like the data is split correctly into the four columns that we need. All that's missing is the custom column that will hold the state name.

27. Add a State column to the PostOffice_CT query and rename the Added Custom step to Added State Column. The State column should appear as the last table column.

28. Select the State column and drag it by its title bar to the left, positioning it before the ZIP Code column. The APPLIED STEPS area should now have an additional step—Reordered Columns.

29. In the Query Settings Properties Name box, enter CT as the new name for your query. Click the Rename button in the pop-up box to confirm.

Figure 21.20 shows the Power Query Editor after the changes made in Steps 27–30.

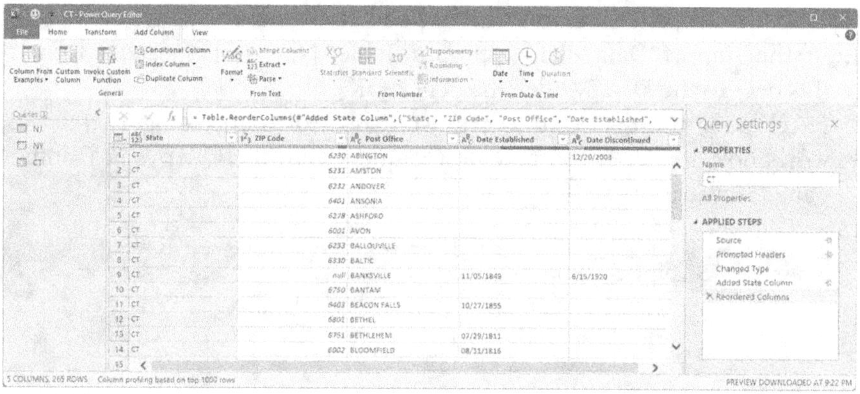

FIGURE 21.20.   The CT query after modification.

30. Choose File | Close & Load.

The `PQ_NJ_NY.xlsx` workbook now contains three queries. Let's save it under a new name so we can include CT.

31. Choose File | Save As, enter `PQ_NJ_NY_CT.xlsx` for the filename, and click Save.

Now that we've got three separate queries with the same structure, let's combine them into one.

Step 4: Combining Data Using Append Query

32. To combine the data for all three states, right-click the NJ query name and choose Append.

An Append dialog appears with two drop-down lists. The Primary table is already specified as NJ.

33. Click the Three or more tables radio button.

Excel displays the Append dialog where you can select the tables you need to combine.

34. Enter the data as shown in Figure 21.21 and click OK to close the Append dialog.

Excel opens the Append1 - Power Query Editor window. You can see the result of the Append operation with one step in the APPLIED STEPS section and a query named Append1.

If the Query Settings pane is not visible, you can turn it back on by choosing View | Query Settings.

Append

Concatenate rows from three or more tables into a single table.

○ Two tables   ● Three or more tables

Available tables

| NJ |
| NY |
| CT |

Add >>

Tables to append

| NJ |
| NY |
| CT |

OK   Cancel

FIGURE 21.21. The Append query allows you to create a single table that includes the rows from multiple tables. In our project, the NJ, NY, and CT data will be merged into one table.

35. Change the query name to `Tri-State Data`, as shown in Figure 21.22.

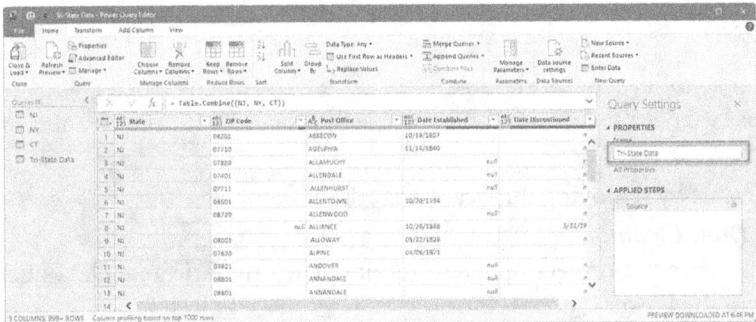

FIGURE 21.22. The Power Query Editor displays the result of the Append operation. We've also changed the name of the combined query to Tri-State Data.

Notice that the formula bar shows the statement that combines three datasets:

```
= Table.Combine({NJ, NY, CT})
```

You could edit the `Table.Combine` formula to include more items if needed without having to open the Append dialog.

36. Choose File | Close & Load.

The Tri-State Data query now appears in the Queries & Connections pane and the active sheet (Tri-State Data) displays an Excel table with combined data for post offices in all three states (NJ, NY, and CT), as shown in Figure 21.23.

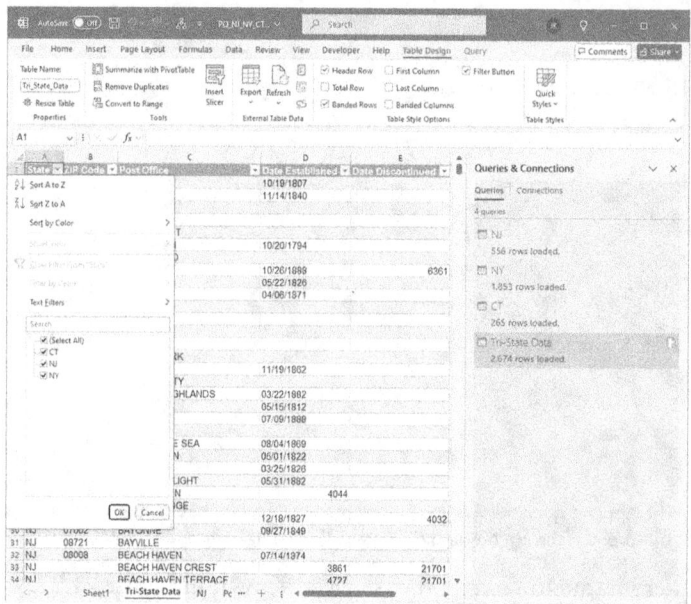

FIGURE 21.23.   An Excel table on the Tri-State Data worksheet displays the data from all three post office queries.

37. Press Ctrl+S to save the workbook.

Now that we have the data in one place, let's proceed with the cleanup of this data.

We will start by removing duplicate records.

Step 5: Data Cleanup

38. In the Queries & Connections pane, double-click the Tri-State Data query to open it in the Power Query Editor.

Select the Post Office column by clicking on its title bar and choose Home | Remove Rows | Remove Duplicates.

A Remove Duplicates step is added to the APPLIED STEPS area in the Query Settings pane.

There are many blank entries in the Date Discontinued column. For consistency's sake, let's replace all blanks with `null` values.

NOTE	*Empty cells are shown as null in power queries.*

39. Select the Date Discontinued column and choose Home | Replace Values.

Excel displays the Replace Values dialog.

40. Leave the Value to Find text box empty, enter `null` in the Replace with text box, and click OK.

A Replaced Value step is added to the APPLIED STEPS pane, as shown in Figure 21.24.

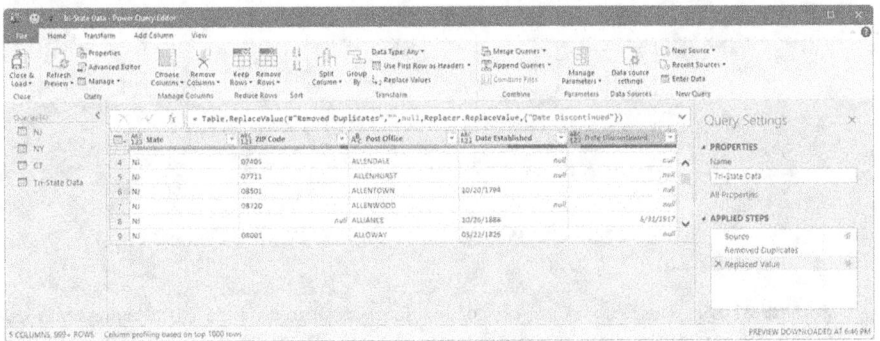

FIGURE 21.24. The Power Query Editor with a Replaced Value step.

Before we can shape the data into its final output, we need to include a bit of logic in our query. To display the count of active and discontinued post offices, let's add a custom column to hold the category we need.

41. Choose Add Column | Add Custom Column and complete the dialog as shown in Figure 21.25. Press OK when done.

Custom Column

Add a column that is computed from the other columns.

New column name

Category

Custom column formula ⓘ

```
= if [Date Discontinued] = null then "Active" else
    "Discontinued"
```

Available columns

State
ZIP Code
Post Office
Date Established
Date Discontinued

<< Insert

Learn about Power Query formulas

✓ No syntax errors have been detected.   OK   Cancel

FIGURE 21.25. Entering a logical expression in a custom column formula. Here, we are specifying the value to be displayed when the post office is discontinued or active.

The Power Query Editor now shows the Category column with Active and Discontinued values (Figure 21.26).

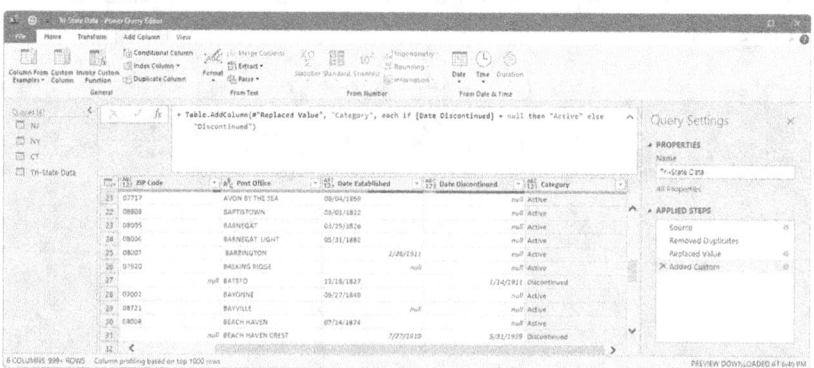

FIGURE 21.26.   The custom Category column data was created via the logical expression shown in Figure 21.25.

Conditional Logic

The M language used in Power Query does not have a `Case` statement like some other programming languages. Instead, M uses conditional logic with the `if...then...else` construct to handle multiple conditions. To evaluate multiple conditions, use the following example:

```
if [Condition1] then [Result1]
else if [Condition2] then [Result2]
else [ResultDefault]
```

---

42. Rename the Added Custom step to `Added Category Column`.

Step 6: Shaping Data into the Final Output

To produce the final report, we will need only two columns: `State` and `Category`. There are various ways to remove columns from a table. You can choose the Remove Columns option to remove the selected columns or the Remove Other Columns option to remove all but the selected columns. There is also a way to remove columns using the Choose Columns button on the Home tab of the Power Query Editor toolbar.

43. Select Home | Choose Columns, select State and Category, and click OK (Figure 21.27).

Our table is now reduced to two columns. Let's shape the data into the final output by using the Group By option.

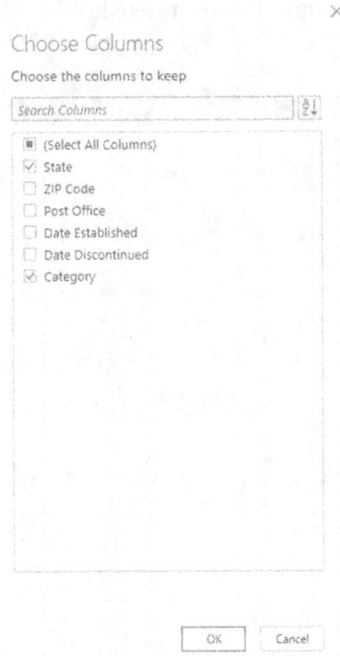

FIGURE 21.27.   Choosing the columns to keep. Our project requires only two columns: State and Category.

44. Choose Transform | Group By and complete the Group By dialog as shown in Figure 21.28. Click OK when done.

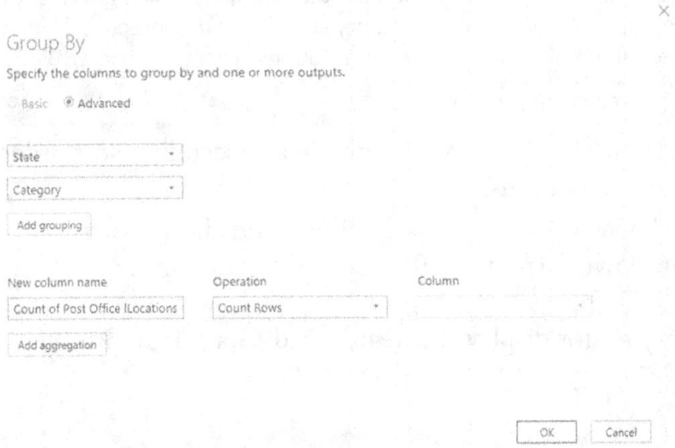

FIGURE 21.28.   Specifying the grouping criteria.

The Power Query Editor displays the grouping output, as shown in Figure 21.29.

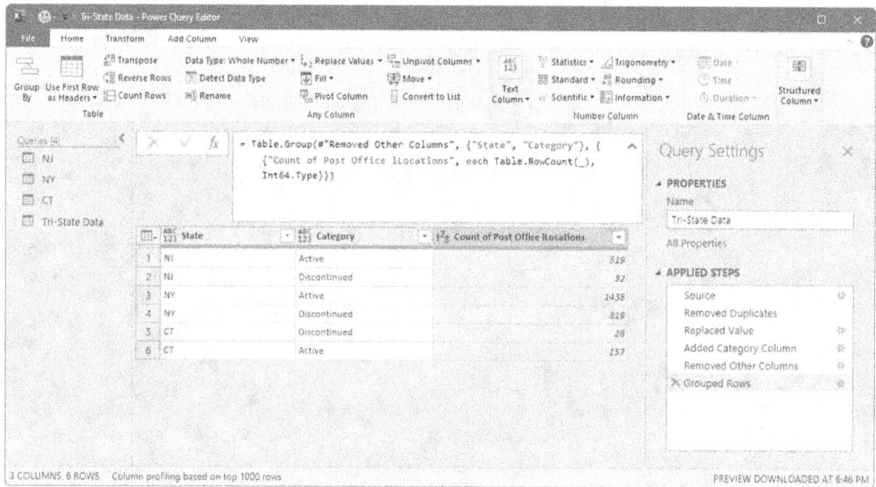

FIGURE 21.29.   The Group By output for this project.

Aggregating Data

To aggregate and group data in the Power Query Editor, select the column you want to group by and click the Group By button in the Home tab. In the Group by window (Figure 21.28), you can group by multiple columns using the Advanced radio button. The Operation drop-down box lets you choose from various aggregate functions such as Count Rows, Average, Min, Max, Sum, and so on. If you require more than one grouping, click the Add grouping button. To remove a group, point to the group name, and notice the three-dot button to the right of the name. Click this button to reveal menu options: Delete, Move Up, and Move Down.

Let's add a final touch to the data by applying an ascending sort order to the `State` and `Category` columns.

45. Click the drop-down arrow in the State column and choose Sort ascending.

46. Click the drop-down arrow in the Category column and choose Sort ascending.

The Power Query Editor displays the rearranged data (Figure 21.30).

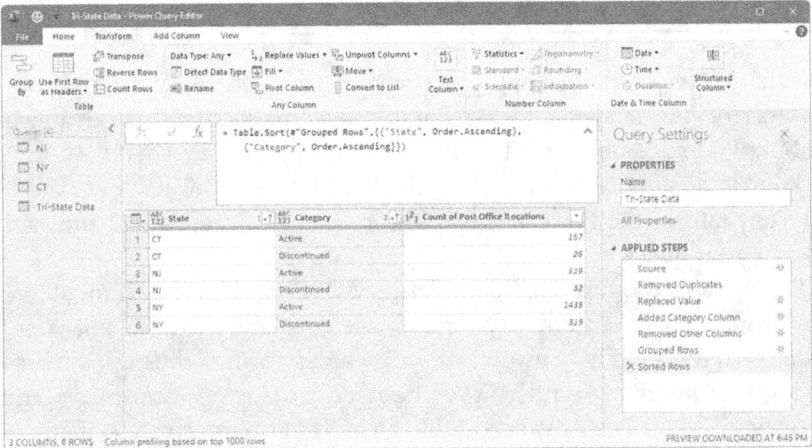

FIGURE 21.30.   The Power Query Editor window showing the final output from the data manipulations performed in this chapter's project.

47. Choose File | Close & Load to close the Power Query Editor and load the data into the worksheet.
Figure 21.31 shows all the queries that you created and their output.

48. Save the `PQ_NJ_NY_CT.xlsx` workbook.

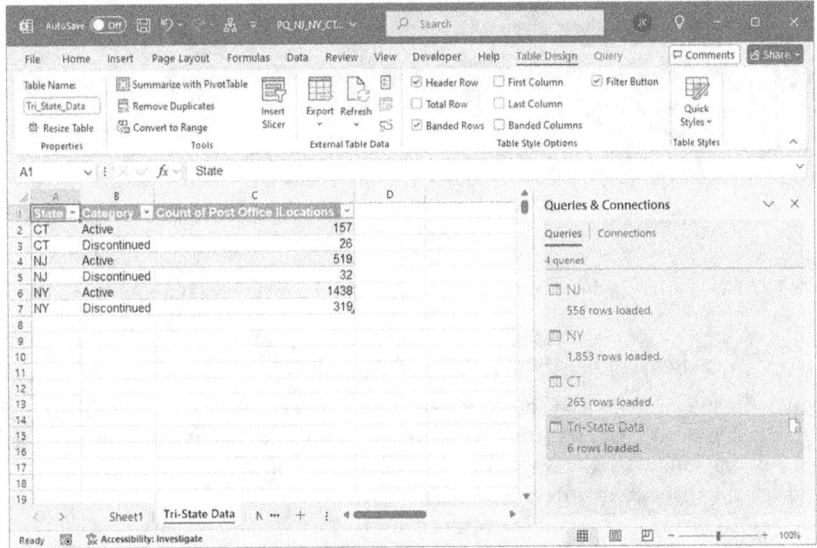

FIGURE 21.31.   An Excel workbook showing the final output of data loading and transformation using the Power Query feature.

USING THE ADVANCED EDITOR

Now that you've created a couple of power queries, let's look at the M code that was created. You can get a full view of the code by accessing the Advanced Editor. Simply double-click the Tri-State Data query in your `PQ_NJ_NY_CT.xlsx` workbook and choose Advanced Editor. This button is available on the Home and View tabs in the Power Query Editor window. Figure 21.32 shows the M code for your query.

The Advanced Editor includes a Display Options drop-down menu. Within this menu, you can customize your code viewing experience to suit your individual preferences. This customization includes various settings, such as displaying line numbers, rendering whitespace, displaying mini-maps, and enabling word wrap. The Advanced Editor window also informs you if any syntax errors are detected in the M code.

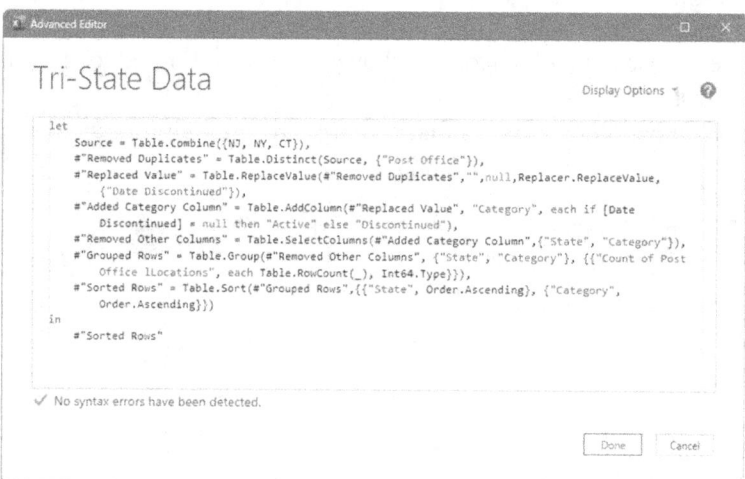

FIGURE 21.32.   Reviewing the M code for the Tri-State Data power query in the Advanced Editor.

Reusing the Output of One Power Query in Another

One power query can serve as a source for another power query. If you right-click a power query in the Queries & Connections pane in your workbook's Queries pane and select Reference on the Query tab, a new power query will be created and will produce the same output as the original power query.

For instance, if `Query1` already holds the data you require, you can use it as a source in another power query. By doing so, Power Query will automatically generate the necessary

M code for `Query2`. This allows you to seamlessly reference the results of `Query1` within `Query2`, making your data transformation process more efficient. Here's how the M code for `Query2` might look:

```
let
    Source = Query1
in
    Source
```

In this example, the `let` expression defines `Source` as the output of `Query1`, and the `in` expression returns the result of `Source`. This allows `Query2` to use and build upon the data transformations already defined in `Query1`. You can then add specific steps to `Query2` to transform the data obtained from `Query1`.

POWER QUERY VS. EXCEL FORMULA LANGUAGE AND EXCEL VBA

While the M language used in Power Query formulas is quite different from Excel formulas and VBA language, it is essential to become proficient with it if you intend to create complex data solutions using the Get Data (Power Query) interface in Excel. The following link provides a reference to the Power Query M formula language (informally known as M):

https://learn.microsoft.com/en-us/powerquery-m/

Table 21.1 shows examples of the M language formulas:

TABLE 21.1.   Examples of M language functions.

Description	Functions
Create a new custom column in Excel.	`= Excel.Workbook([Content])`
Create a new custom column in a CSV file.	`= Csv.Document([Content])`
Create a new column by concatenating values from two existing columns.	`= [Column1Name] & " "` `    & [Column2Name]`
Remove the first two characters from the column entry.	`= Text.Range([ColumnName], 2)`
Place a value in a column based on a condition.	`= if[ColumnName]= "V" then` `"Vacation" else "Other"`
Rename the column to `Employee`.	`= Table.RenameColumns(Source,` `{{"oldColumnName", "Employee"}})`
Get a list of headers used in your table.	`= Table.ColumnNames(Source)`
Get the name of the second column. ***Note:*** Lists in the query are 0-based. 1 will get the second column.	`= Table.ColumnNames(Source){1}`

LEARNING ABOUT VARIOUS M LANGUAGE FUNCTIONS

You can learn about various M language functions in the Power Query Editor. Simply type a function in the formula bar and press Enter to see the syntax and examples of the function usage.

For example, to find out how to use `Table.Combine`, type `=Table.Combine` in the formula bar. After pressing Enter, you should see the output shown in Figure 21.33.

The M language is case-sensitive; therefore, to avoid errors caused by case sensitivity, pay attention to lowercase and uppercase letters in the expressions.

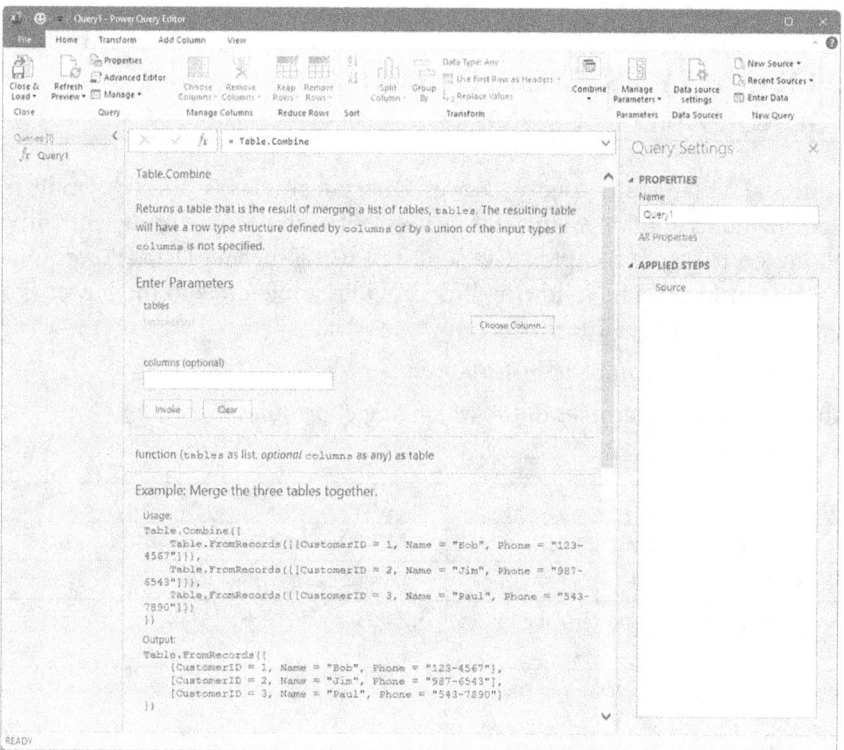

FIGURE 21.33. You can look up information about M language functions by using the formula bar in the Power Query Editor.

Let's take a couple of minutes now to find information about some of the functions shown in Table 21.1. While you can use the formula bar in an existing power query, in Hands-On 21.1, you will use a blank power query for this purpose.

⊙ Hands-On 21.1   Using a Blank Query for Trying Out M Language Functions

1. Close the Power Query Editor window if open and close the PQ_NJ_NY_ CT.xlsx workbook.
2. Open a new blank Excel workbook.
3. In the Excel ribbon, choose Data | Get Data | From Other Sources | Blank query.

   The Power Query Editor appears with the insertion point in the formula bar ready for your input. The APPLIED STEPS area displays one step named Source, which is currently empty as there is no formula for this step. If you open the Advanced Editor (Home | Advanced Editor), you will see the following M language script:

```
let
    Source = ""
in
    Source
```

4. In the Power Query Editor's formula bar, type = Text.Range and press Enter. The Data pane displays the information about the requested function. There is also an Invoke button below the function example that allows you to try it out for yourself.
5. Provide the parameters for this function and click the Invoke button. You can use the Hello World text string and extract the substring from that text starting at index 6, as shown in the function description (see Figure 21.34), or use your own text.

   After you click the Invoke button, the Power Query Editor formula shows the result of running this function (see Figure 21.35). Notice that the Name text box in the Properties area of the Query Settings pane lists a property named Invoked Function and the formula bar displays the following code:

```
= Query1("Hello World", 6, null)
```

   When activated, the Advanced Editor displays the following:

```
let
    Source = Query1("Hello World", 6, null)
in
    Source
```

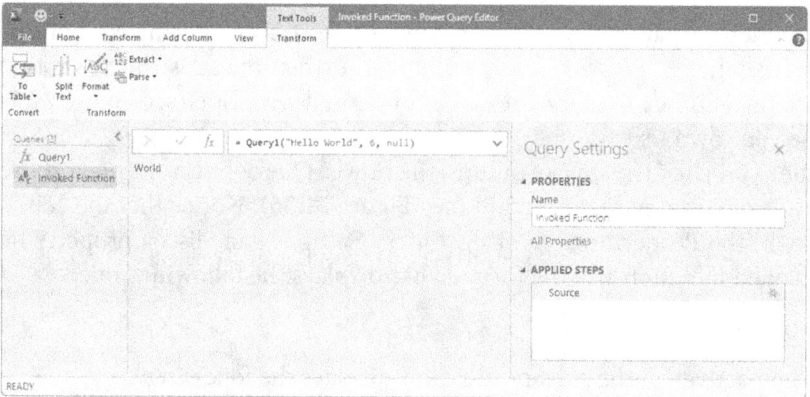

FIGURE 21.34. Entering parameters for the Text.Range M language function.

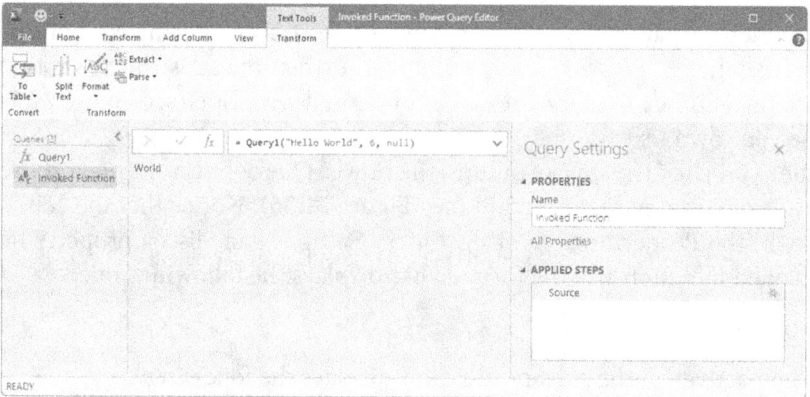

FIGURE 21.35. The result of invoking the Text.Range function in the Power Query Editor.

You can continue testing other functions in the formula bar and invoke them to view the result. The blank query can be used as a scratch pad like the Immediate window in the VBE screen.

6. Select the blank query name (Query1) in the Queries pane to the left of the formula bar and change the Name property of your query in the Query Settings pane to `TestFunctions`.

7. Choose File | Close & Load to save your changes.
Excel creates a `TestFunctions` connection, as shown in Figure 21.36.

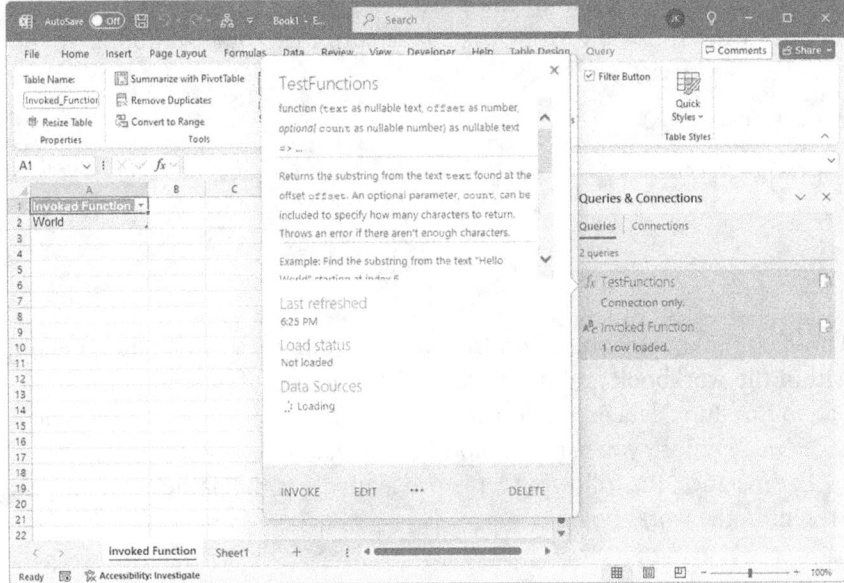

FIGURE 21.36. The blank power query that you created has no data, so Excel created just a connection reference. This way, the query will be saved with your workbook, and you can edit it anytime. Also, you can reference this query from other queries.

8. Save the workbook as `PowerQueries_Scratchpad.xlsx`.

CREATING A QUERY FROM A TABLE

In addition to creating queries from external data sources, you can use the Get Data button to work with a table of data in the currently open workbook. Simply choose Data | Get Data | From Other Sources | From Table / Range command to create a query linked to a selected Excel table. If a selected range is not part of a table, it will be converted into a table.

THE GET DATA AND VBA SUPPORT

To enhance the functionality of the Get Data feature in the Get & Transform Data section of the ribbon's Data tab, the VBA object model introduces the `Queries` and `WorkbookQuery` objects. These objects come with a variety of properties and methods designed to help users efficiently manage and manipulate their data. By using the `Queries` object, you can access a collection of all the queries present in the workbook.

You can try out some of the properties of the `Queries` object in the Immediate window, as shown below. Make sure that your workbook `PQ_NJ_NY_CT.xlsx` is active:

```
?ThisWorkbook.Queries.Count
 4
?ThisWorkbook.Queries.Parent.Name
PQ_NJ_NY_CT.xlsx

?ThisWorkbook.Queries.Item(1).Name
NJ
```

The `WorkbookQuery` object, on the other hand, represents an individual query within the workbook and provides methods to modify its properties, allowing you to fine-tune data extraction and transformation processes. Using the `WorkbookQuery` object, you can list, modify, delete, and refresh queries.

For instance, the following VBA procedure lists the names of all the queries present in the workbook:

```
Sub ListAllQueries()
    Dim wbQuery As WorkbookQuery
    Dim ws As Worksheet
    Dim i As Integer

    ' Create a new worksheet to display the query names
    Set ws = ThisWorkbook.Worksheets.Add
    ws.Name = "Query List"

    i = 1
    For Each wbQuery In ThisWorkbook.Queries
        ws.Cells(i, 1).Value = wbQuery.Name
        i = i + 1
    Next wbQuery
End Sub
```

The following VBA procedure will refresh a specific query to ensure it retrieves the latest data:

```
Sub RefreshQuery()
    Dim wbQuery As WorkbookQuery

    ' Reference the query by its name
    Set wbQuery = ThisWorkbook.Queries("Tri-State Data")
```

```
            ' Refresh the query
            wbQuery.Refresh
End Sub
```

You can also utilize the macro recording feature in Excel to automate the process of creating and refreshing your queries. You cannot, however, record the actions performed in the Power Query Editor.

Hands-On 21.2 walks you through the process of recording three queries obtained from a Web data source.

⦿ Hands-On 21.2   Automating the Creation of Queries

1. Choose File | New | Blank workbook.
2. Choose Developer | Record Macro.
3. In the Record Macro dialog, type `CreateQuery` as the macro name, choose ThisWorkbook in the Store macro in dropdown, and then click OK.
4. Choose Data | Get Data | From Other Sources | From Web.
5. In the URL field, enter: `http://livingwage.mit.edu/states/06` and click OK.
6. In the Navigator, choose Select multiple items and choose all the tables as shown in Figure 21.37.

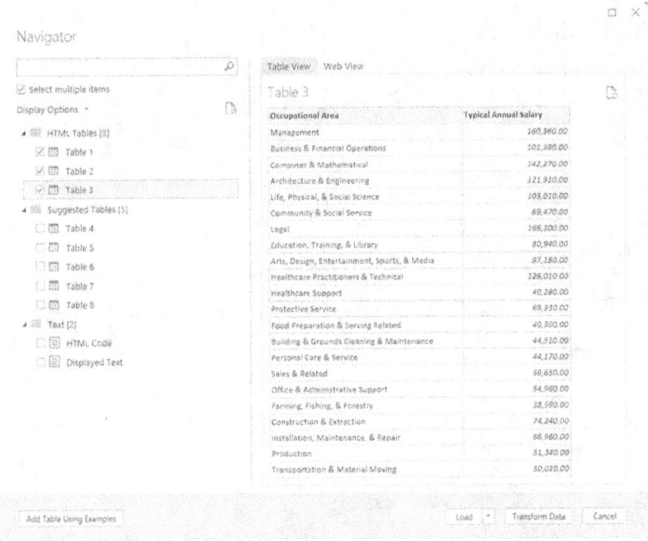

FIGURE 21.37.   Selecting the data provided by the Web resource for a power query.

7. Select Load | Load To.
8. In the Import Data dialog, choose Table, uncheck Add this data to the Data Model, then click OK.

All three tables are loaded into the workbook, as shown in Figure 21.38.

9. Choose Developer | Stop Recording to end the macro recording session.

10. Save the workbook as `RecordedQueries.xlsm`.

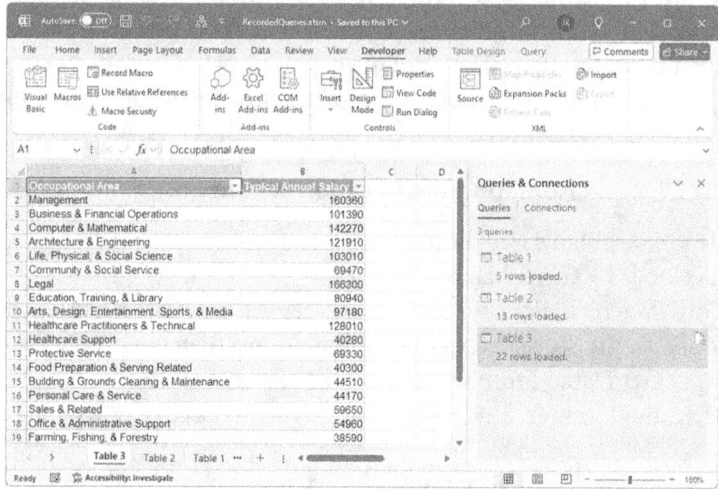

FIGURE 21.38. Three power queries were generated after the selection process shown in Figure 21.37.

Let's open the VBE Code window and examine the code that was recorded in Hands-On 21.2 (see Figure 21.39).

FIGURE 21.39. A partial listing of the recorded macro that retrieves three tables from a Web source and creates corresponding power queries. The snippet illustrates both the beginning of the macro and a section further down in the code, highlighting how the macro is structured in a split VBE Code window.

Notice that the `Queries.Add` method is used to add a new `WorkbookQuery` object to the `Queries` collection. This method requires the name of the query and the M formula for the query. You can also provide a description of the query, but this is optional. The `Web.BrowserContents` function is used for downloading data from the Web. This function expects the URL of the Web site as a text string. When recording a power query from a Web source, the `Table` function is often used to handle various data transformation operations on the retrieved data. For example, `Table.TransformColumnTypes` is used to change the data types of columns in a table.

An OLE DB query connection uses the `Microsoft.Mashup.OleDb.1` provider. In addition to the provider's name, the connection string includes `$Workbook` as the `Data Source` attribute and the name of the query in the `Location` attribute. The `QueryTable.Refresh` method updates the `QueryTable`. The optional argument `BackgroundQuery` specifies whether the query should be updated in the background. The `False` setting returns the control to the procedure only after all data has been fetched to the worksheet. With the `True` setting, the control is returned to the procedure as soon as a database connection is made, and the query is submitted.

By recording queries using different options, you can discover other functions helpful in automating queries in Excel.

ADDITIONAL LEARNING RESOURCES

This chapter demonstrated only a few of the basic capabilities for getting and transforming data in Excel. To delve into the more advanced features, you might need an entirely separate book dedicated to those topics. If you'd like more hands-on experience with building power queries, be sure to check out the following tutorials from Microsoft:

- *Power Query 101*—This tutorial will teach you how to connect to a Web data source, select tables for import, replace and filter values, and load the query into a worksheet:

  https://support.office.com/en-us/article/Power-Query-101-008B3F46-5B14-4F8B-9A07-D3DA689091B5

- *Combine Data from Multiple Data Sources*—This tutorial will teach you how to combine a local Excel file with an OData feed and perform aggregations to produce a Total Sales per Product and Year report:

https://support.office.com/en-us/article/Combine-data-from-multiple-data-sources-Power-Query-70cfe661-5a2a-4d9d-a4fe-586cc7878c7d

SUMMARY

This chapter provided a brief introduction to Excel's data import and transformation feature known as Power Query, which was first introduced in Excel 2016 as Get & Transform Data.

After importing and combing data from an Excel workbook and a CSV text file, you learned how to transform the raw data through a series of steps to produce a final aggregated Excel table. You explored the Power Query Editor, using its command buttons on the ribbon to dynamically generate and edit query steps. The formula bar in Power Query Editor was demonstrated as a tool for obtaining information about and testing M code expressions and functions. Additionally, the Advanced Editor allowed you to view and edit the entire query script. Throughout the chapter, you discovered several M language features, including case sensitivity, data types, and logical expressions. Lastly, you learned how macro recording and the Excel object model can help you write VBA procedures to automate the creation and refreshing of queries.

The next chapter focuses on programming VBE.

Part VI

TAKING CHARGE OF THE PROGRAMMING ENVIRONMENT

While VBA provides a very comprehensive object model for automating worksheet tasks, some of the processes and operations that you may need to program are an integral part of the Windows operating system and cannot be controlled with VBA. In this part of the book, we start by learning how to programmatically work with VBA projects, modules, and procedures. Next, you are introduced to the Windows API library of functions that will come to your rescue when you need to overcome the limitations of the native VBA library.

Chapter 22 PROGRAMMING THE VISUAL BASIC EDITOR (VBE)

Having worked through previous chapters of this book, you have already acquired a working knowledge of many of the tools available in the VBE to create, modify, and troubleshoot VBA procedures. VBA also allows you to program its own development environment known as the Visual Basic Integrated Design Environment (VBIDE).

For instance, you can:

- Control Visual Basic projects by getting and setting project properties and adding or removing individual components.
- Control Visual Basic code—add, delete, and modify code, save code to a file, or insert code from a file and search for specific information in the VBA code.
- Control UserForms by programmatically designing a UserForm and dynamically adding or removing controls from a form.
- Work with references—add a reference to an external object library and check for broken references.
- Control the VBIDE user interface by controlling various windows and adding or changing menus and toolbars.

THE VISUAL BASIC EDITOR OBJECT MODEL

To program and manipulate the VBA environment, you need to utilize the objects and functionalities provided by the `Microsoft Visual Basic for Applications Extensibility 5.3` library. This library offers a set of tools to automate and customize your VBA projects by accessing various elements within the VBA Editor.

To ensure that you can run the procedures in this chapter, perform the steps in Hands-On 22.1.

NOTE	*Please note that the files for the "Hands-On" projects can be found in the companion files.*

Hands-On 22.1 Trusting Access to the VBA Project Object Model

1. Open a new workbook and save it as `C:\VBAExcel2024_ByExample\Chap22.xlsm`.
2. To trust the Visual Basic project, click the Developer tab and then choose Macro Security. Excel displays the Trust Center dialog box with various macro settings.
3. Select the Trust access to the VBA project object model checkbox and click OK (see Figure 22.1).

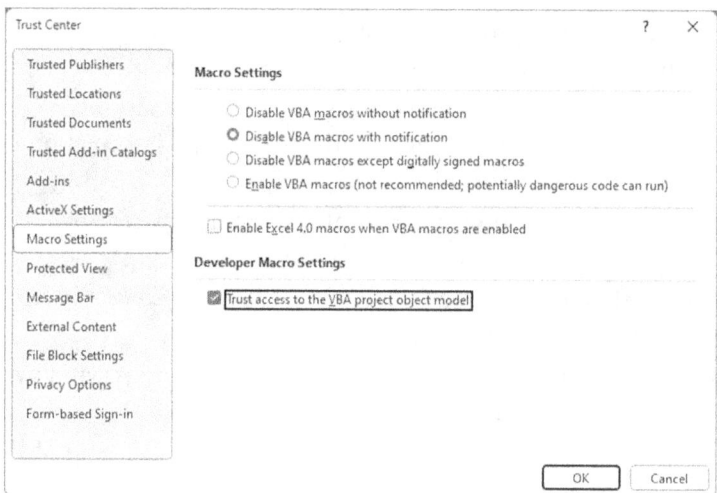

FIGURE 22.1.   You must set access to the VBA project object model to allow for programming the VBE environment.

NOTE	*If access to the VBA project object model is not enabled, an attempt to run a VBA procedure that accesses objects from the Microsoft Visual Basic for Applications Extensibility 5.3 library results in the following runtime error message: "Programmatic access to Visual Basic Project is not trusted."*

4. Choose the Visual Basic button on the Developer tab to activate the VBE window.

   In the Project Explorer window, select VBAProject (Chap22.xlsm) and use the Properties window to rename it `Chap22_SourceCode`.

5. To create a reference to the Microsoft Visual Basic for Applications Extensibility 5.3 library, choose Tools | References. Check the Microsoft Visual Basic for Applications Extensibility 5.3 reference, as shown in Figure 22.2, and click OK.

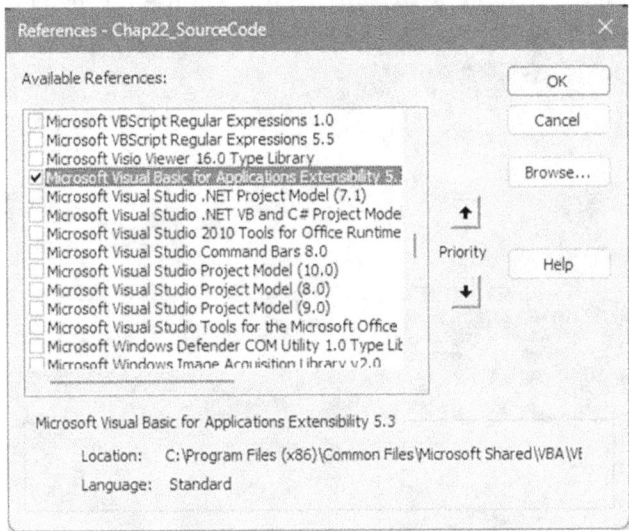

FIGURE 22.2.   Setting a reference to the Microsoft Visual Basic for Applications Extensibility 5.3 library.

UNDERSTANDING THE VBE OBJECTS

Some of the key objects you'll be interacting with include:

- `VBProject`—Represents a VBA project within the VBE screen
- `VBComponent`—Represents a component (e.g., module, class, user form) within the VBA project

- CodeModule—Represents the code contained in a VBComponent
- VBIDE—The root object representing the entire VBA development environment

By utilizing these objects, you can perform a variety of tasks, such as:

- *Creating and deleting modules*—You can add new modules, classes, or user forms to a VBA project, as well as delete existing ones.
- *Editing code*—You can programmatically add, remove, or modify code within a module.
- *Manipulating project properties*—You can modify the properties of VBA projects, components, and modules.

In the Object Browser (Figure 22.3), the Microsoft Visual Basic for Applications Extensibility 5.3 library is referred to as VBIDE. You will use this name when referencing this library in code.

FIGURE 22.3. The Object Explorer lists all the members of the VBIDE object model.

The VBE object model contains five collections of objects, as follows:

- `VBProjects` collection—This collection contains each `VBProject` object that is currently open in the development environment. Use the `VBProject` object to set properties for the project. The `VBProject` object also allows you to access the `VBComponents` collection and the `References` collection.

- `VBComponents` collection—Use this collection to access, add, or remove components in a project. A component can be a form, a standard module, or a class module contained in a project.

- `References` collection—Use this collection to add or remove references in the VBA project. Each VBA project can reference one or more libraries or projects. Use the `Reference` object to find out what references are currently selected in the References dialog box for the specific VBA project.

- `AddIns` collection—Use this collection to access the `AddIn` objects. Add-ins are programs that add extended capabilities and features to Microsoft Excel or other Microsoft 365 applications.

- `Windows` collection—Use this collection to access window objects such as the Project Explorer window, Properties window, or currently open Code windows.

- `CodePanes` collection—Use this collection to access the open code panes in a project. A Code window can contain one or more code panes. A code pane contained in a Code window is used for entering and editing code.

- `CommandBars` collection—This collection contains all of the command bars in a project, including command bars that support shortcut menus.

ACCESSING THE VBA PROJECT

The only way to determine the current setting for Trust access to the VBA project object model in the Trust Center (see Figure 22.1) is by using error trapping. The following procedure displays a message if access to the VBA project is not trusted. Instructions on how to change the security settings are then displayed in a text box placed in a new workbook.

⊙ Hands-On 22.2   Checking Access to the VBA Project Using VBA

> **NOTE**   *All VBA code presented in this chapter will fail unless you followed the steps in Hands-On 22.1.*

1. Switch to the VBE window and insert a new module into `Chap22_SourceCode(Chap22.xlsm)`.

2. In the Code window, enter the `AccessToVBProj` procedure as shown below:

```
Sub AccessToVBProj()
Dim objVBProject As VBProject
Dim strMsg1 As String
Dim strMsg2 As String
Dim response As Integer

On Error Resume Next

If Application.Version >= "16.0" Then
    Set objVBProject = ActiveWorkbook.VBProject

    strMsg2 = "The access to the VBA "
    strMsg2 = strMsg2 + " project must be trusted for this "
    strMsg2 = strMsg2 + "procedure to work."
    strMsg2 = strMsg2 + vbCrLf + vbCrLf
    strMsg2 = strMsg2 + " Click 'OK' to view instructions,"
    strMsg2 = strMsg2 + "  or click 'Cancel' to exit."

    If Err.Number <> 0 Then
        strMsg1 = "Please change the security settings to "
        strMsg1 = strMsg1 & "allow access to the VBA project:"
        strMsg1 = strMsg1 & Chr(10) & "1. "
        strMsg1 = strMsg1 & "Choose Developer | Macro Security."
        strMsg1 = strMsg1 & Chr(10) & "2. "
        strMsg1 = strMsg1 & "Check the 'Trust access" _
            & " to the VBA project object model'. "
        strMsg1 = strMsg1 & Chr(10) & "3. Click OK."

        response = MsgBox(strMsg2, vbCritical + vbOKCancel, _
                "Access to VB Project is not trusted")

            If response = 1 Then
                Workbooks.Add
                With ActiveSheet
                  .Shapes.AddTextbox _
                  (msoTextOrientationHorizontal, _
```

```
                    Left:=0, Top:=0, Width:=300, _
                        Height:=100).Select
                    Selection.Characters.Text = strMsg1
                    .Shapes(1).Fill.PresetTextured _
                        PresetTexture:=msoTextureBlueTissuePaper
                    .Shapes(1).Shadow.Type = msoShadow6
                End With
            End If
        Exit Sub
    End If

    MsgBox "There are " & objVBProject.References.Count & _
        " project references in " & objVBProject.Name & "."
    End If
End Sub
```

The `AccessToVBProj` procedure begins by checking the application version that is currently in use. If the Trust access to the VBA project object model setting is turned off in the Trust Center dialog box's Developer Macro Settings area, the attempt to set the object variable `objVBProject` will cause an error. The procedure traps this error with the `On Error Resume Next` statement. If an error occurs, the `Err` object will return a non-zero value. At this point, we can tell the user that security settings must be adjusted for the procedure to run. Instead of simply displaying the instructions in the message box, we print them to a worksheet so users can follow them easily while accessing the necessary options. They can also print them out if they want to make this change later (see Figure 22.4).

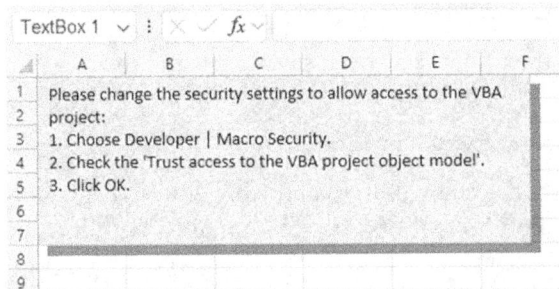

FIGURE 22.4. Instructions for allowing access to the VBA project object model are generated by the example procedure.

3. Run the `AccessToVBProj` procedure.

If you followed the instructions in Hands-On 22.1, you should get a message that displays the number of references that are set in the current `VBProject`.

Otherwise, you will get a message prompting you to click OK to view instructions on how to make the change. The instructions are then placed in a text box, as shown in Figure 22.4.

FINDING INFORMATION ABOUT A VBA PROJECT

As you already know, each new workbook in the Microsoft Excel user interface has a corresponding workbook project named VBAProject. You can change the project name to a more meaningful name by entering a new value for the Name property in the Properties window or by accessing the VBAProject properties dialog box using the Tools menu. You can also perform this change programmatically.

If the project you want to edit is currently highlighted in the Project Explorer window, simply type the following statement in the Immediate window to replace the default VBA project name with your own:

```
Application.VBE.ActiveVBProject.Name = "Chap22_SourceCode"
```

If the VBA project you want to change is not active, the following statement can be used:

```
Workbooks("Chap22.xlsm").VBProject.Name = "Chap22_SourceCode"
```

To change the description of the VBProject object, type the following statement on one line in the Immediate window:

```
Workbooks("Chap22.xlsm").VBProject.Description =
    "Programming Visual Basic Editor"
```

You can find out whether the project has been saved by using the Saved property of the VBProject object:

```
MsgBox Application.VBE.ActiveVBProject.Saved
```

Visual Basic returns False if the changes to the project have not been saved.

To find out how many component objects are contained within a specific VBA project, use this code:

```
MsgBox Workbooks("Chap22.xlsm").VBProject.VBComponents.Count
```

To find out the name of the currently selected VBComponent object, type the following lines of code in the Immediate window, pressing Enter after each statement:

```
Set objVBComp = Application.VBE.SelectedVBComponent
    MsgBox objVBComp.Name
```

You can also quickly find out the number of references defined in the References dialog box by typing the following statement in the Immediate window:

```
?Application.VBE.ActiveVBProject.References.Count
```

VBA PROJECT PROTECTION

To prevent users from viewing code, you can lock each VBA project.

To lock a VBA project, perform the following tasks:

1. In the Project Explorer, right-click the project you want to protect, and then click [ProjectName] Properties on the shortcut menu. You can also choose the [ProjectName] Properties option from the Tools menu.
2. In the Project Properties dialog box, click the Protection tab and select the Lock project for viewing checkbox. Enter and confirm the password, and then click OK (see Figure 22.5).

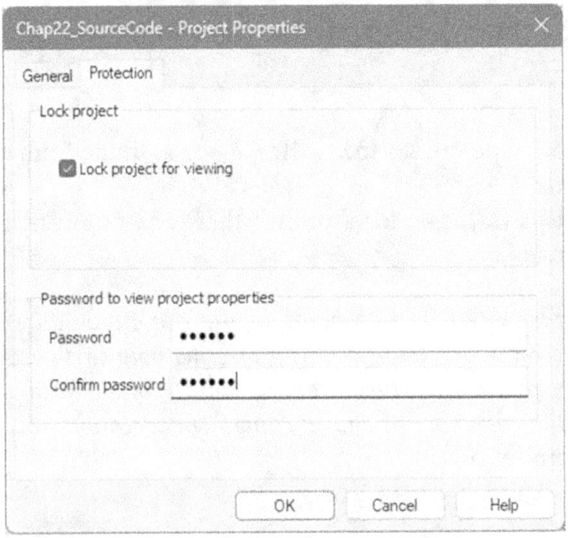

FIGURE 22.5.   Protecting access to the VBA project.

The next time you open the workbook file, you will be prompted to enter the password when attempting to view the code in the project.

There is no way to programmatically specify a password for a locked VBA project. You should check whether the project is protected before you attempt to edit the project or run code that accesses information about the project's components.

To determine whether a VBA project is locked, check the Protection property of the VBProject object.

The following function procedure demonstrates how to check the Protection property of the VBA project in an Excel workbook.

 Hands-On 22.3   Using VBA to Determine Whether the VBA Project Is Protected

1. In the same module where you entered the procedure in Hands-On 22.2, enter the IsProjProtected function procedure as shown below:

```
Function IsProjProtected() As Boolean
    Dim objVBProj As VBProject

    Set objVBProj = ActiveWorkbook.VBProject

    If objVBProj.Protection = vbext_pp_locked Then
        IsProjProtected = True
    Else
        IsProjProtected = False
    End If
End Function
```

2. To test the above function, type MsgBox IsProjProtected in the Immediate window and press Enter.

The MsgBox function displays False for a project that is not protected and True if protection is turned on.

NOTE	*If a project is protected and you attempt to run a procedure that needs to access information in this project without first checking whether the protection is turned on, runtime error 50289 appears with the following description: Can't perform operation since the project is protected.*

WORKING WITH MODULES

All standard modules, class modules, code modules located behind worksheets and workbooks, and UserForms are members of the VBComponents collection of a VBProject object.

To determine the type of the component object, use the Type property of the VBComponent object, as shown in Figure 22.6.

The **Type** property settings for the VBComponent object are described in the following table.

⟦ ⟧ **Expand table**

Constant	Value	Description
vbext_ct_StdModule	1	Standard module
vbext_ct_ClassModule	2	Class module
vbext_ct_MSForm	3	Microsoft Form
vbext_ct_ActiveXDesigner	11	ActiveX Designer
vbext_ct_Document	100	Document Module

FIGURE 22.6. Type property settings for the VBComponent object as shown in the online help at *https://learn.microsoft.com/en-us/office/vba/Language/Reference/Visual-Basic-Add-in-Model/properties-visual-basic-add-in-model - type*.

The code of each `VBComponent` is stored in a `CodeModule`. The `UserForm` component has a graphical development interface called ActiveX Designer.

The following sections demonstrate several procedures that access the `VB-Components` collection to perform the following tasks:

- Listing all modules in a workbook
- Adding a module
- Removing a module
- Removing all code from a module
- Removing empty modules
- Copying (exporting and importing) modules

Listing All Modules in a Workbook

The procedure below generates a list of all modules contained in `Chap22.xlsm` workbook. The name of each module and the description of the module type are placed in a two-dimensional array and then written to a worksheet. Because the `Type` property of the `VBComponent` object returns a constant or a numeric value containing the type of object (see Figure 22.6), the procedure uses a function to show the corresponding description of the object.

⊙ Hands-On 22.4   Listing All Workbook Modules

1. Insert a new module into the current VBA project in `Chap22.xlsm` workbook.

2. In the Code window, enter the following procedure and function:

```
Sub ModuleList()
    Dim objVBComp As VBComponent
    Dim listArray()
    Dim i As Integer


    If ThisWorkbook.VBProject.Protection = vbext_pp_locked Then
        MsgBox "Please unprotect the project to run this " & _
            "procedure."
        Exit Sub
    End If

    i = 2

    For Each objVBComp In ThisWorkbook.VBProject.VBComponents
        ReDim Preserve listArray(1 To 2, 1 To i - 1)
        listArray(1, i - 1) = objVBComp.Name
        listArray(2, i - 1) = GetModuleType(objVBComp)
        i = i + 1
    Next

    With ActiveSheet
        .Cells(1, 1).Resize(1, 2).Value = Array("Module Name", _
            "Module Type")
        .Cells(2, 1).Resize(UBound(listArray, 2), _
            UBound(listArray, 1)).Value = Application._
            Transpose(listArray)
        .Columns("A:B").AutoFit
    End With

    Set objVBComp = Nothing
End Sub

Function GetModuleType(comp As VBComponent)
    Select Case comp.Type
        Case vbext_ct_StdModule
            GetModuleType = "Standard module"
        Case vbext_ct_ClassModule
            GetModuleType = "Class module"
        Case vbext_ct_MSForm
            GetModuleType = "Microsoft Form"
        Case vbext_ct_ActiveXDesigner
            GetModuleType = "ActiveX Designer"
        Case vbext_ct_Document
```

```
            GetModuleType = "Document module"
        Case Else
            GetModuleType = "Unknown"
    End Select
End Function
```

3. Run the `ModuleList` procedure and then switch to the Microsoft Excel application window to view the results.

Figure 22.7 shows the result of running the `ModuleList` procedure.

⊿	A	B
1	Module Name	Module Type
2	ThisWorkbook	Document module
3	Sheet1	Document module
4	Module1	Standard module
5	Module2	Standard module
6		

FIGURE 22.7.   Module names and module types are retrieved into a worksheet in Hands-On 22.4.

> **NOTE**   *If the VBA project is protected when you execute the* `ModuleList` *procedure, you will see a warning message.*

Adding a Module to a Workbook

Use the `Add` method of the `VBComponents` collection to add a new module to `ThisWorkbook`. The `CreateModule` procedure in the next Hands-On exercise prompts the user for the module name and the type of module. When this information has been provided, the `AddModule` procedure is called.

◉ Hands-On 22.5   Adding a Module to a Workbook

1. Insert a new module into the current VBA project in `Chap22.xlsm` workbook.
2. In the Code window, enter the following procedures:

```
Sub CreateModule()
    Dim modType As Integer
    Dim strName As String
    Dim strPrompt As String

    strPrompt = "Enter a number representing the type of module:"
    strPrompt = strPrompt & vbCr & "1 (Standard Module)"
    strPrompt = strPrompt & vbCr & "2 (Class Module)"
    modType = Val(InputBox(prompt:=strPrompt, _
        Title:="Insert Module"))
```

```
        If modType = 0 Then Exit Sub
        strName = InputBox("Enter the name you want to assign to " & _
                "new module", "Module Name")
        If strName = "" Then Exit Sub
        AddModule modType, strName
    End Sub


    Sub AddModule(modType As Integer, strName As String)
        Dim objVBProj As VBProject
        Dim objVBComp As VBComponent

        If InStr(1, "1, 2", modType) = 0 Then Exit Sub

        Set objVBProj = ThisWorkbook.VBProject
        Set objVBComp = objVBProj.VBComponents.Add(modType)
        objVBComp.Name = strName

        Application.Visible = True

        Set objVBComp = Nothing
        Set objVBProj = Nothing
    End Sub
```

3. Run the `CreateModule` procedure. Enter `1` for a standard module when prompted and click OK.
4. Enter `TestModule` as the module name and click OK.
   When the procedure is completed, you should see a new module named TestModule in the Project Explorer window. Do not delete this module, as you will need it for the next Hands-On.

Removing a Module

Use the following procedure to delete the `TestModule` module you added in Hands-On 22.5.

ⓞ **Hands-On 22.6   Removing a Module from the Workbook**

1. Insert a new module into the current VBA project in the `Chap22.xlsm` workbook.
2. In the Code window, enter the following procedure:

```
Sub DeleteModule(strName As String)
    Dim objVBProj As VBProject
    Dim objVBComp As VBComponent
```

```
    Set objVBProj = ThisWorkbook.VBProject

    Set objVBComp = objVBProj.VBComponents(strName)

    objVBProj.VBComponents.Remove objVBComp

    Set objVBComp = Nothing
    Set objVBProj = Nothing
End Sub
```

3. Run the `DeleteModule` procedure by typing the following statement in the Immediate window and pressing Enter to execute:

```
DeleteModule "TestModule"
```

At this point, `Chap22_SourceCode` `(Chap22.xlsm)` should contain the four standard modules created so far in this chapter.

Deleting All Code from a Module

Use the `CountOfLines` property of the `CodeModule` object to return the number of lines of code in a code module. Use the `DeleteLines` method of the `CodeModule` object to delete a single line or a specified number of lines.

The `DeleteLines` method can use two arguments. The first required argument specifies the first line you want to delete. The second argument is optional and specifies the total number of lines you want to delete. If you don't specify how many lines you want to delete, only one line will be deleted. The following procedure deletes all code from the specified module.

⦿ Hands-On 22.7   Deleting a Module's Code

1. Insert a new module into the current VBA project in the `Chap22.xlsm` workbook.

2. In the Code window, enter the following procedure:

```
Sub DeleteModuleCode(strName As String)
    Dim objVBProj As VBProject
    Dim objVBCode As CodeModule
    Dim firstLn As Long
    Dim totLn As Long

    Set objVBProj = ThisWorkbook.VBProject
    Set objVBCode = objVBProj.VBComponents(strName).CodeModule
    With objVBCode
        firstLn = 1
```

```
        totLn = .CountOfLines
        .DeleteLines firstLn, totLn
    End With

    Set objVBProj = Nothing
    Set objVBCode = Nothing
End Sub
```

3. Insert a new module in the current VBA project and rename it `DeleteTest`.
4. Copy the first procedure you created in this chapter (in Hands-On 22.2) into the `DeleteTest` module.
5. In the Immediate window, enter the following statement:

```
DeleteModuleCode "DeleteTest"
```

When you press Enter, all the code in the `DeleteTest` module is removed. Do not delete the empty `DeleteTest` module. You will remove it programmatically in the next example.

Deleting Empty Modules

In the course of writing your VBA procedures, you may have inserted a number of new modules in your VBA project. While most of these modules contain valid code, there are probably a couple of empty modules that were left behind. You can remove all the unwanted empty modules in one sweep with a VBA procedure.

To remove a module, use the `Remove` method of the `VBComponents` collection. This method requires that you specify the type of component you want to remove. Use the enumerated constants shown in Figure 22.6 earlier in this chapter to indicate the type of component.

The following procedure iterates through the `VBComponents` collection and checks whether the retrieved component is a standard module or a class module. If the module contains less than three lines, we assume that the module is empty and okay to delete. We write the information about the deleted modules into the Immediate window.

(◉) Hands-On 22.8   Deleting an Empty Module

1. Insert a new module into the current VBA project in the `Chap22.xlsm` workbook.
2. In the Code window, enter the following procedure:

```
Sub DeleteEmptyModules()
```

```
    Dim objVBComp As VBComponent

    Const vbext_ct_StdModule As Long = 1
    Const vbext_ct_ClassModule As Long = 2

  For Each objVBComp In ActiveWorkbook.VBProject.VBComponents
     Select Case objVBComp.Type
       Case vbext_ct_StdModule, vbext_ct_ClassModule
         If objVBComp.CodeModule.CountOfLines < 3 Then
           Debug.Print "(deleted) " & objVBComp.Name & vbTab & _
               "declarations: " & objVBComp.CodeModule. _
            CountOfDeclarationLines & vbTab & _
               "Total code Lines: " & _
               objVBComp.CodeModule.CountOfLines
            ActiveWorkbook.VBProject.VBComponents. _
               Remove objVBComp
         End If
     End Select
   Next
   Set objVBComp = Nothing
End Sub
```

3. Run the `DeleteEmptyModules` procedure.

The `DeleteTest` module that we created in Hands-On 22.7 should now be removed from the current VBA project. Check the Immediate window for information about the deleted module.

Copying (Exporting/Importing) a Module

Sometimes, you may want to copy modules between VBA projects. There is no single method to perform this task. To copy a module, you must perform the following two steps:

1. Export a module to an external text file.

The `Export` method of the `VBComponent` object saves the component as a separate text file. You must specify the name of the file to which you want to export the component. The filename must be unique, or an error will occur.

2. Import a module from an external text file.

The `Import` method of the `VBComponent` object adds the component to a project from a file. You must specify the path and filename of the file from which you want to import the component. The workbook file that will receive the imported component must be open.

Let's assume that you want to copy `Module1` from the `Chap22.xlsm` workbook to another workbook file named `Chap22b.xlsm`. The procedure that follows requires three arguments for the copy operation: the name of the workbook containing the module you want to copy, the name of the workbook that will receive the copied module, and the name of the module you will be copying.

(•) Hands-On 22.9   Exporting/Importing a Module

1. Insert a new module into the current VBA project in the `Chap22.xlsm` workbook.
2. In the Code window, enter the following procedure:

```
Sub CopyAModule(wkbFrom As String, wkbTo As String, _
            strFromMod As String)
    Dim wkb As Workbook
    Dim strFile As String

    Set wkb = Workbooks(wkbFrom)

    strFile = wkb.Path & "\vbCode.bas"
    wkb.VBProject.VBComponents(strFromMod).Export strFile

    On Error Resume Next
    Set wkb = Workbooks(wkbTo)
    If Err.Number <> 0 Then
        Workbooks.Open wkbTo
        Set wkb = Workbooks(wkbTo)
    End If

    wkb.VBProject.VBComponents.Import strFile
    wkb.Save

    Set wkb = Nothing
End Sub
```

3. In the Immediate window, execute the following two lines of code that create a new workbook and save it as `Chap22b.xlsm` in your `VBAExcel2024_ByExample` folder:

```
Set newWorkbook = Workbooks.Add
newWorkbook.SaveAs "C:\VBAExcel2024_ByExample\Chap22b.xlsm", _
    xlOpenXMLWorkbookMacroEnabled
```

4. In the Immediate window, type the following statement and press Enter:

```
CopyAModule "Chap22.xlsm", "Chap22b.xlsm", "Module1"
```

When you execute the above statement, the `CopyAModule` procedure exports `Module1` from `Chap22.xlsm` to a file named `vbCode.bas`. Next, the `vbCode.bas` file is imported into the `Chap22b.xlsm` workbook and the workbook is saved. You may want to add an additional statement to this procedure to remove the `vbCode.bas` file from your computer (use the `Kill` statement you learned earlier in this book).

5. Activate `VBAProject(Chap22b.xlsm)` and notice `Module1` in the `Modules` folder. `Module1` should contain the same procedure as `Module1` in the `Chap22.xlsm` workbook.

NOTE	*When you copy a module to another workbook, you may also need to add the necessary references to the VBA project to ensure that the code can be compiled and then run without errors. Please refer to Hands-On 22.18 and 22.19 to learn how you can add a reference programmatically.*

Copying (Exporting/Importing) All Modules

Sometimes, you may want to transfer all your VBA code from one project to another. The procedure shown below exports all the modules in the specified workbook file to an external text file and then imports them into another workbook. An error occurs if the receiving workbook cannot be activated. The procedure traps this error by executing the code that opens the required workbook. If the text file with the same name already exists in the same folder, the procedure ensures that the file is deleted before the specified project modules are exported.

⊙ Hands-On 22.10   Exporting/Importing All Modules

This Hands-On assumes that the `Chap22b.xlsm` workbook created in the previous Hands-On is open:

1. Insert a new module into the current VBA project in the `Chap22.xlsm` workbook.

2. In the Code window, enter the following procedure:

```
Sub CopyAllModules(wkbFrom As String, _
                   wkbTo As String)

    Dim objVBComp As VBComponent
    Dim wkb As Workbook
    Dim strFile As String

    Set wkb = Workbooks(wkbFrom)
```

```
On Error Resume Next
Workbooks(wkbTo).Activate
If Err.Number <> 0 Then Workbooks.Open wkbTo

strFile = wkb.Path & "\vbCode.bas"
If Dir(strFile) <> "" Then Kill strFile

For Each objVBComp In wkb.VBProject.VBComponents
    If objVBComp.Type <> vbext_ct_Document Then
        objVBComp.Export strFile
      Workbooks(wkbTo).VBProject.VBComponents.Import strFile
    End If
Next

Set objVBComp = Nothing
Set wkb = Nothing
End Sub
```

3. In the Immediate window, type the following statement and press Enter to run the procedure:

```
CopyAllModules "Chap22.xlsm", "Chap22b.xlsm"
```

When you execute the above statement, the CopyModules procedure will copy all the modules from the Chap22.xlsm workbook to the Chap22b.xlsm workbook that was created in Hands-On 22.9. Observe that VBAProject(Chap22b.xlsm) now includes an additional module named Module11. This occurred because we previously copied a module to this workbook. When we executed the procedure, Excel appended a 1 to the original Module1 name.

NOTE	*When you copy a module whose name is the same as the name of a module in the receiving workbook, Excel assigns a new name to the inserted module following its default naming conventions.*

WORKING WITH PROCEDURES

Code modules contain procedures, and at times, you may want to:

- List all the procedures contained in a module or in all modules.
- Programmatically add or remove a procedure from a module.
- Programmatically create an event procedure.

The following sections demonstrate how to perform the above tasks.

Listing All Procedures in All Modules

Code modules contain declaration lines and code lines. You can obtain the number of lines in the declaration section of a module with the `CountOfDeclarationLines` property of the `CodeModule` object. Use the `CountOfLines` property of the `CodeModule` object to get the number of code lines in a module. Each code line belongs to a specific procedure. Use the `ProcOfLine` property of the `CodeModule` object to return the name of the procedure in which the specified line is located. You must provide the line number you want to check and the constant that specifies the type of procedure to locate. All subprocedures and function procedures use the `vbext_pk_Proc` constant. The following procedure prints to the Immediate window a list of all modules and all procedures within each module in the current VBA project:

```
Sub ListAllProc()
    Dim objVBProj As VBProject
    Dim objVBComp As VBComponent
    Dim objVBCode As CodeModule
    Dim strCurrent As String
    Dim strPrevious As String

    Dim x As Integer

    Set objVBProj = ThisWorkbook.VBProject

    For Each objVBComp In objVBProj.VBComponents
    If InStr(1, "1, 2", objVBComp.Type) Then
        Set objVBCode = objVBComp.CodeModule
        Debug.Print objVBComp.Name

        For x = objVBCode.CountOfDeclarationLines + 1 To _
                objVBCode.CountOfLines
            strCurrent = objVBCode.ProcOfLine(x, vbext_pk_Proc)

            If strCurrent <> strPrevious Then
                Debug.Print vbTab & objVBCode.ProcOfLine( _
                        x, vbext_pk_Proc)
                strPrevious = strCurrent
            End If
        Next
    End If
    Next

    Set objVBCode = Nothing
```

```
    Set objVBComp = Nothing
    Set objVBProj = Nothing
End Sub
```

When this procedure finishes executing, the list can be seen in the Immediate window (see Figure 22.8).

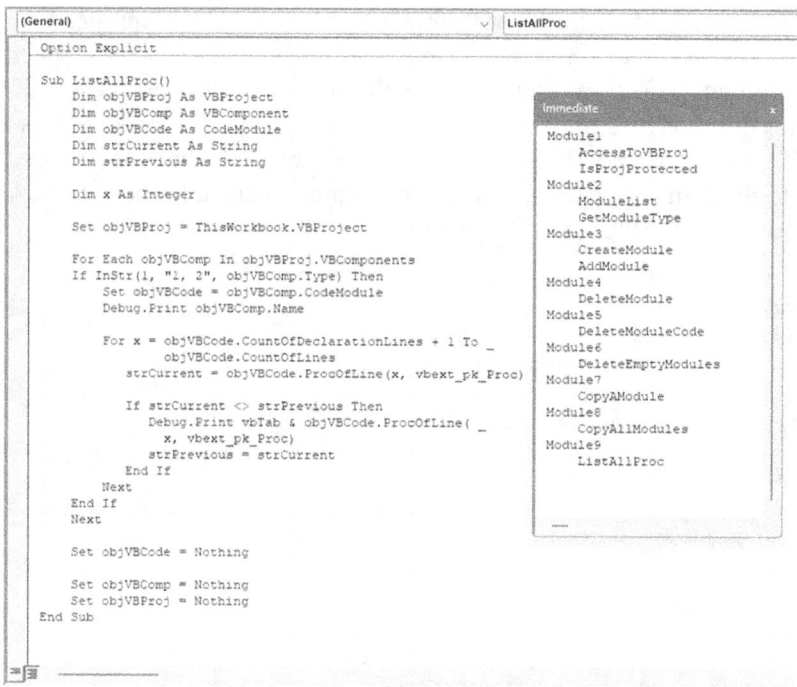

FIGURE 22.8. Listing all module names and their procedures in the Immediate window.

Adding a Procedure

It is fairly easy to write procedure code into a module. Use the `InsertLines` method of the `CodeModule` object to insert a line or lines of code at a specified location in a block of code. The `InsertLines` method requires two arguments: the line number at which you want to insert the code and the string containing the code you want to insert. The following example writes a simple procedure that opens a new workbook and renames the active sheet. The procedure is inserted at the end of the specified module code.

Hands-On 22.11 Adding a Procedure to a Module Using VBA

1. Insert a new module into the VBA project in the `Chap22.xlsm` workbook.

2. In the Code window, enter the following procedure:

```
Sub AddNewProc(strModName As String)
    Dim objVBCode As CodeModule
    Dim objVBProj As VBProject
    Dim strProc As String

    Set objVBProj = ThisWorkbook.VBProject

    Set objVBCode = objVBProj.VBComponents( _
        strModName).CodeModule

    strProc = "Sub CreateWorkBook()" & Chr(13)
    strProc = strProc & Chr(9) & "Workbooks.Add" & Chr(13)
    strProc = strProc & Chr(9)
    strProc = strProc & "ActiveSheet.Name = ""Test""" & Chr(13)
    strProc = strProc & "End Sub"

    Debug.Print strProc

    With objVBCode
        .InsertLines .CountOfLines + 1, strProc
    End With

    Set objVBCode = Nothing
    Set objVBProj = Nothing
End Sub
```

3. In the Immediate window, enter the following statement to run the above procedure and press Enter:

```
AddNewProc "Module6"
```

When the procedure finishes executing, `Module6` will contain a new procedure named `CreateWorkBook`. The Immediate window will display the procedure code.

Deleting a Procedure

Use the `DeleteLines` method of the `CodeModule` object to delete a single line or a specified number of lines. The `DeleteLines` method has two arguments; one is required and the other is optional. You must specify the first line you want to delete. Specifying the total number of lines you want to delete is optional. Before deleting an entire procedure, locate the line at which the specified procedure begins. This is done with the `ProcStartLine` property of the `CodeModule` object. This property requires two arguments: a string containing the name of the procedure and the kind of procedure to delete. Use the `vbext_pk_Proc` constant to delete a subprocedure or a function procedure.

The following procedure deletes the `CreateWorkBook` procedure created in the previous Hands-On exercise.

Hands-On 22.12   Deleting a Procedure from a Module Using VBA

1. Insert a new module into the current `VBAProject` in the `Chap22.xlsm` workbook.
2. In the Code window, enter the following procedure and functions:

```vba
Sub DeleteProc(strModName As String, strProcName As String)
    Dim objVBProj As VBProject
    Dim objVBCode As CodeModule
    Dim firstLine As Long
    Dim totalLines As Long

    On Error GoTo ErrorHandler

    ' Validate inputs
    If strModName = "" Or strProcName = "" Then
        MsgBox "Module or procedure name cannot be empty.", _
            vbExclamation
        Exit Sub
    End If

    ' Use the reusable function to get the VBProject object
    Set objVBProj = GetVBProject()

    If objVBProj Is Nothing Then
        MsgBox "Failed to retrieve VBProject object.", vbCritical
        Exit Sub
    End If

    ' Check if the procedure exists
    If ProcExists(strModName, strProcName) Then
      Set objVBCode = objVBProj.VBComponents(strModName).CodeModule
        With objVBCode
            firstLine = .ProcStartLine(strProcName, vbext_pk_Proc)
            totalLines = .ProcCountLines(strProcName, vbext_pk_Proc)
            .DeleteLines firstLine, totalLines
        End With
    Else
```

```vb
        MsgBox "Procedure not found in the specified module.", _
            vbInformation
    End If

Cleanup:
    ' Release objects
    Set objVBCode = Nothing
    Set objVBProj = Nothing
    Exit Sub

ErrorHandler:
    MsgBox "An error occurred: " & Err.Description, vbCritical
    Resume Cleanup
End Sub


Function GetVBProject() As VBProject
    On Error GoTo ErrorHandler

    ' Return the VBProject object of the current workbook
    Set GetVBProject = ThisWorkbook.VBProject
    Exit Function

ErrorHandler:
    MsgBox "An error occurred while accessing the VBProject: " & _
        Err.Description, vbCritical
    Set GetVBProject = Nothing
End Function


Function ProcExists(strModName As String, _
                    strProcName As String) As Boolean

    Dim objVBProj As VBProject

    Set objVBProj = ThisWorkbook.VBProject

    On Error Resume Next

    ' first find out if the specified module exists
    If ModuleExists(strModName) = True Then
        ProcExists = objVBProj.VBComponents _
        (strModName).CodeModule.ProcStartLine _
        (strProcName, vbext_pk_Proc) <> 0
    End If
End Function
```

```
Function ModuleExists(strModName As String) As Boolean
  Dim objVBProj As VBProject

  Set objVBProj = ThisWorkbook.VBProject

  On Error Resume Next

  ModuleExists = Len(objVBProj.VBComponents(strModName).Name) <> 0
End Function
```

3. In the Immediate window, enter the following statement to run the DeleteProc procedure and press Enter:

```
DeleteProc "Module6", "CreateWorkBook"
```

When the procedure finishes executing, Module6 will no longer contain the CreateWorkBook procedure. Before attempting to delete a procedure from a specified module, it is recommended that you check whether the module and the procedure with the specified name exist. Therefore, we have created two functions to handle these checks: ProcExists and ModuleExists. You can call these functions whenever you need to test for the existence of a module or a procedure. Additionally, to improve code reusability, we have added the GetVBProject function to retrieve the VBProject object. This way we don't need to repeatedly include the same setup code for obtaining the VBProject object.

Creating an Event Procedure

While you could create an event procedure programmatically by using the InsertLines method of the CodeModule object, as you did earlier in the Adding a Procedure section, there is an easier way. Because event procedures have a specific structure and usually require a number of parameters, Visual Basic offers a special method to handle this task. The CreateEventProc method of the CodeModule object creates an event procedure with the required procedure declaration and parameters. All you need to do is specify the name of the event you want to add and the name of the object that is a source of the event. The CreateEventProc method returns the line at which the body of the event procedure starts. Use the InsertLines method of the CodeModule object to insert the code in the body of the event procedure.

The following procedure adds a new worksheet to the current workbook and writes the Worksheet_SelectionChange event procedure to its code module.

⊙ Hands-On 22.13   Creating an Event Procedure with VBA

1. Insert a new module into the VBA project in the `Chap22.xlsm` workbook.

2. In the Code window, enter the following procedure:

```
Sub CreateWorkSelChangeEvent()
    Dim objVBCode As CodeModule
    Dim wks As Worksheet
    Dim firstLine As Long

    ' Add a new worksheet
    Set wks = ActiveWorkbook.Worksheets.Add

    ' create a reference to the code module of
    ' the inserted sheet
    Set objVBCode = wks.Parent.VBProject.VBComponents( _
        wks.Name).CodeModule

    ' create an event procedure and return the line at
    ' which the body of the event procedure begins

    firstLine = objVBCode.CreateEventProc( _
        "SelectionChange", "Worksheet")

    Debug.Print "Procedure first line: " & firstLine

    ' proceed to add code to the body of the event procedure
    objVBCode.InsertLines firstLine + 1, Chr(9) & _
        "Dim myRange As Range"
    objVBCode.InsertLines firstLine + 2, Chr(9) & _
        "On Error Resume Next"
    objVBCode.InsertLines firstLine + 3, Chr(9) & _
        "Set myRange = Intersect(Range(""A1:A10""),Target)"
    objVBCode.InsertLines firstLine + 4, _
        Chr(9) & "If Not myRange Is Nothing Then"
    objVBCode.InsertLines firstLine + 5, _
        Chr(9) & Chr(9) & _
        "MsgBox ""Data entry or edits are not permitted."""
    objVBCode.InsertLines firstLine + 6, _
        Chr(9) & "End If"

    Set objVBCode = Nothing
    Set wks = Nothing
End Sub
```

3. Run the `CreateWorkSelChangeEvent` procedure.

As soon as the procedure finishes executing, the newly inserted sheet module is activated and displays the following event procedure code:

```
Private Sub Worksheet_SelectionChange(ByVal Target As Range)
    Dim myRange As Range
    On Error Resume Next
    Set myRange = Intersect(Range("A1:A10"), Target)
    If Not myRange Is Nothing Then
        MsgBox "Data entry or edits are not permitted."
    End If
End Sub
```

4. To test the newly inserted event procedure, switch to the sheet where the above procedure is located and click on any cell in the A1:A10 range. The `Worksheet_SelectionChange` event procedure will be triggered, causing a message to appear.

WORKING WITH USERFORMS

In Chapter 16, you learned how to create and work with UserForms. Creating UserForms is done most easily by utilizing the manual method; however, at times, you may find it necessary to use VBA to create a quick form on the fly and display it correctly on the user's screen.

To programmatically add a UserForm to the active project, use the `Add` method of the `VBComponents` collection and specify `vbext_ct_MSForm` for the type of component to add:

```
Dim objVBComp As VBComponent
Set objVBComp = Application.VBE.ActiveVBProject. _
    VBComponents.Add(vbext_ct_MSForm)
```

To change the name of the UserForm, use the `Name` property of the `VBComponent` object. To change other properties of the UserForm, use the `VBComponent`'s `Properties` collection. For example, to change the name and caption of the UserForm, use the following statement block:

```
With objVBComp
    .Name = "ReportGenerator"
    .Properties("Caption") = "My Report Form"
End With
```

Here's the complete procedure:

```
Sub ReportGeneratorForm()
    Dim objVBComp As VBComponent

    Set objVBComp = Application.VBE.ActiveVBProject. _
        VBComponents.Add(vbext_ct_MSForm)
    With objVBComp
        .Name = "ReportGenerator"
        .Properties("Caption") = "My Report Form"
    End With
    Set objVBComp = Nothing
End Sub
```

To delete the UserForm from the project, use the Remove method of the VBComponents collection:

```
Set objVBComp = Application.VBE.ActiveVBProject. _
    VBComponents("ReportGenerator")
Application.VBE.ActiveVBProject.VBComponents.Remove objVBComp
```

Creating and Manipulating UserForms

The procedure in Hands-On 22.14 creates a simple UserForm, as shown in Figure 22.9, and writes procedures for each of the form's controls.

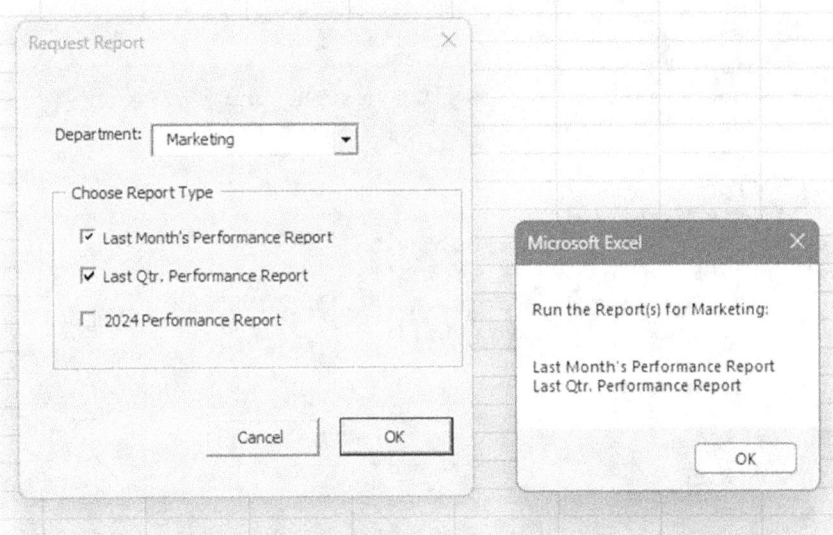

FIGURE 22.9.   The UserForm, as well as all the controls and procedures used by this form, were created entirely with VBA (see Hands-On 22.14).

⊙ Hands-On 22.14   Creating a Custom UserForm with VBA

1. Insert a new module into the VBA project in the Chap22.xlsm workbook.
2. In the Code window, enter the following procedure:

```
Sub AddUserForm()
Dim objVBProj As VBProject
Dim objVBComp As VBComponent
Dim objVBFrm As UserForm
Dim objChkBox As Object
Dim x As Integer
Dim sVBA As String

Set objVBProj = Application.VBE.ActiveVBProject
Set objVBComp = objVBProj.VBComponents.Add(vbext_ct_MSForm)

With objVBComp
' read form's name and other properties
    Debug.Print "Default Name " & .Name
    Debug.Print "Caption: " & .DesignerWindow.Caption
    Debug.Print "Form is open in the Designer window: " & _
        .HasOpenDesigner
    Debug.Print "Form Name " & .Name
    Debug.Print "Default Width " & .Properties("Width")
    Debug.Print "Default Height " & .Properties("Height")

' set form's name, caption and size
    .Name = "RequestReport"
    .Properties("Caption") = "Request Report"
    .Properties("Width") = 250
    .Properties("Height") = 250
End With

Set objVBFrm = objVBComp.Designer
With objVBFrm
    With .Controls.Add("Forms.Label.1", "lbName")
        .Caption = "Department:"
        .AutoSize = True
        .Width = 48
        .Top = 30
        .Left = 20
    End With

    With .Controls.Add("Forms.Combobox.1", "cboDept")
        .Width = 110
        .Top = 30
```

```vba
        .Left = 70
End With

' add frame control
With .Controls.Add("Forms.Frame.1", "frReports")
    .Caption = "Choose Report Type"
    .Top = 60
    .Left = 18
    .Height = 96
End With

' add three check boxes
Set objChkBox = .frReports.Controls.Add("Forms.CheckBox.1")
With objChkBox
    .Name = "chk1"
    .Caption = "Last Month's Performance Report"
    .WordWrap = False
    .Left = 12
    .Top = 12
    .Height = 20
    .Width = 186
End With

Set objChkBox = .frReports.Controls.Add("Forms.CheckBox.1")
With objChkBox
    .Name = "chk2"
    .Caption = "Last Qtr. Performance Report"
    .WordWrap = False
    .Left = 12
    .Top = 32
    .Height = 20
    .Width = 186
End With

Set objChkBox = .frReports.Controls.Add("Forms.CheckBox.1")
With objChkBox
    .Name = "chk3"
    .Caption = Year(Now) - 1 & " Performance Report"
    .WordWrap = False
    .Left = 12
    .Top = 54
    .Height = 20
    .Width = 186
End With
```

```
      ' Add and position OK and Cancel buttons
      With .Controls.Add("Forms.CommandButton.1", "cmdOK")
            .Caption = "OK"
            .Default = "True"
            .Height = 20
            .Width = 60
            .Top = objVBFrm.InsideHeight - .Height - 20
            .Left = objVBFrm.InsideWidth - .Width - 10
      End With

      With .Controls.Add("Forms.CommandButton.1", "cmdCancel")
            .Caption = "Cancel"
            .Height = 20
            .Width = 60
            .Top = objVBFrm.InsideHeight - .Height - 20
            .Left = objVBFrm.InsideWidth - .Width - 80
      End With
End With

' populate the combo box
With objVBComp.CodeModule
      x = .CountOfLines
      .InsertLines x + 1, "Sub UserForm_Initialize()"
      .InsertLines x + 2, vbTab & "With Me.cboDept"
      .InsertLines x + 3, vbTab & vbTab & ".addItem ""Marketing"""
      .InsertLines x + 4, vbTab & vbTab & ".addItem ""Sales"""
      .InsertLines x + 5, vbTab & vbTab & ".addItem ""Finance"""
      .InsertLines x + 6, vbTab & vbTab & _
          ".addItem ""Research & Development"""
      .InsertLines x + 7, vbTab & vbTab & _
          ".addItem ""Human Resources"""
      .InsertLines x + 8, vbTab & "End With"
      .InsertLines x + 9, "End Sub"

      ' write a procedure to handle the Cancel button

      Dim firstLine As Long
      With objVBComp.CodeModule
          firstLine = .CreateEventProc("Click", "cmdCancel")
          .InsertLines firstLine + 1, "    Unload Me"
      End With

      ' write a procedure to handle OK button
      sVBA = "Private Sub cmdOK_Click()" & vbCrLf
      sVBA = sVBA & " Dim ctrl As Control" & vbCrLf
```

```
sVBA = sVBA & " Dim chkflag As Integer" & vbCrLf
sVBA = sVBA & " Dim strMsg As String" & vbCrLf
sVBA = sVBA & " If Me.cboDept.Value = """" Then " & vbCrLf
sVBA = sVBA & "   MsgBox ""Select the Department.""" & _
                   vbCrLf
sVBA = sVBA & "    Me.cboDept.SetFocus " & vbCrLf
sVBA = sVBA & "     Exit Sub" & vbCrLf
sVBA = sVBA & "  End If" & vbCrLf
sVBA = sVBA & "  For Each ctrl In Me.Controls " & vbCrLf
sVBA = sVBA & "  Select Case ctrl.Name" & vbCrLf
sVBA = sVBA & "       Case ""chk1"", ""chk2"", ""chk3""" _
                   & vbCrLf
sVBA = sVBA & "  If ctrl.Value = True Then" & vbCrLf
sVBA = sVBA & "  strMsg = strMsg & vbCrLf & ctrl.Caption " _
                   & Chr(13) & vbCrLf
sVBA = sVBA & "     chkflag = 1" & vbCrLf
sVBA = sVBA & "    End If" & vbCrLf
sVBA = sVBA & "   End Select" & vbCrLf
sVBA = sVBA & " Next" & vbCrLf
sVBA = sVBA & " If chkflag = 1 Then" & vbCrLf
sVBA = sVBA & "  MsgBox ""Run the Report(s) for """
sVBA = sVBA & " & Me.cboDept.Value & "":"""
sVBA = sVBA & " & Chr(13) & Chr(13) & strMsg" & vbCrLf
sVBA = sVBA & " Else" & vbCrLf
sVBA = sVBA & "  MsgBox ""Please select Report type.""" _
                   & vbCrLf
sVBA = sVBA & " End If" & vbCrLf
sVBA = sVBA & "End Sub"

    .AddFromString sVBA
End With
Set objVBComp = Nothing
End Sub
```

In the `AddUserForm` procedure, the following statement creates a blank UserForm:

```
Set objVBComp = objVBProj.VBComponents.Add(vbext_ct_MSForm)
```

Next, the form's default name and other properties (`Caption`, `Width`, and `Height`) are written to the Immediate window and then reset with new values.

Before we can access the content of the UserForm, we need a reference to the `VBComponent`'s `Designer` object, like this:

```
Set objVBFrm = objVBComp.Designer
```

Several With...End With statement blocks are used to add controls (labels, combo boxes, frames, checkboxes, and command buttons) to the blank UserForm and position them within the form by using the Top and Left properties. The InsideHeight and InsideWidth properties are used to move the OK and Cancel buttons to the bottom of the UserForm. These properties return the height and width, in points, of the space that's available inside the form.

The remaining code in the procedure creates various event procedures for the UserForm and its controls. The first one is the UserForm_Initialize() procedure, which will populate the combo box control with department names before the form is displayed on the user's screen. Next, the event procedures for command buttons (OK and Cancel) are created. The cmdCancel_Click() event procedure unloads the form, and the cmdOK_Click() procedure displays a message box with information about the types of reports selected via the checkboxes.

Writing Code for Event Procedures

There are several techniques for adding code to event procedures:

❏ Using the InsertLines method—This involves directly inserting lines of code into the CodeModule.

❏ Using the AddFromString method—Create a string containing the desired code, and then add it to the CodeModule using this method. The added code will be placed immediately before the first procedure in the module.

❏ Using the AddFromFile method—This allows you to insert code from an external text file directly into the module.

NOTE	*The* CreateEventProc *method automatically adds the* Private Sub *and the* End Sub *statements and a space between these lines. All you need to do is add the actual code using the* InsertLines *statements. The* CreateEventProc *method returns the number of the line in the module where the* Private Sub *statement was added.*

3. Run the AddUserForm procedure.
   When the procedure finishes its execution, the VBE screen will display the form shown earlier in Figure 22.9.

4. Choose View | Code or press F7 and review the procedures that were programmatically created for the form by the AddUserForm procedure.

5. Choose Run | Run Sub/UserForm or press F5 to display and work with the form.

If you'd like to display your custom UserForm in a specific location on the screen, consider adding the following event procedure to the `AddUserForm` procedure. You must code this procedure by using one of the techniques described earlier.

```
Private Sub UserForm_Activate()
  With ReportSelector
    .Top = 100
    .Left = 25
  End With
End Sub
```

Copying UserForms Programmatically

If you need to add an existing UserForm to another workbook, you can simply export the form to disk by choosing File | Export File in the VBE screen. Excel will create a form file (identified with a `.frm` extension) that you can then import to another VBA project by choosing the File | Import File command in the VBE screen.

You can also automate the export/import process of UserForms by writing VBA code. The following example procedure exports the form created in the previous Hands-On. After the form is imported, a procedure is written to a standard module of the `Chap22b.xlsm` file (created earlier in this chapter) to display the form.

Hands-On 22.15   Copying a UserForm with VBA

1. Insert a new module into the VBA project in the `Chap22.xlsm` workbook.
2. In the Code window, enter the following procedure:

```
Sub UserFormCopy(strFileName As String)
    Dim objVBComp As VBComponent
    Dim wkb As Workbook

    On Error Resume Next
    Set wkb = Workbooks(strFileName)
     If Err.Number <> 0 Then
        Workbooks.Open ActiveWorkbook.Path & "\" & strFileName
         Set wkb = Workbooks(strFileName)
     End If

    For Each objVBComp In ThisWorkbook.VBProject.VBComponents
```

```
           If objVBComp.Type = 3 Then   ' this is a UserForm
               ' export the UserForm to disk
               objVBComp.Export Filename:=objVBComp.Name
               ' import the UserForm to a specific workbook
               wkb.VBProject.VBComponents.Import _
                   Filename:=objVBComp.Name
               ' delete two form files created by the Export method
               Kill objVBComp.Name
               Kill objVBComp.Name & ".frx"
           End If
       Next

       ' add a standard module to the workbook
       ' and write code to show the UserForm
       Set objVBComp = wkb.VBProject.VBComponents. _
           Add(vbext_ct_StdModule)

   objVBComp.CodeModule.AddFromString _
       "Sub ShowRequestReport()" & vbCrLf & _
       "    RequestReport.Show" & vbCrLf & _
       "End Sub" & vbCrLf

   ' close the Code pane
       objVBComp.CodeModule.CodePane.Window.Close

   ' run the ShowReportSelector procedure to display the form
       Application.Run wkb.Name & "!ShowRequestReport"

       Set objVBComp = Nothing
       Set wkb = Nothing
   End Sub
```

3. Run the `UserFormCopy` procedure by entering the following statement in the Immediate window:

```
UserFormCopy "Chap22b.xlsm"
```

When you execute the above statement, the UserForm is imported into `Chap22b.xlsm` workbook and displayed on the user's screen.

WORKING WITH REFERENCES

When you write VBA procedures, you often need to access objects that are stored in external object libraries. For example, so far in this chapter, you have used objects defined in the Microsoft Visual Basic for Applications Extensibility

5.3 library. In other chapters of this book, you have worked with objects exposed by the Microsoft Word 16.0 object library, Microsoft Outlook 16.0 object library, Microsoft Access 16.0 object library, Microsoft ActiveX Data Objects 6.1 library, and so on.

Understanding Early Binding and Late Binding

There are two ways to expose an object model to your Excel application: *early binding* and *late binding*. Binding means exposing the client object model to the host application, in this case, Microsoft Excel:

- You use early binding when you expose the object model at design time. This is done by choosing Tools | References in the VBE screen. The References dialog box lists files with which you can bind. To manipulate a specific application in your Excel VBA project, you must select the checkbox next to the name of the library you want to use.

- You perform late binding when you bind the object library in your VBA code at runtime. Instead of using the References dialog box, you use the `GetObject` or `CreateObject` functions.

By adding a reference to the external object library using the Tools | References dialog box (early binding), you are able to get on-the-fly programming assistance for the objects you need to include in your VBA code, consequently avoiding many syntax errors. You can also view the application's object model via the Object Browser and have access to the application's built-in constants. Additionally, your code runs faster because the references to the external libraries are checked and compiled at design time. Problems arise, however, when you move your code to other computers that do not have the external libraries installed. The procedures that ran perfectly well on your computer suddenly begin to display compile-time errors that cannot be trapped using standard error-handling techniques. To ensure that the end users have the required references and object libraries, you must write code that checks not only whether these libraries are present but also that they are the correct version.

The following sections will show you how to:

- List references to the external object libraries that have been selected in the References dialog box.
- Add a reference to a specific library on the fly.
- Remove missing library references.
- Check for broken references.

Creating a List of References

The `Reference` object in the `References` collection represents a reference to a type library or a VBA project. You can use various properties of the `Reference` object to:

- Find out whether the reference is built in or added by a developer (`BuiltIn` property).
- Determine whether the reference is broken (`IsBroken` property).
- Find out the reference version number (`Major` and `Minor` properties).
- Get the description of the reference as it appears in the Object Browser (`Description` property).
- Return the full path to the workbook, DLL, OCX, TLD, or OLB file that is a source of the reference (`FullPath` property).
- Return the globally unique identifier for reference (`Guid` property).
- Determine the reference type (`Type` property).

The following procedure prints to the Immediate window the names of all VBA projects, the names and full paths of selected references for each VBA project, and the names of each project's components.

⊙ Hands-On 22.16   Listing VBA Project References and Components Using VBA

1. Insert a new module into the VBA project in the Chap22.xlsm workbook.
2. In the Code window, enter the following procedure:

```
Sub ListPrjCompRef()
        Dim objVBPrj As VBIDE.VBProject
        Dim objVBCom As VBIDE.VBComponent
        Dim vbRef As VBIDE.Reference

    ' list VBA projects as well as references and
    ' component names they contain
        For Each objVBPrj In Application.VBE.VBProjects
        Debug.Print objVBPrj.Name
        Debug.Print vbTab & "References"
        For Each vbRef In objVBPrj.References
            With vbRef
            Debug.Print vbTab & vbTab & .Name & "---" & .FullPath
            End With
        Next
        Debug.Print vbTab & "Components"
```

```
      For Each objVBCom In objVBPrj.VBComponents
            Debug.Print vbTab & vbTab & objVBCom.Name
      Next
   Next
   Set vbRef = Nothing
   Set objVBCom = Nothing
   Set objVBPrj = Nothing
End Sub
```

3. Run the `ListPrjCompRef` procedure.

When the procedure finishes executing, the Immediate window displays information about all the VBA projects that are currently open in Excel.

Adding a Reference

The `AddFromFile` method of the `References` collection is used to add a reference to a project from a file. You must specify the project library filename, including its path. The following procedure adds a reference to the Microsoft Scripting Runtime library, which is stored in the `scrrun.dll` (dynamic link library) file.

⊙ Hands-On 22.17   Adding a Project Reference with VBA

1. Insert a new module into the VBA project in the `Chap22.xlsm` workbook.

2. In the Code window, enter the following procedure:

```
Sub AddRef()
    Dim objVBProj As VBProject

    Set objVBProj = ThisWorkbook.VBProject

    On Error GoTo ErrorHandle
    objVBProj.References.AddFromFile _
        "C:\Windows\System32\scrrun.dll"
    MsgBox "The reference to the Microsoft Scripting " _
        & "Runtime was set."
    Application.SendKeys "%tr"

ExitHere:
    Set objVBProj = Nothing
    Exit Sub
ErrorHandle:
    MsgBox "The reference to the Microsoft Scripting " & _
        " Runtime already exists."
    GoTo ExitHere
End Sub
```

3. Run the `AddRef` procedure. When a message box appears, click OK.
   If the reference to the Microsoft Scripting Runtime was set during the procedure execution, the References dialog box will appear with a checkmark next to Microsoft Scripting Runtime.

4. Close the References dialog box if it is open.

What is GUID?

Every type library has an associated Globally Unique Identifier (GUID) that is stored in the Windows registry. If you know the GUID of the reference, you can add a reference by using the `AddFromGuid` method. This method requires three arguments: a string expression representing the GUID of the reference, the major version number of the reference, and the minor version number of the reference. The `AddFromGuid` method searches the registry to find the reference you want to add.

The following procedure prints to the Immediate window the names, GUIDs, and version numbers of the libraries that are already installed in the active workbook's VBA project. The procedure also adds a reference to the Microsoft ActiveX Data Objects 6.1 library.

⊚ Hands-On 22.18 Obtaining Information About Installed VBA Libraries from the Registry

1. In the Code window where you entered the previous procedure, enter the `AddRef_FromGuid` procedure as shown below:

```
Sub AddRef_FromGuid()
   Dim objVBProj As VBProject
   Dim i As Integer
   Dim strName As String
   Dim strGuid As String
   Dim strMajor As Long
   Dim strMinor As Long

   On Error Resume Next

   Set objVBProj = ActiveWorkbook.VBProject

   ' reference the Microsoft ActiveX Data Objects 6.1 library
   ThisWorkbook.VBProject.References.AddFromGuid _
      "{B691E011-1797-432E-907A-4D8C69339129}", 6, 1

   ' Find out what libraries are already installed
   For i = 1 To objVBProj.References.Count
```

```
            strName = objVBProj.References(i).Name
            strGuid = objVBProj.References(i).GUID
            strMajor = objVBProj.References(i).Major
            strMinor = objVBProj.References(i).Minor
            Debug.Print strName & " - " & strGuid & _
                ", " & strMajor & ", " & strMinor
      Next i
End Sub
```

2. Run the `AddRef_FromGuid` procedure.

The procedure produces the following list of references in the Immediate window:

```
VBA - {000204EF-0000-0000-C000-000000000046}, 4, 2
Excel - {00020813-0000-0000-C000-000000000046}, 1, 9
stdole - {00020430-0000-0000-C000-000000000046}, 2, 0
Office - {2DF8D04C-5BFA-101B-BDE5-00AA0044DE52}, 2, 8
VBIDE - {0002E157-0000-0000-C000-000000000046}, 5, 3
MSForms - {0D452EE1-E08F-101A-852E-02608C4D0BB4}, 2, 0
Scripting - {420B2830-E718-11CF-893D-00A0C9054228}, 1, 0
ADODB - {B691E011-1797-432E-907A-4D8C69339129}, 6, 1
```

If some of the references are missing from your list, ensure that `Chap22b.xlsm` is closed and run the procedure again.

Removing a Reference

To remove an unwanted reference from the VBA project, use the `Remove` method of the `References` collection. The following procedure removes the reference to the Microsoft ActiveX Data Objects 6.1 object library that was added by the `AddRef_FromGuid` procedure in the previous Hands-On.

(◉) **Hands-On 22.19   Removing a Reference Using VBA**

1. In the Code window where you entered the previous procedure, enter the `RemoveRef` procedure as shown below:

```
Sub RemoveRef()
    Dim objVBProj As VBProject
    Dim objRef As Reference
    Dim sRefFile As String

    Set objVBProj = ActiveWorkbook.VBProject

    ' Loop through the references and delete
    ' the reference to ADODB library
    For Each objRef In objVBProj.References
```

```
        If InStr(1, objRef.Description, "Data Objects 6.1") > 0 Then
            objVBProj.References.Remove objRef
            Exit For
        End If
    Next objRef
End Sub
```

2. Run the `RemoveRef` procedure. When the procedure finishes executing, open the References dialog box to verify that the reference to the Microsoft ActiveX Data Objects 6.1 object library is no longer selected.

In addition to removing references to external object libraries, you can remove any existing references to other VBA projects. This is done by checking the `BuiltIn` property of the `Reference` object and removing the reference when the `BuiltIn` property is not `True`:

```
For Each objRef in objVBProjReferences
    If Not objRef.BuiltIn Then objVBProj.References.Remove objRef
Next objRef
```

The `BuiltIn` property of the `Reference` object returns `False` if a reference isn't a default reference. When a reference is not built in, it can be removed. Default references cannot be removed.

Checking for Broken References

If the required object libraries are not installed on a user's computer or aren't the correct version, the bad references are marked as "missing" in the References dialog box. You can use the `IsBroken` property to find these invalid references. The `IsBroken` property returns a Boolean value `True` if the `Reference` object no longer points to a valid reference in the registry. If the reference is valid, `False` is returned. The code to check for broken references should be included or called from the `Workbook_Open` event procedure before attempting to add any new references via code.

The following example procedure checks for broken references.

(◉) **Hands-On 22.20   Checking for Broken References in a VBA Project**

1. In the `ThisWorkbook` module of `Chap22_SourceCode` (`Chap22.xlsm`), enter the following `Workbook_Open` event procedure:

```
Private Sub Workbook_Open()
    Dim objVBProj As VBProject
    Dim objRef As Reference
    Dim refBroken As Boolean
```

```
        Set objVBProj = ThisWorkbook.VBProject

        ' Loop through the selected references in
        ' the References dialog box
        For Each objRef In objVBProj.References
        ' If the reference is broken, get its name and its GUID
            If objRef.IsBroken Then
                Debug.Print objRef.Name
                Debug.Print objRef.GUID
                refBroken = True
            End If
        Next
        If refBroken = False Then
                Debug.Print "All references are valid."
        End If
    End Sub
```

2. Save and close the `Chap22.xlsm` workbook. Do not exit Microsoft Excel.
3. Reopen the `Chap22.xlsm` workbook.
   When the workbook opens, Excel executes the code in the `Workbook_Open` event procedure.
4. Switch to the VBE window and activate the Immediate window. If broken references are found in the active project, you will see the reference name and its GUID; otherwise, a message is displayed that all references are valid.

If you'd like to test whether a specific reference is valid, insert a new module into the active VBA project and write a function procedure like this:

```
Function IsBrokenRef(strRef As String) As Boolean
    Dim objVBProj As VBProject
    Dim objRef As Reference

    Set objVBProj = ThisWorkbook.VBProject

    For Each objRef In objVBProj.References

    If strRef = objRef.Name And objRef.IsBroken Then
        IsBrokenRef = True
        Exit Function
    End If
    Next

    IsBrokenRef = False
End Function
```

To test the above function, you could enter the following statements in the Immediate window:

```
ref = IsBrokenRef("OLE Automation")
?ref
```

If `True`, the reference is invalid; if `False`, it is valid.

WORKING WITH WINDOWS

The VBE screen contains numerous windows. Each window (VBE main window, Project Explorer, Properties window, Immediate and Watch windows, Code window, Designer windows, and so on) is represented by the `Window` object. Each `Window` object is a member of the `VBIDE.Windows` collection. Use the `Type` property of the `Window` object to determine the window type. Available window types are listed in Table 22.1.

TABLE 22.1.   Window types available in the VBA project.

Window Description	Constant	Value
Code window	vbext_wt_CodeWindow	0
Designer	vbext_wt_Designer	1
Object Browser	vbext_wt_Browser	2
Watch window	vbext_wt_Watch	3
Locals window	vbext_wt_Locals	4
Immediate window	vbext_wt_Immediate	5
Project Explorer window	vbext_wt_ProjectWindow	6
Properties window	vbext_wt_PropertyWindow	7
Find dialog box	vbext_wt_Find	8
Search and Replace dialog box	vbext_wt_FindReplace	9
Toolbox	vbext_wt_Toolbox	10
Linked window frame	vbext_wt_LinkedWindowFrame	11
Main window	vbext_wt_MainWindow	12
Tool window	vbext_wt_ToolWindow	15

The following procedure loops through all the open windows in the VBE, closes the Immediate window, and displays a dialog box with the names of open windows.

⊙ **Hands-On 22.21   Working with Windows in the VBE Screen**

1. Insert a new module into the VBA project in the Chap22.xlsm workbook.
2. In the Code window, enter the following procedure:

```
Sub ListOpenWindows()
    Dim objWin As VBIDE.Window
    Dim strOpenWindows As String

    strOpenWindows = "The following windows are open:" & _
            vbCrLf & vbCrLf

    For Each objWin In Application.VBE.Windows
        Select Case objWin.Type
            Case vbext_wt_Immediate
                MsgBox objWin.Caption & " window was closed."
                objWin.Close
            Case Else
                strOpenWindows = strOpenWindows & _
                    objWin.Caption & vbCrLf
        End Select
    Next
    MsgBox strOpenWindows
    Set objWin = Nothing
End Sub
```

3. Run the ListOpenWindows procedure.
4. You should see the list of all open windows in a message box.

WORKING WITH VBE MENUS AND TOOLBARS

In Chapter 18, you wrote VBA code to create or modify shortcut menus. Using the same CommandBar object that you are already familiar with, you can now customize menus and toolbars in the VBE. To work with the CommandBars collection, you need to ensure that a reference to the Microsoft Office 16.0 object library is set in the References dialog box. If the reference to this library is not set, Excel displays a "User-defined type not defined" error message when the code attempts to access the CommandBars collection.

Generating a Listing of VBE CommandBar Objects and Controls

The following procedure lists all the CommandBar objects that can be found on the VBE screen. Each command bar is defined as a menu bar, toolbar, or pop-up menu via the `Type` property of the `CommandBar` object. Each `CommandBar` object has a number of controls assigned to it. The procedure lists all these controls for each `CommandBar`, including the control IDs.

⊙ Hands-On 22.22   Listing VBE CommandBar Objects and Controls

1. Insert a new module into the VBA project in the `Chap22.xlsm` workbook.
2. In the Code window, enter the following procedure:

```
Sub ListVBECmdBars()
    Dim objCmdBar As CommandBar
    Dim strCmdType As String
    Dim c As Variant

    Workbooks.Add
    Range("A1").Select

    With ActiveCell
        .Offset(0, 0) = "CommandBar Name"
        .Offset(0, 1) = "Control Caption"
        .Offset(0, 2) = "Control ID"
    End With

    For Each objCmdBar In Application.VBE.CommandBars
        Select Case objCmdBar.Type
            Case 0
                strCmdType = "toolbar"
            Case 1
                strCmdType = "menu bar"
            Case 2
                strCmdType = "popup menu"
        End Select

        ActiveCell.Offset(1, 0) = objCmdBar.Name & _
            " (" & strCmdType & ")"

        For Each c In objCmdBar.Controls
            ActiveCell.Offset(1, 0).Select
            With ActiveCell
                .Offset(0, 1) = c.Caption
                .Offset(0, 2) = c.ID
            End With
        Next
    Next
```

```
       Columns("A:C").AutoFit

       Set objCmdBar = Nothing
   End Sub
```

3. Run the `ListVBECmdBars` procedure.

The procedure creates a new workbook with the names of `CommandBar` objects and controls found in the VBE (see the partial listing in Figure 22.10).

	A	B	C	D
1	CommandBar Name	Control Caption	Control ID	
2	Menu Bar (menu bar)	&File	30002	
3		&Edit	30003	
4		&View	30004	
5		&Insert	30005	
6		F&ormat	30006	
7		&Debug	30165	
8		&Run	30012	
9		&Tools	30007	
10		&Add-Ins	30038	
11		&Window	30009	
12		&Help	30010	
13	Standard (toolbar)	Microsoft Excel	106	
14		Insert Object	32806	
15		&Save Book2	3	
16		Cu&t	21	
17		&Copy	19	
18		&Paste	22	
19		&Find...	141	
20		Can't Undo	128	
21		Can't Redo	129	
22		&Continue	186	

FIGURE 22.10. You can list all the CommandBar objects available in the VBE by running the custom ListVBECmdBars procedure, as demonstrated in this section.

Adding a CommandBar Button to the VBE

The following procedure adds a new command button to the end of the Tools menu in the VBE screen.

Hands-On 22.23   Modifying the VBE Tools Menu

1. In the same module where you entered the `ListVBECmdBars` procedure (see the previous section), type the following procedure:

```
Sub AddCmdButton_ToVBE()
    Dim objCmdBar As CommandBar
    Dim objCmdBtn As CommandBarButton

    ' get the reference to the Tools menu in the VBE
    Set objCmdBar = Application.VBE.CommandBars.FindControl _
        (ID:=30007).CommandBar
```

```
' add a button to the Tools menu
Set objCmdBtn = objCmdBar.Controls.Add(msoControlButton)

' set the new button's properties
With objCmdBtn
    .Caption = "List VBE menus and toolbars"
    .onAction = "ListVBECmdBars"
End With
End Sub
```

Please do not run this procedure yet, as it is not complete. To run a custom procedure assigned to any VBE menu item, you need to raise the Click event of the CommandBarButton. Use the CommandBarEvents object to trigger the Click event when a control on the CommandBar is clicked. This is done in a class module.

2. Choose Insert | Class Module.
3. In the Properties window, change the name of the Class1 module to clsCmdBarEvents.

This is a very important step that can't be skipped.

4. In the clsCmdBarEvents module Code window, enter the following code:

```
        Public WithEvents cmdBtnEvents As CommandBarButton

Private Sub cmdBtnEvents_Click( _
  ByVal Ctrl As Office.CommandBarButton, _
  CancelDefault As Boolean)
On Error Resume Next
MsgBox "inside class module"
' run the procedure specified in the onAction property
Application.Run Ctrl.OnAction

' specify that we already handled this event
CancelDefault = True
End Sub
```

Notice that the first statement in the class module uses the WithEvents keyword to declare an object called cmdBtnEvents of the type CommandBarButton, whose events we want to handle. Next, we specify that this object (cmdBtnEvents) is to receive the Click event when the menu item is selected. The first statement in the cmdBtnEvents_Click event procedure will prevent an error message from appearing in case the procedure specified in the onAction property doesn't exist. The next statement will run the procedure specified in the control's onAction property. Since the onAction property of the controls on the VBE CommandBars object does not directly execute the specified procedure, you must invoke the required procedure using the Run method of the Application object.

Now, let's connect the class module with the standard module containing the code of the AddCmdButton_ToVBE procedure that you created in Step 1.

5. Enter the following declaration line at the very top of the module that contains the AddCmdBtn_ToVBE procedure:

```
Dim myClickEvent As clsCmdBarEvents
```

In the previous declaration statement, myClickEvent is a module-level variable defined by the class clsCmdBarEvents. This variable will serve as a link between the menu item and the clsCmdBarEvents class module.

The final step requires adding additional code to the AddCmdButton_ToVBE procedure so that Visual Basic knows that it needs to handle the Click event for the menu item.

6. Enter the following code at the very end of the AddCmdButton_ToVBE procedure:

```
' create an instance of the clsCmdEvents class
Set myClickEvent = New clsCmdBarEvents

' hook up the class instance to the newly added button
Set myClickEvent.cmdBtnEvents = objCmdBtn

Set objCmdBtn = Nothing
Set objCmdBar = Nothing
```

The modified AddCmdButton_ToVBE procedure should look as follows:

```
Sub AddCmdButton_ToVBE()
    Dim objCmdBar As CommandBar
    Dim objCmdBtn As CommandBarButton

    ' get the reference to the Tools menu in the VBE
    Set objCmdBar = Application.VBE.CommandBars.FindControl _
            (ID:=30007).CommandBar

    ' add a button to the Tools menu
    Set objCmdBtn = objCmdBar.Controls.Add(msoControlButton)

    ' set the new button's properties
    With objCmdBtn
        .Caption = "List VBE menus and toolbars"
        .onAction = "ListVBECmdBars"
    End With
```

```
        ' create an instance of the clsCmdEvents class
        Set myClickEvent = New clsCmdBarEvents

        ' hook up the class instance to the newly added button
        Set myClickEvent.cmdBtnEvents = objCmdBtn

        Set objCmdBtn = Nothing
        Set objCmdBar = Nothing
    End Sub
```

7. Run the `AddCmdButton_ToVBE` procedure.

The procedure places a new menu item on the Tools menu (see Figure 22.11) and connects this item with the event handler located in the class module.

8. Choose Tools | List VBE menus and toolbars.

Visual Basic triggers the `Click` event of the selected menu item and runs the procedure code specified in the `onAction` property. When you switch to the Microsoft Excel application window, you should see a new workbook with a complete listing of the VBE `CommandBars` and their controls.

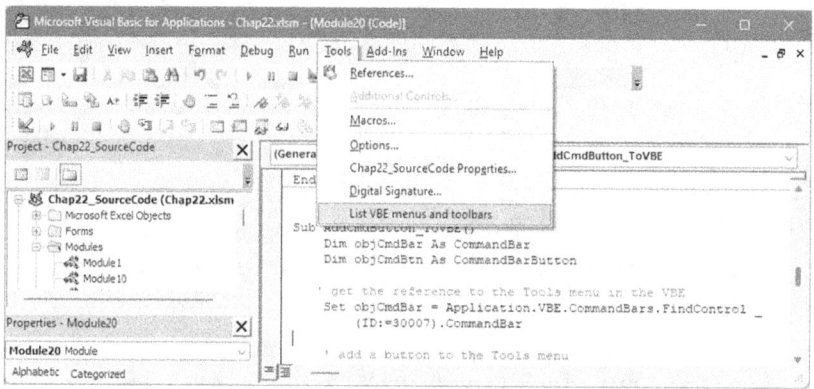

FIGURE 22.11. A custom menu option was added to the VBE Tools menu by a VBA procedure.

Removing a CommandBar Button from the VBE

The following procedure removes the custom menu item that was added to the Tools menu by the procedure in the previous section.

(⊙) **Hands-On 22.24   Removing a Custom Option from the VBE Menu**

1. In the same module where you entered the `AddCmdButton_ToVBE` procedure (see the previous section), type the following procedure:

```
Sub RemoveCmdButton_FromVBE()
    Dim objCmdBar As CommandBar
    Dim objCmdBarCtrl As CommandBarControl

    ' get the reference to the Tools menu in the VBE
    Set objCmdBar = Application.VBE.CommandBars("Tools")

    ' loop through the Tools menu controls
    ' and delete the control with the matching caption
    For Each objCmdBarCtrl In objCmdBar.Controls
      If objCmdBarCtrl.Caption = "List VBE menus and toolbars" Then
            objCmdBarCtrl.Delete
        End If
    Next

    Set objCmdBarCtrl = Nothing
    Set objCmdBar = Nothing
End Sub
```

2. Run the `RemoveCmdButton_FromVBE` procedure.
   Upon the procedure's completion, the Tools menu in the VBE screen no longer displays our custom item "List VBE menus and toolbars."

SUMMARY

In this chapter, you used numerous objects, properties, and methods from the Microsoft Visual Basic for Applications Extensibility 5.3 object library to control the VBE.

In the next chapter, you will learn how to take advantage of the Windows API functions when programming VBA.

Chapter **23**

CALLING WINDOWS API FUNCTIONS FROM VBA

While programming your Excel VBA applications, you may encounter a situation where VBA does not offer a method or property for performing a specific programming task, such as obtaining the user's screen resolution setting or changing the appearance of the UserForm. This is hardly a reason to give up. To ensure that all your program's specifications are implemented, look no further than the Windows operating system itself. A feature that is not directly supported by VBA might be supported by one of the thousands of functions that are exposed by the Windows Application Programming Interface (API). Therefore, overcoming many limitations of VBA boils down to learning how to locate the required Windows API function and then using it in your VBA procedure. This chapter shows you how functions found within the Windows API can help you extend your VBA procedures in areas where VBA does not provide the desired functionality.

UNDERSTANDING THE WINDOWS API LIBRARY FILES

The Windows API is a collection of subroutines and functions located in files called Dynamic Link Libraries (DLLs). These files have the file extension .dll and are located in the Windows System32 or SysWOW64 folder on every PC running Windows. The most popular DLLs are listed in Table 23.1.

The Windows API functions are written in the C language and can be accessed from VBA by utilizing the `Declare` statement, as discussed in the next section.

NOTE	*The following link provides a list of Windows API functions organized by category:* *http://msdn.microsoft.com/en-us/library/Aa383686.*

TABLE 23.1.   Windows API library files. The first three files are known as the main (core) DLLs. The remaining files are known as extension DLLs.

API Library File	Description
`user32.dll`	This library file contains numerous functions that can be called upon whenever your VBA program needs to manage the Windows environment. For example, here, you can find functions that relate to the use of windows such as setting or returning a window position, size, and state, determining whether the window is active, or whether it's a parent window or a child window. Functions found in this library will also allow you to handle messages between various windows and dialog boxes, as well as manage menus, cursors, the keyboard, and the clipboard.
`kernel32.dll`	This library file contains functions that manage the low-level operating system functions, such as memory management, resource management, computer drives, and file and folder management, as well as reading and writing to the Windows registry.
`gdi32.dll`	This library file has functions that will allow you to manage output to the screen. For example, you can manipulate fonts, drawings, graphics, bitmap images, and various display functions.
`comctl32.dll`	This library file provides common GUI controls such as `TreeView` or `ToolBar` controls.
`mapi32.dll`	This library file includes functions for working with electronic mail.
`netapi32. dll`	This library file provides functions for accessing and controlling networks.
`odbc32.dll`	This library file provides functions that allow applications to work with databases that are compliant with the Open Database Connectivity (ODBC).
`tapi32.dll`	This library file provides telephony functions used in managing voice mail and automated attendant phone systems.
`winmm.dll`	This library file allows access to multimedia capabilities such as playing sounds and managing audio devices.

HOW TO DECLARE A WINDOWS API FUNCTION

To access Windows API functions from VBA, you must use the `Declare` statement. This statement lets your program know where the function is located. The `Declare` statement must be added to the general declaration section of a standard module or a `UserForm` module. When added to the standard module, the API function can be called from anywhere within your VBA application. If added to the general declaration section of the `UserForm`, the function will have a local scope, available only to the procedures of that form. To indicate that the function is local to the form, you must precede the `Declare` statement with the `Private` keyword.

The syntax of the `Declare` statement depends on whether or not the procedure returns a value:

- If the procedure returns a value, declare it as a function, as shown in the syntax below:

```
[Public | Private] Declare Function name Lib "libname"
    [Alias "aliasname"] [[ByVal | ByRef] argument [As Type]
    [,[ByVal | ByRef] argument [As Type]…] [As Type]
```

- If the procedure does not return a value, declare it as a subroutine, as shown in the syntax below:

```
[Public | Private] Declare Sub name Lib "libname"
    [Alias  "aliasname"] [[ByVal | ByRef] argument [As Type]
    [,[ByVal | ByRef]argument [As Type]…]
```

The `Public` and `Private` keywords define the scope of the function or subroutine. Recall that a scope of `Private` will not allow the procedure to be used outside of the module in which it is declared.

The `Declare` statement must be followed by the `Function` or `Sub` keyword, depending on whether or not the procedure returns a value.

The name of the function or subroutine is followed by the name of the DLL library in which the function or subroutine is located.

Some functions and subroutines have an *alias* that indicates that the function has another name in the library. Aliases make it possible to call the function or subroutine by any name you want while providing a reference to the actual name of the function. An alias is particularly useful for those API functions that use characters that are illegal in VBA.

For example, many Windows API functions have names that begin with the underscore character (_). Consider the Windows API function `_lopen`, which opens an existing file. To have Excel VBA recognize this function's name as legal, use the `alias` keyword like this:

```
Declare Function lopen Lib "kernel32.dll" Alias _
    "_lopen" (ByVal lpPathName As String, _
    ByVal iReadWrite As Long) As Long
```

You should also alias the DLL function when its name conflicts with the VBA function of the same name. For example, use the following declaration when calling the `GetObject` function to let Excel know that you want to use the Windows API `GetObject` function and not the Excel version of this function:

```
Public Declare Function GetObjectAPI Lib "gdi32.dll" _
    Alias "GetObject" (ByVal hObject As Long, _
    ByVal nCount As Long, lpObject As Any) As Long
```

Sometimes, arguments need to be passed to functions and subroutines. Arguments can be passed by reference (default) or by value, as explained in the next section.

When declaring functions, you will need to define the type of value that is returned by including the `As Type` construct. Most API functions return a long integer that can be indicated by appending `As Long` at the end of the function declaration, or you can use a shortcut by specifying the return type in the name of the function. For example, if the function returns a long integer, append the ampersand (`&`) to the function name, as in the following:

```
Public Declare Function GetObjectAPI& Lib "gdi32.dll" _
Alias "GetObject" (ByVal hObject As Long, _
    ByVal nCount As Long, lpObject As Any)
```

The `GetObjectAPI&` function indicates that the `GetObject` function, which is located in the `gdi32.dll` library, returns a long integer.

NOTE	*The `Declare` statements that we've just discussed are for 32-bit operating systems. If used on a 64-bit system, they will fail. To ensure compatibility with 64-bit systems, you need to adjust the `Declare` statements in VBA to include the `PtrSafe` keyword. Additionally, replace data types that are pointer-dependent, such as `Long` with `LongPtr`. These changes allow the same code to work on both 32-bit and 64-bit systems seamlessly. This topic is covered in detail later in this chapter.*

Passing Arguments to the API Functions

When calling API functions, you must know what type of arguments a function expects to receive. Windows API functions expect the arguments to be passed by reference or by value. Passing an argument by reference means that the function passes a 32-bit or 64-bit pointer to the memory address where the value of the argument is stored. When the argument is passed by reference, it is possible for the function to change the value of the argument because it works with the actual memory address where the argument is stored. You can use the ByRef keyword when passing arguments by reference or omit this keyword entirely from the Declare statement, as passing arguments to function procedures by reference is the default in VBA.

When passing arguments by value, only a copy of the argument is sent to the function; therefore, the function cannot change the original value of the argument. If the API function expects to receive an argument by value and instead receives the argument by reference, the function will not work properly. Strings are always passed to API functions by value (ByVal). Always use the ByVal keyword when passing arguments by value.

Understanding the API Data Types

Windows API functions are written in C, and many of these functions rely on pointers, handles, and specific memory structures that require precise type declarations. When working with Windows API functions in VBA, it's essential to understand how C data types used by the API map to their VBA equivalents. C and VBA differ significantly in their handling of data types, so accurately declaring these in VBA ensures compatibility and prevents runtime errors.

The C data types and their VBA equivalents are listed in Table 23.2.

TABLE 23.2.   C data types vs. VBA data types.

C Data Type	VBA Equivalent (32-Bit)	VBA Equivalent (64-Bit)	Description
BOOL	Long	Long	Boolean value (nonzero = TRUE, 0 = FALSE)
CHAR / BYTE	Byte	Byte	8-bit unsigned integer
SHORT	Integer	Integer	16-bit signed integer
INT / LONG	Long	Long	32-bit signed integer
DWORD	Long	Long	32-bit unsigned integer

(Contd.)

C Data Type	VBA Equivalent (32-Bit)	VBA Equivalent (64-Bit)	Description
HANDLE / LPVOID	Long	LongPtr	Pointer/handle to objects or memory addresses
LPSTR / LPCSTR	String (ANSI)	String (ANSI)	Null-terminated string (ANSI)
LPWSTR / LPCWSTR	String (Unicode)	String (Unicode)	Null-terminated string (Unicode)
FLOAT / DOUBLE	Single / Double	Single / Double	32-bit or 64-bit floating-point value
SIZE_T	Long	LongPtr	Integer size that varies based on the operating system
STRUCT	User-Defined Type	User-Defined Type	Custom structures (mapped as Type in VBA)

Examples of Declaring Arguments

If the API function expects the Integer data type, you should pass it by value (ByVal) using the following syntax:

```
ByVal argumentname As Integer
```

The above Integer type declaration can also be written as:

```
ByVal argumentname%
```

The Long data type is the most common data type in the Windows API functions. If the API function expects the Long data type, you should pass it by value using the following syntax:

```
ByVal argumentname As Long
```

The above Long type declaration can also be written as:

```
ByVal argumentname&
```

The Windows API functions expect string arguments to be passed in the LPSTR format. In the C language, the LPSTR data type is a memory pointer to an array of characters. Because VBA stores strings in a different way than the API functions expect, to handle textual data, your VBA procedure must create a string buffer (a string filled with spaces or null characters) before calling the API function. In addition, because C uses 0-terminated strings and VBA does not, you must strip the 0 terminator from the end of the string returned by the API function (see Hands-On 23.1 later in this chapter). Strings are always passed to

API functions by value (ByVal) even when the API function updates the string. Use the following syntax when declaring arguments with the String data type:

```
ByVal argumentname As String
```

The above String type declaration can also be written as:

```
ByVal argumentname$
```

In the C language, the STRUCT data type represents multiple variables and is the equivalent of a user-defined data type (UDT) in VBA. Before declaring and calling a Windows API function that uses a Structure argument, you must first define the structure using a Type...End Type construct.

Let's look at the following example. The API function called GetCursorPos retrieves the mouse cursor's position in screen coordinates, This function requires the lpPoint variable in the form of the POINTAPI structure:

```
Public Declare PtrSafe Function GetCursorPos Lib _
    "user32.dll" (ByRef lpPoint As POINTAPI) As Long
```

Adding the PtrSafe keyword after the Declare keyword will allow the above API declaration to work with 64-bit Excel. When writing VBA code in 32-bit Excel, this keyword should not be used.

Use the Type...End Type construct to declare the POINTAPI structure, as shown below:

```
Private Type POINTAPI
    x as Long
    y as Long
End Type
```

By convention, the user-defined data type name is written in uppercase. The x and y are the coordinates of the cursor (relative to the screen). Once the structure is declared as a user-defined data type in VBA, you must declare a variable of that type to use when you call the Windows API function:

```
Dim cPos As POINTAPI
```

You can give the variable any name you choose as long as the name does not conflict with any VBA reserved keywords. Next, to retrieve the coordinates of the mouse, you need to write a VBA procedure that calls the GetCursorPos API function, like this:

```
Sub getMouseCoordinates()
    Dim cPos As POINTAPI
```

```
GetCursorPos cPos
    Debug.Print "x coordinate:" & cPos.x
    Debug.Print "y coordinate:" & cPos.y
End Sub
```

UDT arguments are passed to Windows API functions by reference (`ByRef`), as shown in Figure 23.1.

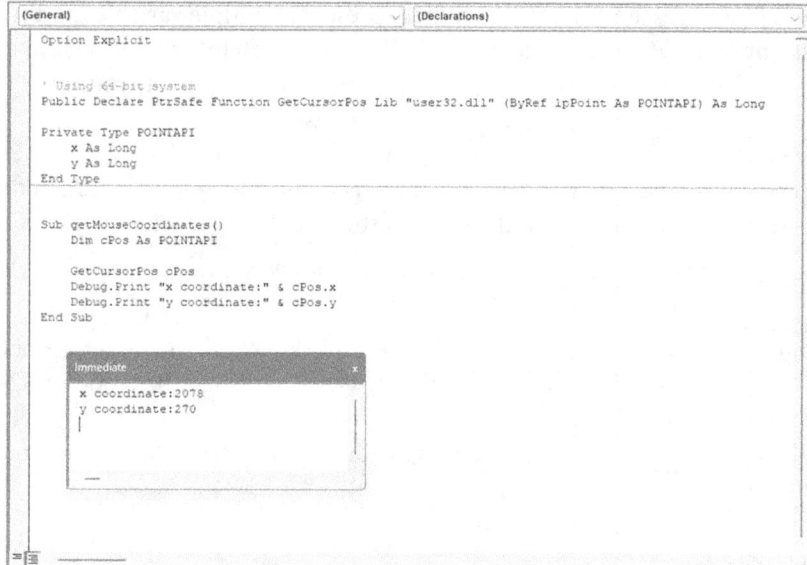

FIGURE 23.1.   Passing the user-defined data type to a Windows API function.

The `Any` data type argument is used for those API functions that can accept more than one data type for the same argument. The `Any` data type is passed by reference (`ByRef`) and its syntax is:

```
argumentname As Any
```

Using Constants with Windows API Functions

Most of the Windows API functions rely on a number of predefined constants that have to be passed to them in specific arguments. Some functions can work with more than a dozen constants, depending on what type of information you want to retrieve. For example, the following `GetSystemMetrics` function expects the `nIndex` argument that specifies the type of metric you want to return:

```
Public Declare PtrSafe Function GetSystemMetrics Lib "user32.dll" _
    (ByVal nIndex As Long) As Long
```

The `nIndex` argument must be passed to the `GetSystemMetrics` function as one of the number of predefined constants whose name begins with the letters SM followed by the underscore (_).

For example, to return the width and height of the screen of the primary display monitor in pixels, you must pass to the function the SM_CXSCREEN and SM_CYSCREEN constants as the `nIndex` argument. To do this, follow these steps:

1. Start by entering in a module the following constant declaration:
```
Public Const SM_CXSCREEN = 0   'defines the screen width
Public Const SM_CYSCREEN = 1   'defines the screen height
```
2. Enter the `Declare` statement to tell Excel that the external function is available:

```
Public Declare PtrSafe Function GetSystemMetrics Lib "user32.dll" _
    (ByVal nIndex As Long) As Long
```

3. Write a VBA procedure that calls the above function to retrieve the screen's dimensions:

```
Sub GetScreenResolution()
    Dim xSM As Long
    Dim ySM As Long

    xSM = GetSystemMetrics(SM_CXSCREEN)
    ySM = GetSystemMetrics(SM_CYSCREEN)

    Debug.Print "Your screen resolution is: " & _
    xSM & " x " & ySM
End Sub
```

4. Save the workbook file. You may also want to choose Debug | Compile VBA Project to ensure that there are no errors in the code you've entered.
5. Run the `GetScreenResolution` procedure. Assuming there were no runtime errors during the execution of your code, you should see your screen's resolution information in the Immediate window. The data retrieved should match the current setting of your screen resolution as displayed in the Control Panel. Figure 23.2 displays the code as entered in the Visual Basic module. The Immediate window shows the current screen resolution of the primary monitor.

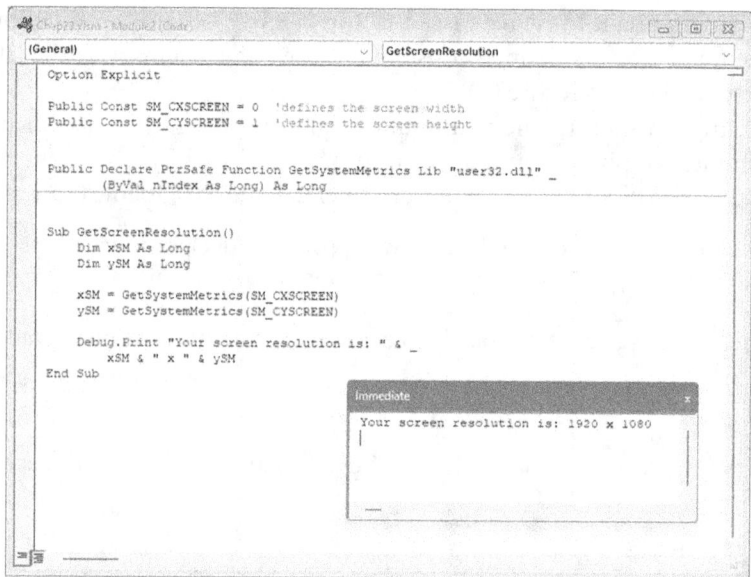

FIGURE 23.2. Retrieving information about screen resolution from a Windows API function using predefined constants.

When your application includes numerous calls to API functions and a great number of constants need to be declared in support of these functions, it is often difficult to know which constants are used with which function. To keep track of your constants and to make your code easier to write and comprehend, after declaring your constants with the `Const` keyword, you can wrap them using an enumeration.

What Is an Enumeration?

An *enumeration* (often abbreviated to `enum`) is a distinct data type that consists of a set of named constants, typically representing related values. Enumerations are widely used in programming to make code more readable and maintainable by replacing numeric values with meaningful names. For example, to keep the screen resolution constants together, use the `Enum` statement like this:

```
Public Enum SysMetConst
    SM_CXSCREEN = 0 ' Screen width
    SM_CYSCREEN = 1 ' Screen height
End Enum
```

The `Enum` statement declares a type for an enumeration. The `Enum` type is used to hold a collection of constants and make it easier to work with programs. The `Enum` statement can appear only at a module level. You can use any name for the `Enum` as long as it does not conflict with any of the VBA reserved keywords.

After defining an enumeration in VBA, you can use it to pass meaningful constants as parameters to a Windows API function. To simplify the API call, you can create a wrapper function in VBA that will accept the Enum type as the parameter.

What Is a Wrapper Function?

A wrapper function is a helper function that "wraps" around another function (often a complex function such as a Windows API function) to make it easier, safer, and more convenient to use. Its purpose is to simplify the interface between your code and the wrapped function, handle repetitive tasks, and provide added functionality such as error checking.

So, if you want to use the Windows API function GetSystemMetrics to retrieve system metrics, instead of providing hard to remember, numeric codes for each metric, you can simplify things by writing a wrapper function:

```
' Wrapper function
    Public Function ScreenRes(ByVal metric As SysMetConst) As Long
        ScreenRes = GetSystemMetrics(metric)
    End Function
```

Once created, your wrapper function can be used throughout your project.

A VBA procedure can be written to display the result of the function to the user:

```
Sub WhatIsMyScreenResolution()
    Dim screenWidth As Long
    Dim screenHeight As Long

    MsgBox ScreenRes(screenWidth) & " x " & ScreenRes(screenHeight)
End Sub
```

The Enum and the Wrapper function technique of calling an API function are depicted in Figure 23.3.

While encapsulating constants in an enumeration and writing wrapper functions may seem at first like much more coding to do, using this method will save you time in the long run. Also, the constants that are encapsulated in the Enum statement will be available in the IntelliSense drop-down box, saving you from typing them manually and possibly introducing errors into your code.

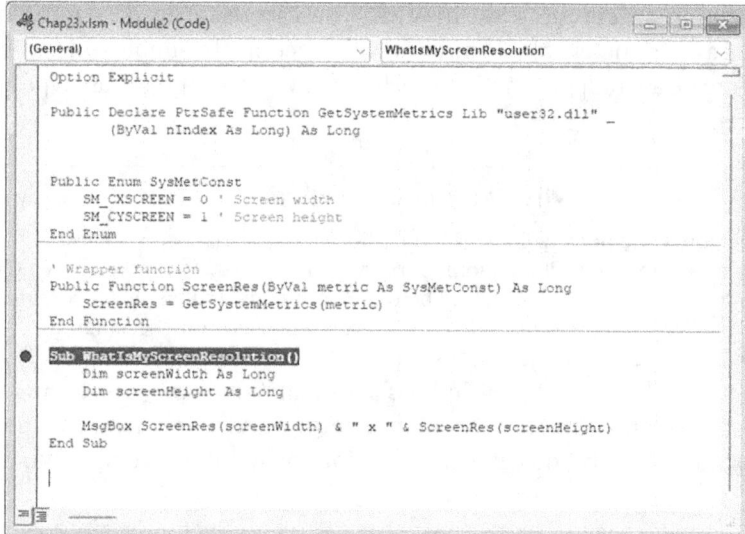

FIGURE 23.3.   Retrieving information about screen resolution from a Windows API function using the Enum keyword and a wrapper function.

EXCEL 64-BIT AND THE WINDOWS API

As mentioned earlier, to use the API functions with 64-bit Excel, you must modify the `Declare` statements to differentiate between 32-bit and 64-bit calls. To help avoid system crashes, truncation, and overflow errors when using API calls in 64-bit systems, Microsoft has introduced the `PtrSafe` keyword and two special data types, `LongLong` and `LongPtr`:

- The `PtrSafe` keyword indicates that the `Declare` statement is compatible with 64-bit systems. This keyword is mandatory on 64-bit systems.
- The `LongLong` data type is an 8-byte data type that is only available in 64-bit versions of standalone Office 2024 and Microsoft 365.
- `LongPtr` is a variable data type that is a 4-byte data type on 32-bit systems and an 8-byte data type on 64-bit versions of Microsoft 365 applications.

When making a call to an API function, you must include the `PtrSafe` keyword in the `Declare` statement and also use `LongPtr` where the `Long` data type was used to return a handle or pointer (see the sidebar).

> ## Understanding Pointers and Handles
>
> A *pointer* is a reference to a specific location in physical memory where the application stores data or programming instructions. Versions of VBA prior to 2013 did not have a pointer data type (LongPtr). Pointers and handles were stored using 32-bit variables declared as the Long data type. The Long data type cannot be used for pointers and handles on 64-bit systems because it is not large enough to store the 64-bit values returned by API functions.
>
> A *handle* is a unique identifier that Windows assigns to each window, dialog box, and control so that it can reference them. In 32-bit systems, handles were declared as the Long data type. In 64-bit systems, they must be declared using the LongPtr data type.

To locate a window on a 64-bit system, declare the API FindWindow function, as shown below:

```
Declare PtrSafe Function FindWindow Lib "user32.dll" Alias _
    "FindWindowA" (ByVal lpClassName As String, _
    ByVal lpWindowName As String) As LongPtr
```

Notice that the Declare statement is followed by the PtrSafe keyword and the function's result is of the LongPtr data type. The Microsoft documentation states that Declare statements without the PtrSafe keyword are assumed not to be compatible with the 64-bit applications. Another important note specifies that the data types in the Declare statement will have to be updated to use LongPtr if they refer to handles and pointers. The above FindWindow API function returns a handle to the window; therefore, its return value is declared As LongPtr.

In comparison, the FindWindow function on a 32-bit system is declared like this:

```
Declare Function FindWindow Lib "user32.dll" Alias _
    "FindWindowA" (ByVal lpClassName As String, _
    ByVal lpWindowName As String) As Long
```

If the function requires a parameter that represents a handle or a pointer, in 64-bit Excel, this parameter data type will also need to be replaced with the LongPtr data type. For example, the following API function is used to retrieve a window's dimensions:

```
Declare PtrSafe Function GetWindowRect Lib "user32.dll" _
    (ByVal hwnd As LongPtr, lpRect As RECT) As Long
```

In the above function, the first parameter is a handle and is therefore declared using the LongPtr data type.

Similar changes need to be made to the Structure data type (UDT) if any member variable refers to a pointer or a handle. For example, the following API function allows the user to select a folder in 64-bit Excel:

```
Declare PtrSafe Function SHBrowseForFolder Lib "shell32.dll" _
    Alias "SHBrowseForFolderA" (ByRef lpBrowseInfo As BROWSEINFO) _
    As Long
```

To use the SHBrowseForFolder function, you need to define the BROWSEINFO structure (user-defined data type) in VBA:

```
Public Type BROWSEINFO
    hWndOwner As LongPtr          ' Handle to the owner window
    pIDLRoot As LongPtr           ' Pointer to the root item
    pszDisplayName As String      ' Pointer to the display name buffer
    lpszTitle As String           ' Pointer to the title of the dialog
    ulFlags As Long               ' Flags specifying options
    lpfn As LongPtr               ' Pointer to a callback function
    lParam As LongPtr             ' Application-defined value
    iImage As Long                ' Index of the image associated with
                                    the folder
End Type
```

In the above user-defined data type declaration, some member variables reference handles or pointers and, therefore, are represented by the LongPtr data type.

If your VBA application utilizes Windows API functions and needs to be run in Excel 2024 or earlier, you should use the VBA7 conditional compilation constant.

The VBA7 compilation constant allows you to determine the version of VBA being used so you can run the appropriate code segment for that version. For example:

```
#if VBA7 Then
    ' declare API function with the PtrSafe keyword

#else
    ' declare API function without the PtrSafe keyword

#end if
```

Use the Win64 compilation constant when you need to provide code for the 32-bit version of Excel as well as for the 64-bit version:

```
#if Win64 Then
```

```
' declare API function using the Declare statement with the _
    PtrSafe keyword

#else

  ' declare API function without the PtrSafe keyword

#end if
```

NOTE	*For more information on compatibility between the 32-bit and 64-bit versions of Microsoft Office, please refer to the Microsoft documentation at http://msdn.microsoft.com/en-us/library/ee691831.aspx.*

Using Conditional Compilation

When you run a procedure for the first time, Visual Basic converts the VBA statements you used into the machine code understood by the computer. This process is called *compiling*. You can also perform the compilation of your entire VBA project manually before you run your procedure. To do this, simply choose Debug | Compile (your VBA project name) in the VBE window. You can tell Visual Basic to include or ignore certain blocks of code when compiling and running by using so-called *conditional compilation*.

To enable conditional compilation, use special expressions called *directives*.

Use the #Const directive to declare a Boolean (True, False) constant. Next, check this constant inside the #If...Then...#Else directive. The portion of code that you want to compile conditionally must be surrounded by these directives. Notice that the If and Else keywords are preceded by a number sign (#). If a portion of code is to be run, the value of the conditional constant has to be set to True (-1). Otherwise, the value of this constant should be set to False (0).

Declare the conditional constant in the declaration section of the module like this:

```
#Const User = True
```

This declares the conditional constant named User.

Conditional compilation can be used to compile an application that will be run on different platforms (Windows or Macintosh, Win32-, Win64-bit). It is also useful in localizing an application for different languages or excluding certain debugging statements before the VBA application is sent off for distribution. The program code excluded during the conditional compilation is omitted from the final file; thus, it has no effect on the size or performance of the program.

ACCESSING WINDOWS API DOCUMENTATION

Most of the Windows API Declare statements are rather long and require many parameters. Therefore, to save keystrokes and to avoid errors, many programmers prefer to resort to cutting and pasting the `Declare` statements from the `Win32API_PtrSafe.txt` file that contains Windows API declarations and constants for Visual Basic, updated for the 64-bit version of Office. This file replaces the original `Win32API.txt` file. For more information on how to download and install the `Win32API_PtrSafe.txt` file, see the following link or search for the name of the text file in your browser: *https://www.microsoft.com/en-us/download/details.aspx?id=9970&msockid=06180f8fe3806dce338a1bfee2286c9e.*

You can find the Windows API reference online at *http://msdn.microsoft.com/en-us/library/aa383749(VS.85).aspx.*

USING WINDOWS API FUNCTIONS IN EXCEL

Now that we've discussed the steps involved in writing API functions and looked at the data types and parameters these functions expect, let's leave the theory behind and do some practical programming. The procedures in Hands-On 23.1 will teach you how you can make calls to API functions from a VBA code module. This particular example will introduce four API functions that you can use to return the following information:

- The current version number of Windows and information about the operating system platform (`GetVersionEx`)
- The path of your Windows directory (`GetWindowsDirectory`)
- The name of the user currently logged on (`GetUserName`)
- The user's computer name (`GetComputerName`)

NOTE	*Please note that the files for the "Hands-On" projects can be found in the companion files.*

⊙ Hands-On 23.1 Retrieving Information About the Computer/User

NOTE	*This Hands-On uses API function calls for the 64-bit Excel installation. If you are working with a 32-bit system, please modify the* `Declare` *statement as described earlier in this chapter.*

Open a new workbook in Excel and save it in a macro-enabled file format as `C:\VBAExcel2024_ByExample\Chap23.xlsm`.

1. Switch to the VBE window and insert a new module.

2. In the module's Code window, enter the following four declarations of API functions:

```
Public Declare PtrSafe Function GetVersionEx Lib _
    "kernel32.dll" Alias "GetVersionExA" _
    (ByRef lpVersionInformation As OSVERSIONINFO) _
    As Long

Public Declare PtrSafe Function GetWindowsDirectory _
    Lib "kernel32.dll" Alias "GetWindowsDirectoryA" _
    (ByVal lpBuffer As String, ByVal nSize As Long) _
    As Long


Public Declare PtrSafe Function GetUserName _
    Lib "advapi32.dll" Alias "GetUserNameA" _
    (ByVal lpBuffer As String, nSize As Long) As Long

Declare PtrSafe Function GetComputerName _
    Lib "kernel32.dll" Alias "GetComputerNameA" _
    (ByVal lpBuffer As String, nSize As Long) As Long
```

3. Below the last `Declare` statement, enter the following UDT definition:

```
Type OSVERSIONINFO
    dwOSVersionInfoSize As Long
    dwMajorVersion As Long
    dwMinorVersion As Long
    dwBuildNumber As Long
    dwPlatformId As Long
    szCSDVersion As String * 128
End Type
```

The `OSVERSIONINFO` data structure contains operating system information. As mentioned earlier in this chapter, `Structure` data types are handled in VBA by user-defined data types.

The first member variable, `dwOSVersionInfoSize`, specifies the size, in bytes, of the data structure. Before calling the `GetVersionEx` function, it is necessary to set the size of `dwOSVersionInfoSize` to the size of the `OSVERSIONINFO` structure by using the `Len` function (see the procedure code in Step 5 below).

The `dwMajorVersion` and `dwMinorVersion` variables identify the major and minor version numbers of the operating system. The `dwBuildNumber` variable identifies the build number of the operating system. The `szCSDVersion` variable contains a null-terminated string that can provide additional information about the operating system, such as the version of the service pack.

4. Enter the following `OpSysInfo` procedure that calls the `GetVersionEx` API function to retrieve information about the Windows platform:

```
Sub OpSysInfo()
    Dim os As OSVERSIONINFO
    Dim osVer As String

    os.dwOSVersionInfoSize = Len(os)
    GetVersionEx os
    osVer = os.dwMajorVersion & "." & os.dwMinorVersion
    Debug.Print "Windows Version = " & osVer
    Debug.Print "Windows Build Number = " & os.dwBuildNumber
    Debug.Print "Windows Platform ID = " & os.dwPlatformId
    Debug.Print "Additional info = " & os.szCSDVersion
End Sub
```

5. Save the changes to the `Chap23.xlsm` workbook and run the `OpSysInfo` procedure.

After executing the above procedure on the machine running the Windows 10/11 Pro, the following information appears in the Immediate window:

```
Windows Version = 10.0
Windows Build Number = 22631
Windows Platform ID = 2
Additional info =
```

6. Enter the `PathToWinDir` procedure in the Code window just below the `OpSysInfo` procedure:

```
Sub PathToWinDir()
    Dim strWinDir As String
    Dim lngLen As Long

    strWinDir = String(255, 0)
    lngLen = GetWindowsDirectory(strWinDir, Len(strWinDir))
    strWinDir = Left(strWinDir, lngLen)
    MsgBox "Windows folder: " & strWinDir
End Sub
```

In the above procedure, we need to obtain textual data from the API function. This requires that you first create a string filled with spaces or `null` characters and then pass it to the function:

```
strWinDir = String(255, 0)
```

The VBA String function is used to return a string containing a repeating character string of the specified length. In this example, the string will be filled with null characters. You could also use the Space function to fill the string with 255 spaces like this:

```
strWinDir = Space(255)
```

In the next statement, we call the API function named GetWindowsDirectory, passing to it the required two arguments: the receiving string buffer (strWinDir) that we previously defined and the buffer's length (Len(strWinDir)). If the API function succeeds, it will return the length of the string copied to the buffer. If the function fails, the return value will be zero. Once we have the length of the returned string, we use the VBA Left function to extract the specified number of characters from the beginning of the buffer:

```
strWinDir = Left(strWinDir, lngLen)
```

7. Save the changes to the Chap23.xlsm workbook and run the PathToWinDir procedure in step mode by pressing F8. By stepping through the code, you can investigate the values of the individual variables that are passed to the API function to obtain the Windows path.

8. Enter the LoggedOnUserName function procedure in the Code window just below the PathToWinDir procedure:

```
Function LoggedOnUserName() As String
    Dim strBuffer As String * 255
    Dim strLen As Long

    strLen = Len(strBuffer)
    GetUserName strBuffer, strLen

    If strLen > 0 Then
        LoggedOnUserName = Left$(strBuffer, strLen - 1)
    End If

    MsgBox LoggedOnUserName
End Function
```

The above function procedure begins by declaring a strBuffer variable of fixed size (255 characters). This variable will be filled by the API function with the name of the logged-on user. The procedure also defines the strLen variable to hold the length of the buffer. The length of the buffer is set to the length of the strBuffer variable, which is initially 255 characters long. Next, to get the logged-on username, the procedure calls the API function `GetUserName`, passing it two arguments as required by the function's `Declare` statement located at the top of the code module. If the length of the returned string is greater than zero (`0`), the function succeeded and the `strBuffer` variable should contain the name of the logged-on user. Before we return the username to the VBA function, however, we need to extract the specified number of characters from the beginning of the `strBuffer` and strip off the terminating `null` character, which is also returned by the API function:

```
LoggedOnUserName = Left$(strBuffer, strLen - 1)
```

NOTE	*As mentioned earlier, the C language uses 0-terminated strings that are not recognized by VBA. Therefore, when calling a Windows API function that returns a string value, you must strip the 0 terminator from the end of the C string so VBA can recognize it correctly.*

9. Save the changes to the `Chap23.xlsm` workbook and run the `LoggedOnUserName` function by typing its name in the Immediate window and then pressing Enter. When the function completes, you should see a message box with the name you used to log on to your computer.

10. Enter the `GetUserComputerName` procedure below the last procedure code:

```
Sub GetUserComputerName()
    Dim strCompName As String
    Dim retval As Long

    strCompName = Space(255)
    retval = GetComputerName(strCompName, 255)
    strCompName = Left(strCompName, _
        InStr(strCompName, vbNullChar) - 1)
    Debug.Print "Your computer name is: " _
        & strCompName
End Sub
```

11. Save the changes to the `Chap23.xlsm` workbook and run the `GetUserComputerName` procedure.

Warnings and Precautions

When writing VBA procedures that call Windows API functions, follow these guidelines:

- Be sure to check all variable types, constants, and values that are required by the API function by checking the function documentation available online.
- It is always better to specify the type of the variable explicitly rather than relying on the Any variable type.
- Be sure to pass strings to API functions by using the ByVal keyword.
- Before running the VBA procedure that calls the Windows API function, always save any changes made to the code module. Unexpected errors in the code may crash your system, and you will lose any unsaved work.

In the next Hands-On, you will work with the Excel VBA UserForm and learn how API functions can be used to change the appearance of the default UserForm. In particular, you will add an icon to the title bar, as well as include the missing Maximize and Minimize buttons. Next, you will make the form resizable and transparent. Finally, you will add the UserForm to the Windows task list. Figure 23.4 displays how the UserForm will look after you've completed Hands-On 23.2.

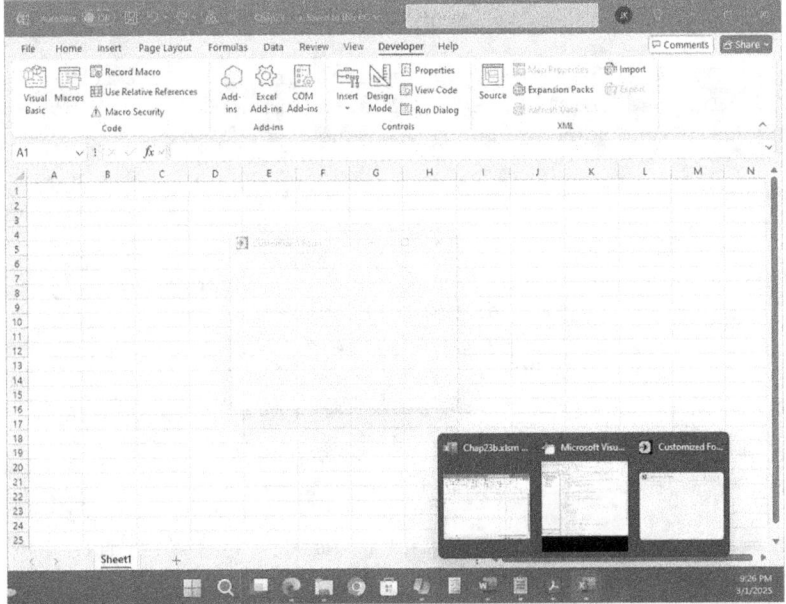

FIGURE 23.4. The Microsoft Excel UserForm can be customized by calling Windows API functions. In the Hands-On 23.2 exercise, we made the UserForm transparent, added an icon to the title bar, and included the Maximize and Minimize buttons.

⊙ Hands-On 23.2 Enhancing a VBA UserForm by Calling Windows API Functions

NOTE	*This Hands-On uses API function calls for the 64-bit Excel installation. If you are working with a 32-bit system, please modify the* Declare *statement as described earlier in this chapter.*

1. Create a new workbook and save it in a macro-enabled file format as `C:\ VBAExcel2024_ByExample\Chap23b.xlsm`.
2. In the VBE window, insert a new module into the `VBAProject (Chap23b. xlsm)` workbook.
3. In the module's Code window, enter the following API function, variable, and constant declarations:

```
' API FUNCTIONS DECLARATIONS

Declare PtrSafe Function FindWindow Lib "user32.dll" Alias _
    "FindWindowA" (ByVal lpClassName As String, _
    ByVal lpWindowName As String) As LongPtr

Declare PtrSafe Function SendMessageA Lib "user32.dll" _
    (ByVal hwnd As LongPtr, ByVal wMsg As Long, _
    ByVal wParam As LongPtr, ByVal lParam As LongPtr) As LongPtr

Declare PtrSafe Function ExtractIconA Lib "shell32.dll" _
(ByVal hInst As LongPtr, ByVal lpszExeFileName As String, _
    ByVal nIconIndex As Long) As Long

Declare PtrSafe Function GetActiveWindow Lib "user32.dll" _
    () As Long

Declare PtrSafe Function SetWindowPos Lib "user32.dll" _
                (ByVal hwnd As LongPtr, _
                ByVal hWndInsertAfter As LongPtr, _
                ByVal x As Long, _
                ByVal y As Long, _
                ByVal cx As Long, _
                ByVal cy As Long, _
                ByVal wFlags As Long) As Long

Declare PtrSafe Function GetWindowLong Lib "user32.dll" Alias _
    "GetWindowLongA" (ByVal hwnd As LongPtr, _
    ByVal nIndex As Long) As Long

Declare PtrSafe Function SetWindowLong Lib "user32.dll" Alias _
```

```
    "SetWindowLongA" (ByVal hwnd As LongPtr, _
    ByVal nIndex As Long, _
    ByVal dwNewLong As Long) As LongPtr

Declare PtrSafe Function SetLayeredWindowAttributes Lib _
    "user32.dll"
    (ByVal hwnd As LongPtr, ByVal crey As Byte, _
    ByVal bAlpha As Byte, ByVal dwFlags As Long) As LongPtr

Declare PtrSafe Function FlashWindow Lib "user32.dll" _
    (ByVal hwnd As LongPtr, ByVal bInvert As Long) As Long

Declare PtrSafe Sub Sleep Lib "kernel32.dll" _
    (ByVal dwMilliseconds As Long)

' variable declarations

Public hwnd As Long  ' handle to the active window

' Constant declarations
Public Const GWL_EXSTYLE = (-20)
Public Const GWL_STYLE = (-16)
Public Const WS_EX_LAYERED = &H80000
Public Const WS_EX_APPWINDOW = &H40000
Public Const WS_MINIMIZEBOX = &H20000
Public Const WS_MAXIMIZEBOX = &H10000
Public Const WS_THICKFRAME = &H40000
Public Const SWP_NOMOVE = &H2
Public Const SWP_NOSIZE = &H1
Public Const SWP_NOACTIVATE = &H10
Public Const SWP_HIDEWINDOW = &H80
Public Const SWP_SHOWWINDOW = &H40
Public Const SW_SHOW = 5
Public Const HWND_TOP = 0
Public Const LWA_ALPHA = &H2&
Public Const WM_SETICON = &H80
```

4. Choose Insert | UserForm to add a new form to `VBAProject`.
5. Right-click the UserForm and choose View Code.
6. In the UserForm1 Code window, enter the following three procedures:

```
Private Sub UserForm_Activate()
    CustomizeForm
End Sub

Private Sub UserForm_Initialize()
```

```
    With Me
        .Caption = "Customizing Form...Please Wait"
        .BackColor = RGB(255, 255, 51)
    End With
End Sub


Private Sub AddIcon_OnTitleBar(strIconBmpFile As String)
    Dim fLen As Long

    If Len(Dir(strIconBmpFile)) <> 0 Then
        'fLen = ExtractIconA(0, strIconBmpFile, 0)
        fLen = ExtractIconA(0, strIconBmpFile, 0)
        SendMessageA FindWindow(vbNullString, Me.Caption), _
            WM_SETICON, False, fLen

    Else
        Exit Sub
    End If
End Sub

Private Sub CustomizeForm()
    Dim wStyle As Long
    Dim xStyle As Long
    Dim bOpacity As Byte

    'get the handle of the active window
    hwnd = GetActiveWindow

    ' get user attention by flashing the window
    Call FlashThisWindow(hwnd)

    ' begin customization process
    AddIcon_OnTitleBar _
        "C:\VBAExcel2024_ByExample\Images\arrow.bmp"

    bOpacity = 120 ' set opacity

    'retrieve the active window's styles
    wStyle = GetWindowLong(hwnd, GWL_STYLE)

    ' modify the window style settings
    wStyle = wStyle Or WS_MINIMIZEBOX   'add the minimize button
    wStyle = wStyle Or WS_MAXIMIZEBOX   'add the maximize button
    wStyle = wStyle Or WS_THICKFRAME    'add a sizing border
```

```
        ' apply the revised style
        Call SetWindowLong(hwnd, GWL_STYLE, wStyle)

        ' retrieve the active window's extended styles
        xStyle = GetWindowLong(hwnd, GWL_EXSTYLE)

        ' modify the window extended style settings
        xStyle = xStyle Or WS_EX_LAYERED     ' change opacity
        xStyle = xStyle Or WS_EX_APPWINDOW   ' add window to _
                                                the task bar

        ' apply the revised extended style
        Call SetWindowLong(hwnd, GWL_EXSTYLE, xStyle)

    Call SetLayeredWindowAttributes(hwnd, 0, bOpacity, LWA_ALPHA)

    Call SetWindowPos(hwnd, HWND_TOP, 0, 0, 0, 0, _
                        SWP_NOMOVE Or _
                        SWP_NOSIZE Or _
                        SWP_NOACTIVATE Or _
                        SWP_HIDEWINDOW)

    Call SetWindowPos(hwnd, HWND_TOP, 0, 0, 0, 0, _
                        SWP_NOMOVE Or _
                        SWP_NOSIZE Or _
                        SWP_NOACTIVATE Or _
                        SWP_SHOWWINDOW)

End Sub


Sub FlashThisWindow(myForm As Long)
    Dim counter As Integer
    ' declare return value used for flashing the window
    Dim retval As LongPtr

    For counter = 1 To 10
        ' toggle the look of the window
        retval = FlashWindow(myForm, 1)
        Sleep 500  ' wait for 5 seconds
    Next counter
    retval = FlashWindow(myForm, 0)
    UserForm1.Caption = "Customized Form"
End Sub
```

When you run the Sub/UserForm, Excel will look for the UserForm_Initialize procedure and will execute the code found therein. This code tells Excel to replace the default caption with the specified text and change the form background color to a shade of yellow represented by the RGB(255, 255, 51) function. Next, the UserForm_Activate event will be triggered. Here we make a call to the CustomizeForm VBA procedure, which begins by obtaining a handle to the active window via the call to the GetActiveWindow API function. You'll need this handle for all the API functions used in this solution, hence the hwnd variable is declared with the Public keyword in the standard module. After obtaining a window reference, we want to get the user's attention by flashing the form. This is done within the code of the FlashThisWindow procedure, which makes a direct call to the following Windows API functions:

```
Declare PtrSafe Function FlashWindow Lib "user32.dll" _
    (ByVal hwnd As LongPtr, ByVal bInvert As Long) As Long


Declare PtrSafe Sub Sleep Lib "kernel32.dll" _
    (ByVal dwMilliseconds As Long)
```

The FlashWindow function returns 0 if the window's look was inactive before flashing, or 1 if its look was active. The hwnd parameter is the handle to the window. The bInvert parameter specifies how to flash. Non-zero will switch the title bar from active to inactive, or vice versa. Zero will restore the window to its normal look.

The Sleep procedure pauses program execution for a specified amount of time. Sleep does not return any value. It requires that you specify the number of milliseconds to hold program execution for (see the dwMilliseconds parameter).

After flashing the UserForm1 window on and off, we again change the form's title using the Caption property and go on to perform form customizations. To put an icon on the form's title bar, we call the AddIcon_OnTitleBar procedure and pass to it the name and location of the bitmap image we want to use.

Before you can do anything with the image, you must ensure that it is indeed found in the specified pathname. Therefore, the AddIcon_OnTitleBar procedure begins by checking whether the specified file exists, and if it does, we call upon three API functions that will allow us to extract the icon from the file (ExtractIconA) and then display it on the form's title bar (SendMessageA and FindWindow).

To find `UserForm1`, you can use the `FindWindow` API function like this:

```
FindWindow(vbNullString, Me.Caption)
```

The `FindWindow` function finds a window handle based on the exact window title. A *window handle* is a unique identifier that Windows assigns to each window, dialog box, or control. Once you have a window handle, you can pass it to the API functions for all operations involving windows. In the above code snippet, the VBA statement `Me.Caption` will provide the title of `UserForm1` by using the `Caption` property of the form so that the `FindWindow` function will be able to locate the `UserForm1` window and return the handle, which is a `Long` (4-byte) value.

> **NOTE**
> *If you want to find the handle of Microsoft Excel, keep in mind that VBA provides a special built-in property named `hWnd` for the Excel `Application` object; thus, it is not necessary to call the API function to retrieve the handle.*

The result of the `FindWindow` function (in this case, the handle to `UserForm`) is then fed along with other arguments to the `SendMessageA` function, which tells Windows to perform the requested action: placing the icon image on the found form's title bar. If the icon image is not found, the procedure does not do anything; you will end up with a UserForm that does not have the icon.

Before you can make modifications to the UserForm's title bar, it is necessary to get the current style information of the active window. This is done by accessing the window's configuration memory using the `GetWindowLong` API function and passing it the `GWL_STYLE` constant that you defined in the standard code module. Once you have the `GWL_STYLE` setting in the `wStyle` variable, you can modify the setting by using the bitwise `OR` operator, as shown below:

```
wStyle = wStyle Or WS_MINIMIZEBOX    'add the minimize button
wStyle = wStyle Or WS_MAXIMIZEBOX    'add the maximize button
wStyle = wStyle Or WS_THICKFRAME     'add a sizing border
```

To have Windows actually apply the new settings, the procedure goes on to call the `SetWindowLong` API function:

```
Call SetWindowLong(hWnd, GWL_STYLE, wStyle)
```

The parameters passed to the `SetWindowLong` API function inform Windows which window should be modified (`hWnd`), what settings need to be changed (`GWL_STYLE`), and what the replacement settings (`wStyle`) are.

You can follow the same logic to make modifications to the extended window styles in order to make your form transparent and visible on the taskbar:

```
' retrieve the active window's extended styles
 xStyle = GetWindowLong(hWnd, GWL_EXSTYLE)

' modify the window extended style settings
 xStyle = xStyle Or WS_EX_LAYERED    ' change opacity
 xStyle = xStyle Or WS_EX_APPWINDOW  ' add window to the task
bar

' apply the revised extended style
 Call SetWindowLong(hWnd, GWL_EXSTYLE, xStyle)
```

Next, you need to tell Windows the amount of transparency you want to apply to the form. This is done by calling the following function:

```
Call SetLayeredWindowAttributes(hWnd, 0, bOpacity, LWA_ALPHA)
```

The second argument in the above function, 0, is the transparent color of the window; this argument is not used. The third argument is the amount of transparency you want. Zero (0) will make the form completely transparent and 255 will make it completely opaque.

The procedure uses the setting of 120 stored in the bOpacity variable. The last parameter, LWA_ALPHA, tells the window that the form should be transparent.

The last two API function calls in the CustomizeForm procedure ensure that the name of the UserForm1 window appears in the top position on the window's taskbar and the entry is removed when the form is closed.

7. Run UserForm1 to examine the result of the applied customizations (see Figure 23.4 earlier in this chapter).

Notice the presence of the icon on the title bar as well as the Minimize and Maximize buttons to the left of the default Close (X) button. The Minimize button will not be visible if the form is too small. Resize the form, dragging its borders to see all buttons. When you click on the icon in the upper-left corner, you should see a menu with options allowing you to move, size, minimize, maximize, and close the form (see Figure 23.5). When you minimize the form, it will jump into the taskbar. You can activate the form directly from the taskbar by right-clicking it and choosing Restore.

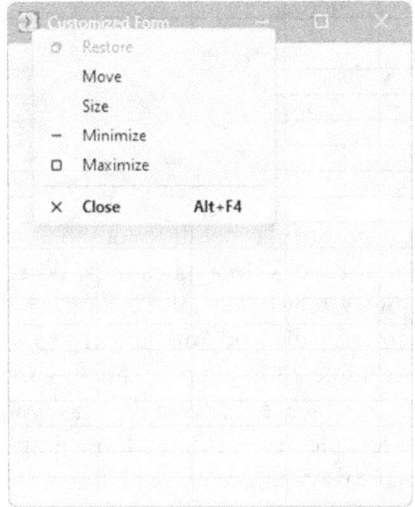

FIGURE 23.5. The Microsoft Excel UserForm was customized in the Hands-On 23.2 exercise. Clicking the icon placed in the title bar reveals a menu. The UserForm can also be easily resized.

The `Winmm.dll` library offers many multimedia functions for audio, video, and timer management. For example, you can check for audio devices on your system using the `waveOutGetNumDevs function` like this:

```
Private Declare PtrSafe Function waveOutGetNumDevs Lib "winmm.
dll" () As Long

Sub CheckAudioDevices()
   Dim numDevices As Long
   numDevices = waveOutGetNumDevs()
   MsgBox "Number of audio output devices: " & numDevices
End Sub
```

You can also retrieve the names of your audio devices to your Excel worksheet as shown in the completed `Chap23b.xlsm` workbook in the companion files.

Another function from the `Winmm.dll` library, `PlaySound`, can be used to play a sound:

```
Private Declare PtrSafe Function _
    PlaySound Lib "winmm.dll" Alias "PlaySoundA" ( _
    ByVal lpszName As String, _
    ByVal hModule As LongPtr, _
    ByVal dwFlags As Long) As Long

Private Const SND_ASYNC As Long = &H1
```

```
Private Const SND_FILENAME As Long = &H20000

Sub PlaySoundExample()
    Dim soundFile As String
    soundFile = "C:\Windows\Media\chimes.wav" ' Path to a .wav file
    PlaySound soundFile, 0, SND_ASYNC Or SND_FILENAME
End Sub
```

These examples highlight just a few of the powerful capabilities of the `Winmm.dll` library that you can use to enhance your Excel VBA applications. Whether you're playing sound effects or retrieving system audio information, calling upon the `Winmm.dll` library can add multimedia and interactivity to your projects. Each Windows API library (see Table 23.1) offers access to specific system functionalities, including user interface enhancements (as you've seen in this chapter's UserForm example), file system operations, graphics rendering, and more. To further explore the possibilities of API libraries, refer to the Microsoft official documentation provided in the links included in this chapter. You can also browse online coding forms, such as Stack Overflow, for additional examples and practical scenarios. Another resource is the Windows SDK documentation, which provides comprehensive details on APIs, data types, and constants.

SUMMARY

While VBA offers programmers a huge library of objects, properties, and methods, there are certain interface features supported by the Windows operating system that cannot be accessed by calling the native VBA library. The good news is that some of the VBA limitations can be directly dealt with by having your VBA procedure call a function from the Windows API. This chapter introduced you to the basic Windows API concepts and showed you how you can call API functions from VBA. It also pointed you to resources you can use to find and research functions that should help you extend your VBA programs by providing more functionality via the Windows API.

In the next chapter, we will move into the next part of this book, which will cover more advanced concepts in Excel VBA. You will learn about creating classes and programming in standalone class modules.

Part VII
ADVANCED CONCEPTS IN EXCEL VBA

Throughout this book, you have explored various ways to accomplish tasks in Excel and developed numerous procedures and functions containing programming code. This code may be useful for other Excel applications. As your Excel VBA procedures grow more complex, however, managing scattered and disorganized code becomes increasingly challenging. Copying code across projects and constantly updating procedures to accommodate new requirements can lead to a tangled, unmanageable mess. To prevent this, it's important to know about Excel's class module feature, which enables you to create self-contained, reusable code.

Chapter 24   Creating Classes in VBA

Chapter 24 CREATING CLASSES IN VBA

A class helps you organize your code into more manageable objects that you can easily reuse and adjust when necessary. Classes also make it easier to share your programming code with others. The class hides its inner workings from the rest of the program. Any programmer can use the class without knowing how that class was put together. It's like driving a car. A driver does not need to know the intricate details of a car manufacturing process.

By learning about classes, you can transition from procedural programming, which you already know, to Object-Oriented Programming (OOP), where you create and work with objects. These objects consist of data, referred to as *attributes* or *properties*, and code, known as *procedures* or *methods*. Objects are created by adding class modules to your Excel VBA projects. Think of a class as a cookie cutter: once defined, it enables you to produce an unlimited number of cookies. In Excel, these cookies represent the custom objects generated from the class within a class module.

IMPORTANT TERMINOLOGY

In this chapter, you will work with advanced VBA concepts: VBA classes, class objects, and collection classes. Before diving into the theory and this chapter's hands-on examples, let's review the following terms:

- *Object*—A logical representation of a thing. This thing can be a tangible, physical entity, such as a person, a car, a customer or an employee, or a logical entity, such as an order, a transaction, or a report. It can also be something related to your specific Excel application, such as a process that simplifies data validation or manipulation of data. In other words, an object is basically anything you need to define.

- *Collection*—An object that contains a set of related objects.

- *Class*—A definition of an object that includes its name, properties, methods, and events. The class acts as a sort of object template from which an instance of an object is created at runtime.

- *Instance*—A specific object that belongs to a class is referred to as an *instance of the class*. When you create an instance, you create a new object that has the properties and methods defined by the class.

- *Class module*—A module that contains the definition of a class, including its property and method definitions.

- *Module*—A module containing sub procedures and function procedures that are available to other VBA procedures and are not related to any object in particular.

- *Form module*—A module that contains the VBA code for all event procedures triggered by events occurring in a UserForm or its controls. A form module is a type of class module.

- *Event*—An action recognized by an object, such as a mouse click or a keypress, for which you can define a response. Events can be caused by a user action or a VBA statement or can be triggered by the system.

- *Event procedure*—A procedure that is automatically executed in response to an event initiated by the user or program code or triggered by the system.

CREATING AND USING CUSTOM OBJECTS

The VBE Insert menu has two module options: Module and Class Module. So far, you've used modules to create subroutine and function procedures. You'll use the class module for the first time in this chapter to create a custom class named CAsset and learn how to define its properties and methods.

To create custom objects, it's essential to first understand what a class is. In the preceding terminology section, we described a class as an object template, and before that, we compared it to a cookie cutter—a common analogy. Just as a cookie cutter shapes the appearance of a cookie, a class defines the structure and behavior of an object. Before using a class, you need to create an instance of it. These object instances are the cookies, each possessing the attributes (properties) and actions (methods) defined by the class. Similar to cutting multiple cookies with the same cutter, you can generate numerous instances of a class. Additionally, the properties of each instance can be modified independently, ensuring they remain distinct from one another.

A class module lets you define your own custom classes, complete with custom properties and methods. Recall that a property is an attribute of an object that defines one of its characteristics, such as shape, position, color, title, and so on. You can create the properties for your custom objects by writing property procedures in a class module.

There are three types of property procedures (`Property Get`, `Property Let`, and `Property Set`). You will learn how to work with property procedures in Custom Project 24.1.

A *method* is an action that the object can perform. The object methods are also created in a class module by writing subroutines or function procedures.

Custom Project 24.1 will introduce you to the process of creating a custom object named `CAsset`. This object will contain information about a single computer hardware asset. It will have four properties to hold the information about: `AssetType`, `Manufacturer`, `Model`, and `Price`. It will also have a method that will allow you to modify the price. The asset information for this project is provided in the `AssetInfo.txt` file in the companion files and depicted in Figure 24.1.

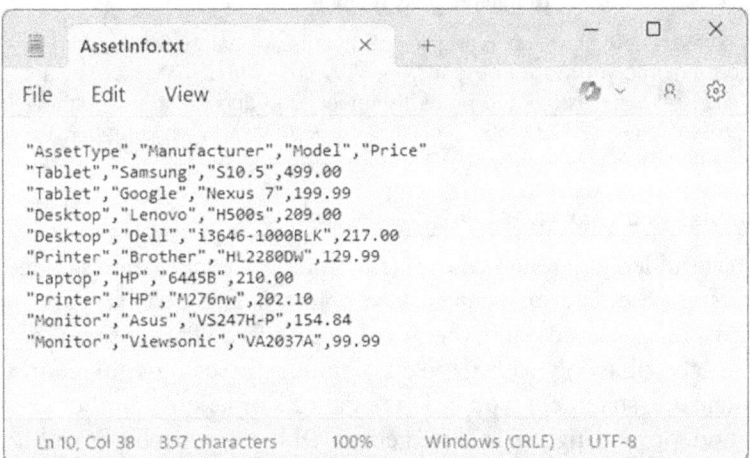

FIGURE 24.1. This text file (AssetInfo.txt) provides the data for the custom CAsset object class.

As you can see in Figure 24.1, the `AssetInfo` data file contains several lines (records). The data between the quotes is treated as a single field. Fields are delimited by a comma (`,`). This type of text file is often called a *comma-delimited file* or a *sequential access file*. You may recall from an earlier chapter that in sequential access files, the data is retrieved in the same order as it is stored. Sequential access files can be opened in `Input`, `Output`, or `Append` mode.

In this project, you will use the `Input` mode to read the data from the file into your custom object's properties. Because this file contains data for several assets, you will also reinforce your understanding of collections by reading the data from the text file into a collection of `CAsset` objects and then manipulating these objects. So, let's get started.

NOTE	*Please note that the figures and files for the hands-on projects can be found in the companion files.*

⦿ Custom Project 24.1a   Creating a Class Module

1. Open a blank Excel workbook and save it as `C:\VBAExcel2024_ByExample\Chap24.xlsm`.
2. Switch to the VBE screen.
3. Select VBAProject (Chap24.xlsm) in the Project Explorer window and choose Insert | Class Module.
4. Highlight the Class 1 module in the Project Explorer window and use the Properties window to rename the class module `CAsset`.

Naming a Class Module

Each time you create a new class module, give it a meaningful name. Set the name of the class module to the name you want to use in your VBA procedures that use the class. The name you choose for your class should be easily understood and identify the "thing" the object class represents. As a rule, the object class name is prefaced with an uppercase `C`.

Member Variables in a Class Module

Once you have added and renamed the class module, the next step is to declare variables to store the data for your custom object. These variables, known as *data members,* are declared using the `Private` keyword. This ensures that the variables are accessible only within the class module. By using `Private` instead of the familiar `Dim` statement, you hide the data members from other parts of the application, preventing external references. Only the procedures within the same class module where the variables are defined can modify their values.

Since a variable's name also serves as the property name for the object, it's important to use meaningful, descriptive names for your data members. It's a common convention to preface the variable names with m_ to indicate they belong to a class.

Let's continue with our project by declaring data members for our CAsset class.

⊙ Custom Project 24.1b   Declaring Data Members of the CAsset Class

Type the following declaration lines at the top of the CAsset class module:

```
' declarations
Private m_AssetType As String
Private m_Manufacturer As String
Private m_Model As String
Private m_Price As Currency
```

Notice that the name of each data member variable begins with the prefix m_.

Introduction to Property Procedures

Declaring the variables with the Private keyword ensures that the variables are not directly accessed from outside the object. This means that the VBA procedures from outside the class module will not be able to set or read data stored in those variables. To enable other parts of your VBA application to set or retrieve the asset data, you must add special property procedures to the CAsset class module.

There are three types of property procedures:

- Property Let allows other parts of the application to set the value of a property.
- Property Get allows other parts of the application to get or read the value of a property.
- Property Set is used instead of Property Let when setting the reference to an object.

Property procedures are executed when an object property needs to be set or retrieved. The Property Get procedure can have the same name as the Property Let procedure.

You should create property procedures for each property of the object that can be accessed by another part of your VBA application.

The easiest of the three types of property statements to understand is the `Property Get` procedure. Let's examine the syntax of the property procedures by taking a closer look at the `Property Get AssetType` procedure. As a rule, the property procedures contain the following parts:

- A procedure declaration line that specifies the name of the property and its data type:

```
Property Get AssetType() As String
```

`AssetType` is the name of the property and `As String` determines the data type of the property's return value.

- An assignment statement like the one used in a function procedure:

```
AssetType = m_AssetType
```

`AssetType` is the name of the property and `m_AssetType` is the data member variable that holds the value of the property you want to retrieve or set. The `m_AssetType` variable should be defined with the `Private` keyword at the top of the class module.

- The `End Property` keywords that specify the end of the property procedure:

```
Property Get AssetType() As String
    AssetType = m_AssetType
End Property
```

Writing Property Procedures

The `CAsset` class has four properties (`AssetType`, `Manufacturer`, `Model`, and `Price`) that need to be exposed to a VBA procedure that you will write later. Because this procedure will read a data file and then write it into a collection of `CAsset` objects, our next step involves writing the required `Property Get` and `Property Let` procedures.

⊚ **Custom Project 24.1c   Writing Property Procedures for the CAsset Class**

Type the following `Property Get` and `Property Let` procedures in the `CAsset` class module, just below the declaration section:

```
' Property procedures

Property Get AssetType() As String
```

```
        AssetType = m_AssetType
End Property

Property Let AssetType(ByVal aType As String)
        m_AssetType = aType
End Property

Property Get Manufacturer() As String
        Manufacturer = m_Manufacturer
End Property

Property Let Manufacturer(ByVal aMake As String)
        m_Manufacturer = aMake
End Property

Property Get Model() As String
        Model = m_Model
End Property

Property Let Model(ByVal aModel As String)
        m_Model = aModel
End Property

Property Get Price() As Currency
        Price = m_Price
End Property

Property Let Price(ByVal aPrice As Currency)
        m_Price = aPrice
End Property
```

Notice that each type of required asset information needs a separate `Property Get` procedure. Each of the `Property Get` procedures returns the current value of the property. The `Property Get` procedure is like a function procedure. Like function procedures, it contains an assignment statement. As you recall from Chapter 4, to return a value from a function procedure, you must assign it to the function's name.

Immediate Exit from Property Procedures

Just like the `Exit Sub` and `Exit Function` keywords allow you to exit early from a subroutine or a function procedure, the `Exit Property` keywords give you a way to immediately exit from a property procedure. Program execution will continue with the statements following the statement that called the `Property Get`, `Property Let`, or `Property Set` procedure.

In addition to retrieving values stored in data members (private variables) with `Property Get` procedures, you write corresponding `Property Let` procedures to allow other parts of the application to change the values of these variables as needed. You can make a property read-only by NOT writing a corresponding `Property Let` procedure.

The `Property Let` procedures require at least one parameter that specifies the value you want to assign to the property. This parameter can be passed by value (see the `ByVal` keyword in the `Property Let Price` procedure shown earlier) or by reference (`ByRef` is the default). If you need a refresher on the meaning of these keywords, see the section titled "Passing Arguments by Reference and Value" in Chapter 4.

The data type of the parameter passed to the `Property Let` procedure must have the same data type as the value returned from the `Property Get` procedure with the same name. Notice that the `Property Let` procedures have the same name as the `Property Get` procedures prepared in the preceding section.

Defining the Scope of Property Procedures

You can place the `Public`, `Private`, or `Static` keyword before the name of a property procedure to define its scope. For example, to indicate that the `Property Get` procedure is accessible to other procedures in all modules, use the following statement format:

```
Public Property Get AssetType() As String
```

To make the `Property Get` procedure accessible only to other procedures in the module where it is declared, use the following statement format:

```
Private Property Get Model() As String
```

To preserve the `Property Get` procedure's local variables between procedure calls, use the following statement format:

```
Static Property Get Manufacturer() As String
```

If not explicitly specified using either `Public` or `Private`, property procedures are public by default. Also, if the `Static` keyword is not used, the values of local variables are not preserved between the procedure calls.

Writing Class Methods

Apart from properties, objects usually have one or more methods. A *method* is an action that the object can perform. Methods allow you to manipulate data stored in a class object. Methods are created using sub or function procedures. To make a method available outside the class module, use the `Public` keyword in front of the `Sub` or `Function` definition.

The CAsset class that you create in this project has one method that allows you to calculate the new price. Assume that the asset's price can be decreased by a specific percentage or amount. Let's continue with our project by writing a class method that calculates the new price.

Custom Project 24.1d   Writing a Method for the CAsset Class

Type the following NewPrice function procedure in the CAsset class module:

```vba
' function to calculate new price
Public Function NewPrice(discountType As Integer, _
                    currentPrice As Currency, _
                    amount As Long) As Currency

    If amount >= currentPrice Then
        NewPrice = currentPrice
        Exit Function
    End If
    Select Case discountType
      Case 1 ' by percent
        If amount > 50 Then
           amount = 50
        End If
          NewPrice = currentPrice - ((currentPrice * _
            amount) / 100)

        Case 2 ' by amount
            NewPrice = currentPrice - amount
    End Select
End Function
```

The NewPrice function defined with the Public keyword in a class module serves as a method for the CAsset class. To calculate a new price, a VBA procedure from outside the class module must pass three arguments: discountType, currentPrice, and amount. The discountType argument specifies the type of the calculation. Suppose you want to decrease the asset price by 5% or by $5. The first option will decrease the price by the specified percentage, and the second option will subtract the specified amount from the current price. The currentPrice argument is the current price figure for an asset, and amount determines the value by which the price should be changed. The other assumptions in the new price calculation that you might want to include can be specified with the conditional statements.

Creating an Instance of a Class

You have now completed the definition of the CAsset class. You have declared member variables and wrote property procedures and a method for the class CAsset. Every time you define a class; you must do this in a class module. In VBA, you can define only one class in a class module. The name of the class is the name of the module.

After defining the class, you can create objects based on that class. This process takes place in a standard module. You start by declaring an object variable. If the name of the class module is CAsset, declare a variable of type CAsset and set that variable to a new instance of the class, like this:

```
Dim asset As CAsset
Set asset = New CAsset
```

It is also possible to combine the two statements in a single statement, like this:

```
Dim asset As New CAsset
```

The asset variable represents a reference to an object of the CAsset class. You can name your object variable anything you want, except you cannot use any of the VBA reserved words. All the properties and methods defined in the CAsset class will now be available in the asset variable. When you declare the object variable with the New keyword, VBA creates the object and allocates memory for it; however, the object isn't instanced until you refer to it in your procedure code by assigning a value to its property or running one of its methods.

Let's continue our hands-on project by writing the VBA procedure that reads the data from the text file into a collection of CAsset objects.

(◉) Custom Project 24.1e   Writing Code

1. Copy the AssetInfo.txt file from the companion files to your C:\ VBAExcel2024_ByExample folder.
2. In the VBE screen, choose Insert | Module to add a standard module to the current VBA project.
3. In the Properties window, rename the module AssetInfo.
4. In the Project Explorer window, double-click the AssetInfo module to activate the Code window.
5. In the AssetInfo Code window, enter the Retrieve_AssetInfo procedure as shown here:

```
Sub Retrieve_AssetInfo()
    ' declare two object variables
```

```vba
' one for the object and the other
' for the collection of objects
Dim asset As CAsset
Dim AssetsColl As Collection

' declare variables for reading the data file
Dim strAssetType As String
Dim strMake As String
Dim strModel As String
Dim itemPrice As String

' declare a variable to specify discount type
' in the calculation of new asset price
Dim intDiscount As Integer

' declare variables used by the MsgBox function
Dim strTitle As String
Dim strPrompt As String

' declare variables to facilitate data
' entry in a worksheet and the Immediate window
Dim strFilePath As String
Dim strRecord As String
Dim wRow As Integer

' declare variables used for collection purpose
Dim counter As Integer
Dim aKey As String

' declare variables for accessing an object
' in a collection via a key
Dim assetKey As String
'Dim m As Object

' if error occurs go to the next statement
On Error Resume Next

' initialize various variables
strFilePath = "C:\VBAExcel2024_ByExample\AssetInfo.txt"
counter = 0

wRow = 1

strPrompt = "Enter 1 for the percent discount or "
strPrompt = strPrompt + " 2 for the amount discount"
```

```vba
strTitle = "Price Discount Type"

' create an instance of the collection object
Set AssetsColl = New Collection

' open the text file for reading
Open strFilePath For Input As #1

'check is the file is available
If Err.Number <> 0 Then
    MsgBox "File not found!", vbCritical, "File Error"
    Exit Sub
End If

' ask the user the type of discount to apply
intDiscount = CInt(InputBox(strPrompt, strTitle, 1))

' add a new empty worksheet
ActiveWorkbook.Worksheets.Add

' --------------------------------------
' loop until end of file is encountered
' --------------------------------------
Do While Not EOF(1)

    'read data from the text file into four variables
    Input #1, strAssetType, strMake, strModel, itemPrice

    If strAssetType = "AssetType" Then
    ' ----------------------------------------------
    ' enter column headings in the worksheet 1st row
    ' 5th column is for new price calculation
    ' ----------------------------------------------
        With ActiveSheet
            .Cells(1, 1) = strAssetType
            .Cells(1, 2) = strMake
            .Cells(1, 3) = strModel
            .Cells(1, 4) = itemPrice
            .Cells(1, 5) = "New " & itemPrice
        End With
        ' skip lines of code following the if statement
        GoTo Label_SkipHeading
    End If

    '--------------------------------------
    ' create an instance of the CAsset class
```

```vba
        '---------------------------------------
        Set asset = New CAsset

        counter = counter + 1
        aKey = "record" & counter

        '---------------------------------
        ' set properties of the asset object
        '---------------------------------
        asset.AssetType = strAssetType
        asset.Manufacturer = strMake
        asset.Model = strModel
        asset.Price = itemPrice

        '----------------------------------------------
        ' add asset object to the AssetsColl collection
        ' and assign a custom key for that object
        '----------------------------------------------
        AssetsColl.Add asset, aKey

        Set asset = Nothing

Label_SkipHeading:
        Resume Next
    Loop

    'Close the text file
    Close #1

    ' display informational message
    MsgBox "Asset Collection contains " & _
        AssetsColl.Count & " items.", _
        vbInformation, "Total Items"

    '------------------------------------------
    ' iterate through the collection and access
    ' each instance of the CAsset class
    ' printing the data to the Immediate window
    '------------------------------------------
    For Each asset In AssetsColl
       Debug.Print asset.AssetType & vbTab & _
       asset.Manufacturer & vbTab & _
       asset.Model & vbTab & FormatNumber(asset.Price, 2)
    Next asset
```

```
'----------------------------------------------
' iterate through the collection to access
' each instance of the CAsset class
' this time entering data the active worksheet
'----------------------------------------------
For Each asset In AssetsColl
 'set next row in the worksheet
    wRow = wRow + 1
    'write record to the worksheet
    With ActiveSheet
        .Cells(wRow, 1) = asset.AssetType
        .Cells(wRow, 2) = asset.Manufacturer
        .Cells(wRow, 3) = asset.Model
        .Cells(wRow, 4) = asset.Price
        ' calculate the discount
        .Cells(wRow, 5) = asset.NewPrice(intDiscount, _
                            asset.Price, 100)
    End With
Next asset
Selection.CurrentRegion.Columns.AutoFit

'retrieve the asset from a collection by a key
assetKey = InputBox("Enter key", "Retrieval", "record1")

Set asset = AssetsColl.Item(assetKey)
strRecord = "Asset Type" & vbTab & asset.AssetType & _
            vbCrLf
strRecord = strRecord & "Manufacturer" & vbTab & _
            asset.Manufacturer & vbCrLf
strRecord = strRecord & "Model" & vbTab & vbTab & _
            asset.Model & vbCrLf
strRecord = strRecord & "Price" & vbTab & vbTab & _
            Format(asset.Price, "Currency")

MsgBox strRecord, vbInformation + vbOKOnly, _
        "Retrieving " & assetKey

End Sub
```

6. Run the `Retrieve_AssetInfo` procedure. Reply to all the procedure prompts by accepting the default values.
   After running the procedure, you should see the asset data entered into a worksheet and in the Immediate window, as shown in Figures 24.2 and 24.3.

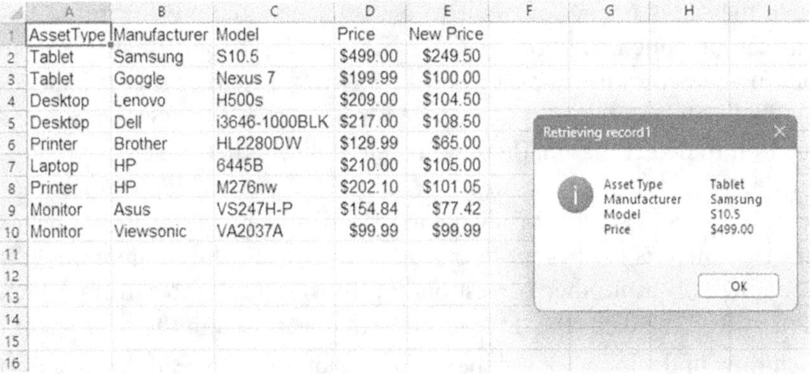

	A	B	C	D	E
1	AssetType	Manufacturer	Model	Price	New Price
2	Tablet	Samsung	S10.5	$499.00	$249.50
3	Tablet	Google	Nexus 7	$199.99	$100.00
4	Desktop	Lenovo	H500s	$209.00	$104.50
5	Desktop	Dell	i3646-1000BLK	$217.00	$108.50
6	Printer	Brother	HL2280DW	$129.99	$65.00
7	Laptop	HP	6445B	$210.00	$105.00
8	Printer	HP	M276nw	$202.10	$101.05
9	Monitor	Asus	VS247H-P	$154.84	$77.42
10	Monitor	Viewsonic	VA2037A	$99.99	$99.99

FIGURE 24.2. The asset data (see Figure 24.1) is stored in a collection of objects and written to the worksheet. The New Price column does not exist in the original file and was added by the VBA procedure to demonstrate the use of class methods.

The retrieval input box demonstrates how a key can be used for accessing objects in a collection. The asset details for the specified record are displayed in the message box above.

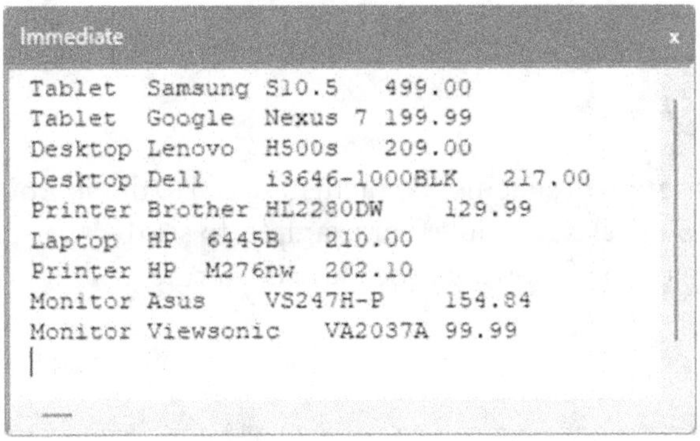

FIGURE 24.3. The asset data in the provided text file (see Figure 24.1) is written to the Immediate window.

The `Retrieve_AssetInfo` procedure starts off by declaring and initializing numerous variables that will be used by various sections of the code. Because you are dealing with an external file, you want to make sure that if the file cannot be found, a message is displayed and the procedure ends. The `Number` property of the VBA `Err` object will return a number other than zero if some problem is encountered while opening the file. To read the file, you must open it in `Input` mode using the following statement:

```
Open strFilePath For Input As #1
```

Once the file is open, you want to read it sequentially from top to bottom. This can be done using the `Do While` or `Do Until` loop. Text files contain a special character known as an End-of-File (`EOF`) marker that is appended to the file by the operating system. When reading the file, you can use the `EOF` function to detect that marker and thus know whether the end of the file was reached. The statement, `Do While Not EOF(1)` means that you want to keep executing the statements inside the loop until all data in the file has been read. This statement is equivalent to `Do Until EOF(1)`. The number between the parentheses is a number corresponding to the file number from which you want to read the data (the same number that was used in the `Open` statement).

Each time in the loop, we use the `Input #` statement to read the data from the file into four variables:

```
Input #1, strAssetType, strMake, strModel, itemPrice
```

After writing out the column names into the worksheet, we create our asset object and set its four properties (`AssetType`, `Manufacturer`, `Model`, and `Price`), using the values stored in the variables:

```
asset.AssetType = strAssetType
asset.Manufacturer = strMake
asset.Model = strModel
asset.Price = itemPrice
```

Each of the preceding assignment statements is actually a call to the appropriate `Let` procedure in the `CAsset` class module. For example, to set the `AssetType` property of the asset object, the following procedure is executed:

```
Property Let AssetType(ByVal aType As String)
    m_AssetType = aType
End Property
```

You can execute the procedure line by line using the F8 key to gain a better understanding of what's going on when these statements are being executed.

At this point, the asset object contains the first record data, which is the second line in our text file. Before handling the next record's data, we use the `Add` method to add the asset object to the `AssetsColl` collection:

```
AssetsColl.Add asset, aKey
```

Each object in the collection is identified by a key that we created by concatenating a number and the word `record`, obtaining `record1`, `record2`, `record3`, and so on.

After adding the asset object to the collection, we release the memory by setting it to `Nothing` and continue to the next record, executing the statements within the loop, skipping only those that were used for the preparation of the column headings. A new object is created, its properties are set, and the object is added to the collection. The same process repeats until the `EOF` is reached. When we are done looping, we close the file using the `Close#1` statement. We should now have nine asset objects in the `AssetsColl` collection.

The remaining code in the procedure iterates through the collection of objects and prints the data to the Immediate window and to the worksheet. When we retrieve the objects from the collection, VBA goes on to execute the `Property Get` procedures that you wrote in the `CAsset` class module. When writing the new price to the worksheet, we call upon the `NewPrice` method. This method uses the `intDiscount` variable whose value was obtained from the user earlier in the procedure. If you accept the default value in the input box, the price is reduced by the specified percentage. The last parameter of the `NewPrice` method, which denotes `amount`, is hardcoded. Based on the amount that is entered, the `IF` statements included in the `NewPrice` method will execute or will be skipped. When entering prices, it is often necessary to appropriately format the data. The `Retrieve_AssetInfo` procedure uses the `FormatNumber` function to format the `Price` data in the Immediate window:

```
FormatNumber(asset.Price, 2)
```

The second argument of the `FormatNumber` function specifies how many places to the right of the decimal are displayed. To format the number as a `Currency` data type, change the preceding statement to:

```
FormatCurrency(asset.Price, 2)
```

CREATING A CUSTOM APPLICATION

The next project in this chapter demonstrates the process of creating and working with a custom object called `CEmployee`. This object will represent an employee and will have properties such as `ID`, `FirstName`, `LastName`, and `Salary`. It will also have a method for modifying the current salary. Be sure to perform all the exercises. Again, we'll start by creating a class module.

Custom Project 24.2a   Creating a Class Module

1. In the Project Explorer window of VBAProject(Chap24.xlsm), choose Insert | Class Module.
2. Highlight the Class 1 module in the Project Explorer window and use the Properties window to rename the class module CEmployee.

Custom Project 24.2b   Declaring Members of the CEmployee Class

Type the following declaration lines at the top of the CEmployee class module:

```
' declarations
Private m_LastName As String
Private m_FirstName As String
Private m_Salary As Currency
Private m_ID As String
```

Notice that the name of each data member variable begins with the prefix m_.

Custom Project 24.2c   Writing Property Get Procedures for the CEmployee Class

Type the following Property Get procedures in the CEmployee class module, just below the declaration section:

```
Property Get ID() As String
  Id = m_ID
End Property

Property Get LastName() As String
  LastName = m_LastName
End Property

Property Get FirstName() As String
  FirstName = m_FirstName
End Property

Property Get Salary() As Currency
  Salary = m_Salary
End Property
```

Let's continue with our project and write the required Property Let procedures for our custom CEmployee object. The CEmployee class will only have three properties (LastName, FirstName, and Salary). Each of these properties will require a separate Property Let procedure. To make the ID property read-only, we will not write a Property Let procedure for it.

⊙ **Custom Project 24.2d   Writing Property Let Procedures for the CEmployee Class**

Type the following `Property Let` procedures in the `CEmployee` class module:

```
Property Let LastName(L As String)
  m_LastName = L
End Property

Property Let FirstName(F As String)
  m_FirstName = F
End Property

Property Let Salary(ByVal dollar As Currency)
  m_Salary = dollar
End Property
```

The `CEmployee` object will have one method to calculate the new salary. Assume that the employee's salary can be increased or decreased by a specific percentage or amount.

Let's continue with our project by writing the required class method.

⊙ **Custom Project 24.2e   Writing Methods for the CEmployee Class**

1. Type the following `CalcNewSalary` function procedure in the `CEmployee` class module:

```
Public Function CalcNewSalary(choice As Integer, _
  curSalary As Currency, amount As Long) As Currency
  Select Case choice
    Case 1 ' by percent
      CalcNewSalary = curSalary + ((curSalary + amount)/100)
    Case 2 ' by amount
      CalcNewSalary = curSalary + amount
  End Select
End Function
```

The `CalcNewSalary` function defined with the `Public` keyword in a class module serves as a method for the `CEmployee` class. To calculate a new salary, a VBA procedure from outside the class module must pass three arguments: `choice`, `curSalary`, and `amount`. The `choice` argument specifies the type of the calculation. Suppose you want to increase the employee salary by 5% or by $5. The first option will increase the salary by the specified percentage, and the second option will add the specified amount to the current salary. The `curSalary` argument is the current salary figure for an employee, and `amount` determines the value by which the salary should be changed.

2. Enter the following SetID method in the CEmployee class module:

```
' Method to set the ID internally within the class
Friend Sub SetID(Value As String)
    m_ID = Value
End Sub
```

Recall that the ID field in the CEmployee class has Property Get but no Property Let, meaning it is read-only. A Friend Sub method SetID is included to internally set the ID directly from the worksheet data. The SetID method will be used in the UserForm_Initialize procedure.

What Is the Friend Method in Excel VBA?

The Friend method (or modifier) in Excel VBA is used to define members such as procedures, functions, or properties in a class module that are accessible only within the same VBA project. This means that while Private restricts access to the containing class/module and Public allows access from anywhere (including other projects), Friend strikes a balance, making the member accessible to other modules or class modules within the same project.

Friend members are hidden from outside projects. In our custom project, the SetID method is declared as Friend because it allows controlled assignment of the ID property by code within the same project but keeps it inaccessible to external applications or users of the class.

USING EVENT PROCEDURES IN THE CLASS MODULE

As you already know, an event is an action recognized by an object. Custom classes recognize only two events: Initialize and Terminate. These events are triggered when an instance of the class is created and destroyed, respectively. The Initialize event is generated when an object is created from a class (see the preceding section on creating an instance of a class). In the CEmployee class example, the Initialize event will also fire the first time that you use the emp variable in the code. Because the statements included inside the Initialize event are the first ones to be executed for the object before any properties are set or any methods are executed, the Initialize event is a good place to perform the initialization of the objects created from the class.

As you recall, the ID is read-only in our CEmployee class. You can use the Initialize event to assign a unique five-digit number to the m_ID variable.

Let's continue with our project by writing the Initialize event procedure in our CEmployee class.

Custom Project 24.2f   Writing the Initialize Event Procedure for the CEmployee Class

In the `CEmployee` class module, enter the following `Class_Initialize` procedure:

```
Private Sub Class_Initialize()
    Randomize
    m_ID = Int((99999 - 10000) * Rnd + 10000)
End Sub
```

The `Class_Initialize` procedure initializes the `CEmployee` object by assigning a unique five-digit number to the variable `m_ID`. To generate a random integer between two given integers where `ending_number` = 99999 and `beginning_number` = 10000, the following formula is used:

```
=Int((ending_number - beginning_number) * Rnd + beginning_number)
```

The `Class_Initialize` procedure also uses the `Randomize` statement to reinitialize the random number generator. For more information on using the `Rnd` and `Int` functions, as well as the `Randomize` statement, see the online help.

The `Terminate` event occurs when all references to an object have been released. This is a good place to perform any necessary cleanup tasks. The `Class_Terminate` procedure uses the following syntax:

```
Private Sub Class_Terminate()
    [cleanup code goes here]
End Sub
```

To release an object variable from an object, use the following syntax:

```
Set objectVariable = Nothing
```

When you set the object variable to `Nothing`, the `Terminate` event is generated. If you have written a `Terminate` event procedure, any code in this event is executed then.

CREATING A FORM FOR DATA COLLECTION

Implementing your custom `CEmployee` object requires that you design a custom form. We will create the form shown in Figure 24.4 and later connect it with our custom `CEmployee` object.

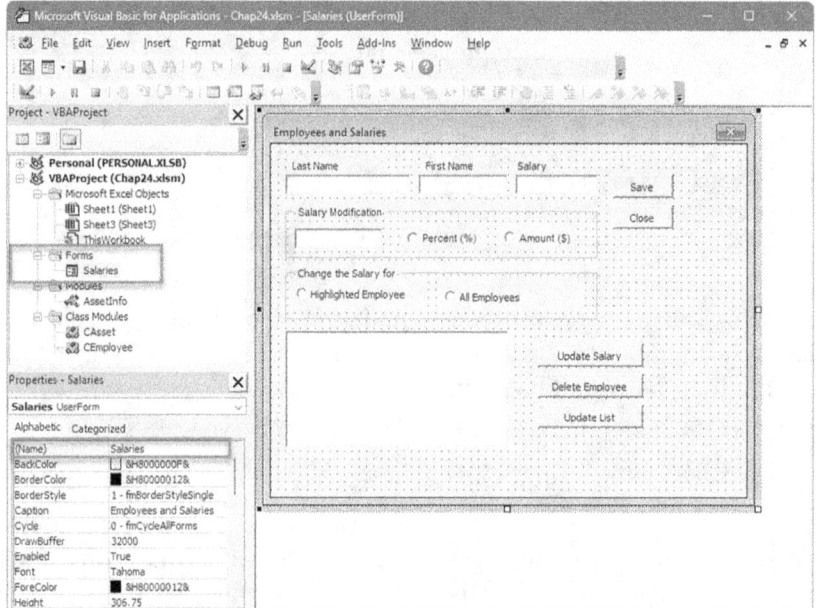

FIGURE 24.4.   This form demonstrates the use of the CEmployee custom object.

⊙ Custom Project 24.2g   Creating a User Form for the CEmployee Object

1. Highlight the current VBA project in the Project Explorer window and choose Insert | UserForm.
   A blank form should appear with UserForm1 in its title bar.

2. Make the form larger by positioning the mouse on the selection handle in the bottom right-hand corner and dragging it to the right and down. Make the form big enough to hold all the controls and labels shown in Figure 24.4.

3. Choose View | Toolbox.
   A small Toolbox should appear on the screen with standard Visual Basic buttons that can be placed on the UserForm.

4. Click the Text Box control in the Toolbox and drag it to the form. Drop it at the position shown in Figure 24.4. Use the same technique to add all the remaining controls and labels to the form.

5. Now that all the controls and labels are placed on the form, set the form and control properties listed in Table 24.1. To set each of these properties:

   i. On the form, click on the object listed in the Object column in Table 24.1.

ii. In the VBE Properties window, locate the object's property name listed in the Property column in Table 24.1.

iii. In the VBE Properties window, replace the property setting in the second column with the value stated in the Settings column in Table 24.1.

TABLE 24.1. Setting the properties for the custom form.

Object	Property	Settings
UserForm1	Name Caption	Salaries Employees and Salaries
label1	Caption	Last Name
Text box below the Last Name label	Name	txtLastName
label2	Caption	First Name
Text box below the First Name label	Name	txtFirstName
label3	Caption	Salary
Text box below the Salary label	Name	txtSalary
frame1	Caption	Salary Modification
Text box in the frame titled Salary Modification	Name	txtRaise
Option button 1	Name Caption	optPercent Percent (%)
Option button 2	Name Caption	optAmount Amount ($)
frame2	Caption	Change the Salary for
Option button 3	Name Caption	optHighlighted Highlighted Employee
Option button 4	Name Caption	optAll All Employees
List box	Name Height Width	lboxPeople 91.45 180.75
Command button 1	Name Caption	cmdSave Save
Command button 2	Name Caption	cmdClose Close
Command button 3	Name Caption	cmdUpdate Update Salary
Command button 4	Name Caption	cmdDelete Delete Employee
Command button 5	Name Caption	cmdEmployeeList Update List

CREATING A WORKSHEET FOR DATA OUTPUT

Now that the UserForm (`Employees and Salaries`) is ready, we need to prepare a worksheet for data entry (see Figure 24.5).

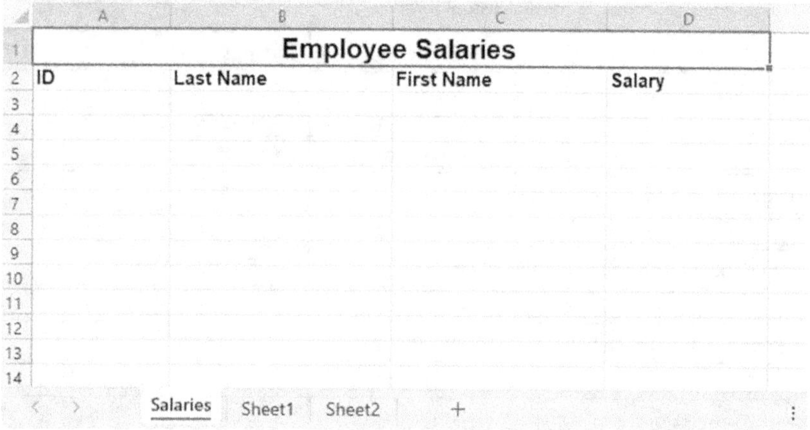

	A	B	C	D
1		Employee Salaries		
2	ID	Last Name	First Name	Salary
3				
4				
5				
6				
7				
8				
9				
10				
11				
12				
13				
14				

Salaries   Sheet1   Sheet2   +

FIGURE 24.5. Data entered on the Employees and Salaries form will be transferred to the Salaries worksheet.

Custom Project 24.2h   Creating a Worksheet for Data Transfer

1. Insert a new worksheet into the current workbook and rename it `Salaries`.
2. Prepare the `Salaries` data entry worksheet as shown in Figure 24.5.

WRITING CODE BEHIND THE USERFORM

With the form and worksheet ready, all that is left to do for the implementation of `CEmployee` is to write the necessary procedures in the form class module. We will need an event procedure to initialize the controls on the form and several click event procedures that will run when a user clicks the form's buttons. We also need to write a function to locate employee data in a worksheet.

Custom Project 24.2i   Writing Code for the UserForm

1. Switch back to the VBE window and double-click the form background to activate the form module. You can also click the View Code button in the Project Explorer window when the form is selected.

2. Enter the following variable declarations at the top of the Code window of the Salaries form module just below the Option Explicit statement.

```
' Global declarations
Private CEmployees As Collection
    Dim emp As New CEmployee
    Dim index As Integer
    Dim ws As Worksheet
    Dim extract As String
    Dim cell As Range
    Dim lastRow As Integer
    Dim empLoc As Integer
    Dim startRow As Integer
    Dim endRow As Integer
    Dim choice As Integer
    Dim amount As Long
```

The CEmployees collection serves as a container for storing employee data. The variable emp represents a new instance of the CEmployee class. Other variables defined in this context will be utilized by VBA procedures linked to various controls on the form.

3. Enter the following procedures in the Salaries Code module:

```
Private Sub UserForm_Initialize()
    Dim ws As Worksheet
    Dim lastRow As Long
    Dim i As Long
    Dim emp As CEmployee
    Dim dataLine As String

    ' Form preparation
    txtLastName.SetFocus
    cmdEmployeeList.Visible = False
    lboxPeople.Enabled = False
    Frame1.Enabled = False

    txtRaise.Value = ""
    optPercent.Value = False
    optAmount.Value = False

    txtRaise.Enabled = False
    optPercent.Enabled = False
    optAmount.Enabled = False

    Frame2.Enabled = False
    optHighlighted.Enabled = False
    optAll.Enabled = False
```

```
    cmdUpdate.Enabled = False
    cmdDelete.Enabled = False

    ' Initialize the Employee collection
    Set CEmployees = New Collection

    ' Reference the Salaries worksheet
    Set ws = ThisWorkbook.Worksheets("Salaries")

    ' Find the last row with data in column A (ID column)
    lastRow = ws.Cells(ws.Rows.Count, 1).End(xlUp).Row

    ' Ensure there is data beyond the header row
    If lastRow > 2 Then
        ' Loop through each row of data
        For i = 3 To lastRow
            ' Create a new Employee instance
            Set emp = New CEmployee

            ' Set the read-only ID using the Friend method
            emp.SetID ws.Cells(i, 1).Value

            ' Populate other Employee properties
            emp.LastName = ws.Cells(i, 2).Value
            emp.FirstName = ws.Cells(i, 3).Value
            emp.Salary = ws.Cells(i, 4).Value

            ' Add the Employee to the collection
            CEmployees.Add emp

            ' Create a comma-delimited string for the ListBox
            dataLine = emp.ID & ", " & emp.LastName & ", " & _
                emp.FirstName & ", " & emp.Salary

            ' Add the dataLine to the ListBox
            Me.lboxPeople.AddItem dataLine

        Next i
        EnableHiddenControls
        Me.lboxPeople.Enabled = True
    End If
End Sub

Sub EnableHiddenControls()
    'enable hidden controls
    cmdEmployeeList.Value = True
```

```
      cmdUpdate.Enabled = True
      cmdDelete.Enabled = True
      Frame1.Enabled = True
      txtRaise.Enabled = True
      optPercent.Enabled = True
      optAmount.Enabled = True
      Frame2.Enabled = True
      optHighlighted.Enabled = True
      optAll.Enabled = True
      txtLastName.SetFocus
End Sub
```

The `UserForm_Initialize` procedure is responsible for populating the list box control on the form when employee data is available on the `Salaries` worksheet. If no data is present, the procedure ensures that only the controls required for the initial data entry are enabled. During form initialization, certain controls may need to be shown or hidden, as well as enabled or disabled. To streamline repeated operations in your project, it's advisable to create a separate procedure to handle these tasks. The `EnableHiddenControls` procedure fulfills this purpose and is called from the `UserForm_Initialize` procedure after populating the list box control. It is also invoked by the `cmdSave_Click` event procedure when the user clicks the Save button after entering employee data.

Figure 24.6 shows the result of the statements inside the `UserForm_Initialize` procedure that enable only the desired controls when the form is first loaded and there is no existing employee data on the `Salaries` worksheet.

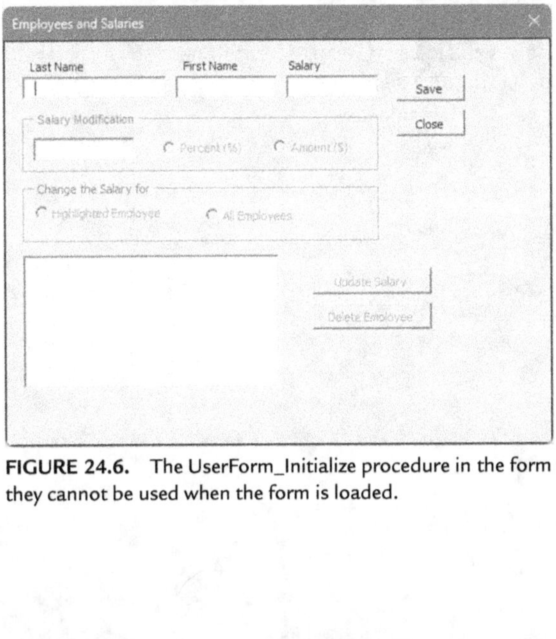

FIGURE 24.6. The UserForm_Initialize procedure in the form module disables certain controls so they cannot be used when the form is loaded.

4. Enter the following `cmdSave_Click` procedure to transfer the data entered on the form to the spreadsheet:

```
'transfer form data to the worksheet

Private Sub cmdSave_Click()
If txtLastName.Value = "" Or txtFirstName.Value = "" Or _
   txtSalary.Value = "" Then
    MsgBox "Enter Last Name, First Name and Salary."
    txtLastName.SetFocus
    Exit Sub
End If
If Not IsNumeric(txtSalary) Then
    MsgBox "You must enter a value for the Salary."
txtSalary.SetFocus
    Exit Sub
End If
If txtSalary < 0 Then
    MsgBox "Salary cannot be a negative number."
    Exit Sub
End If
Worksheets("Salaries").Select
index = ActiveSheet.UsedRange.Rows.Count + 1
lboxPeople.Enabled = True
'set and enter data into the CEmployees collection
With emp
    Cells(index, 1).Formula = emp.ID
    .LastName = txtLastName
    Cells(index, 2).Formula = emp.LastName
    .FirstName = txtFirstName
    Cells(index, 3).Formula = emp.FirstName
    .Salary = CCur(txtSalary)
    If .Salary = 0 Then Exit Sub
    Cells(index, 4).Formula = emp.Salary
    CEmployees.Add emp
End With
'delete data from text boxes
txtLastName = ""
txtFirstName = ""
txtSalary = ""
EnableHiddenControls
End Sub
```

The `cmdSave_Click` procedure starts off with validating the user's input in the Last Name, First Name, and Salary text boxes. If data is provided, VBA assigns to the variable `index` the number of the first empty row on the `Salaries` worksheet. The next statement enables the form's list box control. When the program reaches the `With emp ...` construct, a new instance of the `CEmployee` class is created. The `LastName`, `FirstName`, and `Salary` properties are set based on the data entered in the corresponding text boxes, and the `ID` property is set with the number generated by the `Class_Initialize` event procedure. Each time VBA sees the reference to the instanced `emp` object, it will call the appropriate `Property Let` procedure in the class module.

The last section of this chapter demonstrates how to walk through this procedure step by step to see exactly when the property procedures are executed. After setting the object property values, VBA transfers the employee data to the `Salaries` worksheet. The last statement inside the `With emp...` construct adds the user-defined object `emp` to a custom collection called `CEmployees`.

Next, Visual Basic clears the form's text boxes and enables command buttons that were turned off by the `UserForm_Initialize` procedure.

5. Enter the `cmdEmployeeList_Click` procedure, as shown here:

```
Private Sub cmdEmployeeList_Click()
   ' add new employee data to list box
   lboxPeople.Clear
   For Each emp In CEmployees
     lboxPeople.AddItem emp.ID & ", " & _
     emp.LastName & ", " & emp.FirstName & ", $" & _
     Format(emp.Salary, "0.00")
  Next emp
End Sub
```

The `cmdEmployeeList_Click` procedure is attached to the Update List command button. The `cmdEmployeeList_Click` procedure begins by clearing the contents of the list box and then populating it with the items stored in the custom collection `CEmployees`, as shown in Figure 24.7.

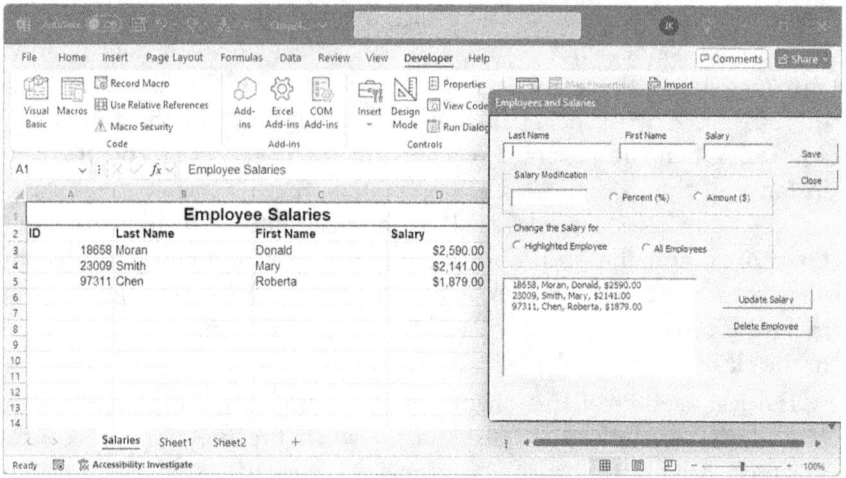

FIGURE 24.7. The list box control displays employee data as entered in the custom collection CEmployees.

6. Type the following `cmdClose_Click` procedure:

```
Private Sub cmdClose_Click()
  ' remove the user form
  Unload Me
End Sub
```

The `cmdClose_Click` procedure allows you to remove the user form from the screen and finish working with the custom collection of employees. When you run the form again, the employees you enter will become members of a new `CEmployees` collection.

7. Type the following `cmdDelete_Click` procedure:

```
Private Sub cmdDelete_Click()
  ' remove an employee from the CEmployees custom collection
  ' make sure that an employee is highlighted in the
  ' list control

  If lboxPeople.ListIndex > -1 Then
    MsgBox "Selected item number: " & lboxPeople.ListIndex
    extract = CEmployees.Item(lboxPeople.ListIndex + 1).ID
    MsgBox extract
    Call FindId
    MsgBox empLoc
    Range("A" & empLoc).Delete (3)
    MsgBox "There are " & CEmployees.Count & _
        " items in the CEmployees collection. "
```

```
    CEmployees.Remove lboxPeople.ListIndex + 1
    MsgBox "The CEmployees collection has now " & _
        CEmployees.Count & " items."
    cmdEmployeeList.Value = True
    If CEmployees.Count = 0 Then
      Call UserForm_Initialize
    End If
  Else
    MsgBox "Click the item you want to remove."
  End If
End Sub
```

The `cmdDelete_Click` procedure lets you remove an employee from the `CEmployees` custom collection. To delete an employee, you must click the appropriate item in the list box. When you click a list item, the `cmdEmployeeList_Click` procedure is automatically executed. This procedure makes sure that the list box contents are refreshed. The employee is removed from both the collection and the list box. If the list box contains only one employee, VBA calls the `UserForm_Initialize` procedure to disable certain form controls after removing the last employee from the collection.

The `cmdDelete_Click` procedure contains several `MsgBox` statements that allow you to examine the contents of the list box control as you make deletions. In addition to removing the employee from the custom collection, the `cmdDelete_Click` procedure must also remove the corresponding row of employee information from the worksheet. Locating the employee data in the worksheet is handled by the `FindId` function. (See Step 8 for the procedure code.) This function returns to the `cmdDelete_Click` procedure the row number that must be deleted.

8. Type the following function procedure:

```
Private Function FindId()
  ' return number of list box selected row
  Set ws= ActiveWorkbook.Sheets("Salaries")
  startRow = ActiveSheet.UsedRange.Rows.Count + _
    1 - CEmployees.Count
  endRow = ActiveSheet.UsedRange.Rows.Count
  For Each cell In ws.Range(Cells(startRow, 1), _
    Cells(endRow, 1))
    If cell.Value = extract Then
      empLoc = cell.Row
      FindId = empLoc
      Exit Function
    End If
```

```
       Next
End Function
```

The `FindId` function procedure returns to the calling procedure the row number of the employee selected in the form's list box. The search for the data in the worksheet is based on the contents of the variable `extract`, which stores the unique employee number. The search for the employee ID is limited to the first worksheet column and begins with the row containing the first collection item. This approach makes the search faster. You don't want to search the entire worksheet.

9. Type the following `cmdUpdate_Click` procedure:

```
Private Sub cmdUpdate_Click()
 If optHighlighted.Value = False And optAll.Value = False Then
    MsgBox "Click the 'Highlighted Employee' or " _
       & " 'All Employees' option button."
    Exit Sub
  End If
  If Not IsNumeric(txtRaise) Then
    MsgBox "This field requires a number."
    txtRaise.SetFocus
    Exit Sub
  End If
  If optHighlighted.Value = True And _
    lboxPeople.ListIndex = -1 Then
      MsgBox "Click the name of the employee."
      Exit Sub
  End If
  If lboxPeople.ListIndex <> -1 And _
    optHighlighted.Value = True And _
    optAmount.Value = True And _
    txtRaise.Value <> "" Then
      extract = CEmployees.Item(lboxPeople.ListIndex + 1).ID
      MsgBox extract
      Call FindId
      MsgBox empLoc
      choice = 2
      amount = txtRaise
      CEmployees.Item(lboxPeople.ListIndex + 1).Salary = _
        emp.CalcNewSalary(choice, _
      CEmployees.Item(lboxPeople.ListIndex + 1).Salary, amount)
      Range("D" & empLoc).Formula = CEmployees. _
        Item(lboxPeople.ListIndex + 1).Salary
      cmdEmployeeList.Value = True
  ElseIf lboxPeople.ListIndex <> -1 And _
```

```vba
      optHighlighted.Value = True And _
      optPercent.Value = True And _
      txtRaise.Value <> "" Then
        extract = CEmployees.Item(lboxPeople.ListIndex + 1).ID
        MsgBox extract
        Call FindId
        MsgBox empLoc
        CEmployees.Item(lboxPeople.ListIndex + 1).Salary = _
          CEmployees.Item(lboxPeople.ListIndex + 1).Salary _
          + (CEmployees.Item(lboxPeople.ListIndex + _
          1).Salary * txtRaise / 100)
        Range("D" & empLoc).Formula = CEmployees. _
          Item(lboxPeople.ListIndex + 1).Salary
        cmdEmployeeList.Value = True
    ElseIf optAll.Value = True And _
      optPercent.Value = True And _
      txtRaise.Value <> "" Then
        For Each emp In CEmployees
          emp.Salary = emp.Salary + ((emp.Salary * _
              txtRaise) / 100)
          extract = emp.ID
          MsgBox extract
          Call FindId
          MsgBox empLoc
          Range("D" & empLoc).Formula = emp.Salary
        Next emp
        cmdEmployeeList.Value = True
    ElseIf optAll.Value = True And _
      optAmount.Value = True And _
      txtRaise.Value <> "" Then
        For Each emp In CEmployees
          emp.Salary = emp.Salary + txtRaise
          extract = emp.ID
          MsgBox extract
          Call FindId
          MsgBox empLoc
          Range("D" & empLoc).Formula = emp.Salary
        Next emp
        cmdEmployeeList.Value = True
    Else
      MsgBox "Enter data or select an option."
    End If
End Sub
```

With the `cmdUpdate_Click` procedure, you can modify the salary by the specified percentage or amount. The update can be done for the selected employee or all the employees listed in the list box control and collection. The cmdUpdate_Click procedure checks whether the user selected the appropriate option buttons and entered the increased value in the text box. Depending on which options are specified, the Salary amount is updated for one employee or all the employees, either by a percentage or amount. The salary modification is also reflected in the worksheet.

Figure 24.8 displays a new value for the salary of Roberta Chen, which has been increased by 10%. By entering a negative number in the text box, you can decrease the salary by the specified percentage or dollar amount. Updating the employee information in the form also updates the corresponding row in the Salaries worksheet.

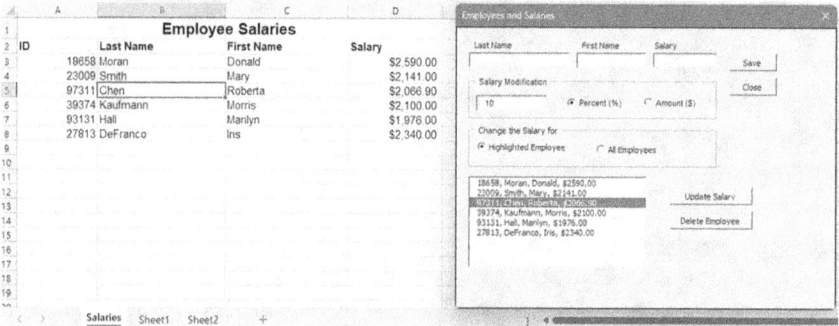

FIGURE 24.8. The employee salary can be increased or decreased by the specified percentage or dollar amount. When you update the employee information in the form, the relevant data is also updated in the worksheet.

WORKING WITH THE CUSTOM CEMPLOYEE CLASS

The application using a custom CEmployee class is now completed. To make it easy to work with the Salaries user form, we'll write a one-line procedure in the standard module to display the form.

◉ Custom Project 24.2j   Writing a Procedure to Display the User Form

1. Insert a standard module into the current project by choosing Insert | Module. Rename this module WorkAndPay.
2. Type the following procedure in the WorkAndPay module:

```
Sub ClassDemo()
   Salaries.Show
End Sub
```

> **NOTE**
> *You can view the* Salaries *form by clicking the form's background and pressing F5, or you can place a button in the* Salaries *worksheet and assign the* ClassDemo *procedure to it.*

WATCHING THE EXECUTION OF YOUR CUSTOM APPLICATION

To help you understand what's going on when your code runs and how your custom object works, let's step through the cmdSave_Click procedure.

Custom Project 24.2k   Walking Through the CEmployee Application Code

1. In the Project Explorer window, right-click the Salaries form and choose View Code.
2. When the Salaries (Code) window appears, select the cmdSave procedure from the combo box at the top left-hand side of the window.
3. Set a breakpoint by clicking in the left margin next to the following line of code, as shown in Figure 24.9:

```
If txtLastName.Value = "" Or txtFirstName.Value = "" Or _
   txtSalary.Value = "" Then
```

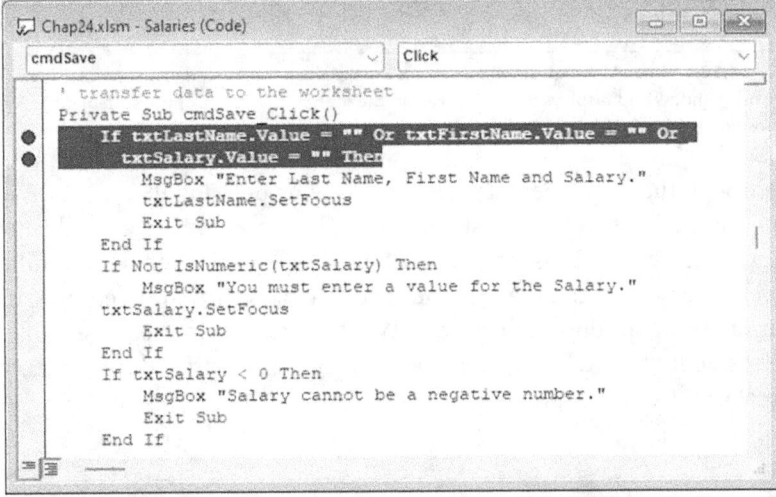

FIGURE 24.9. A red circle in the margin indicates a breakpoint. When VBA encounters the statement with a breakpoint, it automatically switches to the Code window and displays the text of the line as white on a red background.

4. In the Project Explorer window, highlight the WorkAndPay module and press F7 to view the code.

5. Place the cursor anywhere inside the ClassDemo procedure (in the WorkAndPlay module) and press F5 or choose Run | Run Sub/UserForm.

6. When the form appears, enter data in the Last Name, First Name, and Salary text boxes, and click the Save button on the form. Visual Basic should now switch to the Code window because it encountered a breakpoint in the first line of the cmdSave_Click procedure, as shown in Figure 24.10.

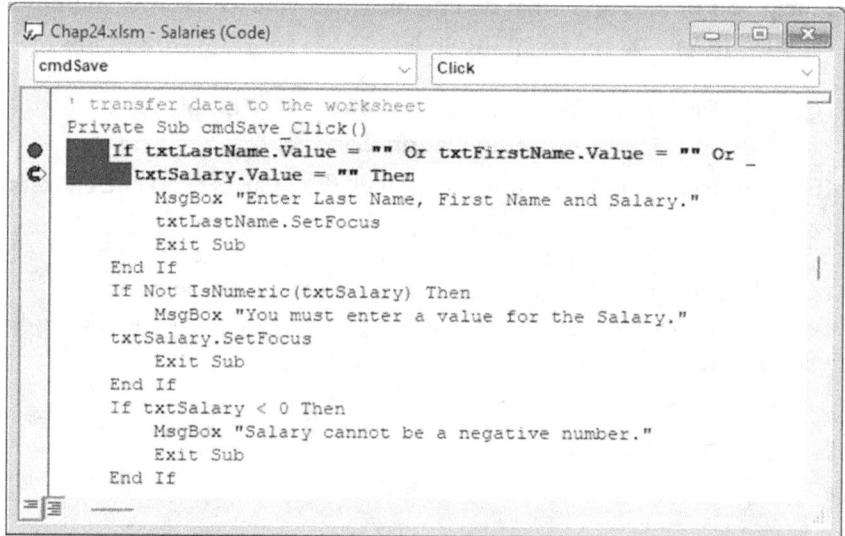

FIGURE 24.10.   When Visual Basic encounters a breakpoint while running a procedure, it switches to the Code window and displays a yellow arrow in the margin to the left of the statement at which the procedure is suspended and highlights the statement in yellow.

7. Step through the code one statement at a time by pressing F8.
   Visual Basic runs the current statement and then automatically advances to the next statement and suspends execution. The current statement is indicated by a yellow arrow in the margin and a yellow background. Keep pressing F8 to execute the procedure step by step. When Visual Basic encounters the With emp statement, it switches to the Class_Initialize procedure, as shown in Figure 24.11. Continue pressing F8.

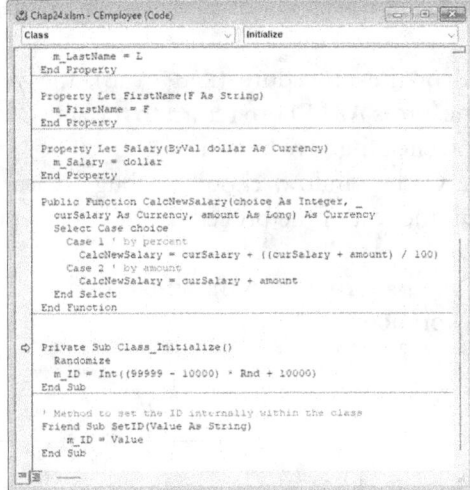

FIGURE 24.11. When Visual Basic encounters the reference to the object variable emp, it goes out to execute the Class_Initialize procedure. After executing the statements inside this procedure, VBA returns to the cmdSave_Click procedure.

When Visual Basic encounters the statement Cells(Index, 1).Formula = emp.ID, it executes the Property Get ID procedure in the CEmployee class module, as shown in Figure 24.12.

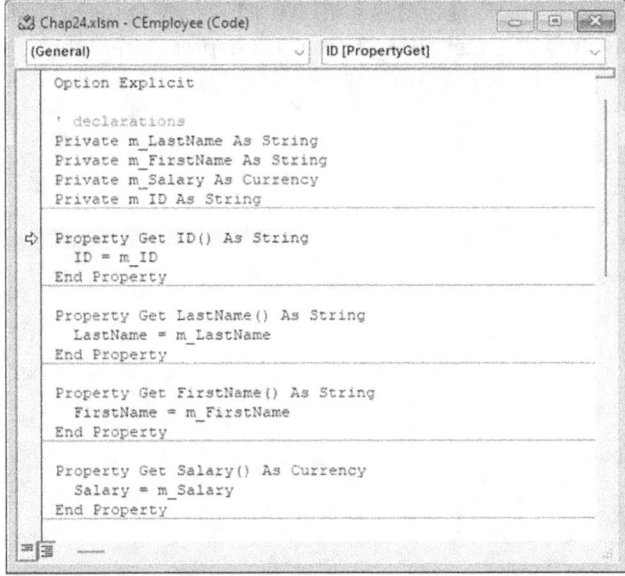

FIGURE 24.12. Reading properties of your custom object is accomplished through the Property Get procedures.

8. Using the F8 key, trace the execution of the `cmdSave_Click` procedure to the end.

9. When VBA encounters the end of the procedure (`End Sub`), the yellow highlight is turned off. At this time, press Alt+F11 to return to the active form. Enter data for a new employee and click the Save button.

10. When Visual Basic displays the Code window, choose Debug | Clear All Breakpoints. Now, press F5 to run the rest of the procedure without stepping through it.

11. Click the Close button on the `Salaries` form to exit the application.

12. Save and close the `Chap24.xlsm` workbook.

SUMMARY

In this chapter, you used class modules in the VBE screen to create two custom objects. You saw how to define your objects' properties by using `Property Get` and `Property Let` procedures. You wrote custom methods for your objects, learned how to create an instance of a class, and were introduced to using event procedures in a class module. Next, you created a custom form for your object, wrote code behind the user form, and learned how to debug your custom Excel VBA application.

This chapter completes Part VII of this book, in which we focused on introducing advanced concepts in Excel VBA. The next chapter covers the topic of using HTML and XML with Excel VBA.

Part **VIII**

Working Together: VBA, HTML, XML, and the REST API

With the Internet and intranets, worksheet data becomes easily accessible and shareable around the clock. Excel not only allows you to capture data from the Web but also enables you to publish it seamlessly.

This section of the book introduces you to integrating Excel VBA with HTML (Hypertext Markup Language) and XML (Extensible Markup Language), as well as leveraging a modern Web service known as the REST API.

Chapter 25 — USING HTML AND XML IN EXCEL 2024

E xcel 2024 continues to be a powerhouse for data analysis and automation. Among its many features is the ability to integrate with Web technologies such as HTML and XML.

HTML enables the creation and presentation of structured information. XML, on the other hand, is a great tool for organizing and transporting data in a format that's both machine-readable and human-readable. Using these technologies in conjunction with Excel VBA opens tremendous opportunities for data interaction and management.

With the integration of HTML and XML into VBA, Excel users can effortlessly harness the power of Web-based data. HTML can be utilized for creating dynamic outputs such as reports or dashboards, while XML can enhance your ability to manage and exchange data. This chapter focuses on practical ways of using HTML and XML with VBA in Excel 2024.

To maximize your benefit from this chapter, you should have a connection to the Internet.

CREATING HYPERLINKS USING VBA

Excel, like other applications in Microsoft 365, allows you to create hyperlinks in your worksheets. You can use hyperlinks to link to other sheets, workbooks, Web sites, or specific file locations manually via the Insert Hyperlink dialog box (see Figure 25.1) or by writing just a few lines of VBA code.

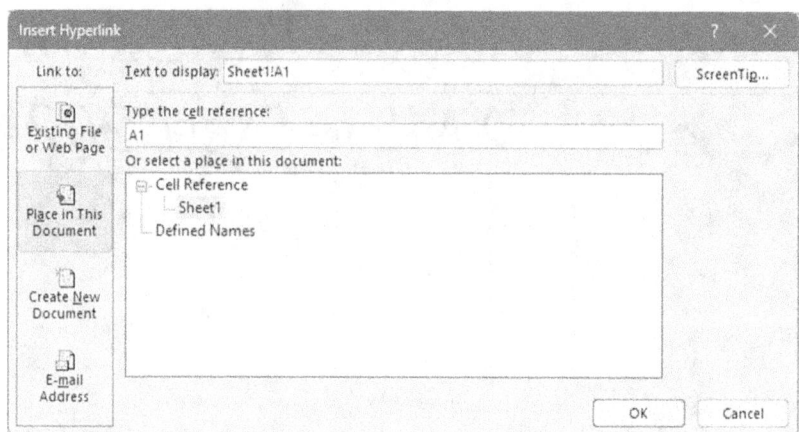

FIGURE 25.1. The Insert Hyperlink dialog (Insert | Link | Insert Link) is used to insert a hyperlink in a Microsoft Excel worksheet.

In VBA, each hyperlink is represented by a `Hyperlink` object. To create a hyperlink to a Web page, use the `Add` method of the `Hyperlinks` collection with the following syntax:

```
Expression.Hyperlinks.Add(Anchor, Address, [SubAddress],
    [ScreenTip], [TextToDisplay])
```

The arguments in square brackets are optional.

`Expression` denotes a worksheet or range of cells where you want to place the hyperlink. `Anchor` is an object to be clicked. This can be either a `Range` or `Shape` object.

`Address` points to a local network or a Web page. `SubAddress` is the name of a range in the Excel file. `ScreenTip` allows the display of a screen label. `TextToDisplay` is the name that you'd like to display in a spreadsheet cell for a specific hyperlink.

In Hands-On 25.1, we will place a hyperlink in a worksheet cell. This hyperlink should take you to the Yahoo!® site when clicked.

NOTE	*Please note that the files for the "Hands-On" projects can be found in the companion files.*

Hands-On 25.1 Using VBA to Place a Hyperlink in a Worksheet Cell

1. Open a new workbook and save it as `C:\VBAExcel2024_ByExample\Chap25.xlsm`.

2. Switch to the VBE screen and insert a new module into `VBAProject (Chap25.xlsm)`.

3. In the Code window, enter the code of the `FollowMe` procedure:

```
Sub FollowMe()
    Dim myRange As Range
    Set myRange = Sheets(1).Range("A1")

    myRange.Hyperlinks.Add Anchor:=myRange, _
        Address:="http://search.yahoo.com/", _
        ScreenTip:="Search Yahoo", _
        TextToDisplay:="Click here"
End Sub
```

4. Run the `FollowMe` procedure.

After running the `FollowMe` procedure, cell A1 in the first worksheet will contain a hyperlink titled "Click here" with the screen tip "Search Yahoo" (see Figure 25.2). If you are connected to the Internet, clicking on this hyperlink will activate your browser and load the Yahoo! search engine.

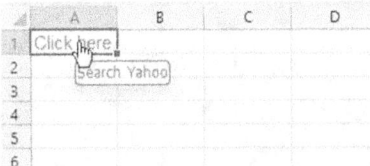

FIGURE 25.2.   This hyperlink was placed in a worksheet by a VBA procedure.

If you'd rather not place hyperlinks in a worksheet but want your users to reach the required Internet pages directly from an Excel worksheet, use the `Follow-Hyperlink` method. This method syntax is shown below:

```
Expression.FollowHyperlink(Address, [SubAddress], [NewWindow],
    [AddHistory], [ExtraInfo], [Method], [HeaderInfo])
```

`Expression` returns a `Workbook` object. `Address` is the address of the Web page that you want to activate. `SubAddress` is a fragment of the object to which the hyperlink address points. This can be a range of cells in an Excel worksheet.

`NewWindow` indicates whether you want to display the document or page in a new window; the default setting is `False`. `AddHistory`, is not currently used and is reserved for future use.

`ExtraInfo` gives additional information that allows jumping to the specific location in a document or on a Web page. For example, here, you can specify the text for which you want to search.

`Method` specifies the method in which the additional information (`ExtraInfo`) is attached. This can be one of the following constants: `msoMethodGet` or `msoMethodPost`. When you use `msoMethodGet`, `ExtraInfo` is a string that's appended to the URL address. When using `msoMethodPost`, `ExtraInfo` is posted as a string or byte array.

`HeaderInfo` is a string that specifies header information for the HTTP request. The default value is an empty string.

Let's use the `FollowHyperlink` method in a VBA procedure We'll use the Microsoft Bing search engine to find pages containing the text entered in a worksheet cell.

(◉) Hands-On 25.2   Using a Search Engine to Find Text Entered in a Worksheet Cell

1. Insert a new sheet into the current workbook.
2. Switch to the VBE screen and double-click the Sheet2 (Sheet2) object in the Microsoft Excel Objects folder located in `VBAProject (Chap25.xlsm)`.
3. In the Sheet2 Code window, enter the `Worksheet_BeforeDoubleClick` event procedure:

```
Private Sub Worksheet_BeforeDoubleClick(ByVal Target As Range, _
        Cancel As Boolean)
    Dim strSearch As String

    strSearch = Sheets(2).Range("C3").Formula
    If Target = Range("C3") Then
        Cancel = True
        ActiveWorkbook.FollowHyperlink _
        Address:="https://www.bing.com/search?q=" & strSearch
    End If
End Sub
```

4. Switch to the Microsoft Excel application window. In cell C3 on Sheet2, enter any word or term about which you want to find information (see Figure 25.3).

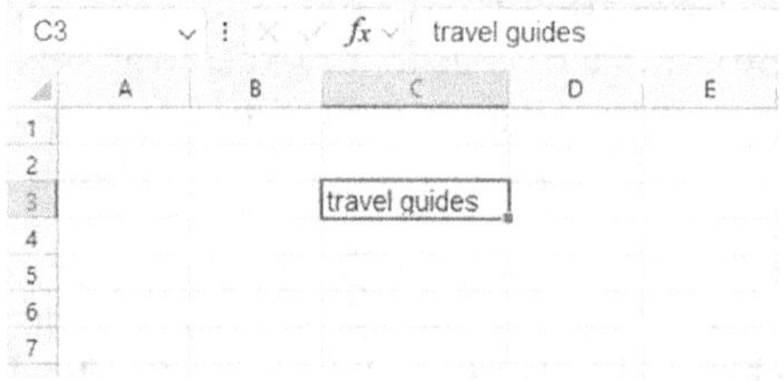

FIGURE 25.3. A Microsoft Excel worksheet can be used to send search parameters to any search engine on the Internet.

5. Make sure that you are connected to the Internet. Double-click cell C3. This will cause the text entered in cell C3 to be sent to Microsoft Bing. The screen should show the index to the found topics with the specified criteria (see Figure 25.4).

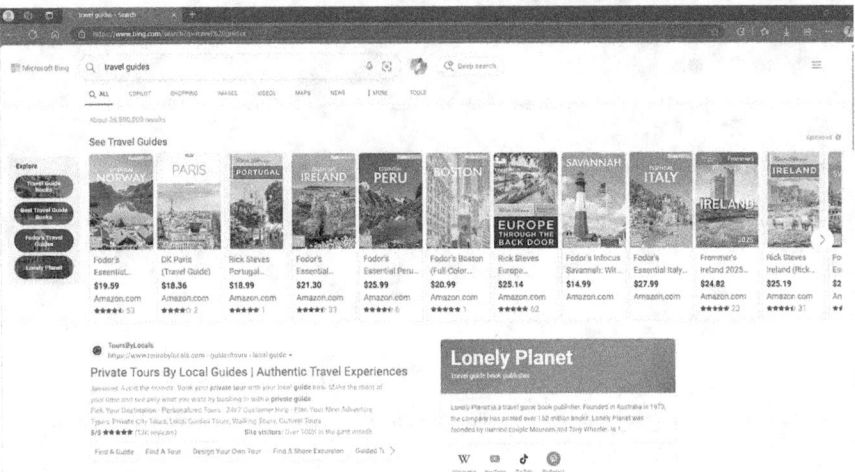

FIGURE 25.4. A Web page opened from a Microsoft Excel worksheet lists topics that were found based on the criteria entered in a worksheet cell (see Figure 25.3).

Removing Hyperlinks

To clear hyperlinks, you can use the `Hyperlinks.Delete` method:

```
Sub RemoveHyperlink()
    Worksheets("Sheet1").Range("A1").Hyperlinks.Delete
End Sub
```

CREATING AND PUBLISHING HTML FILES USING VBA

Excel allows you to save files in the HTML format using `.htm` or `.html` file extensions. When you save an Excel file as `.html`, you can view your worksheets using your favorite Web browser. You can save your HTML files directly to the Web server, a network server, or a local computer. To do this, click File | Save As. Select the folder for the file location, and in the Save as type drop-down box, choose Web Page. Working with nothing other than the user interface (see Figures 25.5 and 25.6), you can place an entire workbook or selected sheets on the Web page. Detailed instructions on how to go about placing an entire workbook or any worksheet (or its elements, such as charts or PivotTables) on a Web page can be found in the Excel online help. Because this book is about programming, we will focus only on the way these tasks are performed by writing VBA code.

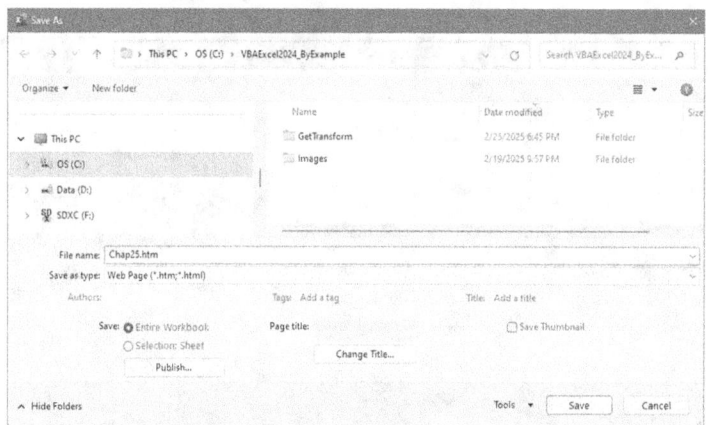

FIGURE 25.5.   The Save As dialog box allows users to save their workbook as a Web page.

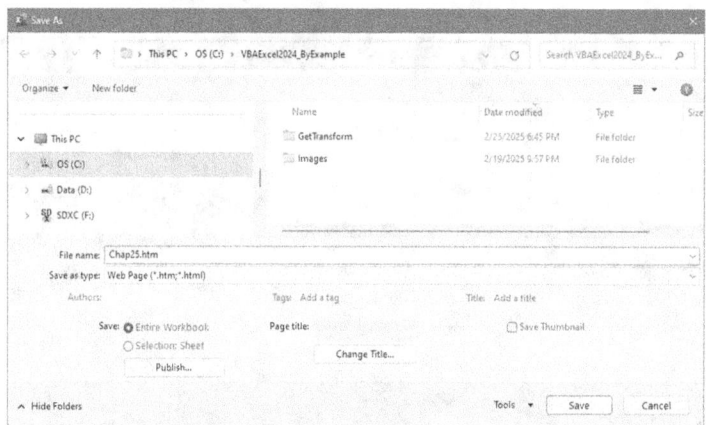

FIGURE 25.6.   The Publish as Web Page dialog box appears after clicking the Publish button on the Save As dialog box (see Figure 25.5).

The VBA object library offers objects for publishing worksheets on Web pages. To programmatically create and publish Excel files in the HTML format, you should become familiar with the `PublishObject` object and the `PublishObjects` collection.

PublishObject represents a worksheet element that was saved on a Web page, while PublishObjects is a collection of all the PublishObject objects of a specific workbook. To add a worksheet element to the PublishObjects collection, use its Add method. This method will create an object representing a specific worksheet element that was saved as a Web page.

The format of the Add method is as follows:

```
Expression.Add(SourceType, Filename, Sheet, Source, HtmlType,
    [DivID], [Title])
```

The arguments in square brackets are optional. Expression returns an object that belongs to the PublishObjects collection. SourceType specifies the source object using one of the constants shown in Table 25.1.

TABLE 25.1.   SourceType constants used with the PublishObjects.Add method.

SourceType Constants	Value	Description
xlSourceAutoFilter	3	An AutoFilter range
xlSourceChart	5	A chart
xlSourcePivotTable	6	A PivotTable report
xlSourcePrintArea	2	A range of cells selected for printing
xlSourceQuery	7	A query table (an external data range)
xlSourceRange	4	A range of cells
xlSourceSheet	1	An entire worksheet
xlSourceWorkbook	0	A workbook

Filename is a string specifying the location where the source object (SourceType) was saved. This can be a Uniform Resource Locator (URL) or the path to a local or network file. Sheet is the name of the worksheet that was saved as a Web page. Source is a unique name that identifies a source object. This argument depends on the SourceType argument.

Source is a range of cells or a name applied to a range of cells when the SourceType argument is the xlSourceRange constant. If the SourceType argument is a constant such as xlSourceChart, xlSourcePivotTable, or xlSourceQuery, Source specifies the name of a chart, PivotTable report, or query table.

HtmlType specifies whether the selected worksheet element is saved as static (non-interactive) HTML or an interactive Microsoft Office Web Component (this feature was depreciated in Excel 2007 and is used only for backward compatibility. See the side note for more information.). HtmlType constants are listed in Table 25.2.

TABLE 25.2. HTMLType constants used with the PublishObjects.Add method.

HTMLType Constants	Description
`xlHTMLCalc` (depreciated)	Use the Spreadsheet component. This component makes it possible to view, analyze, and calculate spreadsheet data directly in an Internet browser. This component also has options that allow you to change the formatting of fonts, cells, rows, and columns.
`xlHTMLChart` (depreciated)	Use the Chart component. This component allows you to create interactive charts in the browser.
`xlHTMLList` (depreciated)	Use the PivotTable component. This component allows you to rearrange, filter, and summarize information in a browser. This component is also able to display data from a spreadsheet or a database (for instance, Microsoft Access, SQL Server, or OLAP servers).
`XlHTMLStatic` (default)	Use static (non-interactive) HTML for viewing only. The data published in an HTML document does not change.

NOTE	*Office Web Components (OWCs) are ActiveX controls that provide four components: Spreadsheet, Chart, PivotTable, and Data Source Control (DSC). In the Office releases prior to 2007 (XP/2003/2000), these components made it possible to use Excel analytical options in an Internet browser. OWCs were discontinued in Office 2007. If you need OWCs to support older applications, you will need to reinstall these components or allow users to download and install them on the fly when the document that requires their use is opened in a browser.*

`DivID` is a unique identifier used in the HTML `DIV` tag to identify the item on the Web page. `Title` is the title of the Web page.

Before we look at how you can use the `Add` method from a VBA procedure, you also need to learn how to use the `Publish` method of the `PublishObject` object. This method allows publishing an element or a collection of elements in a particular document on the Web page. This method is quite simple and looks like this:

```
Expression.Publish([Create])
```

`Expression` is an expression that returns a `PublishObject` object or `PublishObjects` collection. The optional argument, `Create`, is used only with a `PublishObject` object. If the HTML file already exists, setting this argument to `True` will overwrite the file. Setting this argument to `False` inserts the item or items at the end of the file. If the file does not yet exist, a new HTML file is created, regardless of the value of the `Create` argument.

Now that you've been introduced to VBA objects and methods used for creating and publishing an Excel workbook in HTML format, let's get back to programming. In the following Hands-On, you will create an Excel worksheet with an embedded chart and publish it as static HTML.

Hands-On 25.3   Creating and Publishing an Excel Worksheet with an Embedded Chart

1. Create a new workbook and save it as `C:\VBAExcel2024_ByExample\ PublishExample.xlsm`.
2. Right-click Sheet1 and choose Rename. Type `Help Desk` and press Enter.
3. In the `Help Desk` worksheet, enter data as shown in Figure 25.7.
4. To create the chart, select cells A1:B10 and choose the Insert tab. In the Charts group, click the Column drop-down and select 2-D Column. Add Data labels to columns in the plot area and set other chart elements as depicted in Figure 25.7.

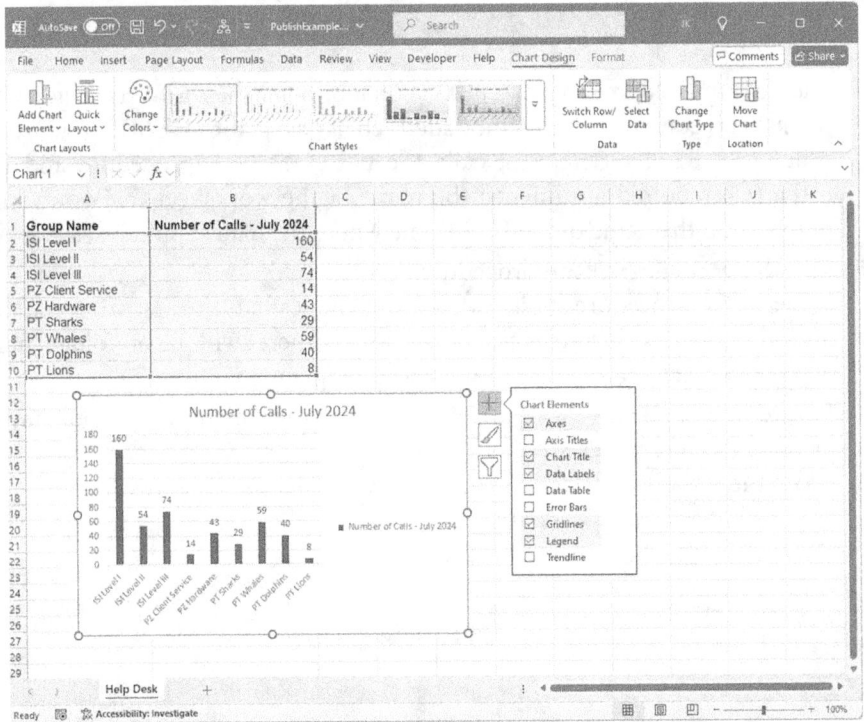

FIGURE 25.7.   A worksheet with an embedded chart can be placed on a Web page by using Save As and choosing Web Page as the file format or via a VBA procedure.

5. Activate the VBE screen, insert a new module into VBAProject (PublishExample.xlsm), and enter the following two procedures:

```
' The procedure below will publish a worksheet
' with an embedded chart as static HTML

Sub PublishOnWeb(strSheetName As String, _
                 strFileName As String)

    Dim objPub As Excel.PublishObject
    Set objPub = ThisWorkbook.PublishObjects.Add( _
        SourceType:=xlSourceSheet, _
        Filename:=strFileName, Sheet:=strSheetName, _
        HtmlType:=xlHtmlStatic, Title:="Calls Analysis")
    objPub.Publish True
End Sub

Sub CreateHTMLFile()
  Call PublishOnWeb("Help Desk", _
    "C:\VBAExcel2024_ByExample\WorksheetWithChart.htm")
End Sub
```

The first procedure above, PublishOnWeb, publishes a Web page with a worksheet containing an embedded chart as static HTML. The second procedure, CreateHTMLFile, calls the PublishOnWeb procedure and feeds it the two required arguments: the name of the worksheet that you want to publish and the name of the HTML file where the data should be saved.

6. Run the CreateHTMLFile procedure.
   When this procedure finishes, a new file called WorksheetWithChart. htm is created on your hard drive. Also, there will be a folder named WorksheetWithChart_files for storing supplemental files.

7. Locate the WorksheetWithChart.htm file and open it in your favorite browser (see Figure 25.8).

8. Close the WorkhseetWithChart.htm file.

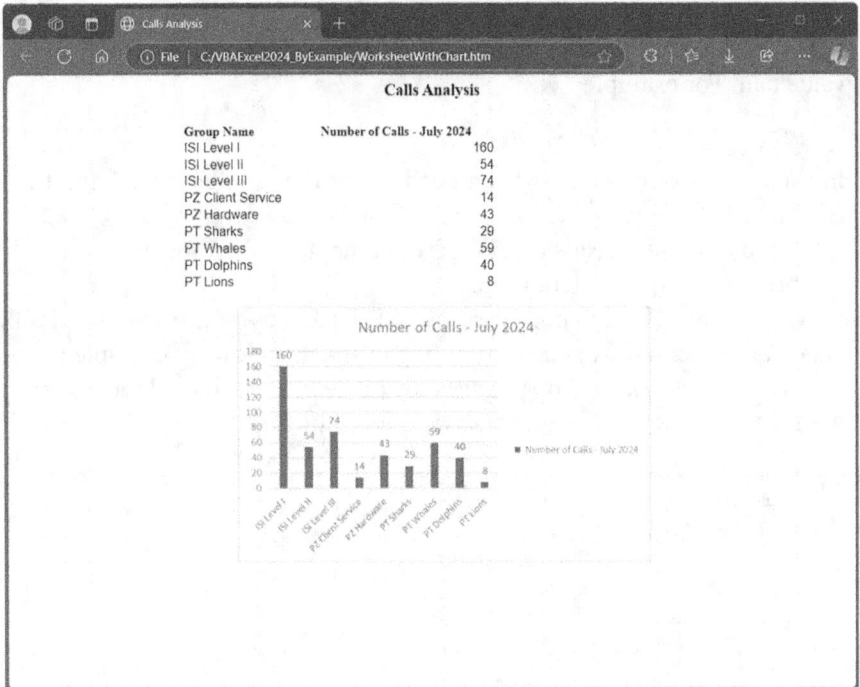

FIGURE 25.8.   An Excel worksheet published as a static (non-interactive) Web page.

CREATING AN HTML FILE FROM AN EXCEL RANGE

To build HTML documents in code, you must become familiar with HTML tags, which define the structure and content of a Web page. HTML tags act as instructions that tell the browser how to display and format text, images, links, and other elements. Tags are enclosed in angle brackets (< >) and typically come in pairs: an opening tag and a closing tag. For example, the `<p>` tag is an opening tag for a paragraph element and the `</p>` tag is the closing tag, indicated by a forward slash (/):

```
<p>This is an example paragraph. </p>
```

Some tags do not require a closing tag because they are self-contained. For example:

```
<img src="image.png" alt="Image description" />
```

`<img>` is a self-closing tag used to display an image.

Tags can include attributes that provide additional information or specify behavior. Attributes are written within the opening tag and consist of a name-value pair. For example:

```
<a href="https://www.DeGruyter.com">Visit This Publisher</a>
```

In the above example, `href` is an attribute of the `<a>` tag, specifying the link's destination.

When a browser reads an HTML document, it interprets the tags and their attributes to render content visually.

VBA enables the dynamic creation of HTML files by extracting and utilizing data from specified worksheet cell ranges. In the following example procedure (`CreateHTMLFile`), we loop through the range `A1:B10` to read the data from the `PublishExample.xlsm` workbook we created in Hands-On 25.3:

```vba
Sub CreateHTMLFile()
    Dim FilePath As String
    Dim FileNum As Integer
    Dim Row As Long, Col As Long
    Dim strHTML As String
    Dim DataRange As Range
    Dim ws As Worksheet

    ' Set worksheet and data range
    Set ws = ThisWorkbook.ActiveSheet
    Set DataRange = ws.Range("A1:B10")

    ' Initialize HTML structure with enhanced table styling
    strHTML = "<html>" & vbCrLf
    strHTML = strHTML & _
        "<head><title>Excel Data</title></head>" & vbCrLf
    strHTML = strHTML & "<body>" & vbCrLf
    strHTML = strHTML & "<h1>Excel Data Table</h1>" & vbCrLf
    strHTML = strHTML & _
    "<table style='border-collapse: collapse; " & _
        "border: 1px solid black;'>" & vbCrLf

    ' Add title row with shading
    strHTML = strHTML & _
        "<tr style='background-color: #f2f2f2;'>" & vbCrLf
    For Col = 1 To DataRange.Columns.Count
        strHTML = strHTML & _
          "<th style='border: 1px solid black; padding: 5px;'>" & _
            DataRange.Cells(1, Col).Value & "</th>" & vbCrLf
    Next Col
```

```
    strHTML = strHTML & "</tr>" & vbCrLf

    ' Add remaining rows and columns with cell borders
    For Row = 2 To DataRange.Rows.Count
        strHTML = strHTML & "<tr>" & vbCrLf
        For Col = 1 To DataRange.Columns.Count
            strHTML = strHTML & _
          "<td style='border: 1px solid black; padding: 5px;'>" & _
            DataRange.Cells(Row, Col).Value & "</td>" & vbCrLf
        Next Col
        strHTML = strHTML & "</tr>" & vbCrLf
    Next Row

    ' Close HTML table and body
    strHTML = strHTML & "</table>" & vbCrLf
    strHTML = strHTML & "</body>" & vbCrLf
    strHTML = strHTML & "</html>"

    ' Define file path
    FilePath = ThisWorkbook.Path & "\EnhancedExcelData.html"

    ' Save HTML content to file
    FileNum = FreeFile
    Open FilePath For Output As FileNum
    Print #FileNum, strHTML
    Close FileNum

    MsgBox "Enhanced HTML file created at: " & FilePath
End Sub
```

The `CreateHTMLFile` procedure, which you can enter in the VBE screen of the current `PublishExample.xlsm` workbook, reads the specified range of cells, constructs an HTML table, and saves it as an `.html` file. The HTML string is created with rows and cells corresponding to the data. Several key HTML elements are used to construct the table:

1. `<html>` and `<body>`:

These tags form the basic structure of the HMTL document.

- `<html>` wraps all the content of the file, specifying that it is an HTML document.
- `<body>` contains the visible content of the Web page, including the title, headings, and table.

2. `<head>` and `<title>`

   - The `<head>` section contains metadata and non-visible elements, such as the `<title>`tag.
   - The `<title>` tag specifies the title of the Web page, which appears in the browser's title bar.

3. `<h1>`

   - The `<h1>` tag is used for the main heading of the page, typically the largest and most prominent heading.

4. `<table>`

   - The `<table>` element defines the table itself. In the procedure, it includes inline CSS (Cascading Style Sheets) for styling.
   - The `style` attribute specifies visual properties, such as `border-collapse: collapse;` for merging borders and `border: 1px solid black;` for single-line borders around cells.

5. `<tr>`

   - The `<tr>` tab defines a table row. Each row in the table is wrapped in `<tr>` tags. In the example procedure, the title row (first row) is styled separately with a light gray background using the `style` attribute.

6. `<th>`

   - The `<th>` tag is used for table headers, typically displayed in bold and centered by default.
   - In the example procedure, these are used for the cells in the title row (the first row of the table).

7. `<td>`

   - The `<td>` tag represents a standard data cell in a table. These cells hold the actual data extracted from the worksheet. Inline styling, such as borders and padding, is applied for consistent appearance.

8. CSS styling (via the `style` attribute)
   Inline CSS is used extensively to format the table. For example:

   - `border: 1px solid black;` creates a thin black border around cells.
   - `padding: 5px;` adds space inside cells for better readability.
   - `background-color: #f2f2f2;` shades the title row for emphasis.

What Is CSS?

CSS is a style sheet language used to describe the visual presentation of HTML or XML documents. It controls the layout, design, and appearance of Web pages, making it easier to style and format content. CSS separates content (HTML) from design, allowing for consistent styling across multiple Web pages and efficient updates. The key aspects of CSS are selectors, properties, and values. The selectors define which HTML elements the style applies to (e.g., body, h1, .class, or #id). The properties specify the style attributes such as color, font-name, and margin. Values set the desired outcome for the property (e.g., red, 20px, or 5px). The following is a simple style sheet example code that makes all <h1> elements blue, sets their font size to 24 pixels, and centers the text:

```
h1 {
color: blue;
font-size: 24px;
text-align: center;
}
```

In the CreatedHTMLFile procedure, instead of using an external style sheet, we used inline CSS to style specific table elements, such as borders, padding, and background colors. Inline CSS applies styles directly to individual HTML elements using the style attribute. In the procedure example, the first row of the table (title row) is styled with a light gray background using style='background-color: #f2f2f2;'. Style borders are applied to all cells using style='border: 1px solid black;'. Padding is added for better spacing using padding: 5px;.

Inline CSS makes the code lengthy and harder to maintain because styles need to be applied repeatedly for each element. It works quite well, however, for small tasks such as dynamic HTML generation in VBA, where embedding all styles directly in the HTML ensures simplicity and portability.

When you run the CreatedHTMLFile procedure, you'll find a new HTML file named ExcelEnhancedData.html in your working directory. Opening this file in a browser displays the data in a simple HTML table, as depicted in Figure 25.9.

Excel Data Table

Group Name	Number of Calls - July 2024
ISI Level I	160
ISI Level II	54
ISI Level III	74
PZ Client Service	14
PZ Hardware	43
PT Sharks	29
PT Whales	59
PT Dolphins	40
PT Lions	8

FIGURE 25.9.   This HTML file was dynamically generated from data found in an Excel worksheet range.

The remaining sections of this chapter introduce you to XML. The XML functionality is not new; it was added to Excel in version 2000 and has been much improved since. We will look at the XML file format in detail later on in this chapter after we've learned what XML is and how it is used in Excel.

WHAT IS XML?

XML is a standard that provides a mechanism for designing your own custom markup language and using that language for describing the data in your own documents. Although XML was designed specifically for delivering information over the World Wide Web, it is being utilized in other areas, such as storing, sharing, and exchanging data. Like HTML, XML is a markup language; however, HTML and XML serve different functions. HTML describes the Web page layout by using a set of fixed non-customizable tags, while XML lets you describe data content using custom tags.

The main goal of XML is the separation of content from presentation. Because XML documents are text files, XML is independent of the operating system platform, software vendor, and natural or programming language. XML makes it easy to describe any data (structured or unstructured) and send it anywhere across the Web using common protocols, such as HTTP or FTP. As long as any two organizations can agree on the XML tag set to be used to represent the data being exchanged, the data can be interpreted and easily exchanged no matter what back-end systems these organizations run or what databases they use. Although anyone can describe the data by creating a set of custom tags, the representatives of many industry groups have defined and published XML schemas that dictate how XML documents are formatted to represent data for their industry. XML schemas define the structure and data types that are allowed within an XML document and enforce the document's conformity to the rules.

Let's take a look at the following XML document (`Courses.xml`) that was created using Windows Notepad:

```
<?xml version = "1.0"?>
<Courses>
    <Course ID = "VBA1EX">
        <Title>Beginning VBA in Excel</Title>
        <Startdate>3/4/2024</Startdate>
        <Sessions>6</Sessions>
    </Course>
    <Course ID = "VBA2EX">
        <Title>Intermediate VBA in Excel</Title>
```

```
        <Startdate>4/13/2024</Startdate>
        <Sessions>8</Sessions>
    </Course>
    <Course ID = "VBA3EX">
        <Title>Advanced VBA in Excel</Title>
        <Startdate>9/7/2024</Startdate>
        <Sessions>12</Sessions>
    </Course>
</Courses>
```

The first line of an XML document is called the XML declaration:

```
<?xml version = "1.0"?>
```

The above instruction identifies the file as an XML file. Recent versions of Excel can easily open the structured data file even if you omit this instruction. The declaration line is also known as a processing instruction. This instruction begins and ends with a question mark (?) and contains the name of the application (in this example, xml) to which the instruction is directed, as well as additional information that needs to be passed to the XML processor, such as the version number and optional encoding and standalone attributes:

```
<?xml version = "1.0" encoding = "UTF-8" standalone = "Yes"?>
```

The encoding attribute specifies the character style to be applied. When the standalone attribute is set to Yes, it tells the XML processor that the document does not reference an external file.

You can include other processing instructions in the XML file if you need the processing application to take a specific action. For example, you can specify that the file be opened by Excel by adding the following instruction to the above XML markup:

```
<?mso-application progid = "Excel.Sheet"?>
```

Notice the simple structure of XML. Similar to HTML, XML uses tags for data markup. Unlike HTML, however, XML tags are not predefined. You can change the name of the tag to anything you want. For example, <Courses> and </Courses> can become <Classes> and </Classes>. You can create custom tags that best describe data in your document. To do this correctly, you'll need to follow some simple XML rules so that your document is well formed. This is explained in the next section of this chapter.

An XML document contains one or more elements, data attributes, and text. The top element (in this example, the element marked with the <Courses> tag) is called a root node. Every XML file must have a start root node and an end root node. There can be only one root node in the file. The start tag is represented by left and right-angle brackets (< >), and the end tag has a left angle bracket, forward slash, and a right-angle bracket (< / >). The names of the tags are case-sensitive. The name of the start tag and the name of the corresponding end tag must match exactly.

Elements may contain text and other elements. For example, the `<Courses>` element is defined to contain one or more `<Course>` elements. Notice that in the previous example, the data for the `<Course>` element is provided by the `ID` attribute:

```
<Course ID = "VBA1EX">
```

The values of attributes must be surrounded by double or single quotation marks.

Notice that the `<Course>` element has three other elements: `<Title>`, `<Startdate>`, and `<Sessions>`. The second and third `<Course>` elements have exactly the same structure. The structure of the XML document is very logical and easy to follow. You can quickly add more data to the file by following the same pattern.

Character Encodings in XML

When you type an XML document into Notepad and save it, you can choose from one of several supported character encodings, including ANSI, Unicode (UTF-16), Unicode (Big Endian), or UTF-8. The encoding declaration in the XML document identifies which encoding is used to represent the characters in the document. UTF-8 encoding allows the use of non-ASCII characters, regardless of the language of the user's operating system and browser or the language version of Microsoft 365. When you use UTF-8 or UTF-16 character encoding, an encoding declaration is optional. XML parsers can determine automatically whether a document uses UTF-8 or UTF-16 Unicode encoding.

WELL-FORMED XML DOCUMENTS

When you create or modify an XML document, you must make sure that your XML file is well formed. Here's what makes an XML document well formed:

- An XML document must have one root element. In HTML, the root element is always `<HTML>`, but in an XML document, you can name your root element anything you want. Element names must begin with a letter or underscore character. The root element must enclose all other elements.

Elements must be properly nested. The XML data must be hierarchical; the beginning and ending tags cannot overlap:

```
<Employee>
  <Employee ID>090909</Employee ID>
</Employee>
```

- All element tags must be closed. A start tag must be followed by an end tag:

```
<Sessions>5</Sessions>
```

- You can use shortcuts, such as a single slash (/), to end the tag so you don't have to type the full tag name. For example, if the current `<Sessions>` element is empty (does not have a value), you could use the following tag:

```
<Sessions />
```

- Tag names are case-sensitive. The tags `<Title>` and `</Title>` aren't equivalent to `<TITLE>` and `</TITLE>`. For example, the following line:

```
<Title>Beginning VBA Programming</Title>
```

- is not the same as:

```
<TITLE>Beginning VBA Programming</TITLE>
```

- All attributes must be in quotation marks:

```
<Course Id = "VBAEX1"/>
```

- You cannot have more than one attribute with the same name within the same element.
- If the `<Course>` element has two `ID` attributes, they must be written separately, as shown below:

```
<Course ID = "VBAEX1"/>
<Course ID = "VBAEX2"/>
```

Checking that an XML document is well formed is similar to syntax checking in VBA. When you try to open an XML file in Excel that is not well formed, you will receive an error message similar to the one in Figure 25.10. You can force this error by removing the ending "s" from the tag `<Courses>` in the `Courses.xml` file while it is open in Notepad. When you do this, the beginning `<Course>` tag will not match the ending `</Courses>` tag; thus, an error will occur when you try to open the file with Excel. Notice that the error message specifies the type of error that was found and the name of the source file. To help you troubleshoot the error, the XML Import Error dialog box includes the Details button. To investigate the error, select the error in the XML Import Error dialog box (Figure 25.10) and click the Details button. This brings up a dialog box that details the error. You must fix all the errors before you can successfully open the XML file in Excel.

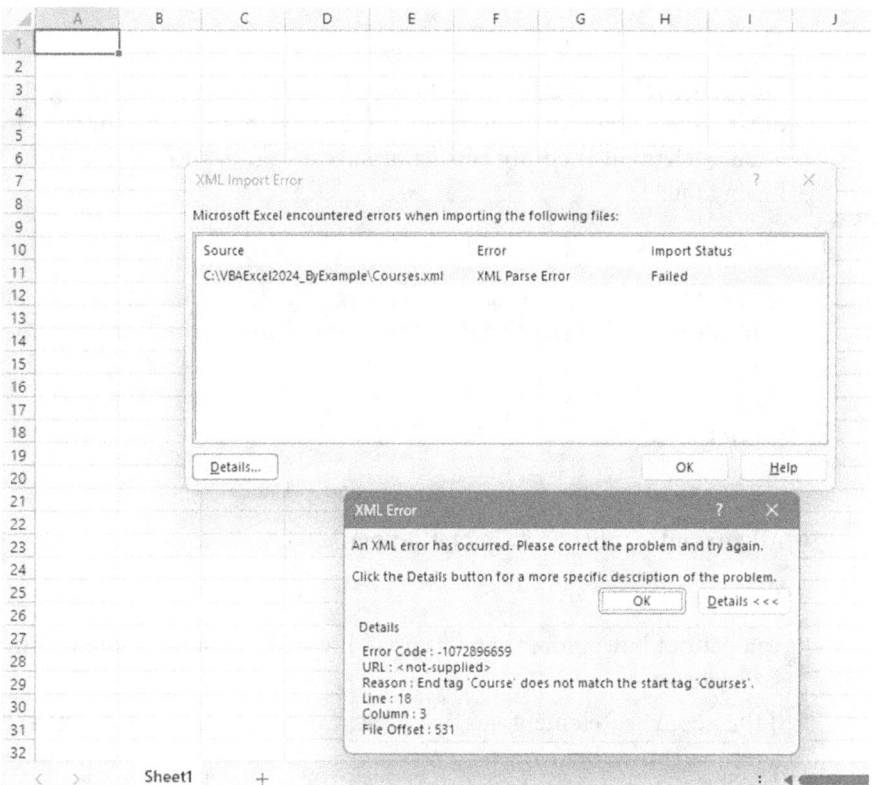

FIGURE 25.10.   When the XML document is not well formed, Excel displays an XML Import Error dialog box when you try to open the file. The Details button shows the explanation of the error.

You do not need to wait for Excel to discover errors in XML files. To verify that the document is well formed, it's a good idea to open it in the browser before attempting this task with Excel. Double-click the XML filename, and it should open up in your default browser (Figure 25.11).

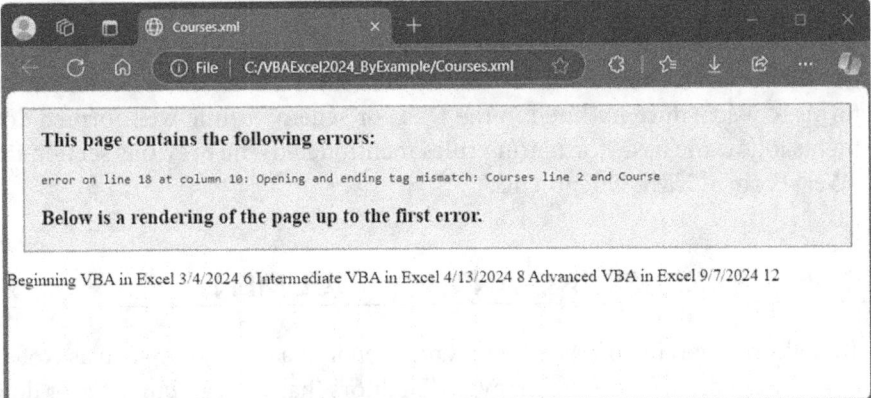

FIGURE 25.11.   The Microsoft Edge browser depicted here notifies you about the error in the XML file.

What Is a Parser?

If you want to read, update, create, or manipulate any XML document, you will need an XML parser. A *parser* is a software engine, usually a Dynamic Link Library (DLL), that can read and extract data from XML. Microsoft Internet Explorer and Edge browsers have a built-in XML parser (`MSXML*.DLL`) that can read and detect all non-well-formed documents. `MSXML` has its own object model, known as DOM (Document Object Model), which you can use from VBA to quickly and easily extract information from an XML document (see "The XML Document Object Model" section later in this chapter).

VALIDATING XML DOCUMENTS

There are two types of validation in XML. One is checking whether the document is well formed (see the previous section). The other type of validation requires that you create a Document Type Definition (DTD) or a set of rules known as a *schema* to determine the type of elements and attributes an XML document should contain, how these elements and attributes should be named, and how the elements should be related.

Creating a DTD or schema for an XML document is optional. Create either one only if you are planning to validate data. In XML, data validation is accomplished by comparing the document with the DTD or schema. When you open the XML document in a parser, the parser compares the DTD to the data and raises an error if the data is invalid. This book does not explore the creation and use of DTDs or schemas. These topics alone would require a separate chapter. What you should remember from this section is that a valid XML document is not the same as a well-formed XML document. A valid XML document conforms to a structure outlined in the DTD or schema, while well-formed documents follow the basic formatting rules mentioned in the previous section titled "Well-Formed XML Documents."

EDITING AND VIEWING AN XML DOCUMENT

To make changes in an XML document, open it in a text editor such as Notepad or an XML editor. There are many XML editors that you can purchase or download free from the Internet. Figure 25.12 shows an XML file open in Notepad ++ and in the XML Copy Editor Store Edition downloaded from Microsoft Store. The advantage of using XML editors is that they come with special features that organize your XML data into an easy-to-read format and allow you to create well-formed documents. They also provide validation features.

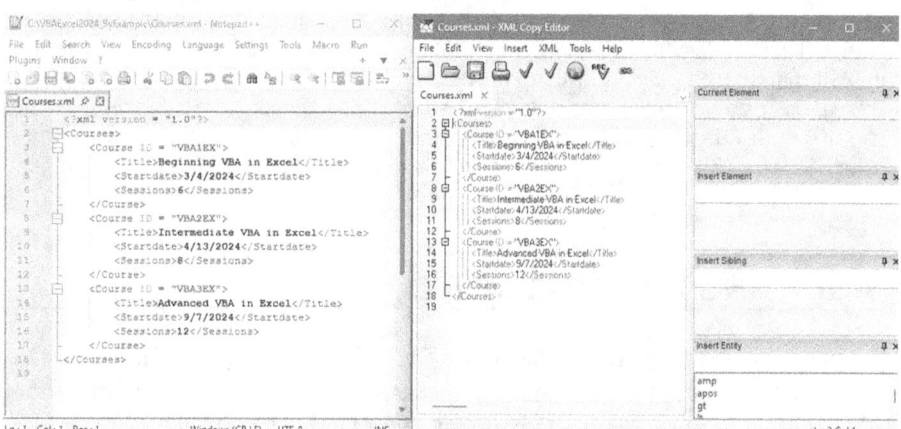

FIGURE 25.12.   The example Courses.xml file opened in two different editors.

You can make your XML documents legible and clear by using comments. The XML processor ignores all commented text. A comment begins with the `<! --` characters and ends with the `-->` characters. Within your comment, you can use any characters except for a double hyphen (`--`). A comment can be placed anywhere within an XML document provided that it's outside (not within) other markup tags. Let's add a comment to the `Courses.xml` document that we discussed earlier.

⊙ Hands-On 25.4 Adding a Comment to an XML Document

1. Create a new folder named VBAExcel2024_XML.
2. Copy the Courses.xml file from the companion files to your VBAExcel2024_XML folder.
3. Right-click the Courses.xml file, choose Open with, and then select Notepad.
   If Notepad is not listed, select Choose Program, and then locate and select Notepad in the list.
4. Type the following comment between the `<Courses>` and `<Course ID = "VBA1EX">` tags:

   ```
   <!-- You can add more courses to this list -->
   ```

   The beginning of the file should now look like this:

   ```
   <?xml version = "1.0"?>
   <Courses>
   <!-- You can add more courses to this list -->
   <Course ID = "VBA1EX">
   ```

5. Save the file as Courses1.xml and exit Notepad.

Comments can also be used to disable a particular processing instruction or an XML node. For example, you could prevent the display of information about a specific course by commenting out the section of XML markup like this:

```
<!--
  <Course ID = "VBA2EX">
    <Title>Intermediate VBA in Excel</Title>
      <Startdate>4/13/2024</Startdate>
      <Sessions>8</Sessions>
  </Course>
-->
```

Now that you've modified the file, let's open it in the browser to ensure that you have a well-formed XML document.

⊙ Hands-On 25.5 Viewing an XML Document in the Internet Browser

1. Open the Courses1.xml file in your browser.
   Figure 25.13 shows the file open in the Microsoft Edge browser. Here, you can see the hierarchical layout of an XML document very clearly. Microsoft Edge automatically places a down arrow to the left of each element so you can expand and collapse the XML data layout.

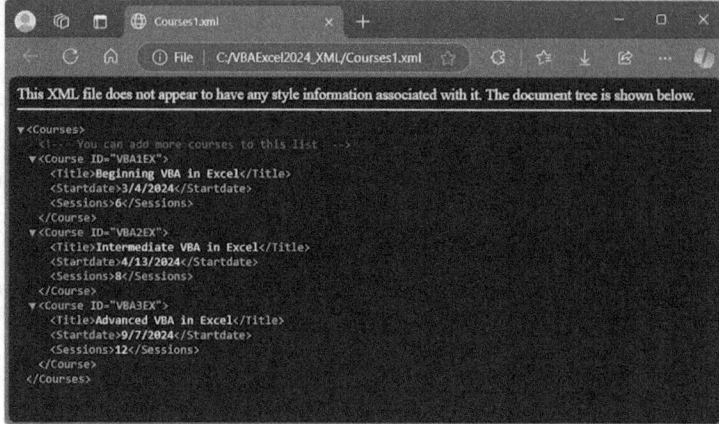

FIGURE 25.13. An XML data file opened in the Microsoft Edge browser.

Edge did not display any errors; therefore, we can assume that the document is well formed.

2. Close the browser.

OPENING AN XML DOCUMENT IN EXCEL

Once you've checked that you have a well-formed XML document by opening it with your browser, let's open it in Excel.

⊙ Hands-On 25.6 Opening an XML Document in Excel

1. Start Excel and open the `Courses1.xml` file.
Excel displays the Open XML dialog box, as shown in Figure 25.14.

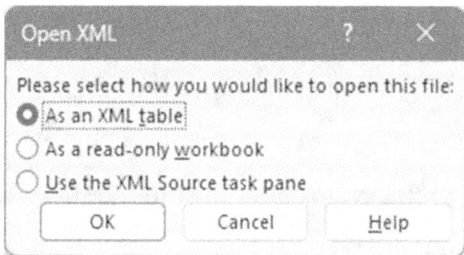

FIGURE 25.14. Excel displays the Open XML dialog box when you open an XML document that does not have a style sheet associated with it.

2. In the Open XML dialog box, select the As an XML table option button and click OK.

Excel tells you that it could not find the schema for the XML document (see Figure 25.15). A schema will be automatically created for you when you click OK.

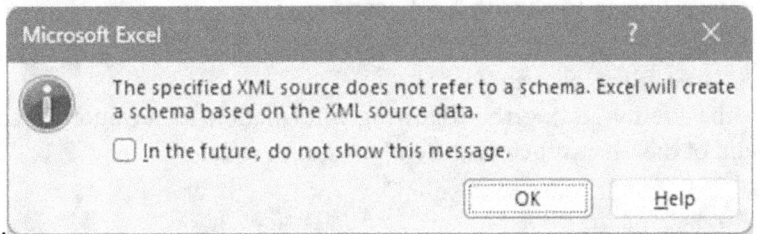

FIGURE 25.15. A schema file provides the rules for the XML document. If it is missing, Excel will infer the schema from the XML data file.

NOTE	*If you are trying to open a very complex XML document, a schema file created by Excel may be incorrect or insufficient for your needs. If this is the case, you will need to create your own XML Schema Description (XSD) file or have someone else create it for you.*

3. Click OK to have Excel create a schema and open the file.

Excel imports the contents of the XML document into an XML table. The cells in the worksheets are mapped to the XML elements in the source file and can be refreshed at any time by clicking the Refresh button on the Table Design tab (Figure 25.16).

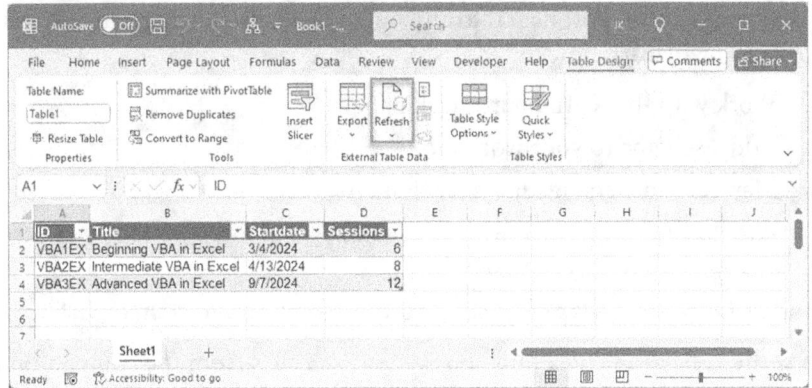

FIGURE 25.16. An XML document opened in Excel.

When Excel creates a schema based on the contents of your XML document, your XML source file becomes read-only. This means that you cannot make changes to the file by editing the XML table in Excel. Excel refers to the schema files it creates as XML maps. Only by creating your own XML map can you write back to your XML document from Excel. The next section of this chapter demonstrates how to work with XML maps.

4. Leave Excel open with the data as shown in Figure 25.16 and open the `Courses1.xml` file in Notepad.

5. Modify the file by adding the following information about another course to the end of the file just before the end `</Courses>` tag:

```
<Course ID = "VBA1Outlook">
  <Title>Beginning VBA in Outlook</Title>
  <Startdate>10/10/2024</Startdate>
  <Sessions>6</Sessions>
</Course>
```

6. Save the file and close Notepad.

7. In Excel, click the Refresh button on the Table Design tab.
   Notice that Excel has added a new row of data to the XML table.

8. Save the workbook as `C:\VBAExcel2024_ByExample\Chap25b.xlsm` in the macro-enabled format and then close it.

WORKING WITH XML MAPS

XML schemas in Excel are called *XML maps*. You can associate one or more schemas with a workbook and then map all or some of the schema elements to various cells or ranges on a worksheet. Using XML mapping makes it relatively easy to import and export data into and out of Excel. In the following Hands-On project, you will learn how to:

- Work with the XML Source task pane
- Add a schema to your workbook
- Map cells to elements in an XML map
- Populate the XML map with XML data

⊙ **Hands-On 25.7   Mapping Schema Elements to Worksheet Cells**

1. Copy the `Employees.xml` and `Employees.xsd` files from the companion files into your `VBAExcel2024_XML` folder.

2. Open a new workbook in Microsoft Excel.

3. Click the Source button in the XML group on the Developer tab.
   Excel displays the XML Source task pane, as shown in Figure 25.17.

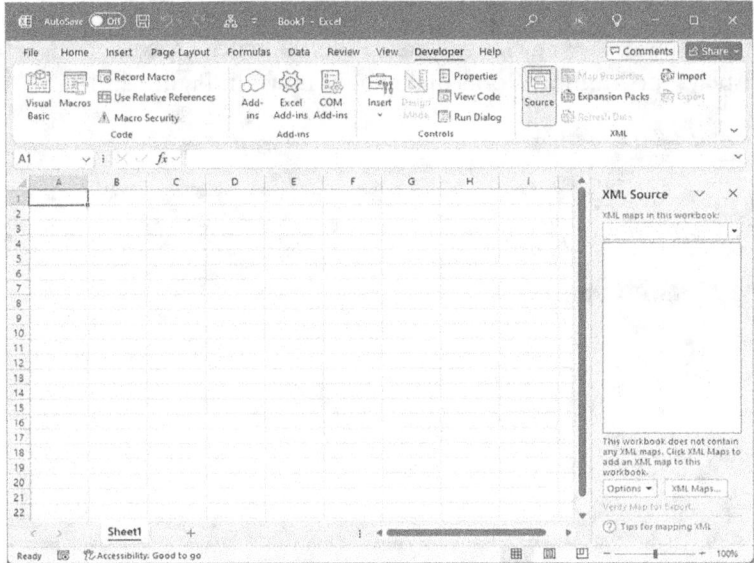

FIGURE 25.17. The XML Source task pane.

The XML Source task pane is used for displaying XML maps found in the XML data or schema documents, and mapping XML elements to cells or ranges on a worksheet. If the current worksheet doesn't have any XML maps associated with it, the XML Source task pane is blank. The XML Source task pane includes two buttons (Options and XML Maps) and one hyperlink (Verify Map for Export).

4. In the XML Source task pane, click the XML Maps button.
   Excel displays the XML Maps dialog box shown in Figure 25.18.

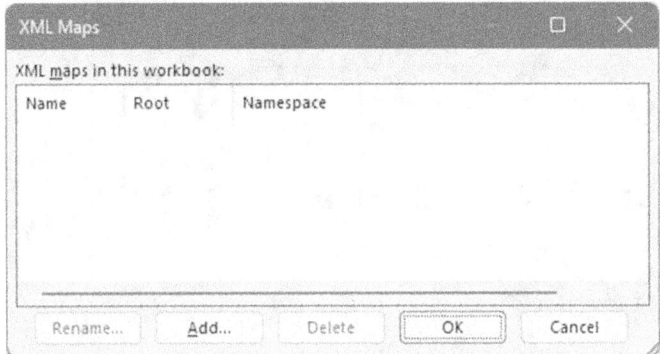

FIGURE 25.18. Use the XML Maps dialog box to add, delete, or rename an XML map associated with the workbook.

5. Click the Add button in the XML Maps dialog box.

6. In the Select XML Source dialog box, switch to the `VBAExcel2024_XML` folder, select the `Employees.xsd` schema file, and click Open.
Excel displays the Multiple Roots dialog box shown in Figure 25.19.

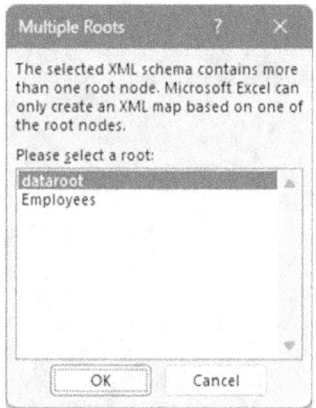

FIGURE 25.19.   If the XML data or schema file contains more than one root node, you must indicate which root node should be used.

7. In the Multiple Roots dialog box, select dataroot and click OK.
Excel displays the XML map name in the XML Maps dialog box shown in Figure 25.20. The name of the map consists of the schema's root element followed by an underscore and the word "Map." You can change the map name by clicking the Rename button.

   You cannot update an existing XML map. Excel only allows you to create new maps or delete existing ones using the XML Maps dialog box. Because of this, you must recreate the XML table created from an XML map any time the source XML schema changes.

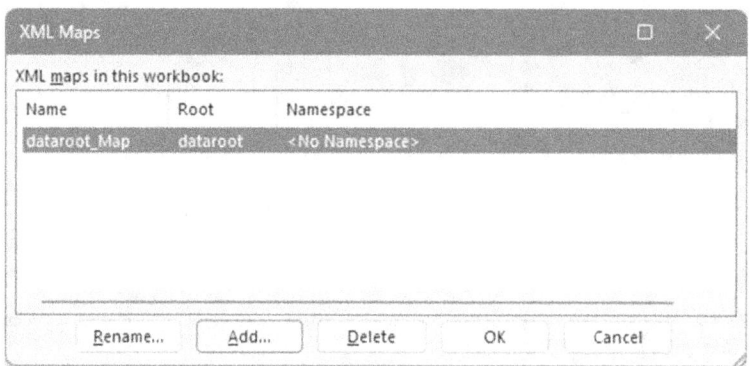

FIGURE 25.20.   The XML Maps dialog box now displays the XML map (dataroot_Map) that was added to the workbook.

8. Click OK to close the XML Maps dialog box and return to Excel.

The XML Source task pane now displays the structure of the XML map, as shown in Figure 25.21.

Notice that the name of the XML map appears in the list box at the top of the XML Source task pane. If the workbook contains more than one XML map, use this list box to select the map you want to work with. Excel obtains the map information from the schema that the XML file references; when the schema is not available, the map is generated based on the content of the XML data file, as you saw earlier in this chapter while opening the Courses1.xml document.

The XML map is displayed as a tree and can be expanded or collapsed by clicking the plus and minus buttons to the left of the element names. Elements

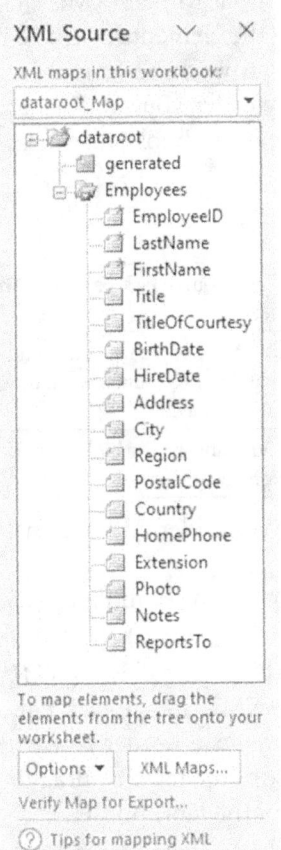

FIGURE 25.21.   The XML Source task pane displays the XML map generated from the XML schema file (Employees.xsd).

in the tree are represented by different icons. For example, in Figure 25.21, the folder icon with a red asterisk in front of the `dataroot` element represents a required parent element. An icon that looks like a sheet of paper with a corner folded down in front of the element name indicates a child element. The child element labeled `generated` stores the date the schema was generated. The folder icon in front of the `Employees` element tells us that this is the repeating parent element with children. The elements below the `Employees` element are child and required child elements. Required elements have a red asterisk in the icon image. To get the list and images of all the icons that can appear in the XML map, click on Tips for mapping XML at the bottom of the XML Source task pane.

Now that you've got the XML map, you can use it to map XML elements to your worksheet. Mapping is done by selecting the elements or entire nodes in the XML map and then dragging them onto a worksheet. You can drag mapped cells anywhere on the worksheet in any order you require. You can only map one schema element to one location in a workbook at a time.

TABLE 25.3. XML mapping guidelines.

XML Mapping	Follow This Procedure...
Single element	Drag the desired element from the XML Source task pane and drop it in the desired location on a worksheet.
Multiple elements	Select the first desired element in the XML Source task pane and hold down the Ctrl key while selecting other elements. Next, drag the selection to the desired location on a worksheet.
Entire node	Click on the parent node. All the child items will be highlighted. Drag the selection to the desired location on a worksheet.

9. In the XML Source task pane, select the Employees folder and drag it to cell A1 on the worksheet.

Excel maps XML elements to a range of cells, as you can see in Figure 25.22.

Notice that the XML elements are laid out in the order in which they appear in the XML Source task pane. Excel generates a structure called an *XML table* when you drag the repeating elements from the XML Source task pane to a worksheet. At this point, the generated table contains a header row with the AutoFilter option enabled. You can adjust the size of the table by dragging the resizer handle found in the bottom-right corner of the table border.

In this example, we have placed all of the XML elements on the worksheet by dragging them from the XML Source task pane and dropping them at a specific cell. When you don't require all the elements, simply drag those you

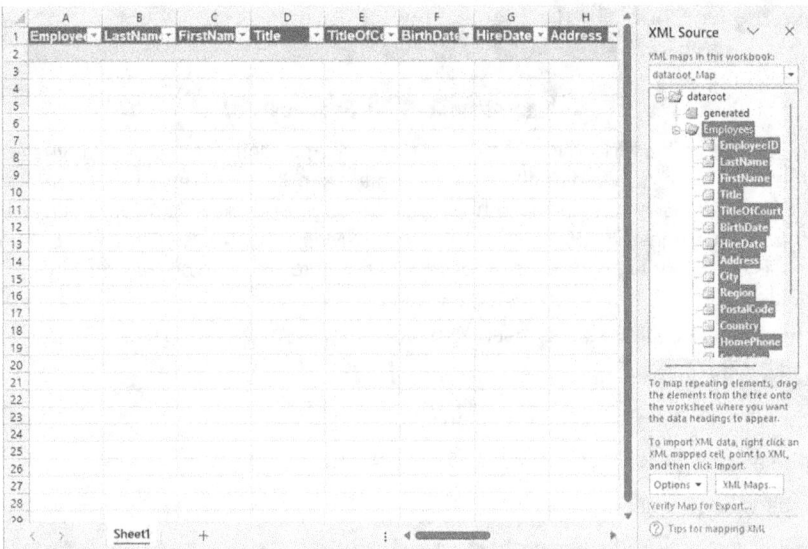

FIGURE 25.22. Mapping XML elements to cells in a worksheet.

need and leave out those you do not need. Mapped elements appear in bold type in the XML Source task pane.

> **NOTE** *Recall that you've already been introduced to the table feature in Chapter 19. XML tables are described in the next section.*

10. To populate the table with data, right-click anywhere within the table and choose XML | Import (or click the Import button on the Developer tab).

11. In the Import XML dialog box, select Employees.xml and click Import. The table on the worksheet is now populated with the data from the selected XML document, as shown in Figure 25.23.

EmployeeID	LastName	FirstName	Title	TitleOfCourtesy	BirthDate	HireDate	Address	City
1	Davolio	Nancy	Sales Representative	Ms.	12/8/1968 0:00	5/1/1992 0:00	507 - 20th Ave. E. Apt. 2A	Seattle
2	Fuller	Andrew	Vice President, Sales	Dr.	2/19/1952 0:00	8/14/1992 0:00	908 W. Capital Way	Tacoma
3	Leverling	Janet	Sales Representative	Ms.	8/30/1963 0:00	4/1/1992 0:00	722 Moss Bay Blvd.	Kirkland
4	Peacock	Margaret	Sales Representative	Mrs.	9/19/1958 0:00	5/3/1993 0:00	4110 Old Redmond Rd.	Redmond
5	Buchanan	Steven	Sales Manager	Mr.	3/4/1955 0:00	10/17/1993 0:00	14 Garrett Hill	London
6	Suyama	Michael	Sales Representative	Mr.	7/2/1963 0:00	10/17/1993 0:00	Coventry House Miner Rd	London
7	King	Robert	Sales Representative	Mr.	5/29/1960 0:00	1/2/1994 0:00	Edgeham Hollow Winchester Way	London
8	Callahan	Laura	Inside Sales Coordinator	Ms.	1/9/1958 0:00	3/5/1994 0:00	4726 - 11th Ave. N.E.	Seattle
9	Dodsworth	Anne	Sales Representative	Ms.	7/2/1969 0:00	11/15/1994 0:00	7 Houndstooth Rd.	London

FIGURE 25.23. A table populated with data from the XML document.

12. Save the workbook in the macro-enabled format as `C:\VBAExcel2024_XML\Employees.xlsm`.

Understanding XML Schemas

Schema files describe XML data using the XML Schema Definition (XSD) language and allow the XML parser to validate the XML document. An XML document that conforms to the structure of the schema is said to be valid. The `Employees.xsd` schema file that we worked with in Hands-On 25.7 was generated in Microsoft Access using built-in menu options. Listed below are some examples of types of information you can find in an XML schema file:

❏ Elements that are allowed in a given XML document

❏ Data types of allowed elements

❏ Number of occurrences of a given element that are allowed

❏ Attributes that can be associated with a given element

❏ Default values for attributes

❏ Elements that are child elements of other elements

❏ Sequence and number of child elements

If you open the `Employees.xsd` file in Notepad, you will notice a number of declarations and commands that begin with the `<xsd>` tag followed by a colon and the name of the command. You will also notice the names of the elements and attributes that are allowed in the `Employees.xml` file as well as the data types for each element. The names of the data types are preceded by the od prefix followed by a colon. For example:

`od:jetType ="text"`	Defines the Jet data type for an element
`od:sqlSType ="nvarchar"`	Defines the Microsoft SQL Server data type for an element
`od:autounique = "yes"`	Defines a Boolean data type for an auto-incremented identity column
`od:nonNullable ="yes"`	Indicates whether or not a column can contain a `null` value

The schema file also specifies the number of times an element can be used in a document based on the schema. This is done via the `minOccurs` and `maxOccurs` attributes.

WORKING WITH XML TABLES

An XML table is a table in Excel that has been mapped to one or more XML elements. In other words, each column in the XML table represents an element in your XML document. In this chapter, you've already created two XML tables based on the `Courses1.xml` and `Employees.xml` documents. After placing your XML data in an XML table in a workbook, you can work with this data just like any other Excel workbook file. This means you can add new columns and rows to your data, include formulas and functions, create charts, and perform various formatting tasks. You can even change the column headings that were automatically created from the XML element names. It is important to keep in mind that even when you change the column headings in the worksheet, the original XML element names will be used to export data from the mapped cells.

The changes you make to the data in the XML table will not affect the XML data that is stored in the original XML data file. Once you are done working with the XML table, you can save it as a standard Excel workbook (`.xlsx`) or in any other file format that is available in the Save As dialog box. You can also export the contents of mapped cells. The XML export feature is explained in the next section.

If the original XML data file has changed, you can update the data in your XML table by clicking the Refresh Data button on the Developer tab. When you use the Refresh command, the data is read from the original XML document into the mapped locations on the worksheet.

If you have another XML file that uses the same mapping, you can import the data from that file into your XML table by clicking the Import button on the Developer tab. Put simply, refreshing updates the XML table with the most current data from the original XML file, while importing gets the data from another XML file that follows the same schema.

When refreshing or importing data you can:

- Overwrite existing data with new data
- Append new data to an existing XML table

These options can be specified via the XML Map Properties dialog box, as shown in Figure 25.24.

The XML Map Properties dialog box allows you to set certain properties that relate to working with XML maps. This dialog can be accessed using any of the techniques listed below:

- Click the Map Properties button on the Developer tab.
- Right-click anywhere in the XML table and choose XML | XML Map Properties.

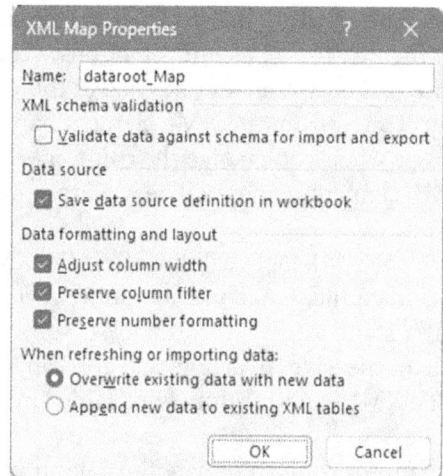

FIGURE 25.24. The XML Map Properties dialog box.

Each XML table in a workbook can be independently manipulated via the XML Map Properties dialog. Table 25.4 lists the XML properties that can be set.

TABLE 25.4. XML properties that can be set using the XML Map Properties dialog box.

XML Property	Description
Name	The name of the active XML map.
Validate data against schema for import and export	Excel will validate XML data against its schema while importing and exporting.
Save data source definition in workbook	Specifies whether your table is dynamic or static. If selected, the XML data is linked to the XML file and can be refreshed. If not selected, the data is static and cannot be refreshed.
Adjust column width	Excel will automatically adjust the width of table columns to fit the data.
Preserve column filter	Excel will preserve the selected column sorting, filtering, and layout.
Preserve number formatting	If selected, Excel will preserve the specified formatting of numbers in the table.
Overwrite existing data with new data	New data from the XML file will replace old data during a refresh or import.
Append new data to existing XML tables	New data from the XML file will be added at the bottom of the XML table during a refresh or import.

Exporting an XML Table

You can preserve the data in your XML table in two ways:

- Save your data to an XML data file.
- To do this, click File | Save As. In the File name box, type a name for the XML data file. In the Save As type list, select XML Data (*.xml) and click Save. Before proceeding with the saving operation, Excel will display the message shown in Figure 25.25.

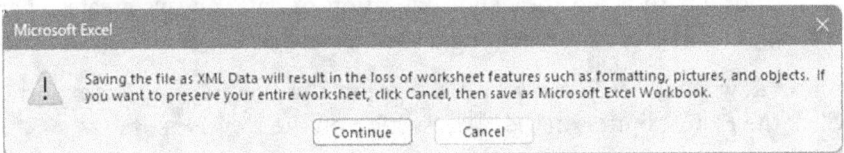

FIGURE 25.25. Excel displays a message about the loss of certain worksheet features prior to saving data in an XML data file.

- To save the data as an XML document, click Continue. If you keep working with this file and make any data and formatting changes, only the data will be saved during subsequent saving operations.
- Save your data by exporting it through the XML map.
- We will see how this feature is used in the following Hands-On. You should be working with the XML table that was created in Figure 25.23 earlier in this chapter.

◉ **Hands-On 25.8   Exporting XML Data in Mapped Worksheet Cells**

Make sure that the Employees.xlsm file you created in Hands-On 25.4 is open in Microsoft Excel. The XML Source task pane should be visible on the right side of the worksheet. If it is missing, click the Source button on the Developer tab.

1. Click the Verify Map for Export hyperlink at the bottom of the XML Source task pane.
If the map is valid for export, Excel displays the message that the map is exportable. If the map is invalid for export, a message is displayed with information about why the map isn't exportable. A map is invalid for export when:

- It contains more than one level of data. Although Excel can import data using multilevel maps, only single-level maps can be exported, such as `dataroot_Map` in Figure 25.21, which you've worked with in previous sections.

- It is denormalized. A map becomes denormalized when non-repeating items from an XML map are included in the XML table on the worksheet. Denormalized elements appear multiple times on the worksheet. If the user changes a non-repeating item in one row, that item will become inconsistent with other rows that should be showing the same data. Because Excel does not know how to reconcile the differences, the table can't be exported. To avoid the denormalization of data, always create separate XML tables for non-repeating and repeating nodes.

2. Click OK when Excel displays the message that the `dataroot_Map` is exportable.
3. Click the Export button in the XML section on the Developer tab.
   If the workbook contains more than one XML map, you will be prompted to select the map to use. You can export data using only one XML map at a time. Excel proceeds to display the Export XML dialog box.
4. In the Export XML dialog box, specify the name of your XML file and the folder where it should be saved. Select your VBAExcel2024_XML folder and enter `Northwind_Employees.xml` in the File name box. Click the Export button to complete the export operation.
5. Open the `Northwind_Employees.xml` document in Notepad to view its contents.
6. Close Notepad.

NOTE	*After exporting XML data in mapped cells to an XML data file, the name of your active workbook does not change. You can continue working with the data in this workbook. If you make changes to existing data or add new rows of data, however, you should re-export the data to the `Northwind_Employees.xml` file.*

XML Export Precautions

When exporting data, be aware of the fact that only the data included in the XML table will be saved; XML elements that were not mapped will not be exported. If you don't want to lose any content during export, always place all the elements from the XML map on the worksheet.

If the XML table contains a formula, the result of the formula (and not the formula itself) will be exported with the other data in the XML table. Formulas that you place in an XML table column must reference XML data elements that contain either a number, time, or date value.

If you add a new column to your XML table and then export the data, the data from this new unmapped column will not be saved. The reason for this is that Excel exports a table as XML using the schema stored in the workbook. The generated XML file must match the XML source file from which the XML table was created. Because the added column is not in the XML source file, Excel cannot save it. Therefore, if you need to add data to the existing XML table, do the following:

1. Open the appropriate schema file in Notepad and add a new element with the name of your new column.
2. Save the modified schema file and close Notepad.
3. Because the XML schema has changed and Excel does not allow you to modify an existing XML map, you will need to create a new XML map and drag the required XML elements to your worksheet. You are already familiar with this process, as it was a part of the Hands-On exercise in the section titled "Working with XML Maps" earlier in this chapter. After mapping your XML elements to cells in a worksheet, simply refresh your XML table. There will be no data in the optional column that you've added to the schema file. You can now proceed to enter the data or formula you need in this empty column. When you export your data to the XML file, the data in the new column will be exported together with the other data in your XML table.

VALIDATING XML DATA

To have Excel validate XML data upon import or export, you need to follow these steps:

1. Select any cell within your XML table on the worksheet and click the Map Properties button on the Developer tab.
   Excel displays the XML Map Properties dialog box shown earlier in Figure 25.24.
2. Select Validate data against schema for import and export.
3. Click OK to close the dialog box.

If you enter an invalid value in any column of your XML table, Excel will not automatically validate your entry. All of the entries will be validated, however, when you click the Export button on the Developer tab to export the data. If any data is found to be invalid, Excel displays an error message similar to the one shown in Figure 25.26.

Note that the error in validating the data does not prevent Excel from saving or exporting. The Details section in the error message dialog box will give you a hint of why the data is invalid so that you can correct the data and re-export it. You may want to define your own data validation rules that comply with the XML schema by using the Data Validation button on the Data tab. Then, Excel will validate your data as you work on the worksheet.

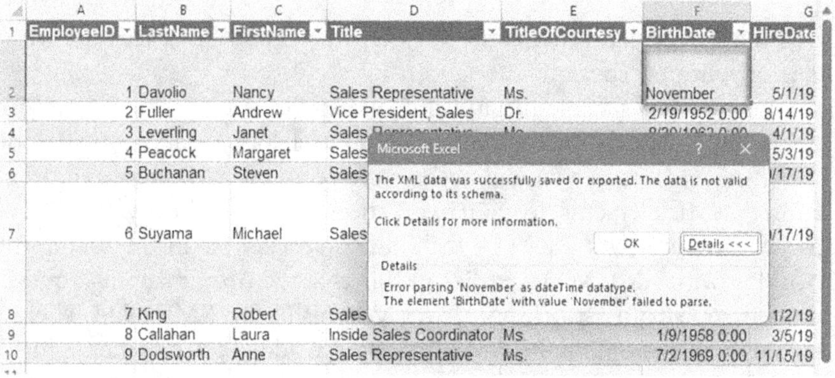

FIGURE 25.26.   Excel displays a message when data is found to be invalid according to its schema during the export or a refresh operation.

PROGRAMMING XML MAPS

Earlier in this chapter, you learned that a workbook can contain more than one XML map. These maps can be from the same schema or different schemas. When mapping XML elements to cells and ranges on the worksheet, keep in mind that mapped cell ranges cannot overlap.

In this section, we will add another XML map to the current workbook, but instead of using a manual method, we'll perform this task programmatically. Excel provides specific objects to deal with programming its XML features, such as the XmlMap object in the XmlMaps collection and the XmlNamespace object in the XmlNamespaces collection.

The XmlMaps collection contains the XmlMap object, which can be used to perform the programming tasks described in the following subsections. You can try out the example code in the Immediate window. Be sure that you have the Employees.xlsm workbook open and your active worksheet contains the

XML table displaying Northwind employees. This table was created earlier in this chapter from the `dataroot_Map` based on the `Employees.xsd` schema (see Hands-On 25.7).

Adding an XML Map to a Workbook

You can add an XML map to a workbook using the `Add` method of the `Xml-Maps` collection. This method requires the location of an XML schema file. If the schema file is not available, you can specify the XML source data file and Excel will create a schema based on that source data.

Earlier in this chapter, you worked with the `Courses1.xml` document. To create an XML map using this file, press Alt+F11 to switch to the VBE screen and type the following statement in the Immediate window:

```
ActiveWorkbook.XmlMaps.Add("C:\VBAExcel2024_XML\Courses1.xml")
```

When you press Enter, Excel will display the message shown earlier in Figure 25.15. Click OK on the message. When you switch back to the Excel application window, you will notice that the `Courses_Map` map is added to the XML maps in the drop-down list at the top of the XML Source task pane (see Figure 25.27).

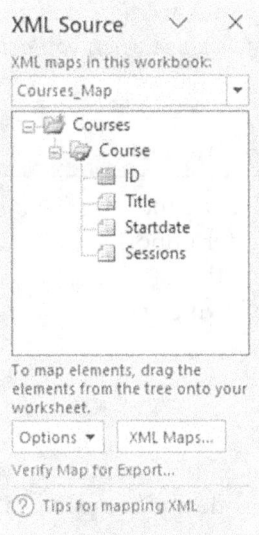

FIGURE 25.27.   Adding another XML map to a workbook.

When Excel creates a new map, it uses the name of the root node for its name, followed by an underscore and the word `Map`. Sometimes, a newly added map may have the same root node name as an existing map. To differentiate one map from another, Excel will add a number following the word Map. So, if you already have a dataroot_Map map in the workbook and you are adding another map whose root node is also named dataroot, Excel will assign the name dataroot_Map2 to the new map.

Deleting Existing XML Maps

To delete an existing XML map from the workbook, use the Delete method of the XmlMap object. The Delete method requires the name of the map to be deleted. Let's delete the Courses_Map map that you added in the previous section. Type the following statement in the Immediate window:

```
ActiveWorkbook.XmlMaps("Courses_Map").Delete
```

When you press Enter, Excel will delete the specified map. This map's name no longer appears in the XML Source task pane's drop-down list.

NOTE	*When you delete the map using the Delete button in the XML Maps dialog box, shown in Figure 25.20 earlier, Excel displays a message informing you "If you delete the specified XML map, you will no longer be able to import or export XML data using this XML map." You do not get this warning message when you delete the XML map programmatically.*

Exporting and Importing Data via an XML Map

Use the XmlMap object to export and import XML data. Use the XmlMap object's Export method for exporting and the Import method for importing.

For example, to export the XML table data through the dataroot_Map that the Employees XML table is mapped to, type the following statement on one line in the Immediate window:

```
ActiveWorkbook.XmlMaps("dataroot_Map").Export
    "C:\VBAExcel2024_XML\InternalContacts.xml"
```

When you press Enter, Excel creates the specified XML document in your VBAExcel2024_XML folder. If the file with the same name already exists in the destination folder, Excel displays an Automation error.

Excel also offers the ExportXML method for those situations when you'd rather export your XML data to a String variable instead of to a file, as is done with the simple Export method. The following procedure demonstrates this:

```
Sub ExportToString()
  Dim strEmpData As String

  ActiveWorkbook.XmlMaps("dataroot_Map").ExportXml _
    Data: = strEmpData
  Debug.Print strEmpData
End Sub
```

To import new XML data into an XML map, copy the `Davolio.xml` document from the companion files to your `VBAExcel2024_XML` folder, and then, in the Immediate window, enter the following statement on one line:

```
ActiveWorkbook.XmlMaps("dataroot_Map").Import
    URL: = "C:\VBAExcel2024_XML\Davolio.xml", Overwrite: = True
```

The `Overwrite` parameter specifies whether or not the newly imported data should overwrite existing data. The `Davolio.xml` file holds data for only one Northwind employee named Nancy Davolio. After running the above statement, the XML table in the `Employees.xlsm` workbook will contain only one record.

Binding an XML Map to an XML Data Source

Each XML map is bound to an XML data source. Use the `DataBinding` property of the `XMLMap` object to find out the name of the data source that is used in the XML map.

For example, when you type the following statement in the Immediate window:

```
Debug.Print ActiveWorkbook.XmlMaps("dataroot_Map").DataBinding
```

Excel returns the following data source: `C:\VBAExcel2024_XML\Davolio.xml`. If you haven't run the statement in the previous section, you should see `C:\VBA-Excel2024_XML\Employees.xml` as the data source.

It is possible to change the data source used by the XML map with the `Load-Settings` method of the `DataBinding` property, as shown below. Be sure to enter this on one line in the Immediate window:

```
ActiveWorkbook.XmlMaps("dataroot_Map").DataBinding.LoadSettings
    ("C:\VBAExcel2024_XML\Employees.xml")
```

After changing the data source used by the XML map, you should refresh the data in your XML table (`Employees.xlsm` workbook) either via the user interface by clicking the Refresh Data button on the Developer tab, Refresh All button on the Data tab, or from code using the `Refresh` method (as shown in the next section).

Refreshing XML Tables from an XML Data Source

Use the `Refresh` method of the `DataBinding` property of the `XmlMap` object to refresh the XML table in your worksheet. The following statement can be used:

```
ActiveWorkbook.XmlMaps("dataroot_Map").DataBinding.Refresh
```

After running the above statement, the XML table in the worksheet should display all employee records.

VIEWING THE XML SCHEMA

To see the schema that is used by an XML map, use the `Schemas` collection of the `XmlMap` object. The `Schemas` property of the `XmlMap` object is used to return the `XmlSchemas` collection. The `XmlSchemas` collection contains `XmlSchema` objects. By using the `XML` property of the `XmlSchema` object, it is possible to return the string representing the content of the specified schema. Try out this code in the Immediate window, pressing Enter after each statement:

```
Set objMap = ActiveWorkbook.XmlMaps(1)
Debug.Print objMap.Name
Debug.Print objMap.Schemas(1).Xml
```

If you'd like to use the above code fragment inside a VBA procedure, don't forget to declare the `objMap` variable with the following statement:

```
Dim objMap As XmlMap
```

NOTE	*By saving the text of the generated schema in a file, you can create a schema file for future use. To do this, open Notepad and paste the data returned by the* `Debug.Print objMap.Schemas(1).Xml` *statement. Next, save the Notepad file using any name you wish, but be sure to use the* `.xsd` *file extension.*

Now that you've acquired a useful vocabulary for programming tasks related to XML maps, let's write a full-fledged VBA procedure that will add an XML map to the current workbook, perform the mapping, and refresh the data. You can work with the current `Employees.xlsm` workbook that already has the `dataroot_Map`.

⊙ Hands-On 25.9   Using VBA to Program XML Maps

1. In the VBE screen, insert a new module into `VBAProject (Employees.xlsm)`.

2. In the module's Code window, enter the `AddNew_XMLMap` procedure as shown below:

```
Sub AddNew_XMLMap()
    Dim lstCourses As ListObject
    Dim lstCol As ListColumn
    Dim objMap As XmlMap
    Dim mapName As String
    Dim strXPath As String

    On Error GoTo ErrorHandler

    ' Create a new XML map
    ActiveWorkbook.XmlMaps.Add _
        ("C:\VBAExcel2024_XML\Courses1.xml ", _
        "Courses").Name = "Courses_Map"

    'location for the new XML table
    Set objMap = ActiveWorkbook.XmlMaps("Courses_Map")
    Range("B20").Select

    ' Create a new List object
    Set lstCourses = ActiveSheet.ListObjects.Add

    ' Bind the first XML element to the first table column
    strXPath = "/Courses/Course/@ID"
    With lstCourses.ListColumns(1)
        .XPath.SetValue objMap, strXPath
        .Name = "ID"
    End With

    ' Add a column to the table
    ' and bind it to an XML node
    Set lstCol = lstCourses.ListColumns.Add
    strXPath = "/Courses/Course/Title"
    With lstCol
        .XPath.SetValue objMap, strXPath
        .Name = "Title"
    End With

    ' Add a column to the table
    ' and bind it to an XML node
    Set lstCol = lstCourses.ListColumns.Add

    strXPath = "/Courses/Course/Startdate"
```

```
    With lstCol
        .XPath.SetValue objMap, strXPath
        .Name = "Start Date"
    End With

    ' Add a column to the table
    ' and bind it to an XML node
    Set lstCol = lstCourses.ListColumns.Add

    strXPath = "/Courses/Course/Sessions"
    With lstCol
        .XPath.SetValue objMap, strXPath
        .Name = "Sessions"
    End With

    ' Set some XML properties
    With ActiveWorkbook.XmlMaps("Courses_Map")
        .ShowImportExportValidationErrors = False
        .AdjustColumnWidth = True
        .PreserveColumnFilter = True
        .PreserveNumberFormatting = True
        .AppendOnImport = False
    End With

    ' Refresh the XML table in the worksheet
    ActiveWorkbook.XmlMaps("Courses_Map").DataBinding.Refresh
Exit Sub

ErrorHandler:
    MsgBox "The following error has occurred: " & vbCrLf _
    & Err.Description
End Sub
```

The AddNew_XMLMap procedure begins by creating the XML map named Courses_Map using the Courses1.xml data file. Next, a new XML table is created in a worksheet. At this time, the table will contain just one column with the default name Column1. We bind this column with the first item in the XML map—ID. The XPath object's SetValue method is used to bind data from an XML map to a table column. This method has two required arguments, Map and XPath. Map is the XML map that has been added to the workbook. In this example, it's the object variable named objMap.

XPath is the XPath statement in the form of a String variable (strXPath) that specifies the XML map data you want to bind to the specified table column. Because ID is an attribute, you must precede it with the @ character. Once ID is mapped to the table column, we replace the default column name with our own (ID), using the Name property of the ListColumn object:

```
strXPath = "/Courses/Course/@ID"
With lstCourses.ListColumns(1)
    .XPath.SetValue objMap, strXPath
    .Name = "ID"
End With
```

Next, we proceed to add another column to the table using the Add method of the ListColumns collection:

```
Set lstCol = lstCourses.ListColumns.Add
```

This column is then bound to the next item in the XML map—Title. Again, we use the SetValue method of the XPath object to do the binding:

```
strXPath = "/Courses/Course/Title"
With lstCol
    .XPath.SetValue objMap, strXPath
    .Name = "Title"
End With
```

In the same manner, we add two more columns to our table and bind each column to the remaining elements in the XML map. Next, we set some XML map properties and proceed to refresh the list. The empty table is now populated with data from the source XML file (Courses1.xml).

3. Switch to the Microsoft Excel application window and choose Developer | Macros to open the Macro dialog box.

4. In the Macro dialog box, select AddNew_XMLMap and click the Run button.
   Excel displays a message informing you that the specified XML source document does not have a schema and Excel will create one on the fly using the XML source data.

5. Click OK on the message.
   Excel adds the specified columns and performs the required data mappings; however, it stops and displays the message about incompatible formatting shown in Figure 25.28 when it gets to the mapping of the <Sessions> element.

   When Excel determines that the cell formatting is not compatible with the data type specified in the XSD for the requested data element, you will receive a warning message, as shown in Figure 25.28. This dialog box contains the following buttons:

Use existing formatting	Click this button to ignore the data type in the XSD file.
Match element data type	Click this button to change the cell formatting to the appropriate type.
Cancel	Click this button to cancel the mapping of this data element.

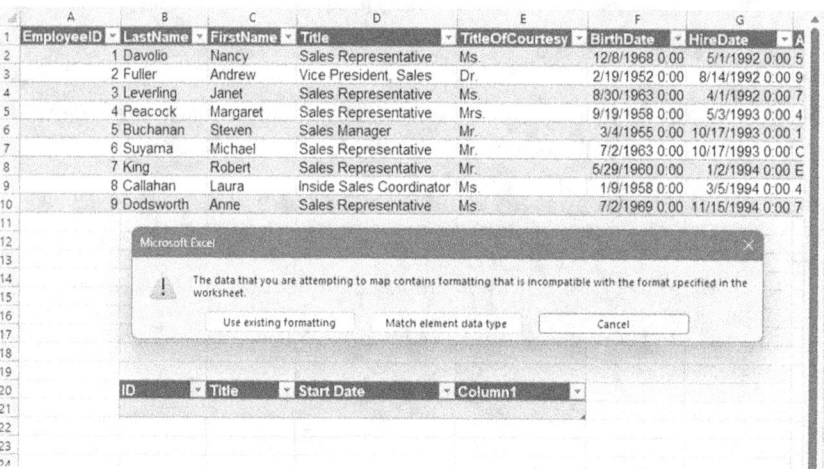

FIGURE 25.28. This warning message appears when the data type of the data being mapped is not compatible with the cell formatting.

6. Click the Match element data type button to proceed with the data mapping. The resulting XML table and XML map are shown in Figure 25.29.

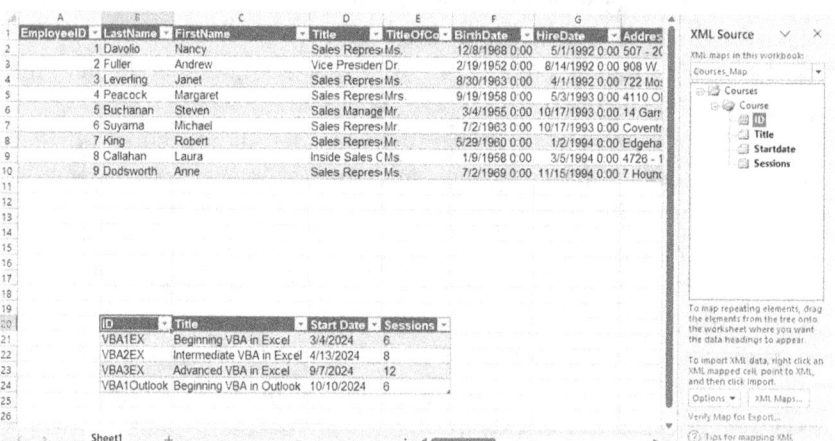

FIGURE 25.29. The Excel worksheet with two XML tables. The upper table was created via the user interface; the one at the bottom was generated programmatically. The XML Source task pane displays the Courses_Map with mapped data elements.

What Is XPath?

XML Path Language (XPath) is a query language used to create expressions for finding data in the XML file. These expressions can manipulate strings, numbers, and Boolean values (`true` or `false`). They can also be used to navigate an XML tree structure and process its elements with XSL Transformations (XSLT) instructions. With XPath expressions, you can easily identify and extract from the XML document specific elements (nodes) based on their type, name, values, or the relationship of a node to other nodes (this is covered later in this chapter).

CREATING XML SCHEMA FILES

When you ask Excel to create an XML map from a specified XML data file, Excel notifies you that the file lacks the schema. Consequently, Excel will generate a schema based on the XML source data. To obtain the schema information that Excel has generated during the XML mapping process, perform the following steps:

1. Open the Immediate window and type the following statement:

```
? ThisWorkbook.XMLMaps(1).Schemas(1).xml
```

When you press Enter, the content of the schema appears in the Immediate window in the form of a very long string.
2. Highlight the retrieved schema text in the Immediate window, and press Ctrl+C to copy it to the clipboard.
3. Open Windows Notepad and press Ctrl+V to paste the data from the clipboard. You may want to format the data as shown in Figure 25.30 to make it easier to understand.
4. Save the file using any name but be sure to specify the `.xsd` file extension.
5. Close Notepad.

```
CoursesSchema.xsd            ×    +

File   Edit   View

<xsd:schema xmlns:xsd="http://www.w3.org/2001/XMLSchema">
<xsd:element nillable="true" name="Courses">
<xsd:complexType>
<xsd:sequence minOccurs="0">
<xsd:element minOccurs="0" maxOccurs="unbounded" nillable="true" name="Course" form="unqualified">
<xsd:complexType><xsd:sequence minOccurs="0">
<xsd:element minOccurs="0" nillable="true" type="xsd:string" name="Title" form="unqualified">
</xsd:element>
<xsd:element minOccurs="0" nillable="true" type="xsd:string" name="Startdate" form="unqualified">
</xsd:element>
<xsd:element minOccurs="0" nillable="true" type="xsd:integer" name="Sessions" form="unqualified">
</xsd:element>
</xsd:sequence><xsd:attribute name="ID" form="unqualified" type="xsd:string">
</xsd:attribute>
</xsd:complexType>
</xsd:element>
</xsd:sequence>
</xsd:complexType>
</xsd:element>
</xsd:schema>

Ln 20, Col 1   822 characters              100%    Windows (CRLF)         UTF-8
```

FIGURE 25.30. This XML schema was generated by Excel during the XML mapping of the XML data file.

USING XML EVENTS

Chapter 15 of this book focused on using event-driven programming in Excel. This section expands your knowledge of Excel events by introducing you to events that occur before and after data is exported, imported, or refreshed via the XML map.

The `Workbook` object provides the following events: `AfterXMLExport`, `AfterXMLImport`, `BeforeXMLExport`, and `BeforeXMLImport`. By writing code for these events in the `ThisWorkbook` code module, you can fully control what happens before and after import, export, and refresh operations.

Event Name: `AfterXMLExport`

This event applies to the `Workbook` object. It occurs after Microsoft Excel saves or exports XML data from the specified workbook. The following parameters are required:

- `Map`—The schema map that was used to save or export data.
- `Url`—The location of the XML file that was exported.

- Result—A constant indicating the result of the save or export operation. Use one of the following xlXmlExportResult constants:

  - xlXmlExportSuccess—Specifies that the XML data file was successfully exported

  - xlXmlExportValidationFailed—Specifies that the content of the XML data file does not match the specified schema map

```
Private Sub Workbook_AfterXMLExport _
    (ByVal Map As XmlMap, _
    ByVal Url As String, _
    ByVal Result As XlXmlExportResult)
    If Result = xlXmlExportSuccess Then
        MsgBox ("XML export succeeded.")
    Else
        MsgBox ("XML export failed.")
    End If
End Sub
```

Event Name: AfterXMLImport

This event applies to the Workbook object. It occurs after an existing XML data connection is refreshed or after new XML data is imported into the specified Microsoft Excel workbook.

The following parameters are required:

- Map—The XML map that will be used to import data.

- IsRefresh—A Boolean value (True/False): True if the event was triggered by refreshing an existing connection to XML data; False if the event was triggered by importing from a different data source.

- Result—A constant indicating the result of the refresh or import operation. Use one of the following xlXmlImportResult constants:

  - xlXmlImportElementsTruncated—Specifies that the content of the specified XML data file has been truncated because the XML data file is too large for the worksheet

  - xlXmlImportSuccess—Specifies that the XML data file was successfully imported

  - xlXmlImportValidationFailed—Specifies that the content of the XML data file does not match the specified schema map

```
Private Sub Workbook_AfterXMLImport _
    (ByVal Map As XmlMap, _
    ByVal IsRefresh As Boolean, _
```

```
     ByVal Result As XlXmlImportResult)
     If Result = xlXmlImportSuccess Then
         MsgBox ("XML import succeeded.")
          ActiveSheet.ListObjects(1).Range.Select
      Selection.Interior.ColorIndex = 35
      ActiveCell.Select
     Else
         MsgBox ("XML import failed.")
     End If
End Sub
```

Event Name: BeforeXMLExport

This event applies to the Workbook object. It occurs before Microsoft Excel saves or exports XML data from the specified workbook. This occurs only when saving to an XML data file format; it does not occur when you are saving to the XML spreadsheet file format.

The following parameters are required:

- Map—The XML map that will be used to save or export data.

- Url—The location where you want to export the resulting XML file.

- Cancel—A Boolean value (True/False). Set to True to cancel the saving or export operation.

```
Private Sub Workbook_BeforeXMLExport _
    (ByVal Map As XmlMap, _
    ByVal Url As String, _
    Cancel As Boolean)

    If (Map.IsExportable) Then
        If MsgBox("Excel is about" & _
        " to export XML from the" & _
        Map.Name & "." & vbCrLf & "Do" & _
        " you want to continue?", _
        vbYesNo + vbQuestion, _
        "XML Export Process") = 7 Then
        Cancel = True
        End If
    End If
End Sub
```

Event Name: BeforeXMLImport

This event applies to the Workbook object. It occurs before an existing XML data connection is refreshed or before new XML data is imported into a Microsoft Excel workbook. The following parameters are required:

- `Map`—The XML map that will be used to import data.
- `Url`—The location of the XML file to be imported.
- `IsRefresh`—A Boolean value (`True`/`False`): `True` if the event was triggered by refreshing an existing connection to XML data; `False` if the event was triggered by importing from a different data source.
- `Cancel`—A Boolean value (`True`/`False`). Set to `True` to cancel the import or refresh operation.

```
Private Sub Workbook_BeforeXMLImport _
    (ByVal Map As XmlMap, _
    ByVal Url As String, _
    ByVal IsRefresh As Boolean, _
    Cancel As Boolean)

    If MsgBox("Excel is about " & _
        " to import XML into the" & _
        " workbook. Continue with" & _
        " importing?", _
        vbYesNo + vbQuestion, _
        "XML Import Process") = 7 Then
        Cancel = True
    End If
End Sub
```

The XML events are also available for the `Application` object. These events are listed below. Recall from Chapter 15 that event procedures for the `Application` object require that you create a new object using the `WithEvents` keyword in a class module:

- `WorkbookBeforeXmlExport`—This occurs before Microsoft Excel saves or exports XML data from the specified workbook. Use this event if you want to capture XML data that is being exported or saved from a particular workbook.
- `WorkbookAfterXmlExport`—This occurs after Microsoft Excel saves or exports XML data from the specified workbook. Use this event if you want to perform an operation after XML data has been exported from a particular workbook.
- `WorkbookBeforeXmlImport`—This occurs before an existing XML data connection is refreshed or new XML data is imported into any open Microsoft Excel workbook. Use this event if you want to capture XML data that is being imported or refreshed to a particular workbook.

⬤ `WorkbookAfterXmlImport`—This occurs after an existing XML data connection is refreshed or new XML data is imported into any open Microsoft Excel workbook. Use this event if you want to perform an operation after XML data has been imported into a particular workbook.

THE XML DOCUMENT OBJECT MODEL

You can create, access, and manipulate XML documents programmatically via the XML Document Object Model (DOM). The DOM has properties and methods for interacting with XML documents. The XML DOM is supplied free with the browser. To use the XML DOM from your VBA procedures, you need to set up a reference to the Microsoft XML object library.

⊙ Hands-On 25.10   Setting Up a Reference to the DOM

1. In the VBE screen of `VBAProject  (Empoyees.xlsm)`, choose Tools | References.
2. In the References dialog box, locate and select Microsoft XML 6.0, as shown in Figure 25.31.
3. Click OK to close the References dialog box.
4. Now that you have the reference set, open the Object Browser and examine the XML DOM's objects, methods, and properties, as shown in Figure 25.32.
5. Close the Object Browser.

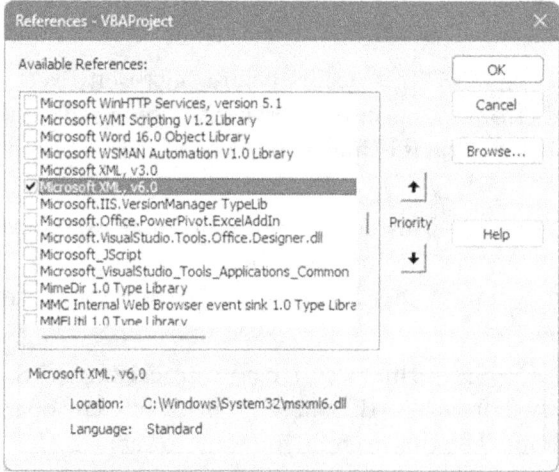

FIGURE 25.31.   To work with XML documents programmatically, you need to establish a reference to the Microsoft XML, v6.0 object library.

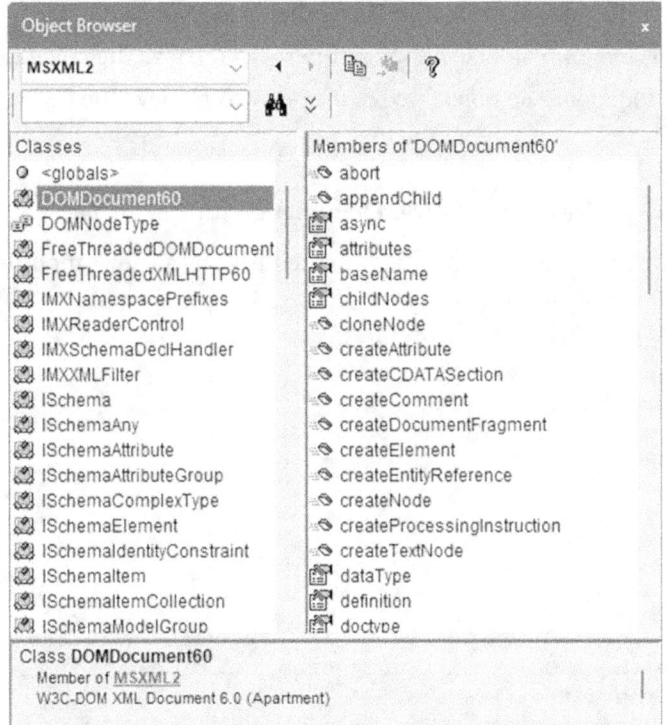

FIGURE 25.32. To see objects, properties, and methods exposed by the DOM, open the Object Browser after setting up a reference to the Microsoft XML, v6.0 object library, as shown in Figure 25.31.

The DOMDocument60 object is the top level of the XML DOM hierarchy. This object represents a tree structure composed of nodes. You can navigate through this tree structure and manipulate the data contained in the nodes by using various methods and properties. The DOMDocument60 object is the parent for all other elements in the DOM hierarchy. Because every XML object is created and accessed from the document, the DOMDocument60 object must be created first.

To work with an XML document, you must create an instance of the DOM-Document60 object, as in the following example:

```
Dim myXMLDoc As MSXM2.DOMDocument60
Set myXMLDoc = New MSXML2.DOMDocument60
```

To make the instantiated DOMDocument60 object useful, you should load it with some data. The following Hands-On demonstrates how to get started with the XML DOM. You will perform the following tasks:

- Create an instance of the DOMDocument60.

- Load XML information from a file using the `Load` method.
- Use the `DOMDocument60` object's `XML` property to retrieve the raw data.
- Use the `DOMDocument60` object's `Text` property to retrieve the text stored in nodes.

(•) Hands-On 25.11   Reading an XML Document with the DOM

1. Enter the following `Load_ReadXMLDoc` procedure in a new module of VBAProject (Employees.xlsm):

```
Sub Load_ReadXMLDoc()
    Dim xmldoc As MSXML2.DOMDocument60

    ' Create an instance of the DOMDocument60
    Set xmldoc = New MSXML2.DOMDocument60

    ' Disable asynchronous loading
    xmldoc.async = False

    ' Load XML information from a file
    If xmldoc.Load("C:\VBAExcel2024_XML\Courses1.xml") Then
        ' Use the DOMDocument60 object's XML property to
        ' retrieve the raw data
        Debug.Print xmldoc.XML
        ' Use the DOMDocument60 object's Text property to
        ' retrieve the actual text stored in nodes
        Sheets.Add
        ActiveSheet.Range("A1").Value = xmldoc.Text
    End If
End Sub
```

The XML DOM has two methods for loading XML information: `Load` and `LoadXML`. Use the `Load` method to load XML information from a text file. Use the `LoadXML` method when loading from a string in memory.

What Is Asynchronous Loading?

By default, MSXML uses an *asynchronous* loading mechanism for working with documents. Asynchronous loading allows you to perform other tasks during long database operations, such as providing feedback to the user as MSXML parses the XML file or giving the user the chance to cancel the operation. Before calling the `Load` method, however, it's a good idea to set the `Async` property of the `DOMDocument60` object to `False` to ensure that when the load returns, the entire document has finished loading. The `Load` method returns `True` if it successfully loaded the data and `False` otherwise.

Having loaded the data into a `DOMDocument60` object, you can use the `XML` property to retrieve the raw data or use the `Text` property to obtain the text stored in document nodes.

2. Run the `Load_ReadXMLDoc` procedure and examine its results in the Immediate window and `Sheet2` of the `Employees.xlsm` workbook. Cell A1 in `Sheet2` should contain the entire string of data.

WORKING WITH XML DOCUMENT NODES

As you already know, the XML DOM represents a tree-based hierarchy of nodes. An XML document can contain nodes of different types. Some nodes represent comments and processing instructions in the XML document, and others hold the text content of a tag. To determine the type of node, use the `nodeType` property of the `IXMLDOMNode` object. Node types are identified by either a text string or a constant. For example, the node representing an element can be referred to as `NODE_ELEMENT` or `1`, while the node representing the comment is named `NODE_COMMENT` or `8`. See the `MSXML2` library in the Object Browser shown in Figure 25.32 in the previous section for the names of other node types.

In addition to node types, nodes can have parent, child, and sibling nodes. The `hasChildNodes` method lets you determine whether a `DOMDocument60` object has child nodes. There's also a `childNodes` property for retrieving a collection of child nodes. Before you start looping through the collection of child nodes, it's a good idea to use the `Length` property of `IXMLDOMNode` to determine how many elements the collection contains.

The `LearnAboutNodes` procedure shown below will get you working with nodes programmatically in no time. The following example demonstrates how to experiment with XML document nodes.

(◉) Hands-On 25.12   Working with XML Document Nodes

1. In the VBE window of `VBAProject (Employees.xlsm)`, insert a new module and enter the `LearnAboutNodes` procedure, as shown below:

```
Sub LearnAboutNodes()
    Dim xmldoc As MSXML2.DOMDocument60
    Dim xmlNode As MSXML2.IXMLDOMNode

    ' Create an instance of the DOMDocument60
    Set xmldoc = New MSXML2.DOMDocument60
```

```
        xmldoc.async = False

        ' Load XML information from a file
        xmldoc.Load ("C:\VBAExcel2024_XML\Courses1.xml")

        ' find out the number of child nodes in the document
        If xmldoc.hasChildNodes Then
            Debug.Print "Number of Child Nodes: " & _
            xmldoc.childNodes.Length

            ' iterate through the child nodes to gather information
            For Each xmlNode In xmldoc.childNodes
                Debug.Print "Node Name: " & xmlNode.nodeName
                Debug.Print vbTab & "Type: " & _
                xmlNode.nodeTypeString & _
                "(" & xmlNode.nodeType & ")"
                Debug.Print vbTab & "Text: " & xmlNode.Text
            Next xmlNode
        End If
    End Sub
```

2. Run the `LearnAboutNodes` procedure in step mode by pressing F8.

The `LearnAboutNodes` procedure prints to the Immediate window the information about child nodes found in the `Courses1.xml` document. Notice that the `Text` property of a node returns all the text from all the node's children in one string (see the text for the `Courses` node below):

```
Number of Child Nodes: 2
Node Name: xml
    Type: processinginstruction(7)
    Text: version="1.0"
Node Name: Courses
    Type: element(1)
    Text: Beginning VBA in Excel 3/4/2024 6 Intermediate VBA
    in Excel 4/13/2024 8 Advanced VBA in Excel 9/7/2024 12
    Beginning VBA in Outlook 10/10/2024 6
```

RETRIEVING INFORMATION FROM ELEMENT NODES

Let's assume that you want to read only the information from the text element nodes and place it in an Excel worksheet. Use the `getElementsByTagName` method of the `DOMDocument60` object to retrieve an `IXMLDOMNodeList` object containing all the element nodes.

The `getElementsByTagName` method takes one argument specifying the tag name to search for. Use `"*"` as the tag to search for all the element nodes, as illustrated in Hands-On 25.13. The following example demonstrates how to obtain data from an XML document's element nodes.

(◉) Hands-On 25.13   Obtaining Data from Element Nodes

1. In the VBE screen of `VBAProject (Employees.xlsm)`, insert a new module and enter the `IterateThruElements` procedure, as shown below:

```
Sub IterateThruElements()
    Dim xmldoc As MSXML2.DOMDocument60
    Dim xmlNodeList As MSXML2.IXMLDOMNodeList
    Dim xmlNode As MSXML2.IXMLDOMNode
    Dim myNode As MSXML2.IXMLDOMNode

    ' Create an instance of the DOMDocument60
    Set xmldoc = New MSXML2.DOMDocument60
    xmldoc.async = False

    ' Load XML information from a file
    xmldoc.Load ("C:\VBAExcel2024_XML\Courses1.xml")

    ' Find out the number of child nodes in the document
    Set xmlNodeList = xmldoc.getElementsByTagName("*")

    ' Open a new workbook and paste the data
    Workbooks.Add
    Range("A1:B1").Formula = Array("Element Name", "Text")
    For Each xmlNode In xmlNodeList
        For Each myNode In xmlNode.ChildNodes
            If myNode.nodeType = NODE_TEXT Then
              ActiveCell.Offset(0, 0).Formula = xmlNode.nodeName
                ActiveCell.Offset(0, 1).Formula = xmlNode.Text
            End If
        Next myNode
        ActiveCell.Offset(1, 0).Select
    Next xmlNode
    Columns("A:B").AutoFit
End Sub
```

2. Run the above procedure in step mode by pressing F8.
   The `IterateThruElements` procedure fills in two worksheet columns with the XML element name and the corresponding text for all the text elements from the `Courses1.xml` document. The procedure result is shown in Figure 25.33.

Notice that this procedure uses two For Each...Next loops. The first one (the outer For Each...Next loop) iterates through the entire collection of element nodes. The second one (the inner For Each...Next loop) uses the nodeType property to find only those element nodes that contain a single text node.

	A	B	C
1	Element Name	Text	
2			
3	Title	Beginning VBA in Excel	
4	Startdate	3/4/2024	
5	Sessions	6	
6			
7	Title	Intermediate VBA in Excel	
8	Startdate	4/13/2024	
9	Sessions	8	
10			
11	Title	Advanced VBA in Excel	
12	Startdate	9/7/2024	
13	Sessions	12	
14			
15	Title	Beginning VBA in Outlook	
16	Startdate	10/10/2024	
17	Sessions	6	
18			
19			

FIGURE 25.33.   You can programmatically retrieve information about element nodes from the XML document. The IterateThruElements procedure was used to create this worksheet.

To list all the nodes that match a specified criterion, use the selectNodes method. In the next example, you will see how you can return to the Immediate window the text found in all Title nodes in the Courses1.xml file.

Hands-On 25.14   Obtaining Data from an Element Node Based on a Condition

1. In the VBE screen of VBAProject (Employees.xlsm), insert a new module and enter the SelectNodes_SpecifyCriterion procedure, as shown below:

```
Sub SelectNodes_SpecifyCriterion()
    Dim xmldoc As MSXML2.DOMDocument60
    Dim xmlNodeList As MSXML2.IXMLDOMNodeList
    Dim myNode As Variant

    ' Create an instance of the DOMDocument60
    Set xmldoc = New MSXML2.DOMDocument60
```

```
    xmldoc.async = False

    ' Load XML information from a file
    xmldoc.Load ("C:\VBAExcel2024_XML\Courses1.xml")

    ' Retrieve all the nodes that match the specified criterion
    Set xmlNodeList = xmldoc.selectNodes("//Title")
    If Not (xmlNodeList Is Nothing) Then
        For Each myNode In xmlNodeList
            Debug.Print myNode.Text
        Next myNode
    End If
End Sub
```

In the `SelectNodes_SpecifyCriterion` procedure, the `"//Title"` criterion of the `selectNodes` method looks for the element named `Title` at any level within the tree structure of the nodes.

2. Run the above procedure in step mode by pressing F8.
   Excel prints to the Immediate window only the names of the courses:

```
Beginning VBA in Excel
Intermediate VBA in Excel
Advanced VBA in Excel
Beginning VBA in Outlook
```

The criterion in the `selectNodes` method can be more complex. Let's assume that you are only interested in the title for the `Course` element with an ID of VBA2EX. To retrieve this information, use the following statement:

```
Set xmlNodeList = xmldoc.selectNodes("//Course[@ID =
    'VBA2EX']//Title")
```

The above statement tells the XML processor to search for an element named `Course` at any level within the tree structure of nodes, find only the course element whose ID attribute contains the value of VBA2EX, and return the `Title` element. To retrieve only the first node that meets the specified criterion, use the `selectSingleNode` method of the XML document. As the argument of this method, specify the string representing the node that you'd like to find. In the next example, you will find the first node that matches the criterion `"/Title"` in the `Courses1.xml` document.

⊙ Hands-On 25.15   Finding a Specific Node

1. In the VBE screen of `VBAProject (Employees.xlsm)`, insert a new module and enter the `Select_SingleNode` procedure, as shown below:

```
Sub Select_SingleNode()
    Dim xmldoc As MSXML2.DOMDocument60
    Dim xmlSingleN As MSXML2.IXMLDOMNode

    ' Create an instance of the DOMDocument60
    Set xmldoc = New MSXML2.DOMDocument60
    xmldoc.async = False

    ' Load XML information from a file
    xmldoc.Load ("C:\VBAExcel2024_XML\Courses1.xml")

    ' Retrieve the reference to a particular node
    Set xmlSingleN = xmldoc.selectSingleNode("//Title")
    Debug.Print xmlSingleN.Text
End Sub
```

2. Run the above procedure in step mode by pressing F8.
   The result of this procedure is the text `Beginning VBA in Excel` written to the Immediate Window. The following statements will retrieve the first `Course` node with the `ID` attribute:

```
Set xmlSingleN = xmldoc.selectSingleNode("//Course//@ID")
Debug.Print xmlSingleN.Text
```

If you replace the last two lines in the `Select_SingleNode` procedure with the above statements and run the procedure again, you should see the text VBA1EX in the Immediate window. Once you find the correct node to work with, you can easily modify its value. For example, to change the text of the first `Course` element with the `ID` attribute, use the following lines of code:

```
Set xmlSingleN = xmldoc.selectSingleNode("//Course//@ID")
xmlSingleN.Text = "VBA1EX2010"
xmldoc.Save "C:\VBAExcel2024_XML\Courses1.xml"
```

Notice that to make a permanent change in the XML document, you must save it using the `Save` method.

When using the `selectSingleNode` method, you should also use the `Is Nothing` conditional expression to determine whether a matching element was found in the loaded XML document, as demonstrated in the next example.

⊙ Hands-On 25.16   Using a Conditional Expression with an Element Node

1. In the VBE screen of VBAProject (Employees.xlsm), insert a new module and enter the Select_SingleNode_2 procedure, as shown below:

```
Sub Select_SingleNode_2()
    Dim xmldoc As MSXML2.DOMDocument60
    Dim xmlSingleN As MSXML2.IXMLDOMNode

    ' Create an instance of the DOMDocument60
    Set xmldoc = New MSXML2.DOMDocument60
    xmldoc.async = False

    ' Load XML information from a file
    xmldoc.Load ("C:\VBAExcel2024_XML\Courses1.xml")

    ' Retrieve the reference to a particular node
    Set xmlSingleN = xmldoc.SelectSingleNode("//Course//@ID")
    If xmlSingleN Is Nothing Then
        Debug.Print "No nodes selected."
    Else
        Debug.Print xmlSingleN.Text
        xmlSingleN.Text = "VBA1EX2024"
        Debug.Print xmlSingleN.Text
        xmldoc.Save "C:\VBAExcel2024_XML\Courses1.xml"
    End If
End Sub
```

2. Run the procedure in step mode by pressing F8.
3. Excel prints to the Immediate window the text of the node before and after modification.
4. Replace the XPath expression "//Course//@ID" with "//Cours//@ID" and run the procedure again.
   You should see the text "No nodes selected" in the Immediate window.

Using the FilterXML Function to Retrieve Data from XML

If you are planning to work a lot with XML, you will be thrilled to find out that Excel offers a powerful function called FilterXML. This function returns specific data from XML content by using the specified XPath expression:

```
FilterXML(xml, xpath)
```

xml is a string in valid XML format. If xml is not valid, FilterXML will return a #VALUE! error. xpath is a string in standard XPath format.

The `FilterXML` function can be entered directly in a worksheet, or it can be called from VBA using the `WorksheetFunction` property of the `Application` object. To quickly learn how this function works, insert a new module in the `Employees.xlsm` VBE screen and enter the following procedure:

```
Sub Load_ReadXMLDoc_FilterXML()
    Dim xmlDoc As MSXML2.DOMDocument60
    Dim retval As String

    ' Create an instance of the DOMDocument60
    Set xmlDoc = New MSXML2.DOMDocument60

    ' Disable asynchronous loading
    xmlDoc.async = False

    ' Load XML information from a file
    If xmlDoc.Load("C:\VBAExcel2024_XML\Courses1.xml") Then
        ' Use the DOMDocument60 object's XML property to
        ' retrieve the raw data to the worksheet
        Sheets.Add
        ActiveSheet.Range("A1").Value = xmlDoc.XML
        Columns("A:A").ColumnWidth = 65
        ' Use the built-in function FilterXML to
        ' retrieve data stored in a specific node
        ActiveSheet.Range("A4").Value = _
          WorksheetFunction.FilterXML( _
          Range("A1").Value, "//Course[@ID='VBA2EX']//Title")
    End If
End Sub
```

When you run this procedure, Excel loads the `Courses1.xml` file and adds a new worksheet to the `Employees.xlsm` workbook. The entire XML string is then placed in cell A1. Next, the specific data is retrieved from XML using the `FilterXML` function. Notice that the first argument of this function points to cell A1 containing the XML data and the second argument specifies that we want to retrieve the `title` element with the `Course ID` set to `VBA2Ex`. Cell A4 in the added worksheet should now contain the text "Intermediate VBA in Excel."

NOTE	*Excel limits the number of characters per cell to 32,767. While that seems like a lot, some XML files you may need to work with will exceed this limit and you will see the `#Value!` error value when you attempt to retrieve the data. You can overcome this limitation by writing additional procedures or functions to clean unnecessary characters from the XML to avoid exceeding the 32K character limit.*

> **XML DOM Additional Learning Resources**
>
> The XML DOM provides a number of other methods that make it possible to programmatically add or delete elements. Covering all of the details of the XML DOM is beyond the scope of this chapter. When you are ready for more information on this subject, please visit the following Web sites:
>
> ❏ *DOM Living Standard: http://www.w3.org/DOM/*
>
> ❏ *Extensible Markup Language: http://www.w3.org/XML/*

XML VIA ADO

Earlier in this book, you learned how to retrieve external data using ActiveX Data Objects (ADO). This section focuses on working with XML and ADO.

Saving an ADO Recordset to Disk as XML

To save an ADO recordset as XML to a file, use the `Save` method of the `Recordset` object with the `adPersistXML` constant. The following example procedure demonstrates how to create XML files from ADO recordsets.

⊙ **Hands-On 25.17   Saving an ADO Recordset as an XML Document**

1. In the VBE screen of VBAProject (Employees.xlsm), insert a new module.
2. Choose Tools | References. In the References dialog box, find and select the reference to the Microsoft ActiveX Data Objects 6.1 Library or earlier.
3. Click OK to close the References dialog box.
4. In the Code window of the new module you added in Step 1, enter the SaveRst_ADO procedure, as shown below:

```
Sub SaveRst_ADO()
    Dim rst As ADODB.Recordset
    Dim conn As New ADODB.Connection
    Const strConn = "Provider = Microsoft.ACE.OLEDB.12.0;" _
    & "Data Source = C:\VBAExcel2024_ByExample\Northwind.mdb"

    ' Open a connection to the database
    conn.Open strConn

    ' Execute a select SQL statement against the database
    Set rst = conn.Execute("SELECT * FROM Products")
```

```
    ' Delete the file if it exists
    On Error Resume Next
    Kill "C:\VBAExcel2024_XML\Products.xml"

    ' Save the recordset as an XML file
    rst.Save "C:\VBAExcel2024_XML\Products.xml", adPersistXML

    rst.Close
    conn.Close
End Sub
```

The `SaveRst_ADO` procedure establishes a connection to the sample `Northwind.mdb` database using the ADO `Connection` object. Next, it executes a `select` SQL statement against the database to retrieve all of the records from the `Products` table. Once the records are placed in a recordset, the `Save` method is called to store the recordset to a disk file. If the disk file already exists, the procedure deletes the existing file using the VBA `Kill` statement. The `On Error Resume Next` statement allows bypassing the `Kill` statement if the file that you are going to create does not yet exist.

5. Run the `SaveRst_ADO` procedure.
6. Use Notepad to open the `C:\VBAExcel2024_XML\Products.xml` file created by the `SaveRst_ADO` procedure.
The file content is depicted in Figure 25.34.

XML files can be element-based or attribute-based. The XML files produced by ADO 2.5 or higher are all attribute-based. XML files generated by ADO are self-describing objects that contain data and metadata (information about the data). If you take a look at the `Products.xml` file in Figure 25.34, you will notice that below the XML document's root tag, there are two child nodes: `<s:Schema>` and `<rs:data>`. The schema node describes the structure of the recordset, while the data node holds the actual data.

Between the `<s:Schema id ="RowsetSchema">` and `</s:Schema>` tags, ADO places information about each column, including field name, position, data type and length, nullability, and whether the column is writable.

Notice that each field is represented by the `<s:AttributeType>` element. The value of the `name` attribute is the field name. The `<s:AttributeType>` element also has a child element, `<s:datatype>`, which holds information about its data type (`integer`, `number`, `string`, etc.) and the maximum field length.

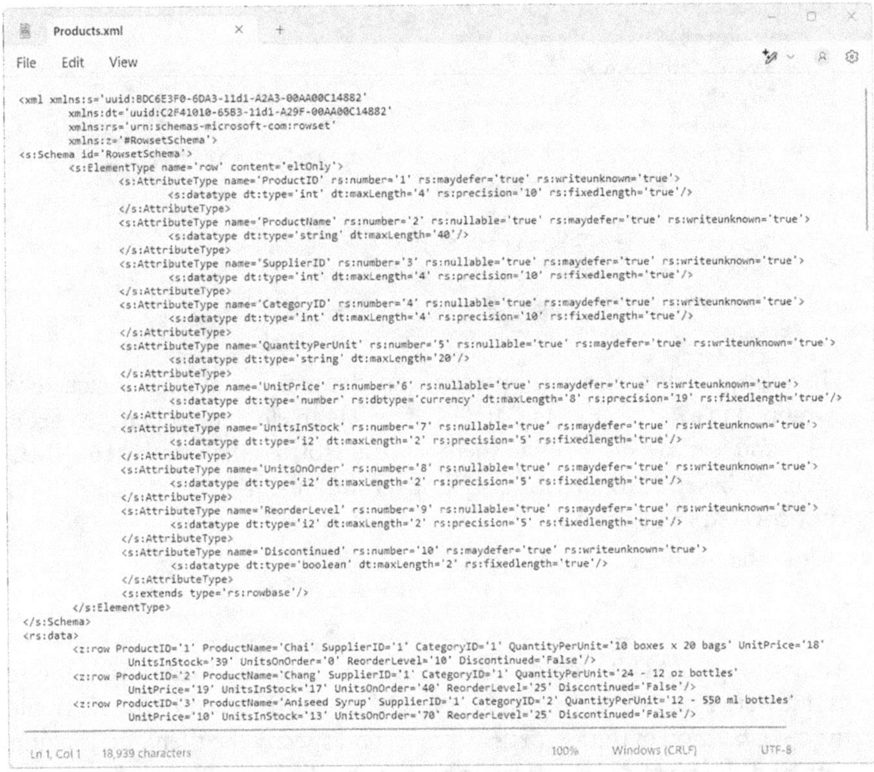


```
<xml xmlns:s='uuid:BDC6E3F0-6DA3-11d1-A2A3-00AA00C14882'
     xmlns:dt='uuid:C2F41010-65B3-11d1-A29F-00AA00C14882'
     xmlns:rs='urn:schemas-microsoft-com:rowset'
     xmlns:z='#RowsetSchema'>
<s:Schema id='RowsetSchema'>
     <s:ElementType name='row' content='eltOnly'>
          <s:AttributeType name='ProductID' rs:number='1' rs:maydefer='true' rs:writeunknown='true'>
               <s:datatype dt:type='int' dt:maxLength='4' rs:precision='10' rs:fixedlength='true'/>
          </s:AttributeType>
          <s:AttributeType name='ProductName' rs:number='2' rs:nullable='true' rs:maydefer='true' rs:writeunknown='true'>
               <s:datatype dt:type='string' dt:maxLength='40'/>
          </s:AttributeType>
          <s:AttributeType name='SupplierID' rs:number='3' rs:nullable='true' rs:maydefer='true' rs:writeunknown='true'>
               <s:datatype dt:type='int' dt:maxLength='4' rs:precision='10' rs:fixedlength='true'/>
          </s:AttributeType>
          <s:AttributeType name='CategoryID' rs:number='4' rs:nullable='true' rs:maydefer='true' rs:writeunknown='true'>
               <s:datatype dt:type='int' dt:maxLength='4' rs:precision='10' rs:fixedlength='true'/>
          </s:AttributeType>
          <s:AttributeType name='QuantityPerUnit' rs:number='5' rs:nullable='true' rs:maydefer='true' rs:writeunknown='true'>
               <s:datatype dt:type='string' dt:maxLength='20'/>
          </s:AttributeType>
          <s:AttributeType name='UnitPrice' rs:number='6' rs:nullable='true' rs:maydefer='true' rs:writeunknown='true'>
               <s:datatype dt:type='number' rs:dbtype='currency' dt:maxLength='8' rs:precision='19' rs:fixedlength='true'/>
          </s:AttributeType>
          <s:AttributeType name='UnitsInStock' rs:number='7' rs:nullable='true' rs:maydefer='true' rs:writeunknown='true'>
               <s:datatype dt:type='i2' dt:maxLength='2' rs:precision='5' rs:fixedlength='true'/>
          </s:AttributeType>
          <s:AttributeType name='UnitsOnOrder' rs:number='8' rs:nullable='true' rs:maydefer='true' rs:writeunknown='true'>
               <s:datatype dt:type='i2' dt:maxLength='2' rs:precision='5' rs:fixedlength='true'/>
          </s:AttributeType>
          <s:AttributeType name='ReorderLevel' rs:number='9' rs:nullable='true' rs:maydefer='true' rs:writeunknown='true'>
               <s:datatype dt:type='i2' dt:maxLength='2' rs:precision='5' rs:fixedlength='true'/>
          </s:AttributeType>
          <s:AttributeType name='Discontinued' rs:number='10' rs:maydefer='true' rs:writeunknown='true'>
               <s:datatype dt:type='boolean' dt:maxLength='2' rs:fixedlength='true'/>
          </s:AttributeType>
          <s:extends type='rs:rowbase'/>
     </s:ElementType>
</s:Schema>
<rs:data>
     <z:row ProductID='1' ProductName='Chai' SupplierID='1' CategoryID='1' QuantityPerUnit='10 boxes x 20 bags' UnitPrice='18'
          UnitsInStock='39' UnitsOnOrder='0' ReorderLevel='10' Discontinued='False'/>
     <z:row ProductID='2' ProductName='Chang' SupplierID='1' CategoryID='1' QuantityPerUnit='24 - 12 oz bottles'
          UnitPrice='19' UnitsInStock='17' UnitsOnOrder='40' ReorderLevel='25' Discontinued='False'/>
     <z:row ProductID='3' ProductName='Aniseed Syrup' SupplierID='1' CategoryID='2' QuantityPerUnit='12 - 550 ml bottles'
          UnitPrice='10' UnitsInStock='13' UnitsOnOrder='70' ReorderLevel='25' Discontinued='False'/>
```


FIGURE 25.34. Saving a recordset to an XML file with ADO produces an attribute-based XML file.

Below the schema definition, you will find the actual data. The ADO schema represents each record using the `<z:row>` tag. The fields in a record are expressed as attributes of the `<z:row>` element. Every XML attribute is assigned a value that is enclosed in a pair of single or double quotation marks; however, if the value of a field in a record is `NULL`, the attribute on the `<z:row>` is not created.

Notice that each record is written out in the following format:

```
<z:row ProductID='1' ProductName='Chai' SupplierID='1'
CategoryID='1'
QuantityPerUnit='10 boxes x 20 bags' UnitPrice='18'
UnitsInStock='39'
UnitsOnOrder='0' ReorderLevel='10' Discontinued='False'/>
```

The above fragment is an attribute-based XML document; however, you may want to have each record written out as follows:

```
<Product>
   <ProductID>1</ProductID>
   <ProductName>Chai</ProductName>
   <SupplierID>1</SupplierID>
   <CategoryID>1</CategoryID>
   <QuantityPerUnit>10 boxes x 20 bags</QuantityPerUnit>
   <UnitPrice>18</UnitPrice>
   <UnitsInStock>39</UnitsInStock>
   <UnitsOnOrder>0</UnitsOnOrder>
   <ReorderLevel>10</ReorderLevel>
   <Discontinued>False</Discontinued>
</Product>
```

The above code fragment represents an element-based XML. Each record is wrapped in a `<Product>` tag, and each field is an element under the `<Product>` tag. You can write a style sheet to transform attribute-based XML into element-based XML. Writing style sheets and using XSL transformations are not covered in this book.

7. Close the `Products.xml` file and exit Notepad.

Loading an ADO Recordset

After saving an ADO recordset to an XML file, you can load it back and read it as if it were a database. To gain access to the records saved in the XML file, use the `Open` method of the `Recordset` object and specify the filename, including its path and the persisted recordset service provider, as `"Provider=MSPersist"`. Let's look at an example that demonstrates opening a persisted recordset.

⦿ Hands-On 25.18 Opening a Persisted Recordset with XML Data

1. In the VBE screen of `VBAProject (Employees.xlsm)`, insert a new module and enter the `OpenAdoFile` procedure, as shown below:

```
Sub OpenAdoFile()
    Dim rst As ADODB.Recordset
    Dim StartRange As Range
    Dim h As Integer

    ' Create a recordset and fill it with
    ' the data from the XML file
    Set rst = New ADODB.Recordset
    rst.Open "C:\VBAExcel2024_XML\Products.xml", _
        "Provider=MSPersist"
```

```
' Display the number of records
MsgBox rst.RecordCount

' Open a new workbook
Workbooks.Add

' Copy field names as headings to the first row
' of the worksheet
For h = 1 To rst.fields.Count
   ActiveSheet.Cells(1, h).Value = rst.fields(h - 1).Name
Next

' Specify the cell range to receive the data (A2)
Set StartRange = ActiveSheet.Cells(2, 1)

' Copy the records from the recordset
' beginning in cell A2
StartRange.CopyFromRecordset rst

' Autofit the columns to make the data fit
Range("A1").CurrentRegion.Select
Columns.AutoFit

' Close the workbook and save the file
ActiveWorkbook.Close SaveChanges:=True, _
   Filename:="C:\VBAExcel2024_ByExample\Products.xlsx"
End Sub
```

The `OpenAdoFile` procedure shown above creates a `Recordset` object and fills it with the data from the `Products.xml` file. After displaying the number of records in the file, the procedure opens a new workbook and fills the first worksheet row with field names. Next, the `CopyFromRecordset` method is used to retrieve all the records into the worksheet. After adjusting the size of the columns to fit the data, the workbook is saved using the default Excel 2024 file format (`.xlsx`).

2. Run the `OpenAdoFile` procedure.

3. Open the `Products.xlsx` file that was created by the `OpenAdoFile` procedure in your `VBAExcel2024_ByExample` folder.
You should see all of the records from the `Products.xml` document nicely arranged in rows and columns and, therefore, easy to analyze and make changes to.

4. Close the `Products.xlsx` file. Do not close the `Employees.xlsm` workbook, as we will continue to use it in the next section.

Saving an ADO Recordset into a DOMDocument60 Object

You can save an ADO recordset directly into an XML DOMDocument60 object using the following code:

```
Set xmlDoc = New MSXML2.DOMDocument60
rst.Save xmlDoc, adPersistXML
```

The next Hands-On exercise demonstrates how to use the DOM to modify XML data in the recordset generated by the ADO Save method.

⊚   Hands-On 25.19   Modifying a Recordset Saved into an XML DOM-Document Object

1. In the VBE screen of VBAProject (Employees.xlsm), insert a new module and enter the SaveToDOM procedure, as shown below:

```
Sub SaveToDOM()
    Dim conn As ADODB.Connection
    Dim rst As ADODB.Recordset
    Dim xmlDoc As MSXML2.DOMDocument60
    Dim myNode As IXMLDOMNode
    Dim strCurValue As String

    ' Declare constant used as database connection string
    Const strConn = "Provider=Microsoft.ACE.OLEDB.12.0;" _
    & "Data Source=C:\VBAExcel2024_ByExample\Northwind.mdb"

    ' Open a connection to the database
    Set conn = New ADODB.Connection
    conn.Open strConn

    ' Open the Shippers table
    Set rst = New ADODB.Recordset
    rst.Open "Shippers", conn, adOpenStatic, adLockOptimistic

    ' Create a new XML DOMDocument60 object
    Set xmlDoc = New MSXML2.DOMDocument60

    ' Add the default namespace declaration
    ' to the Namespace names of the DOMDocument60 object
    ' using the setProperty method of the DOMDocument60 object

    xmlDoc.setProperty "SelectionNamespaces", _
    "xmlns:rs='urn:schemas-microsoft-com:rowset'" & _
    " xmlns:z='#RowsetSchema'"
```

```
' Save the recordset directly into
' the XML DOMDocument60 object
rst.Save xmlDoc, adPersistXML
Debug.Print xmlDoc.XML

' Modify shipper's phone
Set myNode = xmlDoc.selectSingleNode( _
"//z:row[@CompanyName='Speedy Express']/@Phone")
strCurValue = myNode.Text
Debug.Print strCurValue
myNode.Text = "(508)" & Right(strCurValue, 9)
Debug.Print myNode.Text

xmlDoc.Save "C:\VBAExcel2024_XML\Shippers_Modified.xml"

' Cleanup
Set xmlDoc = Nothing
Set conn = Nothing
Set rst = Nothing
Set myNode = Nothing
End Sub
```

After saving the recordset into the XML DOMDocument60 object, the procedure locates a node matching a specified search string by using the selectSingleNode method. Notice that the XPath expression used as an argument of this method searches for the Phone attribute in the z:row element nodes that have a CompanyName attribute set to Speedy Express:

```
Set myNode = xmlDoc.selectSingleNode( _
    "//z:row[@CompanyName='Speedy Express']/@Phone")
```

Once the required phone number is located, the procedure modifies the area code, as follows:

```
strCurValue = myNode.Text
myNode.Text = "(508)" & Right(strCurValue, 9)
```

If you'd rather remove the Phone entry completely, you could use the following code:

```
Set myNode = xmlDoc.selectSingleNode( _
    "//z:row[@CompanyName='Speedy Express']")
myNode.Attributes.removeNamedItem "Phone"
```

The removeNamedItem method removes an attribute from the attributes of a given node. This method requires one parameter: a string specifying the name of the attribute to remove from the collection.

2. Run the procedure in step mode by pressing F8. Make sure the Immediate window is open so you can see the results of various `Debug.Print` statements at once.

3. Use Notepad to open the `C:\VBAExcel2024_XML\Shippers_Modified.xml` file created by the `SaveToDOM` procedure.
Notice the modified phone number for the `Speedy Express` record.

4. Close the `Shippers_Modified.xml` file and exit Notepad.

5. Close the `Employees.xlsm` workbook, saving changes when prompted.

UNDERSTANDING NAMESPACES

XML, a markup language that uses custom tags, allows you to define unique tag names to describe your data. Conflicts can arise, however, when combining multiple XML documents, as identical tag names might represent different meanings. For instance, the `<TABLE>` tag in an Excel XML document would differ in meaning from the `<TABLE>` tag in a furniture catalog.

To resolve such conflicts, the XML namespaces specification ensures that element and attribute names remain unique within their respective sets (namespaces). A *namespace* is a collection of unique names, identified by a Uniform Resource Identifier (URI), which could be a URL or a Uniform Resource Name (URN). Typically, the namespace declaration appears at the start of the XML document. Interestingly, the URI doesn't need to be valid or conform to any specific format. Most namespaces use URIs for their names because URIs inherently guarantee uniqueness.

Take a look at the following lines in the `Shippers_Modified.xml` file that was created in Hands-On 25.19:

```
<xml xmlns:s="uuid:BDC6E3F0-6DA3-11d1-A2A3-00AA00C14882"
    xmlns:dt="uuid:C2F41010-65B3-11d1-A29F-00AA00C14882"
    xmlns:rs="urn:schemas-microsoft-com:rowset"
    xmlns:z="#RowsetSchema">
```

You can declare a namespace in any element by using the `xmlns` attribute. A namespace whose `xmlns` attribute is not followed by a prefix is referred to as a *default namespace*. Therefore, elements or attributes with no prefix will be assumed to be part of the default namespace. In the above example, there are four namespaces, each of which is associated with a particular prefix (`s`, `dt`, `rs`, and `z`). In the XML document, these prefixes are used in front of element and attribute names to indicate which namespace they are referencing. In other words, anything with an "s" in front of it applies to the `uuid:BDC6E3F0-6DA3-11d1-A2A3-00AA00C14882` namespace, and anything marked with the "z" prefix references the `RowsetSchema` namespace.

The key takeaway from this section is that namespaces are merely arbitrary identifiers. Their purpose is to differentiate between tags with identical names that require distinct processing. By doing so, namespaces prevent naming conflicts within XML documents.

UNDERSTANDING OPEN XML FILES

Since the release of Excel 2007, all workbook files are saved by default in an XML file format. This file format known as *Open XML* uses four-letter file extensions (`.xlsx`, `.xlsm`, `.xltx`, `.xltm`, and `.xlam`).

The first two letters of the file extension refer to the application, in this case, `xl` represents Excel. The third letter (`s/t/a`) indicates the specific file type: `s` represents a spreadsheet, `t` represents a template, and `a` represents an add-in. The last letter (`x/m`) specifies whether the file format supports macros: `x` is a macro-free file, and `m` is a macro-enabled file.

The Open XML file is actually a compressed ZIP file. A ZIP file contains one or more files that have been compressed to reduce their file size. By changing the Excel file extension to `.zip`, you can take a look inside the ZIP container using WinZip or another ZIP-aware tool, or use the built-in compressed folders feature in Windows.

The Open XML file format gives developers the ability to directly edit the workbook without the need to open Excel. This means that you can work with the file content without having an Excel application installed on your computer. The same applies to Word and PowerPoint documents that follow the same Open Packaging Conventions (OPC) specification. You can easily insert new data, edit existing data, modify document properties, and add or remove specific XML parts.

This section takes a detailed look inside the compressed file, known as a *package*. Figure 25.35 shows the contents of the `Chap25b.xlsm` workbook file created in this chapter.

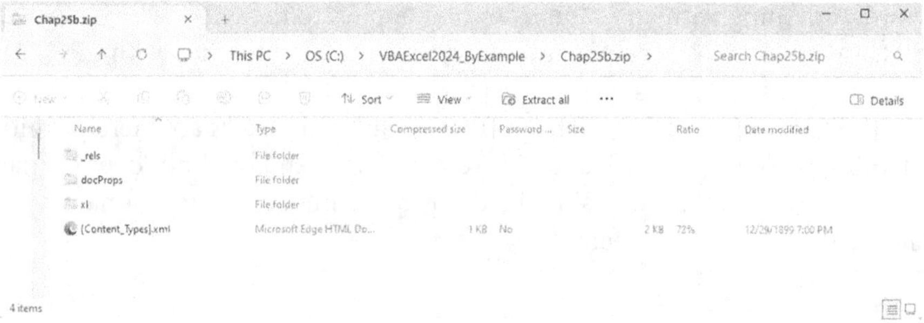

FIGURE 25.35. A sample Excel 2024 workbook is shown here after renaming the file with a .zip extension and opening it with compressed folders in File Explorer.

The package file contains a number of documents called *parts* grouped into various folders. Every part has a defined content type that describes whether it's a worksheet, image, sound, or other binary object. Some types of parts are shared across all Microsoft 365 applications; others are unique to the application. For example, a worksheet part can only be found in an Excel file. While most parts are XML documents, some parts (such as images, VBA projects, or embedded OLE objects) are stored in their native format as binary files. Every part within a container package is connected to at least one other part using a special part referred to as a *relationship*. A relationship file is an XML document with a .rels extension.

At the root level in Figure 25.35, you will notice three folders named _rels, docProps, and xl, and an XML file called [Content_Types].xml:

- The [Content_Types].xml file—This XML file lists the types of files that are included in the package.

- The _rels folder—The parts listed in the Excel package are linked together via relationships. The .rels file in the _rels folder defines the package relationships. You will see here the relationships between the document properties files, docProps/app.xml and docProps/core.xml, and the xl/workbook.xml file. Parts that are related to other parts contain a _rels subfolder. Within this subfolder, you will find a .rels file that describes the relationships. The name of the relationship consists of the filename of the original part and the .rels extension. For example, for the Workbook.xml file listed in the xl folder, there is a relationship file named Workbook.xml.rels in the xl\rels folder (see Figure 25.36).

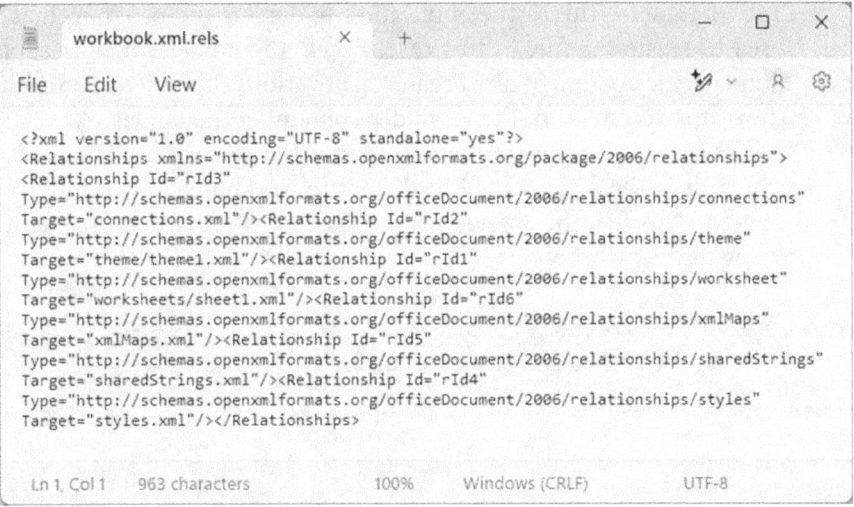

```
workbook.xml.rels                    ×      +

File    Edit    View

<?xml version="1.0" encoding="UTF-8" standalone="yes"?>
<Relationships xmlns="http://schemas.openxmlformats.org/package/2006/relationships">
<Relationship Id="rId3"
Type="http://schemas.openxmlformats.org/officeDocument/2006/relationships/connections"
Target="connections.xml"/><Relationship Id="rId2"
Type="http://schemas.openxmlformats.org/officeDocument/2006/relationships/theme"
Target="theme/theme1.xml"/><Relationship Id="rId1"
Type="http://schemas.openxmlformats.org/officeDocument/2006/relationships/worksheet"
Target="worksheets/sheet1.xml"/><Relationship Id="rId6"
Type="http://schemas.openxmlformats.org/officeDocument/2006/relationships/xmlMaps"
Target="xmlMaps.xml"/><Relationship Id="rId5"
Type="http://schemas.openxmlformats.org/officeDocument/2006/relationships/sharedStrings"
Target="sharedStrings.xml"/><Relationship Id="rId4"
Type="http://schemas.openxmlformats.org/officeDocument/2006/relationships/styles"
Target="styles.xml"/></Relationships>

Ln 1, Col 1    963 characters          100%      Windows (CRLF)        UTF-8
```

FIGURE 25.36.   This relationship file defines how different parts of an Excel workbook are connected.

The `Workbook.xml.rels` file uses XML syntax to declare relationships between the workbook and its associated components, such as worksheets, shared strings, styles, and other elements. Each relationship is defined with attributes such as:

- `ID`—A unique identifier for the relationship

- `Type`—Specifies the type of content (e.g., worksheet, shared strings, etc.)

- `Target`—Indicates the target file or part that the relationship points to

The `Workbook.xml.rels` file plays a crucial role in ensuring that all parts of the workbook are correctly linked and function as intended. It is not meant to be manually edited, however, as changes could corrupt the workbook.

- The `docProps` folder—This folder contains two document properties files that were referenced in the `.rels` file: `app.xml` and `core.xml`. These property files store information that you enter in Excel when you click the File tab and choose Info and then Properties. The `core.xml` part consists of properties such as the document title, subject, and author. The `app.xml` part stores application-specific properties such as the name and the version of the application, company name, and others.

- The `xl` folder—This is the application folder for the program that was used to create the file, in this case, Excel. This folder contains application-specific document files organized in various subfolders. Figure 25.37 shows the root level of the `xl` folder containing the `workbook`, `shared-Strings`, `vbaProject`, and `styles` parts. The `sharedStrings.xml` part stores all of the strings used in the entire workbook. If you modify the string in this file, the change will be applied to every occurrence of the string in your workbook.

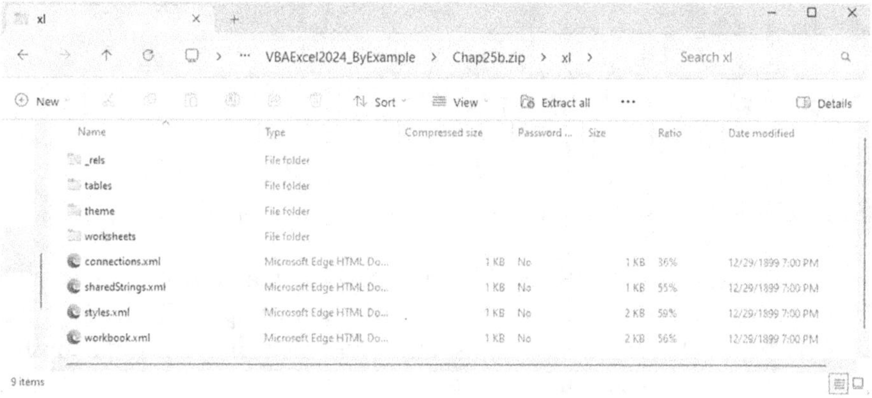

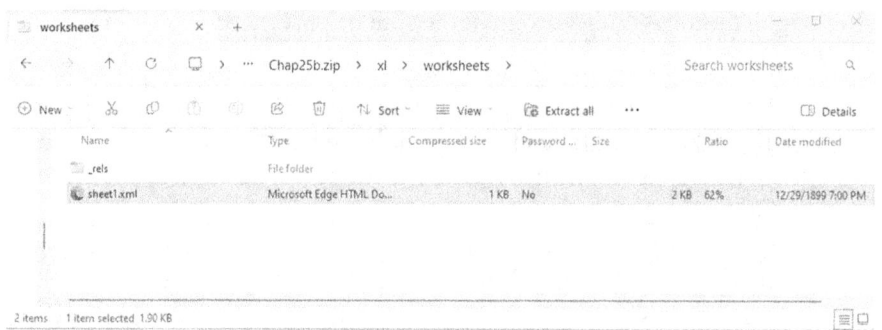

FIGURE 25.37.   The contents of the xl folder.

The `worksheets` folder within the `xl` folder contains a separate XML part for every worksheet, in this case, `sheet1.xml` (Figure 25.38).

FIGURE 25.38.   The contents of the worksheets folder.

Figure 25.39 shows the contents of the `sheet1.xml` part.

```
▼<worksheet xmlns="http://schemas.openxmlformats.org/spreadsheetml/2006/main"
  xmlns:r="http://schemas.openxmlformats.org/officeDocument/2006/relationships"
  xmlns:mc="http://schemas.openxmlformats.org/markup-compatibility/2006"
  xmlns:x14ac="http://schemas.microsoft.com/office/spreadsheetml/2009/9/ac"
  xmlns:xr="http://schemas.microsoft.com/office/spreadsheetml/2014/revision"
  xmlns:xr2="http://schemas.microsoft.com/office/spreadsheetml/2015/revision2"
  xmlns:xr3="http://schemas.microsoft.com/office/spreadsheetml/2016/revision3"
  mc:Ignorable="x14ac xr xr2 xr3" xr:uid="{00000000-0001-0000-0000-000000000000}">
  <dimension ref="A1:D5"/>
 ▼<sheetViews>
    <sheetView tabSelected="1" workbookViewId="0"/>
  </sheetViews>
  <sheetFormatPr defaultRowHeight="14.25"/>
 ▼<cols>
    <col min="1" max="1" width="11.625" bestFit="1" customWidth="1"/>
    <col min="2" max="2" width="22.125" bestFit="1" customWidth="1"/>
    <col min="3" max="4" width="11" bestFit="1" customWidth="1"/>
  </cols>
 ▼<sheetData>
  ▼<row r="1" spans="1:4">
   ▼<c r="A1" t="s">
      <v>0</v>
    </c>
   ▼<c r="B1" t="s">
      <v>1</v>
    </c>
   ▼<c r="C1" t="s">
      <v>2</v>
    </c>
   ▼<c r="D1" t="s">
      <v>3</v>
    </c>
  </row>
  ▼<row r="2" spans="1:4">
   ▼<c r="A2" s="1" t="s">
      <v>4</v>
    </c>
```

FIGURE 25.39.   The contents of the sheet1.xml file in Chap25b.xlsm workbook.

Notice that all the sheet data is contained within the `<sheetData>` element. Each data row has its own `<row>` element and an index (`r` attribute). Rows use a `spans` attribute to indicate the number of cells occupied. Other attributes may be used to indicate row style or custom formatting. Cell values are stored in the `c` element. The `r` attribute holds the cell address using the `A1` reference style notation (e.g., `A2`, `B2`); the `s` attribute indicates which style was used. The numbers used in the `style` attribute are described in the `xl/styles.xml` part. The `t` attribute indicates a data type (`String`, `Number`, or `Boolean`). For example, `t="s"` denotes that the underlying value is a string, not a number. String values are not stored in cells unless they are the result of a calculation. They are stored in the `sharedStrings.xml` part. Each unique text string found within a workbook is listed only once in this part. This prevents duplication of information, saves space, and improves the speed of loading and saving workbooks. If the cell value is textual, then the numeric value inside the `v` element is an index to a particular string in the `sharedStrings.xml` document.

You can easily replace the `sharedStrings` file in the package with a file containing strings from another language, thus providing multiple language support for your worksheet users.

After this short overview of the internals of the Open XML file format, let's spend some time putting this newfound knowledge to practical use. The next section will introduce you to working with XML document parts programmatically.

MANIPULATING OPEN XML FILES WITH VBA

Earlier in this chapter, you worked with XML document nodes using the XML DOM. In this section, you will use the DOM objects, properties, functions, and XPath expressions to augment some of the XML parts found in the Excel workbook. To work with the XML document parts, we need to know how to zip and unzip Excel workbook files with VBA.

NOTE	*The next two Hands-On exercises require 7-Zip, an external archiving tool, to be installed on your computer. You can download 7-Zip directly from https://7-zip.org.* *Choose the version that matches your system (64-bit x64 or 32-bit x86), and click the download link for the appropriate version. Once downloaded, run the installer and follow the on-screen instructions to complete the installation.*

(⊙) **Hands-On 25.20   Unzipping an Excel Workbook with VBA**

1. Copy the `SupportedEquipment.xlsx` workbook from the companion files to your `VBAExcel2024_ByExample` folder.
2. Open a new Microsoft Excel workbook and save it in a macro-enabled format as `C:\VBAExcel2024_ByExample\ManipulateXMLParts.xlsm`.
3. Press Alt+F11 to activate the VBE and select VBAProject (ManipulateXML-Parts.xlsm) in the Project Explorer window.
4. Choose Insert | Module.
5. Choose Tools | References. In the Available References list box, select the Microsoft XML, v6.0 object library, and click OK to exit the References dialog box.

6. In the Code window of `VBAProject (ManipulateXMLParts.xlsm)`, enter the variable declaration and the `UnizpExcelFile` procedure code as follows:

```
' Declare a module-level variable
Public blnIsFileSelected As Boolean

Sub UnzipExcelFile()
    Dim strSourceFile As Variant
    Dim strSourceFileName As String
    Dim strZipFolder As String
    Dim strZipFile As String

    ' Prompt the user to select an Excel file
    strSourceFile = Application.GetOpenFilename( _
    FileFilter:="Excel Files (*.xlsx; *.xlsm), *.xlsx; *.xlsm", _
    Title:="Select Excel file to copy and convert to ZIP")

    Debug.Print strSourceFile & " - sourceFile"

    ' Exit if no file was selected
    If strSourceFile = False Then
        blnIsFileSelected = False
        Exit Sub
    End If

    blnIsFileSelected = True

    ' Extract the file name from the selected file path
    strSourceFileName = Mid(strSourceFile, _
        InStrRev(strSourceFile, "\") + 1)
    Debug.Print strSourceFileName & " - sourceFileName"

    strZipFile = Left(strSourceFileName, _
        Len(strSourceFileName) - 5) & ".zip"

    Debug.Print strZipFile & " - Zip File"

    ' Copy the file to another one using .zip extension
    FileCopy strSourceFile, strZipFile

    ' Create the "ZipPackage" folder if it doesn't already exist
    On Error Resume Next
    strZipFolder = "C:\VBAExcel2024_ByExample\ZipPackage"
    MkDir strZipFolder
    Debug.Print strZipFolder & " - Zip Folder"
```

```
        ExtractWith7Zip strZipFile, strZipFolder

        ' Notify the user
        MsgBox "Unzipping complete! Files are extracted to: " & _
            strZipFolder, vbInformation

        ' Delete the zip file
        Kill strZipFile

        ' Activate Windows Explorer
        Shell "Explorer.exe /e," & strZipFolder, vbNormalFocus
End Sub

Sub ExtractWith7Zip(sourceZip As String, _
    destinationFolder As String)
    Dim sevenZipPath As String
    Dim shellObj As Object
    Dim returnCode As Variant


    ' Adjust based on your 7-Zip installation path
    sevenZipPath = "C:\Program Files\7-Zip\7z.exe"

    ' Execute 7-Zip command
    Set shellObj = CreateObject("WScript.Shell")
    returnCode = shellObj.Run(Chr(34) & sevenZipPath & _
        Chr(34) & " x " & Chr(34) & _
        sourceZip & Chr(34) & " -o" & Chr(34) & _
        destinationFolder & Chr(34), 0, True)

    MsgBox "Extraction complete!", vbInformation

    shellObj.Quit
    Set shellObj = Nothing
End Sub
```

The `UnzipExcelFile` procedure prompts the user to select an Excel file to unzip using the `GetOpenFilename` method of the `Application` object. This method returns either the file path as a string or `False` if no file is selected. By declaring `strSourceFile` as `Variant`, the program can handle both the string (file path) and the `False` value without causing a type mismatch error. The procedure then extracts the filename from the selected file path and modifies it by removing the last five characters (the dot and the four characters of the Excel file extension) and appending `.zip` to generate the zip filename. Using the `FileCopy` statement, the source file is copied with the new `.zip` extension. Next, the `MkDir` statement creates a destination folder named `ZipPackage` for the extracted contents. The `ExtractWith7Zip` subroutine is then called with two arguments, `strZipFile` and `strZipFolder`, to begin the unzipping process. The `7-Zip` executable is invoked, and a zip extraction command is executed via the `Run` method of the `WScript.Shell` object. After all files are extracted to the specified folder, the `Shell` object is closed, and the procedure notifies the user of the completion. The `.zip` file is subsequently deleted using the VBA `Kill` statement. Finally, the `Shell` function is called with the `/e` parameter to display the contents of the `ZipPackage` folder in File Explorer, showing the files in list view.

7. Run the `UnzipExcelFile` procedure. When prompted to select the file to unzip, choose the `SupportedEquipment.xlsx` workbook from your `C:\VBAExcel2024_ByExample` folder.

The File Explorer window will pop up automatically when the unzip process is complete (see Figure 25.40).

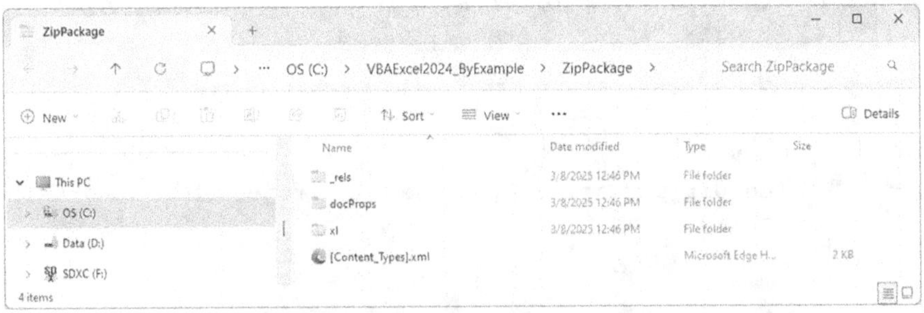

FIGURE 25.40. The contents of the compressed (zipped) folder named ZipPackage created in Hands-On 25.20. When you change an Excel file extension to .zip, you reveal its underlying structure because Excel files (starting from Excel 2007) are essentially ZIP archives containing multiple XML files and other resources. The Excel file, when unzipped, splits into various folders and files, as discussed earlier in this chapter.

After unzipping an Excel file, you can explore the workbook structure, inspect the content of individual sheets (stored as XML), or analyze other components such as styles, themes, and macros. You can modify certain elements (such as styles or XML data) and re-zip the contents. Once you rename it back to `.xlsx or .xlsm`, you can open it in Excel with the changes applied. This is helpful for troubleshooting corrupted files, analyzing structure, or tweaking workbook contents.

Before we make changes to the XML parts (as shown in subsequent Hands-On exercises), let's create another VBA procedure that will allow us to zip the files back into the Excel container.

Hands-On 25.21 Zipping Files to Create an Excel 2024 Package Container

1. In the same module where you entered the UnzipExcelFile procedure in the previous Hands-On, enter the following ZipToExcel procedure:

```
Sub ZipToExcel()
    Dim strZipFile As String
    Dim strZipFolder As String
    Dim strExcelFile As String
    Dim mFlag As Boolean
    Dim shellObj As Object
    Dim command As String
    Dim sevenZipPath As String

    strZipFolder = "C:\VBAExcel2024_ByExample\ZipPackage"
    strZipFile = "C:\VBAExcel2024_ByExample\PackageModified.zip"
    mFlag = False

    'check if folder is empty
    If Len(Dir(strZipFolder & "\*.*")) < 1 Then
        MsgBox "There are no files to zip."
        Exit Sub
    End If

    ' check if a VBA project exists
    If Len(Dir(strZipFolder & "\xl\vbaProject.bin")) > 0 Then
        mFlag = True
    End If

    ' Path to 7-Zip executable (update this if needed)
    sevenZipPath = "C:\Program Files\7-Zip\7z.exe"

    ' Construct the 7-Zip command to include
    ' all files and subfolders in the folder
    command = Chr(34) & sevenZipPath & Chr(34) & " a -tzip " & _
        Chr(34) & strZipFile & Chr(34) & " " & Chr(34) & _
        strZipFolder & "\*.*" & Chr(34) & " -r"

    Debug.Print command

    ' Create a shell object to execute the command
    Set shellObj = CreateObject("WScript.Shell")
    shellObj.Run command, 0, True
```

```
    MsgBox "Folder has been zipped successfully!", _
        vbInformation, "Zipping Completed."

    'Create Excel file name
    If mFlag Then
        strExcelFile = Replace(strZipFile, ".zip", ".xlsm")
    Else
        strExcelFile = Replace(strZipFile, ".zip", ".xlsx")
    End If


    ' Check if the destination file exists
    If Dir(strExcelFile) <> "" Then
        ' Delete the existing file or handle it as needed
        Kill strExcelFile
    End If

    ' Rename the strZipFile
    Name strZipFile As strExcelFile

    Set shellObj = Nothing
    MsgBox "File was renamed successfully."
End Sub
```

The `ZipToExcel` procedure begins by setting `ZipPackage` as the name of the ZIP folder and `PackageModified.zip` as the target ZIP file. Before proceeding with the copy operation, two checks are performed. First, the procedure exits if there are no files in the `ZipPackage` folder. Second, it checks for the presence of the `vbaProject.bin` file within the `xl` folder of the ZIP folder. This check determines whether the destination file will be assigned a macro-free or macro-enabled Excel format when the ZIP archive is renamed. The `mFlag` variable is set to `true` if the `vbaProject.bin` file exists. If files are present in the `ZipPackage` folder, the procedure constructs and executes the `7-Zip` command to compress all files and subfolders in the folder. Following this, the appropriate Excel file extension is chosen based on the value of the `mFlag` variable. Finally, the selected extension is used to replace the `.zip` extension, completing the procedure.

2. Run the `ZipToExcel` procedure.
When the procedure has been executed, you should see the `PackageModified.xlsx` workbook file in your `VBAExcel2024_ByExample` folder. Because we have not yet made any changes to the XML parts contained in the

`SupportedEquipment.xlsx` file that was unzipped in Hands-On 25.20, the `PackageModified.xlsx` file should contain the same content as this file.

3. Open the `PackageModified.xlsx` file in Microsoft Excel to confirm its integrity and ensure it is not corrupt. If a message indicating file corruption appears, adjust the process by modifying the `ZipToExcel` procedure to introduce a slightly longer pause, allowing sufficient time for each file to be fully compressed and saved.

Introducing a Pause in VBA Procedures

A great way to introduce a delay in your VBA procedure is by using the `Application.Wait` method, which pauses the procedure without freezing the Excel interface, keeping the application responsive:

```
Application.Wait Now + TimeValue("00:00:05") ' Pauses for 5 seconds
```

As an alternative, you can create a pause with a loop and the `Timer` function:

```
Dim PauseTime As Single
PauseTime = Timer + 5  ' 5-second pause
Do While Timer < PauseTime
    DoEvents ' Keeps the application responsive
Loop
' Your remaining code...
```

---

4. Close the `PackageModified.xlsx` workbook.

5. Delete the `PackageModified.xlsx` workbook from your `VBAExcel2024_ByExample` folder.

In the next three Hands-On examples, we will utilize both the zip and unzip procedures from Hands-On 25.20 and 25.21 to modify some XML parts in the Excel ZIP archive. The procedure in Hands-On 25.22 demonstrates how to retrieve to a worksheet the unique text values that are stored in the `sharedStrings.xml` part shown in Figure 25.41.

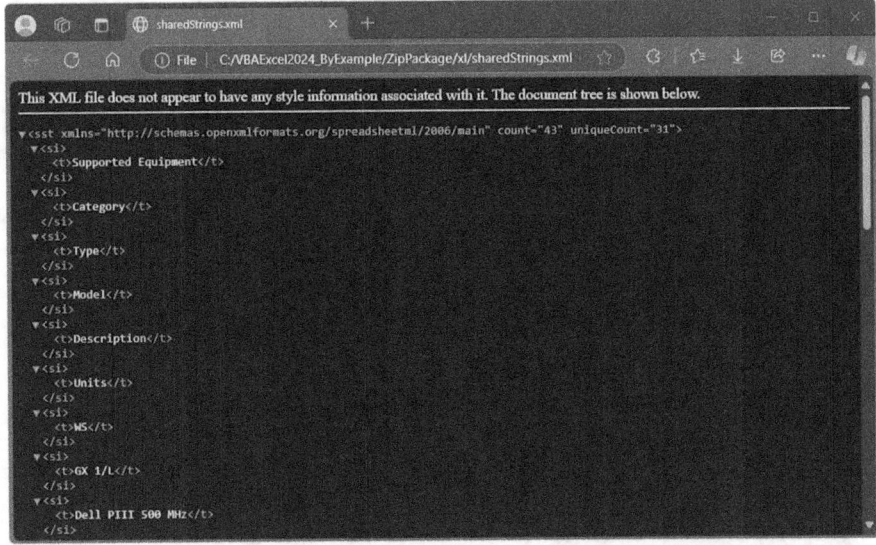

FIGURE 25.41.   Partial content of the sharedStrings.xml part in the SupportedEquipment.xlsx workbook.

Hands-On 25.22   Retrieving Unique Text Values from the shared-Strings XML File

1. Insert a new module in VBAProject (ManipulateXMLParts.xlsm), and in the Code window, enter the following ListUniqueValues procedure:

```
Sub ListUniqueValues()
    Dim xmlDoc As MSXML2.DOMDocument60
    Dim myNodeList As MSXML2.IXMLDOMNodeList
    Dim i As Integer
    Dim strFile As String
    Dim strFolder As String
  Dim xNode As Variant
    Dim cNode As Variant

    i = 1
    strFolder = "C:\VBAExcel2024_ByExample\ZipPackage\xl\"
    strFile = "sharedStrings.xml"

    Set xmlDoc = New MSXML2.DOMDocument60
    xmlDoc.async = False
    xmlDoc.Load strFolder & strFile
    Set myNodeList = xmlDoc.ChildNodes
```

```
        Worksheets(1).Activate
        'Iterate over the elements and print their Text property
        For Each xNode In xmlDoc.ChildNodes
        '    only look at type=NODE_ELEMENT
            If xNode.NodeType = 1 Then
                For Each cNode In xNode.ChildNodes
                    ActiveSheet.Cells(i, 1).Value = cNode.Text
                    i = i + 1
                Next
            End If
        Next
        Columns("A").AutoFit
        Set myNodeList = Nothing
        Set xmlDoc = Nothing
End Sub
```

The `ListUniqueValues` procedure uses the `Load` method of the `DOMDocument60` object to open the `sharedStrings.xml` file. The procedure retrieves the text property of all the child nodes and writes them into a worksheet.

2. Ensure that the `VBAExcel2024_ByExample` folder contains the folder named `ZipPackage` with XML parts. If this folder is missing, run the procedure in Hands-On 25.20.

3. Run the `ListUniqueValues` procedure.

    The result of the procedure is shown in Figure 25.42.

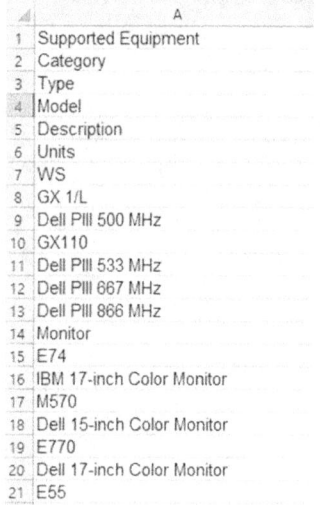

FIGURE 25.42. Partial contents of the sharedStrings.xml part for the SupportedEquipment.xlsx workbook now retrieved into an Excel worksheet.

Now that you retrieved text values stored in the `sharedStrings.xml` part, let's proceed to edit this file. In the next Hands-On, we will replace the worksheet's text entry `Monitor` with `Flat Panel Monitor` without opening the Excel application.

⊙ **Hands-On 25.23   Modifying the sharedString XML File**

1. Insert a new module in `VBAProject (ManipulateXMLParts.xlsm)`, and in the Code window, enter the following `Text_Replace` procedure:

```
Sub Text_Replace()
    Dim xmlDoc As MSXML2.DOMDocument60
    Dim myNode As MSXML2.IXMLDOMNode
    Dim srchStr As String
    Dim newStr As String
    Dim strFile As String
    Dim strFolder As String
    Dim strFileToEdit As String

    strFolder = "C:\VBAExcel2024_ByExample\ZipPackage\xl\"
    strFile = "sharedStrings.xml"
    strFileToEdit = strFolder & strFile

    ' Check if the folder does not exist
    If Dir(strFolder, vbDirectory) = "" Then
        Call UnzipExcelFile
        If blnIsFileSelected = False Then
            MsgBox "The process was abandoned."
            Exit Sub
        End If
    End If

    Set xmlDoc = New DOMDocument60
    xmlDoc.async = False
    xmlDoc.validateOnParse = False
    xmlDoc.SetProperty "SelectionNamespaces", _
    "xmlns:a='http://schemas.openxmlformats.org/" & _
        "spreadsheetml/2006/main'"
    xmlDoc.SetProperty "SelectionLanguage", "XPath"
    xmlDoc.Load (strFileToEdit)

    srchStr = InputBox("Please enter the string to find:", _
            "Search for String")

    If srchStr <> "" Then
        ' find the text that needs to be replaced
```

```
    Set myNode = xmlDoc.SelectSingleNode("//a:t[text()='" + _
        srchStr + "']")
    If myNode Is Nothing Then GoTo ExitHere
Else
    GoTo ExitHere
End If

' replace text
newStr = InputBox("Please enter the replacement string for " _
    & srchStr, "Replace with String")
If newStr <> "" Then
    myNode.Text = newStr
    xmlDoc.Save strFileToEdit
Else
    Exit Sub
End If
ExitHere:
' zip the files in the package
Call ZipToExcel
Set xmlDoc = Nothing
Set myNode = Nothing
End Sub
```

In the above procedure, we use the `InputBox` function to prompt the user to enter the string to search for. If `text` was specified, we use the following statement to find the node with the specified text entry:

```
Set myNode = xmlDoc.SelectSingleNode("//a:t[text()='" + _
        srchStr + "']")
```

The XPath `text()` function is used to retrieve the text value of a node. The XPath expression tells the XML parser to look at the single element node and select the `t` node, shown earlier in Figure 25.41, where the text content is equal to the value of the `srchStr` variable. Instead of using the XPath `text()` function, you can examine the `t` node using the dot operator, like this:

```
Set myNode = xmlDoc.SelectSingleNode("//a:t[.='" + srchStr + "']")
```

Note that the `a:` string before the node name is the alias that was assigned to the namespace using the `SetProperty` property of the `XMLDocument` object.

If the node with the text entry was not found, we go to the `Exit Here` label and perform the final tasks before exiting the procedure. If the node was found, we prompt the user for the replacement text. When we get the new string, we write it to the node and save the file:

```
myNode.Text = newStr
xmlDoc.Save strFileToEdit
```

Next, we need to zip the files back into an Excel container, so we call the `ZipToExcel` procedure that we created earlier.

2. In the File Explorer, delete the `ZipPackage` folder that was created in Hands-On 25.20.

3. Run the `Text_Replace` procedure. When prompted for the filename to unzip, select `SupportedEquipment.xlsx`. When prompted for the string to find, enter `Monitor` and click OK. When prompted for the new text, enter `Flat Panel Monitor` and click OK.

4. When the procedure is completed, open the `PackageModified.xlsx` file in Excel. Each cell entry that previously had Monitor as the underlying text value in column B should now show Flat Panel Monitor.

5. Close the `PackageModified.xlsx` workbook.

6. Delete the `PackageModified.xlsx` workbook from your `VBAExcel2024_ByExample` folder.

The next Hands-On demonstrates how to change the size of the left margin and remove the entire `pageSetup` node from the `sheet1.xml` part.

⊙ Hands-On 25.24   Changing and Removing Elements in an XML Part

1. Insert a new module in `VBAProject (ManipulateXMLParts.xlsm)`, and in the Code window, enter the following `ChangeLeftMargin_RemovePageSetup` procedure:

```
Sub ChangeLeftMargin_RemovePageSetup()
    Dim xmlDoc As DOMDocument60
    Dim myNode As MSXML2.IXMLDOMNode
    Dim strSrchNode As String
    Dim strFolder As String


    strFolder = "C:\VBAExcel2024_ByExample\ZipPackage\"

    ' Check if the folder does not exist
    If Dir(strFolder, vbDirectory) = "" Then
        Call UnzipExcelFile
        If blnIsFileSelected = False Then
            MsgBox "The process was abandoned."
            Exit Sub
        End If
    End If
```

```
    Set xmlDoc = New DOMDocument60
    xmlDoc.async = False
    xmlDoc.validateOnParse = False
    xmlDoc.Load (strFolder & "xl\worksheets\Sheet1.XML")


' xmlDoc.Load ("C:\VBAExcel2024_ByExample\ZipPackage\xl\" _
    & "worksheets\Sheet1.XML")

    xmlDoc.SetProperty "SelectionNamespaces", _
        "xmlns:x14ac='http://schemas.openxmlformats.org/" & _
            "spreadsheetml/2006/main'"

    strSrchNode = "/x14ac:worksheet/x14ac:pageMargins/@left"

    Set myNode = xmlDoc.SelectSingleNode(strSrchNode)
    Debug.Print "previous left margin = " & myNode.Text

    myNode.Text = "0.50"

    Set myNode = xmlDoc.SelectSingleNode("//x14ac:pageSetup")

    On Error Resume Next
    myNode.ParentNode.RemoveChild myNode

    xmlDoc.Save ("C:\VBAExcel2024_ByExample\ZipPackage\xl\" _
        & "worksheets\Sheet1.XML ")

    Set myNode = Nothing
    Set xmlDoc = Nothing
End Sub
```

We begin the `ChangeLeftMargin_RemovePageSetup` procedure by loading the `sheet1.xml` part into the `DOMDocument60` object. To modify the left margin, we need to access the value of the `left` attribute within the `<pageMargins>` element, which is nested under the `<worksheet>` element. Before constructing the appropriate XPath expression, it's essential to determine whether the XML elements belong to any namespaces. If you examine the `Sheet1.xml` document, you'll notice that the root `<worksheet>` element includes four namespace declarations, as illustrated in Figure 25.43. Since the elements we wish to access are not in the default namespace (referred to as `no namespace`) but rather in the namespace `http://schemas.microsoft.com/office/spreadsheetml/2009/9/ac`, we must associate a prefix (e.g., `x14ac`) with this namespace URI. This prefix is then used in our XPath expression to accurately target the desired elements. In MSXML, binding a prefix to a namespace URI is straightforward using the `setProperty` method of the `XMLDocument60` object:

```
xmlDoc.setProperty "SelectionNamespaces", _
        "xmlns:x14ac='http://schemas.openxmlformats.org/" & _
            "spreadsheetml/2006/main'"
```

Having specified the namespace prefix, we can now use it in the XPath expression and successfully access the required node:

```
strSrchNode = "/x14ac:worksheet/x14ac:pageMargins/@left"
Set myNode = xmlDoc.selectSingleNode(strSrchNode)
```

```
▼<worksheet xmlns="http://schemas.openxmlformats.org/spreadsheetml/2006/main"
  xmlns:r="http://schemas.openxmlformats.org/officeDocument/2006/relationships"
  xmlns:mc="http://schemas.openxmlformats.org/markup-compatibility/2006"
  xmlns:x14ac="http://schemas.microsoft.com/office/spreadsheetml/2009/9/ac"
  xmlns:xr="http://schemas.microsoft.com/office/spreadsheetml/2014/revision"
  xmlns:xr2="http://schemas.microsoft.com/office/spreadsheetml/2015/revision2"
  xmlns:xr3="http://schemas.microsoft.com/office/spreadsheetml/2016/revision3"
  mc:Ignorable="x14ac xr xr2 xr3" xr:uid="{7CAC10C3-3F8E-42B6-8844-6D2BDBB35CBF}">
```

FIGURE 25.43.   Examining namespace declarations in the sheet1.xml document.

Once we have located the node with the specified attribute, we set the node's text to the new value:

```
myNode.Text = "0.50"
```

The remaining part of the procedure locates the `<pageSetup>` element inside the `sheet1.xml` part and uses the `RemoveChild` method to remove this node:

```
Set myNode = xmlDoc.SelectSingleNode("//x14ac:pageSetup")
On Error Resume Next
myNode.ParentNode.RemoveChild myNode
```

In the case that the node is not found (for example, you may have mistyped the element name in the XPath expression), `On Error Resume Next` will skip over the node removal statement. The last statement will save the changes made in the `sheet1.xml` part.

2. Run the `ChangeLeftMargin_RemovePageSetup` procedure.
3. After running the procedure, open the `C:\VBAExcel2024_ByExample\ZipPackage\xl\worksheets\sheet1.xml` file in your browser and verify that the left margin is now set to `0.50` and the file no longer contains the `<pageSetup>` element (see Figure 25.44).

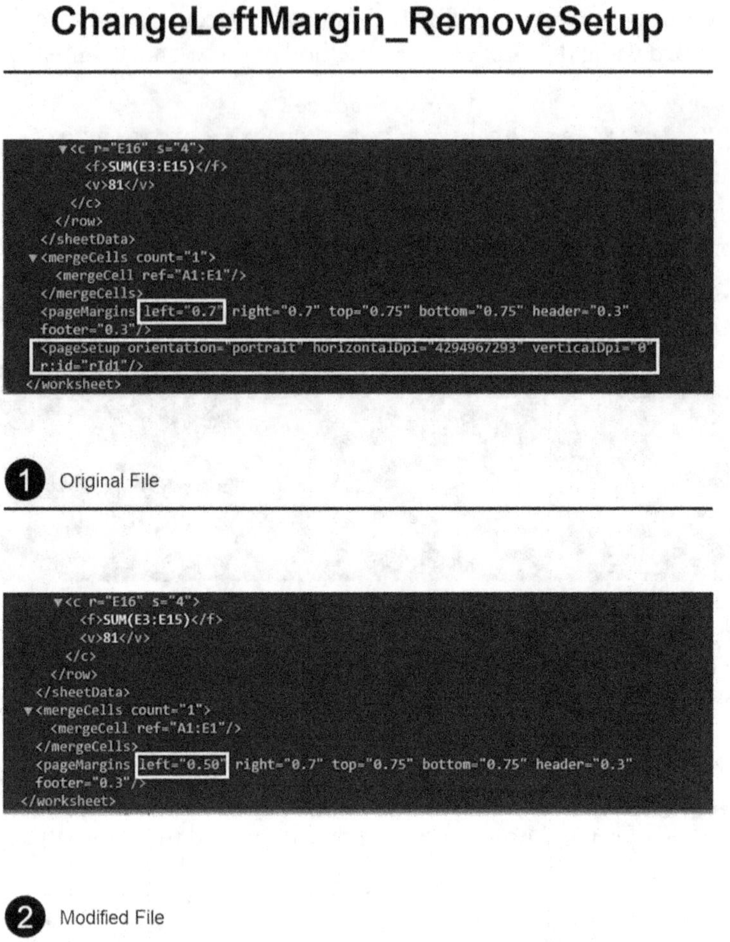

ChangeLeftMargin_RemoveSetup

```
▼<c r="E16" s="4">
    <f>SUM(E3:E15)</f>
    <v>81</v>
  </c>
  </row>
</sheetData>
▼<mergeCells count="1">
  <mergeCell ref="A1:E1"/>
</mergeCells>
<pageMargins left="0.7" right="0.7" top="0.75" bottom="0.75" header="0.3"
footer="0.3"/>
<pageSetup orientation="portrait" horizontalDpi="4294967293" verticalDpi="0"
r:id="rId1"/>
</worksheet>
```

❶ Original File

```
▼<c r="E16" s="4">
    <f>SUM(E3:E15)</f>
    <v>81</v>
  </c>
  </row>
</sheetData>
▼<mergeCells count="1">
  <mergeCell ref="A1:E1"/>
</mergeCells>
<pageMargins left="0.50" right="0.7" top="0.75" bottom="0.75" header="0.3"
footer="0.3"/>
</worksheet>
```

❷ Modified File

FIGURE 25.44. The content of Sheet1.xml before and after modification by the ChangeLeftMargin_ RemoveSetup procedure.

4. Close the `sheet1.xml` file.
5. Delete the `ZipPackage` folder from the `VBAExcel2024_ByExample` folder.
6. Save and close the `ManipulateXMLParts.xlsm` workbook.

SUMMARY

This chapter explored working with HTML and XML in Excel VBA. You learned how to insert hyperlinks in worksheet cells to access Web sites and used the

`FollowHyperlink` method to find Web pages based on keywords. Additionally, you discovered how to create and publish HTML files by building HTML documents in code using data from Excel ranges, while gaining familiarity with essential HTML tags.

The chapter delved into XML, highlighting its flexibility compared to HTML. You studied XML's structure and learned how it allows you to define custom tags to describe data. Practical applications included viewing and editing XML documents in Excel, and using XML maps, tables, and schemas, including programming these features with VBA. You also worked with the DOM to write VBA procedures to find and retrieve information from element nodes. You learned how to save an ADO recordset to disk as an XML file. You explored namespaces and read and manipulated Open XML files using VBA.

In the next chapter, you will expand your VBA expertise by working with a VBA `Dictionary` object, using regular expressions, and calling a Web service known as the REST API. The latter will require some of the XML skills you've acquired in this chapter.

Chapter **26** ***EXCEL AND THE REST API***

We've almost reached the end of this book. In this last chapter, we will focus on expanding your VBA skill set by covering topics such as working with a VBA `Dictionary` object, using regular expressions, and calling a new type of Web service known as the REST API.

INTRODUCTION TO THE VBA DICTIONARY OBJECT

After you've learned how arrays and collections can help you store values while your program is running, you may be surprised to learn that there is still another way to manipulate your data. An object known as the *VBA dictionary* operates like a collection object but offers more flexibility and speed. It works in a similar way to a normal dictionary, allowing you to look up values based on a key you provide. A dictionary is a collection of key-value pairs, and each key must be unique. Unlike an array, its size does not need to be predefined. Its data type is `Variant` so you can enter any type of data you want to keep track of (text, numbers, dates, arrays, or objects).

The dictionary object is not part of standard VBA. It is a part of the Microsoft Scripting Runtime library (`scrrun.dll`).

Accessing the VBA Dictionary

There are two ways in which you can access the VBA dictionary in your VBA code. One is referred to as *early binding* and the other is called *late binding*. You should already be familiar with these terms from previous chapters:

- *Early binding*—To use early binding, you need to add a reference to the *Microsoft Scripting Runtime* library, as described in the next section. With early binding, your code is compiled before it runs, so your procedures can run much faster. Once the reference is added to the required library, you can write the following code to define your `Dictionary` object:

```
Dim myDict As New Scripting.Dictionary
```

With early binding, you can count on *Intellisense* to help you with syntax and programming assistance.

- *Late binding*—With late binding, your object will be compiled while your code runs, therefore, it will be slower. You are not required to set up a library reference. To define the `Dictionary` object, specify `Scripting.Dictionary` using the `CreateObject` method, as shown here:

```
Dim myDict As Object
Set myDict = CreateObject("Scripting.Dictionary")
```

Adding a Reference to the Microsoft Scripting Runtime Library

You add the reference to the Microsoft Scripting Runtime library in the same way you added references to other libraries that were introduced in this book. Select Tools | References in the VBE screen and scroll down in the References window until you locate the library. Select it and click OK. Figure 26.1 shows this selection.

After you've added the library reference, it's a good idea to examine the content of this library by using the Object Browser. In the VBE screen, press F2 or choose View | Object Browser. Choose Scripting in the first dropdown, as shown in Figure 26.2. Notice the available types of objects, one of them being Dictionary. By clicking on each `Dictionary` member, you can find out the type of the member (method, property) and its function. For example, the `Item` property is the default member of `Scripting.Dictionary` and is used to set or get the item for a given key, while `Add` is a method used to add a new key and item to the dictionary.

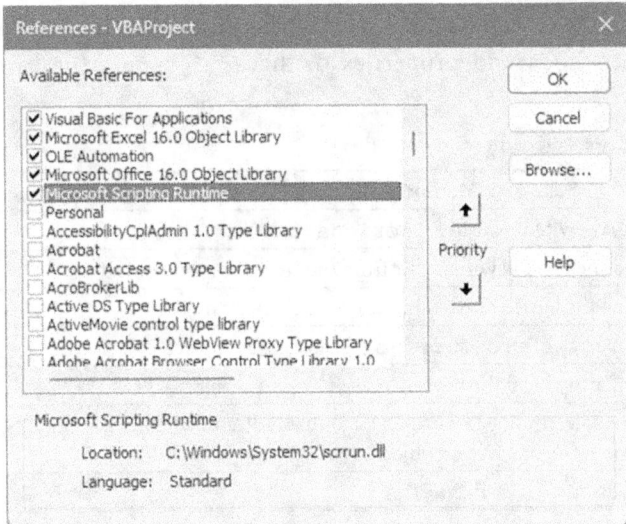

FIGURE 26.1.   Adding a reference to the Microsoft Scripting Runtime library.

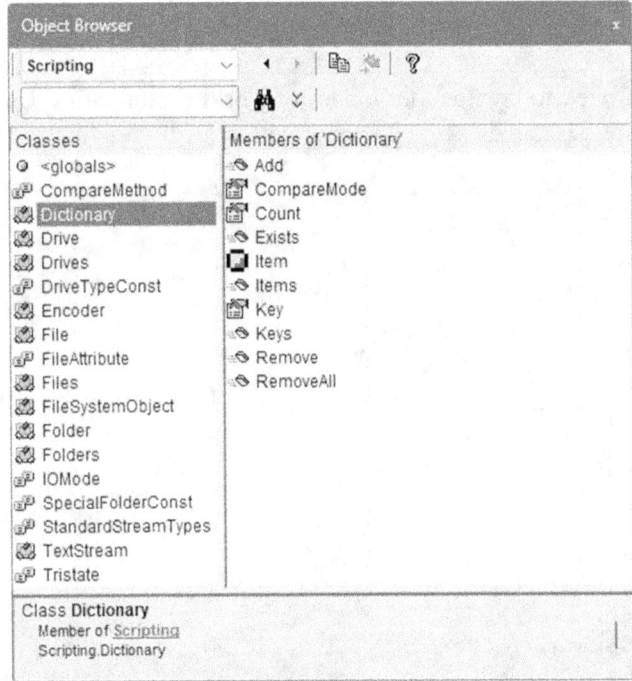

FIGURE 26.2.   Examining the content of the Microsoft Scripting Runtime library.

Working with the Dictionary Object's Properties and Methods

Table 26.1 lists the methods and properties of the `Scripting.Dictionary` object.

TABLE 26.1. The Dictionary object's properties and methods.

Methods	
Add	Adds a new key/item pair to a `Dictionary` object
Remove	Removes a specified key/item pair from the `Dictionary` object
RemoveAll	Removes all the key/item pairs in the `Dictionary` object
Items	Returns an array of all the items in a `Dictionary` object
Keys	Returns an array of all the keys in a `Dictionary` object
Exists	Returns a Boolean value (`true`/`false`) that indicates whether a specified key exists in a `Dictionary` object
Properties	
Key	Sets a new key value for an existing key value in the `Dictionary` object
Item	Sets or returns the value of an item in a `Dictionary` object
Count	Returns the number of key/value pairs in a `Dictionary` object
CompareMode	Sets or returns the comparison mode for comparing keys in a `Dictionary` object

Let's assume that you need to create a dictionary of world capital cities. Using the `Add` method of the `Scripting.Dictionary`, you can fill your dictionary like this:

```
Sub FillDictionary()
    ' declare a dictionary object
    ' uses early binding
    ' requires the reference to the Microsoft
    ' Scripting Runtime Library
    Dim objDict As New Scripting.Dictionary

    'add items to the dictionary
    objDict("USA") = "Washington D.C."
    objDict("Canada") = "Ottawa"
    objDict("France") = "Paris"
    objDict("England") = "London"
    objDict("Hungary") = "Budapest"
    objDict("Italy") = "Rome"
    objDict("Japan") = "Tokyo"
    objDict.Add "Germany", "Berlin"
    objDict.Add Key:="China", Item:="Beijing"
End Sub
```

In the code snippet above, you will notice three different ways to add an item to a dictionary. Use whatever method feels more comfortable to you.

After filling in the dictionary, you will want to read its keys/values. You can use the Keys method to return an array of all the keys like this:

```
' iterate through the dictionary to read its keys
Dim i As Long
For i = LBound(objDict.Keys) To UBound(objDict.Keys)
    Debug.Print i & "--> " & objDict.Keys(i)
Next i
```

In this code, we utilize the upper and lower bound methods (UBound and LBound, respectively) of an array to list the keys. The following statement retrieves a specific key:

```
' retrieve 3rd key
Debug.Print "3rd key=" & objDict.Keys()(2)
```

Recall that arrays are zero-based. You can place all keys into an array like this:

```
Dim aKeys() As Variant
aKeys = objDict.Keys
Debug.Print "Dictionary contains " & UBound(aKeys) + 1 & " keys."
Debug.Print "The first key is " & aKeys(0)
```

It is easy to remove a key from a dictionary, but before removal, always check whether the key exists using the Exists method. This method is what makes the Dictionary object easier to manipulate than collections and arrays. The keys are case-sensitive, so pay attention to the case when checking for the key existence:

```
' remove the key if it exists
If objDict.Exists("France") Then
    objDict.Remove "France"
Else
    Debug.Print "This key does not exist"
End If
```

The Count property returns the total items in the dictionary and the Item property is used to set the value of a dictionary item. In the following example, we replace the value for Germany:

```
' count the items in the dictionary
Debug.Print "Now the dictionary contains " & objDict.Count & " keys."
objDict.item("Germany") = "Frankfurt"
Debug.Print objDict("Germany") & " is a city in Germany."
```

Using the `For Each` loop, you can list all the items or keys in the dictionary, like this:

```
' list all the items in the dictionary
Dim item As Variant
For Each item In objDict.Items
    Debug.Print item
Next
' List all keys in the dictionary
Dim key As Variant
For Each key In objDict.Keys
    Debug.Print key & " --> " & objDict.item(key)
Next
```

To clear your dictionary of all items, use the `RemoveAll` method:

```
' remove all key/item pairs
objDict.RemoveAll
' verify that dictionary is empty
Debug.Print objDict.Count
End Sub
```

> **NOTE** *The complete* `FillDictionary` *procedure can be found in the* `Chap26.xlsm` *workbook in the companion files.*

Recall that key searches are case-sensitive by default. You can, however, use the `CompareMode` property to change this behavior. The mode change must be specified right after the creation of the `Dictionary` object, before adding any data to the dictionary. Let's look at some code:

```
Sub CaseNotSensitive()
    Dim objDict As New Scripting.Dictionary
    objDict.CompareMode = TextCompare
    objDict.Add "Math", 89
    objDict.Add "English", 70
    objDict.Add "Chemistry", 90
    Debug.Print objDict.Exists("math")
End Sub
```

In this example, we set `CompareMode` to `TextCompare` to allow for searches that are not case-sensitive. So instead of typing `Math` for the key name, we can simply enter `math` in lowercase and the `Exists` method is able to locate the key. The `CompareMode` can also be set to two other settings: `BinaryCompare` and `DatabaseCompare`. `BinaryCompare` performs binary comparisons; it is case-sensitive, so `EventName` is not the same as `eventName` or `eventname`. The `DatabaseCompare` performs a comparison based on information in the database.

The latter only exists in Microsoft Access (it's not available in Excel and other VBA-capable Office applications).

Dictionary Versus Collection

Table 26.2 shows several advantages of the `Dictionary` object over the native VBA `Collection` object. In your programming endeavors, you should always pick the object that is most suitable for a particular task.

TABLE 26.2.   Dictionary object versus Collection object.

Dictionary	Collection
Retrieving items in a dictionary is faster than in a collection.	Retrieving items is slower than in a dictionary.
It is easier to search for a given item.	It is harder to search for a given item.
The `Item` value can be changed directly.	The `Item` value must be first removed and then the changed item can be added back.
It uses keys to locate a particular item; keys can be checked for existence.	The keys are used to look up data but cannot be retrieved. Uses index values, which are harder to work with.
It offers `CompareMode` for changing case sensitivity.	Collections are case-sensitive, and the sensitivity cannot be changed.
Key values can be any data type.	Key values must be strings.
Items can be easily removed using the `Re-moveAll` method.	Removing items from a collection requires re-defining the `Collection` object.
It cannot store a reference to a custom collection.	It can store a reference to a custom collection.
It requires a reference to the Microsoft Scripting Runtime library or an object created using late binding.	`Collection` is an object available in the VBA library.

INTRODUCTION TO REGULAR EXPRESSIONS

As you know, VBA has many useful functions that apply to strings. You can use the `Len` function to return the number of characters in a string; for example, `?Len("Today is Sunday")` returns the number 15.

You can use the `Left`, `Right`, and `Mid` functions to return a portion of a string. For example, `?Left("Today is Sunday", 5)` returns the string `Today`. Replace `Left` with `Right`, and you return the five rightmost characters from that string. If you need to return a string starting from a particular character in the string, you can use the `Mid` function to extract the required portion of the string. For example, let's extract `is` from our test string:

```
?Mid("Today is Sunday", 7, 2)
```

This returns the following string: `is`.

If you instead want to find the position of the first occurrence of one string within another string, you can call the `InStr` function like this:

```
?InStr("Today is Sunday and it's a holiday", "Sunday")
```

This will display the number `10`, indicating that the string was found starting at character 10. If you don't need to search from the beginning of the string, specify the position to search from in the first argument, like this:

```
?InStr(5, "Today is Sunday and it's a holiday", "o")
```

This searches for the letter `o`, starting from the fifth character in the specified string. The letter `o` is found in the 29th position.

As you can see, using string manipulation functions in VBA is quite straightforward; however, there will be many situations in your VBA programs where these methods will not be enough. You may need to perform complex text searches, replacements, and validations and match a certain pattern of characters in various expressions. A pattern match can involve a character, a word, a group of words, or an entire sentence. This is where a working knowledge of regular expressions can come to your rescue. Simply put, a *regular expression* (often referred to as *regex* or *RegExp*) is a sequence of characters that specifies a search pattern in text. Regular expressions are commonly used in search engines, Find/Replace dialogs in text editors, and word processing applications. Integrating regular expressions with Excel VBA allows for efficient data cleansing, validation, and extraction tasks. Before you try out some pattern matching in VBA, let's go over a few basic concepts.

Character Matching in RegExp Patterns

A *pattern* is basically a schema that you put together to match character combinations in strings. Patterns are composed of simple characters, such as `/abc/`, or a combination of simple and special characters, such as `/ab*c/`. The latter indicates that we want to match a single `a` followed by zero or more `b` characters followed by `c`. The `*` character after `b` means zero or more occurrences of the preceding item. Table 26.3 shows some examples of special characters used in regular expression patterns. For more detailed information, you will need to look at other sources. This section supplies just a bare minimum of the regex knowledge for you to complete a specific VBA programming task, which will be introduced later in this chapter.

TABLE 26.3. Special characters used in regular expression patterns.

Pattern	Description	Example	Found Matches
. (a single dot)	Matches any single character except vbNewLine	d.g	deg, dig, dog
[characters]	Matches any single character between brackets [ ]	[jv]	Would match j and v in java
[^characters]	Matches any single character that is not between brackets [ ]	[^jv]	Would match a in java
[start-end]	Matches any character that is part of the range in brackets [ ]	[0-9] [A-Z]	Matches any number in the range 0 to 9 Matches any character in the range A to Z
\	Escapes special characters so that special characters can be searched for	\[	Escaped character. Matches a [character.
\n	New line	\n	Matches a new line (vbNew-Line)
\r	Carriage return	\r	Matches a carriage return (vbCr)
\t	Tab	\t	Matches a tab character (vbTab)
\w	Matches any word character alphanumeric and underscore.	\w	Would match morning in morning
\W	Matches any non-alphanumeric characters and the underscore	\W	Would match @ in your email address
\s	Matches any whitespace character (spaces, tabs, line breaks)	\s	Would match the space in good morning
\S	Matches any non-whitespace character	\S	Would match good and morning in good morning
\d	Matches any decimal digit	\d	Would match 7 in 7Eleven
\D	Matches any character that is not a digit character (0–9)	\D	Would match Eleven in 7Eleven

Quantifiers in RegExp Patterns

Regular expression patterns can include quantifiers that allow you to control how many times a match occurs (see Table 26.4).

TABLE 26.4. Quantifiers used in regular expression patterns.

Quantifier	Description	Example	Found Matches
*	Matches zero or more of the preceding characters/digits	b\w*	Matches a b character. Matches any word character (alphanumeric and under-score). Matches zero or more of the preceding items.
+	Matches one or more of the preceding characters/digits	b\w+	Same as above, but matches one or more
?	Matches zero or one	colou?r	Matches color, colour
{n}	Matches "n" many times	c{2}	Matches a c character twice. Case-sensitive. Will match the second and third letter c in Cecilia is sick.
{n,}	Matches at least "n" occur-rences of the preceding item	a{3,}	Matches all of the a characters in baaarber and baaaar-ber
{n,m}	Matches the specified quantity of the previous character/digit	a{1,3}	Will match one to three of the previous items. It will match the a in barber, the two a's in baarber, and the three a's in baaarber.

You can use parentheses, (), to group multiple items in your pattern together. You can also indicate the beginning and ending of lines and words and other patterns indicating that a match is possible. These topics are beyond the scope of this book. Numerous books have been devoted to the subject of regular expression pattern-matching operations. Also, hundreds of Web sites offer useful tools that will help you decipher an unknown regex pattern and help you understand its building elements. While regex can take some time to learn, once mastered, these skills can be reused in many other programming languages for validating, replacing, and extracting data from strings.

Using the RegExp Object in VBA

Many programming languages provide built-in support for working with regexes. To use regexes in your VBA programs and take advantage of the built-in programming assistance (Intellisense), you will need to add a reference to an external library. In your VBE screen, choose Tools | References, scroll down in the available references list, select Microsoft VBScript Regular Expressions 5.5

(see Figure 26.3), then click OK. Once you've added this library to your VBA project, the Object Browser will display all available properties and methods of the `RegExp` object (see Figure 26.4).

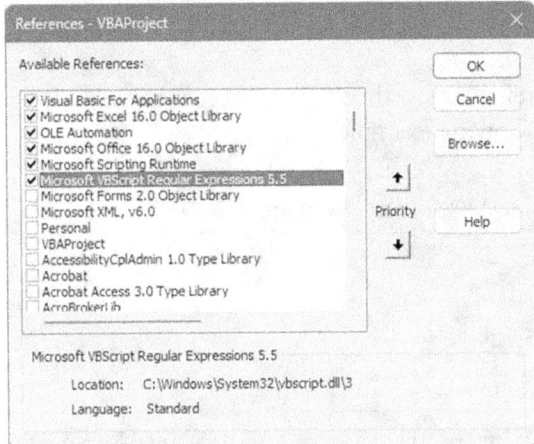

FIGURE 26.3.   Setting a reference to the Microsoft VBScript Regular Expressions 5.5 object library

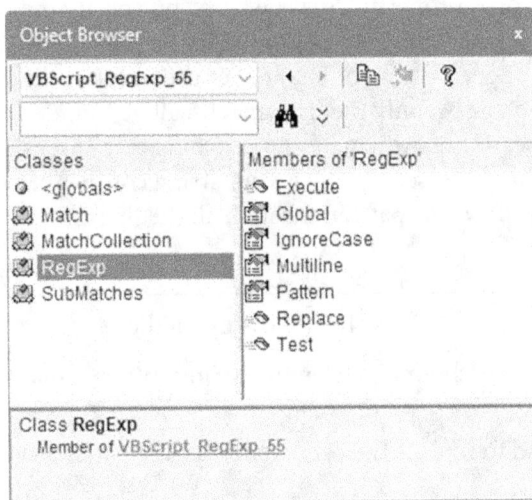

FIGURE 26.4.   Regular expressions can be used via the VBScript Regular Expressions 5.5 library displayed as VBScript_RegExp_55 in the Object Browser.

The RegExp Object Declaration

Regular expressions in VBA are implemented through the `RegExp` object, which provides methods and properties for defining and using regex patterns.

After setting a reference to the Microsoft VBScript Regular Expressions 5.5 library, you can declare the `RegExp` object using the following early-binding syntax:

```
Dim oRegExp As RegExp
Set oRegExp = New RegExp
```

If you don't want to add the reference to the `RegExp` object via the References dialog box, use either of the following examples:

```
Dim oRegExp As Object
Set oRegExp = CreateObject("VBScript.RegExp")
```

or

```
Dim oRegExp As Object
Set oRegExp = New VBScript_RegExp_55.RegExp
```

The RegExp Object's Properties

The `RegExp` object offers four properties that are used to specify the pattern to be matched and various options that should be turned on or off. The most important property is the `Pattern` property. This is where you specify the pattern that you are going to use for matching against the string.

The `Global` property can be set to `True` or `False`. Set it to `True` to find all matches in the pattern. If set to `False`, only the first match will be found.

The `IgnoreCase` property set to `True` will make the matching case-insensitive.

The `Multiline` property should be set to `True` if your string consists of multiple lines and you want to run the same pattern through all the lines.

The RegExp Object's Methods

To work with the `RegExp` object, you can use the following methods:

- Use the `Test` method to search for a pattern in a string. When a match is found, `True` is returned.
- Use the `Replace` method to replace the occurrences of the pattern with a replacement string.
- Use the `Execute` method to return all the matches of the pattern that were found in the provided string.

Writing VBA Procedures Using the RegExp Object

Let's create example procedures that use some of the methods and properties of the `RegExp` object. The first procedure will use the `Test` method to check whether the supplied text string contains a correctly formatted phone number. Assume we want to ensure that phone numbers are in the following US format: (999) 999-9999.

Please note that all the code files and figures for the hands-on projects can be found in the companion files.

⊙ Hands-On 26.1   Testing for a Pattern Match

1. Copy the `Chap26.xlsm` workbook from the companion files to your `C:\ VBAExcel2024_ByExample` folder.
2. In the VBE screen, add a new module and rename it `RegExpressions`.
3. Enter the following VBA procedure and then run it:

```
Sub RegExp_TestDemo()
    Dim oRegExp As RegExp
    Dim strToSearch As String
    Set oRegExp = New RegExp
    strToSearch = "Customer phone number: (201) 234-7899."
    ' match US Phone numbers in the format: (999) 999-9999
    oRegExp.Pattern = "\(\d{3}\) \d{3}-\d{4}"
    MsgBox oRegExp.Test(strToSearch)
End Sub
```

When this procedure executes, you should see True in the message box.

4. Remove the parentheses surrounding the area code and run the procedure again.

The phone number will no longer match the pattern we defined, so the `Test` method will return `False`.

Let's look at the pattern expression we defined for the phone: `\(\d{3}\) \d{3}-\d{4}`

`\ (`	Escaped character. Matches a "(" character.
`\d`	Matches any digit character (0–9).
`{3}`	Quantifier. Matches three of the preceding digits.
`\ )`	Escaped character. Matches a ")" character.
	Matches a space character (empty space).
`\d`	Matches any digit character (0–9).
`-`	Dash character. Matches a "-" character.
`{4}`	Quantifier. Matches four of the preceding digits.

Let's write a procedure that will search for all occurrences of characters provided in the pattern and replace them with nothing, meaning that they will be removed from the search string.

Hands-On 26.2 Replacing All Occurrences of the Pattern Match

1. In the VBE window, enter the code of the procedure, as shown in Figure 26.5.

```
Sub RegExp_ReplaceDemo()
    Dim oRegExp As RegExp
    Set oRegExp = New RegExp

    Dim strToSearch As String
    strToSearch = "{""userId"":1,""Title"":""Account Manager"",""TaskCompleted"":false}"

    'Find all matches in the pattern
    oRegExp.Global = True

    'Match any character in this set
    oRegExp.Pattern = "[\{\}""]"
    MsgBox oRegExp.Replace(strToSearch, "")
End Sub
```

FIGURE 26.5. This VBA procedure uses the Replace method of the RegExp object to remove characters specified in the pattern from the searched string.

2. Execute the `RegExp_ReplaceDemo` procedure.
Figure 26.6 displays the message box with the resulting string.

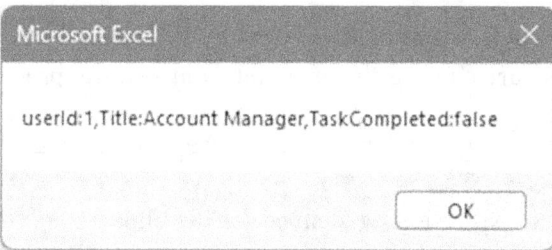

Microsoft Excel ✕

userId:1,Title:Account Manager,TaskCompleted:false

OK

FIGURE 26.6. The Replace method of the RegExp object has successfully removed all occurrences of { } "" characters from the original string.

Note that prior to defining the pattern, we used the `Global` property of the `RegExp` object to ensure that all the characters specified in the pattern are removed, not just their first occurrence.

3. Comment out the line of code that sets the `Global` property to `True` and run the procedure again.
This time, the message box should display a partially cleaned-up string; only the first occurrence of each character in the pattern was removed, giving you the following text string:

```
"userId":1,"title":"Account Manager","TaskCompleted":false}
```

Finally, let's create a procedure that uses the `Execute` method of the `RegExp` object to display all the matches for the pattern that was found in the search string. We will use the same pattern and search string as in the previous procedure.

⊙ Hands-On 26.3   Obtaining a List of All the Matches in the Pattern

1. Create a copy of the procedure from Hands-On 26.2 and rename it `RegExp_ExecuteDemo`.

2. Add the following two declaration statements:

```
Dim matches As Object
Dim itm As Variant
```

3. Uncomment the statement that sets the `Global` property to `True`.

4. Comment out the statement that displays the message box. We will not be making any replacements in this procedure.

5. Add the following code before the `End Stub` statement:

```
Set matches = oRegExp.Execute(strToSearch)
For Each itm In matches
    Debug.Print itm
Next
```

6. Run the completed `RegExp_ExecuteDemo` procedure.
   After running the procedure, you should see all the matches that were found, listed on separate lines in the Immediate window.

Now, you've learned the basics of using the `RegExp` object in your VBA procedures. We will revisit this object in a later Hands-On project when we need to locate and extract specific strings from a JSON response obtained from an external resource. If you've never heard of JSON, read on. The next section introduces you to communicating with Web servers: sending requests and processing responses.

INTRODUCTION TO THE REST API

Over the past few years, the method that developers use to connect to external resources and share information between various computer systems frequently includes a mysterious initialism: API. This vastly popular term is formed from the initial letters of *Application Programming Interface*.

APIs allow programs and scripts to communicate with each other. These programming interfaces expose certain data, services, and functionality of an application so other developers can use them. This allows one product to interact with other products. In other words, the APIs are specifically built to be consumed by another application programmatically. APIs allow you to automate many tasks and create user-friendly dashboards and client applications both for mobile and Web use. Furthermore, they allow you to extend your product functionality by grabbing the required resources from somewhere else. Basically, APIs make things easier. Think of how many times you have seen a Google map embedded in some Web site. That map came from the Google Maps API! Many popular Web sites, such as YouTube, X/Twitter, and Meta (Facebook), as well as a multitude of commercial and government Web sites, provide APIs that allow you to get and update their data. To use these APIs, you don't need to know how they were created internally. All you need is to get acquainted with their documentation and find out how to ask for what you need and how to process the response. Some APIs are free, others require that you pay for the service. Some will ask you to create a developer account to obtain an API key for authentication purposes. This key consists of a set of letters and numbers that uniquely identify you to the application. An API key is like a password; it is important to keep it secure.

There are different types of APIs. The one that everyone is using and talking about right now is called Representational State Transfer (REST). The REST API, also referred to as a RESTful API, was created by computer scientist Roy Fielding. REST is a set of rules (also known as an architectural style) that developers must follow to create programs on a server that allow communication with various client applications. In a RESTful system, a client application sends a request to the server, usually over the HTTP protocol. This request might be to fetch data (GET request), alter the state of the data (PUT request), create data (POST request), delete some data (DELETE request), or modify some details about the resource (PATCH request). POST, GET, PUT, and DELETE are special verbs that specify a CRUD (Create, Read, Update, Delete) action that needs to be performed on the server. After the server completes the action, it sends a response back to the client application, often in the form of a representation of the requested resource. A very important fact in the REST API is that all client requests are stateless. This means that each request must contain all the necessary information for the server to process the request. In other words, the client cannot depend on the server to remember prior requests. Server responses can be formatted in plain text/HTML, XML, or JSON.

In this section, you will learn how to use VBA to implement a GET method to fetch data from some free APIs that you can access without applying for a developer account or using authentication. There is a lot to know about the REST API and, unfortunately, there isn't enough room in this chapter to cover all the methods for using it. Even the simple GET method can get quite complex when you need to pass parameters to the service and authenticate yourself. Thus, for simplicity's sake, we will focus on the basic syntax, which should provide you with enough understanding of the topic so you are ready to dive deeper into it the next time you encounter it.

Accessing REST APIs with VBA

VBA does not have a special method for accessing REST APIs. There is, however, a special XMLHTTPRequest object that allows you to use external API from your VBA code. You must declare the XMLHttpRequest object by using early or late binding.

As mentioned earlier in this chapter, to use early binding, you must set up a reference to an external library. In this case, the library you need is called *Microsoft XML, v6.0*. This is the same library we used in Chapter 25 when we were learning about using XML with Excel. Figure 26.7 shows the selection of this library in the References dialog box. You will need to scroll down in the Available References list to find it.

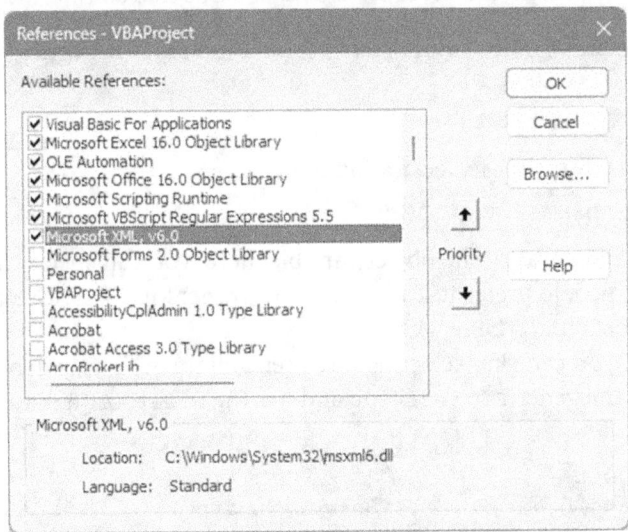

FIGURE 26.7. Activating the Microsoft XML, v.60 object library.

Once the library is selected in the References dialog box, you can view the object's properties and methods (see Figure 26.8) using the Object Browser (View | Object Browser, or press F2).

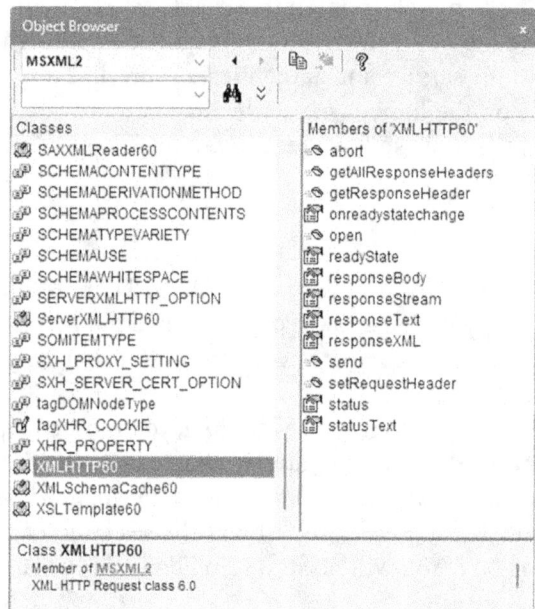

FIGURE 26.8. Exploring the Microsoft XML, v.60 object library using the Object Browser.

To use the XMLHTTPRequest object, you must declare it in your VBA code:

```
Dim httpReq As MSXML2.XMLHTTP60
Set httpReq = New MSXML2.XMLHTTP60
```

Alternatively, put everything in the declaration line:

```
Dim httpReq As New MSXML2.XMLHTTP60
```

Note that httpReq is the name of the object variable; here, you can specify any name you like. Some people prefer to call it xmlhttp; others use xhr. Use the name that you feel most comfortable with.

With late binding, you can skip selecting a reference library. Simply declare the object variable of the generic Object type and use the CreateObject function like this:

```
Dim httpReq As Object
Set httpReq = CreateObject("MSXML2.ServerXMLHttp")
```

In this chapter's examples, we will be using early binding so we can rely on the built-in programming assistance while writing our code.

Methods and Properties of the XMLHTTPRequest Object

Let's look at the methods and properties that we need to be familiar with to make successful requests to any external API.

The most important methods are `Open` and `Send`. Use the `Open` method to initialize a request. You will need to provide the two required arguments. The first argument is a method, which can be either `GET`, `POST`, or `PUT`. The second argument is the URL of the resource you are calling. You may optionally pass a `Boolean` value (`True`/`False`) to indicate whether the API call is meant to be asynchronous (`True` is the default) or synchronous (`False`). If you pass `False`, processing waits until the response is returned from the server.

Using the `httpReq` object variable defined earlier, here is how you would set up a basic `GET` request:

```
httpReq.Open "GET","strURL ", False
```

Here, `strURL` is the URL address of the Web server you want to access.

The actual request to the server is made using the `Send` method. If the request was declared asynchronous (`True`), then this method returns immediately; otherwise, it waits until the response is received. You can pass additional optional arguments with the `Send` method. The simple syntax of this method is as follows:

```
httpReq.send
```

The `Send` method will send the request to the resource you indicated in the second parameter of the `Open` method.

The `Abort` method is handy for aborting the current request. You will use it when you want to stop the request.

Any problems and issues with an API can often be resolved by examining API headers, and we have three methods to deal with headers: `setRequestHeader`, `getResponseHeader`, and `getAllResponseHeaders`. Headers provide extra information about each API call and response. The most common API headers are listed in Table 26.5.

TABLE 26.5. Common API headers.

API Header	Description
`Authorization`	This header contains the authentication credentials for HTTP authentication.
`WWW-Authenticate`	The server may send this header if it needs some form of authentication before sending the response for the requested resource. It may include an error code `401`, which means unauthorized.

(Contd.)

API Header	Description
Accept_Charset	The client may send this header with a request to let the server know which character sets (such as UTF-8, ISO-8859-1, and Windows-1251) are acceptable by the client.
Content-Type	This header tells the client what media type (e.g., `application/xml` and `application/json`) the response is sent in. This helps the client to correctly process the response received from the server.
Cache-Control	This header contains caching directives (instructions) defined by the server for the response. These directives determine how a resource is cached, where it's cached, and its maximum age before expiring. The `no-cache` entry in the `Cache-Control` header indicates that returned responses can't be used for subsequent requests to the same URL before checking whether server responses have changed.

To get all the response headers from the HTTP request, use the `getAllResponseHeaders` method. This method must be used after you've set up the `Send` method to send the request. For example,

```
httpReq.Send
Debug.Print httpReq.getAllResponseHeaders
```

The second statement above should print to the Immediate window the string containing response headers. What you get in the response headers depends on the server. Here is an example of the headers obtained from the resource we'll be querying in Hands-On 26.4:

```
Cache-Control: no-cache
Connection: keep-alive
Date: Mon, 10 Mar 2025 03:36:30 GMT
Pragma: no-cache
Content-Length: 1163241
Content-Type: application/xml
Expires: -1
Server: Microsoft-IIS/10.0
Access-Control-Expose-Headers: Request-Context
Request-Context: appId=cid-v1:39f1cd0a-de7e-435f-bf7d-
39de930d88c6
X-AspNetmvc-Version: 5.2.9
Strict-Transport-Security: max-age=31536000 ; includeSubDomains
; preload
X-Content-Type-Options: nosniff
X-Frame-Options: SAMEORIGIN
```

As you can see from this example, the headers are key/value pairs in text format separated by a colon.

To find out a specific header value you want, use the `getResponseHeader` method, providing it with the argument denoting the specific header value you want. For example, to get the value of the `content-type` header, use the following line of code:

```
Debug.Print HttpReq.getResponseHeader("content-type")
```

The `setRequestHeader` method is used by the client to provide information to the server about the types of content that are acceptable for the response, the acceptable character sets, the list of acceptable encoding, etc. For a full list of standard request fields, see the article at *https://en.wikipedia.org/wiki/List_of_ HTTP_header_fields#Requests*.

It is the responsibility of the server to consider the sent-in client requirements. The `setRequestHeader` method has two required arguments that specify the name of the header and its value. You must call it after calling the `Open` method, but before calling `Send`. Here is an example:

```
httpReq.Open "GET", strURL, False
httpReq.setRequestHeader("Accept", "text/xml")
httpReq.Send
```

The above example will tell the server that the client is looking for a response in `text/xml` format.

NOTE	*Before sending a request to the server, you may want to query the server to find out whether the server is operational and what server resources are available. By sending a* HEAD *request instead of* GET *or* POST, *the server will send back its response headers. Here is a short VBA procedure that demonstrates how to query the server:*
	
	`Sub GetOnlyHeaders()`
	`Dim xhr As New MSXML2.XMLHTTP60`
	`xhr.Open "HEAD", "https://itunes.apple.com/" & _`
	`    "search?term=celine+dion", False`
	`xhr.send`
	`  If xhr.ReadyState = 4 Then`
	`      Debug.Print xhr.getAllResponseHeaders`
	`  End If`
	`End Sub`

With the `XMLHTTPRequest` object's methods covered, let's look at its properties.

The `readyState` property specifies the state of the request. There are five possible values:

- `0` means uninitialized
- `1` means loading
- `2` means loaded
- `3` means interactive
- `4` means complete

Every time the `readyState` changes, the `onreadystatechange` event is fired. When `readyState` is `4` and status is `200`, the response is ready.

The `status` property returns the HTTP status code from the server. The most common status is `200`, which means OK, signifying that the request was successful. The `403` status means that the access is forbidden. The `404` status denotes that the requested page/resource was not found. For a complete list of statuses, go to https://en.wikipedia.org/wiki/List_of_HTTP_status_codes.

The `statusText` property returns the text version of the HTTP status code. It is useful for programming error messages. You will see the example of using `status` and `statusText` in Hands-On 26.4.

Finally, the last four properties (`responseText`, `responseXML`, `response-Stream`, and `responseBody`) specify the form in which the HTTP response can be returned. Use the `responseXML` property if you know that the response body is an XML formatted text. If you need the response body as a string, use the `responseText` property. The `responseStream` property will return the server response in the form of binary-encoded data (such as UTF-8, UCS-2, UCS-3, and Shift_JIS). The `responseBody` property is useful when the response body is binary. This property returns the response content as a byte stream.

Making a Basic GET Request

You may be wondering how exactly you can put this newfound knowledge to use. Here again, the best learning occurs from working through an example.

In the first Hands-On example of calling the REST API, we will access the National Highway Traffic Safety Administration (NHTSA) Product Information Catalog and Vehicle Listing (vPIC) API. The following is a direct link: *https://vpic.nhtsa.dot.gov/api/*.

It describes API methods that can be used to gather information on vehicles and their specifications in various types of data formats (XML, CSV, and JSON). According to the NHTSA documentation, the vPIC dataset is populated using the information submitted by the motor vehicle manufacturers. We will get a list of all the makes available in the vPIC dataset. We will get this list in XML format and will push it into an Excel worksheet. Let's get started.

Hands-On 26.4 Requesting Data from the REST API

1. In the VBE screen, insert a new standard module to your Chap26.xlsm workbook. Use the Properties box to rename the module APIRequest_XML.
2. In the APIRequest_XML module Code window, enter the following procedure:

```
' set up a reference to the following Library
' --->Microsoft XML, v6.0
' using the References dialog box

Sub RequestData_XML_toExcel()

Dim URL As String
Dim httpReq As Object
Dim Resp As New MSXML2.DOMDocument60
Dim wkb As Workbook
Dim ws As Worksheet
Dim Header As Boolean
Dim n As Integer
Dim i As Integer
Dim c As Integer
Dim startTime As Single
Dim endTime As Single
Dim strExcelFile As String

On Error GoTo ErrorHandler

URL = "https://vpic.nhtsa.dot.gov/api/vehicles/
getallmakes?format=xml"

strExcelFile = "C:\VBAExcel2024_ByExample\MakesList.xlsx"

' delete Excel file if it exists
If Dir(strExcelFile) <> "" Then Kill strExcelFile

Set httpReq = CreateObject("MSXML2.ServerXMLHttp")
httpReq.Open "GET", URL, False

httpReq.send
Debug.Print httpReq.getAllResponseHeaders
' check if we sucessfully made the request
If httpReq.status <> 200 Then
```

```vba
        Debug.Print httpReq.status & ":" & httpReq.statusText
        Set httpReq = Nothing
        Exit Sub
End If

' Assume we got through to the API
Debug.Print httpReq.responseText

Resp.LoadXML httpReq.responseText

' set up an Excel workbook for our data
Set wkb = Workbooks.Add
Set ws = wkb.Sheets(1)
ws.Activate

Header = False
i = 1

ws.UsedRange.Clear

Dim NodeList As MSXML2.IXMLDOMNodeList
Dim iNode As MSXML2.IXMLDOMNode

startTime = Timer() ' start the timer

Set NodeList = Resp.getElementsByTagName("AllVehicleMakes")
If NodeList.length <> 0 Then
    For c = 0 To NodeList.length - 1
        For n = 0 To NodeList.item(c).childNodes.length - 1
            Set iNode = NodeList.item(c).childNodes(n)

            If Header = False Then
                ws.[A1].Offset(0, n).Value = iNode.baseName
            End If
            ws.[A1].Offset(i, n).Value = iNode.Text
        Next n
    i = i + 1
    ' this will get only the first 100 items from the response
    ' and will shorten the processing time
    ' comment it out to get all records
    ' (processing may take over 5 minutes)
    If i = 101 Then Exit For
    Next

    'autofit columns
```

```
    ws.Columns.AutoFit

    ' save and close the workbook
    wkb.SaveAs strExcelFile
    wkb.Close

    endTime = Timer() ' end the timer

    Debug.Print "Processing completed in " & _
        Format(Round(endTime - startTime, 2), "#0.00") & " _
            seconds."

    MsgBox "Processing completed in " & _
        Int((endTime - startTime) / 60) & " minutes, " & _
        Round(endTime - startTime, 2) Mod 60 & " seconds.", _
        vbInformation

    Debug.Print "The worksheet with All Vehicle Makes is
        completed."
Else
    Debug.Print "No data to process. Please check for errors."
End If

ExitHere:
' cleanup
    Set httpReq = Nothing
    Set ws = Nothing
    Set wkb = Nothing
    Exit Sub
ErrorHandler:
    Debug.Print Err.Number & ":" & Err.Description
    Resume ExitHere
End Sub
```

Click in the selection bar next to the `httpReq.send` statement in the procedure to put a breakpoint on this line of code, and press F5 to run the procedure.

When the break mode is activated and you see the yellow selection in the code window, keep pressing F8 to debug the code line by line. Keep the Immediate window open while debugging the procedure so you can see the output of the `Debug` statements. Activate the Immediate window by pressing Ctrl+G.

The first `Debug` statement in this procedure should return the contents of the headers as shown earlier in this chapter. Notice that we are asking for the headers after we have sent the request to the server. This statement is for demonstration purposes only. Most of the time, you don't need to look at headers unless problems arise and you need to troubleshoot some issues. After sending the request, the server gives you back a status code: `200` means that the request was successful and the server produced the response. Only then should we proceed with the rest of the code. There is no point in asking for a response when an error code was returned in the server `status` property. If the status is not `200`, we will output to the Immediate window the status code and the equivalent text error message using the `status` and `statusText` properties discussed earlier. If the request is successful, we use the `responseText` property to get the response and print it to the Immediate window.

Debug statements help you understand the output from the server. Now that we have the output, we also want to save it so we can use it for a specific purpose. In this procedure, we use the `LoadXML` method of the `XMLDocument` object to load the server response into an XML document, which we declare at the top of the procedure in the `Resp` object variable. Note that we extensively covered working with XML in Chapter 25. After the XML data is retrieved to the Immediate window, we write some VBA code to read the XML nodes and output their contents to an Excel worksheet in a new workbook.

When you have reached the end of the procedure, open the `MakesList.xlsx` workbook that was created in this Hands-On exercise.

Figure 26.9 illustrates the contents of XML data retrieval in the Immediate window. Figure 26.10 shows the same data written on an Excel worksheet.

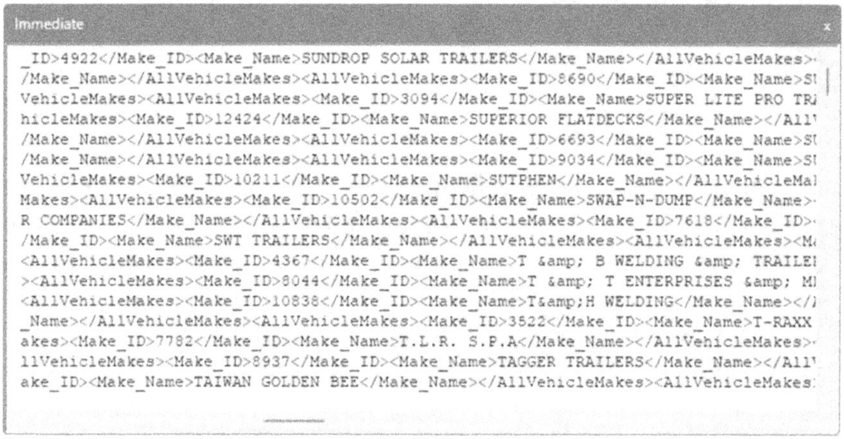

FIGURE 26.9. The AllVehicleMakes list was created from the REST API response and written to the Immediate window. There were too many vehicles listed, however, so we also wrote that list to an Excel worksheet in the MakesList.xlsx workbook.

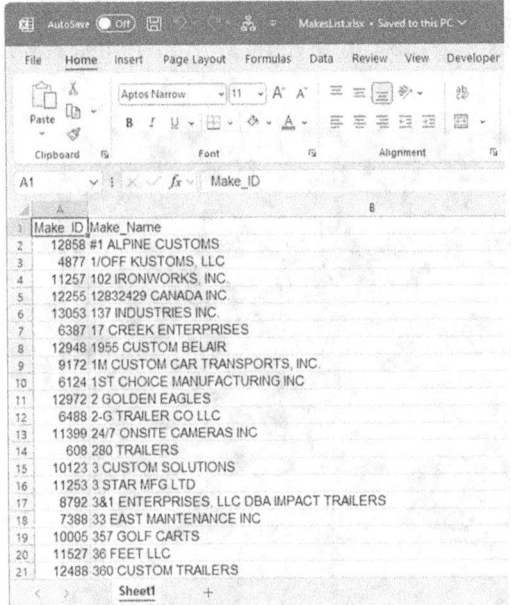

FIGURE 26.10.   The XML data returned from the REST API written to an Excel worksheet.

Using the documentation available for the Vehicle API, with its easy-to-understand examples of API methods and output format, you can modify the `RequestData_XML_toExcel` procedure in a few places to obtain other types of content. For example, to get all the manufacturers, simply pass another URL, change the name of the Excel workbook to save the data, and set `NodeList` to `Manufacturers`, as this is the name of the XML root node for this dataset:

```
URL = "https://vpic.nhtsa.dot.gov/api/vehicles/GetAllManufacture
rs?format=xml&page=2"
strExcelFile = "C:\VBAExcel2024_ByExample\Manufacturers.xlsx"
Set NodeList = Resp.getElementsByTagName("Manufacturers")
```

An Overview of JSON

Using XML with REST API requests and responses requires knowledge of the XML language and a good understanding of the process of generating and modifying XML structures. In recent years, a new text format, known as JSON (pronounced "Jason") has become an alternative to XML. *JSON*, which stands for *JavaScript Object Notation*, is a language-independent data format often used for data interchange between disparate systems due to its lightweight format. The JSON format is a human-readable text that consists of name/value pairs.

Figure 26.11 displays the sample of the JSON file obtained from the free fake API that is available for testing and prototyping (*https://jsonplaceholder. typicode.com/users*).

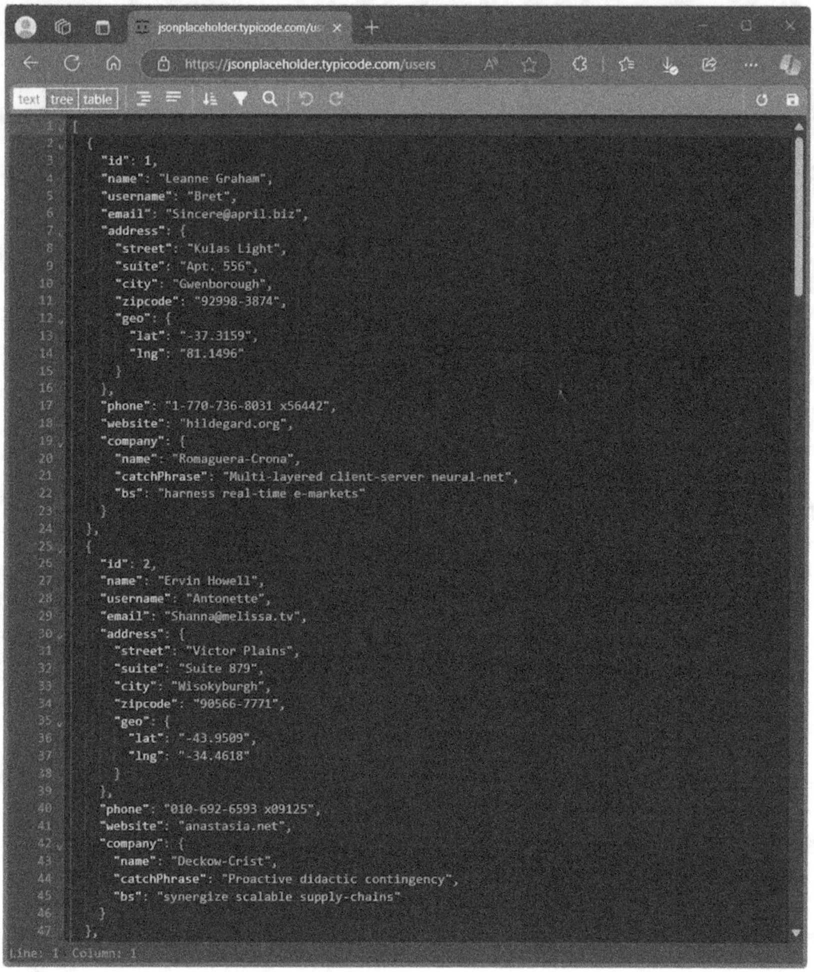

FIGURE 26.11. Example of the JSON file format.

The JSON format can include:

- An object, which is an unordered set of name/value pairs.

  Objects begin with { (left brace) and end with } (right brace). Each name is followed by : (colon), and the name/value pairs are separated by , (comma).

- An array, which is an ordered collection of values.

  Arrays begin with [(left bracket) and end with] (right bracket). Values are separated by , (comma). A value can be a string enclosed in double quotes, a number, a `True` or `False`, an object, or an array. Values can be nested so you can create very elaborate JSON structures to fit all your needs.

- A string, which is a sequence of zero or more Unicode characters, wrapped in double quotes using backslash escapes.

- A number (e.g., digits 1–9, fractions, and exponents).

- Whitespace (e.g., space, linefeed, carriage return, or horizontal tab).

In the following example, the JSON object represents three courses in an array called `courses`. Notice that the array starts and ends with square brackets. Within the array are three objects, one for each course:

```
{
"courses": [
{
"courseID": 1001,
"title": "Access 2024 / Microsoft 365 Programming by Example",
"dateOffered": "September 14",
"location": "virtual"
},
{
"courseID": 1002,
"title": "Excel 2024 / Microsoft 365 Programming by Example",
"dateOffered": "September 18",
"location": "New York, Hilton"
},
{
"courseID": 1003,
"title": "PowerPoint VBA Programming by Example",
"dateOffered": "December 5-7",
"location": "Unspecified"
}
]
}
```

Because JSON is a simple text format, you can use Windows Notepad to create, save, and view JSON files. The above snippet can be found in the companion files as `Courses.json`.

The JSON format is supported by virtually all modern browsers, and because it is smaller than the XML encoding, JSON can provide bigger performance gains when sending large amounts of data over a network. The format you should use for data exchange (plain text/HTML, XML, or JSON) will depend on your specific situation and your project requirements.

When working with JSON data, you may come across another term, *JSONP* or *Jason with Padding*. JSONP allows you to get JSON data from a server in a different domain. This will help you get around the cross-domain security policy that modern browsers implement.

A more detailed description and examples of JSON formatting can be found at www.json.org.

Loading JSON Data into Excel

In the next Hands-On example, we will access another free REST API resource, but this time, we will receive the response formatted as JSON. The direct link to that resource is *https://api.zippopotam.us.*

The Web site shows how the API can be used to autocomplete the City and State text boxes based on the zip code you enter. We will build a similar form in Excel (see Figure 26.12) that obtains the zip code data through that API.

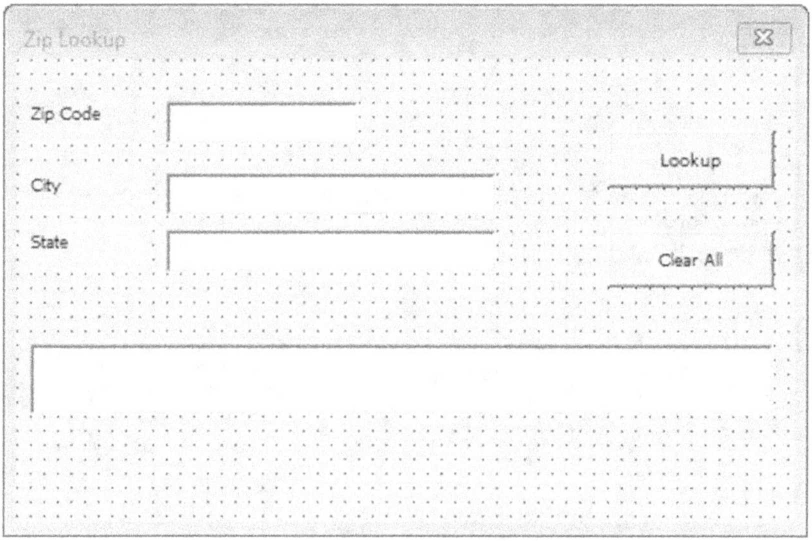

FIGURE 26.12. Excel form used in Zip Code Demo REST API – JSON

If you enter `api.zippopotam.us/us/90210` in your favorite browser, you should see the JSON response shown in Figure 26.13.

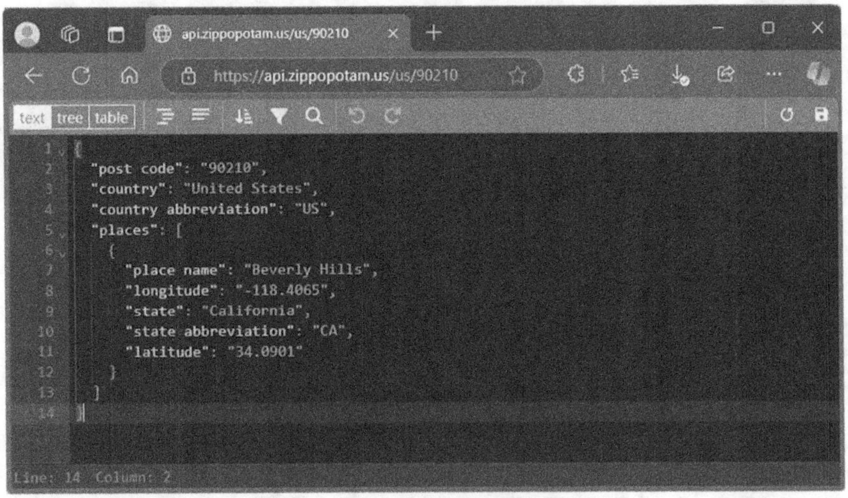

FIGURE 26.13. JSON structure returned from the Zippopotam API.

To work with the JSON format in VBA, we don't have a method that can help us easily extract individual key values from the JSON string. For simple structures such as the one shown in Figure 26.13, we can use the basic string functions discussed earlier in this chapter, or we can use what we've learned so far about regular expressions. Let's take the latter route to gain more experience.

⊙ Hands-On 26.5   Requesting Data from the REST API (JSON Example)

1. In the `Chap26.xlsm` workbook, choose Insert | UserForm to create a new form, as shown in Figure 26.12 earlier.
   This form should include the following:
   - Three unbound text boxes with caption properties set to `Zip Code`, `City`, and `State` and name properties set to `txtZip`, `txtCity`, and `txtState`.
   - Two command buttons with caption properties set to `Lookup` and `Clear All`, and name properties set to `cmdRequest` and `cmdClearAll`.
   - One unbound text box without a label. The name property of the text box is `txtStatus`. This box is used to return any errors received during the lookup process.
2. Rename `UserForm1` to `ZipLookup`.
3. Double-click the form's Lookup button.
4. In the `ZipLookup` Code module, enter the following VBA procedure, which will run when the Lookup button is clicked

```
Private Sub cmdRequest_Click()
    ClearFields

    If Me.txtZip = "" Or IsNull(Me.txtZip) Then
        With Me
            .txtStatus.Visible = True
            .txtStatus = "Please enter Zip Code"
            .txtZip.SetFocus
        End With
        Exit Sub
    End If
    If Len(Me.txtZip) <> 5 Or Not IsNumeric(Me.txtZip) Then
        Me.txtStatus.Visible = True
        Me.txtStatus = "Zip code must be 5-digit long."
        Exit Sub
    End If
    RequestData Me.txtZip
End Sub
```

This procedure will make sure that the Zip Code text box is not empty and is entered in the correct format. In the first line of this procedure, we will call a `ClearFields` procedure that will clear the existing entries in the City, State, and Status text boxes and will set the visibility of the Status box to `False` when it is empty. In the last procedure statement, we will call the `RequestData` procedure and pass it the current value from the Zip text box. The `RequestData` procedure will make a call to the REST API. We will enter it later in the standard code module.

5. Enter the following two procedures in the same `ZipLookup` Code module. The first of these procedures will handle the form's Clear All button click:

```
Private Sub cmdClearAll_Click()
    Me.txtZip = ""
    ClearFields
    Me.txtZip.SetFocus
End Sub
Sub ClearFields()
    With Me
        .txtCity = ""
        .txtState = ""
        .txtStatus = ""
        .txtStatus.Visible = False
    End With
End Sub
```

6. Save the changes you've made in the `ZipLookup` Code module.

7. Choose Insert | Module to add a new standard module to `VBAProject` (`Chap26.xlsm`). Use the Properties window to rename it `APIRequest_JSON`.

8. In the `APIRequest_JSON` module, enter the code of the `RequestData` procedure as follows:

```
Sub RequestData(ByVal postcode As String)
Dim httpReq As MSXML2.XMLHTTP60
Dim oRegExp As Object
Dim rec As Variant
Dim aRecords As Variant
Dim aRecord As Variant
Dim fld As Variant
Dim fldName As String
Dim fldContent As String
Dim strAObj As String
Dim webResponse As String

Set httpReq = New MSXML2.XMLHTTP60
httpReq.Open "GET", _
    "https://api.zippopotam.us/us/" + postcode, False
httpReq.send
If httpReq.status <> "200" Then
    ZipLookup.txtStatus.Visible = True
    ZipLookup.txtStatus = "Error Code: " & _
        httpReq.status & " - " & httpReq.statusText
    Exit Sub
End If
'get the entire JSON string
webResponse = httpReq.responseText

Debug.Print "Below is raw web response json" & vbCrLf
Debug.Print webResponse
' convert json string to an array
Set oRegExp = New RegExp

oRegExp.Global = True
oRegExp.Pattern = "[\[\]\{\}""]+"
strAObj = "[{"

If InStr(1, webResponse, strAObj) > 0 Then
    webResponse = Replace(webResponse, strAObj, ", ")
End If

Debug.Print webResponse
```

```
webResponse = oRegExp.Replace(webResponse, "")
Debug.Print webResponse

aRecords = Split(webResponse, ", ")
For Each rec In aRecords
    aRecord = Split(rec, ", ")
        For Each fld In aRecord
            fldName = Split(fld, ":")(0)
                fldContent = Split(fld, ":")(1)
                 If fldName = "place name" Then
                     ZipLookup.txtCity = fldContent
                 End If
                 If fldName = "state" Then
                     ZipLookup.txtState = fldContent
                 End If
        Next
    Debug.Print fldName & vbTab & vbTab & fldContent
Next
Set httpReq = Nothing
End Sub
```

9. Save the changes in the module.

10. In the Project Explorer, select the ZipLookup form and choose Run | Sub/
UserForm.

11. Click the Lookup button. You should see the error message shown in Figure
26.14.

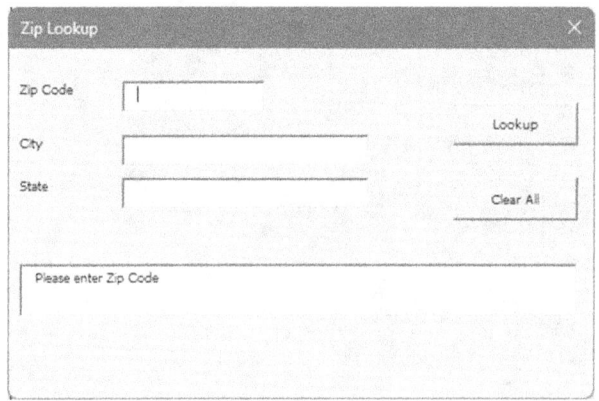

FIGURE 26.14.   The Excel form used in Zip Code Demo REST API – JSON displays the message in a text box when the Lookup button is clicked without providing the zip code.

12. Click the Clear All button. The error message box should disappear.

13. Enter your zip code or any valid zip code you can recall and click the Lookup button.

If the zip code exists, you should see both the City and State text boxes populated (see Figure 26.15).

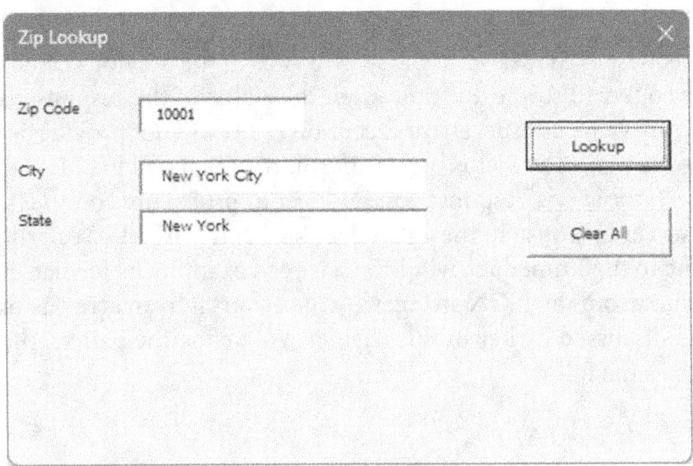

FIGURE 26.15.   The Excel form used in Zip Code Demo REST API – JSON displays zip code data from the Zippopotam API.

If the zip code is invalid, for example, you've entered `11345`, you should see "Error Code: 404 -", as shown in Figure 26.16.

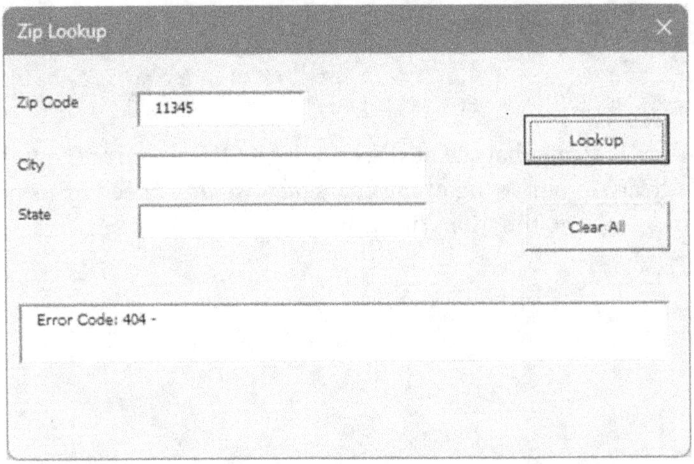

FIGURE 26.16.   The Excel form used in Zip Code Demo REST API – JSON displays an error code when the zip code is invalid.

To retrieve the zip code data, we pass the value from the Zip Code text box to the `RequestData` procedure in the string type variable called `postcode`. Notice that we then add this value to the URL string like this:

```
httpReq.Open "GET", "https://api.zippopotam.us/us/" + _
    postcode, False
```

We send the request to the server in the same way as in the previous Hands-On exercise. Our code will stop executing when the status of the response is not equal to `200`. Any value but successful execution (`200`) will not provide us with a response, so we make the Status box visible on the form and load it with the error message. If the server response code is `200`, our procedure continues, and we store the server response in the `webResponse` string variable. We print the response string to the Immediate window so we can examine its format. To get individual values from the JSON string, we will convert it to an array using the `RegExp` object discussed earlier in this chapter. We define the pattern that we want to match as the following:

```
oRegExp.Pattern = "[\[\]\{\}""""]+"
```

The `Global` property of the `oRegExp` object will ensure that all occurrences of the characters specified in our pattern will be removed when we apply the `Remove` method. Prior to that, we want to find and replace the character string `[{` that follows the "`places`" key in the returned JSON string. This is how the string appears in the Immediate window:

```
{"post code": "10001", "country": "United States", "country
abbreviation": "US", "places": [{"place name": "New York
City", "longitude": "-73.9967", "state": "New York", "state
abbreviation": "NY", "latitude": "40.7484"}]}
```

The `[{` characters indicate that the next section of JSON is an array containing an object. We clean up our string of the characters we don't need by using the VBA `InStr` and `Replace` functions like this:

```
strAObj = "[{"
If InStr(1, webResponse, strAObj) > 0 Then
    webResponse = Replace(webResponse, strAObj, ", ")
End If
```

After this code is completed, we print the revised string to the Immediate window:

```
{"post code": "10001", "country": "United States", "country
abbreviation": "US", "places": , "place name": "New York
City", "longitude": "-73.9967", "state": "New York", "state
abbreviation": "NY", "latitude": "40.7484"}]}
```

Notice that `"places":` is now followed by a space and a comma.

Next, we use pattern matching and replace with a comma all the braces and brackets that are remaining in the previous string. As a result, we get the following string:

```
post code: 10001, country: United States, country abbreviation:
US, places: , place name: New York City, longitude: -73.9967,
state: New York, state abbreviation: NY, latitude: 40.7484
```

Now, all that's left to do is split the above string into the name/value pairs so we can get access to individual items. We can use the VBA `Split` function to break the string each time we encounter a comma and a space:

```
aRecords = Split(webResponse, ", ")
```

The `aRecords` variable is an array and we can count the number of items we have in it using the `UBound` function. If you enter `?Ubound(aRecords)` in the Immediate window during the break mode (while your code is running), you should see 8 as a return value. If you type `?aRecords(0)`, you will get back the first name/value pair like this:

```
?aRecords(0)
post code: 10001
```

We use the `For Each` loop to iterate through the array. To make the process easier, we split each of the name/value pairs into individual fields using the colon separating the name from the value. The first (0) item will be a field name and the second (1) item will be the field content (value). We use the `If` statement to find the value for the place name field that we put in the `txtCity` text box on our form, and then look for the state to get the state. The looping process continues until we find the values we are searching for.

To gain more understanding of how this code performs its magic trick of parsing the JSON string using the VBA string functions and regular expressions, set a breakpoint on the first `Debug` statement in the `RequestData` procedure, then go back to your form, enter a valid zip code, and click the Lookup button.

When VBA encounters the break statement and the code window appears, you can use Step Into (F8) to step through the entire code line by line, asking questions and checking responses in the Immediate window. You can also try to set up some watch expressions, which were introduced in Chapter 8.

Sending JSON Data to REST APIs

You can send data to REST APIs using POST, PUT, or DELETE requests. The following procedure demonstrates how to use the PUT method to update a resource. We will be using the fake API so the resource will not really be updated on the server, but it will be faked as if it is:

```
Sub PutDataToAPI()
    Dim http As Object
    Dim putData As String
    Dim response As String

    Set http = CreateObject("MSXML2.XMLHTTP")

    ' JSON data

    putData = "{""Id"":""1""," _
        & """title"":""Tester""," _
        & """body"":""Project API Test""}"

    http.Open "PUT", _
        "https://jsonplaceholder.typicode.com/posts/1", False
    http.setRequestHeader "Content-Type", "application/json"
    http.send putData
    response = http.responseText

    MsgBox "Response: " & response
End Sub
```

Formatting JSON data for server communication can be challenging. While it's possible to write the entire string on a single line, this approach often makes the code harder to read and maintain. To improve readability and manageability, you can break a long JSON string into multiple lines. Use the concatenation operator (&) and double up the quotation marks to ensure valid syntax.

When you run the PutDataToAPI procedure, you get the response shown in Figure 26.17.

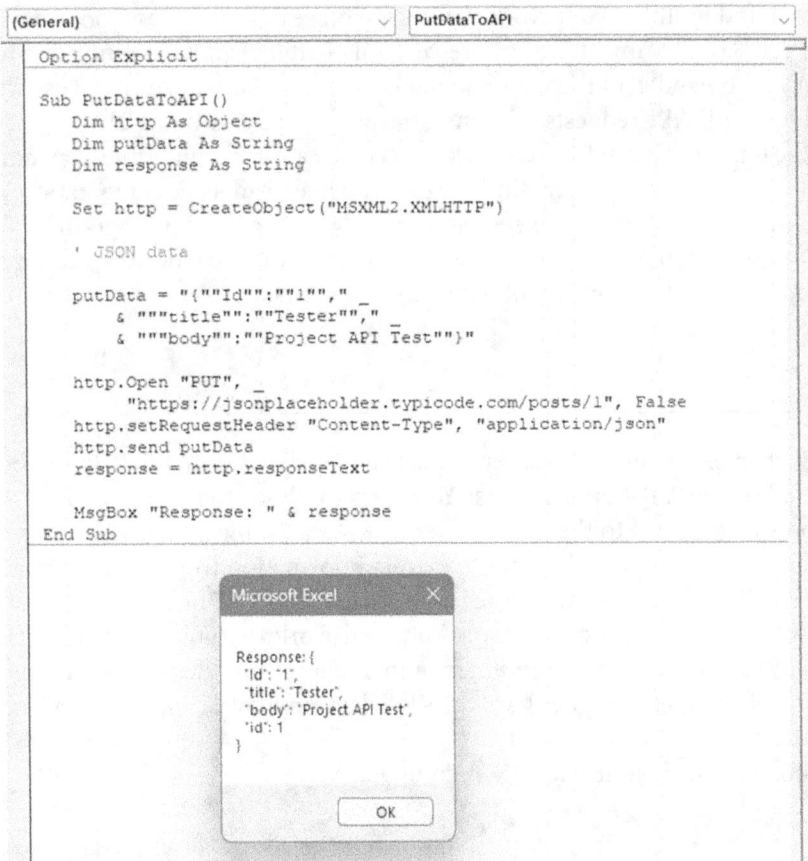

FIGURE 26.17.   PUT HTTP request to Web service returned as a JSON-formatted string.

Parsing JSON with Third-Party Libraries

Parsing REST API responses, especially complex ones, will not be as straightforward as the examples you've tried here. You may need a custom VBA library written specifically to help you handle the intricate JSON format your response can be returned in. Search the Web for this topic and you're bound to find a great set of VBA-JSON tools created by Tim Hall and available via *https://github.com/VBA-tools*. It contains the JSON conversion and parsing for VBA (Windows/Mac, Excel, Access, and other Microsoft 365 applications). I have successfully used it in several of my own VBA projects and it made the task of dealing with the JSON output so much easier. There are other tools that you may also like and find useful in your work if you keep on searching. These tools may not work 100% of the time, but they are certainly better than having nothing built into VBA to handle all the data parsing issues that you will come across while making REST API requests. Getting a response and not being able to turn it into the format you can further work with can be a painful and time-consuming effort. Do your own research; find the tool you like, read its documentation, and be ready to use it in your project. Now that you know how to program in VBA (you've learned about classes, dictionaries, arrays, and collections), you can create your own tool or improve one that already exists.

SUMMARY

In this chapter, you explored several external libraries that should help you build more advanced VBA applications. You learned about the `Dictionary` object and how it compares to the native VBA `Collection` object. You saw how regular expressions can make it easier to extract information from JSON-formatted data. Finally, you learned the basics of making a REST API request and parsing both XML and JSON response data. Your real-world assignments will be more complex than the examples presented in this chapter, but these simple examples should help you reach higher levels of VBA development quickly and with more confidence.

Good luck with your Excel VBA coding!

INDEX

www.ingramcontent.com/pod-product-compliance
Lightning Source LLC
Chambersburg PA
CBHW080229200526

45165CB00025B/2284